Principles of Chemistry

Principles of Chemistry

Editor
Donald R. Franceschetti, PhD
The University of Memphis

SALEM PRESS

A Division of EBSCO Information Services, Inc.
Ipswich, Massachusetts

GREY HOUSE PUBLISHING

∞ The paper used in these volumes conforms to the American National Standard for Permanence of Paper for Printed Library Materials, Z39.48-1992 (R1997)

Publisher's Cataloging-In-Publication Data
(Prepared by The Donohue Group, Inc.)

Names: Franceschetti, Donald R., 1947- editor.
Title: Principles of chemistry / editor, Donald R. Franceschetti, Ph.D., the University of Memphis.
Description: [First edition]. | Ipswich, Massachusetts : Salem Press, a division of EBSCO Information Services, Inc. ; Amenia, NY : Grey House Publishing, [2016] | Includes bibliographical references and index.
Identifiers: ISBN 978-1-61925-501-2 (hardcover)
Subjects: LCSH: Chemistry.
Classification: LCC QD31.3 .P75 2016 | DDC 540--dc23

PRINTED IN THE UNITED STATES OF AMERICA

CONTENTS

Publisher's Note

Salem Press is pleased to add *Principles of Chemistry* to its collection, the first in a new *Principles of* series. Physics, Astronomy, and Computer Science will follow later this year. This new resource introduces students and researchers to the fundamentals of chemistry using easy-to-understand language, giving readers a solid start and deeper understanding and appreciation of this complex subject.

The 128 entries range from Acid anhydrides to Zone refining and are arranged in an A to Z order, making it easy to find the topic of interest. Each entry includes the following:

- Related fields of study to illustrate the connections between the various branches of chemistry, including biochemistry, molecular biology, organic and inorganic chemistry, nuclear and physical chemistry, geochemistry, and metallurgy;
- A brief, concrete summary of the topic and how the entry is organized;
- Principal terms that are fundamental to the discussion and to understanding the concepts presented;
- Illustrations that clarify difficult concepts via models, diagrams, and charts of such key topics as the relationship between acids and bases, molecular structures, bonds, and reactions;
- Equations that demonstrate how chemical formulas are written, how equations are balanced, and other important functions;
- Photographs of significant contributors to the study of chemistry;
- Sample problems that further demonstrate the concept presented;
- Further reading list that relates to the entry.

This reference work begins with a comprehensive introduction to the field, written by editor Donald R. Franceschetti, PhD, starting with man's earliest understanding of the primary substance of life and concluding with a discussion of chemistry as the "central science" and touching on the exciting contemporary research being conducted around such fascinating materials as graphene and buckyballs.

The book's backmatter is another valuable resource and includes:

- Periodic Table of Elements, including an introduction and history of the table and a second table that with additional information about atomic weights;
- Nobel Notes that explain the significance of the prizes to the study of science and their interdisciplinary nature;
- Nobel Prize winners in the area of chemistry from the first awards in 1901, given to Jacobus Henricus van't Hoff "in recognition of the extraordinary services he has rendered by the discovery of the laws of chemical dynamics and osmotic pressure in solutions" to the prize awarded in 2015 to Tomas Lindahl, Paul Modrich, and Aziz Sancar "for mechanistic studies of DNA repair";
- Glossary;
- General bibliography;
- Subject index.

Grey House Publishing extends its appreciation to all involved in the development and production of this work. The entries have been written by experts in the field. Their names and affiliations follow the Editor's Introduction.

Principles of Chemistry, as well as all Salem Press titles, is available in print and as an e-book. Please visit www.salempress.com for more information.

Editor's Introduction

The Science of Matter and its Transformations

Humans have always looked for explanations about the world in which they live. From the earliest speculations on the nature of matter and its transformations to today, a science of chemistry has developed. Our understanding has grown over the centuries, slowly at first and with increasing speed since the Scientific Revolution of the seventeenth century ushered in the scientific method. While philosophers pondered the nature of chemical transformation, more pragmatic individuals recorded their observations, sometimes in code to protect them from the eyes of the unworthy.

The word *chemistry* is derived from the Arabic *al-chemy* and possibly from the Egyptian *khem* referring to the fertile black mud produced by the annual overflowing of the Nile. The ancient Egyptians knew something of chemical reactions, but subsequent writers often attributed their work to even more ancient and even mythical individuals, so the situation is unclear.

Matter is Made Up of Atoms

Chemistry got its real start only when atoms were first proposed and studied. The atomic hypothesis, which states that matter was composed of indestructible atoms, was put forth by Leucippus (c. 480–420 BCE) and his disciple Democritus (c. 460–370 BCE). The atomic hypothesis met resistance since it suggested that there was no "soul" and that, therefore, life necessarily ended completely at death. The pagan religions varied in their reaction to this but to the Christians and some Jews, the concept of life as finite was anathema. It became even less acceptable when the medieval Catholic theologians "baptized" Aristotle and began talking about transubstantiation.

The ancient Greek philosopher Empedocles (492–432 BCE) argued for four elements: air, earth, fire and water, to which Plato (428–347 BCE) and Aristotle (384–322 BCE) added a fifth element known as *aether* or *quintessence* (which translates as "fifth element") to differentiate celestial objects from terrestrial ones. In his *Timaeus,* Plato considered the five elements to correspond to the five perfect solids.

The alchemists took a rather different view of the composition of matter. Alchemy had its origins in the Hellenistic period (300 B.C.E. to 300 C.E.), but, as an accumulation of experimental results, it began in earnest around 1300 C.E. Over time as more was slowly learned about transformations that could be achieved by heating substances, the elements of Empedocles were put aside in favor of a trio of principles: mercury, sulfur and salt. The Renaissance physician, alchemist, and prolific writer, Philippus Aureolus Theophrastus Bombastus von Hohenheim, (1493–1541), who assumed the name Paracelsus, held this view and introduced alchemical ideas into medical practice. Paracelsus, like Copernicus, Martin Luther, and even Leonardo da Vinci, was a product of the Renaissance, part mystic and part rational.

Views began to change when Robert Boyle published *The Skeptical Chymist* in 1661, calling into doubt not only the ancient Greek theory of four elements—air, earth, fire and water—but also the *tria prima* of the alchemists: salt, sulfur and mercury. Chemistry then proceeded by the slow accumulation of facts until 1789, when Antoine Laurent Lavoisier, assisted by his wife, Marie, published the *Traité élémentaire de chimie* in1789, which included the first modern list of elements. Lavoisier's beheading in 1794 was a great loss to science.

After about 70 more years of trying to find a system in which to organize the elements, the Russian Mendeleev produced the first modern table of elements, published in the German journal, *Zeitschrift für Chemie,* in 1869, highlighting the periodic variation of chemical properties.

While the development of physical theory can be traced to the work of Galileo and Newton in the seventeenth century, chemical theory had to wait until the nineteenth and early twentieth centuries to see comparable development. In retrospect, the reasons for the delay are not hard to see. The modern chemist takes the existence of atoms for granted. Atoms have well defined atomic masses and combine to form molecules held together by chemical bonds, the essence of which are the quantum rules for the behavior of electrons, particles much smaller than the smallest atom. The atomic nature of matter and the electron theory of bonding are now so well established that no serious scientist doubts them. but only 110 years ago, the very existence of molecules in the liquid and solid states was a matter of debate. Some scientists and philosophers saw no point in describing the behavior of particles that could never be seen. Only the accumulation of overwhelming evidence allowed progress to be made.

One source of confusion was the existence of subatomic particles. Once chemists had established that matter could be broken down into elements, and that all the compounds known could be expressed as compounds of less than 90 elements, there was actually considerable resistance to accepting the electron, which weighed about 1/1840th as much as a hydrogen atom, as a distinct particle. Instead the cathode rays (beams of electrons) that physicists could produce when applying an electric field to heated metals in a vacuum were thought of as waves in the luminiferous ether, a mysterious medium assumed to permeate all space, whose vibrations were light waves.

Einstein's special theory of relativity did away with the need for the ether and scientists generally came to accept the atom as composed of sub-atomic parts. This new realization prompted a revolution in physics as well, as it was realized that all the positive charge in the atom was contained in the very tiny nucleus, about which the electrons moved in paths that could not be visualized but rather had some of the characteristics of waves.

With the formulation of the Schrödinger equation and the discovery of electron spin in the 1920s, it became possible, in principle at least, to describe with some accuracy the structure of each molecule. Even then, exact solutions were generally impossible to achieve and the description of molecules or ions with even a few electrons was not possible until the advent of modern high-speed computers.

Traditional Branches of Chemistry

Traditionally chemistry has been divided into four main branches—analytical chemistry, physical chemistry, organic chemistry and inorganic chemistry—and is bounded by the interdisciplinary fields of biochemistry and chemical physics.

Analytical chemistry, broken up into qualitative and quantitative analysis, deals with identifying the elements in a chemical sample (qualitative analysis) and their relative proportions (quantitative analysis). As technology has improved, analytical chemistry has advanced to new levels of sophistication. It is now possible to isolate a few dozen argon atoms within a tankful of cleaning fluid, for example. Less extreme, perhaps, is the ability of chemists to find impurities in bodies of water at the level of parts per billion or parts per trillion. The increasing sensitivity of analytical methods can present a challenge to the regulators of air and water pollution. Many would argue that regulations have actually gotten away from the need to protect the public and become a regime unto themselves. If a substance is lethal at 1 ppt (part per thousand) but it can be detected at 0.01 ppb (part per billion) in the water supply, how much effort and cost is justified to get rid of it?

Physical chemistry involves the physically measurable properties of substances and their interaction with electromagnetic radiation. Physical chemistry and chemical physics are essentially two sides of the same coin. Physical chemists tend to be more concerned with the mechanism of chemical reactions, and the measurement of thermodynamic properties of different compounds while chemical physicists are more interested in the electronic structure of molecules.

Organic chemistry had its origin at a time it was generally believed that there was a class of compounds that could be produced only by living organisms. While this vitalist hypothesis was disproved by the synthesis of urea (clearly an organic compound) from ammonium cyanate (generally considered inorganic) by Friedrich Wohler (1800–1882) in 1828, organic chemistry continues as the field of chemistry devoted to carbon compounds, because of the unique versatility of the carbon atom. Indeed over 90% of known compounds are carbon compounds.

The character of organic chemistry owes a lot to William Henry Perkin (1838–1907), who isolated the first of the coal-tar dyes—mauve. At first these dyes were valued because they could be used to stain cloth fibers in a variety of different colors, leading to a revolution in fashion. It was soon found, however, that coal-tar dyes could selectively stain living tissue, leading to an even more far reaching revolution in microscopy. Moreover, they could be used to combine with living organisms, destroying some disease carrying bacteria and ushering in the earliest form of chemotherapy. The academic culture of the German universities, which idealized the research specialist, was particularly suited to producing generations of organic chemists who advanced the German chemical industry to new heights. This development did, alas, have its dark side as well. A group of German chemical companies formed the *Interessengemeinshaft der Deutschen Teerfarbenfabriken*, a conglomerate of dye color interest groups that became known as IG Farben. Although the scientists working as part of IG Farben made important contributions to all areas of chemistry, the group's role as the supplier of poison gas used in Holocaust gas chambers led to the group's dissolution following World War II. Despite these dark episodes, overall, chemical discovery has benefitted mankind far more than harmed it.

The rest of chemistry traditionally gets lumped together under the heading inorganic chemistry, where the focus is often on transition metal compounds and the properties of the lanthanides and actinides. New research areas abound in inorganic chemistry. Consider graphene, a layer of carbon that is the thickness of a single atom with intriguing electrical properties; macromolecular chemistry, which deals with enormous molecules; nanotechnology, which finds new phenomena in the behavior of materials that are just a few atomic layers thick; the chemistry of surfaces, where molecules behave differently than in the interior of homogeneous phases; and materials science, which finds important effects when defects are introduced into bulk materials.

ISOMERS AND STEREOCHEMISTRY

The arrangement of atoms in a molecule has become a source of concern to chemists. For simple compounds like NaCl, the empirical formula provides an adequate description for many purposes–but not all. In certain cases, more than one compound could have the same empirical formula. The compound, C_3H_8O, for instance, comes in three forms or isomers: two alcohols–$CH_3CH_2CH_2OH$ (n-propanol), with the OH group attached to an end carbon, and $CH_3CHOHCH_3$ (isopropanol), with an OH group attached to the central carbon—and methyl ethyl ketone, in which the oxygen atom connects a methyl group, CH_3, with an ethyl group CH_3 CH_2. As the number of carbon atoms increases, the number of isomers increases dramatically.

Yet another type of isomerism—stereoisomerism—occurs in organic compounds and crystals, a fact discovered by the young Louis Pasteur (1822–1895), who made many contributions to chemistry and physics, as well as those for which he is justly famous in the field of microbiology. As Pasteur showed, a molecule with four different substituent groups attached to a carbon atom will generally not be superimposable on its mirror image. In its pure form, the molecule and its mirror image will have identical melting points and the molecule and its mirror image may crystallize in mirror image forms. Such pairs of molecules are known as optical isomers as they rotate the plane of polarized light by equal amounts in opposite directions. Accounting for the spatial arrangement of such molecules would require advances in both typography and chemical nomenclature.

Optical isomerism is typical of the products of living organisms. Usually only one form of the molecule will be found in living cells. Naturally occurring proteins are long chain polymers of about 20 amino acids. The amino acids all have the general formula HCR (NH_2) (COOH) where different R groups correspond to different amino acids. With the exception of glycine, for which the R group is another hydrogen, all the amino acids are asymmetric. Speculations as to the reasons this should be so abound. Possibly all living beings are descendants of a single reproducing molecule. Possibly the first living things involved molecules that would selectively adsorb on an optically active form of quartz. We just don't know.

THE CENTRAL SCIENCE AND ITS BORDERS: CHEMISTRY AND PHYSICS

The American Chemical Society has taken to referring to chemistry as the central science, noting its great importance to physics, on one side, and biology, on the other.

Since the discovery of electrons in cathode rays and the elucidation of the Schrödinger equation with its ability to predict the optical spectra of atoms, the electron theory of matter has become the dominant paradigm in chemistry and chemical physics.

In the mid-twentieth century, it was determined that within the atomic nucleus the protons and neutrons occupied energy levels in a manner somewhat analogous to those of electrons in the atom. Some forms of radioactive decay can be understood as a rebalancing of protons and neutrons among energy levels, resulting in either beta emission, producing an atom with one more proton and one less neutron, or positron emission, producing an atom with one more neutron and one less proton. The nuclear weak force, which is responsible for such transition, is sometimes called the cosmic alchemist for this reason. The alpha particle ejected in the radioactive decay of some elements is identical to the helium nucleus and is a particularly stable combination of two protons and two neutrons. Any helium found on earth is a product of such decay, as the mass of the helium atom is too small to be gravitationally bound to the earth for long. And, of course, the induced fission of large nuclei to form two smaller nuclei occurs in nuclear bombs and power plants.

An effect of research in nuclear physics is that the periodic table has grown longer. Elements up to ununoctium (number 118) have been reported to date. While this development contributes to our understanding of nuclear physics, the effect on chemistry has not been great, as atoms of the new elements have been created only in very small numbers and they decay rapidly.

The impact of chemistry on astronomy has also been great. Now it is appreciated that many molecules have been formed in space, where concentrations are low and collisions infrequent, including many that would not be observed under terrestrial conditions. Confirming the existence of such species presents a continuing challenge to chemists and spectroscopists.

Within the interdisciplinary area of physical chemistry/chemical physics, one now finds theorists, experimentalists, and computationalists, the latter being specialists who use computers to determine molecular electronic structures and to systematize what is known about different related structures. Recent exciting developments include combinatorial chemistry, the use of automation to synthesize large numbers of related compounds that can then be screened for desired properties and quantitative structure activity relationships, or QSARs, which can guide the synthesis of promising pharmaceuticals by suggesting which compounds are the most promising to investigate.

While there is still a popular perception of the chemist as a man in a white coat working at a laboratory bench, nowadays chemists can be of either gender, may work with computers in an office or library on data from space or the deep ocean, or derive equations at a blackboard.

The Central Science and its Borders: Chemistry and Biology

A watershed in the biological sciences was reached when James D. Watson (1938–) and Francis Harry Compton Crick (1916–2004) published the double helix structure of DNA in the journal *Nature* in 1955. It had long been established that DNA was the genetic material of eukaryotic cells, that it somehow encoded the structure of the molecules that made each organism distinct, and that the DNA structure itself could be duplicated. The boundary between biochemistry and molecular biology began to disappear. A number of leading chemistry departments, including at Harvard University, have renamed themselves as departments of chemistry and chemical biology.

Chemistry, International Politics, Schools, and Standards

Chemistry lends itself to school instruction, as there is much that can be learned in a high school or college laboratory, without major expense. In the United States, the notion that a well-educated high school graduate should have a year-long course in biology, followed by a year of chemistry, and another year of physics, was proposed by the "Committee of Ten" appointed by the National Education Association and including the presidents of Harvard University and the University of Michigan. Their report, issued in 1892, has continued to influence American secondary education up to the present time.

As World War II and the Manhattan project led to the first nuclear explosions in 1945, the role of American colleges and universities in scientific research began to change drastically. The ideas set forth by Vannevar Bush, head of the Wartime Office of research and development, in *Science: the Endless Frontier*, prepared at the request of President Franklin D. Roosevelt, led to the establishment of the National Science Foundation in 1950, with the mission of supporting all non-medical scientific research and development. Universities responded by erecting new buildings, hiring additional faculty, and establishing PhD programs and postdoctoral fellowships. For a time it seemed that the United States had become the leading nation in science and technology and would retain that position indefinitely.

Things changed quite suddenly on October 4, 1957, when the Soviet Union launched Sputnik I, the first successful earth-orbiting satellite. The American public was, frankly, frightened that the same space vehicles capable of launching a satellite might be used to deliver a nuclear bomb. To better prepare American science students, the National Science Foundation was assigned the additional mission of upgrading secondary science instruction, by establishing summer courses for science and math teachers, special summer opportunities for gifted students, and a series of fellowships and traineeships that made it possible for students to devote their full time to graduate study. Special curricula were developed in biology, chemistry, and physics, all of which were still influenced by the 1892 report of the Committee of Ten.

At the time of the first Moon landing in 1969, the US had regained its leadership position. By 1976, however, the government found that the apparatus set up to train scientists had worked *too* well; rumors of scientists who could not find a job in the sciences began to spread. Scientific careers lost their luster compared to those in finance. American educators began to question their approach to teaching science and, thus, a quest for science education standards began. Trying to establish standards for science education has proven difficult. Standards proposed by the National Science Teachers' Association have stressed the interrelatedness of the sciences and the scientific method, as opposed to the need to teach specific scientific facts. The latest version of those standards, the *Next Generation Science Standards*, has been our guide for the selection of topics for this volume.

Donald R. Franceschetti, PhD

Further Reading

Brock, W. H. *The Chemical Tree: A History of Chemistry.* New York: Norton, 2000. Print.

Cobb, Cathy, and Harold Goldwhite. *Creations of Fire: Chemistry's Lively History from Alchemy to the Atomic Age.* New York: Plenum, 1995. Print.

Emsley, John. "The Pieces of the Periodic Table Nature's Building Blocks: An A-Z Guide to the Elements." *Journal of Industrial Ecology* 13 (n.d.): n. pag. 154-155. Web.

Hoffmann, Roald. *The Same and Not the Same.* New York: Columbia UP, 1995. Print.

Ihde, Aaron J. *The Development of Modern Chemistry.* New York: Harper & Row, 1964. Print.

National Research Council. *National Science Education Standards.* S.l.: National Academy, 1996. Print.

Next Generation Science Standards: For States, by States. Washington, D.C.: National Academies, 2013. Print.

Quadbeck-Seeger, Hans-Jürgen, and Gunther Schulz. *The Periodic Table through History Who Charted the Elements?* Weinheim: Wiley-VCH, 2007. Print.

Contributors

Donald R. Fransceschetti, PhD
University of Memphis

Marianne M. Madsen, MS
University of Utah

Richard M. Renneboog, MSc
University of Western Ontario

Principles of Chemistry

A

ACID ANHYDRIDES

FIELDS OF STUDY

Organic Chemistry

SUMMARY

The characteristic properties and reactions of acid anhydrides are discussed. Acid anhydrides are useful compounds in organic synthesis reactions, helping to form complex molecular structures from simple starting materials.

PRINCIPAL TERMS

- **acyl group:** a functional group with the formula –RCO, where R is connected by a single bond to the carbon atom of the carbonyl (C=O) group.
- **carboxylic acid:** an organic compound containing a carboxyl functional group and having the general formula RC(=O)OH.
- **functional group:** a specific group of atoms with a characteristic structure and corresponding chemical behavior within a molecule.
- **organic acid:** an acid derived from an organic compound.
- **R (generic placeholder):** a symbol used primarily in organic chemistry to represent a hydrocarbon side chain or other unspecified group of atoms in a molecule; can be used specifically for an alkyl group, with Ar used to represent an aryl group.

The Nature of Acid Anhydrides

The term "anhydride" indicates the absence of either water or its components, oxygen and hydrogen, from a molecular structure. Acid anhydrides are compounds in which a bond has been formed between two acid functional groups by the virtual elimination of –H from one acid group and –OH from the other. Most common acid anhydrides are formed from carboxylic acids, but they can also be formed from any organic acid and many inorganic acids. Mixed acid anhydrides are also possible and are, in fact, essential components of certain biochemical cycles. Acid anhydrides can also be thought of as esters formed from an acyl group. Generally, acid anhydrides are represented as follows:

$$R-C(=O)-O-C(=O)-R \qquad Ar-C(=O)-O-C(=O)-Ar$$

The generic placeholders R– and Ar– indicate an alkyl group substituent and an aryl group substituent, respectively.

Acid anhydrides of linear molecules are generally used to add single substituents to a molecular structure. Acid anhydride compounds with a cyclic structure are also commonly used in synthetic procedures to elaborate a molecular structure. In principle, an anhydride can be formed from any carboxylic acid. However, in practice, very few acid anhydrides are available, and other carboxylic-acid derivatives are more readily obtained for particular purposes. Thus, linear carboxylic acids are seldom prepared or utilized. Dicarboxylic acids, in which the formation of the anhydride produces a five- or six-membered ring structure, are more readily converted to their anhydride form and are generally more useful in synthetic methods.

The most common acid anhydride is acetic anhydride. Because of its relative ease of handling and because it does not release hydrogen chloride (HCl) gas as a product of the reaction, it is the preferred reagent for forming acetate esters. In pure form, acetic acid tends to absorb atmospheric moisture. It is a common laboratory practice to add a small amount of acetic anhydride to pure acetic acid to counteract this effect, as any moisture consumed by the hydrolysis of acetic anhydride simply produces more acetic acid, which would not be a contaminant in the reagent.

ACID ANHYDRIDES

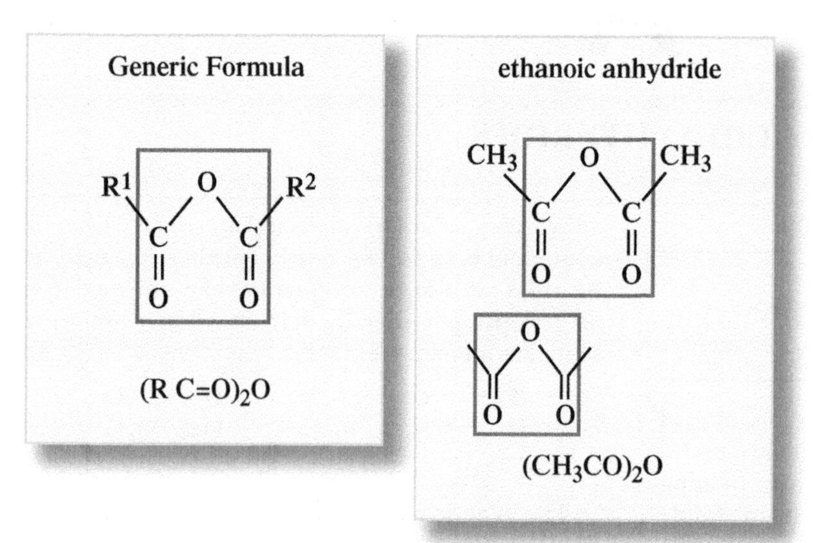

Formation of Acid Anhydrides

Simple acid anhydrides are never formed by dehydration reactions. Acetic anhydride is produced commercially by the reaction of acetic acid with ethenone in the gas phase, according to the following equation:

$$H_3C{-}C({=}O){-}OH + H_2C{=}C{=}O \longrightarrow H_3C{-}C({=}O){-}O{-}C({=}O){-}CH_3$$

The anhydrides of larger linear carboxylic acids are more difficult to prepare, but in some cases, they can be formed by heating the acid in the presence of a strong dehydrating agent, such as phosphorus pentoxide (P_2O_5), and separating the anhydride from the reaction mixture as it forms. Cyclic anhydrides such as maleic anhydride and succinic anhydride form readily when their parent acids are heated. The process is assisted by the removal of water from the reaction as it is formed, which prevents the reverse reaction from occurring. Ring size is vitally important for this process: five- and six-membered rings are well suited to the geometric constraints of the electron orbitals in carbon and oxygen atoms, while seven-membered rings are more difficult to form because of the strain that the bond angles of such a ring impose on the molecule. Adipic acid, or 1,6-hexanedioic acid, does not form the corresponding cyclic anhydride; instead, it eliminates a molecule of both carbon dioxide and water to produce the less strained compound cyclopentanone.

Reactions of Acid Anhydrides

Acid anhydrides are reactive compounds widely used for the formation of esters. By far the most common commercial use of acetic anhydride is in the formation of acetate esters of carbohydrate compounds such as sugars, starches, and celluloses. The carbonyl (C=O) groups of an acid anhydride readily interact with nucleophiles, which are chemical species that are attracted to positively charged species. This happens partly because the electronegative oxygen atoms of the carbonyl groups tend to draw electron density away from the carbon atoms, causing the carbon to behave as though it were positively charged, and partly because the way the bonds are arranged around the carbon atoms makes it easier for nucleophiles to approach and interact. After the nucleophile is added and the leaving group is eliminated, the carbonyl groups re-form in the new molecule. This same basic mechanism permits acid anhydrides to be converted into different acid derivatives. Reaction with water reconverts the anhydride into its acid form, which allows other types of reactions to occur after an anhydride group has been added to a substrate molecule. For example, maleic anhydride is commonly used in a Diels-Alder reaction, which is a type of reaction that results in the addition of a six-membered ring structure to the substrate molecule. Hydrolysis of the resulting molecule produces two carboxylic acid functional groups (–COOH) that can be modified by other reactions to produce a desired product.

Acid anhydrides also react well with ammonia to produce the corresponding amide of one carboxylic acid group and the ammonium salt of the other. The reaction of acetic anhydride with ammonia, for example, would produce acetamide and ammonium acetate:

$$H_3C{-}C({=}O){-}O{-}C({=}O){-}CH_3 + 2NH_3 \longrightarrow H_3C{-}C({=}O){-}NH_2 + H_3C{-}C({=}O){-}O^-NH_4^+$$

A reaction with a substituted amine such as ethylamine would produce the corresponding substituted amide compound.

Anhydrides are also able to act as acylating agents in the Friedel-Crafts acylation of benzene and benzene-like aromatic compounds. One half of the anhydride function bonds chemically to the aromatic ring as an acyl substituent, while the other half is reconverted into the acid. The reaction is typically catalyzed by a Lewis acid, such as aluminum trichloride ($AlCl_3$) or ferric chloride ($FeCl_3$). Phthalic anhydride, for example, can be used to acylate benzene, producing the corresponding benzoic acid derivative, according to the following equation:

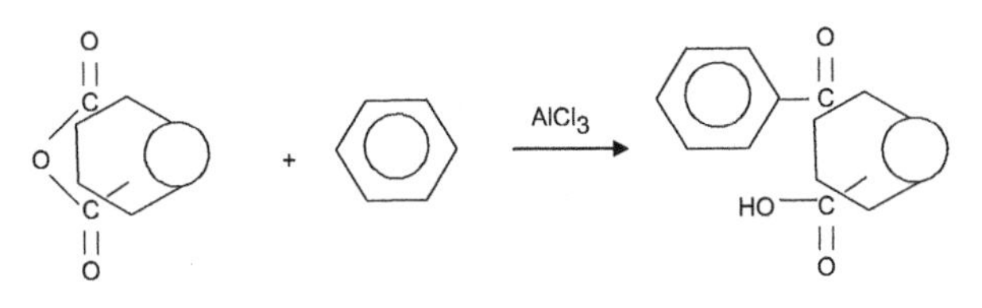

Diels-Alder, Friedel-Crafts, and similar reactions are useful in organic synthetic chemistry because they allow the chemist to easily and quickly build a complex molecular structure from very simple starting materials. The reactions typically produce hydrogen chloride as well, but this is an inconsequential by-product and is seldom, if ever, shown in structural formula equations.

SAMPLE PROBLEM

Given the chemical name isobutyric anhydride, draw the skeletal structure of the molecule.

Answer:

The name indicates that the material is the anhydride of isobutyric acid, which has the structure

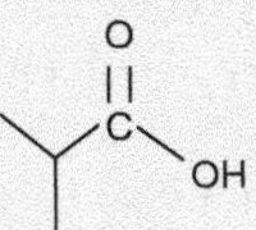

Isobutyric anhydride must therefore have the structure

NOMENCLATURE OF ACID ANHYDRIDES

Acid anhydrides are named according to the parent acid from which they are formed, simply by adding "anhydride" to the name of the parent acid. Accordingly, the anhydride of butyric acid would be named butyric anhydride, while maleic acid and succinic acid readily form maleic anhydride and succinic anhydride. Mixed anhydrides, though uncommon, are also known. For a mixed anhydride, the names of the two parent acids are used, with the "higher" acid being named in second place. For example, the mixed anhydride of acetic acid and phosphoric acid would be named acetic phosphoric anhydride. Such mixed anhydrides sometimes appear in enzyme-mediated biochemical processes.

Richard M. Renneboog, MSc

BIBLIOGRAPHY

Hendrickson, James B., Donald J. Cram, and George S. Hammond. *Organic Chemistry.* 3rd ed. New York: McGraw, 1970. Print.

Jones, Mark Martin. *Chemistry and Society.* 5th ed. Philadelphia: Saunders College Pub., 1987. Print.

Morrison, Robert Thornton, and Robert Neilson. Boyd. *Organic Chemistry.* 7th ed. Englewood Cliffs, N.J.: Prentice Hall, 2003. Print.

Wuts, Peter G. M., and Theodora W. Greene. *Greene's Protective Groups in Organic Synthesis.* 4th ed. Hoboken: Wiley-Interscience, 2007. Print.

ACID CHLORIDES

FIELDS OF STUDY

Organic Chemistry; Inorganic Chemistry

SUMMARY

The characteristic properties and reactions of acid chlorides are discussed. Acid chlorides are useful compounds in both organic and inorganic synthesis reactions for the relatively easy formation of complex molecular structures from simple starting materials.

PRINCIPAL TERMS

- **acid derivative:** a compound formed by modifying the structure of an acid, such as an ester created by the reaction between a carboxylic acid and an alcohol.
- **functional group:** a specific group of atoms having a characteristic structure and corresponding chemical behavior within a molecule.
- **nucleophile:** a chemical species that tends to react with positively charged or electron-poor species.
- **reactivity:** the propensity of a chemical species to undergo a reaction under applied conditions.
- **R (generic placeholder):** a symbol used primarily in organic chemistry to represent a hydrocarbon side chain or other unspecified group of atoms in a molecule; can be used specifically for an alkyl group, with Ar used to represent an aryl group.

THE NATURE OF ACID CHLORIDES

The acid chlorides, also known as acyl chlorides, are characterized by the presence of a carbonyl functional group (C=O) connected by a single bond to a chlorine atom. Acid chlorides are acid derivatives of either alkyl or aryl carboxylic acids, and in fact the acid chloride functional group (–COCl) is essentially a carboxyl group (–COOH) in which the hydroxyl group (–OH) has been replaced by chlorine. The general formula of an acid chloride shows the acid chloride group attached to either an R– or an Ar– placeholder, where R indicates an alkyl group (a basic hydrocarbon) and Ar represents an aryl group (a hydrocarbon derived from an aromatic ring):

Acid chlorides are the most reactive derivatives of carboxylic acids. This relatively high reactivity and the ease with which acid chlorides can be prepared make them extremely versatile reagents in many applications. They are especially prone to reactions with nucleophiles, which typically replace the chlorine atom with another atom or group.

NOMENCLATURE OF ACID CHLORIDES

ACID CHLORIDE

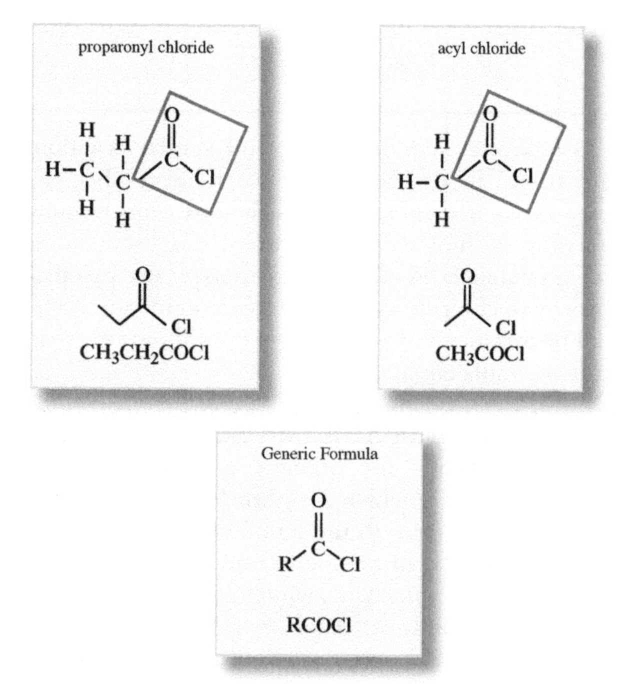

Acid chlorides are named by replacing the "-ic acid" ending of the parent acid name with "-yl chloride." The acid chloride of benzoic acid, for example, is benzoyl chloride, while the acid chloride of oxalic acid is oxalyl chloride. The naming convention for acid chlorides of more complex acids, in which other functional groups take priority, uses the prefix "chlorocarbonyl-" ahead of the proper name of the parent acid.

Preparation of Acid Chlorides

Acid chlorides can be readily prepared from essentially any carboxylic acid. The standard method of preparation is via reaction with thionyl chloride ($SOCl_2$), phosphorus trichloride (PCl_3), or phosphorus pentachloride (PCl_5). The reaction with thionyl chloride is generally preferred, as it is easier to obtain the acid chloride in a pure form using this reactant than with either of the phosphorus-based chlorinating agents. Any volatile thionyl chloride can be easily removed by distillation, whereas the phosphorus-based by-products are reactive solids that are not as readily eliminated from the reaction mixture.

The inorganic chlorinating agents used to prepare acid chlorides are themselves acid chlorides of their corresponding acids. Thionyl chloride is the acid chloride of sulfurous acid (H_2SO_3), while phosphorus trichloride and phosphorus pentachloride are acid chlorides of phosphoric acid (H_3PO_4).

SAMPLE PROBLEM

Given the chemical name 7-methyloctanoyl chloride, draw the skeletal structure of the molecule.

Answer:

The name indicates that the compound is a derivative of 7-methyloctanoic acid, which has the following structure:

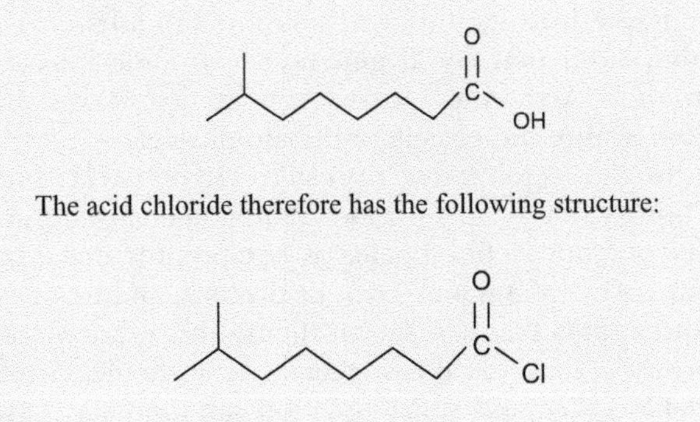

The acid chloride therefore has the following structure:

Reactions of Acid Chlorides

Acid chlorides are the reagents of choice for acylation reactions, which add an acyl group (RCO–) to a compound. The reaction of an alcohol with an acid chloride can replace the hydrogen atom of the alcohol's hydroxyl group with the acyl group to form the corresponding ester (RCOOR'), with hydrogen chloride (HCl) as a by-product. Similarly, the reaction of an acid chloride with ammonia (NH_3) or an amine will readily form the corresponding amide compounds. For example, the reaction of octanoyl chloride ($C_7H_{15}COCl$), a derivative of octanoic acid, with dimethylamine (CH_3NHCH_3) produces *N,N*-dimethyloctanamide ($C_7H_{15}CON(CH_3)_2$), where the *N,N*- prefix indicates that the two methyl groups (–CH_3) are attached to the nitrogen atom.

Perhaps the most important use of acid chlorides is in the formation of ketones by the acylation of benzene and benzene-like aromatic compounds via Friedel-Crafts acylation reactions, which use a Lewis acid such as aluminum trichloride ($AlCl_3$) or ferric chloride ($FeCl_3$) to catalyze the reaction. The reaction between propanoyl chloride (C_2H_5COCl) and benzene (C_6H_6), for example, produces the ketone propiophenone, officially named 1-phenyl-1-propanone and also known as ethyl phenyl ketone, according to the following diagram:

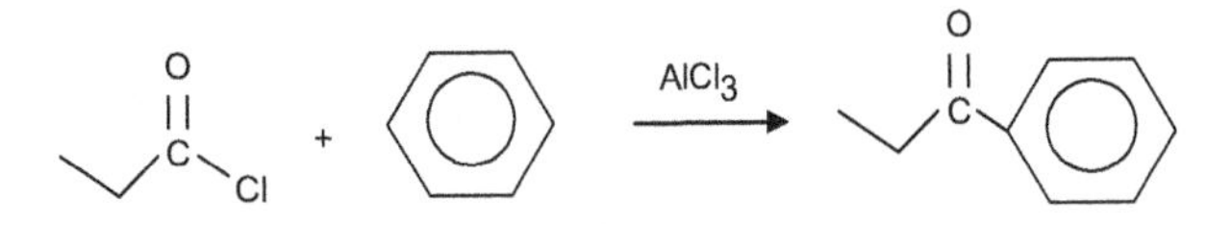

The by-product of this reaction is again hydrogen chloride, which is usually given off as a gas. The by-products of such reactions are not a consideration and are seldom, if ever, included in the equations.

Richard M. Renneboog, MSc

Bibliography

Hendrickson, James B., Donald J. Cram, and George S. Hammond. *Organic Chemistry*. 3rd ed. New York: McGraw, 1970. Print.

Jones, Mark Martin. *Chemistry and Society*. 5th ed. Philadelphia: Saunders College Pub., 1987. Print.

Morrison, Robert Thornton, and Robert Neilson. Boyd. *Organic Chemistry*. 7th ed. Englewood Cliffs, N.J.: Prentice Hall, 2003. Print.

Wuts, Peter G. M., and Theodora W. Greene. *Greene's Protective Groups in Organic Synthesis*. 4th ed. Hoboken: Wiley-Interscience, 2007. Print.

ACIDS AND BASES

FIELDS OF STUDY

Physical Chemistry; Biochemistry; Chemical Engineering

SUMMARY

The characteristics of acids and bases are discussed, and the progressive theories of Svante Arrhenius, Johannes Brønsted and Thomas Lowry, and Gilbert N. Lewis are described. The relationship of pH and acidity or basicity, and the technique of titration are presented in detail.

PRINCIPAL TERMS

- **Brønsted-Lowry acid-base theory:** definitions of acids and bases developed separately in 1923 by Danish chemist Johannes Nicolaus Brønsted and English chemist Martin Lowry; defines an acid as any compound that can release a hydrogen ion and a base as any compound that can accept a hydrogen ion.
- **dissociation:** the separation of a compound into simpler components.
- **equilibrium constant:** a numerical value characteristic of a particular equilibrium reaction, defined as the ratio of the equilibrium concentration of the products to that of the reactants..
- **Lewis acid-base theory:** definitions of acids and bases developed in 1923 by American chemist Gilbert N. Lewis; defines an acid as any chemical species that can accept an electron pair and a base as any chemical species that can donate an electron pair.
- **pH:** a numerical value that represents the acidity or basicity of a solution, with 0 being the most acidic, 14 being the most basic, and 7 being neutral.
- **protonation:** the addition of a proton, in the form of a hydrogen cation (H^+), to an atom, ion, or molecule.

THE NATURE OF ACIDS AND BASES

Acidity and basicity are concepts that refer to both the tactile nature of the respective solutions and the nature of their chemical activity. Acidic solutions, such as vinegar and lemon juice, typically have a distinctly sour taste, an irritating smell, and a "sticky" feel between the fingers. Basic solutions, such as ammonia-based window cleaners and household bleach, typically have a bitter taste, pungent smell, and a "slippery-oily" feel between the fingers. While tasting, inhaling, and touching acidic and basic solutions is most definitely not recommended, historically this is precisely how early chemists experienced different materials before there was any workable theory of atomic structure to explain the properties and reactions of different compounds. In everyday life, some such experiences are unavoidable, especially with regard to foods. Indeed, many food flavors depend on the acidity or basicity of the particular condiment or other foodstuff. One cannot, for example, make pickles without vinegar (acetic acid) or fluffy pancakes without baking soda (sodium bicarbonate).

The essential differences between acidic and basic solutions have been recognized for centuries but could not be explained adequately before an understandable and workable model of atomic structure. In the 1880s, Svante Arrhenius (1859–1927) recognized that salts dissociate into charged ions in solution with water. In 1887, long before subatomic particles such as protons were discovered and identified, he defined an acid as a compound that could generate hydrogen ions in solution and a base as a compound that could generate hydroxide ions in solution. Arrhenius's theory was the first to explain both acidity and basicity at the atomic level.

In 1923, Johannes Brønsted (1879–1947) and Thomas Lowry (1874–1936) simultaneously and independently defined acids as compounds that can donate hydrogen ions (H^+), or protons, and bases as compounds that can accept them. This is known as Brønsted-Lowry acid-base theory. Accordingly, a base that has accepted a hydrogen ion can then act as an acid by protonating another compound, and an acid that has donated a hydrogen ion can act as a base by accepting a hydrogen ion from another compound. This fact is the basis of conjugate acids and bases.

Also in 1923, American chemist Gilbert N. Lewis (1875–1946) published a more comprehensive definition of acids and bases that applied to both organic

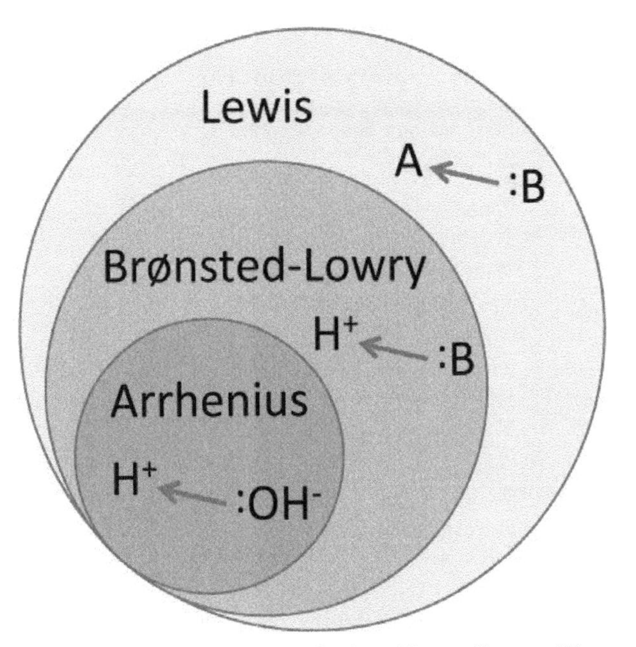

The relationship between Arrhenius, Lewis, and Brønsted-Lowry acid-base theories. By Tem5psu via Wikimedia Commons

and inorganic chemistry. According to Lewis's definition, a base is a compound that has a lone pair of electrons that can be used to complete the valence group of another atom, and an acid is a compound that can use a lone pair of electrons from another atom to complete its own valence group. This Lewis acid-base theory was a direct result of Lewis's earlier work in valence theory and did much to unify the science of chemistry by providing a set of theoretical principles common to both organic and inorganic chemistry.

ACIDS, BASES, WATER, AND PH

Water is unique in its molecular structure, physical and chemical properties, and interaction with other materials. The water molecule has two hydrogen atoms bonded to an oxygen atom. As an element in the second period of group 16 of the periodic table, the oxygen atom has six electrons in its valence shell. The bonds to the two hydrogen atoms give the oxygen atom a filled valence-shell electron configuration of eight. According to the valence-shell electron-pair repulsion (VSEPR) principle, the 2*s* and 2*p* orbitals that contain those eight electrons, arranged in four pairs, hybridize to form four equivalent sp^3 hybrid atomic orbitals oriented to the four apexes of a tetrahedron. This arrangement of atomic orbitals arises from the force of electrostatic repulsion experienced between the pairs of like-charged electrons. Two of these orbitals form the bonds to the hydrogen atoms; the other two orbitals are not involved in bonds, and each contains a pair of free electrons, often called a lone pair. This imparts a high degree of polarity to the water molecule, as the lone pairs create a region of high negative charge. In addition, the electrons from the hydrogen atoms are localized in the oxygen-hydrogen bonds, exposing the positive charge of the hydrogen nuclei. This charge separation in the molecule is known as a dipole moment. Water molecules can therefore act somewhat like bar magnets and stick to each other, positive charge to negative charge. The cohesion between the water molecules that results from dipole-dipole interaction gives water an extraordinarily high boiling point relative to its molecular weight.

In liquid water, water molecules undergo dissociation as described by the autolysis equilibrium equation:

$$2H_2O \rightleftharpoons H_3O^+ + OH^-$$

The concentration of both the acid hydronium ions (H_3O^+) and the base hydroxide ions (OH^-) is exactly equal, as required by the balanced reaction equation, and liquid water is neither acidic nor basic but neutral. Experimental analysis has determined that the concentration of both is 10^{-7} moles per liter, or molars (M). The equilibrium dissociation constant for the reaction is determined as:

$$pK_{eq} = [H_3O^+][OH^-] = (10^{-7})(10^{-7}) = 10^{-14}$$

where [*x*] is the concentration of *x* in molars. Since this is an equilibrium reaction, addition of any material that increases the concentration of one ion must decrease the concentration of the other ion while maintaining the value of the equilibrium constant (K_{eq}) at 10^{-14}. This is the foundational principle of the pH scale.

The pH of an aqueous solution is defined by the following equation:

$$pH = -\log[H_3O^+]$$

(The notation $[H^+]$ is often used instead of $[H_3O^+]$, though the latter form is more accurate.) The pH scale is a logarithmic scale, such that every increase of one unit of pH corresponds to a tenfold change in the concentration of hydronium ions. The use of logarithmic values greatly simplifies calculations of both acid and base concentrations. The logarithm, or log, of a value is the exponent, or power, to which a base number must be raised to produce that value. For example, the log of 10^3 is 3.

The ease of using logs comes from the power rules of mathematics. When multiplying two numbers, their respective powers are added together. For example, 100 multiplied by 1,000 is 100,000; in scientific notation, this is written as $10^2 \times 10^3 = 10^5$. Observe that the sum of the first two exponent values, 2 and 3, is equal to the third exponent value, 5. Conversely, when dividing two values, their exponents are subtracted from each other. Returning to the autolysis equilibrium of water, the logarithmic expression of the equation for the equilibrium constant of the process is as follows:

$$\log(10^{-14}) = \log(10^{-7})(10^{-7}) = (-7) + (-7)$$

By using the p-scale relationship, this becomes:

$$-\log(10^{-14}) = -\log(10^{-7})(10^{-7}) = -[(-7) + (-7)]$$

$$pK_{eq} = p[H_3O^+] + p[OH^-] = 7 + 7 = 14$$

Since the concentration of hydronium and hydroxide ions in water is in equilibrium, the pH and the pOH must always add up to the neutral value of 14. Adding any material that changes the concentration of either ion causes the concentration of the other ion to shift to compensate and restore the equilibrium condition. Thus, adding an acid material such as hydrochloric acid (HCl) increases the hydronium concentration and forces the hydroxide concentration to decrease by the same amount to maintain the equilibrium constant. Similarly, adding sodium hydroxide (NaOH) increases the hydroxide concentration and causes the hydronium concentration to decrease by the same amount.

TITRATION

Titration is an extremely useful technique for determining the concentration of a solution by comparing it to another solution of known concentration. The process entails a single, very controlled reaction that has a well-defined end point. First, a standard reference compound is weighed accurately and made up to a solution of precise volume. This solution, called the titrant, is then used to carefully neutralize a sample of an unknown solution of approximately the same concentration, called the analyte. The amount of titrant used to attain the precise end point of the acid-base neutralization reaction, along with its concentration, reveals the precise concentration of the analyte. Since most acid and base solutions are clear and colorless, the pH of the reacting solution is closely monitored either electronically or with an

SAMPLE PROBLEM

Rank the following aqueous solutions in order of increasing acidity:

- vinegar, pOH = 10.6
- beer, pH = 4.6
- grapefruit juice, pH = 3.2
- 0.05 M oxalic acid, pOH = 12.4
- 1 M sodium hydroxide, pOH = 0
- 1 M hydrochloric acid, pH = 0.1
- 0.01 M potassium hydroxide, $[H_3O^+] = 10^{-12}$ M
- blood plasma, $[H_3O^+] = 3.9 \times 10^{-8}$ M

Answer:

Using the relationship pH + pOH = 14, one can solve for the pH of a substance by subtracting its known pOH from 14. Thus, the pH of vinegar is 3.4, the pH of the oxalic acid solution is 1.6, and the pH of the sodium hydroxide solution is 14.

Using the definition of pH as $-\log[H_3O^+]$, the pH of the potassium hydroxide solution is 12, and the pH of blood plasma is 7.4.

Since the more acidic a solution is, the lower its pH is, then the above solutions are ranked from most acidic to least acidic as:

1. 1 M hydrochloric acid, pH = 0.1
2. 0.05 M oxalic acid, pH = 1.6
3. grapefruit juice, pH = 3.2
4. vinegar, pH = 3.4
5. beer, pH = 4.6
6. blood plasma, pH = 7.4
7. 0.01 M potassium hydroxide, pH = 12
8. 1 M sodium hydroxide, pH = 14

indicator compound that changes color at a specific pH. The titrant is added to the analyte using a burette, a cylinder marked with a precise volumetric scale. The end of the burette is fitted with a stopcock and dropping tip, allowing the user to strictly control the amount of titrant that is meted out.

Richard M. Renneboog, MSc

Bibliography

Askeland, Donald R., Wendelin J. Wright, D. K. Bhattacharya, and Raj P. Chhabra. *The Science and Engineering of Materials.* Boston: Cengage Learning, 2016. Print.

Jones, Mark Martin. *Chemistry and Society.* 5th ed. Philadelphia: Saunders College Pub., 1987. Print.

Laidler, Keith J. *Chemical Kinetics.* 3rd ed. New York: Harper, 1987. Print.

Myers, Richard. *The Basics of Chemistry.* Westport: Greenwood, 2003. Print.

Skoog, Douglas Arvid, Donald M. West, and F. James. Holler. *Skoog and West's Fundamentals of Analytical Chemistry.* 9th ed. Belmont (Calif.): Cengage Learning, 2014. Print.

Vieira, Ernest R. *Elementary Food Science.* 4th ed. Gaithersburg: Aspen, 1999. Print.

ACIDS AND BASES: BRØNSTED-LOWRY THEORY

FIELDS OF STUDY

Physical Chemistry; Inorganic Chemistry; Chemical Engineering

SUMMARY

The basic properties of acids and bases according to the Brønsted-Lowry theory are elaborated. Acidity and basicity (or alkalinity) are measured using the pH scale, which charts the concentration of H^+ ions relative to pure water.

PRINCIPAL TERMS

- **amphoterism:** the ability of a compound to act as either an acid or a base, depending on its environment and the other materials present.
- **conjugate pair:** an acid and the conjugate base that is formed when it donates a proton, or a base and the conjugate acid that is formed when it accepts a proton.
- **hydronium ion:** a polyatomic ion with the formula H_3O^+, formed by the addition of the hydrogen cation (H^+) to a molecule of water; also called oxonium (IUPAC preference) or hydroxonium.
- **proton acceptor:** a compound or part of a chemical compound that has the ability to accept a proton (H^+) from a suitably acidic material in a chemical reaction.
- **proton donor:** a compound or part of a chemical compound that has the ability to relinquish a proton (H^+) to a suitably basic material in a chemical reaction.

Johannes Nicolaus Brønsted, a Danish physical chemist.
Peter Elfelt via Wikimedia Commons

Thomas Martin Lowry
Anonymous via Wikimedia Commons

The Nature of the Brønsted-Lowry Acids and Bases

In 1923, chemists Johannes Nicolaus Brønsted (1879–1947) and Thomas Martin Lowry (1874–1936) defined acids and bases in a straightforward way. Simply stated, an acid, a proton donor, is any compound that can release a proton from its molecular structure:

$$AH \rightarrow A^- + H^+$$

A base is a proton acceptor, any compound that can accept a proton from another compound:

$$B + H^+ \rightarrow BH^+$$

Acids that act in this way are termed "protic" acids, since the H^+ cation is identically a proton. For most practical purposes, this definition is sufficient, as most water-soluble acids are either protic or amphiprotic compounds. (Amphoterism is the ability of a compound to act as either an acid or a base, depending on its environment and the other materials present.)

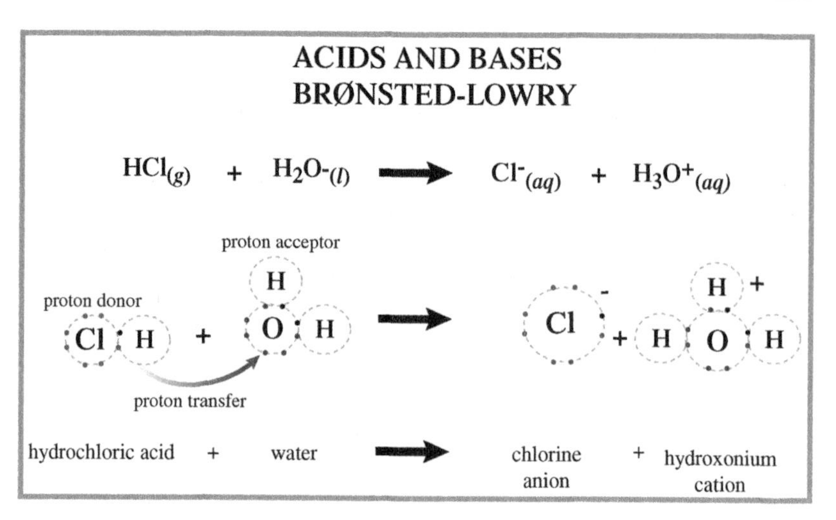

The Brønsted-Lowry definition refines those formulated before the development of the modern atomic theory. In 1838, Justus von Liebig (1803–73) described acids as materials containing hydrogen that could be replaced by a metal, which indeed they are. The acidic hydrogen atom on compounds such as the inorganic sulfuric acid and the organic acetic acid is easily replaced by metal ions such as magnesium to form magnesium sulfate ($MgSO_4$) or sodium to form sodium acetate ($C_2H_3NaO_2$). The corresponding equations are as follows:

$$Mg^{2+} + H_2SO_4 \rightarrow MgSO_4 + 2H^+$$

and

$$Na^+ + CH_3COOH \rightarrow CH_3COONa + H^+$$

Svante Arrhenius (1859–1927) first elucidated the definition of acids and bases as specifically forming hydrogen ions and hydroxide ions in aqueous solutions. This statement seems to limit the identity of acids and bases to protic acids and hydroxide salts. The Brønsted-Lowry definition, however, recognizes that acidic and alkalinic behaviors are complementary actions of the same fundamental molecular structure. That is, a compound that releases a proton in acting as an acid produces a corresponding anion as its conjugate base. An acid and its conjugate base are a conjugate pair and do not exist in isolation from each other in an equilibrium system. Accordingly, both the acid and base equations above should be properly written as follows:

$$AH \rightleftharpoons A^- + H^+$$

and

$$B + H^+ \rightleftharpoons BH^+$$

These equations more clearly demonstrate the conjugate relationship of the various components, and they include such compounds as ammonia, which do not dissociate into ions in solution. When dissolved in water, ammonia accepts a proton from the water molecules, according to the following equation:

$$NH_3 + H_2O \rightleftharpoons NH_4^+ + OH^-$$

and so acts as a base without itself dissociating to release a hydroxide ion, as would be required by the Arrhenius definition.

MEASURING ACIDS AND BASES: THE pH SCALE

Equal molar quantities of different acids do not produce equally acidic solutions. The conjugate nature of the components of an acid dissociation equilibrium is an important factor in determining the acidity of the resulting solution. Hydrogen chloride and nitric acid, for example, are completely dissociated into ions in aqueous solutions, but acetic acid and other organic acids generally do not dissociate completely in solution. The resulting solutions all

SAMPLE PROBLEM

The material corresponding to the chemical formula AlH_3O_3 is an amphoteric compound and can act as either the base aluminum hydroxide, $Al(OH)_3$, or the acid aluminic acid, H_3AlO_3, both of which have the same molecular structure. Write the equations describing both modes of activity and identify the conjugate pairs in each case.

Answer:

As a base, the reaction releases OH^- from the compound by cleaving the Al–O bond, as

$$Al(OH)_3 \rightarrow Al(OH)_2^+ + OH^-$$

The chemical species $Al(OH)_2^+$ is the conjugate acid of the base $Al(OH)_3$.

As an acid, the reaction releases a proton by cleaving an O–H bond rather than an Al–O bond, as

$$Al(OH)_3 \rightarrow {}^-OAl(OH)_2 + H^+$$

The chemical species $^-OAl(OH)_2$ is the conjugate base of the acid $Al(OH)_3$.

have different concentrations of H^+ ions and, thus, different acidities. As the most common material on the planet, water is the standard by which H^+ ion concentrations are measured. The structure of the water molecule allows it to dissociate into its component ions in solution without the participation of a second material (a process known as "autolysis"), according to the following equation:

$$H_2O \rightleftharpoons H^+ + OH^-$$

In pure water, this equilibrium produces equal quantities of both H^+ and OH^-. The quantity of each has been determined experimentally to be 10^{-7} moles per liter (M). Solutions in which the concentration of H^+, $[H^+]$, is greater than 10^{-7}M are acidic, while those in which $[H^+]$ is less than 10^{-7}M are alkaline. The pH scale was developed as a simple means of communicating the $[H^+]$ of an aqueous solution. The pH of an aqueous solution is defined as follows:

$$pH = -\log[H^+]$$

Because the $[H^+]$ of pure water is 10^{-7}M, the pH of pure water is defined as +7. A solution of HCl in water is hydrochloric acid, characterized by the complete dissociation of HCl into H^+ and Cl^- ions. Because hydrochloric acid has an [HCl] of 0.01M, it also has a $[H^+]$ of 0.01M, or 10^{-2}M. Therefore, the corresponding pH is +2. Similarly, a solution in which the $[H^+]$ is 10^{-9}M, for example, has a pH of +9.

A complementary scale, the pOH scale, can be used for alkaline solutions, based on the following identical relationship:

$$pOH = -\log[OH^-]$$

Since the concentration of H_2O in pure water is always a strictly constant value, the relationship $[H^+][OH^-]$ must also be as strictly constant. In pure water, this p-scale value is the equilibrium constant of the autolysis reaction, with the value $(10^{-7})(10^{-7})$, or 10^{-14}. Therefore, in any aqueous solution the product $[H^+][OH^-]$ must also always be 10^{-14}. If the $[H^+]$ is greater than 10^{-7}M, the equilibrium of the autolysis reaction requires that the $[OH^-]$ be reduced by the corresponding amount. (When two numbers are multiplied together, their logarithmic values combine.) Thus, the value of the autolysis equilibrium is 7 + 7 = 14 (the pH and the pOH of a solution must always add up to 14). Normally, aqueous acid solutions have pH between 1 and 7, corresponding to $[H^+]$ in the range of 10^{-1} to 10^{-7}. A pH with a value of 0 or less is also possible. A 10M solution of HCl, for example, has $[H^+]$ of 10^1M and a corresponding pH of –1.

ACID-BASE NEUTRALIZATION REACTIONS

The goal of a neutralization reaction is to bring the pH of a particular solution to the neutral pH of 7. The product of any neutralization reaction is a salt. This is easily seen in the neutralization reaction of hydrochloric acid, HCl, with sodium hydroxide, NaOH:

$$HCl + NaOH \rightarrow H_2O + NaCl$$

By adding hydrochloric acid to a solution of sodium hydroxide to the point at which the pH is 7, one ends up with a solution of plain "salt water," NaCl in H_2O. This process is called "titration" and is a common laboratory method for determining the precise $[H^+]$ of a solution. The technique can also be used to analyze many other materials by determining the end point or equivalence point of a specific reaction. The technique depends on a means of monitoring the change in pH of the solution as acid or base is added. This has traditionally been done by using litmus paper or another indicator material that changes color according to the pH, but this method has been superseded by electronic devices that can measure the pH directly and display the measured values in graphic form.

Richard M. Renneboog, MSc

Bibliography

Daniels, Farrington, and Robert A. Alberty. *Physical Chemistry*. 3rd ed. New York: Wiley, 1966. Print.

Douglas, Bodie E., Darl H. McDaniel, and John J. Alexander. *Concepts and Models of Inorganic Chemistry*. 3rd ed. New York: Wiley, 1994. Print.

Laidler, Keith J. *Chemical Kinetics*. 3rd ed. New York: Harper, 1987. Print.

Myers, Richard. *The Basics of Chemistry*. Westport: Greenwood, 2003. Print.

Skoog, Douglas Arvid, Donald M. West, and F. James. Holler. *Skoog and West's Fundamentals of Analytical Chemistry*. 9th ed. Belmont (Calif.): Cengage Learning, 2014. Print.

ACTINIDES

FIELDS OF STUDY

Inorganic Chemistry; Nuclear Chemistry

SUMMARY

The basic properties and characteristics of the actinide elements are presented. The actinides are the elements with atomic numbers from 89 (actinium) to 103 (lawrencium) and are characterized by having an incomplete 5*f* shell as well as electrons in the 6*d* and 7*s* shells. All actinides are radioactive.

PRINCIPAL TERMS

- **f-block:** the portion of the periodic table containing the elements with incompletely filled orbitals; includes all lanthanides (excluding lutetium) and actinides (excluding lawrencium).
- **oxidation state:** a number that indicates the degree to which an atom or ion in a chemical compound has been oxidized or reduced.
- **pyrophoricity:** a property of some solids and liquids that causes them to spontaneously combust when exposed to air.
- **radioactivity:** the emission of subatomic particles due to the spontaneous decay of an unstable atomic nucleus, the process ending with the formation of a stable atomic nucleus of lower mass.

The Nature of the Actinides

All of the actinides reside in the f-block of the periodic table, though it is a matter of some debate whether one of the actinides, either actinium or lawrencium, should be considered a d-block element instead. The f-block elements are characterized by having their valence electrons in the *f* orbitals, with some of the outer *d*-orbitals being incompletely filled. This arrangement of electrons is responsible for the multivalent nature of actinides, enabling them to form compounds in which they have taken on different oxidation states; in other words, the atoms can lose or gain different numbers of electrons in order to form chemical bonds. Actinides characteristically lose electrons in order to form positively charged ions, or cations.

All of the actinides are scarce metals, with only thorium and uranium, and to a lesser extent plutonium, qualifying as primordial elements, meaning they have existed in the universe since before the earth was formed. All other actinides are trace elements that either occur in nature through radioactive decay or were synthesized in the laboratory through nuclear reactions in high-energy particle accelerators. All of the actinides are characterized by radioactivity. They also all display pyrophoricity, meaning that fine particles or shavings of them ignite on contact with air, appearing as a spark.

ACTINIDES

Atomic Structure of the Actinides

The *f* orbitals are presently the highest atomic orbitals in the known elements of the periodic table. When the atoms ionize, however, the 6*d* and 7*s* electrons are removed first. At this energy level, the 5*f*, 6*d*, and 7*s* levels are very similar in energy, and their distance from the atomic nucleus means that they have relatively low ionization energies and therefore easily form ions. This is because the farther an electron orbital is from the nucleus, the less force the positively charged protons exert on the negatively charged electrons to hold them in place.

The radioactive nature of the actinides and the scarcity of many of them make the study of these elements very difficult, and little is known about their chemical behavior. Of those that have been studied in some detail, the atoms have been observed to be multivalent, with oxidation states ranging between +2 and +7. The +3 oxidation state is most common overall, having been observed in all of the actinides, but it is not always the most stable.

Actinide Compounds

Of the actinide elements, only uranium and thorium occur in sufficient abundance to be economically useful. They are mined and smelted most commonly from the ore known as pitchblende (uraninite). Uranium occurs naturally in the form of various radioactive isotopes, or radioisotopes, with uranium-238 and uranium-235 being the most common. The small difference in atomic mass between the two isotopes is sufficient to permit their separation by physical means. Uranium forms various compounds with fluorine, all of which convert when heated to

SAMPLE PROBLEM

Use the chart below, which shows the order in which electron orbitals are filled, to determine the electron distribution of uranium (atomic number 92) and lawrencium (atomic number 103). Follow the arrows from top to bottom, keeping in mind that *s* orbitals can hold a maximum of two electrons, *p* orbitals hold six electrons, *d* orbitals hold ten electrons, and *f* orbitals hold fourteen electrons. Which electrons are the valence electrons?

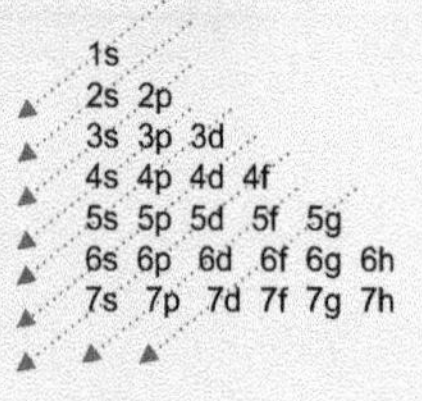

Answer:

The total number of electrons in each element is equal to the atomic number of the element. Thus, uranium has ninety-two electrons, ordered as follows:

$$1s^2 2s^2 2p^6 3s^2 3p^6 4s^2 3d^{10} 4p^6 5s^2 4d^{10} 5p^6 6s^2 4f^{14} 5d^{10} 6p^6 7s^2 5f^3 6d^1$$

In this notation, the exponents tell how many electrons are in each orbital. One would expect the 5*f* orbital to fill completely before any electrons appear in the 6*d* orbital, but this is not the case for many actinides, which is what gives them the ability to take different oxidation states. Because the 5*f* and 6*d* orbitals are so far from the nucleus, at least on a subatomic scale, the difference in energy levels required to occupy each one is much smaller than in lower orbitals, so electrons can easily move between them as needed to make the atom more stable. There is no simple way to determine the exact distribution of valence electrons in these higher orbitals without additional information.

Uranium has six valence electrons, in the 5*f*, 6*d*, and 7*s* orbitals. Atoms form ions in order to attain stable electron configurations, which are represented by the electron distributions of the noble gases (helium, neon, argon, krypton, xenon, and radon). The electron distributions of all elements apart from hydrogen can also be conveyed by putting the symbol of the noble gas immediately preceding the element in brackets, followed by the extra electrons not contained in the noble gas. For example, the last noble gas before uranium in the periodic table is radon, which has an electron configuration of $1s^2 2s^2 2p^6 3s^2 3p^6 4s^2 3d^{10} 4p^6 5s^2 4d^{10} 5p^6 6s^2 4f^{14} 5d^{10} 6p^6$, so the electron configuration of uranium can be written as $[Rn]7s^2 5f^3 6d^1$. For elements in the f-block, the valence electrons are the ones outside the noble gas configuration, as these are the electrons the uranium atom would have to lose to achieve a stable electron configuration.

Lawrencium has 103 electrons, ordered as follows:

$$1s^2 2s^2 2p^6 3s^2 3p^6 4s^2 3d^{10} 4p^6 5s^2 4d^{10} 5p^6 6s^2 4f^{14} 5d^{10} 6p^6 7s^2 5f^{14} 6d^1$$

Unlike uranium, lawrencium follows the expected pattern of filling the 5*f* orbital completely before starting to fill the 6*d* orbital. It has seventeen valence electrons, again in the 5*f*, 6*d*, and 7*s* orbitals.

the compound uranium hexafluoride, UF_6, which becomes a gas at 56.5 degrees Celsius (134 degrees Fahrenheit). Using differential diffusion in gas centrifuges, the uranium hexafluoride becomes enriched with the lighter isotope uranium-235. The enriched metal can then be recovered from the cooled "yellowcake" and used as the fuel of nuclear reactors. Further enrichment to increase the proportion of uranium-235 produces so-called weapons-grade uranium, capable of undergoing the nuclear-fission chain reaction that produces a nuclear explosion.

In a nuclear reactor, a controlled fission process is maintained in which the nuclei of the uranium-235 atoms decay rapidly. The energy released is used indirectly to produce steam to drive turbines for electrical-power generation. The process gradually depletes the amount of uranium-235 in the fuel rods until they are spent. At that point the rods must be replaced. The spent rods are still highly radioactive, however, and safe storage is problematic, as the natural half-life of uranium-235 is more than seven hundred million years.

One particular type of nuclear reactor, called a breeder reactor, uses both uranium-238 and thorium-232 in a fission process that results in the formation of the isotopes plutonium-239 and plutonium-233. The advantage of breeder reactors is that they can generate more fissionable fuel material than they consume. On the other hand, plutonium is far more toxic than uranium, and plutonium-239's half-life of around twenty-four thousand years poses a very serious environmental hazard.

ACTINIDES IN DAILY LIFE

The best-known applications of actinides are in nuclear power and nuclear weapons, which typically use uranium, plutonium, or to a lesser extent thorium. However, some of the rarer elements in this group, although highly radioactive and toxic, have a few practical uses as well. Some actinides are used in alloys and for components in high-end scientific instrumentation. Actinium (atomic number 89) and californium (atomic number 98) have uses in cancer treatment. Americium (atomic number 95) is perhaps the only actinide commonly found in households, as it is a component of some kinds of smoke detectors. Tiny amounts of curium (atomic number 96) have been used to power cardiac pacemakers. Most of the other actinides have no known practical use outside of scientific research.

Richard M. Renneboog, MSc

BIBLIOGRAPHY

Douglas, Bodie E., Darl H. McDaniel, and John J. Alexander. *Concepts and Models of Inorganic Chemistry.* 3rd ed. New York: Wiley, 1994. Print.

Johnson, Rebecca L. *Atomic Structure.* Minneapolis: Lerner, 2008. Print.

Jones, Mark Martin. *Chemistry and Society.* 5th ed. Philadelphia: Saunders College Pub., 1987. Print.

Kean, Sam. *The Disappearing Spoon: And Other True Tales of Madness, Love, and the History of the World from the Periodic Table of the Elements.* New York: Little, Brown, 2010. Print.

Mackay, K. M., R. A. Mackay, and W. Henderson. *Introduction to Modern Inorganic Chemistry.* 6th ed. Cheltenham: Nelson, 2002. Print.

Miessler, Gary L., Paul J. Fischer, and Donald A. Tarr. *Inorganic Chemistry.* 5th ed. Upper Saddle River: Prentice Hall, 2013. Print.

Wehr, M. Russell, James A. Richards Jr., and Thomas W. Adair III. *Physics of the Atom.* 4th ed. Reading: Addison-Wesley, 1984.Print.

Winter, Mark J. *The Orbitron: A Gallery of Atomic Orbitals and Molecular Orbitals.* University of Sheffield, n.d. Web.

ACTIVATION ENERGY

FIELDS OF STUDY

Physical Chemistry; Inorganic Chemistry; Biochemistry

SUMMARY

The activation energy of a process is defined, and its importance in chemical processes is elaborated. Activation energy is a widely variable quantity in different reactions but is nevertheless characteristic of any specific reaction process.

PRINCIPAL TERMS

- **Arrhenius equation:** a mathematical function that relates the rate of a reaction to the energy required to initiate the reaction and the absolute temperature at which it is carried out.
- **catalyst:** a chemical species that initiates or speeds up a chemical reaction but is not itself consumed in the reaction.
- **chemical reaction:** a process in which the molecules of two or more chemical species interact with each other in a way that causes the electrons in the bonds between atoms to be rearranged,

resulting in changes to the chemical identities of the materials.

- **reaction rate:** how much of a particular reaction or reaction step occurs per unit time.
- **transition state:** an unstable structure formed during a chemical reaction at the peak of its potential energy that cannot be isolated and ultimately breaks down, either forming the products of the reaction or reverting back to the original reactants.

ACTIVATION ENERGY IN CHEMICAL REACTIONS

Activation energy can be thought of as a barrier that the reactants in a chemical reaction must overcome if the reaction is to proceed to the formation of products. The molecules that are involved must rearrange to form either a transition state or an intermediate that is higher in energy than the starting materials. An intermediate is a stable chemical structure formed during a reaction process that can often be captured and isolated by chemical means. Once this structure is formed, the reaction process can either progress to form products or revert back to the original reactants.

Activation energy is the energy required for a specific chemical reaction to occur. In a reaction, two reactant molecules contact each other with the energy of their ambient states. (The ambient state of a material can be thought of as its "default" state—the state it takes at one atmosphere of pressure and what is commonly considered to be room temperature.) In the case of a spontaneous reaction, the energy of the collision is sufficient to initiate the formation of the transition state or intermediate. In a nonspontaneous reaction, the energy of the molecular collision is not sufficient, and the two molecules will not interact. The input of some additional energy is required to drive the two molecules together so that the transition state or intermediate is formed and the reaction can proceed. The energy released in the transformation of reactants into products is generally sufficient to drive the reactions of other molecules in the reaction mixture.

Another way to look at activation energy is to think of it as the minimum amount of energy that two interacting molecules must gain in order to weaken bonds between atoms in both molecules so that those bonds can be rearranged. Since chemical reactions are essentially processes of breaking and making bonds, having sufficient energy to overcome the strength of the appropriate bonds is essential if there is to be any reaction between the two molecules.

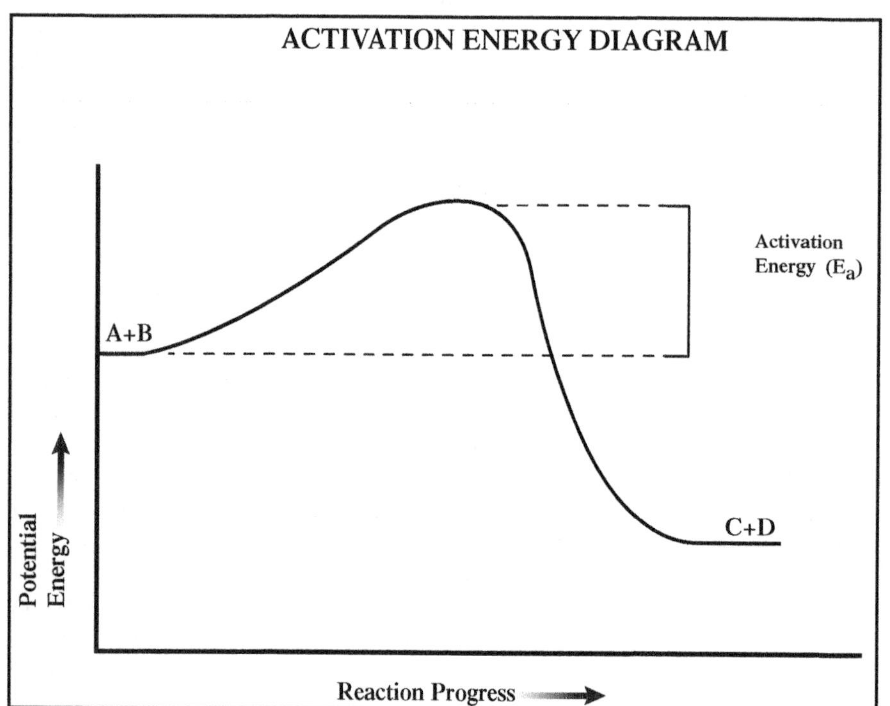

ACTIVATION ENERGY AND REACTION RATES

Reaction rates can be related directly to their activation energies. This relationship is defined by the Arrhenius equation, formulated in 1884 by the Swedish scientist Svante Arrhenius (1859–1927), who received the Nobel Prize in Chemistry in 1903. The Arrhenius equation relates the rate constant of a reaction to its activation energy and the absolute temperature and has the following form:

$$k = Ae^{-E/RT}$$

where k is the rate constant for the reaction or process; A is the pre-exponential factor, also known in some cases as the frequency factor; E (or E_a) is the activation energy for the reaction or process; R is the gas constant; T is the absolute temperature; and the mathematical constant e is the base of the natural logarithm, so that the natural logarithm (ln) of e is equal to 1. The Arrhenius equation has been found to apply not only to chemical reactions but to physical processes as well. The relationship can be most clearly seen by plotting experimentally determined logarithmic values of k against the inverse of the absolute temperature, $1/T$. This results in a straight line plot, from which the activation energy of the reaction or process can be calculated.

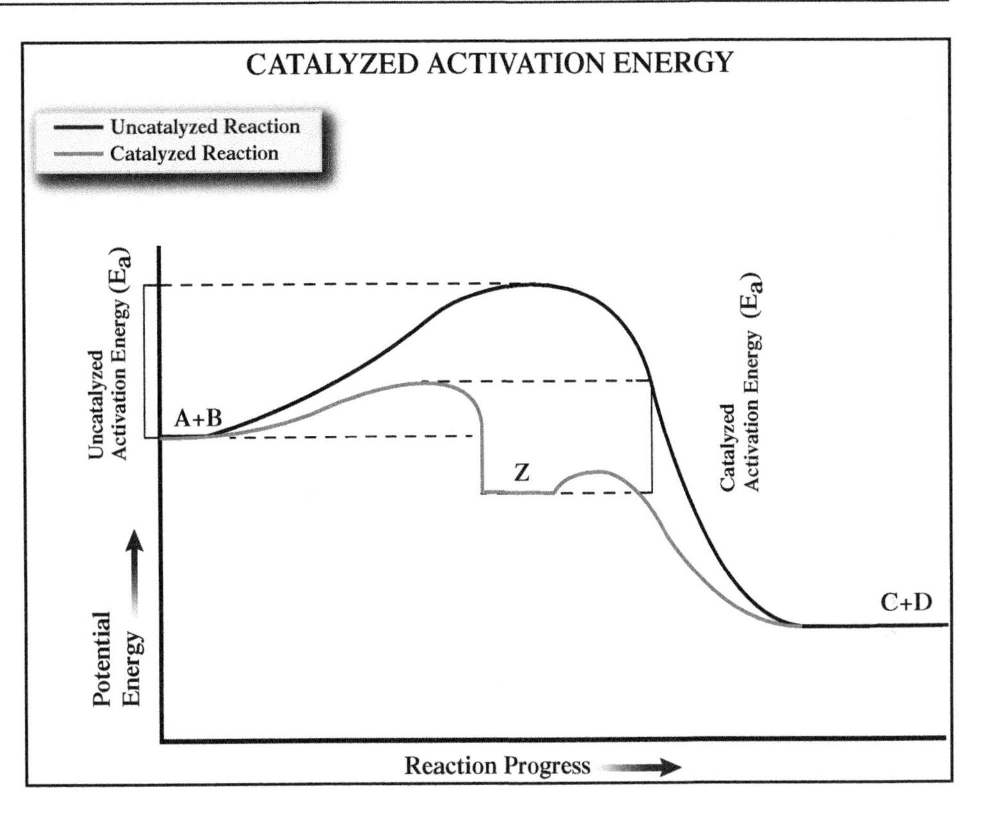

The pre-exponential factor A is identified as the value that the specific rate constant k would have if the activation energy E were zero (a spontaneous reaction). In that special case, the exponent $-E/RT$ would also be equal to zero, making $e^{-E/RT}$ equal 1 and thus causing the value of k to be equal to A. For different specific reactions, the value of A ranges over several orders of magnitude, but the rate constant k is determined almost solely by the value of $e^{-E/RT}$, which can range over several hundred orders of magnitude, depending on the relative values of E and T.

The course of a reaction depends on the relative difference between the energy of the reactants and that of the products. The greater this difference is, the more impetus there is for the reaction to proceed to the formation of products. This is typically illustrated in a plot of energy versus the reaction coordinate, a symbolic representation of the progress of a reaction. In the plot, the energy level of the reactants, on the left, is either higher or lower than that of the products, on the right. In between, a curved line rises from the energy level of the reactants to a maximum value before falling to the energy level of the products. The difference between the energy level of the reactants and this peak energy value represents the

SAMPLE PROBLEM

Use the Arrhenius equation to determine the activation energy at 0°C for a reaction having a specific rate constant (k) of 0.023 moles per liter per second and a pre-exponential factor (A) of 2,303 moles per liter per second. Use the gas constant

$$R = 8.314 \frac{\text{J}}{\text{mol K}}$$

continued on next page

Answer:

Convert the temperature from degrees Celsius to kelvins, given that K = °C + 273.15:

$$K = 0 + 273.15 = 273.15$$

The Arrhenius equation is

$$k = Ae{-}E/RT$$

Rearrange the equation using natural logarithmic (ln) relationships:

$$\ln k = \ln A + \ln(e^{-E/RT})$$

$$\ln k = \ln A - \frac{E}{RT}$$

$$\frac{E}{RT} = \ln A - \ln k$$

$$E = RT(\ln A - \ln k)$$

Substitute in the values of R $(8.314\frac{\text{J}}{\text{mol K}})$, T (temperature), A (pre-exponential factor), and k (rate constant). Calculate, paying attention to the units throughout:

$$E = RT(\ln A - \ln k)$$

$$E = (8.314\frac{\text{J}}{\text{mol K}} \times 273.15\text{ K})(\ln 2303 - \ln 0.023)$$

$$E = (8.314\frac{\text{J}}{\text{mol K}} \times 273.15\text{ K})[7.742 - (-3.772)]$$

$$E = 26147.938\frac{\text{J}}{\text{mol}}$$

The activation energy of the reaction is 26147.938 J/mol, or 26.148 kJ/mol. Note that the logarithmic values of A and k have no units.

activation energy for the reaction, while the difference between the energy levels of the reactants and the products represents the energy released in the reaction, also called its enthalpy.

The specific rate of any individual reaction is determined by its activation energy. However, in mass quantities, the energy differences between reactants and products in the system also play a role. This can be understood by considering the Boltzmann fraction $e{-}E/RT$, which describes the fraction of molecules in the system having energy greater than E. As energy is released from several reactions, the fraction of molecules present at any given time with sufficient energy to react increases, and more reactions can occur in any given time period. Each reaction requires the same activation energy, and the amount of energy that is available in the system to permit reactions to occur may be anything from "barely enough" to "excessive."

The activation energy of a reaction can be greatly reduced by the inclusion of a catalyst, a material that takes part in the reaction mechanism but is not consumed in the reaction. Catalysts function by forming an activated complex with the reactants, typically constraining the reactant molecules in an orientation that they would otherwise have to achieve through collision with each other. This reduces the energy necessary to achieve that particular orientation—that is, the activation energy—so that the reaction can proceed. When the reactant molecules are constrained in the activated complex with the catalyst, bonds between certain atoms are weakened and the orientations of atomic and molecular orbitals that must interact are often brought into the proper alignment, or trajectory, for the new bonds to form between atoms.

Activation Energy in Action

The activation energy of a reaction can range from exceedingly small to very large. Two examples serve to illustrate this point. For the first, consider the addition of two parts hydrogen gas (H_2) to one part oxygen gas (O_2). This is an explosive mixture of gases, yet the two mixed gases are quite happy to coexist quietly in the same container, no matter how much is present. The introduction of an initiator such as an electrical spark, however, results in an almost instantaneous reaction to form water (H_2O), accompanied by the release of a great deal of energy. The activation energy of the reaction between hydrogen and oxygen is very low, and the amount of energy released by one reaction is more than sufficient to drive many instances of the reaction in the gas mixture, with each subsequent occurrence releasing an equal amount of energy as the enthalpy of reaction.

The second example is the so-called thermite reaction, in which iron oxide and aluminum metal react to produce aluminum oxide and iron metal. This is a spectacular reaction often demonstrated for chemistry exhibitions. Because the activation energy of the thermite reaction is very high, the reaction is very difficult to initiate and must typically be ignited by a burning piece of magnesium metal; once it has begun, however, it is essentially impossible to stop it due to the amount of energy that is released. Typically, the iron metal falls out of the reaction mixture as a white-hot liquid.

Activation Energy in Biological Systems

Activation energy applies to biochemical processes as well as to physical processes. The chirping of crickets, for example, is dependent on temperature in a manner that is in complete accord with the Arrhenius equation. In biological systems, the activation energy of processes is made a great deal lower by the catalytic action of protein molecules called enzymes. Enzymes have well-defined three-dimensional structural shapes that allow them to coordinate with other molecules in specific ways, rather like the way a key works with a lock. The coordination normally alters the three-dimensional shape of the substrate molecule or otherwise interacts with it so that specific bonds are weakened and the molecular geometry is changed such that reaction is highly favored.

Richard M. Renneboog, MSc

Bibliography

Arnaut, Luís, Sebastião Formosinho, and Hugh Burrows. *Chemical Kinetics: From Molecular Structure to Chemical Reactivity.* Oxford: Elsevier, 2007. Print.

Bell, Jerry A. *Chemistry: A Project of the American Chemical Society.* New York: Freeman, 2005. Print.

Lafferty, Peter, and Julian Rowe, eds. *The Hutchinson Dictionary of Science.* 2nd ed. Oxford: Helicon, 1998. Print.

Masterton, William L., Cecile N. Hurley, and Edward Neth. *Chemistry: Principles and Reactions.* 7th ed. Belmont: Brooks, 2012. Print.

Raymond, Kenneth W. *General, Organic, and Biological Chemistry: An Integrated Approach.* 4th ed. Hoboken: Wiley, 2014. Print.

Zumdahl, Steven S., and Susan Zumdahl. *Chemistry.* 9th ed. Belmont: Brooks Coles, 2013. Print.

ACTIVE TRANSPORT

FIELDS OF STUDY

Biochemistry; Molecular Biology; Genetics

SUMMARY

The process of active transport is defined, and its importance in biochemical processes is elaborated. Active transport is an essential feature of the biochemistry of living systems and helps maintain the necessary concentrations of various biochemical components and electrolytes for the proper functioning of cellular metabolism.

PRINCIPAL TERMS

- **adenosine triphosphate (ATP):** a molecule consisting of adenine, ribose, and a triphosphate chain that is used to transfer the energy needed to carry out numerous cellular processes.
- **cell membrane:** a biological membrane that forms a semipermeable barrier separating the interior of a cell from the exterior.
- **concentration gradient:** the gradual change in the concentration of solutes in a solution across a specific distance.
- **diffusion:** the process by which different particles, such as atoms and molecules, gradually become intermingled due to random motion caused by thermal energy.
- **passive transport:** the passage of materials through a membrane with no input of energy required.

THE MECHANICS OF ACTIVE TRANSPORT

In living cells, biochemical processes transport materials necessary for a properly functioning metabolism through cell membranes. Passive transport does not require an input of energy to move materials across cell walls because it operates in the same direction as the concentration gradient, moving the materials from an area of high pressure to one of low pressure. Active transport can be thought of as a "shuttle service" for ions and other polar materials that cannot pass through a cell membrane by diffusion, a kind of passive transport. Instead, those entities must be physically transported across membranes by various mechanisms collectively termed pumps. A pump is a type of mediated transport system that functions to conduct ions, amino acids, glucose, and other polar compounds through the nonionic lipid bilayer, the highly nonpolar material that makes up the cell wall. Pumps always work against the concentration gradient to move materials out of regions of low concentration and into regions of higher concentration, using energy derived from biochemical reactions. The transported material is subsequently used in other biochemical reactions that return the energy used during transport.

CELL WALLS AND LIPID BILAYERS

Long-chain fatty acids are organic molecules whose molecular structure consists of a single hydrocarbon chain terminated by a carboxylic acid functional group (–COOH). The carboxyl group is highly polar and hydrophilic, while the hydrocarbon moiety, or portion, of the molecule is very nonpolar and hydrophobic. Carboxylic acids are converted to esters by enzyme-mediated reactions with alcohols. In an ester, the carboxyl functional group retains the highly polar character that it had in its free carboxylic acid form, giving the long-chain esters, called lipids, a polar-nonpolar structure similar to that of the free carboxylic acids. When carboxylic acids are esterified with glycerol, which has three hydroxyl (–OH) functional groups, the resulting triesters are called triglycerides. Lipids and triglycerides are the principal forms in which long-chain fatty acids are found in biological systems.

The hydrocarbon chains and the carboxyl-based portions of fatty acids and their esters do not interact with each other due to their different hydrophilicities—that is, the degrees to which they attract and interact with water and other polar molecules—but they are quite capable of interacting with the corresponding portions of other molecules. The hydrocarbon chains associate preferentially with each other, as do the carboxyl portions. The basic structure of the lipid bilayer results from the hydrocarbon portions of the acids of two layers of such molecules intermingling and essentially dissolving each other. The carboxyl functions on the other ends of the hydrocarbon chains thus form two hydrophilic surfaces, one on either side of the very hydrophobic interior layer. The resulting structure is a lipid bilayer.

The walls of all animal cells are formed of lipid bilayers, allowing them to interact with water-based fluids while isolating the sensitive materials and processes that take place within each cell. The fluid inside of each cell is also water based, which necessitates some means of transporting vital polar materials from the exterior of the cell to the interior and moving extraneous materials and metabolites in the opposite direction for elimination. This movement is accomplished by active transport.

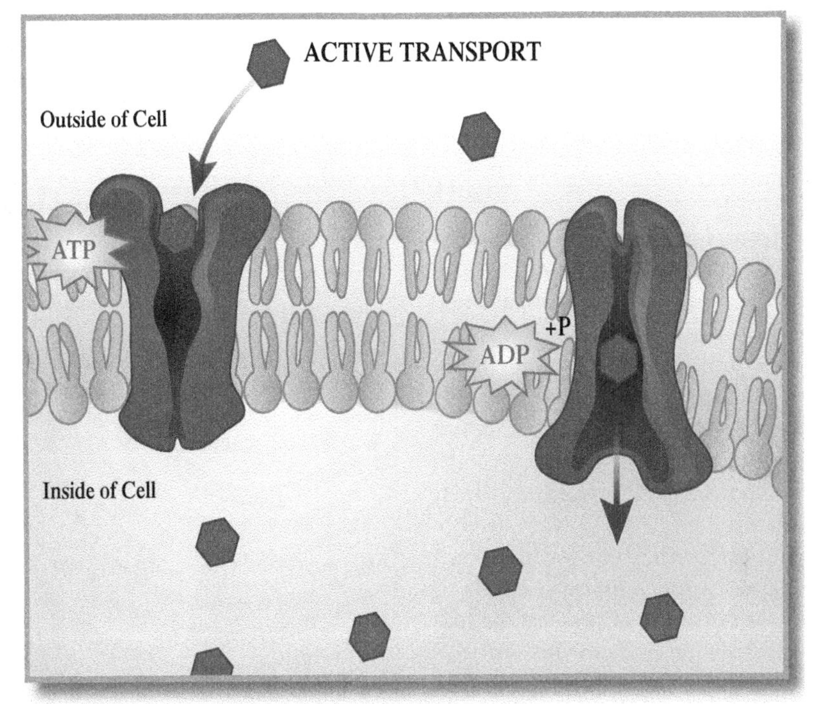

Functions of Active-Transport Systems

Active-transport systems serve a variety of functions in the biochemistry of living systems. Their principal function is to allow the organism to extract "fuels" and other essential materials for use in the metabolic functions that occur within cells. This is a very important function, and the nature of active transport allows cells to retain a relatively high concentration of such materials even when their concentrations outside of the cell are quite low. A second important function of active-transport systems is to regulate and maintain the organism's metabolic steady state, a balanced state in which the material and energy that the organism removes from its environment through living functions is equal to the energy and materials that it returns to the environment through those same functions. The biochemical processes of metabolism use energy and materials taken from the environment. Anabolic processes remove materials from the environment and use energy from reactions involving those materials to build and support the life of the organism. Catabolic processes remove used materials from the organism and return them to the environment, releasing the energy stored in those materials.

Active transport maintains a constant optimal amount of various inorganic elements within the living cells of an organism. Potassium ions, for example, are essential to the proper functioning of many intracellular processes. An active-transport system produces potassium-ion channels in cell walls of nerves and muscles, including the cardiac muscles. Potassium ions are delivered into the cytoplasm of the cell via these channels to replace ejected sodium ions, thus maintaining a constant ionic concentration within the cell. The system maintains a relatively high concentration of potassium ions in most aerobic cells, between 100 and 150 millimolars (mM), whether they are plant, animal, or microbial in nature and regardless of the concentration outside of the cells. (A 1 mM solution has a concentration of 0.001 moles per liter.) The potassium ions that are pumped into the cell also serve to maintain the electric potential across the cell membrane, a factor that affects the free-energy change in reactions involved in active-transport systems.

Active Transport in Action

The transfer of ions across a membrane or against a concentration gradient by active transport is accompanied by a free-energy change (ΔG) that can be calculated by one of two equations. The first equation represents the free-energy change for the transfer of neutral materials against a concentration gradient. This is described by the following equations:

$$\Delta G = RT \ln \frac{c_2}{c_1}$$

$$8.314 \frac{\text{J}}{\text{mol K}}$$

where R is the gas constant and T is the absolute temperature in kelvins, ln is the natural logarithm function, and c_1 and c_2 are concentrations on either side of the membrane in molars, or moles per liter (M), with c_2 being greater than c_1.

The second equation, which represents the free-energy change for the transfer of electrically charged materials, needs to account for the charge on the material being transported and the difference in electric potential across the membrane. The latter is determined by the neutral nature of the lipid bilayer, which causes it to act as a capacitor, or energy-storage device, and the presence of charge as maintained by the potassium ions in the cytosol. The free-energy expression for the transport of charged species across a cell membrane is given by the following equation:

$$\Delta G = RT \ln \frac{c_2}{c_1} + ZF\Delta\Psi$$

where Z is the charge on the ion, F is the Faraday constant (96,485.3365 coulombs per mole, the electric charge on one mole of electrons), and is the difference in electric potential across the membrane in volts.

ATP and Active Transport

The energy used in active-transport systems is obtained through enzyme-mediated reactions of adenosine triphosphate (ATP). ATP molecules consist of a molecule of the nucleobase adenine that is bonded to a molecule of ribose sugar, which in turn is bonded to a triphosphate ion. A magnesium ion coordinates and stabilizes the second and third segments of the triphosphate moiety. Energy is derived from the structure by the enzymatic cleavage of the

SAMPLE PROBLEM

Use the free-energy equation for active transport against a concentration gradient to determine the free energy associated with transporting neutral amino-acid molecules across a membrane from a concentration of 20 μM to one of 43 μM. Assume normal body temperature of 37°C. Use $R = 8.314 \frac{\text{J}}{\text{mol K}}$

Answer:

The materials being transported are electrically neutral. Therefore, use the equation

$$\Delta G = RT \ln \frac{c_2}{c_1}$$

Convert the temperature from °C to K:

$$\text{K} = {}^\circ\text{C} + 273.15$$

$$\text{K} = 37 + 273.15 = 310.15$$

Convert the concentration values from micromolars to molars:

$$c_1 = 20\ \mu\text{M} = 20 \times 10^{-6}\ \text{M} = 0.00002\ \text{M}$$

$$c_2 = 43\ \mu\text{M} = 43 \times 10^{-6}\ \text{M} = 0.000043\ \text{M}$$

Substitute in the values of R, T, c_1, and c_2 and calculate, paying attention to the units throughout:

$$\Delta G = RT \ln \frac{c_2}{c_1}$$

$$\Delta G = (8.314 \frac{\text{J}}{\text{mol K}})(310.15\ \text{K}) \ln \frac{0.000043}{0.00002}$$

$$\Delta G = 1973.8 \frac{\text{J}}{\text{mol}}$$

The free energy of active transport of neutral amino acids across a concentration gradient from 20 μM to 43 μM is 1973.8 joules per mole, or 1.9738 kilojoules per mole.

third phosphate segment from the triphosphate moiety, transforming the molecule into adensosine diphosphate (ADP), and it is restored by concatenating, or joining, a third phosphate ion to ADP to re-form ATP.

The function of muscle cells depends on the active transport of calcium ions and sodium ions, a process termed the calcium ion pump or Ca^{2+} pump. The calcium ion pump works in an organelle of muscle cells called the sarcoplasmic reticulum and is powered by ATP hydrolysis reactions mediated by the enzyme calcium adenosine triphosphatase. This process is critical to the contraction and relaxation of muscle fibers, especially heart muscles. The sarcoplasmic reticulum is a cell structure that stores and releases calcium ions to aid in this contraction and relaxation. In muscle cells, the rapid release of calcium ions from the sarcoplasmic reticulum into the cytosol, the cellular fluid outside of the organelles, triggers contraction of the muscle, while rapid removal of calcium ions from the cytosol and back into the sarcoplasmic reticulum triggers relaxation of the muscle. The normal concentration of free calcium ions in the cytosol is between 0.1 and 0.2 micromolar (μM, or 10^{-6} moles per liter), increasing when the muscle contracts and returning to the normal value when it relaxes.

Richard M. Renneboog, MSc

BIBLIOGRAPHY

Lafferty, Peter, and Julian Rowe, eds. *The Hutchinson Dictionary of Science.* 2nd ed. Oxford: Helicon, 1998. Print.

Lehninger, Albert L. *Biochemistry: The Molecular Basis of Cell Structure and Function.* 2nd ed. New York: Worth, 1975. Print.

Lodish, Harvey, et al. *Molecular Cell Biology.* 7th ed. New York: Freeman, 2013. Print.

Pelczar, Michael J., Jr., E. C. S. Chan, and Noel R. Krieg. *Microbiology: Concepts and Applications.* New York: McGraw, 1993. Print.

Reece, Jane B., et al. *Campbell Biology.* 10th ed. San Francisco: Cummings, 2013. Print.

ACYL CATION

FIELDS OF STUDY

Organic Chemistry; Biochemistry

SUMMARY

The basic structure of acyl cations is defined, and their importance in organic chemical synthesis is described. An acyl cation is formed when the hydroxyl (–OH) group of the carboxylic acid functional group is lost as the hydroxide ion. They are important reagents in Friedel-Crafts acylation reactions.

PRINCIPAL TERMS

- **acylation:** a reaction process by which an acyl group is added to a compound.
- **cation:** any chemical species bearing a net positive electrical charge, which causes it to be drawn toward the negative pole, or cathode, of an electrochemical cell.
- **electrophilic addition:** an addition reaction in which an electrophile, typically a positively charged chemical species, is bonded to a nucleophile, a molecule with a free pair of electrons that can be easily donated; results in the breaking of a multiple bond to form two single bonds.
- **octet rule:** the tendency of atoms when bonding to either accept or donate electrons so as to arrive at eight electrons in the outermost electron shell.
- **reaction mechanism:** the sequence of electron and orbital interactions that occurs during a chemical reaction as chemical bonds are broken, made, and rearranged.

THE NATURE OF THE ACYL CATIONS

An acyl cation (CO^+) is a carbonyl functional group that has a positive charge due to separation from a hydroxide ion (OH^-), usually in a carboxylic acid (a hydrocarbon containing the –RCOOH group). The loss of the hydroxide portion leaves the acyl cation. The term "acyl" originates from acetic acid, a two-carbon carboxylic acid, with which different naturally occurring compounds share a common structural feature. Specifically, the hydroxyl portion (–OH) of the carboxylic acid function is replaced by a bond to another carbon atom. The corresponding structure can be identified for most other carboxylic acids, and the generic term "acyl group" was coined to indicate this structure in a molecule. In acylation, an acyl group is bonded to a substrate molecule in a biochemical system. Studies of its reaction mechanism demonstrate that the process is enzyme-mediated in metabolic processes. A useful laboratory method of modifying a molecular structure is the electrophilic addition reaction known as "Friedel-Crafts acylation," after Charles Friedel (1832–99) and James Mason Crafts (1839–1917).

ELECTRONIC STRUCTURE OF ACYL CATIONS

An acyl cation is an electron-poor species due to the positive charge on the carbon atom of the carbonyl group. The positive charge is stabilized by the adjacent oxygen atom through overlap of the *p* orbitals of the oxygen and carbon atoms, and equivalent molecular structures can be drawn for an acyl cation in which one structure effectively has a triple bond to the carbonyl oxygen atom, which also bears the positive charge:

$$\text{R-C(=O)-OH} \xrightarrow{-\text{OH}^-} \text{R-}\overset{+}{\text{C}}\text{=O} \leftrightarrow \text{R-C}\equiv\text{O}^+$$

The reactive center of an acyl cation is accordingly the carbon atom of the carbonyl group. The charge on the carbonyl carbon in an acyl cation can be described as an extreme case of the charge separation that normally exists in a carbonyl group as a result of the oxygen atom being more electronegative than the carbon. In an acyl cation, the carbon atom bears a full positive charge rather than just the enhanced positive charge character.

ELECTROPHILIC ADDITION AND FRIEDEL-CRAFTS ACYLATION

As the name suggests, electrophilic addition is the addition of an electrophile (a chemical species attracted to electrons) to an electron-rich target molecule, or "substrate." The substrate, which is by definition

ACYL CATION

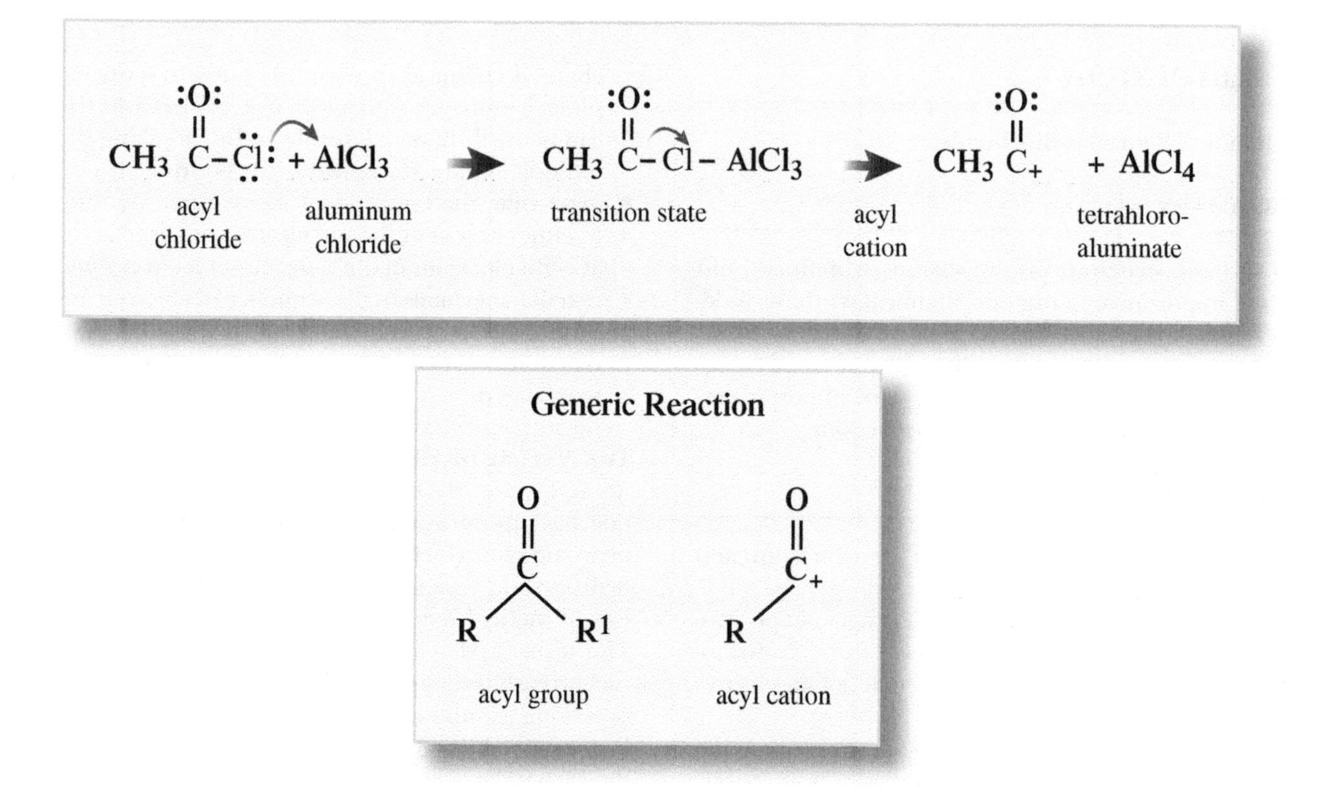

SAMPLE PROBLEM

Given the following compound, identify the reagents that could be used to form it in a Friedel-Crafts acylation reaction using anhydrous ether as the reaction medium:

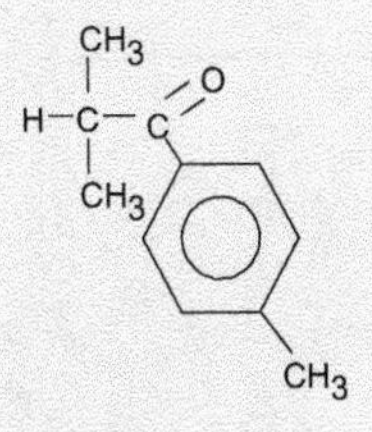

Answer:

The acyl substituent is derived from 2-methylpropanoic acid. The aryl substrate is the compound methylbenzene, or toluene. In anhydrous diethyl ether, the Friedel-Crafts acylation reaction to produce this compound can be carried out using aluminum trichloride ($AlCl_3$) as the Lewis acid catalyst, 2-methylpropanoyl chloride as the acylating agent, and toluene as the substrate molecule.

a nucleophile (an electron donor), typically has a chemical bond system that is able to reorganize electrons to form new bonds. In some cases, additional bonds are formed as the electrons in the molecular orbitals are reorganized. In other systems, the overall effect is a substitution reaction, rather than new bond formation. In the Friedel-Crafts acylation reaction, an acid chloride provides the corresponding acyl cation to replace one of the –H substituents on an aryl system (derived from an aromatic ring). A good Lewis acid, such as aluminum trichloride ($AlCl_3$) or ferric chloride ($FeCl_3$), catalyzes the reaction. In the reaction mechanism, the acyl group and the substrate molecule are coordinated to the metal atom of the Lewis acid, forming an intermediate. This allows the acyl cation to bond to one of the carbon atoms of the aryl group, and recovery of the electrons from the C–H bond reforms the pi-bond system, leaving H^+. The reaction is charge neutral, since the positive charge that was on the acyl cation becomes the positive charge on the H^+ ion.

Acylation is an important feature of biochemical systems. Plant metabolites called "acetogenins" are formed by repeated acylation reactions, and the presence of acetyl coenzyme A (acetyl CoA) is vital to cellular respiration.

NOMENCLATURE

Acyl cations are also called "acylium ions," indicating their positive charge. This is the preferred method of naming acyl cations, and the name is generated by adding *-ium* to the name of the basic structure of the corresponding carboxylic acid. The acyl cation of propanoic acid and benzoic acid, for example, would be known as the propanoylium and benzoylium ions. The names can become unwieldy, however, and it is much more common to refer to the acyl group as the substituent group, as in propanoyl and benzoyl.

Richard M. Renneboog, MSc

BIBLIOGRAPHY

Berg, Jeremy M., John L. Tymoczko, Gregory J. Gatto, and Lubert Strye. *Biochemistry.* 8th ed. New York: W. H. Freeman, 2015. Print.

Gribbin, John. *Science: A History, 1543–2001.* London: Lane, 2002. Print.

Herbert, R. B. *The Biosynthesis of Secondary Metabolites.* 2nd ed. London: Chapman & Hall, 1989. Print.

Johnson, Rebecca L. *Atomic Structure.* Minneapolis: Lerner, 2008. Print.

Mackay, K. M., R. A. Mackay, and W.Henderson. *Introduction to Modern Inorganic Chemistry.* 6th ed. Cheltenham: Nelson, 2002. Print.

Morrison, Robert Thornton, and Robert Neilson. Boyd. *Organic Chemistry.* 7th ed. Englewood Cliffs, N.J.: Prentice Hall, 2003. Print.

Winter, Mark J. *The Orbitron: A Gallery of Atomic Orbitals and Molecular Orbitals.* University of Sheffield, n.d. Web.

ALGORITHMIC CHEMISTRY

FIELDS OF STUDY

Biochemistry, Molecular Evolution, Computer Science

SUMMARY

Algorithmic chemistry attempts to simulate on a computer the evolution of a vessel or pond containing molecules, some of which can catalyze the reactions of other molecules. In the Alchemy program, the molecules are represented by symbol strings and the alphabet of symbols includes those which let a molecule have catalytic properties. The several versions of Alchemy allow one to explore what might have occurred in prebiotic evolution.

PRINCIPAL TERMS

- **algorithm:** A set of procedures for executing a calculation, e. g. the rules for adding two possibly multi-digit numbers with the usual rules for carrying and shifting positions.
- **autocatalysis:** A reaction of the type A+B+C= 2A + B, so that the concentration of A grows in a nonlinear fashion.
- **Belousov-Zhabotinsky reaction:** A complex redox reaction often used to illustrate self organization in an open system removed from equilibrium.
- **catalyst:** a molecule that can participate in a chemical reaction without being consumed.
- **genetic algorithm:** An algorithm that attempts to find a solution to a problem by applying a survival of the fittest rule to a collection of numerical approximations.
- **self-organization:** the tendency of many open systems far from chemical equilibrium to develop significant structure.
- **universal Turing machine:** a model for the programmable digital computer.

The Nature of Algorithmic Chemistry

Anyone who has studied both chemistry and algebra has probably noticed the similarity between chemical equations and algebraic equations. This similarity is even stronger between polymerization equations and some synthetic computer languages. In both cases one has a set of symbol strings, which evolve in time according to a specified rule. Alan Turing (1912-1954), who is often considered the father of the digital computer, showed that any set of rules for changing one symbol string into another could be implemented in what is now known as a universal Turing machine. The lambda calculus, invented by Alonzo Church (1903-1995), is a powerful formalism that accomplishes the same thing. The Alchemy program, short for algorithmic chemistry, developed by Walter Fontana in 1990, simulates the evolution of a collection of symbol strings that undergo random collisions and act on each other.

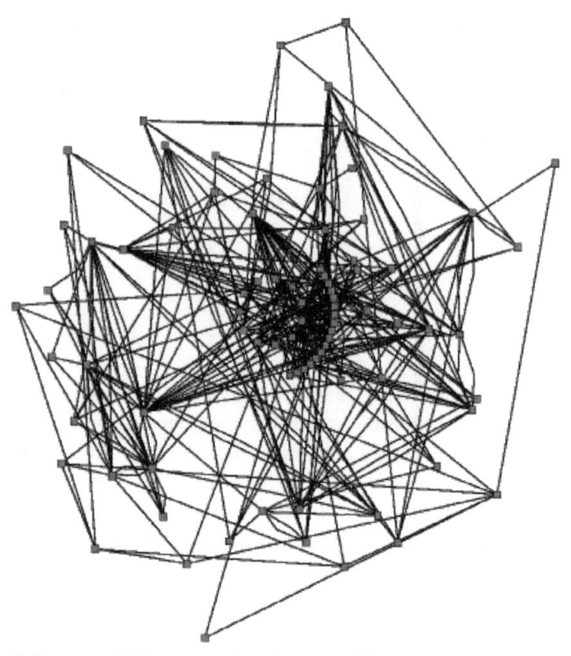

A thaliana metabolic network. TimVickers via Wikimedia Commons

Algorithmic chemistry was devised to identify the formal features of an open chemical reaction vessel so that prebiotic evolution could be modeled. In algorithmic chemistry, the "molecules" are symbol strings. The symbol strings include letters to represent atoms, and operators, which allow the molecules to interact. In a typical Alchemy simulation molecules are selected at random from the set of those present and allowed to interact. Thus an element of

SAMPLE PROBLEM

Given an initial set of monomers {A and B}, list the probable combinations after three iterations. Assume that A-B and A-A bonds are stable, but that B-B bonds cannot form.

We have, assuming for simplicity that AB is the same as BA,

- 1st generation {A, B, AA, AB}, that is to say the monomers and all possible dimers
- 2nd generation {A, B, AA, AB, AAA, AAB, ABA, AAAA, AAAB, AABA), that is the "legal" monomers, dimers, trimmers and tetramers
- 3rd generation will include monomers through 8-mers, a very large number of compounds

randomness is always present and no two simulations yield the same result.

Living creatures have had a profound effect on the Earth's surface and atmosphere. It is believed that the prebiotic earth has a chemically reducing atmosphere and the present atmosphere, characterized by the presence of free oxygen molecules, arose through the metabolism of the first microorganisms. In a famous experiment, current was passed through electrodes in a sealed vessel of a gas mixture simulating the atmosphere of the early earth. When the vessel was opened chemists found concentrations of numerous compounds which might have been incorporated into the first living cells. Alternative scenarios for life's origin involve the freezing of reactive materials and absorption in clays.

Of course, the origin of life on Earth is a question of endless fascination. While the science of thermodynamics tells us that an isolated thermodynamic system will eventually reach a state of maximum randomness, the Nobel Prize winning work of Prigogine and his coworkers has shown that a system, when it is far enough from equilibrium, will exhibit persistent, self-organized structures, provided that the system is open to the exchange of material and energy with the environment, and the system dynamics include an autocatalytic step. These principles can be demonstrated in a number of relatively simple systems. For instance, in the now famous Belousov-Zhabotinsky reaction an initially homogeneous mixture of chemicals in a Petri dish takes on an elaborate pattern as the system evolves. The work of Dyson and Fontana and many others subsequently is an attempt to develop a more detailed model of this phenomenon.

Donald R. Franceschetti, Ph.D.

BIBLIOGRAPHY

Dyson, Freeman. Origins of Life. 2nd ed. New York: Cambridge UP, 1999. Print.

Fontana, W., and L. Buss. ""The Arrival of the Fittest": Toward a Theory of Biological Organization." *Bulletin of Mathematical Biology* 56.1 (1994): 1-64. Web.

Nicolis, G., and I. Prigogine. *Exploring Complexity: An Introduction.* New York: W.H. Freeman, 1989. Print.

Prigogine, I., and Isabelle Stengers. *Order out of Chaos: Man's New Dialogue with Nature.* Toronto: Bantam, 1984. Print.

Stuart A. Kaufmann, *The Origins of Order: Self Organization and Selection in Evolution,* (Oxford, New York, 1993)

Stuart A. Kauffman, "Whispers from Carnot: The Origins of Order and Principles of Adaptation in Complex Non Equilibrium Systems, in *Complexity; Metaphors, Models and Reality,* ed. By G.A. Cowan, D. Pines, D. Meltzer, SFI Studies in the Sciences of Complexity, Vol 19, (Addison-Wesley,1994, New York).

ALKALI METALS

FIELDS OF STUDY

Inorganic Chemistry; Biochemistry; Organic Chemistry

SUMMARY

The basic properties and characteristics of the alkali metal elements are presented. The alkali metals are in group 1 of the periodic table and are highly reactive, easily forming positively charged ions.

PRINCIPAL TERMS

- **electrolysis:** the passage of an electric current through a solution or molten material to induce a chemical reaction, resulting in a reduced chemical species at the negative terminal of the electrolytic cell and an oxidized species at the positive terminal.
- **halogens:** the elements in group 17 of the periodic table (fluorine, chlorine, bromine, iodine, and astatine), all of which are highly electronegative and tend to accept one electron to form a negative ion.
- **malleability:** the ability of a solid material to be deformed by the application of compressive (pushing) force, such as hammering, without breaking or fracturing.
- **reactivity:** the propensity of a chemical species to undergo a reaction under applied conditions.
- **salt:** an ionic compound produced by the reaction of an acid and a base, formed either by combining electron-rich and electron-poor species or by replacing the hydrogen cation (H^+) of the acid with another cation from the base.

THE NATURE OF THE ALKALI METALS

The alkali metals consist of the various elements in the first column, group 1, of the periodic table. Their atoms are characterized as having a single valence electron in the outermost *s* orbital. This arrangement accounts for the ease with which alkali metals form cations with a single positive charge, their high reactivity, and their ability to form numerous salts.

The alkali metal atoms only ever form compounds in which they take on the +1 oxidation state. Hydrogen, the simplest member of group 1, is almost always a gas and, as such, usually does not behave as an alkali metal; under conditions of extremely high pressure and low temperature, however, it is a solid with the same chemical behavior and properties of the other alkali metals. The most common alkali metals—lithium, sodium, and potassium—are well known. Rubidium and cesium are less common and thus much less familiar. The last member of the group, francium, has high radioactivity and is so extremely reactive that it is estimated that no more than fifteen grams of this element exist in the earth's entire crust at any one time.

ATOMIC STRUCTURE OF THE ALKALI METALS

The electronic structure of the alkali metals is exceedingly simple. Its valence shell is composed of just the outermost *s* orbital, containing a single electron. The electronic structure of atoms seems to be most stable when the electron distribution corresponds to completely filled electron shells, with no extra electrons in higher orbitals. Accordingly, each alkali metal is highly electropositive, meaning that it easily donates its single valence electron to form the corresponding positively charged cation (H^+, Li^+, Na^+, K^+, Rb^+, Cs^+, Fr^+). This makes the smallest of them chemically useful as charge carriers and electrolytes (solutes that produce an electrically conductive solution), especially in biochemical systems.

ALKALI METAL COMPOUNDS

The alkali metals commonly form monatomic ions or neutral coordination compounds in which the central alkali metal atom is enclosed within a cage-like neutral organic molecule, as in a fullerene or carbon nanotube. The most common kinds of alkali metal compounds are salts, in which the alkali metal ions are the cations for various anions. A salt is the product of the neutralization reaction between an acid and a base. For example, the neutralization reaction between hydrochloric acid (HCl) and the base sodium hydroxide (NaOH) produces the salt sodium chloride (NaCl) and water (H_2O). Alkali metal atoms are also capable of undergoing reactions with organic compounds and replacing a hydrogen atom in the molecular structure. In small-scale chemical

ALKALI METALS

19
K
potassium
39

1 H hydrogen 1																	2 He helium 4
3 Li lithium 7	4 Be beryllium 9											5 B boron 11	6 C carbon 12	7 N nitrogen 14	8 O oxygen 16	9 F fluorine 19	10 Ne neon 20
11 Na sodium 28	12 Mg magnesium 29											13 Al aluminium 27	14 Si silicon 28	15 P phosphorus 31	16 S sulfur 32	17 Cl chlorine 35	18 Ar argon 40
19 K potassium 39	20 Ca calcium 40	21 Sc scandium 45	22 Ti titanium 48	23 Vi vanadium 51	24 Cr chromium 52	25 Mn manganese 55	26 Fe iron 56	27 Co cobalt 59	28 Ni nickel 59	29 Cu copper 64	30 Zn zinc 65	31 Ga gallium 70	32 Ge germanium 73	33 As arsenic 75	34 Se selenium 79	35 Br bromine 80	36 Kr krypton 84
37 Rb rubidium 85	38 Sr strontium 88	39 Y yttrium 89	40 Zr zirconium 91	41 I niobium 93	42 Mo molybdenum 96	43 Tc technetium 98	44 Ru ruthenium 101	45 Rh rhodium 103	46 Pd palladium 106	47 Ag silver 108	48 Cd cadmium 112	49 In indium 115	50 Sn tin 119	51 Sb antimony 122	52 Te tellerium 128	53 I iodine 127	54 Xe xenon 131
55 Cs caesium 133	56 Ba barium 137	*	72 Hf hafnium 178	73 Ta tantalum 181	74 W tungsten 184	75 Re rhenium 186	76 Os osmium 190	77 Ir iridium 192	78 Pt platinum 195	79 Au gold 197	80 Hg mercury 201	81 Tl thallium 204	82 Pb lead 207	83 Bi bismuth 209	84 Po polonium 209	85 At astatine 210	86 Rn radon 222
87 Fa francium 223	88 Ra radium 226	**	104 Rf rutherfordium 267	105 Db dubnium 268	106 Sg seaborgium 271	107 Bh bohrium 272	108 Hs hassium 270	109 Mt meitnerium 276	110 Ds darmstadtium 281	111 Rg roentgenium 280	112 Cn copernicium 285	113 Uut ununtrium 284	114 Fl flerovium 289	115 Uup ununpentium 288	116 Lv livermorium 293	117 Uus ununseptium 294	118 Uuo ununoctium 289

57 * La lanthanum 139	58 Ce cerium 140	59 Pr praseodymium 141	60 Nd neodymium 144	61 Pm promethium 145	62 Sm samarium 150	63 Eu europium 152	64 Gd gadolinium 157	65 Tb terbium 159	66 Dy dysprosium 163	67 Ho holmium 165	68 Er erbium 167	69 Tm thulium 168	70 Yb ytterbium 173	71 Lu lutetium 175
89 ** Ac actinium 227	90 Th thorium 232	91 Pa protactinium 231	92 U uranium 238	93 Np neptunium 237	94 Pu plutonium 244	95 Am americium 243	96 Cm curium 247	97 Bk berkelium 247	98 Cf californium 251	99 Es einsteinium 252	100 Fm fermium 257	101 Md mendelevium 258	102 No nobelium 259	103 Lr lawrencium 284

synthesis, it is a common practice to use an extremely strong base to remove a specific hydrogen atom from the structure of an organic molecule in order to produce a specific mode of reactivity.

Occurrence of Alkali Metals

Alkali metals are never found in their elemental form in nature, only as compounds or dissolved salts. The most common of these is sodium chloride, which is dissolved in seawater. In industry, both the alkali metal sodium and the halogen chlorine are recovered from molten sodium chloride by electrolysis. Sodium metal reacts strongly with atmospheric moisture, quickly becoming coated with sodium hydroxide, also called caustic soda, and producing hydrogen gas. It is typically stored in inert mineral oil or kerosene to prevent this. Due to its high reactivity with water, sodium metal recovered from electrolysis is mainly converted to sodium hydroxide, which is much easier and safer to handle and store than sodium metal. Potassium metal is recovered in the same way from the electrolysis of potassium chloride (KCl) and must be handled more cautiously since it is more reactive than sodium. All of the alkali metals have high malleability and are typically soft enough to be easily cut with a knife.

Richard M. Renneboog, MSc

SAMPLE PROBLEM

Calculate the amount of sodium metal that can be recovered from 3,500 kilograms of pure sodium chloride.

Answer:

Write out the balanced reaction equation:

$$2NaCl \rightarrow 2Na + Cl_2$$

From this, it can be seen that each mole of NaCl produces one mole of sodium metal. Therefore, the number of moles of sodium chloride in 3,500 kilograms is equal to the number of moles of sodium that would be produced. Calculate the molecular weight of NaCl, adding together the atomic weights of sodium and chlorine:

$$23\frac{g}{mol} + 35.5\frac{g}{mol} = 58.5\frac{g}{mol}$$

Determine the number of moles of NaCl in the sample:

$$3{,}500 \text{ kg} \times 1{,}000\frac{g}{kg} = 3{,}500{,}000 \text{ g}$$

$$\frac{3{,}500{,}000 \text{ g}}{58.5 \text{ g/mol}} = 59{,}829.06 \text{ mol}$$

Because there is only one atom of sodium in one molecule of NaCl, 59,829.06 moles of NaCl contains the same number of moles of sodium. From this, find how many kilograms of sodium metal would be produced:

$$59{,}829.06 \text{ mol} \times 23\frac{g}{mol} = 1{,}376{,}068.38 \text{ g} = 1{,}376.06838 \text{ kg}$$

The reaction would produce 1,376,068.38 grams of sodium metal, or approximately 1,376.07 kilograms.

BIBLIOGRAPHY

Berg, Jeremy M., John L. Tymoczko, Gregory J. Gatto, and Lubert Strye. *Biochemistry.* 8th ed. New York: W. H. Freeman, 2015. Print.

Gribbin, John. *Science: A History, 1543–2001.* London: Lane, 2002. Print.

Johnson, Rebecca L. *Atomic Structure.* Minneapolis: Lerner, 2008. Print.

Kean, Sam. *The Disappearing Spoon: And Other True Tales of Madness, Love, and the History of the World from the Periodic Table of the Elements.* New York: Little, Brown, 2010. Print.

Mackay, K. M., R. A. Mackay, and W.Henderson. *Introduction to Modern Inorganic Chemistry.* 6th ed. Cheltenham: Nelson, 2002. Print..

Miessler, Gary L., Paul J. Fischer, and Donald A. Tarr. *Inorganic Chemistry.* 5th ed. Upper Saddle River: Prentice Hall, 2013. Print.

Winter, Mark J. *The Orbitron: A Gallery of Atomic Orbitals and Molecular Orbitals.* University of Sheffield, n.d. Web.

ALKALINE EARTH METALS

FIELDS OF STUDY

Inorganic Chemistry; Geochemistry; Metallurgy

SUMMARY

The basic properties and characteristics of the alkaline earth metal elements are presented. The alkaline earth metals constitute group 2 of the periodic table and are highly reactive, easily forming ions with a 2+ charge.

PRINCIPAL TERMS

- **halide:** a binary compound consisting of a halogen element (fluorine, chlorine, bromine, iodine, or astatine) bonded to a non-halogen element or organic group; alternatively, an anion of a halogen element.
- **hydroxide:** an anion consisting of one oxygen atom and one hydrogen atom, represented as OH^-; also, an ionic compound in which the OH^- ion is bonded to another element or group.
- **oxidation state:** a number that indicates the degree to which an atom or ion in a chemical compound has been oxidized or reduced.
- **oxide:** a compound formed by the reaction of any element with oxygen, such as carbon dioxide, carbon monoxide, iron oxide, or diphosphorus pentoxide.
- **reactivity:** the propensity of a chemical species to undergo a reaction under applied conditions.

Characteristics of the Alkaline Earth Metals

The alkaline earth metals are the six elements in the second column, group 2, of the periodic table. Their atoms have two valence electrons in the outermost *s* orbital, an arrangement that accounts for the ease with which alkaline earth metals form cations with a 2+ charge, their relatively high reactivity, and their ability to form numerous salts. The alkaline earth metals typically form compounds in which they take on an oxidation state of +2, although the +1 state is also known to occur. Beryllium (Be) is the simplest member of group 2 and is the principal metal ion found in beryl minerals. The other alkaline earth metals are magnesium (Mg), calcium (Ca), strontium (Sr), barium (Ba), and radium (Ra). Radioisotopes of radium are known to occur naturally.

Atomic Structure of the Alkaline Earth Metals

The electronic structure of the alkaline earth metals is relatively simple. Their valence shells are composed of just the outermost *s* orbital, containing a single pair of electrons. Atoms seem to be the most stable when their electron shells are completely filled, with no extra electrons in higher orbitals. Accordingly, each alkaline earth metal is highly electropositive and easily donates its two valence electrons to form the corresponding cation, or positively charged ion (Be^{2+}, Mg^{2+}, Ca^{2+}, Sr^{2+}, Ba^{2+}, Ra^{2+}). This makes alkaline earth metal ions relatively comparable in size to alkali metal ions, but with twice the positive charge density. Because of their size and charge, calcium and magnesium are essential components of biochemical processes.

Alkaline Earth Metal Compounds

The alkaline earth metals commonly form either monatomic (single-atom) ions or electrically neutral coordination compounds in which a central alkaline earth metal atom is surrounded by neutral organic molecules or a cage-like neutral molecule. Alkaline earth metals most commonly bond to halogen anions (negatively charged ions), forming compounds called halides, or other monatomic or complex anions with a charge of 2–.

The alkaline earth oxides and hydroxides are important commercial minerals. Having two valence electrons enables alkaline earth metals to form long chain-like molecular structures in many of the minerals in which they occur. In theory, a corresponding anion balances each metal ion; structurally, however, each metal ion essentially shares its anions with the next metal ion in the chain. The anions thus act as bridges between the metal cations in the molecular structure of the mineral. This feature gives rise to a wide variety of mineral structures having the same chemical formula, some of which are classified as gemstones.

ALKALINE EARTH METALS

38 Sr strontium 88

1 H hydrogen 1																	2 He helium 4
3 Li lithium 7	4 Be beryllium 9											5 B boron 11	6 C carbon 12	7 N nitrogen 14	8 O oxygen 16	9 F fluorine 19	10 Ne neon 20
11 Na sodium 28	12 Mg magnesium 29											13 Al aluminium 27	14 Si silicon 28	15 P phosphorus 31	16 S sulfur 32	17 Cl chlorine 35	18 Ar argon 40
19 K potassium 39	20 Ca calcium 40	21 Sc scandium 45	22 Ti titanium 48	23 Vi vanadium 51	24 Cr chromium 52	25 Mn manganese 55	26 Fe iron 56	27 Co cobalt 59	28 Ni nickel 59	29 Cu copper 64	30 Zn zinc 65	31 Ga gallium 70	32 Ge germanium 73	33 As arsenic 75	34 Se selenium 79	35 Br bromine 80	36 Kr krypton 84
37 Rb rubidium 85	38 Sr strontium 88	39 Y yttrium 89	40 Zr zirconium 91	41 I niobium 93	42 Mo molybdenum 96	43 Tc technetium 98	44 Ru ruthenium 101	45 Rh rhodium 103	46 Pd palladium 106	47 Ag silver 108	48 Cd cadmium 112	49 In indium 115	50 Sn tin 119	51 Sb antimony 122	52 Te tellerium 128	53 I iodine 127	54 Xe xenon 131
55 Cs caesium 133	56 Ba barium 137	*	72 Hf hafnium 178	73 Ta tantalum 181	74 W tungsten 184	75 Re rhenium 186	76 Os osmium 190	77 Ir iridium 192	78 Pt platinum 195	79 Au gold 197	80 Hg mercury 201	81 Tl thallium 204	82 Pb lead 207	83 Bi bismuth 209	84 Po polonium 209	85 At astatine 210	86 Rn radon 222
87 Fa francium 223	88 Ra radium 226	**	104 Rf rutherfordium 267	105 Db dubnium 268	106 Sg seaborgium 271	107 Bh bohrium 272	108 Hs hassium 270	109 Mt meitnerium 276	110 Ds darmstadtium 281	111 Rg roentgenium 280	112 Cn copernicum 285	113 Uut ununtrium 284	114 Fl flerovium 289	115 Uup ununpentium 288	116 Lv livermorium 293	117 Uus ununseptium 294	118 Uuo ununoctium 289

57 * La lanthanum 139	58 Ce cerium 140	59 Pr praseodymium 141	60 Nd neodymium 144	61 Pm promethium 145	62 Sm samarium 150	63 Eu europium 152	64 Gd gadolinium 157	65 Tb terbium 159	66 Dy dysprosium 163	67 Ho holmium 165	68 Er erbium 167	69 Tm thulium 168	70 Yb ytterbium 173	71 Lu lutetium 175
89 ** Ac actinium 227	90 Th thorium 232	91 Pa protactinium 231	92 U uranium 238	93 Np neptunium 237	94 Pu plutonium 244	95 Am americium 243	96 Cm curium 247	97 Bk berkelium 247	98 Cf californium 251	99 Es einsteinium 252	100 Fm fermium 257	101 Md mendelevium 258	102 No nobelium 259	103 Lr lawrendum 284

OCCURRENCE OF ALKALINE EARTH METALS

Alkaline earth metals are never found in their elemental form in nature, only as compounds or dissolved salts. They are normally found as ores and minerals in which they exist as various oxides, sulfides, sulfates, sulfites, silicates, carbonates, and other inorganic compounds. They are typically extracted by mining and smelting operations that release the materials as relatively pure elements.

Magnesium is particularly useful for its strength and is the lightest structural metal known. There are certain dangers associated with magnesium, however. The activation energy of its oxidation reaction is high, but the reaction liberates so much heat energy that, once started, a magnesium fire is extraordinarily difficult to extinguish. Attempting to extinguish a magnesium fire with water only makes it worse, as magnesium reacts with room-temperature water, and it also reacts with nitrogen in air. Because of this, heat operations involving magnesium, such as casting and welding, are carried out in an atmosphere of an inert gas that cannot react with the metal. Magnesium is also an important laboratory reagent since it can form reactive complexes with alkyl and aryl halides that are useful in Grignard reactions. In biochemical systems, the Mg^{2+} ion is a vital component of nerve and enzyme function.

Richard M. Renneboog, MSc

SAMPLE PROBLEM

Determine the electron distribution of magnesium and radium, using the electron distribution chart below if needed. What are their similarities?

1s
2s 2p
3s 3p 3d
4s 4p 4d 4f
5s 5p 5d 5f 5g
6s 6p 6d 6f 6g 6h
7s 7p 7d 7f 7g 7h

Answer:

The atomic number of magnesium is 12, giving it an electron distribution of

$$1s^2 2s^2 2p^6 3s^2$$

The atomic number of radium is 88, giving it an electron distribution of

$$1s^2 2s^2 2p^6 3s^2 3p^6 4s^2 3d^{10} 4p^6 5s^2 4d^{10} 5p^6 6s^2 4f^{14} 5d^{10} 6p^6 7s^2$$

In both magnesium and radium, the outermost orbital is an *s* orbital containing two electrons. Both elements also have a completely filled set of *p* orbitals as the next lowest set.

Bibliography

Berg, Jeremy M., John L. Tymoczko, Gregory J. Gatto, and Lubert Strye. *Biochemistry.* 8th ed. New York: W. H. Freeman, 2015. Print.

Gribbin, John. *Science: A History, 1543–2001.* London: Lane, 2002. Print.

Johnson, Rebecca L. *Atomic Structure.* Minneapolis: Lerner, 2008. Print.

Kean, Sam. *The Disappearing Spoon: And Other True Tales of Madness, Love, and the History of the World from the Periodic Table of the Elements.* New York: Little, Brown, 2010. Print.

Mackay, K. M., R. A. Mackay, and W.Henderson. *Introduction to Modern Inorganic Chemistry.* 6th ed. Cheltenham: Nelson, 2002. Print.

Miessler, Gary L., Paul J. Fischer, and Donald A. Tarr. *Inorganic Chemistry.* 5th ed. Upper Saddle River: Prentice Hall, 2013. Print.

Winter, Mark J. *The Orbitron: A Gallery of Atomic Orbitals and Molecular Orbitals.* University of Sheffield, n.d. Web.

ALKANES

FIELDS OF STUDY

Organic Chemistry

SUMMARY

The characteristic properties and reactions of alkanes are discussed. Alkanes are an infinite series of compounds containing only carbon and hydrogen atoms. Their variety is due to the unique electronic distribution and physical size of the carbon atom. The alkanes are the basic structures of all organic molecules.

PRINCIPAL TERMS

- **functional group:** a specific group of atoms with a characteristic structure and corresponding chemical behavior within a molecule.
- **hydrocarbon:** an organic compound composed solely of carbon and hydrogen atoms.
- **isomer:** one of two or more chemical species that have the same molecular formula but different molecular structures.
- **orbital:** a specific region of space about the nucleus of an atom in which electrons of a given energy level are most likely to be found.
- **saturated:** describes an organic compound in which carbon atoms are attached to other atoms

via single bonds only, allowing the compound to contain the maximum possible number of hydrogen atoms.

- **single bond:** a type of chemical bond in which two adjacent atoms are connected by a single pair of electrons via the direct overlap of their atomic orbitals.

THE NATURE OF THE ALKANES

The alkanes are a series of compounds consisting of only carbon and hydrogen atoms, or hydrocarbons. Because alkanes do not contain any other functional groups in their molecular structure, they are generally unreactive to materials that are not good oxidizing agents or radicals able to extract a hydrogen atom.

The structure of alkanes derives from the geometric arrangement of atomic orbitals on the carbon atom. There are four valence electrons in each carbon atom, two in the 2*s* orbital and one each in two of the three 2*p* orbitals. By combining the 2*s* and 2*p* orbitals, the carbon atom is able to form four equivalent hybrid atomic orbitals, each containing just one of the four valence electrons of the carbon atom. Each of the four orbitals is directed toward an apex of a tetrahedron, with the carbon atom nucleus at its center. Any two of these sp^3 hybrid orbitals ideally form an angle of 109.5 degrees. This angle and the physical diameter of the carbon atom are ideally suited to the formation of innumerable structures by linkage of the carbon atoms in different directions. The electrons in the orbitals of two adjacent carbon atoms readily form a covalent single bond, or sigma (σ) bond, between the two atoms, and each carbon atom can form four such single bonds. Accordingly, organic compounds with structures comprising carbon atoms form the largest and most varied group of chemical compounds known. In alkanes, all of the carbon atoms have four single bonds to either other carbon atoms or to hydrogen atoms, and because they have all of the bonds that they can, they are described as saturated. Thus, another way to describe the alkanes is as the series of saturated hydrocarbons. If a hydrocarbon molecule contains a carbon-carbon C=C double or C≡C triple bond, it is not classed as an alkane, but as an unsaturated hydrocarbon in the alkene or alkyne series, respectively.

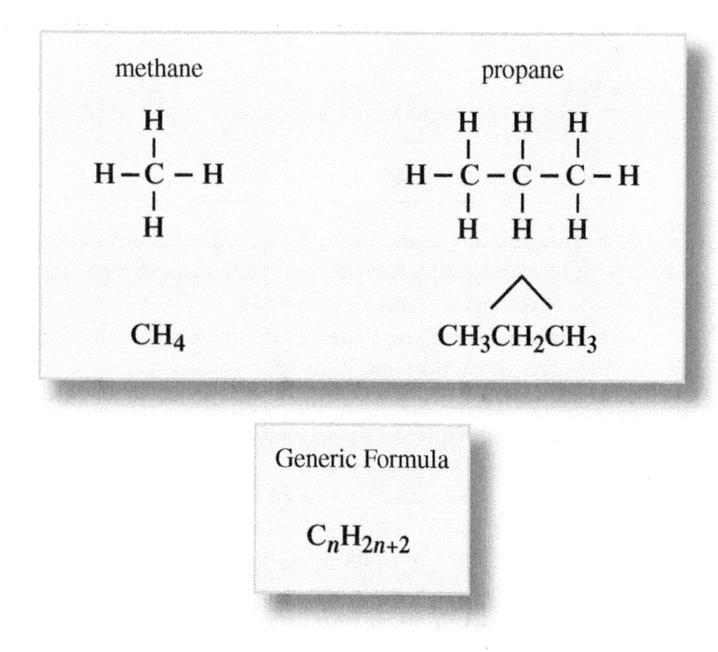

The alkanes have two parallel families of structures, one comprising acyclic molecular structures and the other comprising cyclic molecular structures. The carbon atom skeletons of the acyclic alkane series can be expanded infinitely, at least in principle. The simplest of these is methane (CH_4), having just one carbon atom and four equivalent bonds to as many hydrogen atoms. The next is ethane (C_2H_6), followed by propane (C_3H_8), butane (C_4H_{10}), and so on. All alkanes have the general chemical formula C_nH_{2n+2}.

There are also numerous different arrangements of the same number of carbon and hydrogen atoms, or isomers, beginning with butane. The four carbon atoms of butane can be joined in sequence (linearly) or in a T structure to form the alkane named 2-methylpropane, both of which have the chemical formula C_4H_{10}. With each additional carbon atom, the number of possible isomers multiplies. Four saturated carbon atoms can be arranged in two isomeric forms. With five carbons atoms, there are three isomers, and with six carbon atoms there are five isomers. With seven carbon atoms, the number of isomers begins to outnumber the number of carbon atoms and increase in a dramatic manner, with eight isomers or ten if the optical isomers are included.

Optical isomers are compounds in which a carbon atom is bonded to four different substituent groups—atoms or groups that can be added in place of another, also called "side chains." They are identical in every way except the order in which the bonds to the substituent groups are distributed about that central asymmetric carbon atom. While the physical properties and chemical behavior of two optical isomers are the same, they are nonetheless distinctly different structurally and are therefore separate isomeric forms. With eight carbon atoms, then, the number of isomeric forms jumps to fourteen (eighteen counting optical isomers), and with nine carbon atoms the number of possible isomeric structures numbers more than one hundred, all of which are chemically distinct compounds.

In the cyclic hydrocarbons, the carbon atoms are bonded together in a ring structure, beginning with the three carbon atoms of cyclopropane (C_3H_8). The cyclic series follows the same order as the acyclic alkanes, with cyclobutane (C_4H_8), cyclopentane (C_5H_{10}), and so on. Isomeric structures of the cyclic alkanes also begin with four carbon atoms (methylcyclopropane and cyclobutane), and optical isomers are also possible. The presence of the ring structure adds another type of isomeric structure termed a "geometric isomer." Two side chains on carbon atoms in the ring can either be on the same side or on opposite sides of the virtual plane of the ring.

NOMENCLATURE OF ALKANES

Alkanes are named according to the number of carbon atoms comprising the longest unbranched chain of carbon atoms in the molecular structure, and the positions of side chains are identified accordingly. The name sequence begins with methane, ethane, propane, and butane as common names, and then progresses through names based on the number of carbon atoms: pentane, hexane, heptane, octane, nonane, decane, undecane, and so on. Once the base chain has been determined, the next part of the alkane's name is a number referring to the carbon atom in the base chain to which the side chain is bonded. Carbon atoms in the base chain are numbered in sequence from either end, left or right, so that the side chain branches from the carbon atom with the lowest number in the sequence. The position number is followed by a hyphen and then the name of the substituent. A six-carbon base chain that has methyl groups on the second and third carbon atoms from one end would be named 2,3-dimethylhexane and not 4,5-dimethylhexane.

Substituent alkanes switch the *-ane* suffix for *-yl*, so that methane becomes methyl ($-CH_3$), propane becomes propyl ($-CH_2CH_2CH_3$), butane becomes butyl ($-CH_2CH_2CH_2CH_3$), and so on. Some substituent alkanes are structured in branched chains; these add the prefix *iso-*, as in isopropyl ($-CH_3CHCH_3$, with the second carbon in the side chain linked to the base chain). Isobutyl has $-CH_2$ linked to the base chain and to a central carbon, which is also linked to a hydrogen and two methyls. A prefix such as *di-* is added

SAMPLE PROBLEM

Given the chemical names 4-isopropyl-2,3-dimethylnonane and 1,3-diethyl-5-isopropylcycloheptane, draw the skeletal structures of the molecules.

Answer:

The name of the first compound indicates that the longest chain is nine carbons long (*nonane*). There is an isopropyl side chain on the fourth atom in the parent chain and methyl groups on the second and third carbon atoms in the nonane structure. The structure is therefore

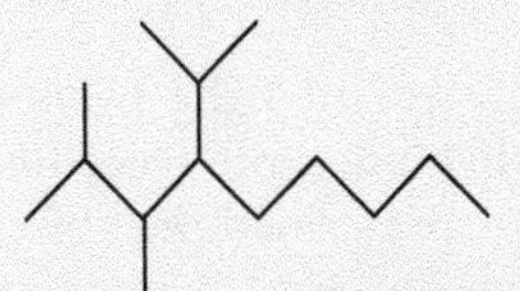

The second alkane compound is composed of seven carbon atoms in a ring (*cycloheptane*). There is an ethyl side chain on one carbon atom and another on the second carbon atom over from the first one. There is also an isopropyl side chain on the fourth carbon atom around from the first. The structure is therefore

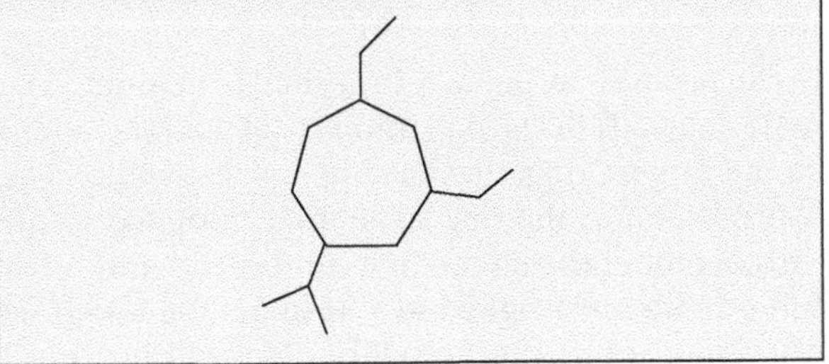

to specify the number of substituents of each type. So a dimethylhexane is a hexane with two methyl side chains. When different side chains are present on a hydrocarbon molecule, they are assigned by the priority of their size, again to attain the lowest numbers, and listed alphabetically.

Alkane structures are typically represented by structural formulas, line drawings that depict the angles between the carbon atoms in such a way that every angle vertex and line end represents a single carbon atom. Such line drawings allow chemists to communicate a large amount of chemical information in a very small space. This is a much more convenient and readily understood representation than strings of *C*s connected by lines, and makes the nature and identity of the side chains immediately apparent. For example, the following structure:

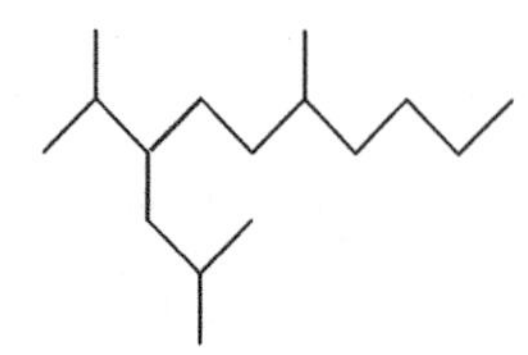

is much more clearly understood than this one:

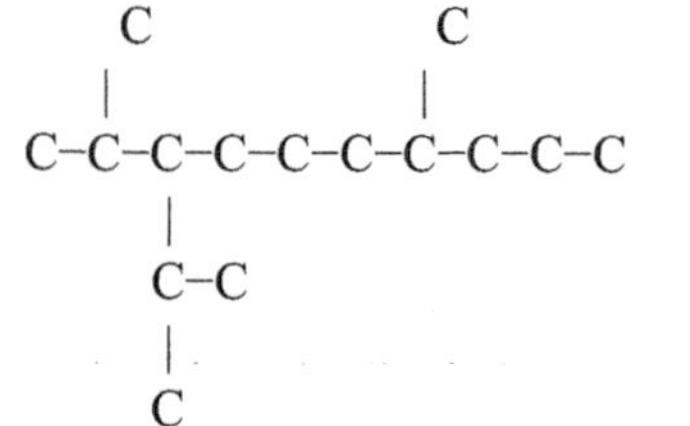

and can be easily assigned the proper name 4-isopropyl-2,7-dimethylundecane, according to the rules of nomenclature prescribed by the International Union of Pure and Applied Chemistry (IUPAC).

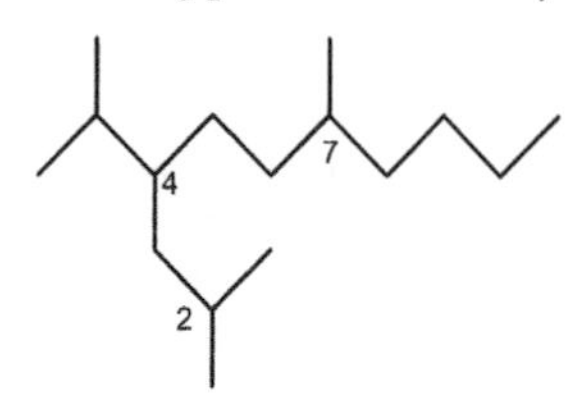

Cyclic alkanes are named in a similar manner. The basic name reflects the number of carbon atoms in the largest ring structure in the molecule. The first position in the ring structure is assigned to the carbon atom bearing the first substituent, and other substituents are assigned so as to attain the lowest set of position numbers. A cyclohexane ring bearing a methyl substituent on one carbon atom and an ethyl substituent on another carbon atom would be named 1-ethyl-3-methylcyclohexane, but not 1-methyl-3-ethylcyclohexane or any other combination of substituent positions. Cyclic alkane structures are slightly harder to draw if there are more than six carbon atoms in the ring, but they are even more readily understood than acyclic alkane structures, especially when cyclic structures are fused together.

FORMATION OF ALKANES

The simple alkanes—methane, ethane, propane, and butane—occur naturally in natural gas, although the quantities of propane and butane are comparatively quite low. These are formed in the same processes that alter the molecular structures of organic materials in the formation of petroleum deposits. Alkanes higher than butane are typically obtained from "cracking," the industrial-scale fractional distillation of crude petroleum and its products. The range of products obtained from the process includes light petroleum ethers that contain essentially all isomers of pentane, hexane, heptane, and octane, as well as gasoline, kerosene, diesel and fuel oils, lubricating oils and greases, waxes, and tars. The residue from the process is normally used to pave highways. The compounds are never obtained from this process as single compounds. Specific alkanes can be prepared by synthetic reactions involving molecules with various functional groups that can undergo substitution, addition, and reduction reactions. However, since alkanes are not themselves reactive, they have little use other than as structural components of other molecules.

REACTIONS OF ALKANES

Alkanes are unreactive compounds with acids and bases. They do not undergo reductions, but are readily and rapidly oxidized by good oxidizing agents, such as molecular oxygen (O_2) and perchloric acid ($HClO_4$). Hence, all hydrocarbons are highly combustible, and reaction with perchloric acid is violent. Due to their nonpolar character and inherent stability or unreactiveness, alkanes typically see use as solvents, though their ability to undergo oxidation to carbon dioxide and water also makes them very useful in quantities as combustion fuels.

Combustion with molecular oxygen occurs by a "radical" process involving electrically neutral single atoms rather than ions. In combustion, electrons in the bonding orbitals of molecules such as oxygen acquire sufficient energy to allow them to transfer to a higher-energy "antibonding" orbital. When this happens, the bond between the two atoms simply ceases to exist, and the individual atoms each have an unpaired electron. This makes such an atom highly reactive. An oxygen atom, for example, is able to abstract a hydrogen atom from an alkane molecule, which begins a radical chain reaction mechanism that ultimately produces water molecules and carbon dioxide molecules. Alkanes essentially undergo no other types of reactions.

Richard M. Renneboog, MSc

BIBLIOGRAPHY

Fenichell, Stephen. *Plastic: The Making of a Synthetic Century*. New York: Harper, 2009. Print.

Haynes, W. H., PhD, David R. Lide, PhD, and Thomas J. Bruno, PhD, eds. *CRC Handbook of Chemistry and Physics,* 96th Edition. New York: CRC/Taylor & Francis Group, 2015. Print.

Hendrickson, James B., Donald J. Cram, and George S. Hammond. *Organic Chemistry*. 3rd ed. New York: McGraw, 1970. Print.

Morrison, Robert Thornton, and Robert Neilson. Boyd. *Organic Chemistry*. 7th ed. Englewood Cliffs, N.J.: Prentice Hall, 2003. Print.

Myers, Richard. *The Basics of Chemistry* Westport: Greenwood, 2003. Print.

ALKENES

FIELDS OF STUDY

Organic Chemistry

SUMMARY

The characteristic properties and reactions of alkenes are discussed. Alkenes are an infinite series of compounds containing only carbon and hydrogen atoms and at least one carbon-carbon double bond. Their variety is due to the unique electronic distribution and physical size of the carbon atom. The alkenes parallel the series of structures of the alkanes.

PRINCIPAL TERMS

- **dehydration reaction:** a chemical reaction in which hydrogen and oxygen atoms are removed from the reactants and combine to form water.
- **double bond:** a type of chemical bond in which two adjacent atoms are connected by four bonding electrons rather than two.
- **functional group:** a specific group of atoms with a characteristic structure and corresponding chemical behavior within a molecule.
- **hydrogenation:** a chemical reaction in which two hydrogen atoms, usually in the form of molecular hydrogen (H_2), are bonded to another molecule, almost always as a result of catalysis.
- **orbital:** a specific region of space about the nucleus of an atom in which electrons of a given energy level are most likely to be found.
- **unsaturated:** describes an organic compound in which carbon atoms are attached to other atoms via double or triple bonds, preventing the compound from containing the maximum possible number of hydrogen atoms.

THE NATURE OF THE ALKENES

The alkenes are a series of compounds consisting of only carbon and hydrogen atoms, or hydrocarbons. Alkenes contain at least one double bond between two adjacent carbon atoms (C=C) as a functional group in their molecular structure and so are generally reactive materials. The structure of alkenes derives from the geometric arrangement of atomic orbitals on the carbon atom. There are four valence electrons in each carbon atom, two in the $2s$ orbital and one each in two of the three $2p$ orbitals. By combining the $2s$ and two of the three $2p$ orbitals the carbon atom is able to form three equivalent sp^2 hybrid atomic orbitals, each containing a single electron. The fourth electron remains in the unhybridized $2p$ orbital. Each of the three hybrid orbitals is directed toward a vertex of a triangle, with the carbon atom nucleus at its center. Any two of the sp^2 hybrid orbitals ideally form an angle of 120 degrees. This angle and the physical diameter of the carbon atom are well suited to the formation of innumerable structures by linkage of the carbon atoms. The sp^2 orbitals of two adjacent carbon atoms readily form a covalent single bond, or sigma (σ) bond, and each carbon atom can form three such single bonds. The unhybridized *p* orbital on the adjacent carbon atoms are able to overlap in a side-by-side orientation to form a pi (ϖ) bond parallel to the sigma bond. Accordingly, alkenes form a large and varied group of chemical compounds in their own right, and the reactivity of the C=C functional group enables the formation of many other compounds. In alkenes, the carbon atoms of the double bonds do not have all of the bonds that they can, and another way to describe the alkenes is as the series of unsaturated hydrocarbons. If a hydrocarbon molecule contains only carbon-carbon single bonds (C–C) or triple bonds (C≡C), it is not classed as an alkene, but as a hydrocarbon in the alkane or alkyne series, respectively.

The alkenes have two parallel families of structures one comprising acyclic molecular structures and the other comprising cyclic molecular structures. The carbon atom skeletons of the acyclic alkene series can be expanded infinitely, at least in principle. The simplest alkene is ethene (C_2H_4), commonly known as ethylene, which has just two carbon atoms and four equivalent bonds to as many hydrogen atoms. The next are propene (C_3H_6), butene (C_4H_8), and so on. All alkenes have the general chemical formula C_nH_{2n}, which is the same as the cyclic alkane series. For alkenes with four or more carbon atoms (butenes or higher), there are also numerous possible isomers, chemical species that have the same molecular formula but different molecular structures. The double bond in butene can be between either the first and second carbon atoms or the second and third carbon atoms in the molecule, both of which have the chemical formula C_4H_8. In addition, the geometry of the C=C bond is fixed and rotation about a C=C bond cannot occur as it can with a C–C bond. This permits the formation of geometric isomers, in which two substituents, or side chains, can be either on the same side of the double bond or on opposite sides. Accordingly, 2-butene has the two isomeric structures:

and the two compounds have both different physical properties and chemical behaviors. The same relationships exist for every C=C bond, within the structural constraints of the molecule.

With each additional carbon atom, the number of possible isomeric forms increases. Four carbon atoms can be arranged in two isomeric alkene forms. With five carbons atoms, there are six isomeric alkene structures, and with six carbon atoms there are sixteen isomeric alkene structures. There are even more isomeric forms when optical isomers are included. Optical isomers are compounds in which a saturated carbon atom has four different substituent groups bonded to the same carbon atom. They are identical in every way except the order which the bonds to substituent groups are distributed around the central asymmetric carbon atom. While the physical properties and chemical behavior of two optical isomers are the same, they are nonetheless distinctly

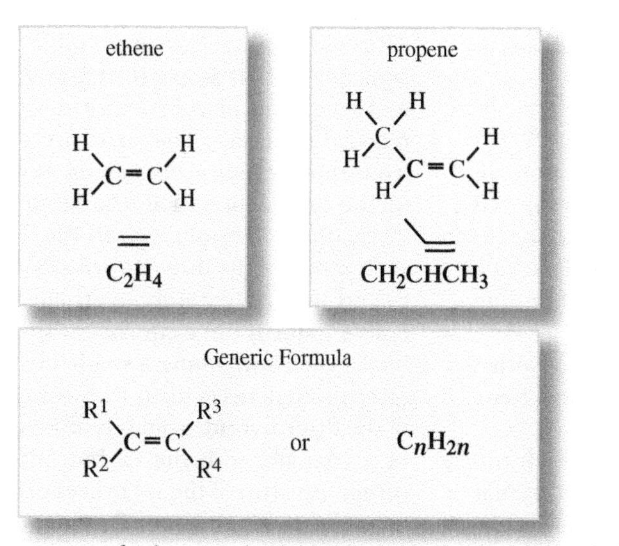

Alkenes can have one or more set of carbon atoms double bonded together. R-groups may be a hydrogen atom, some form of a hydrocarbon, or some other functional group.

different structurally and therefore separate isomeric forms. Optical isomerism is not possible about the carbon atom of a single C=C bond, but the presence of more than one C=C bond in an alkene can satisfy the requirements for optical isomeric structures, in addition to greatly increasing the number of alkene structures.

In the cyclic hydrocarbons, the carbon atoms are bonded together in a ring structure, beginning with the three carbon atoms of cyclopropene, C_3H_4. The cyclic series follows the same order as the acyclic alkenes, with cyclobutene (C_4H_6), cyclopentene (C_5H_8), and so on. Geometric isomers of the cyclic alkenes are not physically possible to achieve for ring structures containing fewer than eight carbon atoms due to bond-angle restrictions.

NOMENCLATURE OF ALKENES

The alkenes are named according to the number of carbon atoms comprising the longest unbranched chain of carbon atoms in the molecular structure, and the positions of side chains are identified accordingly. The name sequence begins with ethylene, propene, and butene as common names, and then progresses through names based on the number of carbon atoms: pentene, hexene, heptene, octene, nonene, decene, and so on. When the base chain has been determined and has three or more carbon atoms, the position of the bond is indicated by the lower of the two position numbers of the carbon atoms in the bond. The position number is inserted into the name of the primary chain immediately before the *-ene* suffix, which identifies the compound as an alkene. (Some naming conventions place the position number before the name of the primary chain.) If no position number is given, it is assumed that the bond is between the first and second carbon atoms. A seven-carbon chain with a C=C bond between the second and third carbon atoms from one end would thus be named hept-2-ene (or 2-heptene), not hept-3-ene or hept-5-ene. The order of the side chains is determined next. The numerical positions of the side chains are assigned as determined by the position assigned to the C=C bond. For example, a six-carbon chain that has methyl groups ($-CH_3$) on the second and third carbon atoms from the end that has the C=C bond would be named 2,3-dimethylhexene. When different side chains are present on a hydrocarbon molecule, they are assigned by the priority of their size, again to attain the lowest position numbers, and named alphabetically. The relative positions of side chains are indicated in the formal name of the compound using the identifiers *cis-* for side chains on the same side of the double bond and *trans-* when the side chains are on opposite sides of the double bond.

Organic structures are typically represented by structural formulas, line drawings that depict the angles between the carbon atoms in such a way that every angle vertex and line end represents a single carbon atom. Such line drawings allow chemists to communicate a large amount of chemical information in a very small space. This is a much more convenient and readily understood representation than strings of *C*s connected by lines and makes the nature and identity of the side chains immediately apparent. For example, the following structure:

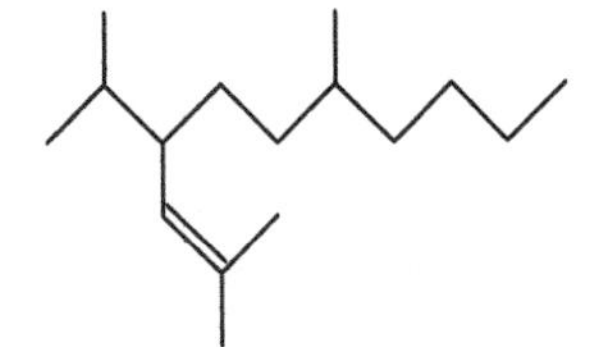

is much more easily and clearly understood as:

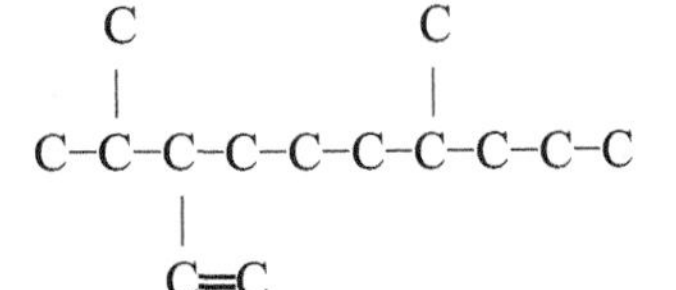

and can be easily assigned the proper name *cis*-4-isopropyl-2,7-dimethylundec-2-ene, according to the rules of nomenclature prescribed by the International Union of Pure and Applied Chemistry (IUPAC).

Cyclic alkenes are named in a similar manner. The base name reflects the number of carbon atoms in the largest ring structure in the molecule. The first position in the ring structure is assigned to the carbon atom bearing the first carbon atom of the C=C bond, and other substituents are assigned so as to attain the lowest set of position numbers. A cyclooctene ring bearing a methyl substituent on the third carbon atom from the beginning of the C=C bond and an ethyl substituent on another carbon atom two positions farther around the ring would be named 5-ethyl-3-methylcyclooctene. Cyclic alkene structures are slightly harder to draw due to the number of carbon atoms in the ring and the geometric constraints imposed by the C=C system, but

SAMPLE PROBLEM

Given the chemical names 4-isopropyl-2,3-dimethylnon-2-ene and 1,3-diethyl-5-isopropylcycloheptene, draw the skeletal structures of the molecules.

Answer:

The name indicates that the longest chain is nine carbons long (*nonene*). The C=C bond is located between the second and third carbon atoms in the chain *(-2-ene)*. There is an isopropyl side chain on the fourth atom in the chain and a methyl group on the second and third carbon atoms in the nonane structure. The structure is therefore

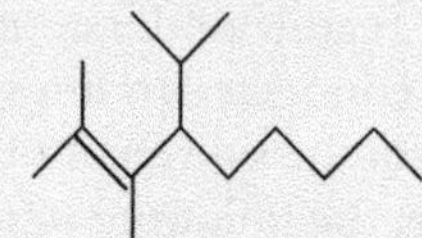

The second alkene compound is composed of seven carbon atoms in a ring (cycloheptene). The C=C bond is the first position in the ring (since no other position is specified in the name). There is an ethyl side chain on one carbon atom and another on the second carbon atom over from the first one. There is also an isopropyl side chain on the fourth carbon atom around from the first. The structure is therefore

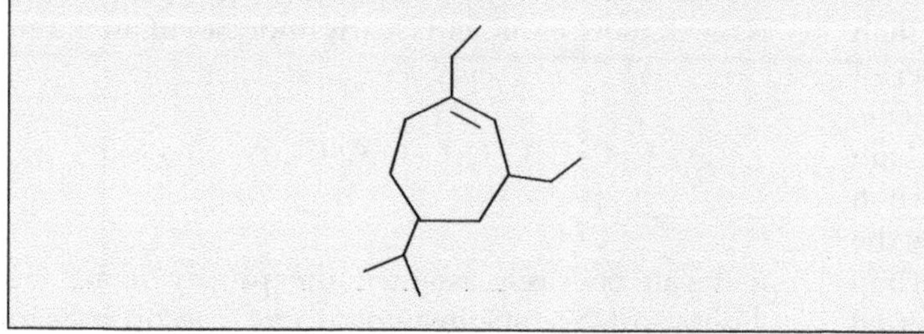

the structural drawings are even more informative than they are for alkane structures.

FORMATION OF ALKENES

The simple alkenes do not occur naturally in appreciable quantities. They are instead produced by dehydration reactions that eliminate the components of water molecules from adjacent carbon atoms in alcohols. Ethanol, for example, can be dehydrated to produce ethylene and water, according to this equation:

$$H_3C\text{–}CH_2\text{–}OH \rightarrow H_2C{=}CH_2 + H_2O$$

Other alkenes are produced in quantity by similar methods. However, the processes are not selective and will generally produce mixtures of alkene isomers rather than single compounds. The processes used are reversible under the conditions used, requiring that the product water and alkenes must be separated from each other as soon as possible. This is usually not a significant problem due to the typically broad difference in boiling points of water and the lower alkenes. In laboratory procedures, specific methods are used to generate C=C bonds in a particular molecular structure. Controlled dehydration reactions can be used to eliminate water molecules from a molecule containing a suitable –OH group. Dehydrohalogenation can be used to eliminate the hydrogen halide molecules when a suitable –X substituent (X = chlorine, bromine, or iodine) is present. Another method is dehalogenation, which eliminates X_2 from molecules containing two halogen atoms on adjacent carbon atoms. Many other specific synthetic methods can also be used in which a leaving group is eliminated from one carbon atom and another substituent from an adjacent atom. These methods are effective when the leaving group and its companion can form a very stable compound that does not react with the C=C bond.

REACTIONS OF ALKENES

Alkenes are reactive compounds. They readily undergo reduction reactions such as hydrogenation, in which a hydrogen atom is bonded to each of the two carbon atoms of the C=C bond. This is typically carried out using a metal catalyst. Alkenes are also readily and rapidly oxidized by good oxidizing agents, such as ozone (O_3) and permanganate ion (MnO_4^-). Electrophilic addition reactions such as halogenation and hydrohalogenation are also facile reactions with alkenes. The C=C bond forms a flattened region in a molecular structure that exposes the electron-rich region, facilitating bond formation with an electrophile (a species attracted to electrons). Simple alkenes are thus able to readily undergo polymerization reactions with ease. Ethylene and stryrene, for example, readily accept an electrophile and subsequently enter into a chain reaction that produces polyethylene and polystyrene, respectively. As hydrocarbons, alkenes are highly combustible, and the extra energy of the C=C bond increases the amount of energy released in combustion and other oxidation reactions. Their reactivity precludes their use as solvents, except when the presence of several C=C bonds in a cyclic structure, as in benzene, makes the compound extremely stable.

Richard M. Renneboog, MSc

Bibliography

Fenichell, Stephen. *Plastic: The Making of a Synthetic Century*. New York: Harper, 2009. Print.

Haynes, W. H., PhD, David R. Lide, PhD, and Thomas J. Bruno, PhD, eds. *CRC Handbook of Chemistry and Physics,* 96th Edition. New York: CRC/Taylor & Francis Group, 2015. Print.

Hendrickson, James B., Donald J. Cram, and George S. Hammond. *Organic Chemistry.* 3rd ed. New York: McGraw, 1970. Print.

Morrison, Robert Thornton, and Robert Neilson. Boyd. *Organic Chemistry.* 7th ed. Englewood Cliffs, N.J.: Prentice Hall, 2003. Print.

Myers, Richard. *The Basics of Chemistry.* Westport: Greenwood, 2003. Print.

ALKYL HALIDES

FIELDS OF STUDY

Organic Chemistry

Summary

The characteristic properties and reactions of alkyl halides are discussed. Alkyl halides are an infinite series of organic compounds in which one or more halogen atoms have replaced hydrogen atoms. Their variety and reactivity are due to the electron distribution and physical size of the carbon and halogen atoms. The alkyl halides parallel the series of structures of the alkanes.

PRINCIPAL TERMS

- **dipole:** the separation of positive and negative charges within a single molecule due to electron density being relatively high in one part of the molecule and relatively low in another.
- **functional group:** a specific group of atoms with a characteristic structure and corresponding chemical behavior within a molecule.
- **halide:** a binary compound consisting of a halogen element (fluorine, chlorine, bromine, iodine, or astatine) bonded to a non-halogen element or organic group; alternatively, an anion of a halogen element.
- **hydrocarbon:** an organic compound composed solely of carbon and hydrogen atoms.
- **substitution reaction:** a chemical reaction in which one component of a compound is replaced by a different atom or group of atoms without altering the basic structure of the molecule.

The Nature of the Alkyl Halides

The alkyl halides are a series of hydrocarbons, in which at least one hydrogen atom has been replaced by a halogen atom. The halogen atoms in alkyl halides act as a functional group, determining the reactivity of the materials. The geometric arrangement of atomic orbitals on the carbon atom and the halogen atom dictate the alkyl halide's overall structure. There are four valence electrons in each carbon atom, two in the 2*s* orbital and one each in two of the three 2*p* orbitals. By combining the 2*s* and the three 2*p* orbitals, the carbon atom is able to form four equivalent *sp3* hybrid atomic orbitals, each containing a single electron. The halogen atoms require only one additional electron to form the electron distribution corresponding to the nearest inert gas atom. Thus, halogens readily form a covalent single bond, or sigma (σ) bond, to carbon atoms.

The alkyl halides have two parallel families of acyclic molecular structures and cyclic molecular structures. The carbon atom skeletons of the acyclic alkyl halide series can be expanded infinitely, at least in principle. Halogen atoms can be substituted for hydrogen atoms to any extent in the alkanes (saturated hydrocarbons) to produce the corresponding alkyl halides. The simplest alkyl halide is fluoromethane, CH_3F, having just one carbon atom and three equivalent bonds to as many hydrogen atoms. The one-carbon methane analogs are difluoromethane (CH_2F_2), trifluoromethane or fluoroform (CHF_3),

ALKYL HALIDES

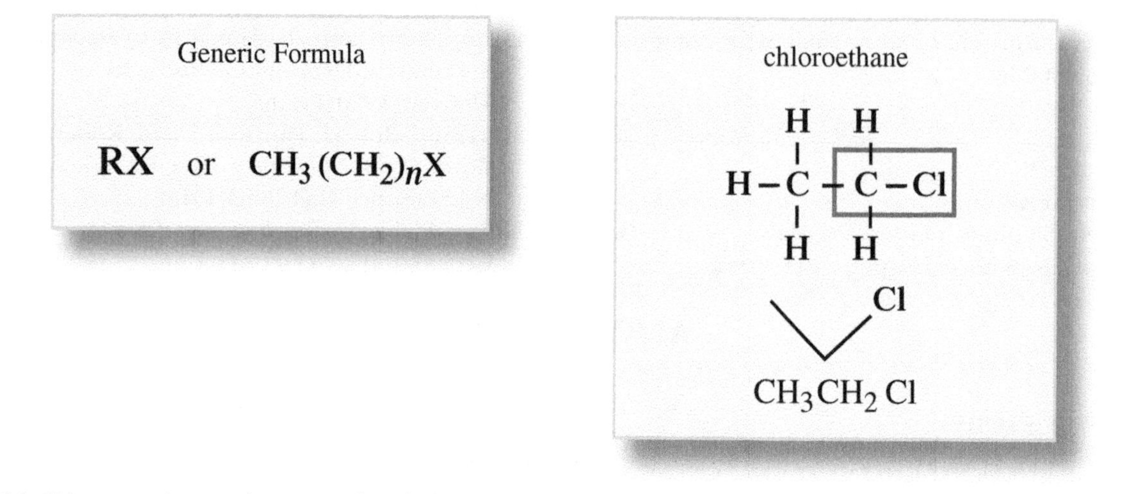

Alkyl halides contain a carbon atom bonded to an atom from the halogen family. The letter X represents any halogen in the chemical formula. R-groups can be any hydrocarbon.

and tetrafluoromethane (CF_4). The corresponding chlorinated analogs CH_3Cl, CH_2Cl_2, $CHCl_3$ and CCl_4 are all commonly known, as are the brominated and iodinated analogs. The two-carbon ethane has a similar series of halogenated derivatives. In principle, all of the various hydrocarbons have their halogenated analogs; in practice, however, they begin to destabilize when two or more halogens are present on adjacent carbon atoms. The larger size of the halogen atoms physically strains the molecular structure, and compounds containing two bromine or iodine atoms tend to give up their halogen components, via dehalogenation and dehydrohalogenation reactions, very easily. Alkanes containing several fluorine atoms, and especially the perfluorinated alkanes, are exceptional. The high electronegativity of the fluorine atom makes C–F bonds very stable, while the relatively small size of the fluorine atom does not physically strain the structure of the molecule.

With each additional carbon atom in the hydrocarbon series, the number of possible isomeric forms increases. An isomer is one of two or more chemical species that have the same molecular formula but different molecular structures. The substitution of halogen atoms for hydrogen atoms vastly increases the number of isomeric forms for each chemical formula, beginning with the three-carbon hydrocarbons. Whereas only one isomer of propane (C_3H_8) is known, there are two isomers in which a halogen atom (X) has been substituted for a hydrogen atom, corresponding to the chemical formula C_3H_7X. When two hydrogen atoms have been substituted ($C_3H_6X_2$), the number of isomers increases to three, then five when three halogens are present ($C_3H_5X_3$), six when four halogens are present, five when five halogens are present, three for six halogens, two for seven halogens, and one perhalogenated compound. A similar pattern of the number of derivative compounds applies for all of the hydrocarbon structures, in which the highest number of isomers for a particular chemical formula occurs when half of the hydrogen atoms have been substituted by halogen atoms. In the cyclic alkyl halides, the carbon atoms are bonded together in a ring structure, but the same general principles of substituting halogen atoms for hydrogen atoms apply.

The presence of a halogen atom in a hydrocarbon structure also creates an electric dipole in the molecule. The higher electronegativity of the halide atom imparts an "electron-withdrawing effect" so that the electron density in the molecule increases near the halogen atom and decreases elsewhere. This makes possible dipole-dipole interactions that affect both the physical properties and the reactivity of the alkyl halides.

Nomenclature of Alkyl Halides

The alkyl halides are named systematically according to the identity of the parent hydrocarbon from which they have been formed. The hydrocarbon structure is named for the number of carbon atoms with the longest unbranched chain in the molecular structure. The formal name sequence begins with methane, ethane, propane, and butane as common names, and then progresses through names based on the number of carbon atoms: pentane, hexane, heptane, octane, nonane, decane, and so on. For alkenes and alkynes (unsaturated hydrocarbons containing one or more carbon-carbon C=C double bonds or carbon-carbon C≡C triple bonds, respectively), the corresponding *-ene* or *-yne* ending of the name is used. When the base chain has been determined, the positions of the substituents (side chains) are assigned so as to have the lowest position numbers in the structure. The position number is inserted into the name of the primary chain immediately before the name of the substituent. Thus, a seven-carbon chain with two chlorine atoms bonded to the second and third carbon atoms from one end would be named 2,3-dichloroheptane, not 5,6-dichloroheptane. When different side chains are present on a hydrocarbon molecule, they are assigned by the priority of their size, again to attain the lowest numbers, and named alphabetically.

Several common names exist for many of the alkyl halides. These continue to be used because they came into common usage before the International Union of Pure and Applied Chemistry (IUPAC) rules of nomenclature were developed. Thus, it is common practice to refer to $CHCl_3$, CHI_3, and CH_2Cl_2 as chloroform, iodoform, and methylene chloride, respectively, instead of the official IUPAC names trichloromethane, tri-iodomethane, and dichloromethane. The compound CCl_4 is commonly referred to as "carbon tet" (an abbreviation of carbon tetrachloride) instead of tetrachloromethane. Another common usage is to refer to the simpler compounds using an alkyl halide name, such as ethyl bromide, rather than bromoethane.

Organic structures are typically represented by structural formulas, line drawings that depict the angles between the carbon atoms in such a way that every angle vertex and line end represents a single carbon atom. Such line drawings allow chemists to communicate a large amount of chemical information in a very small space. This is a convenient and readily understood representation that makes the nature and identity of the side chains immediately apparent. These line drawings use the elemental symbol of any atom other than carbon and hydrogen in its proper location in the molecular structure. The compound 3,3-dichloro-6-butylundecane, for example, is more readily comprehended when depicted as the following structure:

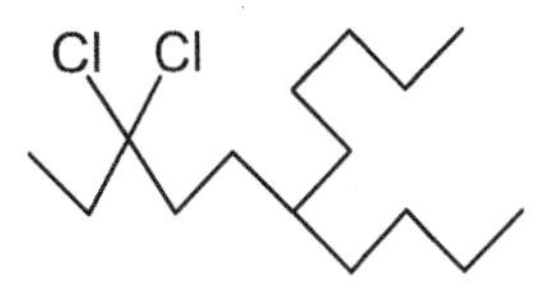

than as this one:

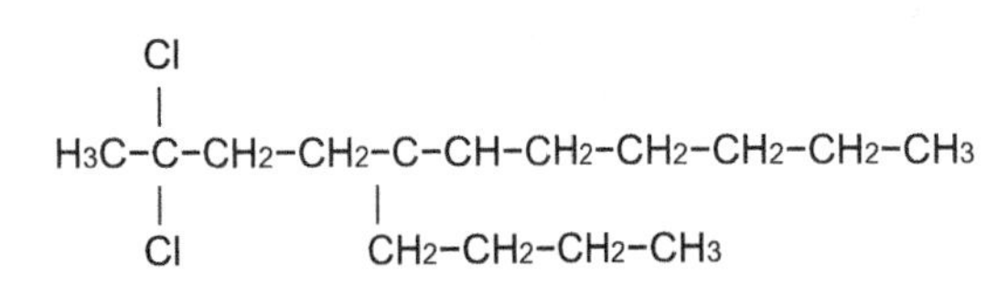

Cyclic alkyl halides are named in a similar manner. The basic name reflects the number of carbon atoms in the molecule's largest ring structure. The first position in the ring structure is assigned to the carbon atom bearing the first substituent of the first carbon atom of a C=C or C≡C bond, and other substituents are assigned so as to attain the lowest set of position numbers. A cyclohexane (C-6) ring with a bromine atom on one carbon atom and an ethyl substituent on another carbon atom two positions farther around the ring would be named 1-bromo-3-ethylcyclohexane. Cyclic alkyl halide structures are as easily drawn as the corresponding hydrocarbon structure.

Formation of Alkyl Halides

Simple alkyl halides can be prepared by a substitution reaction between an alcohol and a hydrogen halide. This normally requires the presence of an inorganic salt, such as zinc (II) dichloride, $ZnCl_2$. More complex alkyl halides require more specific reactions, however, since substitution in acid proceeds through the formation of a carbonium ion intermediate (an ion containing a positively charged carbon atom), which makes many different outcomes possible depending on the structure of the alcohol molecule. A carbonium ion will always rearrange from its original

SAMPLE PROBLEM

Given the chemical name 2,3–dibromo–2,3–dimethylbutane, draw the skeletal structure of the molecule.

Answer:

The name indicates that the material is a derivative of butane, a four-carbon chain of the alkane series. There are two bromine atoms in the structure, on the second and third carbon atoms of the butane skeleton. There are also two methyl groups—also on the second and third carbon atoms. Therefore, the compound has the structure

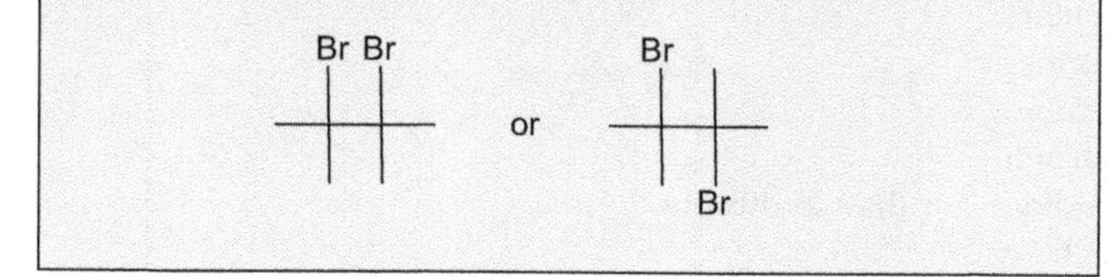

form to a more stable form, which can involve significant rearrangement of the bonds within the molecular structure. Alternatively, a carbonium ion can eliminate a proton to form an alkene, or it can add a nucleophile (an electron-rich species) that may be present in the reaction mixture.

Other methods of forming alkyl halides normally involve the addition of halogens to alkenes. A standard test for the presence of a C=C bond in a molecule is bromination. A small amount of a solution of bromine (Br_2) in carbon tetrachloride (CCl_4) is added to a solution of the suspected alkene or alkyne (an unsaturated hydrocarbon containing at least one C≡C bond). The intense color of the Br_2/CCl_4 solution quickly disappears as the bromine atoms add to the C=C bond to form the corresponding dibromide compound from the C=C bond. As an electrophile (a species attracted to electrons), the C=C bond is also amenable to the addition of hydrogen halides (HX).

Halogen atoms can also be substituted into hydrocarbons by "radical" mechanisms. A radical is an electrically neutral portion of a molecule with an unpaired electron in a bond-forming orbital. Chlorination by a radical mechanism begins with the chlorine molecule separating into two chlorine atoms. One of the chlorine atoms can "abstract" a hydrogen atom from the molecular structure to form hydrochloric acid (HCl), leaving the rest of the molecule as an alkyl radical. This alkyl radical can then form a bond with the other chlorine atom to produce the alkyl chloride. The process is difficult to control, and products can be unpredictable. Chlorination is the most useful radical process. Bromination and iodination do not proceed sufficiently to be useful in producing alkyl bromides and iodides, while fluorination is so aggressive that it will destroy both the carbon-carbon bonds and the carbon-hydrogen bonds in the molecule.

REACTIONS OF ALKYL HALIDES

Alkyl halides are highly reactive compounds under the right conditions. An existing alkyl halide can be converted into another halide or a substituted compound through an appropriate substitution reaction. Their most useful and most used application is as alkylating agents in the synthesis of other compounds. They are the key components of both the Friedel-Crafts and Grignard reactions. The Friedel-Crafts alkylation reaction adds the alkyl group of an alkyl halide to an aromatic system, such as the benzene ring. A Lewis acid compound, such as $AlCl_3$ or $FeCl_3$, catalyzes the reaction. An activated complex forms from the catalyst and the alkyl halide, which are attracted to the electrons of the aromatic system. This process allows the alkyl group to substitute for a hydrogen atom. For example, the Friedel-Crafts alkylation reaction between benzene and *t*-butyl chloride (2-chloro-2-methylpropane), with $AlCl_3$, produces *t*-butyl benzene (2-methyl-2-phenylpropane), according to the following reaction:

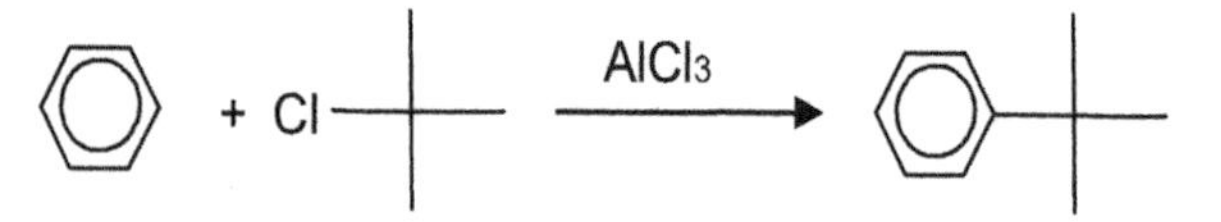

The Grignard reaction takes place between an alkyl halide and magnesium metal. A magnesium atom becomes interposed between the halogen and carbon atoms to which it was bonded, forming the corresponding alkyl magnesium halide compound. The reaction to form the Grignard reagent must be carried out in "dry" solvent that has been treated to remove any traces of water and is also quite vigorous. Once formed, the Grignard reagent can be applied in a wide variety of reactions that form new bonds between carbon atoms. An addition to a carbonyl group (C=O), for example, can turn an ester (RCOOR') into a ketone (RCOR'), an aldehyde (RCOH) into a secondary alcohol (RR'CHOH), or a ketone (RCOR') into a tertiary alcohol (RR'R"COH). (Note that in the general formulas, R, R', and R" represent alkyl groups.) Grignard reactions are perhaps the most generally useful of all synthesis reactions.

Alkyl halides also react with ammonia (NH_3) to produce amines and with amines to produce more highly alkylated amines. These reactions demand careful control, however, as a mixture of all possible products typically results.

Richard M. Renneboog, MSc

Bibliography

Fenichell, Stephen. *Plastic: The Making of a Synthetic Century.* New York: Harper, 2009. Print.

Haynes, W. H., PhD, David R. Lide, PhD, and Thomas J. Bruno, PhD, eds. *CRC Handbook of Chemistry and Physics,* 96th Edition. New York: CRC/Taylor & Francis Group, 2015. Print.

Hendrickson, James B., Donald J. Cram, and George S. Hammond. *Organic Chemistry.* 3rd ed. New York: McGraw, 1970. Print.

Morrison, Robert Thornton, and Robert Neilson. Boyd. *Organic Chemistry.* 7th ed. Englewood Cliffs, N.J.: Prentice Hall, 2003. Print.

Myers, Richard. *The Basics of Chemistry.* Westport: Greenwood, 2003. Print.

ALKYNES

FIELDS OF STUDY

Organic Chemistry

SUMMARY

The characteristic properties and reactions of alkynes are discussed. Alkynes are an infinite series of compounds containing only carbon and hydrogen atoms and at least one carbon-carbon triple bond. Their variety is due to the unique electronic distribution and physical size of the carbon atom. The alkynes parallel the series of structures of the alkanes.

PRINCIPAL TERMS

- **alkylation:** a combination reaction that results in the addition of an alkyl group to a molecule.
- **functional group:** a specific group of atoms with a characteristic structure and corresponding chemical behavior within a molecule.
- **hydrocarbon:** an organic compound composed solely of carbon and hydrogen atoms.
- **hydrogenation:** a chemical reaction in which two hydrogen atoms, usually in the form of molecular hydrogen (H_2), are bonded to another molecule, almost always as a result of catalysis.
- **triple bond:** a type of chemical bond in which two adjacent atoms are connected by six bonding electrons rather than two.

The Nature of the Alkynes

The alkynes are a series of compounds consisting of only carbon and hydrogen atoms, or hydrocarbons. Alkynes contain at least one triple bond between two adjacent carbon atoms (C≡C) as a functional group in their molecular structure and so are highly reactive materials. The structure of alkynes derives from the geometric arrangement of atomic orbitals on the carbon atom. There are four valence electrons in each carbon atom, two in the 2*s* orbital and one each in two of the three 2*p* orbitals. By combining the 2*s* and one of the three 2*p* orbitals, the carbon atom is able to form two equivalent *sp* hybrid atomic orbitals, each containing a single electron. The other two electrons remain in the unhybridized 2*p* orbitals. The two hybrid orbitals are directed in opposite directions, with the carbon atom nucleus at the center. The *sp* hybrid orbitals form an angle of 180 degrees. This angle and the physical diameter of the carbon atom are well suited to the formation of innumerable structures by linkage of the carbon atoms. The *sp* orbitals of two adjacent carbon atoms readily form a covalent single bond, or sigma (σ) bond, and each carbon atom can form three such single bonds. The unhybridized *p* orbitals on the adjacent carbon atoms are able to overlap in a side-by-side orientation to form two pi (ϖ) bonds parallel to the sigma bond. Accordingly, alkynes form a large and varied group of chemical compounds in their own right, and the reactivity of the C≡C functional group enables the formation of many other compounds. In alkynes, the carbon atoms of the triple bonds do not have all of the bonds that they can, and another way to describe the alkynes is as the series of unsaturated hydrocarbons. If a hydrocarbon molecule contains only carbon-carbon single bonds (C–C) or double bonds (C=C), it is not classed as an alkyne, but as a hydrocarbon in the alkane or alkene series, respectively.

The alkynes have two parallel families of structures, one comprising acyclic molecular structures and the other comprising cyclic molecular structures. The carbon atom skeletons of the acyclic alkyne series can be expanded infinitely, at least in principle. The simplest alkyne is ethyne (acetylene), C_2H_2, having just two carbon atoms and two equivalent bonds to as many hydrogen atoms. The next are propyne (C_3H_4), butyne (C_4H_6), and so on. All alkynes have the general chemical formula C_nH_{2n-2}, which is the same as the cyclic alkene series. For alkynes with four or more carbon atoms (butyne or higher), there are also numerous possible isomers, chemical species with the same molecular formula but different molecular structures. The triple bond in butyne can be between either the first and second carbon atoms or the second and third carbon atoms in the molecule, both of which have the chemical formula C_4H_6. In addition, the geometry of the C≡C bond is rigidly fixed and rotation about a C≡C bond cannot occur as it can with a C–C bond. This does not contribute to the formation of geometric isomers, however, as it does in the alkene series. Accordingly, 2-butyne has only the two isomeric structures, represented as follows:

The two compounds have both different physical properties and chemical behaviors.

With each additional carbon atom, the number of possible isomeric forms increases. Four carbon atoms can be arranged in two isomeric alkyne forms. With five carbons atoms, there are three isomeric alkyne structures, and with six carbon atoms, there are seven isomeric alkyne structures. With seven carbon atoms, the number of isomeric alkynes increases to twelve. The presence of additional C≡C and C=C bonds further increases the number of possible skeletal arrangements. There are even more isomeric forms when optical isomers are included. Optical isomers are compounds in which a saturated carbon atom has four different substituent groups, or side chains, bonded to the same carbon atom. They are identical in every way except the order which the bonds to side chains are distributed about that central asymmetric carbon atom. While the physical properties and chemical behavior of two optical isomers are the same, they are nonetheless distinctly different structurally and therefore separate isomeric forms. Optical isomerism is not possible around a carbon atom in a single C≡C bond.

In the cyclic alkynes, the carbon atoms are bonded together in a ring structure, but the restrictions imposed by the linear geometry of the C≡C system prevents the formation of cyclic alkynes of fewer than ten carbon atoms and precludes the possibility of geometric isomers.

NOMENCLATURE OF ALKYNES

Alkynes are named according to the number of carbon atoms comprising the longest unbranched chain of carbon atoms in the molecular structure, and the positions of side chains are identified accordingly. The name sequence begins with ethyne, propyne, and butyne as common names, and then progresses through names based on the number of carbon atoms: pentyne, hexyne, heptyne, octyne, nonyne, decyne, and so on. When the base chain has been determined and has three or more carbon atoms, the position of the C≡C bond is indicated by the lower of the two position numbers of the carbon atom in the bond. The position number is inserted into the name of the primary chain immediately before the portion that identifies the compound as an alkyne. (Some naming conventions place the position number before the name of the chain.) A seven-carbon chain with a C≡C bond between the second and third carbon atoms from one end would thus be named hept-2-yne (or 2-heptyne), not hept-3-yne or hept-5-yne. The order of the side chains is determined next. The numerical positions of the side chains are assigned as determined by the position assigned to the C≡C bond. For example, a six-carbon chain that has methyl groups ($-CH_3$) on the third and fourth carbon atoms from the end that has the C≡C bond would be named 3,4-dimethylhexyne. When different side chains are present on a hydrocarbon molecule, they are assigned by the priority of their size, again to attain the lowest numbers, and named alphabetically.

ALKYNES

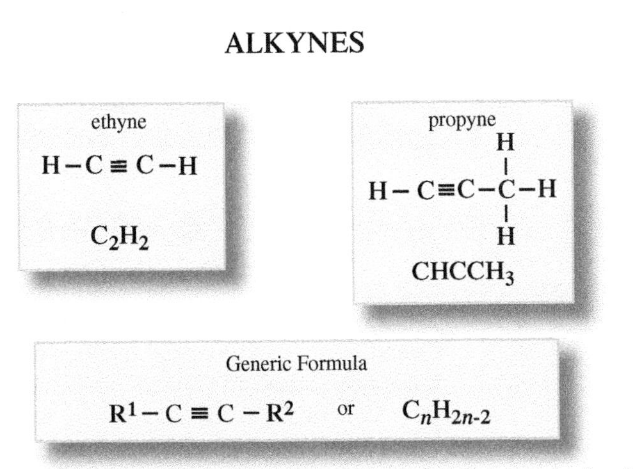

Alkynes can have one or more set of carbon atoms triple bonded together. R-groups may be a hydrogen atom, some form of a hydrocarbon, or some other functional group.

Organic structures are typically represented by structural formulas, line drawings that depict the angles between the carbon atoms in such a way that every angle vertex and line end represents a single carbon atom. Such line drawings allow chemists to communicate a large amount of chemical information in a very small space. This is a much more convenient and readily understood representation than strings of *C*s connected by lines and makes the nature and identity of the side chains immediately apparent. For example, the following structure:

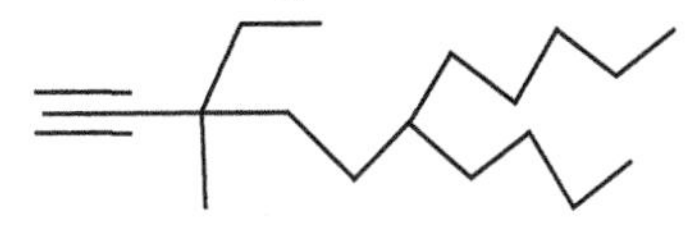

is much more easily and clearly understood than this:

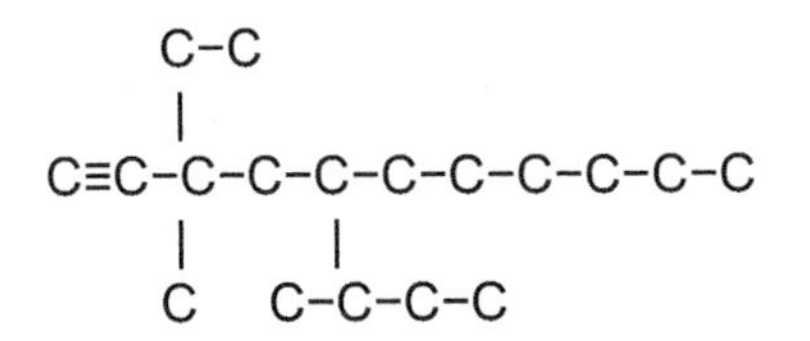

and can be easily assigned the proper name 6-butyl-3-ethyl-3-methylundecyne, according to the rules of nomenclature prescribed by the International Union of Pure and Applied Chemistry (IUPAC).

Cyclic alkynes are named in a similar manner. The basic name reflects the number of carbon atoms in the largest ring structure in the molecule. The first position in the ring structure is assigned to the carbon atom bearing the first carbon atom of the C≡C bond, and other substituents are assigned so as to attain the lowest set of position numbers. A cyclododecyne (C-12) ring bearing a methyl substituent on the third carbon atom from the beginning of the C≡C bond and an ethyl substituent on another carbon atom two positions farther around the ring would be named 5-ethyl-3-methylcyclododecyne. Cyclic alkyne structures are slightly harder to draw due to the number of carbon atoms in the ring and the geometric constraints imposed by the C≡C system, but the structural drawings are highly informative.

FORMATION OF ALKYNES

Alkynes are so highly reactive that they do not occur under normal conditions. Interestingly, compounds containing multiple C≡C bonds have been detected in interstellar space. The C≡C bond can be produced by dehydration reactions, which eliminate the components of water molecules from adjacent carbon atoms in diols, compounds containing two hydroxyl groups. Ethylene glycol, for example, can be dehydrated to produce ethyne and water, according to the following equation:

$$HO\text{–}H_2C\text{–}CH_2\text{–}OH \rightarrow HC{\equiv}CH + 2H_2O$$

Acetylene (ethyne) is never produced in this way, however. It is produced much more efficiently by the reaction of calcium carbide (CaC_2) with water, yielding acetylene and calcium hydroxide, $Ca(OH)_2$. Other alkynes are commonly produced by alkylation reactions in which the ethynyl group is added to other molecular structures as a substituent. Polyhalogenated compounds can be either dehalogenated or dehydrohalogenated to produce the C≡C bond as well. In laboratory procedures, specific methods are used to generate C≡C bonds in a particular molecular structure. Controlled dehydrohalogenation can be used to eliminate hydrogen halide molecules when two suitable –X substituents (X =

SAMPLE PROBLEM

Given the chemical name 4,4-dimethylpent-2-yne, draw the skeletal structure of the molecule.

Answer:

The name indicates that the material is a five-carbon chain (pentyne) with a C≡C bond between the second and third carbon atoms in the chain. There are two methyl groups (dimethyl) on the fourth carbon atom in the chain. The structure is therefore

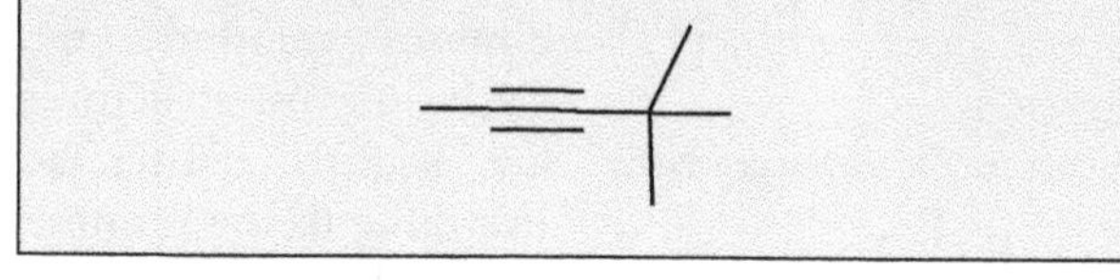

chlorine, bromine, or iodine) are present. Another method is dehalogenation, which eliminates X_2 from molecules containing four halogen atoms on adjacent carbon atoms. Often such reactions produce adjacent C=C bonds instead of the C≡C bond directly, but the resulting allene system can be made to rearrange into the alkyne structure.

Reactions of Alkynes

Alkynes are highly reactive compounds. They readily undergo reduction reactions such as hydrogenation, in which two hydrogen atoms are bonded to each of the two carbon atoms of the C≡C bond. This is typically carried out using a metal catalyst. Alkynes are also readily and rapidly oxidized by good oxidizing agents, such as ozone (O_3) and permanganate ion (MnO_4^-). Electrophilic addition reactions such as halogenation and hydrohalogenation are also facile reactions with alkynes. The C≡C bond forms an open linear region in a molecular structure that exposes the electron-rich region, facilitating bond formation with an electrophile (a species attracted to electrons). Simple alkynes are thus able to readily undergo polymerization reactions with ease. The compound dimethylacetylenedicarboxylate (DMAD) is widely used in Diels-Alder reactions to add functional groups to an existing molecular structure. DMAD typically undergoes electrocyclic self-addition reactions, in which pi bonds are changed to sigma bonds, to form a variety of polycyclic compound:

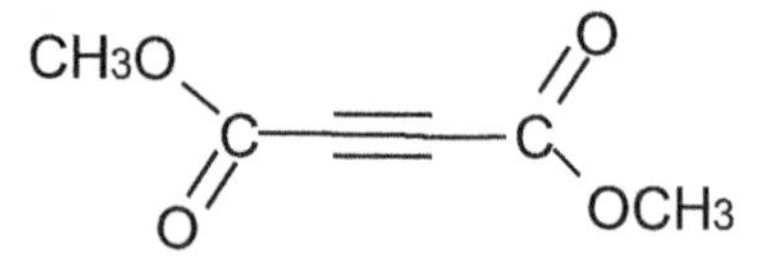

The compounds formed by self-addition are readily reversed by heating to regenerate the material in the desired form.

As hydrocarbons, alkynes are highly combustible, and the extra energy of the C≡C bond increases the amount of energy released in combustion and other oxidation reactions. Because of the high electron density of the C≡C bond, it is sufficiently easy for the hydrogen atom of a terminal alkyne to be removed by a strong base, producing an acetylide anion. This ion can then be used in nucleophilic addition and substitution reactions to add the C≡C functional group to an existing molecular structure, a process called alkylation. Many organometallic complexes are also known in which C≡C groups coordinate to metal atoms as ligands.

Richard M. Renneboog, MSc

Bibliography

Fenichell, Stephen. *Plastic: The Making of a Synthetic Century*. New York: Harper, 2009. Print.

Haynes, W. H., PhD, David R. Lide, PhD, and Thomas J. Bruno, PhD, eds. *CRC Handbook of Chemistry and Physics*, 96th Edition. New York: CRC/Taylor & Francis Group, 2015. Print.

Hendrickson, James B., Donald J. Cram, and George S. Hammond. *Organic Chemistry*. 3rd ed. New York: McGraw, 1970. Print.

Morrison, Robert Thornton, and Robert Neilson. Boyd. *Organic Chemistry*. 7th ed. Englewood Cliffs, N.J.: Prentice Hall, 2003. Print.

Myers, Richard. *The Basics of Chemistry*. Westport: Greenwood, 2003. Print.

ALLOTROPES

FIELDS OF STUDY

Inorganic Chemistry; Geochemistry

SUMMARY

An allotrope is a stable physical form of an element. Many elements can exist in more than one allotropic form. Examples are well-known for carbon, oxygen sulfur, phosphorus and other elements.

PRINCIPAL TERMS

- **allotrope:** one of two or more principal physical forms in which a single pure element occurs, due to differences in chemical bonding or the structural arrangement of the atoms; for example, diamond and graphite are two allotropes of carbon.
- **alloy:** a mixture of a metal and at least one other element, often another metal; also known as a solid solution.
- **amorphous:** a physical form that has no defined or regular structure.
- **buckminsterfullerene:** (C_{60}) an allotrope of carbon consisting of 60 carbon atoms bonded together in the form of a sphere.
- **carbon nanotubes:** an allotrope of carbon consisting of an indeterminate number of carbon atoms bonded together as in graphene and forming a closed tube shape.
- **crystal:** a solid consisting of atoms, molecules, or ions arranged in a regular, periodic pattern in all directions, often resulting in a similarly regular macroscopic appearance.
- **crystal lattice:** the regular geometric arrangement of atoms and molecules in the internal structure of a crystal.
- **fullerenes:** the allotropic form of carbon that includes carbon nanotubes, buckminsterfullerene, and all other such forms of carbon.
- **graphene:** an allotrope of carbon consisting of carbon atoms bonded together in a flat array of six-membered rings.
- **graphite:** an allotrope of carbon consisting of stacked layers of graphene, commonly used for pencil lead.
- **monoclinic:** a crystal shape in which two pairs of opposite faces are in the horizontal and vertical planes, and the third pair of opposite faces is inclined.
- **rhombic:** a crystal shape in which opposite faces are parallel and all are oriented at right angles; also called orthorhombic

ALLOTROPISM AND ALLOTROPES

Allotropism is the ability of an element to exist in more than one stable form. Each form typically has its own unique chemical properties and physical characteristics while retaining its elemental identity. The term allotrope comes from two Greek words meaning "other" and "form." Typically, the difference between allotropic forms of the same element is the manner in which their atoms are bonded to each other or arrayed in a crystal lattice. If the allotrope does not have a regular crystal structure or arrangement of atoms it is described as amorphous. Regular crystal structures are evident by the external shape of the material normally. Allotropes of an element can be thought of as analogous to the isomeric structures of compounds.

CARBON

Carbon has perhaps the most allotropic forms of any element. These include diamond and graphite as well as a number of other forms derived from graphene. Graphene is currently being strongly researched as potentially the most important "new" material known, due to its structure as a flat, two-dimensional sheet of carbon atoms bonded together in conjoined hexagonal rings. It is a curious fact that people have been writing and drawing with this "wonder material" since the first time someone scratched a piece of graphite coal across a flat rock. Graphite, which is composed of layer upon layer of graphene, is the material of choice for pencil lead. The material known as graphene was first identified as a single molecular species after researchers were able to isolate a single layer from a pencil mark by lifting it with a piece of adhesive tape. Graphene has potential as the basic material of transistor structures far smaller than those currently available on silicon-based computer chips, and is the leading candidate for the material of choice for quantum computers. Graphene can be doped with gold or other metal nanoparticles to form quantum dot structures consisting of just a small number of atoms that have very different electronic properties on such a small scale. Other allotropes of carbon are formed by closure of the graphene structure into spherical and tubular shapes. One called buckminsterfullerene, or more colloquially known as a buckyball, consists of sixty carbon atoms bonded together to form a sphere. Recognized at first as something of an oddity formed under special conditions, C_{60} was examined for its special properties. It has since been recognized as a common constituent of wood smoke and so is another example of a new wonder material that is not quite so new. Numerous other C_n molecules similar to C_{60} have since been identified, with the most interesting and potentially useful being the general class of fullerenes that includes carbon nanotubes. These materials are typically structured as a sheet of graphene that has rolled up to form a tube with closed ends, and come in a variety of sizes. Commercially, they are produced by vapor deposition techniques, but they also occur naturally in smoke from carbon sources.

SULFUR

Sulfur is able to exist in a number of equally stable allotropic elemental forms. Some sources claim that more allotropes of sulfur are known than of any other element. The most common of these consists of ring-shaped molecules of eight sulfur atoms. Being a third period element, the valence-shell electrons of the sulfur atom are in a higher quantum level than their counterparts in the carbon atom, and are also much closer in energy to the *d* orbitals. Accordingly, sulfur exhibits very different chemical behavior and its molecular structures give rise to several different allotropic forms. The most common of these consists of rings of eight sulfur atoms forming three distinct crystal structures that are temperature dependent. The amorphous form melts at 107°C (225°F). The most stable at standard temperature is the rhombic form that melts at 113°C (235°F). The monoclinic crystal form is thermodynamically stable only at temperatures above 95°C (203°F) but melts at just 118°C (244°F). As temperature increases, other molecular forms of sulfur are produced as the eight-membered rings open and concatenate into a series of polymeric molecules (S_{16}, S_{24}, S_{32}, etc.). At even higher temperatures, these polymeric forms dissociate into smaller molecules again, ultimately to individual sulfur atoms. The color of the material changes accordingly. When poured into cold water, molten sulfur forms an easily-molded rubber-like mass, and all allotropes slowly reorganize to the stable rhombic form at room temperature.

PHOSPHORUS

Phosphorus is positioned adjacent to sulfur in the periodic table, and so might be expected to have a similar number of allotropic forms. However, only three common stable forms of phosphorus are known. These are designated by color as white, red and

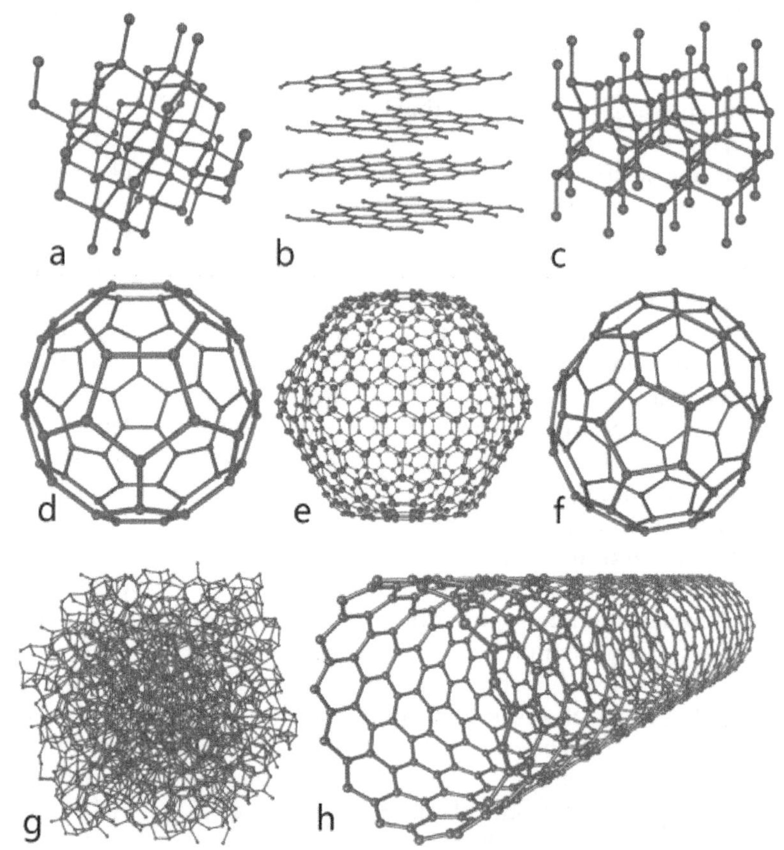

Eight Allotropes of Carbon. Created by Michael Ströck via Wikimedia Commons.

SAMPLE PROBLEM

White phosphorus normally acquires a coating of diphosphorus pentoxide (P2O5) Calculate the weight of P2O5 that can be produced from 100g of white phosphorus.

Answer:

Begin with the atomic and molecular weights of reactants and products. Since white phosphorus is an elemental form, it consists only of P atoms having atomic weight of 31 g.mol-1.

Molecular weight of P2O5 = (2 X 31 + 5 X 16) = 142 g.mol-1

Moles of P in 100 g = (100/31) = 3.22 mole

Moles of P2O5 = (3.22/2) = 1.61 moles

Therefore the weight of P2O5 produced = (1.61 X 142) = 228.6 g

black. The most common allotrope of phosphorus is the white form, which is highly reactive and combusts spontaneously when exposed to air. This feature has made it a fearsome component of munitions. The molecular structure of white phosphorus consists of units of four P atoms bonded together in a tetrahedral arrangement, and two temperature-dependent crystal structures are known. The α form exists at normal temperatures but transforms reversibly to the β form at 195.2 K (-78°C, -172°F). The difference between the two forms is the stacking of the P_4 units within their respective crystal lattices. White phosphorus is normally stored under water to prevent reaction with air, and at high pressures (12,000 atm) and temperatures of 350°C is converted to the thermodynamically most stable allotrope called black phosphorus. The P atoms of black phosphorus are linked together in six-membered rings like the carbon atoms of graphene, and the physical properties of black phosphorus are similar to those of graphite. Heating white phosphorus to 300°C (572°F) anaerobically or exposing it to sunlight causes the structure to break down to an amorphous arrangement of P atoms that is red in color.

THE IMPORTANCE OF ALLOTROPES

An allotrope of an element is almost like having an entirely new element with which to work as it expands the nature of the periodic table. The different allotropes of carbon, for example, have properties that are so different and unique that carbon can seem like five different elements instead of just one. This variable nature enables a wide variety of applications that can have very far-reaching effects and high economic value.

Richard M. Renneboog M.Sc.

BIBLIOGRAPHY

Wiberg, Egon and Wiberg, Nils. *Inorganic Chemistry.* San Diego, CA: Academic Press, 2001. Print.

Greenwood, N.N. and Earnshaw, A. *Chemistry of the Elements* 2nd ed. Burlington, MA: Elsevier Butterworth-Heinemann, 2005. Print.

Kutney, Gerald. *Sulfur. History, Technology, Applications & Industry.* Toronto, ON: ChemTec Publishing, 2007. Print.

House, James E., and Kathleen A. House. *Descriptive Inorganic Chemistry.* Waltham, MA: Academic Press, 2016. Print.

Rodgers, Glen E. *Descriptive Inorganic, Coordination, & Solid-State Chemistry* Toronto, ON: Nelson Education, 2011. Print.

Cobb, Allan B. *The Basics of Nonmetals* New York, NY: Rosen Publishing, 2014. Print.

Corbridge, D.E.C. *Phosphorus. Chemistry, Biochemistry and Technology* 6th ed. Boca Raton, FL: CRC Press, 2013. Print.

Toy, Arthur D.F. *The Chemistry of Phosphorus.* New York, NY: Pergamon Press, 1975. Print.

ALLOYS AND INTERMETALLICS

FIELDS OF STUDY

Metallurgy; Physical Chemistry

SUMMARY

Alloys are typically a combination of two or more metals, but may include any number of other elements as well. The term also applies to any combination of materials that achieves a similar result. Intermetallics are a subset of alloys and are characterized by properties that depend on components present in stoichiometric ratios. Such materials are an important area of ongoing research and development.

PRINCIPAL TERMS

- **alloy:** a mixture of a metal and at least one other element, often another metal; also known as a solid solution.
- **composition:** the identities and relative proportions of different elements or components in a compound, mixture, or other material.
- **compound:** a chemically unique material whose molecules consist of several atoms of two or more different elements.
- **ductility:** the ability of a solid material to be deformed by the application of tensile (pulling) force, such as bending, without breaking or fracturing.
- **element:** a form of matter consisting only of atoms of the same atomic number.
- **homogeneous mixture:** a physical combination of different materials that has a generally uniform distribution of composition, and therefore properties, throughout its mass; called a solution when liquid and an alloy when solid metal.
- **malleability:** the ability of a solid material to be deformed by the application of compressive (pushing) force, such as hammering, without breaking or fracturing.
- **metallic bond:** a type of chemical bond formed by the sharing of delocalized electrons between a number of metal atoms.
- **molecular formula:** a chemical formula that indicates how many atoms of each element are present in one molecule of a substance.
- **phase diagram:** a graphical representation of the physical phases present in systems of two or more different materials with regard to proportional composition and an environmental parameter such as temperature or pressure.
- **solution:** an intimate combination of two or more different materials forming a single physical phase, most typically liquid but may also be gaseous as a mixture of gases or solid as an alloy.
- **stoichiometric ratio:** the exact proportions of different elements or other components required for a balanced chemical reaction or to produce a material having a specific composition.

PHASE DIAGRAMS

Two or more different elements or other materials can be combined in a homogeneous mixture that can best be described as a solution. The composition of the materials can have any relative value, and for a two-component system this can obviously range from 100% of one component and 0% of the second, through to 0% of the first component and 100% of the second. The physical composition of the system can be affected by an environmental condition such as temperature, and each relative composition will exhibit its own specific physical composition according to the environmental condition. For example, a two-component system having a relative composition of 45% of one component and 55% of a second component may be completely solid below a certain temperature. As the temperature increases, however, a liquid phase may form and be in equilibrium with the solid phase, and with a gas phase as well. The same two components in different relative proportions, however, may exhibit a different set of phases. A phase diagram of the system can be drawn using the relative composition as the horizontal axis and the temperatures or pressures at which different phases appear for each composition. For such a graph, generally only one environmental variable (temperature, pressure, etc.) can be changed, and all others kept constant. The phase behavior of a one-component (unary) system can be shown graphically as both

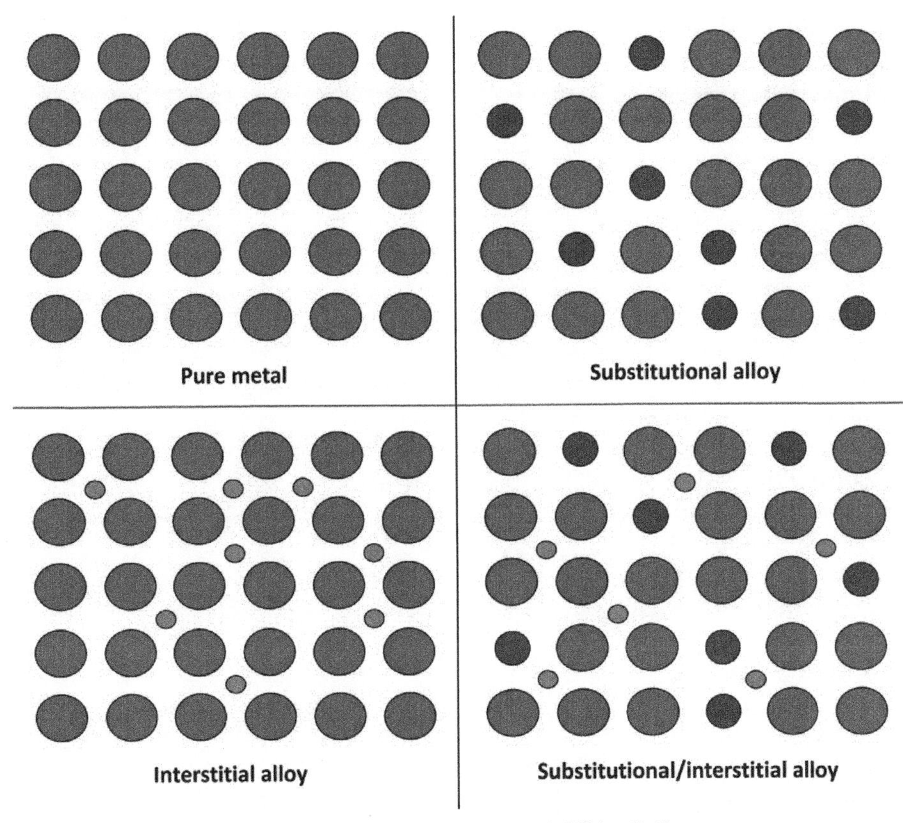

Alloy atomic arrangements showing the different types by Zaereth via Wikimedia Commons

temperature and pressure are changed. A two-component (binary) system presents a more complex scenario, and phase changes under combined temperature-pressure changes can only be shown graphically for a specific composition. Otherwise, temperature-related changes are portrayed at constant pressure, and pressure-related changes are portrayed at constant temperature. Three-component, and perhaps four-component, systems may also be represented using such graphs. However, the phase behavior of such multicomponent systems quickly becomes too complex for simple graphical representation. The number of phases in a system is given by Gibbs phase rule, as follows:

$$P = C - F + 2$$

where P is the number of phases present, C is the number of components and F is the number of degrees of freedom in the system. The number of degrees of freedom is the smallest number of independent variables that must be specified to completely describe the state of the system.

Solutions and Alloys

A solution is typically thought of as a liquid in which another material has dissolved to form a single phase having uniform composition and properties throughout. This description is not rigorous, however, as solutions can also be gaseous and solid. A gaseous solution is typically referred to as a gas mixture, with air being by far the most common example. Solid solutions are also very well known, generally as alloys although the term is applicable as a verb as well as a noun. Any combination of materials that form a single solid phase having uniform properties throughout can be described as an alloy of those materials. It applies equally to compounds and elements, and at this point a solid alloy is indistinguishable from a solution. The term "solid solution" is fully applicable. Metallic alloys have been known for thousands of years, the most well known of these being bronze, a combination of copper and tin. Varying the relative proportions of the two metals produces as many different grades of bronze, each with its own properties of malleability and ductility.

Stoichiometric Ratio

Compounds are formed when atoms share electrons to form interatomic bonds. The result is a fixed three-dimensional structure that corresponds to a specific molecular formula. Because only whole atoms can be involved in the formation of a chemical bond, the ratios of atoms in a molecular formula, and the ratios of reactants and products in a chemical reaction equation, are in whole number ratios called the stoichiometric ratio. The stoichiometric ratio of a compound represents the specific number of each type of atom in the basic structural unit of the material. While it is a hard and fast rule for compounds that the stoichiometric ratio must be stated in whole numbers, this is not necessarily true for blended materials such as alloys. In an alloy, different regions within the material can form combinations having a uniform composition different from the overall composition of the alloy. The important feature of such regions is that they form a different phase in the material, with a specific structural composition and correspondingly different properties. As such they are distinctly different materials and their structure normally forms through an association of an indeterminate number of atoms of each element in the blend. Each identical phase in an alloy must have the same composition and the same internal structure. Phases that meet this requirement are termed stoichiometric alloys, and typically have the sharp melting point that characterizes a pure compound. The difference between different phases that may exist in an alloy is often a difference in the crystal lattice stacking of the atoms in the particular phase. A different stacking pattern can be had for the same molecular formula. One example of how this affects physical properties and chemical behavior can be readily seen in the different allotropic forms of sulfur, in which the crystal structure of one form is monoclinic and another is rhombic, though both are composed of the same S8 molecules. Non-stoichiometric alloys, on the other hand, melt across a broadened range of temperatures, as any impure compound would.

The distribution of atoms throughout an alloy can be compared to the distribution of ions in an electrolytic solution, as is the case in Debye-Hückel theory. In Debye-Hückel theory, dissolved ions are deemed to be unevenly distributed rather than uniformly distributed, due to the various forces of attraction and repulsion that function between them and the effect of the surrounding solvent molecules. The atoms in an alloy are neutral, however, and so do not experience the electrostatic forces that exist between ions. Instead, the atoms in an alloy experience van der Waals forces, and the electronic influences that derive from their respective relative oxidation potentials in the electromotive series. In combination with the effect of atomic orbital energies in the different atoms, these factors introduce granularity into the distribution of atoms throughout the alloy. Each different type of granule that forms represents a different phase in the system. The structure of the granules may simply correspond to the overall composition of the alloy, differing only in the way in which the atoms are stacked into a crystalline array. Other granules, however, may exist only with a very specific stoichiometric ratio of component atoms.

SAMPLE PROBLEM

Calculate the number of phases that can exist in an alloy of iron, carbon, vanadium, chromium and silicon that is affected only by temperature and pressure.

Answer:

The Gibbs phase rule states that $P = C - F + 2$, where P is the number of phases, and F is the number of degrees of freedom.

Only two factors describe the system (temperature and pressure).

There are five components in the system (Fe, C, V, Cr, Si).

Therefore

$$P = C - F + 2$$

$$= (5 - 2 + 2)$$

$$= 5$$

Therefore only five phases should exist in this system according to the Gibbs phase rule.

INTERMETALLICS

Stoichiometric alloys belong to a special class of alloyed materials called intermetallics. An intermetallic material is not so much an alloy as it is a compound, and a great many such materials are known ranging from small organometallic compounds to ceramics and bulk alloys. The central feature of all intermetallic materials is that their identifying structural units have at their core a combination of two or more metal and other elements. Steel, for example, is typically alloyed with a certain small proportion of carbon. In the bulk metal, granules of triiron carbide (Fe3C) form, their presence adding strength characteristics to the material overall. Microscopic examination of the surface of a cross-section of a sample of steel reveals different structural regions due to stoichiometric intermetallic materials such as pearlite, bainesite and martensite. Intermetallic compounds are often synthesized in chemistry laboratories. They are generally referred to as cluster compounds because the center of each molecule is a cluster of one or more metal atoms, often with one or more atoms of another element capable of forming the required electronic interaction with the metal atoms. The interaction between atoms in intermetallic compounds is believed to be a combination of covalent bonding and metal conduction band overlap. These both require that the quantum energy levels of the atoms are sufficiently close in energy that overlap can occur. The metal atoms, typically period three and period four elements, utilize the d orbitals of the atoms to interact with other atoms, and particularly with other metal atoms. In an ordered array of metal atoms, unoccupied quantum levels are sufficiently close in energy to occupied d orbitals that electrons can move relatively freely from one to the other. This is the characteristic of the metal conduction band. In stoichiometric intermetallic materials this condition exists as a minimum energy condition that enables the formation of both true electronic bonding and a regular structure.

The Importance of Alloys and Intermetallics

The vast majority of metals in common use are in fact alloys of base metals that have enhanced properties due to the presence of the other atoms that have been blended in and the intermetallic materials that form within them as a result. Stainless steel, for example, is essentially a blend of iron and a small percentage of chromium. This combination is sufficient to completely prevent rusting of the iron and adds a great deal of toughness and durability to the material. Very few metals actually can be used in their pure form effectively, but become more useful and adaptable when alloyed with other elements. Pure gold, for example, is essentially oblivious to environmental conditions, but is so soft that it can be deformed between one's fingers. All gold constructs are made from harder, tougher alloys of gold. Semiconductor materials (Si, Ge, etc.) are not used in as pure elements, but must be doped by alloying them with other elements such as phosphorus. Ceramics offer a very broad and expanding range of intermetallic materials with increasing importance and applications. Surface chemistry, and particularly catalysis, is an important area of application for alloys and intermetallic materials. Research and development of intermetallic materials for pharmaceutical and other applications is ongoing.

Richard M. Renneboog M.Sc.

Bibliography

Campbell, F.C., ed. *Phase Diagrams. Understanding the Basics Materials.* Park, OH: ASM International, 2012. Print.

Carter, C. Barry, and Grant M. Norton. *Ceramic Materials Science and Engineering.* New York: Springer Science + Business Media, 2007. Print.

Dubois, Jean-Marie and Belin-Ferré, *Esther Complex Metallic Alloys. Fundamentals and Applications.* New York, NY: John Wiley & Sons, 2010. Print.

Hillert, Mats. *Phase Equilibria, Phase Diagrams and Phase Transformations. Their Thermodynamic Basis.* 2nd ed. New York, NY: Cambridge University Press, 2007. Print.

Moiseyev, Valentin N. *Titanium Alloys. Russian Aircraft and Aerospace Applications.* Boca Raton, FL: CRC Press, 2006. Print.

Pöttgen, Rainer, and Dirk Johrendt. *Intermetallics: Synthesis, Structure, Function.* Boston, MA: Walter de Gruyter GmbH, 2014. Print.

Sauthoff, Gerhard. *Intermetallics.* New York, NY: VCH, 1995. Print.

ALLYLIC ALCOHOLS

FIELDS OF STUDY

Organic Chemistry

SUMMARY

The characteristic properties and reactions of allylic alcohols are discussed. Allylic alcohols are very useful compounds in organic synthesis reactions, as they allow for the relatively easy formation of complex molecular structures from simple starting materials.

PRINCIPAL TERMS

- **alcohol:** an organic compound in which a hydroxyl is the primary functional group and is bonded to a saturated carbon atom.
- **double bond:** a type of chemical bond in which two adjacent atoms are connected by four bonding electrons rather than two.
- **functional group:** a specific group of atoms with a characteristic structure and corresponding chemical behavior within a molecule.
- **hydrolysis:** the cleavage of a chemical bond caused by the presence of water.

HYBRID ORBITALS IN ALLYLIC ALCOHOLS

Like other alcohols, allylic alcohols are characterized by the presence of a hydroxyl (–OH) functional group bonded to a saturated carbon atom—that is, a carbon atom that is attached to the maximum possible number of hydrogen atoms via single bonds only. The term "allylic" indicates that the hydroxyl group is bonded to a carbon atom that is adjacent to a carbon atom that is part of a carbon-carbon double bond (C=C). When this is the case, the carbon atom bonded to the hydroxyl group is said to be in the allylic position and can be referred to as an allylic carbon atom, while the hydroxyl group itself can also be called an allylic alcohol group.

An atom is made up of a central nucleus, consisting of positively charged protons and neutral neutrons, surrounded by negatively charged electrons. Electrons are often described as falling within electron shells and inhabiting specific regions about the nucleus known as orbitals. A carbon atom has four electrons in its valence, or outermost, electron shell: two electrons in the 2*s* orbital and two electrons in the 2*p* orbitals. Combining the 2*s* orbital with two of the three 2*p* orbitals results in the formation of three equivalent hybrid sp^2 orbitals. These three orbitals are arranged in a plane, to which the carbon atom's remaining 2*p* orbital is perpendicular. When two adjacent carbon atoms are hybridized in this way, a C=C bond can form between them. The two bonds that make up this double bond are known as the sigma (σ) bond and the pi (ϖ) bond. The sigma bond is formed by the end-to-end overlap of an sp^2 orbital from each atom, while the pi bond is formed by the side-by-side overlap of the non-hybridized 2*p* orbitals. Meanwhile, the allylic carbon atom hybridizes the 2*s* and all three of the 2*p* orbitals to form four equivalent sp^3 hybrid orbitals, meaning that it cannot form a double bond with another carbon atom, unless one of its hybrid orbitals regresses.

NOMENCLATURE OF ALLYLIC ALCOHOLS

Allylic alcohols are not assigned any specific nomenclature under the International Union of Pure and Applied Chemistry (IUPAC) naming conventions. The compounds are named as normal alkenes (organic compounds that include two carbon atoms connected by a double bond) containing a hydroxyl group, the location of which is specified in the name.

It is important to note that the term "allylic alcohol" is not a synonym for the similar-sounding "allyl alcohol." The latter term refers to a specific compound—the simplest of the allylic alcohols, with the molecular formula C_3H_6O—that falls into the broader category of the former, which are compounds characterized by the presence of a hydroxyl functional group bonded to an allylic carbon atom.

REACTIONS OF ALLYLIC ALCOHOLS

Allylic alcohols undergo all of the reactions typical of an alcohol, though they are more active in some ways than others. Substitution of the allylic hydroxyl group with a halide, or halogen anion, can be achieved by a reaction with hydrogen chloride (HCl) or hydrogen bromide (HBr). The halide compound dissociates into a halogen anion and a hydrogen cation (H^+),

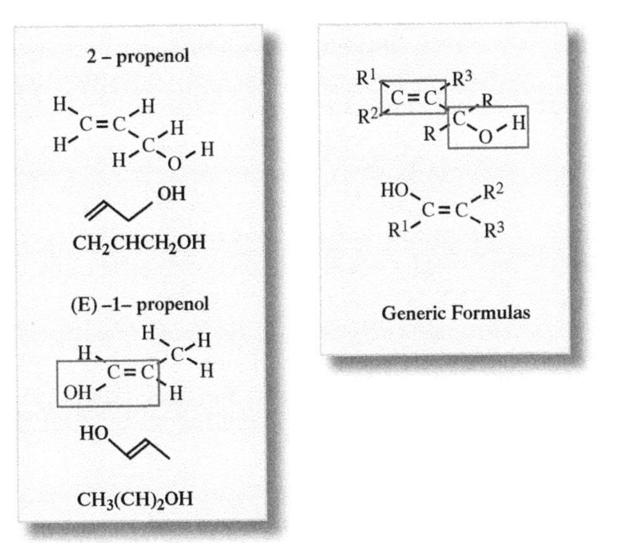

Allylic alcohols have a minimum of one set of carbon atoms double bonded together and an alcohol group. R-groups may be a hydrogen atom, some form of a hydrocarbon, or some other functional group.

and the cation, which is effectively a proton, is added to the -OH group, a process known as protonation. The loss of the resulting H_2O group from the alcohol molecule produces the allylic carbonium ion, which then bonds to the unattached halide ion.

Halogenated compounds, or organic compounds containing a halogen, are also effective in converting an allylic alcohol to the corresponding allylic halide. Allylic halides can be used in a variety of reactions, such as alkylation to form new carbon-carbon single bonds (C–C), allylic amines, and esters. Allyl chloride, for example, can be reacted with ammonia to produce allyl amine. Allyl alcohol is sufficiently acidic that it can be deprotonated—that is, lose a hydrogen cation—by a sufficiently strong base to form the corresponding allylic oxide salt of the metal, which can then be used in a number of different reactions. Deprotonation produces the alkoxide anion, which can then be used as a reagent in substitution reactions for the formation of allyl ethers.

Such reactions can be used to protect an allylic alcohol group in a molecule from undergoing an undesired side reaction in conditions meant to alter another part of a molecule. The allylic alcohol group can be first deprotonated to form the alkoxide anion and then reacted with a compound such as trimethysilyl chloride to produce the corresponding mixed-ether compound. This renders the allylic alcohol group safe from reaction when the other parts of the molecule are altered. The group can then be easily regenerated by a simple hydrolysis reaction that removes the protecting group, leaving the original hydroxyl group behind.

PRODUCTION OF ALLYLIC ALCOHOLS

Allylic alcohols are produced commercially by a variety of methods. Allyl alcohol can be produced by the dehydration of propylene glycol or, alternatively, hydrolysis of allyl chloride with a mild base. Similar processes are used to form an allylic alcohol functional group in another molecular structure. Manganese dioxide (MnO_2) can be used to oxidize a suitable C=C bond to the corresponding diol (a compound containing two hydroxyl groups). Carefully controlled dehydration of the diol product can subsequently produce an allylic alcohol.

SAMPLE PROBLEM

Given the chemical name 2,8-dihydroxy-4,8-dimethylnona-3,6-diene, draw the structure of the molecule and identify any allylic alcohol functional groups.

Answer:

The name 2,8-dihydroxy-4,8-dimethylnona-3,6-diene indicates that the longest carbon atom chain in the molecule consists of nine carbons (nona). There is a C=C bond between the third and fourth carbon atoms in the chain and another one between the sixth and seventh carbon atoms (3,6-diene). The second and eighth carbon atoms are bonded to a hydroxyl group (2,8-dihydroxy), and the fourth and eighth carbon atoms are bonded to a methyl group (4,8-dimethyl). The structure is therefore

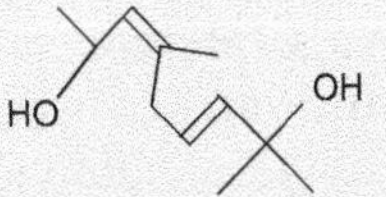

Since each carbon atom bonded to a hydroxyl group is also adjacent to a carbon atom that is part of a C=C bond, both hydroxyl groups are allylic alcohol functional groups.

Further dehydration can produce a mixture of allylic alcohols.

An allylic position can be formed in a molecule in many ways, but allylic alcohols are rather more difficult to achieve directly. The addition of a vinyl Grignard reagent to a ketone—a strictly structured compound consisting of a carbonyl (C=O) group as well as other carbon-containing groups—can result in an allylic alcohols. However, the structure of the ketone must preclude the possibility of dehydration after the allylic alcohol has been formed, as this would alter the substance further.

In industry, allylic alcohols are used primarily to produce substances such as resins and plasticizers, which are in turn involved in the development of safety glass, coatings, and other products. Allylic alcohols are extremely toxic, and their production and use are regulated by various governmental bodies.

Richard M. Renneboog, MSc

BIBLIOGRAPHY

Gooch, Jan W. *Encyclopedic Dictionary of Polymers.* New York: Springer, 2007. Print.

Hendrickson, James B., Donald J. Cram, and George S. Hammond. *Organic Chemistry.* 3rd ed. New York: McGraw, 1970. Print.

Mann, J., et al. *Natural Products: Their Chemistry and Biological Significance.* New York: Wiley, 1994. Print.

Morrison, Robert Thornton, and Robert Neilson. Boyd. *Organic Chemistry.* 7th ed. Englewood Cliffs, N.J.: Prentice Hall, 2003. Print.

Myers, Richard. *The Basics of Chemistry.* Westport: Greenwood, 2003. Print.

Winter, Mark J. *The Orbitron: A Gallery of Atomic Orbitals and Molecular Orbitals.* University of Sheffield, n.d. Web.

Wuts, Peter G. M., and Theodora W. Greene. *Greene's Protective Groups in Organic Synthesis.* 4th ed. Hoboken: Wiley-Interscience, 2007. Print.

AMINES

FIELDS OF STUDY

Organic Chemistry

SUMMARY

The characteristic properties and reactions of amines are discussed. Essentially unlimited in type, amines are useful compounds in organic synthesis reactions, helping form complex heterocyclic molecular structures from simple starting materials.

PRINCIPAL TERMS

- **functional group:** a specific group of atoms with a characteristic structure and corresponding chemical behavior within a molecule.
- **hydrogen bond:** a weak type of chemical bond formed by the attraction of a hydrogen atom to an electronegative atom—an atom with a strong tendency to attract electrons—in the same or another molecule.
- **primary:** describes an organic compound in which one of the hydrogen atoms bonded to a central atom is replaced by another atom or group of atoms, called a substituent.
- **secondary:** describes an organic compound in which two of the hydrogen atoms bonded to a central atom are replaced by other atoms or groups of atoms, called substituents.
- **tertiary:** describes an organic compound in which three of the hydrogen atoms bonded to a central atom are replaced by other atoms or groups of atoms, called substituents.

THE NATURE OF THE AMINES

The term "amine" refers to the class of organic compounds that include one or more nitrogen-based functional groups in their molecular structures. The nitrogen atom has five electrons in its valence electron shell, with two electrons in the 2*s* orbital and three electrons distributed among the three 2*p* orbitals. In order to minimize electron-electron repulsions between the orbitals, the electrons are able to "hybridize" to form C–N, C=N, and C≡N bonds. Compounds containing the C–N bond are called "amines"; compounds containing the C=N bond are termed "imines"; and compounds containing the –C≡N functional group are termed "nitriles."

The amines are divided into four classifications, beginning with the primary amines, which are characterized by the presence of the $-NH_2$ functional group. Generally, they are represented as $R-NH_2$ or $Ar-NH_2$, in which R– and Ar– are generic placeholders indicating an alkyl functional group (derived from a saturated hydrocarbon) and an aryl group (derived from an aromatic ring), respectively. The basic aryl amine is the compound aniline, which consists of a benzene ring with the $-NH_2$ group in place of one of the hydrogen atoms:

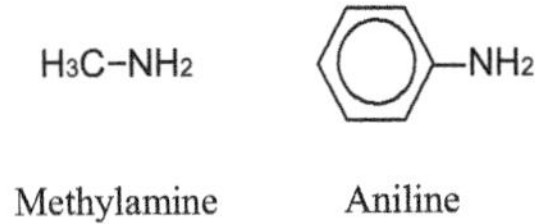

Methylamine Aniline

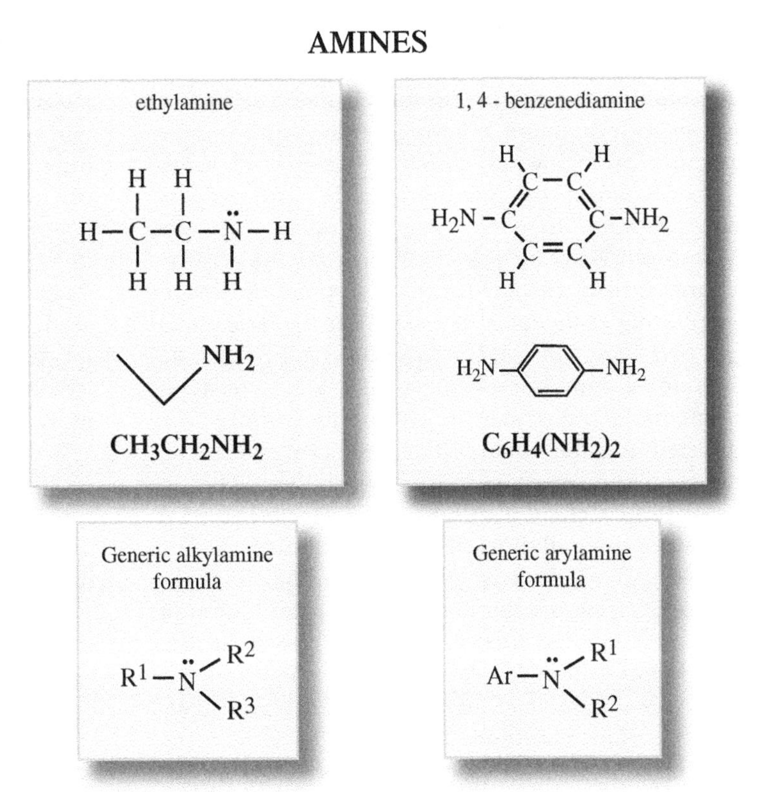

The secondary amines have two substituent groups bonded to the nitrogen atom. Thus, they have the general form R_2-NH, Ar_2-NH, or RAr-NH. Like primary amines, secondary amines are formed when the nitrogen atom is part of the ring structure of the compound. In such heteroatomic compounds, the carbon atoms bonded to the nitrogen atoms are considered to be the alkyl groups of the secondary amine structure. An example of such a compound is pyrrolidine:

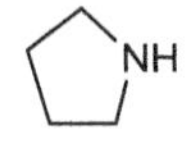

Pyrrolidine

Heterocyclic amines are common in nature and include a wide variety of naturally occurring compounds, such as the alkaloids.

The tertiary amines continue the same pattern, having three substituent groups attached to the nitrogen atom. The simplest tertiary amine is trimethylamine, a pungent alkaline compound with a melting point of –117 degrees Celsius (–178.6 degrees Fahrenheit). The simplest triarylamine is triphenylamine, a crystalline solid with melting point of 127 degrees Celsius (260.6 degrees Fahrenheit).

The lone pair of electrons on the nitrogen atom permits the formation of a chemical bond to a fourth alkyl group, which produces a fourth amine group, as quaternary ammonium salts. The reaction of trimethylamine with chloromethane, for example, produces the ionic compound tetramethylammonium chloride, according to the following equation:

$$H_3C-N(CH_3)_2 + Cl-CH_3 \longrightarrow H_3C-N^+(CH_3)_3\ Cl^-$$

Tetraalkylammonium salts are common additives in soap and shampoo formulations.

The lone pair of electrons on the nitrogen atom affords some special reactivity to amines. Because the *p* orbital holds its two electrons, there is less stability to be gained from hybridizing. When the nitrogen atom is bonded to a C=C bond, the orbital retains all or most of its *p* orbital character. In alkyl amines, the nitrogen atom adopts a tetrahedral arrangement, but in aryl amines, the geometry about the nitrogen atom is essentially a plane.

Nomenclature of Amines

Amines are named systemically according to the identity of the parent hydrocarbon from which they have been formed. The hydrocarbon structure is named for the number of carbon atoms on the longest unbranched chain in the molecular structure; the positions of substituent groups, or side chains, are identified accordingly. The formal name sequence begins with methane, ethane, propane, and butane as common names, and then progresses through names based on the number of carbon atoms: pentane, hexane, heptane, octane, nonane, decane, and so on. For alkene and alkyne compounds (unsaturated hydrocarbons), the corresponding *-ene* or *-yne* ending of the name is used. After the base chain has been determined, the positions of the substituents are assigned according to the lowest position numbers in the structure. The position number is inserted into the name of the primary chain immediately before the name of the substituent. Thus, a seven-carbon chain with two amino groups bonded to the second and third carbon atoms would thus be named 2,3-diaminoheptane and not 5,6-diaminoheptane. Using the standard International Union of Pure and Applied Chemistry (IUPAC) format, the substituent groups are named in alphabetical order in the formal name of the compound, which usually puts the amino substituents first.

Several "informal" names exist for many of the amines, and it is common practice to name a primary amine compound as a substituted amine rather than as an amino-substituted hydrocarbon. For example, 1-aminocycloheptane would typically be referred to as cycloheptylamine. Indeed, many common names for amines have been adopted as the proper IUPAC base name for derivative compounds.

SAMPLE PROBLEM

Given the chemical name (*N*-methyl, *N*-phenyl)cyclohexylamine, draw the skeletal structure of the molecule.

Answer:

The name indicates that the material is the amine of cyclohexane. There is a methyl group and a phenyl group bonded to the nitrogen atom of the amino functional group. The compound therefore has the structure

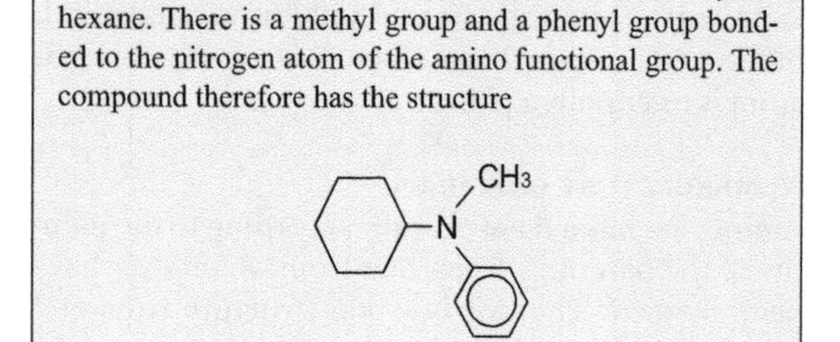

Organic structures are typically represented by line drawings that depict the angles between the carbon atoms in such a way that every angle vertex and line end represents a single carbon atom. Such line drawings allow chemists to communicate a large amount of chemical information in a small space. This is a convenient representation that makes the nature and identity of the side chains readily apparent. These "stick drawings" use the elemental symbol of any atom other than carbon and hydrogen in its proper location in the molecular structure. The compound 2,8-diamino-nona-*cis*-3-*trans*-6-diene, for example, is more readily comprehended when depicted as this:

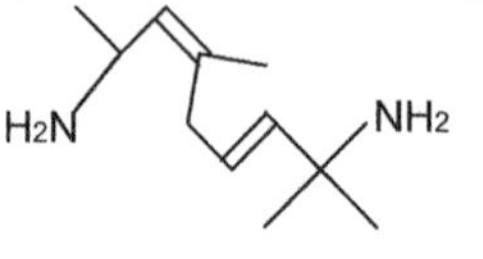

than as this:

```
CH3H CH3
|||
H2N-CH-C=C-CH2
|
CH=CH
|
H3C-C-CH3
|
NH2
```

For amines in which the nitrogen atom of the amino group bears alkyl or aryl substituents, an *N*- is placed before the name of the substituent. For example, a hexane bearing on the third carbon atom of an amino group that has a methyl and an ethyl substituent would be named 3-(*N*-ethyl, *N*-methyl) aminohexane.

Cyclic amines are named in a similar manner. The basic name reflects the number of carbon atoms in the largest ring structure in the molecule. The first position in the ring structure corresponds to either the carbon atom with the first substituent or the first carbon atom of a C=C or other definitive functional group; other substituents are assigned so as to attain the lowest set of position numbers.

FORMATION OF AMINES

Simple amines are formed by the reaction of an alkyl halide with ammonia. Ammonia (NH_3) is produced commercially by the Haber-Bosch process—the high-temperature, high-pressure reaction of nitrogen gas and hydrogen gas—according to this equation:

$$N_2 + 3H_2 \rightarrow 2NH_3$$

As the result of the reaction, ammonia displaces the halogen atom as halide ion, assisted by the elimination of a proton, to form the ammonium salt. Neutralization of the HX product releases the free amine. Aryl amines are normally produced by reducing the corresponding nitro- compound. Nitroarenes are prepared by the reaction of nitric acid (HNO_3) on an appropriate aryl compound, followed by reduction, using hydrogen (H_2) and a catalyst, to convert the $-NO_2$ (nitro) group to the amino group ($-NH_2$). Amines can also be prepared by reacting an ammonia or a primary amine with an appropriate aldehyde (R–CHO) or ketone (R–CO–R') compound, followed by reduction with H_2. In the initial reaction, the nitrogen atom adds to the carbonyl group (C=O) as a nucleophile (electron donor). A proton shifts from the nitrogen atom to the oxygen atom of the carbonyl group, and a molecule of water is subsequently eliminated to form an imine. Reduction of the imine produces the corresponding primary or secondary amine.

The reduction of nitriles can also produce primary amines. Reaction of the –C≡N bond with two molar equivalents of H_2 reduces the nitrile to the corresponding primary amine. When this type of reduction is too powerful, however, a variety of other reducing agents are available. More selectivity in the site of reduction and gentler conditions are the hallmark of reagents such as lithium aluminum hydride and sodium borohydride.

REACTIONS OF AMINES

Amines are reactive compounds widely used to form amides. An amine reacts with an acid chloride to produce the corresponding amide, in which the acyl functional group (–RCO) of the acid chloride substitutes for one of the hydrogen atoms on the nitrogen atom of the amine. For example, acetyl chloride reacts with methylamine to produce *N*-methylacetamide. Similarly, benzoyl chloride (the acid chloride of benzoic acid) can react with diethylamine to produce *N,N*-diethylbenzamide. Reaction with alkyl halides occurs in the same way. For the reaction to be effective, at least one replaceable hydrogen atom must be on the amine nitrogen atom. A similar reaction occurs with acid anhydrides (which contain two acyl groups bonded to a single oxygen atom), as in the reaction between ammonia and acetic anhydride that produces acetamide and ammonium acetate:

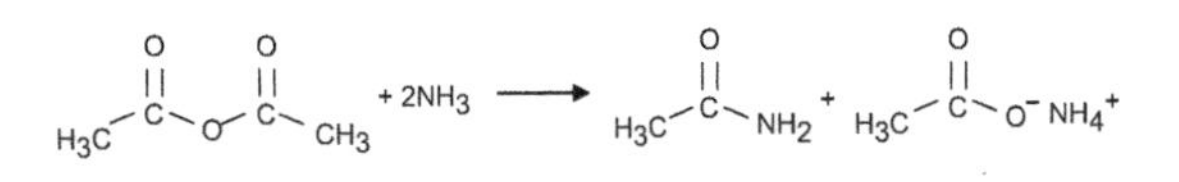

Reaction of a diamine with two equivalents of an acid chloride or an alkyl halide forms the corresponding diamide. However, if the two acid chlorides are in the same molecule, the reaction becomes a polymerization reaction. This is the means by which the series of nylon polymer formulations was discovered, and it remains the major means of their production. From that beginning, a large number of polyamides, including Kevlar, have been developed.

One of the more generally useful reactions of amines is their ability to add to a carbonyl group as a nucleophile. Under the proper conditions, the reaction can be used to convert an ester (R–COO–R') into an amide. More commonly, the reaction is used to form an imine as an intermediate stage in an overall synthesis plan. Once formed, the imine can be easily converted into a variety of other compounds.

Richard M. Renneboog, MSc

BIBLIOGRAPHY

Haynes, W. H., PhD, David R. Lide, PhD, and Thomas J. Bruno, PhD, eds. *CRC Handbook of Chemistry and Physics*, 96th Edition. New York: CRC/Taylor & Francis Group, 2015. Print.

Hendrickson, James B., Donald J. Cram, and George S. Hammond. *Organic Chemistry*. 3rd ed. New York: McGraw, 1970. Print.

Jones, Mark Martin. *Chemistry and Society*. 5th ed. Philadelphia: Saunders College Pub., 1987. Print.

Miessler, Gary L., Paul J. Fischer, and Donald A. Tarr. *Inorganic Chemistry*. 5th ed. Upper Saddle River: Prentice Hall, 2013. Print.

Morrison, Robert Thornton, and Robert Neilson. Boyd. *Organic Chemistry*. 7th ed. Englewood Cliffs, N.J.: Prentice Hall, 2003. Print.

Myers, Richard. *The Basics of Chemistry*. Westport: Greenwood, 2003. Print.

Powell, P., and P. L. Timms. *The Chemistry of the Non-Metals*. London: Chapman, 1974. Print.

Wuts, Peter G. M., and Theodora W. Greene. *Greene's Protective Groups in Organic Synthesis*. 4th ed. Hoboken: Wiley-Interscience, 2007. Print.

AMINO ACIDS

FIELDS OF STUDY

Biochemistry; Organic Chemistry

SUMMARY

The basic structure of amino acids is explained, as is their function in the creation of polypeptides and proteins. The transcription of nucleotides from DNA by RNA to create amino acid chains is also discussed.

PRINCIPAL TERMS

- **amino group:** a functional group containing a nitrogen atom bonded to two hydrogen atoms ($-NH_2$).
- **carboxyl group:** a functional group containing a carbon atom double bonded to an oxygen atom and single bonded to a hydroxyl group (-OH); has the formula CO_2H, typically written -COOH.
- **catalyst:** a chemical species that initiates or speeds up a chemical reaction but is not itself consumed in the reaction.
- **peptide bond:** a covalent bond that links the carboxyl group of one amino acid to the amine group of another, enabling the formation of proteins and other polypeptides.
- **protein:** a biological polymer consisting of one or more long chains of amino acids linked by peptide bonds in a sequence specified by an organism's DNA.

THE NATURE OF AMINO ACIDS

Strictly speaking, an amino acid is any compound whose molecular structure contains both an amino group and a carboxyl group, also called a carboxylic acid group—hence the term "amino acid." The term in general use, however, refers to the specific group of amino acids relevant to the genetic code in the DNA molecule. These twenty (or sometimes twenty-three, depending how they are classified) amino acids are called proteinogenic amino acids, which refers to the fact that they are the only amino acids used in the creation of proteins, enzymes, and other biomolecules. The three disputed amino acids are selenocysteine and pyrrolysine, which are not directly coded for in the genetic code but rather synthesized by other means and incorporated later, and *N*-formylmethionine, which initiates protein creation in some prokaryotes but is typically removed afterward. Selenocysteine is the only one of the three found in eukaryotes.

Of the standard twenty proteinogenic amino acids, nine are deemed "essential" because they are not synthesized in human metabolism but must be acquired through diet. All proteinogenic amino acids are also called α-amino acids, meaning that their amino and carboxyl groups are both bonded to the same carbon atom, known as the α-carbon (alpha carbon). This same carbon atom is also bonded to a hydrogen atom. The fourth atom or group bonded to the α-carbon determines the identity of the amino acid.

Chemically, the amino acids have unique properties due to the presence of both a base and an acid in the same molecule. Self-neutralization, in which the acid transfers a proton to the base, readily takes place to produce a zwitterion, an electrically neutral molecule in which both a positive and a negative charge exist in separate parts of the molecule at the same time. In addition, each amino acid has a unique isoelectric point, which is the specific degree of acidity or basicity (pH) at which the amino acid has no net electrical charge. These particular characteristics are responsible for most, if not all, of the behavior of amino acids and the much larger compounds they form as proteins and enzymes.

FORMATION OF PROTEINS AND ENZYMES

The structures of all proteins are determined by the sequence of amino acids encoded in the DNA molecule. With just one each of the twenty standard amino acids, there are thousands of trillions of possible ways to arrange them in what is called a polypeptide chain, and most proteins and enzymes contain far more than just twenty amino acids.

The synthesis of polypeptides and their formation into proteins is extremely quick, taking as little as six minutes, according to various tracer studies using radioactively labeled amino acids. The process begins with transcription, during which specific enzymes open up the double-stranded structure of a DNA

AMINO ACIDS

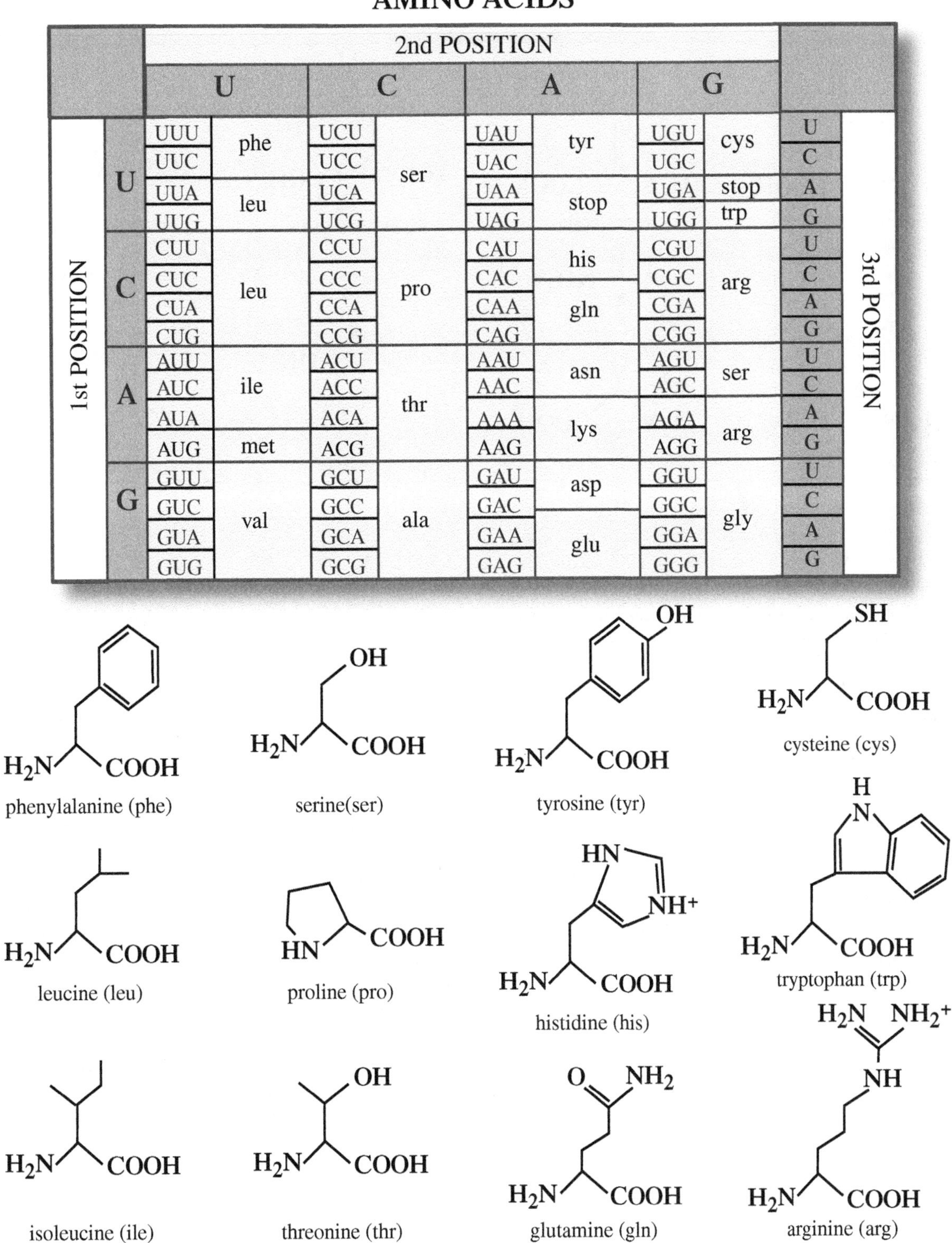

1st POSITION	2nd POSITION U		2nd POSITION C		2nd POSITION A		2nd POSITION G		3rd POSITION
U	UUU	phe	UCU	ser	UAU	tyr	UGU	cys	U
	UUC		UCC		UAC		UGC		C
	UUA	leu	UCA		UAA	stop	UGA	stop	A
	UUG		UCG		UAG		UGG	trp	G
C	CUU	leu	CCU	pro	CAU	his	CGU	arg	U
	CUC		CCC		CAC		CGC		C
	CUA		CCA		CAA	gln	CGA		A
	CUG		CCG		CAG		CGG		G
A	AUU	ile	ACU	thr	AAU	asn	AGU	ser	U
	AUC		ACC		AAC		AGC		C
	AUA		ACA		AAA	lys	AGA	arg	A
	AUG	met	ACG		AAG		AGG		G
G	GUU	val	GCU	ala	GAU	asp	GGU	gly	U
	GUC		GCC		GAC		GGC		C
	GUA		GCA		GAA	glu	GGA		A
	GUG		GCG		GAG		GGG		G

phenylalanine (phe)

serine(ser)

tyrosine (tyr)

cysteine (cys)

leucine (leu)

proline (pro)

histidine (his)

tryptophan (trp)

isoleucine (ile)

threonine (thr)

glutamine (gln)

arginine (arg)

AMINO ACIDS (cont.)

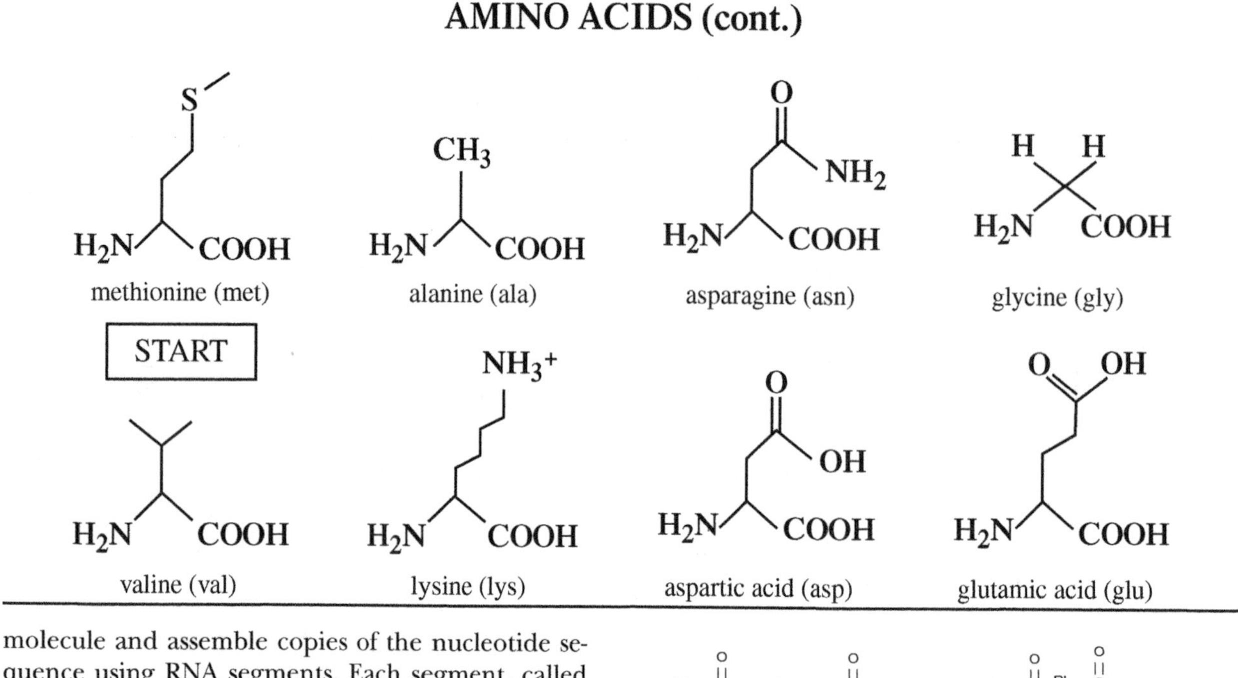

molecule and assemble copies of the nucleotide sequence using RNA segments. Each segment, called messenger RNA (mRNA), carries specific sequences of three nucleotides called codons. During the next step, translation, the mRNA translates the genetic code from DNA to structures called ribosomes, composed of ribosomal RNA (rRNA), where they match up with the rRNA sequence of nucleotides. When this occurs, the codons are exposed. In the cytosol (intracellular fluid) of the cell, a third type of RNA called transfer RNA (tRNA) transfers the specific amino acid corresponding to a particular codon to the mRNA strand in the ribosome. The anticodon on the tRNA segment matches to the codon on the mRNA strand, and specific enzymes there act as a catalyst to form a peptide bond between two neighboring amino acids. A peptide bond is just the normal amide structure that forms between a carboxylic acid and an amine:

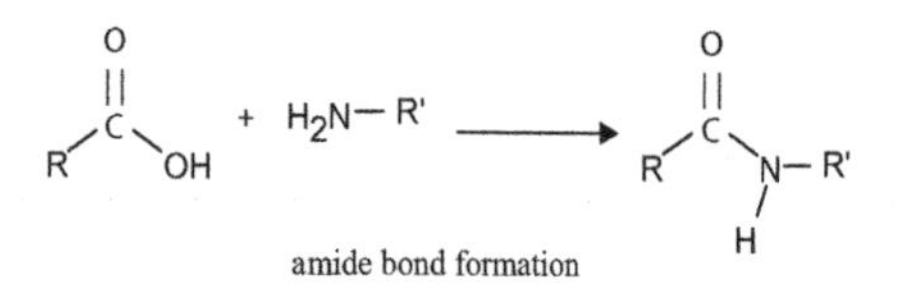

amide bond formation

The term is used to refer specifically to an amide bond formed between amino acids in a polypeptide, as here:

peptide bond formation

TRANSLATING THE GENETIC CODE INTO PROTEINS

The DNA and RNA molecules use only four different nucleotide bases to specify the entire genetic code, yet the number of possible three-nucleotide codons that can be formed from these is more than sufficient to differentiate the twenty standard amino acids. The system is actually quite redundant, with several different codons signifying the same amino acid. There are also specific nucleotide sequences that designate the starting and ending points of a particular sequence of amino acids, and hence the protein structure that derives from that sequence. DNA itself has been known to exist since 1869, but it was not thought to have any relation to genetic information until the connection was unequivocally demonstrated in 1943. Subsequent research eventually revealed the structure and function of DNA and RNA. By preparing synthetic sequences of mRNA codons, researchers were able to determine which codons encoded for each specific amino acid in transcription and translation. The code is translated in the accompanying chart.

Amino Acid	Symbol	mRNA Codon
Alanine	Ala	GCA, GCC, GCG, GCU
Arginine	Arg	AGA, AGG, CGA, CGG, CGC, CGU
Asparagine	Asn	AAC, AAU
Aspartic acid	Asp	GAC, GAU
Cysteine	Cys	UGC, UGU
Glutamic acid	Glu	GAA, GAG
Glutamine	Gln	CAG, CAA
Glycine	Gly	GGA, GGC, GGG, GGU
Histidine	His	CAC, CAU
Isoleucine	Ile	AUA, AUC, AUU
Leucine	Leu	CUA, CUC, CUG, CUU, UUA, UUG
Lysine	Lys	AAA, AAG
Methionine	Met	AUG
Phenylalanine	Phe	UUU, UUC
Proline	Pro	CCA, CCC, CCG, CCU
Serine	Ser	AGC, AGU, UCA, UCG, UCC, UCU
Threonine	Thr	ACA, ACG, ACC, ACU
Tryptophan	Trp	UGG
Tyrosine	Tyr	UAC, UAU
Valine	Val	GUA, GUG, GUC, GUU
Start codon		AUG
Stop codon		UAA, UAG, UGA

SAMPLE PROBLEM

Given the following RNA nucleotide sequence, determine the sequence of amino acids in the protein that may be assembled from its code.

CCGAUGUGGGGGGGCGCUCUUUUUUUGUGCGCU-
CUAUACACGCGGGGCGCGCGAGAUAUAUAGAGC-
GCC

Answer:

The start codon, AUG, begins three units from the left end of the string. No synthesis is carried out until a start codon is encountered. Each subsequent three-unit codon then specifies the next amino acid in the sequence, until a stop codon is encountered.

The sequence begins at AUG and is followed by the codons UGG, GGG, GGC, GCU, CUU, UUU, UUG, UGC, GCU, CUA, UAC, ACG, CGG, GGC, GCG, CGA, GAU, AUA, UAG, AGC, GCC. The stop codon is encountered at UAG, the third codon from the end of the string. Thus, only the amino acids coded for between AUG and UAG will be synthesized.

Based on the chart, the relevant codons produce the following polypeptide string:

Trp-Gly-Gly-Ala-Leu-Phe-Leu-Cys-Ala-Leu-Tyr-Thr-Arg-

Gly-Ala-Arg-Asp-Ile

From this it can be seen that the same codon, AUG, indicates both methionine and the start sequence, while the three stop codons are unique. The context in which the AUG codon appears that determines whether or not it functions as a start codon or the codon for methionine.

AMINO ACIDS AND PROTEINS

The sequence of amino acids in a protein molecule defines its primary structure. Since each amino acid group has a specific geometry dictated by the rules of molecular structure and bond formation, no polypeptide chain or protein can be just a linear molecule. The angles of the bonds at each atom create all sorts of twists and turns along the entire length of the polypeptide molecule, moving the various functional groups on each amino acid into positions that allow them to interact with each other. Some segments of the protein molecule form larger physical shapes, such as spirals or flattened sheets. These constitute the secondary structure of the protein. A third, or tertiary, structure results from the interaction of the various functional groups as they form bonds due to their proximity to each other. A fourth, or quaternary, structure results when two or more protein molecules combine to form a larger reactive complex.

Richard M. Renneboog, MSc

BIBLIOGRAPHY

Berg, Jeremy M., John L. Tymoczko, Gregory J. Gatto, and Lubert Strye. *Biochemistry*. 8th ed. New York: W. H. Freeman, 2015. Print.

Lehninger, Albert L. *Biochemistry: The Molecular Basis of Cell Structure and Function*. 2nd ed. New York: Worth, 1975. Print.

Lodish, Harvey, et al. *Molecular Cell Biology*. 7th ed. New York: Freeman, 2013. Print.

Morrison, Robert Thornton, and Robert Neilson. Boyd. *Organic Chemistry*. 7th ed. Englewood Cliffs, N.J.: Prentice Hall, 2003. Print.

Pine, Stanley H. *Organic Chemistry*. 5th ed. New York: McGraw, 1987. Print.

Reece, Jane B., et al. *Campbell Biology*. 10th ed. San Francisco: Cummings, 2013. Print.

AMMONIUM ION

FIELDS OF STUDY

Organic Chemistry; Inorganic Chemistry; Biochemistry

SUMMARY

The basic structure of the ammonium ion is defined, and its importance in various branches of chemistry is described. An ammonium ion (NH_4^+) forms when a neutral ammonia molecule (NH_3) gains a proton from an acidic material. The NH_4^+ ion behaves in the same manner as any other cation.

PRINCIPAL TERMS

- **ammonia:** an inorganic compound consisting of a nitrogen atom bonded to three hydrogen atoms; exists as a colorless, pungent gas at room temperature.
- **biochemistry:** the chemistry of living organisms and the processes incidental to and characteristic of life.
- **Brønsted-Lowry acid-base theory:** definitions for acids and bases developed separately in 1923 by Danish chemist Johannes Nicolaus Brønsted and English chemist Martin Lowry; defines an acid as any compound that can release a hydrogen ion and a base as any compound that can accept a hydrogen ion.
- **ionization:** the process by which an atom or molecule loses or gains one or more electrons to acquire a net positive or negative electrical charge.
- **reaction mechanism:** the sequence of electron and orbital interactions that occurs during a chemical reaction as chemical bonds are broken, made, and rearranged.

THE NATURE OF THE AMMONIUM ION

Ammonia is a simple compound consisting of a nitrogen atom chemically bonded to three hydrogen atoms; its chemical formula is therefore NH_3. The electron distribution of the nitrogen atom consists of just seven electrons, with two in the 1*s* orbital, two in the 2*s* orbital, and three in the 2*p* orbitals. Energy considerations for the bonds to the three hydrogen atoms make it more favorable for the 2*s* and 2*p* atomic orbitals to "hybridize" into four equivalent sp^3 orbitals, with three of them each containing one electron. These sp^3 orbitals arrange tetrahedrally about the nucleus of the nitrogen atom to form the bonds to the hydrogen atoms. The remaining two electrons exist as a lone pair (that is, not part of a chemical bond). This arrangement exposes a region of high negative-charge density that makes the ammonia molecule highly nucleophilic (prone to donating electrons) and a strong base. Ammonia is therefore susceptible to ionization. It readily accepts a proton (or a hydrogen cation, H^+) from any compound capable of releasing one, in accordance with Brønsted-Lowry acid-base theory. The interaction of the proton's 1*s* orbital with the lone pair of electrons in the sp^3 orbital on the nitrogen atom effectively forms a chemical bond. This results in the formation of the ammonium ion, NH_4^+.

CHEMICAL PROPERTIES OF THE AMMONIUM ION

The ammonium ion forms when the ammonia molecule accepts a proton from a Brønsted-Lowry acid. Because of the chemical bond that forms between the proton and the lone pair of electrons on the nitrogen atom, the ammonium ion is very stable. All four N–H bonds are exactly equivalent, and the

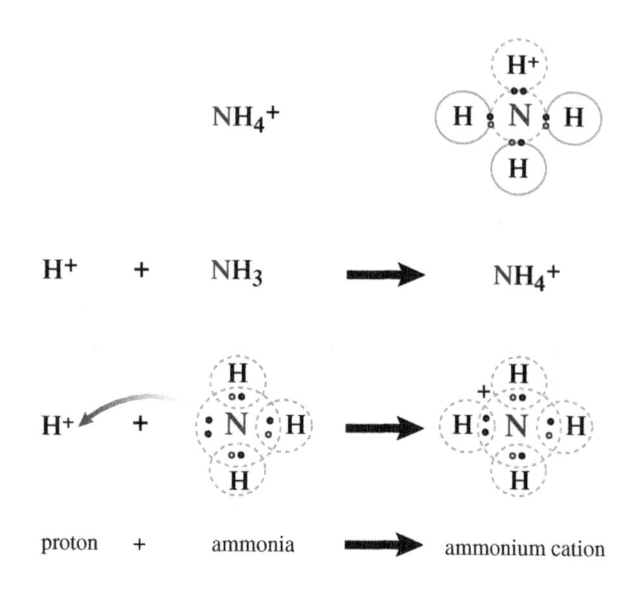

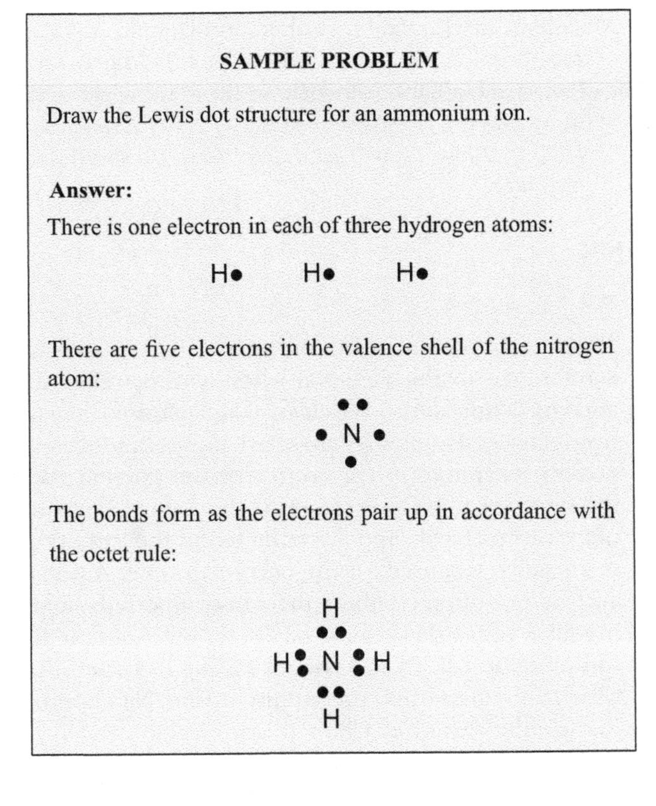

ammonium ion has the same structure and electron distribution as the methane molecule, CH_4. It is also the smallest of complex ions and thus has a positive charge density similar to that of the alkali metal ions, and the ammonium ion is able to form all of the corresponding ionic compounds with anions in the same way. The ammonium ion has a unique property relative to other cations, however, since it is itself the conjugate acid of the base ammonia. In combination with various anions, the resulting compound is able to separate into a neutral ammonia molecule and the corresponding acid of the anion. Ammonium chloride (NH_4Cl), for example, is able to decompose into NH_3 and hydrochloric acid (HCl), as well as dissociate into NH_4^+ and Cl^- ions. The characteristic odor of ammonia is quite noticeable over solid ammonium chloride crystals. Ammonium compounds thus tend to become "self-contaminated" over time due to loss of materials from decomposition.

The Ammonium Ion in Industry and Biochemistry

Nitrogen is an essential component of living systems, required for the formation of amino acids, which are in turn used in metabolic pathways for the synthesis of proteins according to the genetic code of all living things. In plants, amino acids are converted into various other components, such as chlorophyll and various alkaloids. Nitrogen from the air can be "fixed" by soil-borne bacteria that convert N_2 into nitrate ions (NO_3^-), which can be absorbed through the root structure of plants. Alternatively, nitrogen can be added directly to soils as water-soluble nitrate salts, the most common of which is ammonium nitrate, commonly used as a fertilizer. In biochemistry, amino acids and other compounds are also broken down into component fragments through metabolic processes. The amino groups typically end up as highly basic ammonia molecules, which are rapidly neutralized by conversion to highly soluble ammonium ions. The kidneys isolate the ammonium ions and excrete them in urine. The amount of ammonium ions in urine is tightly controlled, and significant variations can indicate various medical conditions.

The other major nitrogen-containing compound produced as a by-product of metabolism is urea ($H_2N–CO–NH_2$), which is produced biochemically by the metabolic breakdown of purines. Urea is of historical relevance to ammonium ion chemistry as the first example of an organic compound synthesized solely from inorganic materials, accomplished in 1828 by German chemist Friedrich Wohler (1800–1882). Urea is also used as a nitrogenous fertilizer and is considered more environmentally friendly than ammonium nitrate:

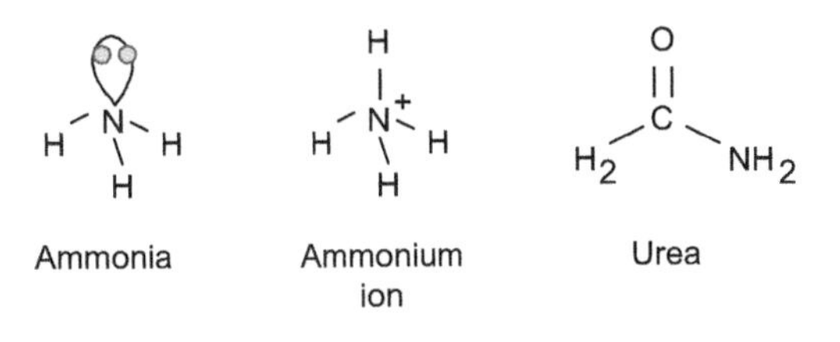

Richard M. Renneboog, MSc

Bibliography

Berg, Jeremy M., John L. Tymoczko, Gregory J. Gatto, and Lubert Strye. *Biochemistry.* 8th ed. New York: W. H. Freeman, 2015. Print.

Gribbin, John. *Science: A History, 1543–2001.* London: Lane, 2002. Print.

Herbert, R.B. *The Biosynthesis of Secondary Metabolites.* 2nd ed. London: Chapman & Hall, 1989. Print.

Johnson, Rebecca L. *Atomic Structure.* Minneapolis: Lerner, 2008. Print.

Mackay, K. M., R. A. Mackay, and W.Henderson. *Introduction to Modern Inorganic Chemistry.* 6th ed. Cheltenham: Nelson, 2002. Print.

Miessler, Gary L., Paul J. Fischer, and Donald A. Tarr. *Inorganic Chemistry.* 5th ed. Upper Saddle River: Prentice Hall, 2013. Print.

Winter, Mark J. *The Orbitron: A Gallery of Atomic Orbitals and Molecular Orbitals. University* of Sheffield, n.d. Web.

ANIONS

FIELDS OF STUDY

Inorganic Chemistry; Geochemistry; Metallurgy

SUMMARY

The basic structure of anions is defined, and the development of the modern theory of atomic structure is elaborated. Electronic structure is fundamental to all chemical behaviors and is responsible for the relationships seen in the periodic table.

PRINCIPAL TERMS

- **cation:** any chemical species bearing a net positive electrical charge, which causes it to be drawn toward the negative pole, or cathode, of an electrochemical cell.
- **ionic bond:** a type of chemical bond formed by mutual attraction between two ions of opposite charges.
- **ionization:** the process by which an atom or molecule loses or gains one or more electrons to acquire a net positive or negative electrical charge.
- **oxoanion:** an ion consisting of one or more central atoms bonded to a number of oxygen atoms and bearing a net negative electrical charge.
- **valence electron:** an electron that occupies the outermost or valence shell of an atom and participates in chemical processes such as bond formation and ionization.

THE NATURE OF ANIONS

An anion is any atom or group of atoms that bears a net negative charge due to the presence of one or more extra valence electrons. The term is a contraction of "anode ions," which is a reference to the fact that when a direct electric current is applied to an electrolytic solution, negatively charged ions are attracted to the anode, or positive terminal, of the source of the current. By the same token, cations, from "cathode ions," have a net positive charge and are attracted to the cathode, or negative terminal, of the current source. Anions and cations often combine to form compounds held together with ionic bonds; one common example is sodium chloride (NaCl), better known as table salt, which is created when the sodium cation, Na^+, bonds to the chloride anion, Cl^-.

The formation of any ion is the result of an atom or molecule gaining or losing one or more valence electrons. This is most apparent in monatomic

ANIONS

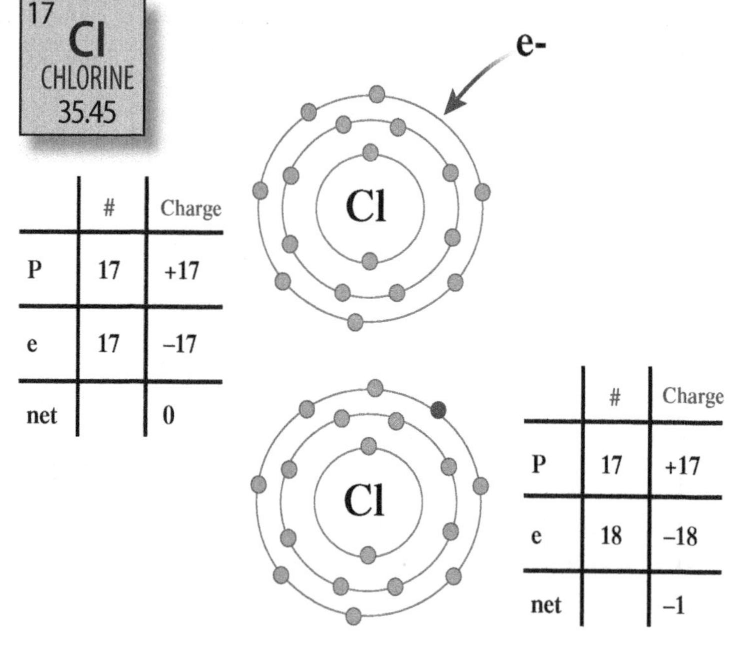

(single-atom) ions, in which the electrical charge is equal to the oxidation state. For example, the halogens—fluorine, chlorine, bromine, iodine, and astatine—are all highly electronegative, meaning that they readily accept an extra valence electron so that their outer electron shell is completely full and therefore stable. The resulting anions are called fluoride, chloride, bromide, iodide, and astatinide, respectively, and they have an electrical charge of 1− because they gained one electron and thus one unit of negative charge. Similarly, the chalcogens oxygen (O) and sulfur (S) readily accept two extra electrons to form the oxide ion, O^{2-}, and the sulfide ion, S^{2-}.

Compound ions in which a central atom is bonded to a number of oxygen atoms are extremely common. These **oxoanions** tend to form very stable compounds and are the basic materials of many minerals.

THE ELECTRONICS OF ANION FORMATION

According to the modern theory of atomic structure, each atom contains a very small, extremely dense nucleus that holds at least 99.98 percent of the atom's mass and all of its positive electrical charge. The nucleus is surrounded by a diffuse and comparatively very large cloud of electrons containing all of the atom's negative electrical charge. These electrons are allowed to possess only very specific energies. This restricts their movement around the nucleus to specific regions called "electron shells." Within each shell are well-defined regions called "orbitals." The strict geometric arrangement of the orbitals regulates the formation of chemical bonds between atoms.

Each shell and orbital is subject to a number of restrictions that dictate how many electrons it can hold. There are four different types of electron orbitals, designated *s*, *p*, *d*, and *f*, each of which can contain a specific number of electrons: *s* orbitals can hold a maximum of two electrons; *p* orbitals, a maximum of six; *d* orbitals, a maximum of ten; and *f* orbitals, a maximum of fourteen. One or more of these orbitals make up an electron shell. The various electron shells are indicated by an integer value known as the "principal quantum number," starting with 1 for the innermost shell, typically referred to as the "n = 1 shell." The standard notation to describe an electron orbital is the principal quantum number, followed by the type of orbital, followed by a superscript number indicating how many electrons it holds. For example, helium has only one *s* orbital, which is completely full, so its electron configuration is represented as $1s^2$. The first *p* orbital appears in the $n = 2$ shell, the first *d* orbital in the $n = 3$ shell, and the first *f* orbital in the $n = 4$ shell. Due to variances in energy levels, electrons usually fill atomic orbitals in the order 1*s*, 2*s*, 2*p*, 3*s*, 3*p*, 4*s*, 3*d*, 4*p*, 5*s*, 4*d*, 5*p*, 6*s*, 4*f*, 5*d*, 6*p*, 7*s* . . . rather than in strict numerical order.

The outermost electron shell is the valence shell, and in all elements, except the noble gases (helium, neon, argon, krypton, xenon, and radon), the valence shell is incompletely occupied. Because of this, the noble gases are often called "inert gases,"

SAMPLE PROBLEM

Draw a simple orbital diagram of the bromine (Br) atom, and identify the electrons that it will lose or gain to become the bromide ion. Use the following chart to fill in the electron orbitals in the proper order, and remember that *s* orbitals hold two electrons, *p* orbitals hold six electrons, *d* orbitals hold ten electrons, and *f* orbitals hold fourteen electrons.

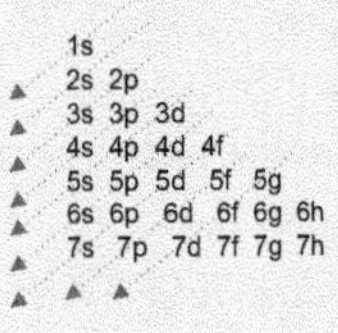

Answer:

The atomic number of bromine is 35, indicating that it has a total of thirty-five electrons, distributed as follows:

$$1s^2 2s^2 2p^6 3s^2 3p^6 4s^2 3d^{10} 4p^5$$

Thus, there are two electrons in the $n = 1$ shell, eight in the $n = 2$ shell, eighteen in the $n = 3$ shell, and seven in the $n = 4$ shell. This can be drawn as

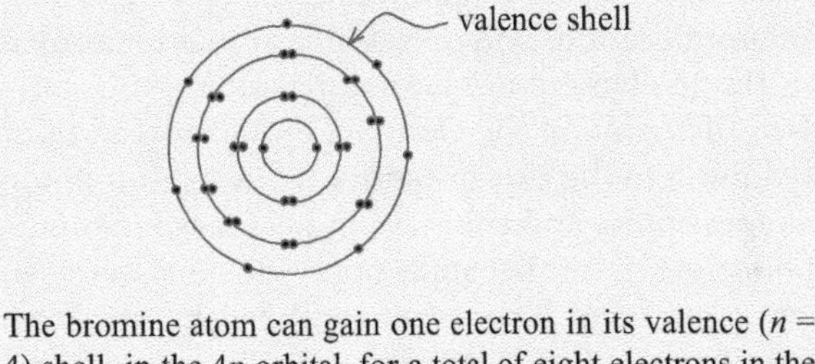

The bromine atom can gain one electron in its valence ($n = 4$) shell, in the 4*p* orbital, for a total of eight electrons in the closed-shell configuration $4s^2p^6$. This creates the bromide ion, Br^-.

reflecting the fact that they are far less chemically reactive than elements whose valence shells are not filled. All noble gases except for helium have eight electrons in their valence shell, as a shell with eight electrons approximates a stable closed-shell configuration of s^2p^6. This is the basis of the octet rule, which states that atoms of lower atomic numbers tend to achieve the greatest stability when they have eight electrons in their valence shells. The closer an element is to having eight valence electrons, the more likely it is to undergo ionization or a chemical reaction in order to achieve a stable configuration, either by gaining enough electrons to complete the octet or by losing all valence electrons so that the next-highest completed shell becomes the outermost shell. Elements with similar electron distributions in their valence shells typically exhibit similar chemical behaviors, which is the basis on which the periodic table of the elements is arranged.

NAMING THE ANIONS

The rules for naming various chemical species are established and standardized by the International Union of Pure and Applied Chemistry (IUPAC). Monatomic anions and polyatomic anions composed of a single element are named by adding the suffix *-ide* to the name of the element, either instead of or in addition to the existing suffix. Thus a sulfur anion becomes sulfide, a xenon anion becomes xenonide, a potassium anion becomes potasside, and so on. In some cases, the suffix is added to the element's Latin name instead of its common name; for example, an anion of silver, which has the Latin name *argentum*, is called argentide.

Oxoanions are generally named for the central atom of the anion and the number of oxygen atoms that surround it, with the charge number given in parentheses at the end. An oxoanion consisting of a sulfur atom surrounded by three oxygen atoms (SO_3^{2-}) has the formal name trioxidosulfate(2–), while one with a sulfur atom and four oxygen atoms (SO_4^{2-}) is called tetraoxidosulfate(2–). The common (nonsystematic) names of these two oxoanions are sulfite and sulfate, respectively, following the convention that the oxoanion with fewer oxygen atoms takes the suffix *-ite* and the one with more oxygen atoms takes the suffix *-ate*. If one element is capable of forming more than two different oxoanions, as is the case with chlorine, the prefixes *hypo-* and *per-* are used as well, so that the common name of ClO^- is hypochlorite, ClO_2^- is chlorite, ClO_3^- is chlorate, and ClO_4^- is perchlorate. Multiple central atoms are indicated by the appropriate numerative prefix; for example, the anion CrO_3^{2-} is chromate, while the anion $Cr_2O_7^{2-}$ is dichromate.

Richard M. Renneboog, MSc

BIBLIOGRAPHY

Haynes, W. H., PhD, *David R. Lide, PhD, and Thomas J. Bruno*, PhD, eds. CRC Handbook of Chemistry and Physics, 96th Edition. New York: CRC/Taylor & Francis Group, 2015. Print.

Johnson, Rebecca L. *Atomic Structure.* Minneapolis: Lerner, 2008. Print.

Mackay, K. M., R. A. Mackay, and W. Henderson. *Introduction to Modern Inorganic Chemistry.* 6th ed. Cheltenham: Nelson, 2002. Print.

Miessler, Gary L., Paul J. Fischer, and Donald A. Tarr. *Inorganic Chemistry.* 5th ed. Upper Saddle River: Prentice Hall, 2013. Print.

Wehr, M. Russell, James A. Richards Jr., and Thomas W. Adair III. *Physics of the Atom.* 4th ed. Reading: Addison-Wesley, 1984. Print.

Winter, Mark J. *The Orbitron: A Gallery of Atomic Orbitals and Molecular Orbitals. University* of Sheffield, n.d. Web.

AROMATICITY

FIELDS OF STUDY

Organic Chemistry

SUMMARY

The electronic property of aromaticity results from the hybridization and overlap of atomic orbitals in the formation of conjugated bond structures. Only those conjugated compounds that conform to the $4n + 2$ rule are aromatic.

PRINCIPAL TERMS

- **benzene:** an organic compound with the molecular formula C_6H_6, consisting of a six-membered carbon ring with a hydrogen atom bonded to each carbon; although sometimes described by alternating single and double carbon-carbon bonds, a more correct description requires resonance between two equivalent structures.
- **conjugation:** the overlap of *p* orbitals between three or more successive carbon atoms in a molecular structure, creating a system of alternating double and single bonds through which electrons can move freely.
- **cyclic delocalization:** a property of some ring molecules, such as benzene, in which the overlap of orbitals between the atoms that make up the ring allows their electrons to move freely about the molecule.
- **hybrid structure:** a representation of molecular structure that averages a number of possible molecular structures that are equivalent in terms of the arrangement of orbitals and electrons within them.
- **pi bond:** a covalent chemical bond formed when parallel *p* orbitals of two adjacent atoms overlap in a side-by-side manner to form two molecular orbitals.
- **resonance:** a method of graphically representing a molecule with both single and multiple covalent bonds whose valence electrons are not associated with one particular atom or bond; two or more diagrams, or resonance structures, depict each possible arrangement of single and multiple bonds, while the true structure of the molecule is somewhere between the different resonance structures, an intermediate form known as the resonance hybrid.

The Nature of Aromaticity

The term "aromatic" does not refer to the effect a compound has on the sense of smell, although the effect was first identified as some smelly compounds and the term generalized. It more accurately describes a characteristic of the electrons in the pi (ϖ) bonds of the molecule. Atoms with their valence electrons in the *p* orbitals, especially carbon atoms, have the unique ability to form a type of extra chemical bond between adjacent atoms. A carbon atom has four electrons in its valence shell, distributed as $2s^22p^2$: two electrons in the *s* orbital and one each in two of the three *p* orbitals. The *s* orbital is spherically symmetric about the atomic nucleus, but the *p* orbitals are figure-eight shaped and at right angles to each other, causing the electrons in these orbitals to experience very asymmetrical forces of repulsion. Because the *s* and *p* orbitals belong to the same electron shell, they are sufficiently close in energy that they are able to combine to form four equivalent hybrid sp^3 orbitals. These typically form four sigma bonds to as many other atoms, characteristic of saturated hydrocarbons:

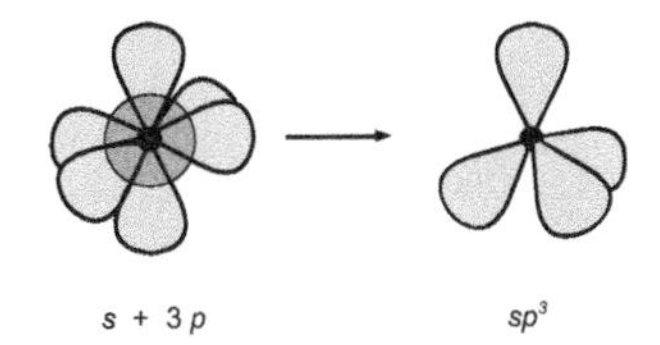

The carbon atom can also hybridize the *s* orbital with just two of the three *p* orbitals to produce three equivalent sp^2 hybrid atomic orbitals, oriented in a plane about the nucleus at 120-degree angles to each other. The third *p* orbital remains a *p* orbital, perpendicular to the plane of the sp^2 orbitals:

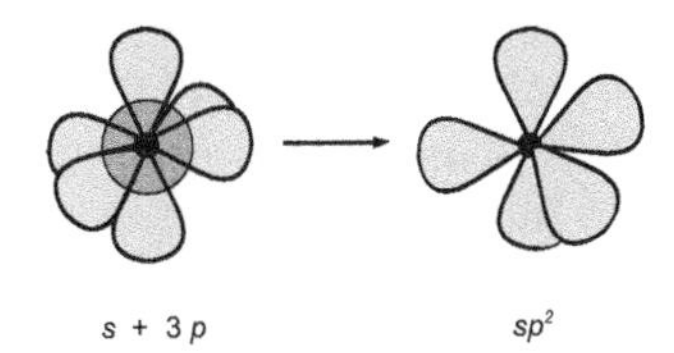

AROMATICITY

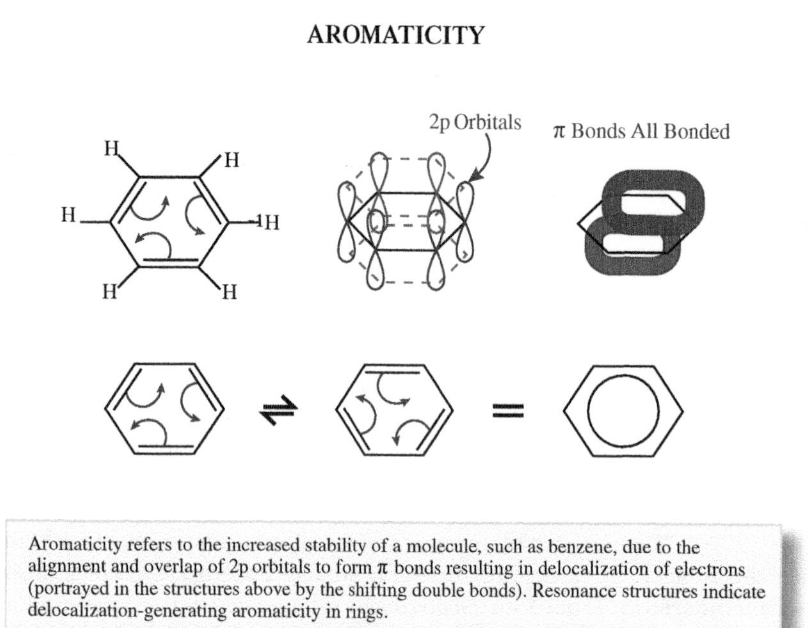

Aromaticity refers to the increased stability of a molecule, such as benzene, due to the alignment and overlap of 2p orbitals to form π bonds resulting in delocalization of electrons (portrayed in the structures above by the shifting double bonds). Resonance structures indicate delocalization-generating aromaticity in rings.

The three sp^2 orbitals form sigma bonds to other atoms by end-to-end overlap of the orbitals on adjacent atoms. The lone *p* orbital does not take part in these bonds, but in order for the carbon atom to achieve its octet of valence electrons, one more electron is required. This in turn requires a second sp^2 hybridized atom next to the first. In this relative position, the lone *p* orbitals on the two atoms can overlap side by side to form a pi-bonding molecular orbital containing two electrons:

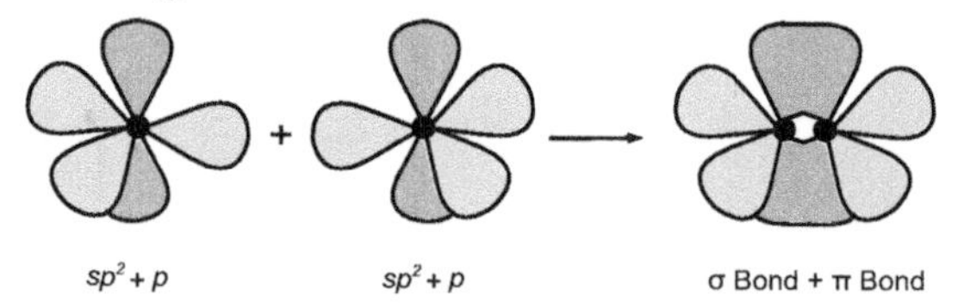

When two pi bonds are adjacent to each other, spanning four carbon atoms, as C=C–C=C, they are said to be in conjugation with each other, or forming a conjugated system. Because all four pi bonds overlap equally well, the four *p* orbitals effectively form a single pi-bonding molecular orbital across all four carbon atoms. The electrons in that molecular orbital can therefore occupy any space along its entire span rather than being restricted to a bond between just two atoms, a phenomenon known as delocalization. In the case of cyclic conjugated systems such as benzene, it is called cyclic delocalization.

BENZENE AND OTHER AROMATIC COMPOUNDS

Benzene, discovered in 1825 by Michael Faraday (1791–1867), was long known to be a type of hydrocarbon compound, but before the development of the modern atomic theory, the structure of the benzene molecule was a mystery. Scientists in the nineteenth century could not reconcile the composition of benzene with the known behaviors of other hydrocarbons. Legend has it that German chemist Friedrich August Kekulé (1829–96), while studying the compound, fell asleep one night before the fireplace in his home and had a dream of a snake eating its own tail; another version of the story has him dreaming about six monkeys holding hands and dancing in a circle. The dream supposedly made Kekulé realize that the structure of benzene was cyclic, not linear. He envisioned the benzene molecule as a ring of six carbon atoms with alternating single (C–C) and double (C=C) carbon-carbon bonds.

Despite this, Kekulé's breakthrough did not solve the mystery of benzene, since the compound simply would not take part in the same kinds of reactions that other carbon-carbon double-bonded compounds underwent with ease. Only when twentieth-century technology enabled close study of the individual bonds in the benzene molecule was the mystery solved. The structure of the benzene molecule had been previously hinted at in the concept of resonance. Kekulé suggested, but could not prove, that the benzene molecule oscillated between two equivalent "resonance structures." It was eventually determined that the molecule exhibits a hybrid structure that is the midpoint between the two resonance structures, with all six carbon-carbon bonds in the molecule being equivalent in length and bond strength rather than alternating between single and double bonds. Molecular orbital theory finally provided a satisfactory description of the benzene molecule as a ring of six carbon atoms in a ring with six sigma bonds and a pi cloud encompassing all

SAMPLE PROBLEM

Using the $4n + 2$ rule, determine if the compound 1,3,5,7-cyclooctatetraene is aromatic.

Answer:

The root of the name, *cycloocta-*, indicates that the hydrocarbon molecule is an eight-membered ring. There are four C=C bonds around the cyclic structure, as indicated by the suffix *-tetraene*, and these alternate with four C–C bonds. Recall that each C=C bond is a sigma bond and a pi bond and that there are two electrons in each pi bond, for a total of eight pi electrons in this compound. Set up the $4n + 2$ equation so that it equals 8 and solve:

$$4n + 2 = 8$$

$$4n = 8 - 2 = 6$$

$$n = \frac{6}{4} = \frac{3}{2}$$

The value of n is not 0 or a positive integer, so the compound does not satisfy the $4n + 2$ rule and thus is not aromatic.

six of the carbon atoms through their respective *p* orbitals:

Kekulé's 'resonance structures' for benzene

The descriptive term "aromatic" was used to indicate the special nature of the conjugated double-bond system of the benzene molecule, perhaps because of benzene's sweet, oily aroma. This bond arrangement provides somewhat more stability than would be provided by single and double bonds alone.

In other compounds with conjugated double-bond systems, some were found to have this aromatic character while others did not. Observations of their chemical behavior resulted in the formulation of the $4n$+2 rule, or Hückel's rule, which states that only conjugated systems in which the number of pi electrons is equal to $4n + 2$, where $n = 0$ or a positive integer, exhibit the property of aromaticity.

REACTIONS OF AROMATIC COMPOUNDS

Aromatic conjugated systems are somewhat more stable and unreactive than nonaromatic compounds, but conjugation does "transmit" electron-withdrawing or -donating influences from other substituents. As electron-rich compounds, benzene and other aromatic compounds are subject to various substitution reactions involving electrophiles (chemical species that attract electrons), such as Friedel-Crafts alkylation and acylation reactions.

Richard M. Renneboog, MSc

BIBLIOGRAPHY

Gribbin, John. *Science: A History, 1543–2001.* London: Lane, 2002. Print.

Hendrickson, James B., Donald J. Cram, and George S. Hammond. *Organic Chemistry.* 3rd ed. New York: McGraw, 1970. Print.

Morrison, Robert Thornton, and Robert Neilson. Boyd. *Organic Chemistry.* 7th ed. Englewood Cliffs, N.J.: Prentice Hall, 2003. Print.

Myers, Richard. *The Basics of Chemistry.* Westport: Greenwood, 2003. Print.

Zumdahl, Steven S., and Susan Zumdahl. *Chemistry.* 9th ed. Belmont: Brooks Coles, 2013. Print.

ATOMIC NUMBER

FIELDS OF STUDY

Physical Chemistry; Organic Chemistry; Inorganic Chemistry

SUMMARY

The atomic number of an element is the number of protons in the nucleus of that element. The atomic number uniquely identifies each element.

PRINCIPAL TERMS

- **atomic mass:** the total mass of the protons, neutrons, and electrons in an individual atom.
- **electrical charge:** a property of subatomic particles that causes them to exert a force on each other, either attractive (if their charges are of opposite signs) or repulsive (if they are of the same sign); by convention, a proton is assigned a charge of 1+ and an electron is assigned a charge of 1–.

- **element:** a form of matter consisting only of atoms of the same atomic number.
- **periodic table:** the chart representing the known elements by atomic number and electron distribution.
- **proton:** a fundamental subatomic particle with a single positive electrical charge, found in the atomic nucleus.

The Nature of the Atomic Number

The atomic number of an element represents the exact number of protons in the nucleus of each atom of that element. Because protons and electrons have opposite electrical charges, the atomic number of an uncharged, neutral atom is also equal to the number of electrons surrounding the atomic nucleus. The number of electrons that surround a neutral atom of an element is a key determinant of that element's specific chemical properties.

In addition to electrons, every atom is composed of a specific number of two subatomic particles, protons and neutrons, that comprise the atomic nucleus. A proton and a neutron are almost equal in mass, with the neutron having only slightly more mass. Between them, an atom's protons and neutrons account for at least 99.98 percent of the atomic mass. Mass is a physical property that is defined relative to an acceptable standard. One atomic mass unit (u) is defined as one-twelfth of the mass of an atom of carbon-12, the most abundant isotope of carbon, which has six protons (according to carbon's atomic number) and six neutrons; therefore, 1 u approximately represents the mass of one nucleon—that is, one proton or one neutron. (An isotope is an atom of a specific element that contains the usual number of protons in its nucleus but a different number of neutrons.)

The atomic weight of an element is determined by calculating the weighted average of the masses of all of that element's isotopes. Because an element's isotopes all have the same number of protons, they exhibit nearly identical chemical properties. Isotopes differ from one another in the number of neutrons in the atomic nucleus, which affects the atomic mass. Due to differences in the atomic mass, heavier isotopes react similarly but more slowly than lighter isotopes of the same element.

Atomic Numbers and the Periodic Table

Prior to the identification of the subatomic particles, scientists attempted to bring some order to their knowledge of the chemical elements. They found that certain elements exhibit similar chemical behaviors. These chemical similarities seemed to have a periodic relationship with the elements' atomic masses; elements with similar chemical properties were found to have atomic masses that are either of nearly the same value or that increase at regular intervals. By arranging the known elements in tables according to the similarities in their chemical behavior, scientists developed the earliest versions of the periodic table. With the discovery of the existence of neutrons, protons, and electrons and the formulation of the octet rule of chemical bonding, the periodic relationships of the elements became more readily understood.

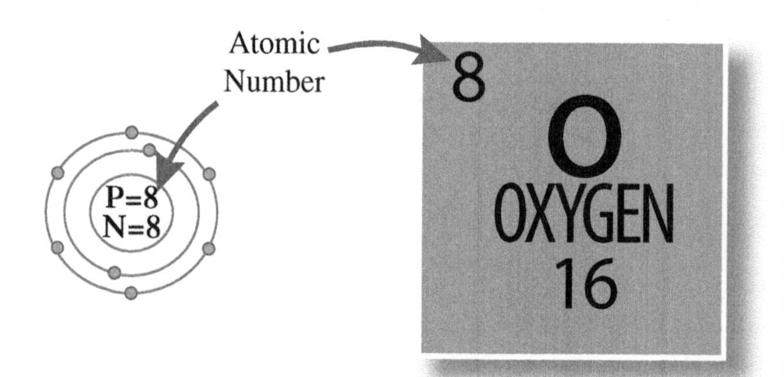

An uncharged atom has an equal number of protons and electrons and therefore has a neutral charge. The electrons occupy a region of space surrounding the atom's nucleus that is about one hundred thousand times larger in diameter than the nucleus. Because chemical reactions and bonds take place at the level of the outermost electrons in an atom (the valence electrons), elements in the same group (the vertical columns of the periodic table) have similar distributions of their outermost electrons and therefore exhibit similar chemical behaviors. For example, all alkali metal elements (group 1) have a single outermost electron, so they all readily give up that electron to form a corresponding alkali metal ion

SAMPLE PROBLEM

Determine the atomic numbers of the chlorine ion Cl^-, the calcium ion Ca^{2+}, an atom of carbon with an atomic mass of 13 u, and an atom of hydrogen with an atomic mass of 3 u. Calculate the number of neutrons in an atom of uranium-235.

Answer:

Because the atomic number is based on the number of protons in an atom's nucleus and ionization is determined by a change in the number of electrons around an atom, ionization does not change the atomic number of an atom. Therefore, the atomic number of a chlorine ion is always 17 and the atomic number of a calcium ion is always 20, according to the number of protons in their respective nuclei.

Furthermore, the elemental identity of an isotope is not changed by its number of neutrons. The isotope carbon-13 has six protons (otherwise it would not be carbon) and seven neutrons instead of the usual six. Radioactive hydrogen-3, known as "tritium," has one proton (otherwise it would not be hydrogen) and two neutrons. The atomic numbers of carbon-13 and hydrogen-3 are therefore 6 and 1, respectively.

The atomic number of uranium is 92, indicating that the nucleus of any uranium isotope contains ninety-two protons, no more and no less. The isotope's mass number represents the total number of protons and neutrons in the isotope's atomic nucleus. Subtracting the element's atomic number from the isotope's mass number yields the number of neutrons in the isotope's atomic nucleus. Thus, uranium-235 has 143 neutrons ($235 - 92 = 143$).

bearing a single net positive charge. As the atomic number of the elements increases, so too do the number and distribution of the outermost electrons that they possess. The periods, the horizontal rows of the periodic table, rank the elements in their respective groups according to the highest energy level of their outermost electrons.

Richard M. Renneboog, MSc

Bibliography

Douglas, Bodie E., Darl H. McDaniel, and John J. Alexander.Concepts and *Models of Inorganic Chemistry. 3rd ed.* New York: Wiley, 1994. Print.

Haynes, W. H., PhD, *David R. Lide, PhD, and Thomas J. Bru*no, PhD, eds. CRC Handbook of Chemistry and Physics, 96th Edition. New York: CRC/Taylor & Francis Group, 2015. Print.

Jones, Mark Martin. Che*mistry and Society.* 5th ed. Philadelphia: Saunders College Pub., 1987. Print.

Lafferty, Peter, and Julian Rowe, eds. *The Hutchinson Dictionary of Science.* 2nd ed. Oxford: Helicon, 1998. Print.

Miessler, Gary L., Paul J. Fischer, and Donald A. Tarr. *Inorganic Chemistry.* 5th ed. Upper Saddle River: Prentice Hall, 2013. Print.

Morrison, Robert Thornton, and Robert Neilson. Boyd. *Organic Chemistry.* 7th ed. Englewood Cliffs, N.J.: Prentice Hall, 2003. Print.

Myers, Richard. *The Basics of Chemistry.* Westport: Greenwood, 2003. Print.

Wehr, M. Russell, James A. Richards Jr., and Thomas W. Adair III. *Physics of the Atom.* 4th ed. Reading: Addison-Wesley, 1984. Print.

ATOMS

FIELDS OF STUDY

Analytical Chemistry; Spectroscopy; Inorganic Chemistry

SUMMARY

The basic structure of atoms is defined, and the development of the modern theory of atomic structure is elaborated. Atomic structure is fundamental to all fields of chemistry, but especially to fields that rely on the intrinsic properties of individual atoms.

PRINCIPAL TERMS

- **atomic mass:** the total mass of the protons, neutrons, and electrons in an individual atom.
- **atomic number:** the number of protons in the nucleus of an atom, used to uniquely identify each element.
- **electron:** a fundamental subatomic particle with a single negative electrical charge, found in a large, diffuse cloud around the nucleus.
- **element:** a form of matter consisting only of atoms of the same atomic number.

- **isotope:** an atom of a specific element that contains the usual number of protons in its nucleus but a different number of neutrons.
- **neutron:** a fundamental subatomic particle in the atomic nucleus that is electrically neutral and about equal in mass to the mass of one proton.
- **nucleus:** the central core of an atom, consisting of specific numbers of protons and neutrons and accounting for at least 99.98 percent of the atomic mass.
- **proton:** a fundamental subatomic particle with a single positive electrical charge, found in the atomic nucleus.

Visualizing the Atom

An atom is the smallest unit of elemental matter that retains and defines the specific properties of that material. The concept can be visualized by imagining a sample of a pure elemental material being repeatedly divided into ever-smaller portions. Eventually a point would be reached at which the material could no longer be subdivided without destroying its identity. The remaining indivisible portion is an atom of the element. Any further subdivision would require breaking the atom into its component protons, neutrons, and electrons.

Historical Theories of Atomic Structure

In their attempts to comprehend the basic nature of the universe, early Greek philosophers, particularly Democritus (ca. 460 BCE–ca. 370 BCE), followed this kind of logical reasoning to the philosophical concept of the *atomos* (indivisible), the fundamental thing from which all matter was made. Greek philosophy, however, was based on thought, not experimentation. Two thousand years later, in the Middle Ages, the practice of alchemy arose alongside the practices of metallurgy, and alchemists sought means of transforming materials into other materials, typically through magic and arcane practices. The refine-ment and working of metals became a very important practical study and gave rise to the first truly scientific book of chemistry, *De re metallica* (On the nature of metals, 1556), by Georgius Agricola (1494–1555). Alchemy eventually gave way to more scientific study of matter through the use of weights and measures, enabling early scientists to recognize the relationships between matter that formed the foundation on which the modern atomic theory has been constructed.

In the nineteenth century, scientists began to reject the philosophical *atomos* in favor of the physical atom as the basic building block of matter. John Dalton (1766–1844) conceived of the atom as though it were a billiard ball: a single, hard, uniform spherical object. He based this view on the behavior of gases, some of which he found he could describe mathematically with the billiard model. Other chemical behaviors required a mechanism to account for electrical charge, however, especially in the case of ionic interactions.

In 1897, J. J. Thomson (1856–1940) discovered that when a polarized electricity source is discharged within a gas-filled tube, the rays emitted from the cathode (the negative electrode), known as cathode rays, obey the mathematics of a stream of charged particles. After calculating the likely mass of the particles, Thomson determined that they were even smaller than atoms, thus proving the existence of subatomic particles. He called these particles "corpuscles" at first, though they soon became known as electrons. Following this discovery, Thomson developed a new atomic model, one in which small, negatively charged particles—electrons—were embedded throughout a positively charged matrix, rather like the plums in a plum pudding. This came to be known as the plum-pudding model.

The Modern Theory of Atomic Structure

In 1909, Ernest Rutherford (1871–1937), aided by Hans Geiger (1882–1945) and Ernest Marsden (1889–1970), conducted the gold-foil experiment, which demonstrated that atoms consist of a nucleus surrounded by primarily empty space. In this experiment, a beam of alpha particles (the nuclei of helium atoms) was directed at a piece of very thin gold foil, which was surrounded by a film strip that would detect alpha particles as they passed through the foil. Rutherford found that while the vast majority of the particles passed directly through the foil as though there was nothing there, some of the particles were deflected in all directions by the foil. This demonstrated that atoms consist of a very small, dense nucleus that can deflect alpha particles, surrounded by a thin, diffuse cloud that is not capable of affecting their movement. Rutherford continued experimenting with alpha particles, and in 1917 he

ATOMS

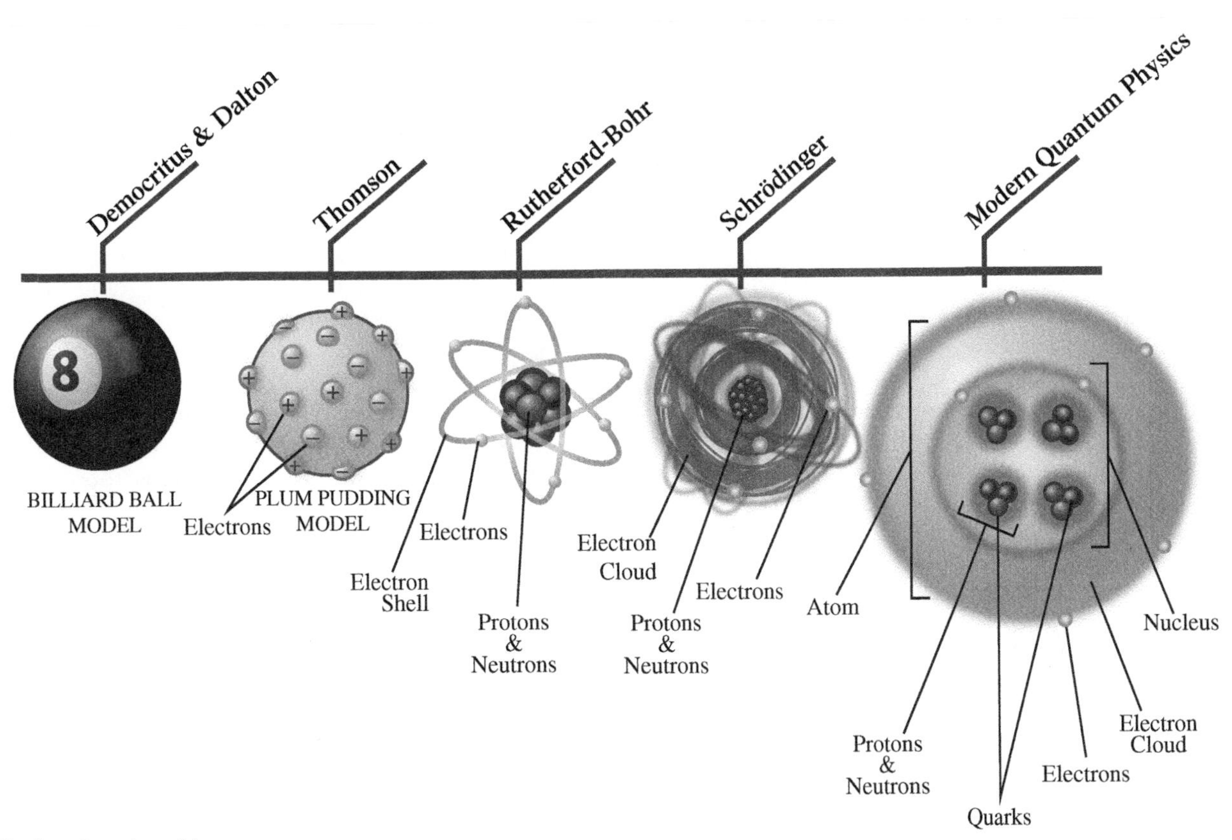

Timeline of atomic models

discovered that the collision of an alpha particle with the nucleus of a nitrogen atom released a particle identical to a hydrogen nucleus—in other words, a proton.

Considerations of the energies associated with electrons in atoms led Niels Bohr (1885–1962) to postulate in 1913 that the electrons must be restricted to specific orbital paths around the nucleus, in accordance with the corresponding wavelengths calculated by Louis de Broglie (1892–1987). Measurement of the wavelengths of various emission and absorption spectra indicated that electrons in atoms can only have certain specific energy levels, which is consistent with the idea that electrons can only follow certain stable orbits around a nucleus.

Because of its electrical neutrality, the neutron remained an elusive theoretical construct until 1932, when an experiment by James Chadwick (1891–1974) demonstrated its existence. Chadwick used alpha particles to bombard a beryllium target, causing it to release radiation that would strike a paraffin target, which would in turn release protons from hydrogen nuclei into an ionization chamber. He determined that the radiation released from the beryllium was actually a neutral particle of approximately the same mass as a proton, now known to be a neutron.

These discoveries about the nature of the various subatomic particles formed the foundation of a new model of atomic structure, based on the principles of quantum mechanics. In this model, the energy levels of electrons can only change by discrete steps, or quanta, and thus electrons can occupy only specific regions of space about the nucleus, which came to be called atomic orbitals. These orbitals have specific shapes and orientations around the nucleus of the atom, which determine how atoms interact to form chemical bonds. The physical orientation of those bonds determines the shapes and properties of both the atoms and the molecules that they form.

SAMPLE PROBLEM

How many neutrons and how many electrons are in an oxygen atom?

Answer:

Locate oxygen (O) in the periodic table to find its atomic number and atomic mass.

- The atomic number of oxygen is 8, meaning that an atom of oxygen has eight protons in its nucleus.
- The atomic mass of oxygen, rounded to two decimal places, is 16.00. The atomic mass is approximately equal to the number of protons plus the num- ber of neutrons. Thus, the total number of protons and neutrons in an atom of oxygen is sixteen.

Calculate the number of neutrons:

There are eight neutrons and eight electrons in an oxygen atom.

atomic mass = protons + neutrons

16 = 8 + neutrons

16 – 8 = neutrons

8 = neutrons

In its elemental and electrically neutral form, every atom has an equal number of protons and electrons. Calculate the number of electrons:

protons = electrons protons = 8

8 = electrons

There are eight neutrons and eight electrons in an oxygen atom

THE PERIODIC TABLE

The periodic table of the elements displays the relationships of the elements according to the quantum mechanical model of atomic structure. Chemists had tried to sequentially order the known elements by their common properties since the late eighteenth century. Dmitri Mendeleev (1834–1907) is credited with developing the first modern periodic table, which he published in 1869, though much information was missing. The periodic table of the elements contains all ninety-eight naturally occurring elements, as well as eighteen more that have been (or are claimed to have been) produced artificially. Each square in the table corresponds to a single element. In a standard periodic table, the internationally accepted symbol of the element is in the center of the square, with its atomic number at the top and its atomic mass at the bottom. Some periodic tables also include the normal oxidation states in which each element is found and its electron configuration ac- cording to the quantum mechanical model of atomic structure.

The identity of each atom is determined by the number of protons in its nucleus, meaning that all atoms with the same number of protons in their respective nuclei are atoms of the same element. However, the number of neutrons and electrons in an atom can change. Neutral atoms must have the exact same number of electrons orbiting the nucleus as there are protons within the nucleus; those atoms with more or fewer electrons than protons are ions, or atoms with a net electrical charge. Atoms with different numbers of neutrons in their nuclei are different isotopes of the same element. The atomic mass of an atom is the total mass of all protons, neutrons, and electrons in the atom; however, because electrons are so small and have such little mass, the atomic mass can be approximated by adding the number of neutrons and protons together. Atomic mass is measured in unified atomic mass units (u), one of which is approximately equal to the mass of a proton or neutron.

Richard M. Renneboog, MSc

BIBLIOGRAPHY

Agricola, Georgius. *De re metallica.* Trans. Herbert Clark Hoover and Lou Henry Hoover. New York: Dover, 1950. Print.

The Britannica Guide to the 100 Most Influential Scientists. Introd. John Gribbin. London: Constable, 2008. Print.

Gribbin, John. *Science: A History, 1543–2001.* London: Lane, 2002. Print.

Johnson, Rebecca L. *Atomic Structure.* Minneapolis: Lerner, 2008. Print.

Kragh, Helge. *Niels Bohr and the Quantum Atom: The Bohr Model of Atomic Structure, 1913–1925.* Oxford: Oxford UP, 2012. Print.

Winter, Mark J. *The Orbitron: A Gallery of Atomic Orbitals and Molecular Orbitals. University* of Sheffield, n.d. Web..

B

BENZENE AND OTHER RINGS

FIELDS OF STUDY

Organic Chemistry

SUMMARY

The characteristic properties and reactions of benzene and other ring systems are discussed, and the special characteristic of aromaticity is described. The benzene structure and ring structures in general are common in natural products and are very useful in electrophilic reactions.

PRINCIPAL TERMS

- **aromatic hydrocarbon:** a hydrocarbon in which the carbon atoms form a ring with alternating double and single bonds, distributed in such a way that all bonds are of equal length and strength; also called an arene.
- **cyclohexane:** a saturated hydrocarbon composed of six methylene bridges (–CH2–) bonded in a six-membered ring structure; has the molecular formula C_6H_{12}.
- **cyclopentane:** a saturated hydrocarbon composed of five methylene bridges (–CH2–) bonded in a five-membered ring structure; has the molecular formula C_5H_{10}.
- **functional group:** a specific group of atoms with a characteristic structure and corresponding chemical behavior within a molecule.
- **pi bond:** a covalent chemical bond formed when parallel *p* orbitals of two adjacent atoms overlap in a side-by-side manner to form two molecular orbitals.

The Nature of Benzene and Other Rings

The term "cyclic" refers to a basic molecular structure of hydrocarbon molecules in which the carbon atoms are bonded together to form a closed loop or ring structure, rather than as a two-ended linear chain. Of these, cyclohexane, cyclopentane, and their derivatives are by far the most common of the saturated hydrocarbon compounds, while benzene and its derivatives form the basis of a vast number of unsaturated compounds.

The cyclic nature of these compounds relates directly to the unique electronic nature of the carbon atom. The carbon atom has four electrons in its valence electron shell, two in the 2*s* orbital and one each in two of the three *p* orbitals in the 2*p* subshell. This arrangement is stabilized at a lower intrinsic energy when the 2*s* and all 2*p* orbitals combine, or hybridize, to form four equivalent sp^3 hybrid atomic orbitals. The sp^3 orbitals are geometrically arranged around the carbon atom nucleus so that they are oriented toward the four apexes of a tetrahedron. This places any two of them at a 109.5-degree angle to each other. As it happens, this particular angle and the physical size of the carbon atom are nearly ideal for the formation of a five- or six-membered ring of carbon atoms. Both configurations are very stable, with essentially no physical strain on the bonds between carbon atoms. Accordingly, a great many compounds are formed of five- and six-membered rings of carbon and other, very similar atoms, such as nitrogen and oxygen. This includes glucose; all of the carbohydrates that derive from glucose and other sugars, including the ribose and deoxyribose that form the structural backbone of RNA and DNA, respectively; and all animal and plant steroids:

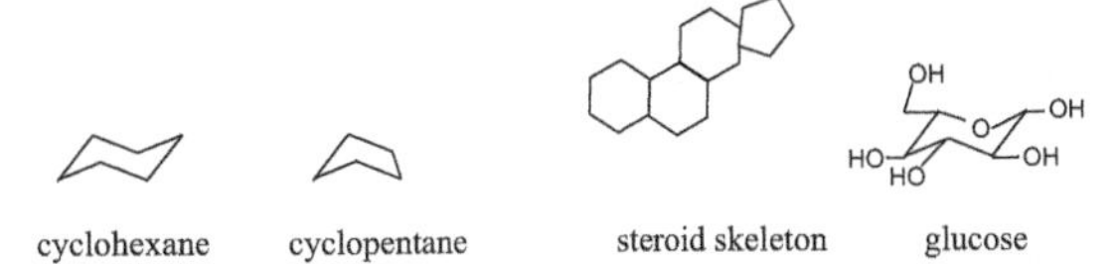

In other compounds, the 2*s* orbital and just two of the three 2*p* orbitals hybridize to form three equivalent sp^2 hybrid atomic orbitals. These are arranged in a plane, radiating outward from the nucleus of the carbon atom. Any two of the three sp^2 orbitals are at an angle of 120 degrees to each other. The third 2*p*

orbital remains as a *p* orbital, oriented perpendicularly to the plane of the sp^2 orbitals. When two carbon atoms with this orbital arrangement are adjacent to each other in a molecule, their remaining unhybridized *p* orbitals are able to overlap side by side, forming a secondary covalent bond called a pi (ϖ) bond. The combination of a pi bond with the standard sigma (σ) bond, in which orbitals overlap head on, is significantly stronger than a sigma bond alone. With regard to molecular structure, the 120-degree bond angle between sp^2 orbitals precisely matches the internal angles of a hexagon. As a result, six sp^2 hybridized carbon atoms in a ring make for a very stable planar molecular structure, characteristic of the compound benzene (C_6H_6). Accordingly, a great many naturally occurring compounds are based on the benzene ring, from the simple amino acid phenylalanine to highly complex plant biopolymers called "lignins." In fact, it would be fair to say that far more naturally occurring compounds incorporate benzene rings in their structures than do not.

Ring structures range in size from three-carbon units and their heteroatomic analogs to very large single rings. In many of these, the benzene ring is present as a phenyl group ($-C_6H_5$). Heteroatomic analogs are also very common. One of the most interesting and potentially most valuable of compounds based on the benzene ring is graphene, perhaps the most advanced material known to modern chemistry. Paradoxically, it is also a naturally occurring substance that people have been writing with for literally hundreds of years. Graphene, for all its technological potential, is the basic structural component of graphite, or pencil lead. The structure of graphene consists of an immense array of carbon atoms bonded together in six-membered rings, each of which is structurally identical to the benzene ring.

Aromaticity

The prevalence of the six-membered benzene ring in nature is due to a special electronic characteristic of its bond structure. Historically, benzene, discovered in 1825 by Michael Faraday (1791–1867), was long known to be a hydrocarbon of the alkene class (unsaturated hydrocarbons with at least one C=C double bond), though it did not exhibit the chemical behavior characteristic of alkenes, and especially not of its homologous six-carbon compounds. Before there was a usable theory of atomic structure, and well before any modern analytical apparatus, scientists were greatly puzzled by the behavior of benzene and at a loss to explain its unexpected stability. Legend has it that German chemist Friedrich August Kekulé (1829–96), while pondering the structure of the benzene molecule, dreamed of a snake swallowing its own tail or of six monkeys holding hands and dancing in a circle. This clue allegedly led to his 1865 proposal that the six carbon atoms of the benzene molecule were joined together in a circle, alternating three double bonds with three C–C single bonds. For many years, scientists questioned whether these were normal single and double bonds or whether there was something special about them.

In 1872, Kekulé further proposed that the C=C and C–C bonds were constantly changing places, causing the molecule to oscillate between two different configurations. While considered outrageous at the time, this later proved to be not far from the truth. Following the development of methods for testing the bonds between atoms, it was discovered that all six carbon-carbon bonds in benzene are identical in length and therefore in bond strength. In fact, the electrons that form the bonds exhibit a form of delocalization, meaning that they are not restricted to a single atom or bond but rather follow orbitals

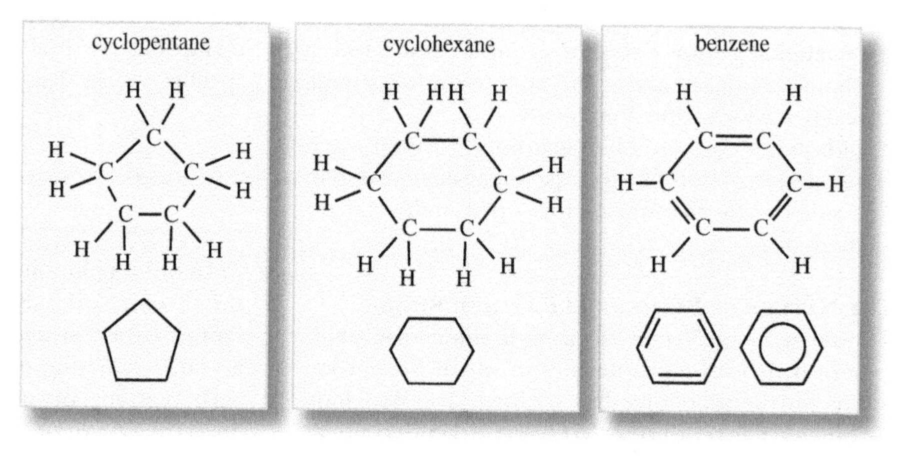

that encompass several adjacent atoms. Thus, while the molecule is best represented with alternating double and single bonds, functionally both types of bond behave more like one-and-a-half bonds (which do not otherwise exist). At low temperatures, as the energy of the molecule decreases, the activity of the electrons in the bonds also decreases, allowing the bonds to differentiate into normal single and double bonds; at normal temperatures, the equalization of the bonds produces an enhanced stability in the molecule, a characteristic known as "aromaticity."

The term "aromatic" may have originally been ascribed for benzene's sweetish, oily aroma, but it has since come to refer to the behavior of electrons in pi-bond systems rather than the compound's effect on the sense of smell. Not all compounds with alternating single and double bonds are aromatic in nature. The compound 1,3,5,7-cyclooctatetraene is a hydrocarbon with alternating single and double bonds throughout its eight-membered ring structure, but it does not exhibit aromaticity. Rather, aromaticity seems to be reserved for compounds that have only certain numbers of pi electrons (electrons engaged in the pi-bond system), in accord with what is known as Hückel's rule, named for German chemist and physicist Erich Hückel (1896–1980). The rule states that compounds only exhibit aromaticity if they have $4n + 2$ pi electrons, where n equals zero or any positive integer.

Benzenes are part of a larger class of aromatic hydrocarbons, also called "arenes." The name "benzene" refers specifically to an aromatic compound with six carbon atoms arranged in a ring and one hydrogen atom bonded to each carbon atom. In benzene derivatives, one or more of the hydrogen atoms have been replaced with substituents. While six-membered arenes are the most common, those with more or fewer carbon atoms exist as well. Aromatic compounds in which one or more of the carbon atoms has been replaced by another atom are called "heterocyclic aromatics."

Nomenclature of Benzene Derivatives

Benzene derivatives are named by first identifying the various substituents (either single atoms or functional groups) bonded to the ring. The International Union of Pure and Applied Chemistry (IUPAC) provides rules for nomenclature of organic compounds that cover all possible combinations of substituents on the benzene molecule. The position of each substituent is specified numerically, with the primary substituent being assigned position 1 on the benzene ring. For example, a benzene ring bearing a cyclohexyl group ($-C_6$) and a neopentyl group [$-CH_2-C(CH_3)_3$] separated by two carbon atoms on the ring would be named 1-cyclohexyl-4-neopentylbenzene.

Because certain substituents are commonly found in many different benzene derivatives, the name of the base derivative can also be used to describe the same derivative with additional substituents. For example, toluene (IUPAC name methylbenzene) is a benzene ring in which one of the hydrogen atoms has been substituted with a methyl group ($-CH_3$), while anisole (methoxybenzene) is a benzene ring with a methoxy functional group ($O-CH_3$). The location of the functional group does not need to be identified in such compounds if it is the only substituent, but if one or more additional substituents are added, their positions in relation to the compound's characteristic

SAMPLE PROBLEM

Given the chemical names 3-*t*-butyl-4-pentyltoluene and *o,m*'-diphenylanisole, draw the skeletal structure of the molecules.

Answer:

The name of the first compound indicates that it is a derivative of toluene, or methylbenzene. There is a *t*-butyl [$-C(CH_3)_3$] group on the third carbon atom in the ring and a pentyl group ($-C_5H_{11}$) on the fourth carbon atom in the ring. The structure is therefore

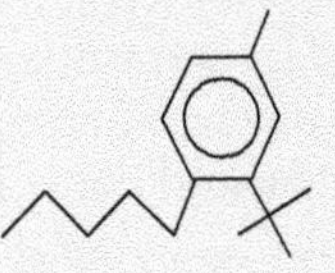

The name of the second compound indicates that it is a derivative of anisole, or methoxybenzene. There is a phenyl group—that is, a benzene ring missing one hydrogen atom—in the ortho position on one side of the molecule and another phenyl group in the meta position on the other side of the molecule. The structure is therefore

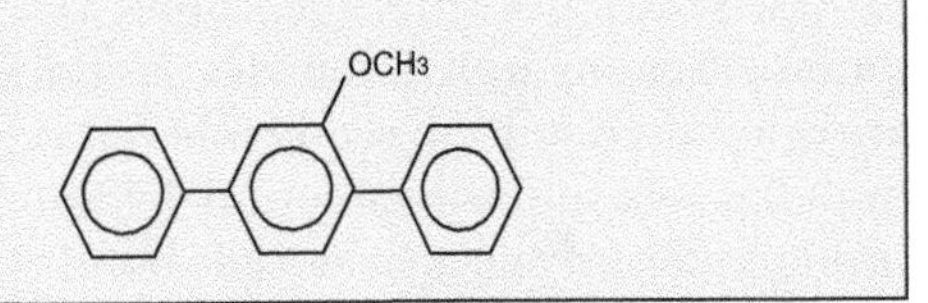

functional group can be identified either by number or by the prefixes *o-* (*ortho-*), *m-* (*meta-*), and *p-* (*para-*). If the characteristic functional group is on the carbon atom designated number 1 (C-1), then the ortho position is one atom away, on the C-2 atom; the meta position is two away, on C-3; and the para position is on C-4, directly across the benzene ring from the original group. Thus, a toluene molecule with a cyclohexyl group directly across the benzene ring from the methyl group can be named *p*-cyclohexyltoluene rather than the more unwieldy 4-cyclohexyl-1-methylbenzene. The ortho/meta/para system can quickly become confusing, however, as it requires the use of *m'-* and *o'-* identifiers for substituents in the C-5 and C-6 positions, respectively. For this reason, the numerical system, which would name the molecule 4-cyclohexyltoluene instead, is generally preferred.

Formation of Benzene Rings

The extra stability associated with the benzene ring makes its formation by dehydration and related reactions very easy. In fact, the ready formation of the benzene-ring structure often leads to undesired products during synthesis reactions. Various condensation reactions with aldehydes, ketones, and esters are often used to form six-membered ring structures in molecules, and subsequent dehydration reactions produce the conjugated double-bond system of the benzene ring.

Faraday originally isolated benzene by distilling coal tar. While this method is still used, the vast majority of benzene is derived from petrochemicals, most commonly via catalytic re-forming. In this method, various straight-chain hydrocarbons are vaporized and mixed with hydrogen gas before being exposed to a catalyst, typically platinum and aluminum oxide, under conditions of high heat and pressure. The process causes the hydrocarbons to re-form into arenes, after which the benzene is separated from the rest of the reaction products, again by distillation.

Applications of Benzene and Other Rings

Benzene is used as a solvent in many applications because of its ability to dissolve nonpolar compounds and its general inertness. However, benzene is also an electron-rich material due to the cloud of pi electrons on each "face" of the molecule, making it attractive to electrophiles (electron acceptors) such as Lewis acids. The prime examples of this electrophilic reactivity are the Friedel-Crafts alkylation and acylation reactions. In the Friedel-Crafts alkylation reaction, the presence of a Lewis acid catalyst can cause an alkyl or aryl halide to give up an alkyl or aryl group (a saturated hydrocarbon or arene derivative, respectively) to replace a hydrogen atom on the benzene ring. A similar substitution occurs in the Friedel-Crafts acylation reaction, with an acyl group (carboxylic acid derivative) from an acyl chloride or an acid anhydride replacing the hydrogen atom.

The electron-rich character of benzene and similar cyclic compounds also permits the formation of organometallic "sandwich" compounds, such as ferrocene and dibenzenechromium. In these compounds, the metal atom coordinates with the pi electrons of the cyclic system in such a way that the metal atom is sandwiched between two planar arenes. One of the more infamous such compounds is methylcyclopentadienyl manganese tricarbonyl (MMT), known as a "half-sandwich" compound because the manganese atom is only bonded to an arene on one side. MMT was used to replace the tetraethyllead in leaded gasoline until it became associated with various health hazards. US regulations still permit low levels of MMT in fuel, but it is no longer commonly used.

Richard M. Renneboog, MSc

Bibliography

Acheson, R.M. *An Introduction to the Chemistry of Heterocyclic Compounds.* 3rd ed. New York: Wiley, 2008. Print.

Hendrickson, James B., Donald J. Cram, and George S. Hammond. *Organic Chemistry.* 3rd ed. New York: McGraw, 1970. Print.

Herbert, R.B. The B*iosynthesis of Secondary Metabolites. 2nd* ed. London: Chapman & Hall, 1989. Print.

Mann, J., et al. *Natural Products: Their Chemistry and Biological Significance.* New York: Wiley, 1994. Print.

Morrison, Robert Thornton, and Robert Neilson. Boyd. *Organic Chemistry.* 7th ed. Englewood Cliffs, N.J.: Prentice Hall, 2003. Print.

O'Neil, Maryadele J., ed. *The Merck Index: An Encyclopedia of Chemicals, Drugs, and Biologicals.* 15th ed. Cambridge: RSC, 2013. Print.

BIOCHEMISTRY

FIELDS OF STUDY

Biochemistry; Organic Chemistry

SUMMARY

The broad field of biochemistry explores the compounds and processes involved in living organisms. Biology and biochemistry are intimately related: biology does not exist without biochemical processes, and biochemistry does not exist without biological systems.

PRINCIPAL TERMS

- **biomolecule:** an organic molecule produced by a living organism.
- **carbohydrate:** an organic compound containing hydroxyl (–OH) and carbonyl (C=O) groups, often with the general formula $C_x(H_2O)_y$; includes sugars, starches, and celluloses.
- **lipid:** a type of biomolecule that is soluble in organic nonpolar solvents and generally insoluble in water; includes fats, waxes, and the major components of organic oils.
- **nucleic acid:** a biopolymer consisting of many different nucleotides bonded together; includes both DNA and RNA.
- **protein:** a biological polymer consisting of one or more long chains of amino acids linked by peptide bonds in a sequence specified by an organism's DNA.

THE BASICS OF BIOCHEMISTRY

Biochemistry is quite literally the chemistry of life. Because they are composed of matter, all living organisms are made of atoms and molecules that obey the same chemical principles as any other atoms and molecules. However, the very complexity of biological systems separates biochemistry from "basic" chemistry. In any one of the trillions of cells that compose the physical body of an animal, tens of thousands of interrelated chemical processes occur almost constantly. These processes are often cyclic in nature and extract energy from matter taken in as nutrition. For example, the process of cellular respiration uses oxygen from the atmosphere and quantities of inorganic phosphate ions to break down glucose into carbon dioxide and water, releasing the energy from the chemical bonds so it can be used to drive other biochemical processes. In green plants, the opposite process occurs: carbon dioxide from the atmosphere and water are combined in the presence of sunlight to produce glucose, which polymerizes to form starches, celluloses, and other carbohydrates. In this way, the sun is the source of all energy that maintains the vast majority of life-forms on the planet.

Deoxyribonucleic acid (DNA) and other nucleic acids are central to the processes of life. DNA molecules are composed of a series of several million nucleotides. Nucleotide sequences are the blueprint for every one of the many thousands of proteins in living beings. The living cell itself exists only as a construct resulting from the hydrophilic ("water loving") and hydrophobic ("water fearing") properties of lipids. All of these biomolecules, and an untold number of others, are the subject of the field of biochemistry.

THE MAJOR BIOCHEMICALS

Carbohydrates

Carbohydrates are compounds composed of carbon atoms, each of which is bonded to the components of a water molecule: two hydrogen atoms and one oxygen atom. Technically, the simplest carbohydrates are methanol (H–CH_2–OH) and ethylene glycol (HO–CH_2CH_2–OH), since the carbon atoms in each of these is bonded to both a hydrogen atom and a hydroxyl (–OH) group. However, the first truly important carbohydrate in biochemical systems is glycerol, an alcohol compound containing three hydroxyl groups. Glycerol is the first product formed during glycolysis, produced by splitting a molecule of glucose into two molecules of glycerol. This compound is also the "backbone" of the phospholipids that form the phospholipid bilayer of cell membranes.

Larger carbohydrate molecules are called saccharides, or sugars, and include fructose, glucose, lactose, and many other similar compounds. The most important of these are ribose and deoxyribose, the essential sugar components of ribonucleic acid (RNA) and DNA, respectively. In plants, photosynthesis

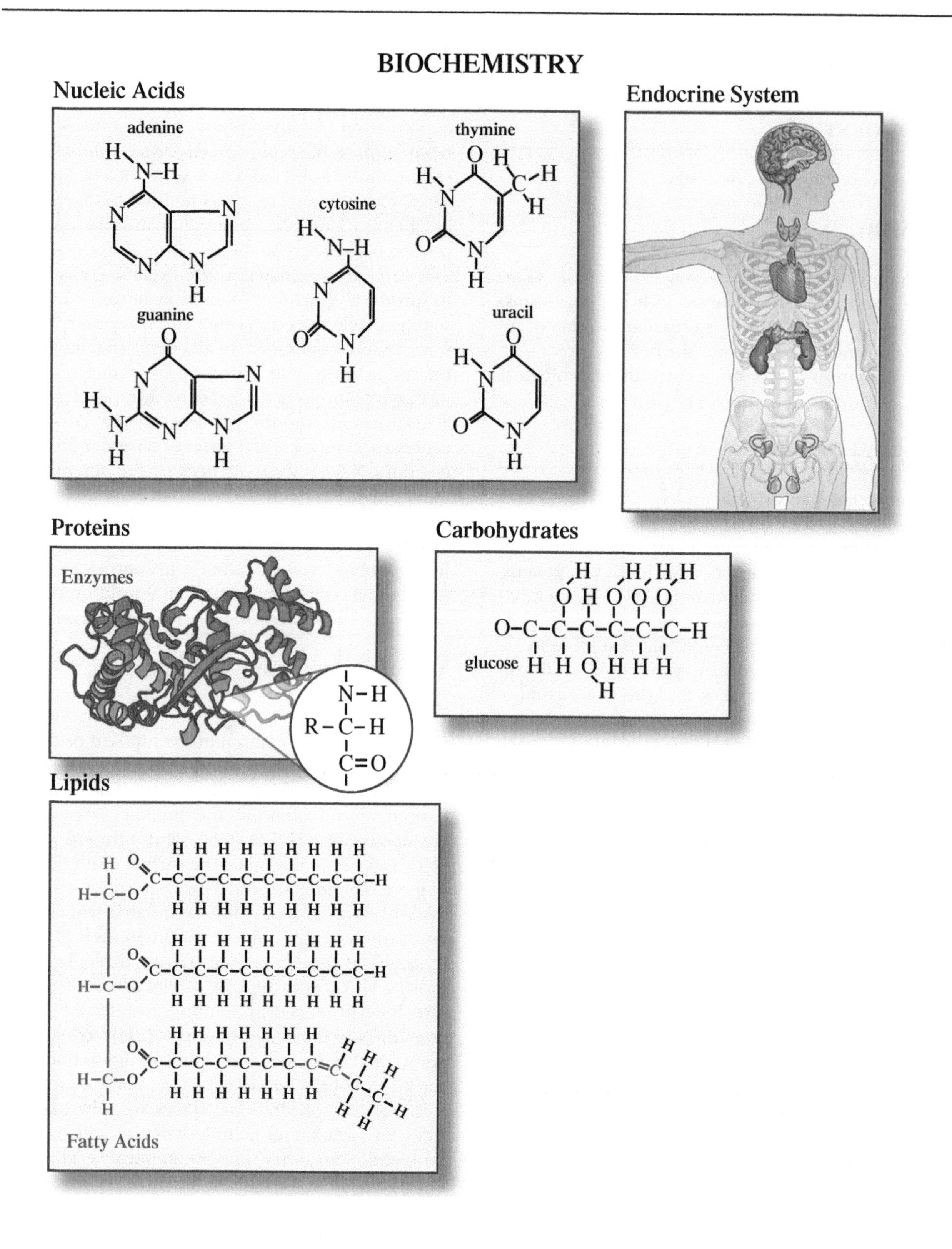
BIOCHEMISTRY
Nucleic Acids
adenine
cytosine
thymine
guanine
uracil
Endocrine System
Proteins
Enzymes
Carbohydrates
glucose
Lipids
Fatty Acids

produces glucose exclusively, mostly because of its stable six-membered ring structure. Fructose and other sugars are produced by plants and animals by the alteration of glucose or some other synthetic process. Glucose polymerizes into various starches and celluloses in plants and into glycogen in animals, providing the primary sources of energy in metabolism.

Proteins

All proteins in living systems are created from a limited assortment of approximately twenty amino acids. Protein molecules can be any length, and the same amino acids can repeat any number of times, so the potential variety of proteins is essentially infinite.

Amino acids bond chemically to each other via amide linkages, which are called peptide bonds when they occur in proteins. Extended chains of amino acids form a protein molecule. Proteins are the principal structural component of the many different types of animal tissues, from heart muscle to hair and fingernails. All of the proteins created and used in living organisms are synthesized by enzyme-mediated processes according to the nucleotide sequences contained in each individual DNA molecule. Therefore, in a way, all living things are simply self-replicating collections of proteins.

Enzymes, which are in fact specialized protein molecules, carry out their functions based on their molecular structure. Each protein molecule has a primary structure consisting of the sequence of amino acids from which it was assembled. The shapes of these individual molecules induce a secondary protein structure, causing some sections to twist in a spiral and others to resemble pleated sheets. A tertiary structure arises as functional groups in different parts of the protein molecule interact with each other. A final, multi-unit quaternary structure forms when two or more independent protein molecules become intertwined but not chemically connected, forming a protein complex or an enzyme complex.

Nucleic Acids

Nucleic acids are the central feature of protein synthesis. DNA carries the genetic code of the corresponding living organism. It was first isolated in 1869 but was not known to bear genetic information until 1943, when Oswald Avery (1877–1955), Colin MacLeod (1909–72), and Maclyn McCarty (1911–2005) demonstrated that introducing DNA from a virulent strain of a bacterium into a nonvirulent strain could produce the virulent strain. In 1953, James D. Watson (1928–) and Francis Crick (1916–2004), with significant assistance from their laboratory manager, Maurice Wilkins (1916–2004), and based on the analytical and theoretical work of Rosalind Franklin (1920–58), first published their discovery that the DNA molecule is in the shape of a double helix.

Since then, methods and techniques for the manipulation and analysis of DNA have advanced, permitting biochemists and geneticists to obtain a much better understanding of the role of genes and chromosomes in the DNA molecule. In February 2001, the journal *Nature* published the first complete analysis of the human genome. Subsequent study of the genome has demonstrated that all modern humans are descended from a very few human populations that originated in Africa in the distant past.

As the genome was deciphered over time, scientists gained an understanding of the mechanisms by which DNA and RNA are synthesized. In DNA replication and protein transcription, certain enzymes separate the dual-strand DNA molecule into two single strands so that RNA segments can assemble against the nucleotide pattern contained in the molecule. These RNA segments are then used to synthesize either proteins or new DNA molecules. DNA and RNA are each formed from only four different nucleotide bases, but the number of possible combinations of those four is sufficient to control the functional process in its entirety.

Lipids

Lipids are superficially related to carbohydrates through the intermediary of the glycerol molecule. The category "lipid" describes several different types of molecule, including the many and various fats, oils, and greases found in the cells of living organisms. Numerous metabolic processes synthesize lipids to carry out a variety of functions. The lipids formed from fatty acids (carboxylic acids attached to long hydrocarbon chains) are the essential material of the cell walls and membranes in animal cells, forming what is called the phospholipid bilayer.

A phospholipid consists of a glycerol molecule that has been esterified by two long-chain fatty-acid molecules and a phosphate ion. The phosphate portion is a "mixed ester," composed of phosphoric acid and one of the hydroxyl groups of the glycerol molecule; the fatty-acid portions form normal carboxylate esters with the other two hydroxyl groups. Thus, the phospholipid molecule has two distinct regions, a phosphoester end that is strongly hydrophilic and a long-chain fatty-acid end that is decidedly hydrophobic. These two different regions do not interact well with each other and are instead attracted to the corresponding sections of other molecules.

When large quantities of phospholipid molecules are present in a water-based environment, the hydrophobic fatty-acid portions are forced to aggregate and effectively blend into each other. Due to the repulsive force between the negatively charged phosphate groups to which they are attached, the molecules automatically form a sandwich-like structure called a bilayer. The two outer surfaces of the bilayer consist of the hydrophilic phosphoester portions of the phospholipid molecules, while sandwiched between them is a thick, hydrophobic layer of intertwined fatty-acid chains. This bilayer fully encloses the interior contents of living animal cells, forming a membrane that allows materials to pass from one side to the other by various mechanisms. Being hydrophilic, the inner and outer surfaces can interact freely with the water-based fluids on either side of the membrane.

Metabolic Processes

The biochemical processes that living systems use to obtain energy and sustain life are collectively called metabolism. These metabolic processes fall into two distinct categories, anabolic and catabolic. Anabolic processes build up the structure of the system; examples include muscle-protein synthesis and bone construction. Catabolic processes deconstruct materials in order to extract energy for anabolic and homeostatic processes; examples include the breakdown of glycogen and fat deposits, the digestion of foods, and the elimination of waste products.

Biochemistry and Organic Chemistry

Since life, as it is known on Earth, is based on carbon compounds, the basic principles of biochemistry are the same as those of organic chemistry. Biochemistry can be thought of as the application of the principles of organic chemistry to aqueous (water-based) substances, which are normally the realm of inorganic chemistry. All of the fluids of living things, such as blood and tree sap, are water based. Accordingly, mineral salts and electrolyte ions are important factors in the maintenance of biological systems, and the field of biochemistry is as concerned with the role of inorganic substances, such as minerals and trace elements, as it is with the organic aspects of biochemical processes.

SAMPLE PROBLEM

Determine whether the following processes are anabolic or catabolic:

- protein synthesis
- protein digestion
- glycolysis
- photosynthesis
- respiration

Answer:

- Protein synthesis is a process for building up the system; it is anabolic.
- Protein digestion breaks down proteins into component amino acids portions; it is catabolic.
- Glycolysis breaks down the glucose molecule into two molecules of glycerol; it is catabolic.
- Photosynthesis assembles glucose from carbon dioxide and water to store energy from the sun; it is anabolic.
- Respiration breaks down the glycerol from glycolysis into carbon dioxide and water to release the energy stored in its bonds; it is catabolic.

Richard M. Renneboog, MSc

Bibliography

Abbott, David. The *Biographical Dictionary of Scientists. New York: P.* Bedrick, 1984. Print.

Acheson, R.M. A*n Introduction to the Chemistry of Heterocyclic Compounds.* 3rd ed. New York: Wiley, 2008. Print.

Berg, Jeremy M., John L. Tymoczko, Gregory J. Gatto, and Lubert Strye. *Biochemistry.* 8th ed. New York: W. H. Freeman, 2015. Print.

Fenichell, Stephen. *Plastic: The Making of a Synthetic Century.* New York: Harper, 2009. Print.

Hendrickson, James B., Donald J. Cram, and George S. Hammond. *Organic Chemistry.* 3rd ed. New York: McGraw, 1970. Print.

Herbert, R. B. *The Biosynthesis of Secondary Metabolites.* 2nd ed. London: Chapman & Hall, 1989. Print.
Jones, Mark Martin. *Chemistry and Society.* 5th ed. Philadelphia: Saunders College Pub., 1987. Print.
Lehninger, Albert L. *Biochemistry: The Molecular Basis of Cell Structure and Function.* 2nd ed. New York: Worth, 1975. Print.
Lodish, Harvey, et al. *Molecular Cell Biology.* 7th ed. New York: Freeman, 2013. Print.
Mann, J., et al. *Natural Products: Their Chemistry and Biological Significance.* New York: Wiley, 1994. Print.
Morrison, Robert Thornton, and Robert Neilson. Boyd. *Organic Chemistry.* 7th ed. Englewood Cliffs, N.J.: Prentice Hall, 2003. Print.
Robbers, James E., Marilyn K. Speedie, and Varro E. Tyler. *Pharmacognosy and Pharmacobiotechnology.* Rev. ed. Baltimore: Williams, 1996. Print.
Vieira, Ernest R. *Elementary Food Science.* 4th ed. Gaithersburg: Aspen, 1999. Print.

BLACK HOLES

FIELDS OF STUDY

Stellar Astronomy; Astronomy; Cosmology

SUMMARY

Black holes are areas of space-time with gravitational fields so strong that nothing can escape. This phenomenon is caused by a mathematical singularity at the center of the black hole. At a black hole, space-time is highly warped. Black holes were first proposed by scientist John Michell in the eighteenth century. However, they were popularized by Albert Einstein as a consequence of his general theory of relativity. While black holes cannot be directly observed, astronomers have found strong evidence for their existence.

PRINCIPAL TERMS

- **event horizon:** the boundary of a black hole beyond which nothing can escape.
- **general relativity:** Albert Einstein's theory that gravity is the result of matter causing space-time to curve. Among other things, this implies that gravity can bend light.
- **supermassive black hole:** a black hole with a mass greater than one hundred thousand times that of the sun.

Discovery of Black Holes

The existence of black holes was first proposed by English clergyman and scientist John Michell (1724–93) in 1783. Michell published a letter stating that certain stars may grow so big that not even light can escape their gravitational pull. This was based on Isaac Newton's (1642–1727) corpuscular theory of light, which essentially stated that light is a particle rather than a wave. A particle could be affected by gravitational pull, while a wave could not. Soon after, a number of experiments seemed to disprove Newton's theory, and as a result Michell's proposal was largely ignored.

This all changed in 1915, when Albert Einstein published his famous theory of general relativity. When tested, Einstein's theories showed that light could be bent by powerful gravitational fields. Thus, if an area of space somehow acquired a large enough gravitational field, light would not be able to escape it. The object would appear as a giant black sphere, effectively invisible in outer space.

Black holes are incredibly difficult to observe directly. However, astronomers can use other methods to determine their locations. First, black holes have a visible effect on the stars around them. When astronomers find a star that appears to be orbiting a large object that neither emits or reflects visible light, that object is probably a black hole. Second, the accretion disks of black holes create so much friction that they heat up and emit electromagnetic radiation, mostly in the x-ray range. Scientists can use x-ray telescopes to detect this radiation and trace it to its source.

Classical Black Holes

Black holes are formed by the collapse of incredibly massive stars. When such a star uses up all its fuel, the delicate balance that allowed it to exist is broken. The gravity from the star's dense core pulls its outer layers inward. As more matter is pulled into the core, the increase in the core's mass and density causes its gravitational pull to grow stronger, which in turn pulls in even more matter. This process increases in speed and power exponentially. Eventually, the area around the core becomes so dense and gravitationally powerful that not even light can escape it. As this area pulls everything around it inward, it continues to grow and increase its gravitational pull. This creates an infinitely dense region of space-time with an infinitely strong gravitational pull. Such a region is called a singularity.

Every singularity has a definitive boundary called an event horizon. An event horizon is the point at which the singularity's escape velocity equals the speed of light, so that no electromagnetic radiation can escape.

Black holes are typically surrounded by accretion disks. An accretion disk is a disk of gas and other matter that orbits the event horizon as it is slowly pulled toward the singularity. The gravity produced by the singularity compresses this matter as it rotates, creating intense amounts of friction. The friction generates so much heat that the accretion disk emits massive amount of radiation, mostly in the form of x-rays.

Supermassive Black Holes

As their name suggests, supermassive black holes are many times larger than normal black holes. The singularities of supermassive black holes are at least one hundred thousand solar masses (one hundred thousand times the mass of the sun). Astronomers theorize that supermassive black holes as large as fifty billion solar masses could exist. The largest known supermassive black hole, SDSS J0100+2802, is twelve billion solar masses.

Many astronomers believe that a supermassive black hole lies at the center of every galaxy. The supermassive black hole at the center of the Milky Way is roughly 27,000 light-years from Earth and more than four million solar masses. Astronomers believe that supermassive black holes play a major role in the formation of galaxies. They have found a direct correlation between the mass of a galaxy's supermassive black hole and the size of the galactic bulge surrounding it.

A supermassive black hole's accretion disk is much brighter, hotter, and larger than that of traditional black holes. Occasionally, an extremely large object passes through a supermassive black hole's event horizon. The incredibly intense gravity near the event horizon crushes most objects, causing a reaction that converts roughly 10 percent of the object's mass into energy. The energy then dramatically shoots away from the black hole. Supermassive black holes sometimes capture objects so large that this reaction creates energy across multiple spectrums, including visible light and x-rays. These impressive events are called quasars.

Scientists are unsure how supermassive black holes grow to their tremendous sizes. While black holes can grow in mass by gradually absorbing more matter, it does not seem possible for one to grow quickly enough to reach such supermassive proportions during its lifetime. Therefore, some scientists theorize that supermassive black holes originated as giant clouds of gas soon after the big bang. This theory bypasses entirely the star phase of a black hole's existence. Proponents claim that some external force, such as large concentrations of dark matter in the early universe, could have compressed the gas quickly enough to form a large black hole, and then continued to push more gas into the singularity at an incredibly fast rate. This would greatly increase the speed of the black hole's growth.

Another theory of the formation of supermassive black holes involves binary black hole systems. A binary black hole system is one in which two nearby black holes orbit a common center of mass. This could happen when two galaxies collide, which would have happened more often early in the universe's history. The combined gravity from the two black holes would then pull their accretion disks out of alignment, which computer simulations show would cause the entirety of the disks to tear and collapse into their singularities. This would greatly increase the size of the black holes in a very short period of time. The two black holes would then merge, creating one black hole with the combined mass and gravitational pull of the entire binary system.

Information Paradox

While black holes are viewed as necessary and mundane products of general relativity, their existence conflicts with a major tenet of quantum mechanics. General relativity deals primarily with the physics of large objects, while quantum mechanics deals with the physics of objects much too small to see. Scientists have been attempting to unify the two theories for many years.

One of the major principles of quantum mechanics is the idea that information can never be truly destroyed. If one has all the pieces of a current state of events, it should theoretically be possible to discover all past states. However, information pulled into the singularity of a black hole is effectively destroyed. It is compressed beyond any form of recognition and may never be able to escape.

British physicist Stephen Hawking (b. 1942) devised a way around this paradox. According to his calculations, due to quantum mechanics, all particles occasionally create temporary copies of themselves. Hawking proposed that black holes emit tiny amounts of radiation, slowly burning off all the matter they have collected over time. According to his models, black holes that fail to absorb new matter will eventually evaporate. Any information collected by a black hole will eventually be released, but not in any form humans can recognize or put back together.

An alternate solution for this paradox involves phenomena called white holes. According to this theory, black holes eventually burst, spewing all the information they have collected in a huge explosion of energy. Proponents of this theory say that this effect actually happens as soon as the black hole forms. However, due to the ways singularities bend time and space, the effect appears to take billions of years to anyone outside the singularity.

Tyler Biscontini

Bibliography

Byrne, Michael. "The Black Holes We See in Space Might Already Be White Holes." *Motherboard*. Vice Media, 21 July 2015. Web. 30 Apr. 2015.

Cain, Fraser. "Never a Star: Did Supermassive Black Holes Form Directly?" *Universe Today*. Author, 7 Sept. 2007. 30 Apr. 2015.

GaBany, R. Jay. "A Singular Place." *Cosmotography*. Author, n.d. Web. 30 Apr. 2015.

Hall, Shannon. "How Do Black Holes Get Super Massive?" *Universe Today*. Fraser Cain, 13 Aug. 2013. Web. 30 Apr. 2015.

Marshall, John. "Introduction: Black Holes." *New Scientist*. Reed Business Information, 6 Jan. 2010. Web. 30 Apr. 2015.

Redd, Nola Taylor. "Black Holes: Facts, Theory, and Definition." *Space.com*. Purch, 9 Apr. 2015. Web. 30 Apr. 2015.

BUCKYBALLS

FIELDS OF STUDY

Molecular Biology; Organic Chemistry

SUMMARY

Buckyballs, or buckminsterfullerene, are spherical fullerene molecules. The name comes from inventor Richard Buckminster Fuller, who designed a geodesic dome that resembles the shape of these molecules—something similar to a soccer ball. *Science* called the buckyball the Molecule of the Year in 1991, and research continues to develop this unexpected molecule with extraordinary characteristics. However, the full potential of buckyballs does not seem to have been reached.

PRINCIPAL TERMS

- **allotrope:** one of two or more principal physical forms in which a single pure element occurs, due to differences in chemical bonding or the structural arrangement of the atoms; for example, diamond and graphite are two allotropes of carbon.
- **aromaticity:** a characteristic of certain ring-shaped molecules in which alternating double and single bonds are distributed in such a way that all bonds are of equal length and strength, giving the molecule greater stability than would otherwise be expected.
- **buckminsterfullerene:** a form of carbon that contains molecules having 60 carbon atoms arranged at the vertices of a polyhedron with hexagonal and pentagonal faces, more commonly know as a buckyball.

- **graphite:** a common mineral, a black or grey soft allotropic form of carbon in hexagonal crystalline form.
- **Hückel's rule:** a theory proposed in 1931 by German chemist and physicist Erich Hückel to determine if a planar ring molecule would have aromatic properties. Hückel's rule states that if a cyclical, planar molecule has 4n + 2(pi) electrons, it will be aromatic.

Buckminsterfullerene, or buckyballs, is an extremely unusual substance. Archimedes described the geometry of solid truncated icosahedrons more than 2,000 years ago. But Leonardo da Vinci first theorized the skeletal, hollow-faced version that is now called buckminsterfullerene in his illustrations in 1494, so its existence has been theorized for hundreds of years. Eiji Osawa of Japan proposed an idea for buckyballs when he observed that the structure of a corannulene molecule, a cyclopentane ring fused with 5 benzene rings, looked like a portion of a soccer ball. He theorized that this could exist in a full-body shape.

Diamonds and graphite, two other allotropes of carbon, have been known for thousands of years. However, it was not until 1985 that buckminsterfullerene was finally discovered, and not until 1990 was it created in sufficient quantities to explore its possible uses. Its discovery completely upset all understanding of the physics and chemistry of carbon, and researchers formed a special branch of chemistry, fullerene science and technology, to study this unique molecule and its relatives.

On September 1, 1985, Harry Kroto arrived at Rice University. He was interested in studying cyanopolyynes and convinced Richard Smalley and Robert Curl to let him use their laser-generated supersonic cluster beam to investigate his ideas. These men, helped by students, vaporized a graphite disk. When it cooled and condensed, they immediately analyzed the resulting material. In addition to finding the carbon snakes that Kroto had postulated, they also found C_{60}—buckyballs. They submitted their findings to *Nature* on September 12, 1985, and 11 years later, the three men were awarded the Nobel Prize.

It is quite surprising that buckyballs were not discovered sooner and without high-tech equipment. Buckyballs are formed, for example, by vaporizing wax molecules in the soot of a burning candle, in geological formations, and in cosmic dust clouds. They are the most common naturally occurring fullerenes.

The most common technique for producing buckyballs is electric arc discharge, where an electric arc is established between two carbon electrodes. The energy from the arc dissipates, breaking carbon from the surface. The carbon then cools in the inert atmosphere and forms buckyballs. Unfortunately, this type of production is not capable of producing commercial quantities of this substance. Another type of apparatus that produces more buckyballs works by vaporizing graphite in a chamber filled with helium. Buckyballs are then extracted in toluene, which is then removed, leaving behind a mixture of mostly C_{60} with small amounts of the larger fullerenes. Another method involves laser ablation.

Chemical Profile

Buckyballs are compounds made up of only carbon. They contain 60 carbon atoms, formed into a cage-like structure. The term can be used slightly more broadly to refer to any closed cage structure consisting only of three-coordinate carbon atoms. They are closely related to buckytubes, or carbon nanotubes, with a cylindrical structure. Other similar structures include C_{70} and C_{76}, but these structures can contain as few as 28 and as many as 600 carbon atoms. The category of fullerenes also includes tubes of carbon that are thousands of times longer than they are wide and have the same icosahedral structure with diameters as small as 2nm. Another type of carbon cage exists called a carbon "onion," in which the carbon cages are all stacked within one another, like Russian dolls. These carbon cages contain millions of atoms and dozens of concentric shells. Scientists have hypothesized carbon cages with 7-sided rings but have not yet actually discovered these.

Unique Structure

Buckyballs have a structure that is unique in all of chemistry. Most molecules are compounds, containing two or more chemical elements. However, buckyballs are made of 60 atoms of only one element—carbon. Each carbon atom occupies an exactly equivalent site and is joined to each of the other atoms in an exactly equivalent way to form a hollow, cage-like structure. This structure has 32 faces, 20 of which are hexagons and 12 of which are pentagons.

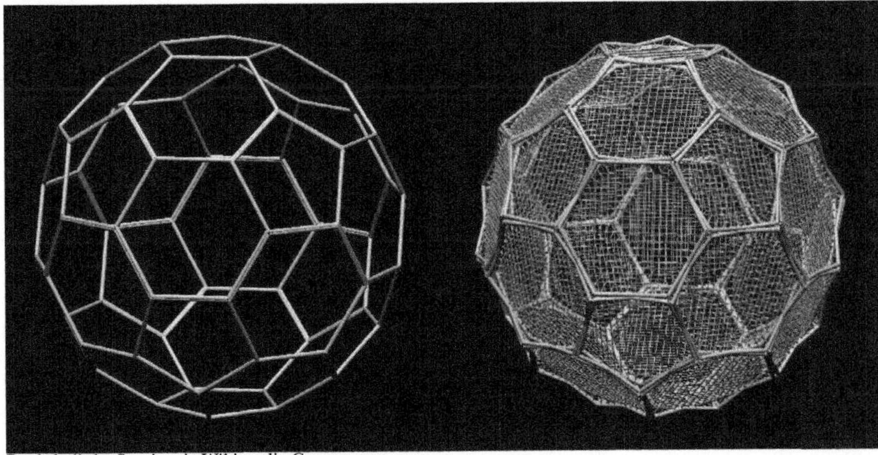
Buckyballs by Segrim via Wikimedia Commons

also become superconductors, handling high currents with no resistance and little loss of energy.

C_{60} and C_{70} have some similar properties, including donor-accepting properties. Oxidation of these substances is irreversible. In studies with mice, buckyballs have shown no toxic effects. Fullerenes with no functional groups can act as antioxidants, but those with functional groups can be highly toxic.

No two pentagons have a common vertex; if they did, this molecule would be much more unstable. The shape and pattern is similar to that of a soccer ball but the proper name for this structure is a truncated icosahedron.

PROPERTIES

Buckyballs are the roundest known molecule, and they are hollow inside. They are hard to break apart, even at extremely high temperatures and pressures, and bounce back when slammed against solid objects. The outside, exposed surface of a buckyball reacts with other molecules while still maintaining its spherical shape. Because it is hollow inside, a buckyball can entrap other smaller molecules, such as helium. However, the inside of the buckyball itself does not react with whatever it encloses. Buckyballs do not actually bond to one another, but they stick together with Van der Waals forces, an attractive force that is not as strong as a chemical bond.

Buckyballs do not follow Hückel's rule which states that if $n = 2(N+1)2$ carbon atoms will have aromaticity in spherical fullerenes. The bonds that hold buckyballs together allow electrons to move among bonds, causing buckyballs to be aromatic and form very stable, inert bonds. However, buckyballs seem to generally avoid having double bonds within the pentagonal rings, making electron delocalization difficult. This means that buckyballs are not superaromatic, as some other substances are. Buckyballs are the only known carbon allotropes that are soluble in common solvents such as benzene, toluene, or chloroform. With treatment, these molecules can

USES

Buckyballs have a wide variety of possible uses that span an unbelievably high number of fields. However, the expense of creating buckyballs inhibits practical use. Research continues on how to make these molecules more cost effectively. The sheer strength of buckyballs make them an exciting possibility in construction materials, and the hollow center excites researchers who are looking for a way to deliver materials intact, for example, directly into a cell. Other researchers are finding ways to use buckyballs and buckytubes as circuit components in nanoelectronic or molecular electronic devices. Buckytubes are one of the strongest materials known and are used in composite materials and as components in scientific instruments. When compressed to 70% of its original size, a buckyball is more than twice as hard as a diamond, one of its close relatives. One interesting possibility is bucky paper, which may be used in fire-resistant materials, in television screens, in heat sinks, and in filter membranes.

Medical possibilities include the following uses:

- Trap free radicals
- Block inflammation
- Fight deterioration of motor function in the body
- Deliver medication directly into cancer cells

Industrial possibilities include the following uses:

- Produce inexpensive, paint-on solar cells
- Store hydrogen as fuel
- Reduce the growth of bacteria in water systems

- Lubricate nearly anything
- Improve wear resistance

The possibilities seem endless for practical applications of this unique substance.

Marianne M. Madsen, MS

BIBLIOGRAPHY

Aldersey-Williams, Hugh. *The Most Beautiful Molecule: The Discovery of the Buckyball.* New York: John Wiley, 1995.

Delgado, Juan L., et al. "Buckyballs," *Polyarenes II.* Topics in Current Chemistry: Volume 350, 2014, pp. 1-64.

Ngo, Christian and Marcel H. Van de Voorde, *Nanotechnology in a Nutshell.* Atlantis Press, 2014, pp. 71-84.

Shanbogh, Pradeep and Nalini G. Sundaram. "Fullerenes Revisited." *Resonance* (February 2015): pp. 123-135.

Thakur, Vijay Kumar and Manju Kumari Thakur, eds. *Chemical Functionalization of Carbon Nanomaterials: Chemistry and Applications.* Boca Raton, FL: Taylor & Francis Group, 2016.

Tomanek, David. *Guide Through the Nanocarbon Jungle: Buckyballs, Nanotubes, Graphene, and Beyond.* Morgan and Claypool (2014).

C

CARBOXYLIC ACIDS

FIELDS OF STUDY

Organic Chemistry

SUMMARY

The characteristic properties and reactions of carboxylic acids are discussed. Carboxylic acids are very useful in organic synthesis reactions and are also essential components of the biochemistry of living systems.

PRINCIPAL TERMS

- **Brønsted-Lowry acid-base theory:** definitions for acids and bases developed separately in 1923 by Danish chemist Johannes Nicolaus Brønsted and English chemist Martin Lowry; defines an acid as any compound that can release a hydrogen ion and a base as any compound that can accept a hydrogen ion.
- **carbonyl group:** a functional group consisting of a carbon atom double bonded to an oxygen atom.
- **functional group:** a specific group of atoms with a characteristic structure and corresponding chemical behavior within a molecule.
- **hydrogen bond:** a weak type of chemical bond formed by the attraction of a hydrogen atom to an electronegative atom—an atom with a strong tendency to attract electrons—in the same or another molecule.
- **hydroxyl group:** a primary functional group consisting of an oxygen atom covalently bonded to a single hydrogen atom.

The Nature of the Carboxylic Acids

According to Brønsted-Lowry acid-base theory, developed separately in 1923 by chemists Johannes Nicolaus Brønsted and Martin Lowry, the term "acid" indicates that a compound is capable of giving up one or more hydrogen ions (H^+), also sometimes known as "protons." The carboxylic acids satisfy this criterion. Whereas ionic acids dissociate completely into positive and negative ions when put into water, carboxylic acids typically do not. When acetic acid (CH_3COOH), for example, is dissolved in water, a small percentage of the acetic acid molecules are dissociated into ions. The vast majority of the material, however, exists in solution as dissolved, but undissociated molecules of acetic acid.

The carboxylic acids are characterized by the presence of the carboxyl, or carboxylic acid, functional group, often represented as –COOH attached to an R– or Ar– placeholder. In this notation, R– and Ar– indicate an alkyl group (a hydrocarbon chain with the general formula C_nH_{2n+1}) or an aryl group (derived from an aromatic ring), respectively. The carboxyl group comprises a carbonyl group (C=O) that is bonded to a hydroxyl group (–OH):

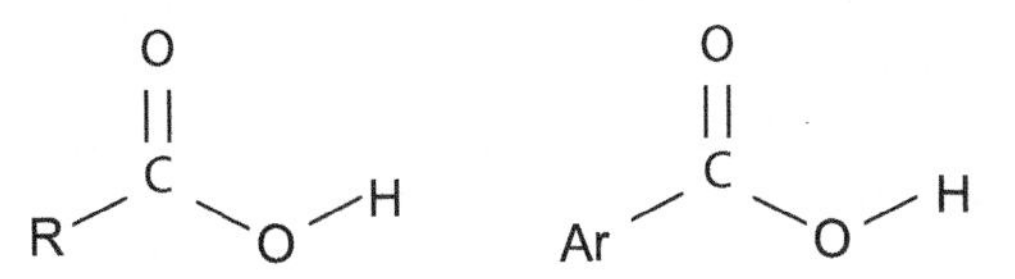

Since the carboxyl function is a substituent functional group, or side chain, rather than a structural component in the molecular structure, the number of compounds that can be classed as carboxylic acids is effectively infinite.

The acidity of carboxylic acids is the result of the bonding between the hydroxyl group and the carbon atom of the carbonyl group. An atomic nucleus is composed of positively charge protons and neutral neutrons and is in turn surrounded by negatively charged electrons. The numbers of protons and electrons are equal in neutral atoms, while atoms with unequal electrons and protons are known as "ions." Electrons can be thought of as existing within specific regions about the nucleus known as "orbitals"; the location of electrons within the orbitals controls how an atom forms bonds and undergoes chemical reactions. In carboxylic acids, the oxygen atom exerts an electron-withdrawing effect through the pi (ϖ) bond (one of two bonds that make up the double bond) with the carbon atom. This effect induces the hydroxyl oxygen atom to draw in the electrons from the O–H bond and hybridize from its normal sp^3 orbital configuration to the sp^2 and p orbital configuration. The p orbital forms an extended pi-system with the carbonyl group, and the extra electron becomes delocalized, or detached from a specific atom or covalent bond. Through delocalization, both of the oxygen atoms become equivalent, and the C–O bond lengths are equalized. The addition of a proton, or hydrogen ion, to an atom or molecule occurs naturally as the reverse step in the following equilibrium process:

$$R\text{–}COOH \leftrightharpoons R\text{–}COO^- + H^+$$

and can revert either one of the oxygen atoms to the hydroxyl group. Over time, the proportion of carbonyl oxygen and hydroxyl oxygen becomes equal. In addition to the interaction of the atoms' orbitals, some of the properties of carboxylic acids are due to those compounds' ability to form hydrogen bonds.

The linear carboxylic acids are essential components of both plant and animal biochemistry. Apart from its popular use as vinegar, acetic acid is the basic material for a vast number of hydrocarbon compounds called "acetogenins," which are produced by plants. It is also the source of the acetyl group of acetyl coenzyme A, which is essential in different stages of respiration and in other biochemical processes. Higher carboxylic acids and dicarboxylic acids such as succinic acid are produced during the breakdown of glucose in cellular respiration and glycolysis. The long-chain fatty acids are the essential component that form the phospholipid bilayer of animal cells. The amino acids, a special class of carboxylic acids, are the material from which all proteins are made, as specified by the genetic code carried by the deoxyribonucleic acid (DNA) molecule.

Nomenclature of Carboxylic Acids

There are common names for many of the simpler carboxylic acids, often based on the source of the compound, which continue to be used despite the development of a standardized system of chemical nomenclature. Methanoic acid (HCOOH) is responsible for the characteristic odor of ants; the common name, formic acid, derives from the Latin word for "ant," *formica.* Next in the series is ethanoic, or acetic, acid (H_3CCOOH), which is responsible for the taste and smell of vinegar. The common name comes from the Latin word for "sour," as does the word "acid." Third in the series is propanoic (propionic) acid, and the fourth is butanoic (butyric) acid. Butyric acid is responsible for the taste and smell of rancid butter, and the name derives from the Latin word *butyrum,* meaning "butter." Pentanoic acid ($C_5H_{10}O_2$) is also known as "valeric acid," and hexanoic acid ($C_6H_{12}O_2$), is known as "caproic acid" (from the Latin word *caper,* meaning "goat"). Octanoic acid ($C_8H_{16}O_2$) and decanoic acid ($C_{10}H_{20}O_2$) were also identified from soured goat's milk and are known as "caprylic acid" and "capric acid," respectively. Other carboxylic acids were isolated from various vegetable oils and have common names that reflect that heritage, including lauric acid (from laurel oil), myristic acid (nutmeg), palmitic acid (palm), stearic acid (beef fat, or stearin), and oleic, linoleic, and linolenic acids (linseed).

The systematic International Union of Pure and Applied Chemistry (IUPAC) names of linear carboxylic acids follow the basic hydrocarbon series, beginning with methanoic and progressing through

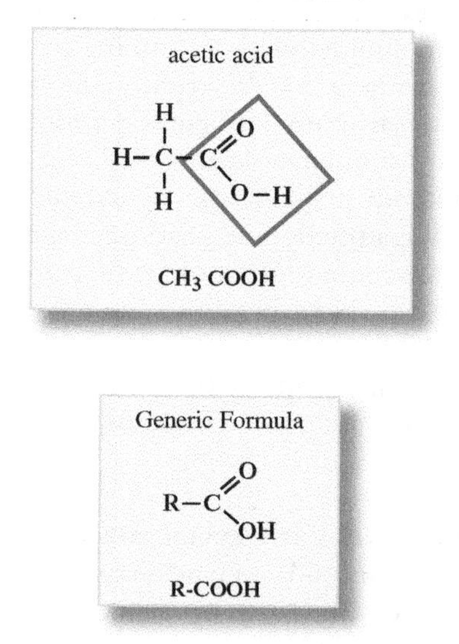

Carboxylic acid, written as COOH in semi-structural formulas, is composed of a carbon atom double bonded to an oxygen atom, single bonded to an alcohol group, and is single bonded to some other R-group.

SAMPLE PROBLEM

Given the chemical name *trans*-γ-(N,N-dimethyl)aminobut-3-enoic acid, draw the structure of the molecule.

Answer:

The root name but-3-enoic acid indicates that the substance is a derivative of the four-carbon carboxylic acid with a C=C bond between the third and fourth carbon atoms in the chain. There is a nitrogen atom bearing two methyl groups bonded to the fourth carbon atom. The *trans-* prefix indicates that the major portions of the molecule are on opposite sides of the C=C bond. The structure of the molecule is therefore

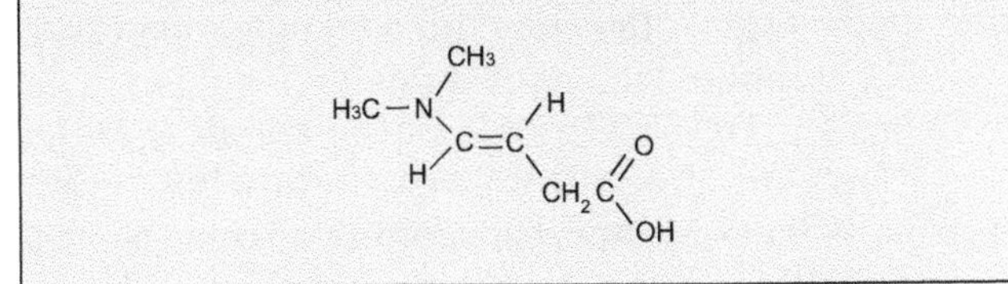

ethanoic, propanoic, and butanoic. The series continues with the numeric names pentanoic, hexanoic, heptanoic, and so on. The name is generated by identifying the longest carbon atom chain in the molecule and adding the suffix *-oic acid*. The carboxyl carbon atom is assigned the first position in the chain. For example, a seven-carbon chain with a C≡C triple bond at the third carbon atom in the chain would thus be named hept-3-ynoic acid. When different side chains are present on a hydrocarbon molecule, they are assigned by the priority of their size to attain the lowest position numbers and are named alphabetically. If a methyl group was on the fifth carbon atom in the seven-carbon chain, the compound would be named systematically as 5-methylhept-3-ynoic acid. In an alkene (an unsaturated hydrocarbon with one or more C=C bonds) and alkene-derived compounds, the relative positions of side chains are indicated in the formal names using the identifiers *cis-* for side chains on the same side of the double bond and *trans-* when the side chains are on opposite sides of the double bond.

A different convention is often used to identify the position of substituents on the chain, using the letters of the Greek alphabet as place markers relative to the carboxyl carbon atom. The adjacent position is assigned as α- (alpha), the next as β- (beta), then γ- (gamma), and so on. The omega fatty acids take their name from this practice, as ω (omega) is the last letter of the Greek alphabet. An omega-3 fatty acid is one that has a C=C bond beginning at the third carbon atom from the ω end of the molecule.

Aryl carboxylic acids are typically named as derivatives of benzoic acid or the corresponding IUPAC name of the base aryl group. Compounds bearing two carboxyl functions are called "dicarboxylic acids." One of the simplest of these is oxalic acid, found in rhubarb and other plants. Tricarboxylic acids are also known, though they are much less common. When it must be named as a substituent of another compound that is not named as a carboxylic acid, the carboxyl functional group is identified as the hydroxycarbonyl substituent.

FORMATION OF CARBOXYLIC ACIDS

Carboxylic acids are typically formed by oxidation of the corresponding alcohol—that is, the alcohol molecule loses one or more electrons. Ethanol, for example, oxidizes to form first acetaldehyde and then acetic acid. Oxidation of the aldehyde (–RCOH) to the carboxylic acid often requires only exposure to air. It is not unusual for bottles of the reagent benzaldehyde, for example, to slowly become bottles of benzoic acid over time, once they have been opened. The process is enhanced and controlled by the use of specific oxidizing agents, eliminating the wait time associated with air oxidation. Other methods of producing acetic acid in bulk quantities include catalytic oxidation of hydrocarbons and the reaction of methanol and carbon monoxide catalyzed by rhodium and iodine. Benzoic acid is typically produced from toluene obtained from the catalytic reforming of petroleum. Similarly, phthalic acid is produced from xylene or naphthalene. Essentially all other carboxylic acids, except those that can be obtained in quantity by hydrolysis of natural fats and oils, are synthesized by elaborating these basic structures through various synthesis reactions.

REACTIONS OF CARBOXYLIC ACIDS

Carboxylic acids form salts and can be titrated or neutralized by a basic compound. When combined, an acid and its corresponding salt dissolved in water form a buffer solution that will maintain a constant pH—a numerical value that represents the acidity or basicity of a solution—even after the addition of limited quantities of an acid or a base. This is a very important feature in biological systems and in laboratory methods.

Carboxylic acids are converted easily to the corresponding acid chloride by reaction with thionyl chloride ($SOCl_2$), phosphorus trichloride (PCl_3), or pentachloride (PCl_5). The acid chloride readily reacts with various nucleophiles (electron-rich chemical species) to produce the corresponding derivatives. Reaction with ammonia (NH_3), primary amines (RNH_2), or secondary amines (R_2NH) produces the corresponding amides ($RCONH_2$). Reaction with an alkoxide or aryloxide anion (RO^- or ArO^-) produces the corresponding ester (RCOOR'). Acid chlorides can also be used to elaborate an aryl compound in the Friedel-Crafts acylation reaction, which results in the addition of substituents to an aromatic ring. Aryl carboxylic acids, such as benzoic acid, also undergo Friedel-Crafts alkylation and acylation reactions, although less actively than other compounds.

Richard M. Renneboog, MSc

BIBLIOGRAPHY

Berg, Jeremy M., John L. Tymoczko, Gregory J. Gatto, and Lubert Strye. *Biochemistry.* 8th ed. New York: W. H. Freeman, 2015. Print.

Hendrickson, James B., Donald J. Cram, and George S. Hammond. *Organic Chemistry.* 3rd ed. New York: McGraw, 1970. Print.

Herbert, R. B. *The Biosynthesis of Secondary Metabolites.* 2nd ed. London: Chapman & Hall, 1989. Print.

Jones, Mark Martin. *Chemistry and Society.* 5th ed. Philadelphia: Saunders College Pub., 1987. Print.

Mann, J., et al. *Natural Products: Their Chemistry and Biological Significance.* New York: Wiley, 1994. Print.

Morrison, Robert Thornton, and Robert Neilson. Boyd. *Organic Chemistry.* 7th ed. Englewood Cliffs, N.J.: Prentice Hall, 2003. Print.

Robinson, Trevor. *The Organic Constituents of Higher Plants.* 6th ed. North Amherst: Cordus, 1991. Print.

Wuts, Peter G. M., and Theodora W. Greene. *Greene's Protective Groups in Organic Synthesis.* 4th ed. Hoboken: Wiley-Interscience, 2007. Print.

CATIONS

FIELDS OF STUDY

Inorganic Chemistry; Geochemistry; Metallurgy

SUMMARY

The basic structure of cations is defined, and the development of the modern theory of atomic structure is elaborated. Electronic structure is fundamental to all chemical behaviors and is responsible for the relationships seen in the periodic table.

PRINCIPAL TERMS

- **anion:** any chemical species bearing a net negative electrical charge, which causes it to be drawn toward the positive pole, or anode, of an electrochemical cell.
- **ionic bond:** a type of chemical bond formed by mutual attraction between two ions of opposite charges.
- **ionization:** the process by which an atom or molecule loses or gains one or more electrons to acquire a net positive or negative electrical charge.
- **multivalent:** describes an atom that has the ability to accept or donate more than one valence electron and thus can exist in more than one oxidation state.
- **valence electron:** an electron that occupies the outermost or valence shell of an atom and participates in chemical processes such as bond formation and ionization.

THE NATURE OF CATIONS

A cation is any atom or group of atoms that bears a net positive charge due to the absence of one or more valence electrons. The term is a contraction of "cathode ion," which is a reference to the fact that when a direct electric current is applied to an electrolytic solution, positively charged ions are attracted to the cathode, or negative terminal, of the source of the current. By the same token, anions, from "anode ions," have a net negative charge and are attracted to the anode, or positive terminal, of the current source. Anions and cations often combine to form compounds held together with ionic bonds; one common example is sodium chloride (NaCl), better known as table salt, which is created when the sodium cation, Na^+, bonds to the chloride anion, Cl^-.

The formation of any ion is the result of an atom or molecule gaining or losing one or more valence electrons. This is most apparent in monatomic (single-atom) ions, in which the electrical charge is equal to the oxidation state. For example, the alkali metals—hydrogen, lithium, sodium, potassium, rubidium, cesium, and francium—are all highly electropositive, meaning that they readily donate their lone valence electron so that their next electron shell becomes the outermost shell, which is ideal because it is full and therefore stable. The resulting cations have an electrical charge of 1+ because they lost one electron and thus one unit of negative electrical charge. Similarly, the alkaline earth metals, such as barium (Ba) and magnesium (Mg), have two electrons in their valence shells, which they readily lose to form cations such as Ba^{2+} and Mg^{2+}. Such cations are said to be "divalent," meaning that they can form single bonds with up to two other ions or molecules. They can also be described as multivalent or polyvalent, which simply means that they can form single bonds with more than one other ion or molecule. Multivalent atoms also have the potential to exist in different oxidation states—that is, they can lose or gain different numbers of electrons to form ions of different electrical charges; phosphorous, for example, commonly forms cations with charges of 5+ and 3+ as well as an anion with a charge of 3–.

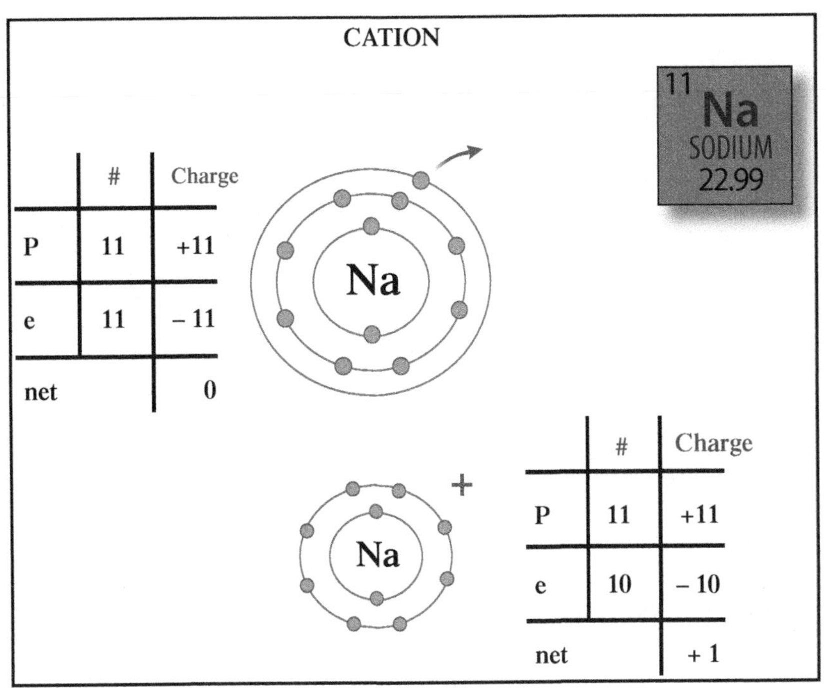

The Electronics of Cation Formation

According to the modern theory of atomic structure, each atom contains a very small, extremely dense nucleus that holds at least 99.98 percent of the atom's mass and all of its positive electrical charge. The nucleus is surrounded by a diffuse and comparatively very large cloud of electrons containing all of the atom's negative electrical charge. These electrons are allowed to possess only very specific energies. This restricts their movement around the nucleus to specific regions called "electron shells." Within each shell are one or more subshells that contain even more well-defined regions called "orbitals." The strict geometric arrangement of the orbitals regulates the formation of chemical bonds between atoms.

Each shell and orbital is subject to a number of restrictions that dictate how many electrons it can hold. There are four different types of electron orbitals, designated *s*, *p*, *d*, and *f*, each of which can contain a specific number of electrons: *s* orbitals can hold a maximum of two electrons; *p* orbitals, a maximum of six; *d* orbitals, a maximum of ten; and *f* orbitals, a maximum of fourteen. One or more of these orbitals make up an electron shell. The various electron shells are indicated by an integer value known as the "principal quantum number," starting with 1 for the innermost shell, typically referred to as the "$n = 1$ shell." The standard notation to describe an electron orbital is the principal quantum number, followed by the type of orbital, followed by a superscript number indicating how many electrons it holds. For example, helium has only one *s* orbital, which is completely full, so its electron configuration is represented as $1s^2$. The first *p* orbital appears in the $n = 2$ shell, the first *d* orbital in the $n = 3$ shell, and the first *f* orbital in the $n = 4$ shell. Due to variances in energy levels, electrons usually fill atomic orbitals in the order 1*s*, 2*s*, 2*p*, 3*s*, 3*p*, 4*s*, 3*d*, 4*p*, 5*s*, 4*d*, 5*p*, 6*s*, 4*f*, 5*d*, 6*p*, 7*s* . . . rather than in strict numerical order.

SAMPLE PROBLEM

Draw a simple orbital diagram of the calcium (Ca) atom, and identify the electrons that it will lose or gain to become the calcium ion. Use the following chart to fill in the electron orbitals in the proper order, and remember that *s* orbitals hold two electrons, *p* orbitals hold six electrons, *d* orbitals hold ten electrons, and *f* orbitals hold fourteen electrons.

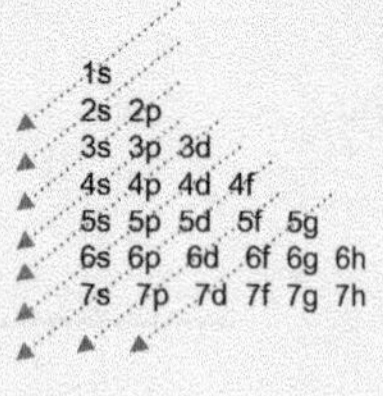

Answer:

The atomic number of calcium is 20, indicating that it has a total of twenty electrons, distributed as follows:

$$1s^22s^22p^63s^23p^64s^2$$

Thus, there are two electrons in the $n = 1$ shell, eight in the $n = 2$ shell, eight in the $n = 3$ shell, and two in the $n = 4$ shell. This can be drawn as

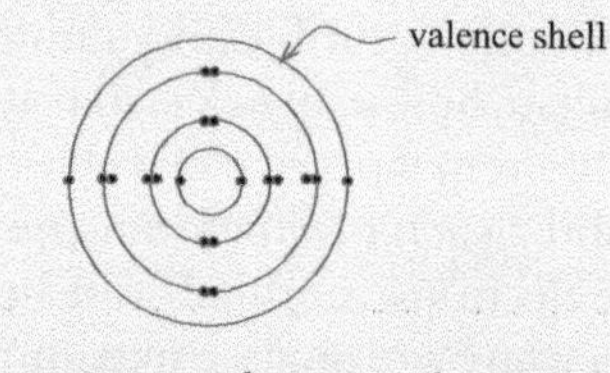

The calcium atom can lose two electrons from its valence ($n = 4$) shell, from the 4*s* orbital, making the $n = 3$ shell the outermost shell, with a closed-shell configuration of $3s^2p^6$. This creates the calcium ion Ca+.

The outermost electron shell is the valence shell, and in all elements, except the noble gases (helium, neon, argon, krypton, xenon, and radon), the valence shell is incompletely occupied. Because of this, the noble gases are often called "inert gases," reflecting the fact that they are far less chemically reactive than elements whose valence shells are not filled. All noble gases except for helium have eight electrons in their valence shell, as a shell with eight electrons approximates a stable closed-shell configuration of s^2p^6. This is the basis of the octet rule, which states that atoms of lower atomic numbers tend to achieve the greatest stability when they have eight electrons in their valence shells. The closer an element is to having eight valence electrons, the more likely it is to undergo ionization or a chemical reaction in order to achieve a stable configuration, either by gaining enough electrons to complete the octet or by losing all valence electrons so that the next-highest completed shell becomes the outermost shell. Elements with similar electron distributions in their valence shells typically exhibit similar chemical behaviors, which is the basis on which the periodic table of the elements is arranged.

NAMING THE CATIONS

The rules for naming various chemical species are established and standardized by the International Union of Pure and Applied Chemistry (IUPAC). Monatomic cations are simply called by the name of the element itself; when it is important to note the charge, this can be given in parentheses immediately following the name, so that a sodium cation, for example, is formally rendered as sodium(1+). If a compound contains a multivalent atom that can form ions of different charges (and thus different oxidation states), the particular ion present can be indicated in the compound's formal name in the same way, as in mercury(2+) chloride ($HgCl_2$), in which the mercury cation has a charge of 2+. Alternatively, the oxidation number can be used in the place of the charge number; this is given as a roman numeral, so that mercury(2+) chloride is also mercury(II) chloride. It is important to remember that only monatomic ions, such as Hg^{2+} in this example, can be assumed to have oxidation numbers that are equal to their charge numbers, as the oxidation number of a polyatomic ion takes into account a number of other factors.

The parenthetical method of indicating charge also applies to polyatomic cations. For example, a molecule consisting of three hydrogen atoms and having a charge of 1+ is called trihydrogen(1+), while the far more complex cation $[Al(POCl_3)_6]^{3+}$ has the IUPAC name hexakis(trichloridooxidophos phorus)aluminum(3+). The latter is an example of a coordination complex, which consists of a central atom or cation, typically metallic, surrounded by various other molecules or anions called "ligands." Such complexes require their own systematic names that indicate the number and identity of the ligands present. The names of these complexes can become quite complicated, requiring careful attention to the IUPAC naming conventions.

If the polyatomic cation is derived from a hydride (a compound containing a hydrogen anion, H^-) and gained its positive charge not by losing an electron but by accepting a hydrogen cation (H^+), a process known as "protonation," it can also be named by adding the suffix *-ium* to the name of the parent hydride. For example, when ammonia (NH_3) accepts a hydrogen cation, it forms ammonium (NH_4^+).

Richard M. Renneboog, MSc

Bibliography

Haynes, W. H., PhD, *David R. Lide, PhD, and Thomas J. Bruno*, PhD, eds. CRC Handbook of Chemistry and Physics, 96th Edition. New York: CRC/Taylor & Francis Group, 2015. Print.

Johnson, Rebecca L. *Atomic Structure.* Minneapolis: Lerner, 2008. Print.

Mackay, K. M., R. A. Mackay, and W. Henderson. *Introduction to Modern Inorganic Chemistry.* 6th ed. Cheltenham: Nelson, 2002. Print.

Miessler, Gary L., Paul J. Fischer, and Donald A. Tarr. *Inorganic Chemistry.* 5th ed. Upper Saddle River: Prentice Hall, 2013. Print.

Wehr, M. Russell, James A. Richards Jr., and Thomas W. Adair III. *Physics of the Atom.* 4th ed. Reading: Addison-Wesley, 1984. Print.

Winter, Mark J. *The Orbitron: A Gallery of Atomic Orbitals and Molecular Orbitals.* University of Sheffield, n.d. Web.

CELL COMMUNICATION

FIELDS OF STUDY

Biochemistry; Molecular Biology; Genetics

SUMMARY

The process of cell communication is defined, and its importance in biochemical processes is elaborated. Cell communication is an essential feature of nerve function and control of the metabolism of living systems, affecting the operation of the autonomic and sympathetic nervous systems.

PRINCIPAL TERMS

- **autocrine signaling:** a type of cell signaling in which the signaling compound is produced within a cell and delivered to receptors on the outside of the same cell.
- **endocrine signaling:** a type of cell signaling in which the signaling compound is produced in one location in the body and transported to a receptor site some distance away.
- **juxtacrine signaling:** a type of cell signaling in which the signaling compound is produced within a cell and delivered to receptors in an adjacent cell via physical contact.
- **paracrine signaling:** a type of cell signaling in which the signaling compound is produced in one location in the body and delivered to receptors in a nearby cell.
- **receptor:** a molecule or molecular structure, typically a protein or enzyme, that interacts only with compounds that have a matching molecular structure; the interaction normally triggers a biochemical response in the cells to which the receptor is attached.
- **transmitter:** a biochemical compound produced to trigger a specific response at a corresponding receptor site.

The Basics of Cell Communication

Cell communication is also known as cell signaling or extracellular signaling. It occurs via a complex multistep process that uses several different pathways. The process begins with the synthesis of a specific signaling compound within a cell or group of cells of the same type. After synthesis, the signaling compound is released by the cell, typically crossing the cell membrane via an active-transport mechanism. The signaling compound is then transported to its target cell. At the target cell, the signaling compound is detected by a specific protein or enzyme receptor site. The detection of the signaling compound triggers a change in cellular metabolism or gene expression in the cell to which the receptor is attached. The final step in the process is the removal of the signaling compound from the receptor and its subsequent elimination from the system.

Cell Communication Functions and Methods

Cell communication or signaling serves a number of purposes, depending on the nature of the living system in which it occurs. In single-cell organisms such as bacteria and protozoa, the production of specific cell-communication compounds coordinates the organisms for cell differentiation or mating, according to which particular reproductive pathway is appropriate. For more complex life-forms, this role is carried out by compounds called pheromones, which can range from relatively simple hydrocarbons to complex proteins.

Pheromones are typically released directly into the environment, where they are detected by other members of the same species. The sense of smell plays a significant role in the detection of pheromones; in many creatures, specific anatomical structures exist for the sole purpose of detecting pheromone molecules. Moths and butterflies, for example, have receptors in their antennae that have evolved to maximize their sensitivity to specific airborne molecules. Almost all other species have a molecule-sensitive structure called the vomeronasal organ as part of their olfactory sense.

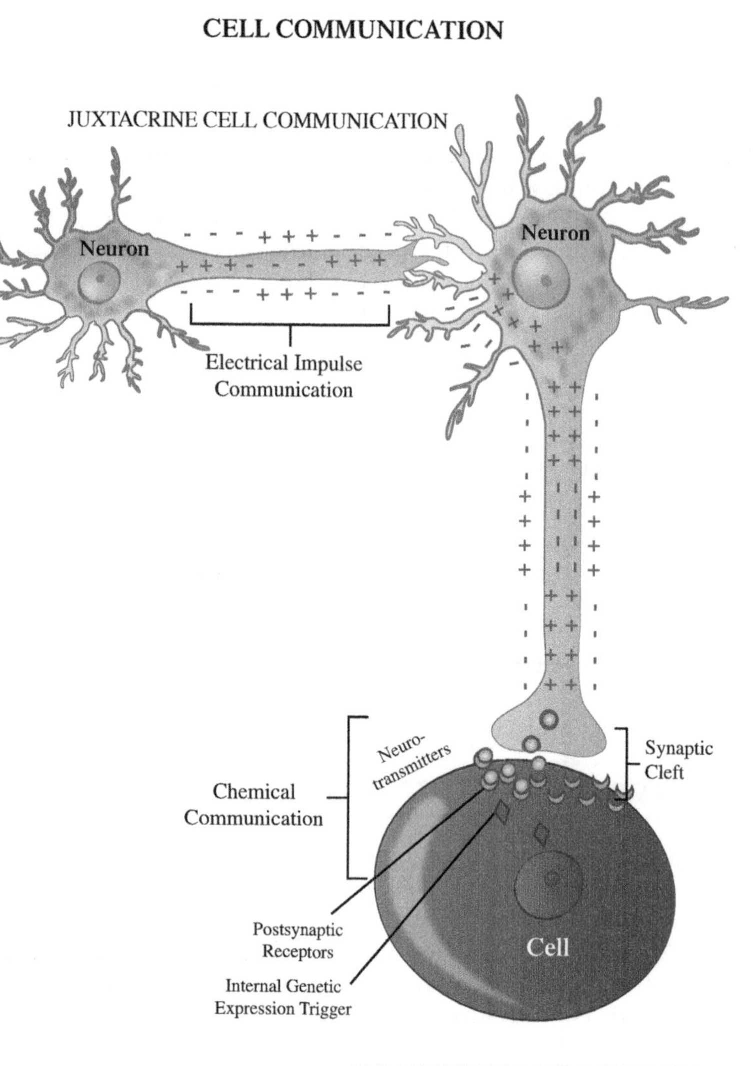

This method of cell communication is also used by plants. The production of aroma compounds and other molecular materials for the attraction of pollinators has long been known. More recent studies have also shown that plants use chemical signals to communicate with other plants and to exert some control over their environment by inhibiting the growth of competitors or promoting the growth of companion plants.

Cell communication within organisms is a more fundamental function and is responsible for a great many essential features of living biochemical systems. Intramolecular cell communication is essential to the control of metabolic processes within the cells of organisms. An extremely complex variety of chemical reactions takes place within cells, involving tens of thousands of different proteins and other biochemicals. All of these many and varied reactions are interconnected by their necessity for maintaining the existence of the organism, and a great many are more intimately related to each other as components of various cyclic biochemical mechanisms.

One function of intracellular communication is to control tissue growth. Signaling by hormones controls anabolic processes, which are processes that extract energy from adenosine triphosphate (ATP) to power the movement of various materials across cell membranes; the construction of protein molecules from amino acids; the movement of calcium

SAMPLE PROBLEM

The rate of lateral movement of a protein molecule in the lipid bilayer of a cell membrane can be described by the equation

$$\text{rate} = \frac{2kT}{3\pi r\eta}$$

where k is the Boltzmann constant, given in kilocalories per kelvin, with a value of $3.3\times10^{-27}\frac{\text{kcal}}{\text{K}}$; T is the absolute temperature in kelvins; r is the effective radius of the protein molecule; and η is the viscosity of the lipid bilayer. If the radius of the protein molecule is estimated to be 0.5 nanometers (nm), what is the rate of lateral movement at a temperature of 37°C? Use $2.4\times10^{-11}\frac{\text{kcal s}}{\text{cm}^3}$ for the value of η.

Answer:

Convert the temperature from °C to °K:

$$\text{K} = °\text{C} + 273.15$$

$$\text{K} = 37 + 273.15$$

$$\text{K} = 310.15$$

Because the viscosity of the lipid bilayer is given in units per cubic centimeter, the radius of the protein molecule also be expressed in centimeters:

$$r = 0.5 \text{ nm} = 0.5 \times 10^{-9} \text{ m} = 0.5 \times 10^{-7} \text{ cm}$$

Substitute in the values of k, T, r and η and calculate, paying attention to the units throughout:

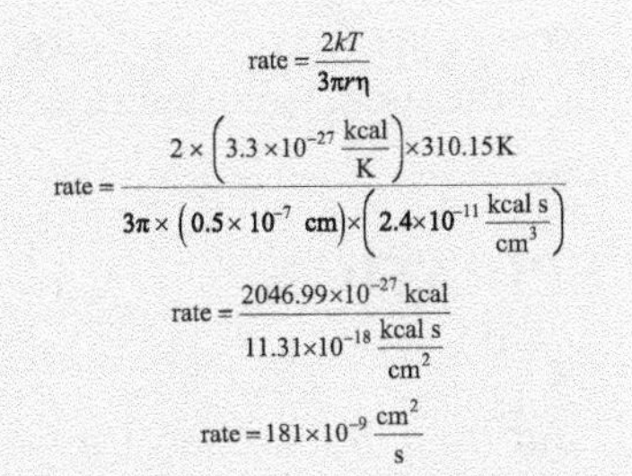

$$\text{rate} = \frac{2kT}{3\pi r\eta}$$

$$\text{rate} = \frac{2\times\left(3.3\times10^{-27}\frac{\text{kcal}}{\text{K}}\right)\times310.15\text{K}}{3\pi\times\left(0.5\times10^{-7}\text{ cm}\right)\times\left(2.4\times10^{-11}\frac{\text{kcal s}}{\text{cm}^3}\right)}$$

$$\text{rate} = \frac{2046.99\times10^{-27}\text{ kcal}}{11.31\times10^{-18}\frac{\text{kcal s}}{\text{cm}^2}}$$

$$\text{rate} = 181\times10^{-9}\frac{\text{cm}^2}{\text{s}}$$

The rate of lateral movement of the protein in a lipid bilayer is 181×10^{-9} square centimeters per second.

into bone structures; and many other processes that build up and maintain the physical structure of the organism. The many biochemical processes that take place within a cell's organelles (organ-like structures within the cell) and cytosol (intracellular fluid surrounding the organelles) extract energy and convert needed materials into unusable product materials that must be eliminated from the organism. Catabolic processes, also controlled by hormone signaling, are the processes that break down and eliminate byproducts of metabolism and materials that are no longer required, such as signaling molecules that have already served their function.

Types of Signaling

Cell communication takes place between cells or within cells. Endocrine signaling is typical of hormonal processes in which the signaling compound is produced in one part of the organism, such as the endocrine glands, and carried in the bloodstream to receptor sites located in other parts of the organism at some distance from the signaling compound's point of origin. Paracrine signaling, which affects receptors located on cells that are in close proximity to the originating cell, is typical of such functions as nerve-signal transmission across neural synapses and is mediated by acetylcholine and other compounds that function as neurotransmitters and neurohormones. Like paracrine signaling, juxtacrine signaling also takes place between adjacent cells; however, juxtacrine signaling relies on physical contact. In autocrine signaling, a cell responds to its own self-synthesized signaling compounds; receptors may be within the cell or located on the outer surface of the cell membrane. This method of cell communication is typical of the action of growth hormones and is especially relevant to the growth of tumors. There is some crossover in the methods of cell communication, since many compounds can signal by more than one method.

Receptors in Cell Communication

All methods of cell communication function by the interaction of a signaling compound with the corresponding receptor. Receptor proteins only bind to specific signaling compounds; their molecular structure also plays a role in their function. Lipophilic (literally, "fat-loving") receptor proteins found on the surfaces of cell and organelle membranes in the cytosol typically interact with fats, oils, and other hydrophobic ("water-fearing") signaling compounds. Hydrophilic ("water-loving") receptor proteins, normally found on cell surfaces, interact with polar or water-soluble signaling compounds. Activation of cell surface receptors often triggers the formation of a secondary signaling compound that delivers the signal into the cell. Secondary signaling compounds formed in response to such triggering include the cyclic forms of adenosine monophosphate (cAMP) and guanisine monophosphate (cGMP), inositol triphosphate (IP_3), and diacylglycerol (DAG).

Receptors are classified in one of four major categories. G-protein-coupled receptors (GPCRs) interact with compounds such as epinephrine, serotonin, and glucagon to activate G proteins that subsequently activate or inhibit a second messenger or ion channel, ultimately bringing about a change in the cell function. Ion-channel-linked receptors, typically activated by acetylcholine, change the conformation of the ion channel to allow the passage of specific ions across the cell membrane. A third class of receptor is the enzyme-linked receptors, which either behave as enzymes themselves or activate associated enzymes. The fourth class of receptor is nuclear receptors, which are found within cells rather than on the cell surface and mainly respond to steroid and thyroid hormones.

Signaling compounds that function in cell communication are many and varied in structure. The neurotransmitter acetylcholine is the principal signaling compound in nerve synapses. Other paracrine signaling compounds include dopamine, serotonin, and gamma-amino butyric acid (GABA). Endocrine signaling typically involves hormones as the signaling compounds. Most hormones are steroid compounds, based on the molecular structure of cholesterol, though the very important thyroid hormone thyroxine is not.

Richard M. Renneboog, MSc

Bibliography

Berg, Jeremy M., John L. Tymoczko, Gregory J. Gatto, and Lubert Strye. *Biochemistry.* 8th ed. New York: W. H. Freeman, 2015. Print.

Lafferty, Peter, and Julian Rowe, eds. *The Hutchinson Dictionary of Science.* 2nd ed. Oxford: Helicon, 1998. Print.

Lehninger, Albert L. *Biochemistry: The Molecular Basis of Cell Structure and Function.* 2nd ed. New York: Worth, 1975. Print.

Lodish, Harvey, et al. *Molecular Cell Biology.* 7th ed. New York: Freeman, 2013. Print.

Pelczar, Michael J., Jr., E. C. S. Chan, and Noel R. Krieg. *Microbiology: Concepts and Applications.* New York: McGraw, 1993. Print.

Reece, Jane B., et al. *Campbell Biology.* 10th ed. San Francisco: Cummings, 2013. Print.

CELLULAR RESPIRATION

FIELDS OF STUDY

Biochemistry; Molecular Biology

SUMMARY

The process of cellular respiration is defined, and its importance in biochemical processes is explained. Cellular respiration controls both the uptake of gases required for maintenance of the living processes and the elimination of by-product gases that would poison the living biochemical system if not removed.

PRINCIPAL TERMS

- **adenosine triphosphate (ATP):** a molecule consisting of adenine, ribose, and a triphosphate chain that is used to transfer the energy needed to carry out numerous cellular processes.

- **aerobic respiration:** a form of cellular respiration that requires oxygen in order to generate energy from glucose.
- **anaerobic respiration:** a form of cellular respiration that does not require oxygen in order to generate energy from glucose.
- **electron transport chain:** a series of oxidation and reduction (redox) reactions in which the electrons released by oxidation are transferred from one molecule to the next, ultimately enabling the production of ATP.
- **Krebs (citric acid) cycle:** a cyclic series of biochemical reactions that completes the conversion of glucose into carbon dioxide, water, and adenosine triphosphate (ATP), consuming and then regenerating citric acid in the process.

UNDERSTANDING CELLULAR RESPIRATION

Cellular respiration is the process that underlies the act of breathing. On the macroscopic scale, living aerobic organisms take in oxygen from the atmosphere through aerobic respiration, normally by inhaling air into the lungs and then exhaling air that has become enriched with carbon dioxide, which is a waste product of metabolic processes. This is not the only means by which organisms take in oxygen, however. Fish extract oxygen from water by passing it over their gills, while some amphibians and insects are able to absorb oxygen from the air through their skin.

By contrast, anaerobic organisms do not require oxygen to maintain their life functions and in fact may cease to live if oxygen is present; rather, they survive by practicing anaerobic respiration. Such organisms include bacteria and deep-ocean dwellers that have evolved to depend on chemosynthetic rather than photosynthetic processes for their energy needs. Plants, especially green plants, also rely on respiration as part of their continuing living processes. Respiration in plants provides the carbon dioxide used in photosynthesis to produce glucose and releases oxygen that is formed as a by-product of photosynthesis.

At the cellular level, during respiration, oxygen being transported in the blood is taken into eukaryotic cells, where it is used to generate adenosine triphosphate (ATP), a substance that carries the energy necessary for many enzyme-mediated biochemical processes.

CELLULAR RESPIRATION FUNCTIONS

Cellular respiration is the fundamental process by which ATP is generated in biological systems. In chlorophyll-bearing plant cells, sunlight is required to begin the process of photosynthesis, which both consumes and produces ATP. Through photosynthesis, carbon dioxide and water are combined to form the simple carbohydrate sugar glucose. The net production of ATP is used to support various aspects of plant metabolism. The photosynthetic reaction also produces oxygen, which is then released into the atmosphere, becoming available to take part in respiration-based biochemical reactions by other organisms. The glucose that is formed through photosynthesis is the principal foodstuff that supports biological systems, whether as a simple sugar or in more complex molecular forms. Once it is ingested, digestive processes break down the more complex forms into the simple sugar, which is subsequently absorbed and used as a high-energy compound. The energy produced by the biochemical breakdown of glucose is extracted via the linked processes of glycolysis and the Krebs cycle, also known as the citric acid cycle.

GLYCOLYSIS

Glycolysis begins when a molecule of glucose is chemically bonded to an inorganic phosphate group by the enzyme hexokinase, producing glucose 6-phosphate (the number indicates the position of the carbon atom to which the phosphate group has been attached). This step consumes one molecule of ATP. The third segment of the triphosphate group on the ATP molecule is cleaved off and transferred to the glucose molecule, leaving adenosine diphosphate (ADP). The conversion of ATP to ADP is the means by which the high energy contained in the phosphate-anhydride bond is released to do work in biochemical systems. Conversely, converting ADP to ATP stores that energy for later use.

The glucose 6-phosphate is converted by another enzyme to fructose 6-phosphate, which is then bonded to a second phosphate group from another ATP molecule. The symmetrical structure of fructose 6-phosphate is then cleaved to produce two molecules of glyceraldehyde 3-phosphate, each of which is then bonded to a second phosphate group to produce 1,3-bisphosphoglycerate. At this point, the glycolysis process begins the net generation of ATP, as the enzyme phosphoglycerate kinase transfers a phosphate group from each of the two 1,3-bisphosphoglycerate molecules to two ADP molecules, thus forming two molecules of ATP. The 3-phosphoglycerate produced by this reaction is isomerized to 2-phosphoglycerate and then converted to phosphoenolpyruvate. Finally, the two molecules of phosphoenolpyruvate undergo reaction with the enzyme pyruvate kinase, which transfers the two remaining phosphate groups to two more molecules of ADP to generate two more molecules of ATP, leaving two units of pyruvate, the anion (negatively charged ion) of pyruvic acid.

The process of glycolysis uses energy obtained from the conversion of two units of ATP to ADP before converting four units of ADP into ATP. In other words, glycolysis doubles the amount of ATP available for use in other energetic processes.

The Krebs Cycle

The Krebs cycle, also known as the citric acid cycle and the tricarboxylic acid cycle, is directly connected to glycolysis through the pyruvate produced in the final step of the process. Each unit of pyruvate loses a molecule of carbon dioxide, leaving an acetyl group that is transferred to a substance called coenzyme A (CoA) to produce acetyl-CoA.

The cycle begins with the interaction of acetyl-CoA with a molecule of oxaloacetate to produce a molecule of citrate. Dehydration converts the citrate into *cis*-aconitate, which is an isomer of aconitic acid from which a hydrogen ion has been removed. The *cis*-aconitate is then rehydrated to produce the isomeric molecule isocitrate. A chemical reaction called decarboxylation takes place, in which a carboxyl group (–COOH) is removed from the isocitrate, thus releasing carbon dioxide. The decarboxylation, mediated by a coenzyme called nicotinamide adenine dinucleotide (NAD+), converts the isocitrate to α-ketoglutarate. A second reaction with NAD+ and CoA decarboxylates the α-ketoglutarate and transfers the succinyl residue to CoA to produce succinyl-CoA.

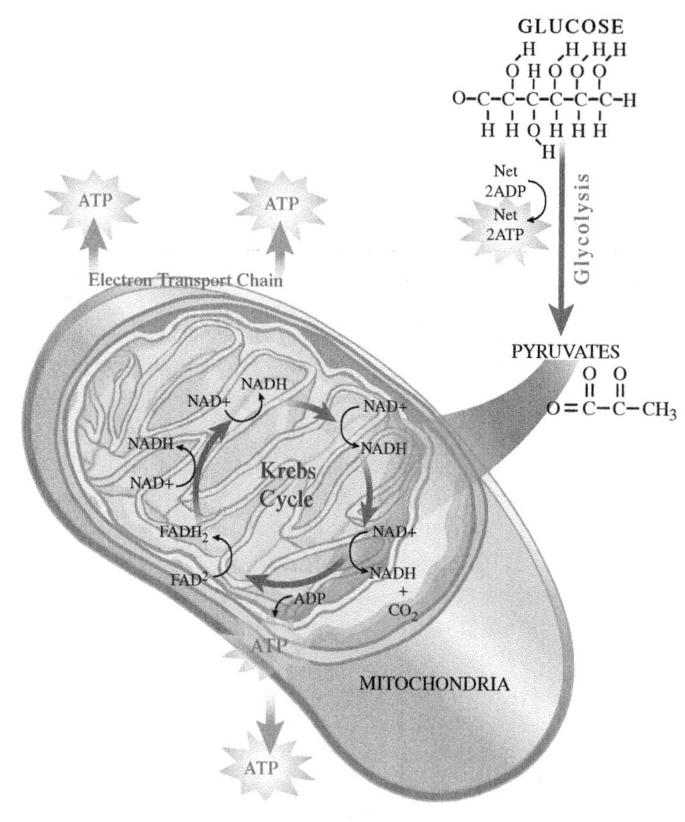

The succinate anion is released from succinyl-CoA through a reaction that converts guanosine diphosphate (GDP) to guanosine triphosphate (GTP). When the succinate is dehydrogenated by flavin adenine dinucleotide (FAD), the product is the fumarate anion, which is then hydrated to the malate structure. Oxidation of malate by NAD+ then produces the oxaloacetate that begins the cycle again.

Electron Transport Chain

Although it is possible for biological systems to extract energy from glucose without the involvement of oxygen, a great deal more energy and ATP is generated by the aerobic decomposition of glucose into carbon dioxide and water. This occurs by a series of processes that includes not only glycolysis and the Krebs cycle but also the electron transport chain.

Through repeated iterations of the Krebs cycle, the six carbon atoms of each glucose molecule are converted to six corresponding molecules of carbon dioxide. Two carbon atoms are eliminated when the two pyruvate units are decarboxylated to form acetyl-CoA, two more are eliminated when isocitrate is converted to α-ketoglutarate,

SAMPLE PROBLEM

What is the yield of ATP when pyruvate, fructose 1,6-biphosphate, and phosphoenolpyruvate are each completely oxidized to carbon dioxide? Assume that glycolysis, the citric acid cycle, and oxidative phosphorylation in the electron transport chain are fully involved.

Answer:

Recall that oxidative phosphorylation generates three ATP for each NADH and two ATP for each $FADH_2$ that is produced in glycolysis and the Krebs cycle. Use a chart of the glycolysis process and the Krebs cycle to identify the necessary steps from the material in question and tabulate the ATP output from that starting point.

Pyruvate:
Pyruvate to acetyl-CoA produces 2 NADH and 6 ATP.
Citric acid to α-ketoglutaric acid produces 2 NADH and 6 ATP.
Succinic acid to fumaric acid produces 2 FADH2 and 4 ATP.
Malic acid to oxaloacetic acid produces 2 NADH and 6 ATP.

$$6 + 6 + 4 + 6 = 22$$

Therefore, a total of twenty-two ATP are produced from oxidation of pyruvate.

Fructose 1,6-Biphosphate:
Fructose 1,6-biphosphate to 1,3-diphosphoglycerate produces 2 NADH and 6 ATP.
1,3-diphosphoglycerate to 3-phosphoglycerate produces 2 ATP.
Phosphoenolpyruvate to pyruvate produces 2 ATP.
Add the ATP produced by the oxidation of pyruvate.

$$6 + 2 + 2 + 22 = 32$$

Therefore, a total of thirty-two ATP are produced from oxidation of fructose 1,6-biphosphate.

Phosphoenolpyruvate:
Phosphoenolpyruvate to pyruvate produces 2 ATP.
Add the ATP produced by the oxidation of pyruvate.

$$2 + 22 = 24$$

Therefore, a total of twenty-four ATP are produced from oxidation of phosphoenolpyruvate.

and the remaining two when α-ketoglutarate is converted to succinyl-CoA.

The electron transport chain is the major generator of ATP during the metabolism of glucose. The process is a complex series of oxidation-reduction (redox) reactions, which involve molecular oxygen and such substances as flavin mononucleotide, ubiquinone, or various cytochromes. An oxidation reaction is one in which a reactant loses one or more electrons, while a reduction reaction is one in which one or more electrons are gained.

In the cytosol, the liquid inside the cells, glycolysis produces a net quantity of two units of ATP directly, in addition to two units of NADH (the reduced form of NAD+). In the mitochondria, enclosed units within eukaryotic cells, the rest of the glycolytic process occurs, beginning with the conversion of pyruvate to acetyl-CoA, which is accompanied by the formation of two more units of NADH. As the Krebs cycle continues, two more units of ATP are produced, as well as six more units of NADH and two of $FADH_2$ (the reduced form of flavin adenine dinucleotide, or FAD). NADH and $FADH_2$ are formed during the electron transport chain through a mitochondrial process known as oxidative phosphorylation. About three units of ATP are produced for each unit of NADH and two for each unit of $FADH_2$. Overall, a total of roughly forty units of ATP are produced, while four are consumed, for a net production of about thirty-six units of ATP per molecule of glucose that is metabolized.

ANAEROBIC RESPIRATION IN ACTION

Anaerobic respiration, a term that has come to mean any type of metabolic process that breaks down molecules into smaller units and uses an electron transport chain to produce ATP, typically extracts energy from glucose via alcoholic fermentation. In the absence of oxygen, fermentation by prokaryotic cells begins with glycolysis to produce two units of pyruvic acid. Two units of ATP and two units of NADH are produced in this process. The pyruvic acid is then decarboxylated to produce two units of carbon dioxide and two units of acetaldehyde. Subsequent reduction of the acetaldehyde by NADH produces two units of ethanol (ethyl alcohol) for a net gain of two units of ATP. Anaerobic respiration by this means is thus an energy-poor method that is generally restricted to a few groups of bacteria and yeasts.

A second form of anaerobic respiration, lactic acid fermentation, is well known to anyone who has "felt the burn" during exercise. As in alcoholic fermentation, the process begins with glycolysis, producing two units each of ATP, NADH, and pyruvic acid. In this process, however, the pyruvic acid is not decarboxylated but rather reduced by the NADH to two units of lactic acid. Again, only two units of ATP are obtained by this process. When insufficient oxygen is present in muscle tissue, lactic acid is produced faster than it can be removed from the muscle tissue, and the familiar burning sensation in the muscle results.

Richard M. Renneboog, MSc

BIBLIOGRAPHY

Berg, Jeremy M., John L. Tymoczko, Gregory J. Gatto, and Lubert Strye. *Biochemistry.* 8th ed. New York: W. H. Freeman, 2015. Print.

Lehninger, Albert L. *Biochemistry: The Molecular Basis of Cell Structure and Function.* 2nd ed. New York: Worth, 1975. Print.

Lodish, Harvey, et al. *Molecular Cell Biology.* 7th ed. New York: Freeman, 2013. Print.

Pelczar, Michael J., Jr., E. C. S. Chan, and Noel R. Krieg. *Microbiology: Concepts and Applications.* New York: McGraw, 1993. Print.

Reece, Jane B., et al. *Campbell Biology.* 10th ed. San Francisco: Cummings, 2013. Print.

CHALCOGENS

FIELDS OF STUDY

Inorganic Chemistry; Geochemistry

SUMMARY

The chalcogens are the group of elements oxygen, sulfur, selenium, tellurium and polonium. Oxygen and sulfur are essential to life on Earth, and trace quantities of selenium are also required. Tellurium is poisonous, and all isotopes of polonium are radioactive.

PRINCIPLE TERMS

- **amphoterism:** the ability of a compound to act as either an acid or a base, depending on its environment and the other materials present.
- **combustion:** a reaction between a fuel and an oxidizing agent that results in new chemical compounds and the release of heat; most often takes place between organic material and molecular oxygen, in which case the products include carbon dioxide and water.
- **conjugation:** the overlap of *p* orbitals between three or more successive carbon atoms in a molecular structure, creating a system of alternating double and single bonds through which electrons can move freely.
- **dipole:** the separation of positive and negative charges within a single molecule due to electron density being relatively high in one part of the molecule and relatively low in another.
- **electronegative:** describes an atom that tends to accept and retain electrons to form a negatively charged ion.
- **mercaptan:** an organic compound characterized by the presence of a sulfhydryl functional group (–SH).
- **nucleophilic:** describes a chemical species that tends to react with positively charged or electron-poor species.
- **organosulfur compound:** an organic compound containing one or more carbon-sulfur bonds.
- **oxidizing agent:** any atom, ion, or molecule that accepts one or more electrons from another atom, ion, or molecule in an oxidation-reduction (redox) reaction and is thus reduced in the process.
- **oxoanion:** an ion consisting of one or more central atoms bonded to a number of oxygen atoms and bearing a net negative electrical charge.
- **proton acceptor:** a compound or part of a chemical compound that has the ability to accept a proton (H+) from a suitably acidic material in a chemical reaction.

The Chalcogen Group

Group VIa, or group 16, of the periodic table is also known as the chalcogen group, and consists of oxygen, sulfur, selenium, tellurium and polonium. Chalcogen is an old term carried over from geology when sulfide and sulfate ores were termed chalcogenides, and has been generalized to refer to elements of the same group. The chalcogens exhibit similar chemical properties due to the arrangement of electrons in their respective valence shells. All members of the chalcogen group have the valence electron distribution *s2p4* and are electronegative. They typically act as oxidizing agents and accept two electrons from other atoms to form nucleophilic chemical species.

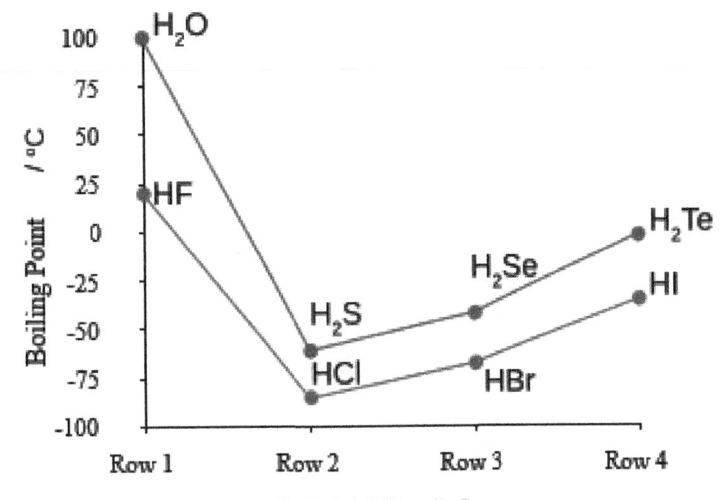

Boiling points of chalcogens by Jkwchui via Wikimedia Commons

Oxygen

Oxygen is the second most abundant element on Earth, comprising about 48% of the mass of the planet. It is found in the gaseous state as molecular oxygen, O_2, and in small quantities as ozone, O_3, in the atmosphere. The remainder is bound in the water molecule, H_2O, and in a multitude of mineral types formed from various oxoanions such as carbonates, sulfates, silicates and oxides. Numerous other anionic species such as hydroxides are also known. In certain compounds oxygen atoms are conducive to the property of amphoterism, by which the compound can act as a base under some conditions or an acid under others. Aluminum hydroxide ($Al(OH)_3$), for example, can act as a base by releasing hydroxide ions (OH-), or as aluminic acid by releasing hydrogen ions (H+), depending on whether the Al-O bonds or the O-H bonds cleave. Oxygen is essential to life through the process of respiration and facilitates the extraction of bioenergy in living cells by the electron transport mechanism. Energy is also released in the process of combustion when materials, especially carbon-based materials, burn. Oxygen has a unique position in the periodic table as a second period element along with carbon and nitrogen. The 2*s* and 2*p* electron orbitals of these elements are able to interact so well that they can produce an essentially infinite series of organic or carbon-based compounds including amino acids and all of the biochemicals that comprise the life processes. Oxygen atoms can form double bonds with carbon and nitrogen atoms and so undergo conjugation with other double bonds in organic compounds. The carbonyl group (C=O) is the most important functional group in this regard, and is enhanced by the electronegativity of the oxygen atom inducing a dipole in the C=O group that facilitates the reactivity of a great many organic compounds.

Sulfur

Considerably less abundant than oxygen, sulfur is able to exist in a number of equally stable allotropic elemental forms. The most common of these consists of ring-shaped molecules of eight sulfur atoms. Being a third period element, the 3*s*3*p* electrons of the sulfur atom are in a higher quantum level than their counterparts in the oxygen atom, and are also much closer in energy to the *d* orbitals. Accordingly, sulfur exhibits very different chemical behavior and does not undergo conjugation in the same way that oxygen does. Nevertheless, their chemical bonding behavior is sufficiently similar that organosulfur compounds are known that are entirely analogous to their oxygen counterparts. Alcohols, for example, are organic molecules containing the -OH functional group. Their organosulfur counterparts contain instead the sulfhydryl -SH group and are generally known as mercaptans or thiols. The presence of sulfur in an organic molecule is indicated by the inclusion of thio- or thia- in the chemical name, for example as thioacetic acid. Sulfur is a major component of many metal and non-metal ores such as iron pyrite (FeS_2, iron sulfide, fool's gold) and galena

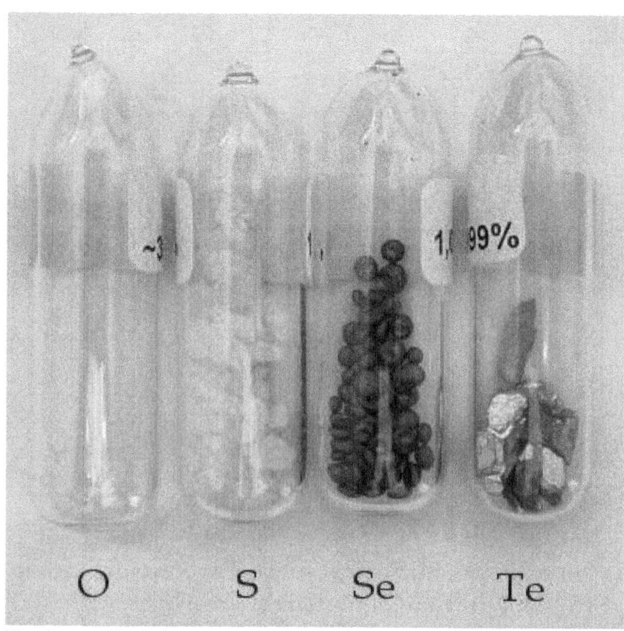

The four stable chalcogens at STP by Tomihahndorf at the German language Wikipedia via Wikimedia Commons

(PbS, lead sulfide). It is also a component of gaseous compounds such as sulfur dioxide (SO_2) and liquids such as carbon disulfide (CS_2). Industrially, sulfur is one of the most important materials known as the central atom of sulfuric acid and an essential additive to rubber and other compounds. Sulfur is also essential to life as a functional center in protein molecules.

Selenium, Tellurium and Polonium

Selenium, and its higher homolog tellurium, are much less abundant the either sulfur or oxygen. Selenium is highly poisonous, but is sufficiently similar to oxygen and sulfur chemically that it is also an essential element in trace quantities for optimum health. Selenium is strongly photoconductive, and becomes a good conductor of electricity when irradiated. This feature was fundamental to the development of the photocopier. Tellurium is only very mildly photoconductive and is a p-type semiconductor. Tellurium compounds are thought to be as toxic as those of selenium, and the main uses of tellurium are in the formation of alloys. Both selenium and tellurium exist in a variety of allotropic forms, much like sulfur. Polonium, the highest of the chalcogens, exists as two metallic allotropes. Both are highly radioactive, with half-lives of 2.9 and 103 years. Other isotopes are known, but have very short half-lives. Polonium releases so much heat energy as it decays that a 0.5g sample quickly reaches a temperature of 500°C and ionizes the surrounding air do that it glows with a blue light called Cerenkov radiation. The amount of energy released as polonium decays has made it a valuable energy source for space exploration.

The Importance of the Chalcogens

The importance of oxygen, sulfur and selenium in regard to life is obvious. However, these are eclipsed by the value of oxygen in the form of ozone. As a component of smog, it is detrimental to air quality and hence to health. The minor percentage that exists in the upper atmosphere, however, is all that stands between the biosphere of Earth and the influx of ultraviolet solar radiation that would quickly destroy all life on its surface and render the planet uninhabitable. Sulfur has high economic value as a component of many different minerals and through its utility as

SAMPLE PROBLEM

Calculate the weight of thiosulfuric acid ($H_2S_2O_3$) that could be produced from 1000g of elemental sulfur (S_8).

Answer:

Begin with the molecular weights of reactants and products. Since only the sulfur based materials are important here, a balanced equation is not necessary. All that is required is to know that molar ratio of sulfur in S_8 and $H_2S_2O_3$ is 4-to-1. Thus one mole of S_8 can produce four moles of $H_2S_2O_3$.

Molecular weight of S_8 = (8 X 32) = 256 g.mol^{-1}
Molecular weight of $H_2S_2O_3$ = (2 + 64 + 48) = 114 g.mol^{-1}

1000 g of S_8 = (1000/256) = 3.906 moles

Therefore the maximum number of moles of $H_2S_2O_3$ possible = (4 X 3.906) = 15.624

Therefore the maximum weight of $H_2S_2O_3$ possible = (15.624 mol X 114 g.mol^{-1})

= 1781.1 g

the basis of many industrial processes. Similarly, tellurium has great value in metallurgical applications. The photoconductivity of selenium may prove extremely valuable in the development of long-term alternative energy sources, while polonium provides a dependable short-term energy source for remote applications such as space exploration.

Richard M. Renneboog M.Sc.

BIBLIOGRAPHY

Bayse, Craig A., and Julia L Brumaghim. *Biochalcogen Chemistry. The Biological Chemistry of Sulfur, Selenium and Tellurium.* American Chemical Society, 2015. Print.

Greenwood, N.N. and Earnshaw, A.. *Chemistry of the Elements* 2nd ed. Burlington, MA: Elsevier Butterworth-Heinemann, 2005. Print.

House, James E., and Kathleen A. House. *Descriptive Inorganic Chemistry.* Waltham, MA: Academic Press, 2016. Print.

Lane, Nick. *Oxygen: The Molecule That Made the World.* Oxford: Oxford UP, 2009. Print.

Kutney, Gerald. *Sulfur. History, Technology, Applications & Industry.* Toronto, ON: ChemTec Publishing, 2007. Print.

Rodgers, Glen E. *Descriptive Inorganic, Coordination, & Solid-State Chemistry.* Toronto, ON: Nelson Education, 2011. Print.

Wiberg, Egon and Wiberg, Nils. *Inorganic Chemistry.* San Diego, CA: Academic Press, 2001. Print.

CHEMICAL BONDING

FIELDS OF STUDY

Organic Chemistry; Physical Chemistry; Molecular Biology

SUMMARY

Chemical bonding, through bond formation and bond dissociation, is the fundamental process by which all chemical reactions occur and depends on a number of factors deriving from the electronic structure of atoms. The primary types of bonds are ionic and covalent.

PRINCIPAL TERMS

- **covalent bond:** a type of chemical bond in which electrons are shared between two adjacent atoms.
- **dipole-dipole:** describes a type of interaction between two molecular dipoles, which are molecules that are polarized due to the greater concentration of electrons in one area.
- **hydrogen bond:** a weak type of chemical bond formed by the attraction of a hydrogen atom to an electronegative atom—an atom with a strong tendency to attract electrons—in the same or another molecule.
- **ionic bond:** a type of chemical bond formed by mutual attraction between two ions of opposite charges.
- **partial charge:** a term used to indicate a degree of charge separation in bonded atoms due to the different electronegativities of the atoms at either end of the bond.

VISUALIZING CHEMICAL BONDING

Chemical bonding can be thought of as "hand shaking" between atoms. A good understanding of chemical bonding requires a sound appreciation of atomic structure, particularly the arrangement of electrons in atoms. The distribution of electrons in atoms is reflected in the arrangement of chemical bonds that can be formed about them. This arrangement determines the three-dimensional structure of molecules and the physical properties of the corresponding compounds. The basic mechanism of chemical bonding is the transfer of electrons between the atoms that have formed the bond to each other. This involves only the outermost or valence electrons in each atom.

ATOMIC STRUCTURE AND ATOMIC ORBITALS

The modern theory of atomic structure is based on the principles of quantum mechanics. According to this mathematical description of the atom, supported by physical observations, atoms contain a very small,

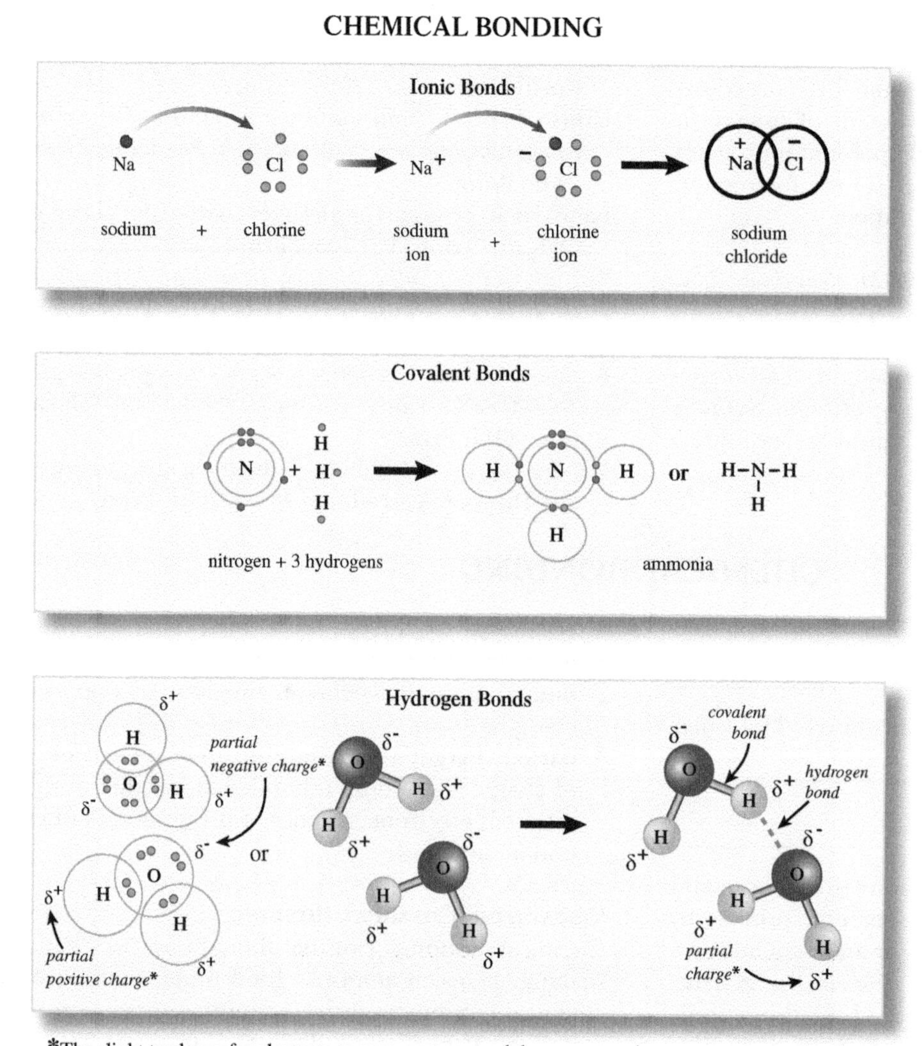

*The slight tendency for electrons to congregate around the more massive oxygen atom rather than the hydrogen atoms causes the oxygen region of the molecule to take on a more electronegative character. This is called a partial negative charge. The hydrogens take on a more electropositive character. Overall, the water molecule remains neutral.

dense nucleus composed of protons and neutrons. Surrounding the nucleus is a much larger, diffuse cloud of electrons, equal in number to the protons in the nucleus. To excellent approximation the wave function of each atom can be written as a product of single electron wavefunctions, describing each of the occupied the electron orbitals.

The rules of quantum mechanics dictate the number and distribution of electrons in the atomic orbitals of any particular atom, and this in turn determines the valence properties of the atom. The lowest level, or *s*, orbitals are spherical about the nucleus. The three *p* orbitals have a figure-eight shape and are oriented at right angles to each other, much like the *x*, *y*, and *z* axes of the Cartesian coordinate system. Under rotations of the xyz coordinates, the orbitals transform into each other. The same can be said of the five *d* orbitals. Conventionally, we draw four of the five *d* orbitals to have a shape like two figure eights crossing each other, while the fifth is shaped like an oversized *p* orbital with a donut about its middle. The *d* orbitals are oriented at angles that place them between the *p* orbitals. These three sets of atomic orbitals are sufficient to describe the most common elements. A set of seven *f* orbitals, with more complex shapes and orientations, is available in higher elements.

THE CHEMICAL BONDING CONTINUUM

The most common types of chemical bonds are ionic and covalent bonds, but it would be incorrect to think that all bonds fit neatly into one of these two categories. The nature of a chemical bond is affected by several factors, including the relative electronegativities of the bonded atoms and their size-related electron densities. These factors determine the extent to which charge separation occurs in the bond. In an ionic bond, one atom has lost electrons to become a positively charged ion, while the other has gained electrons to become a negatively charged ion. The charge separation between them is complete. In a purely covalent bond, the two electrons in the chemical bond between two atoms belong equally to both. When the atoms are dissimilar, however, one may exert a greater influence over the bonding electrons, resulting in an uneven distribution of charge between the two atoms, often referred to as a partial charge. This imparts a certain amount of ionic character to the covalent bond. Thus, chemical bonds actually exist on an ionic-covalent continuum.

IONIC BONDS

Ionic bonds are formed by the attraction between two oppositely charged ions. A simple common example is sodium chloride, in which each sodium ion is surrounded by 6 chloride ions and vice-versa. The attraction between the ions is very strong, and ionic compounds are normally crystalline solids with high melting points. This also covers a broad range of values, as there are ionic compounds with low melting points and ionic compounds with very high melting points. Niobium carbide, for example, melts at 3,500 degrees Celsius, and nickel oxide melts at about 1,990 degrees Celsius. Magnesium chlorate hexahydrate, on the other hand, melts at a mere 35 degrees Celsius. There are higher and lower melting points of ionic solids, but the vast majority fall somewhere within this range.

The strength of an ionic bond is a reflection of the charge density of the ions. The charge density is a measure of the amount of electrical charge contained within a specific volume or region of space. Elements that form ions of small diameter have high charge density and are thus able to fit together in a crystal lattice (a regular array of ions, atoms, or molecules in the structure of a crystal) more compactly than ions with larger diameters. Such compounds tend to have higher melting points and hardness than compounds composed of larger or differently sized ions.

Ionic bonds can be strongly affected by solvents, particularly water. Most dissolve in water, in degrees ranging from sparingly soluble to completely soluble. Sodium chloride and similar salts are prime examples; they dissolve as the water molecules surround and stabilize both the positive and negative ions that make up the solid compound. As long as enough free solvent is available, the material will continue to dissolve, but when the amount of available solvent molecules is not sufficient to stabilize any more ions, dissolution ceases and the solution is said to be saturated. The energy state of the dissolved ions typically becomes lower as their entropy (degree of disorder) increases.

COVALENT BONDS

Covalent bonds are the opposite of ionic bonds, as they represent the mutual sharing of electrons rather than the complete transfer of electrons that defines ions and ionic bonds. A covalent bond forms between two atoms by the overlap of atomic orbitals of corresponding energy. This allows two electrons, one from each atom, to occupy the resulting molecular orbital in a way that satisfies the valence-shell electron distribution of both atoms. The simplest example of this is the covalent bond between the two

SAMPLE PROBLEM

Identify the types of bonds in the compounds $CaSO_4$, F_2, and $[Cr(NH_3)_4][Cr(CN)_6]$. Then explain why liquid water, H_2O, has such a high boiling point compared to other compounds of similar molecular weight.

Answer:

$CaSO_4$: Calcium forms the cation (positively charged ion) Ca^{2+}, while sulfate is an anion (negatively charged ion) with the formula SO_4^{2-}. The bond is formed between two ions and is therefore ionic.

F_2: Fluorine atoms have seven electrons in their valence shell and are strongly electronegative. This means that they attract electrons easily, forming the anion F^-, but do not naturally lose electrons to form a cation. Because of this, the bond cannot be ionic; instead, the two F atoms form a covalent bond by sharing one valence electron from each.

$[Cr(NH_3)_4][Cr(CN)_6]$: This is a material formed from the two complex ions $[Cr(NH_3)_4]^{3+}$ and $[Cr(CN)_6]^{3-}$. Both complex ions are formed by the coordination of ligands, NH_3 in the first case and CN^- in the second, to a central chromium ion, Cr^{3+}. The bonds in them are therefore coordination bonds, while the bond formed between the two complex ions is an ionic bond.

Liquid water consists of molecules of H_2O, in which the covalent bonds between the hydrogen and oxygen atoms leave the positive charge of the hydrogen nuclei exposed. Each oxygen atom has two lone (unshared) pairs of valence electrons, forming a region of high negative charge. When the positively charged hydrogen nuclei are attracted to the negative charge of the oxygen atoms' lone pairs, a hydrogen bond is formed, increasing the energy required to separate the molecules and thus making the boiling point greater than in compounds that do not form hydrogen bonds.

hydrogen atoms of the hydrogen molecule H_2. Each hydrogen atom has just one electron in an *s* atomic orbital, designated 1*s* because it has the lowest possible energy level, but an *s* orbital can contain up to two electrons. By combining their respective 1*s* orbitals and forming a directly overlapping molecular orbital, each hydrogen atom effectively contains two electrons in its 1*s* orbital, thus completely filling its valence shell. The same principle applies to the formation of all covalent bonds, regardless of the atoms involved.

Covalent bonds can display pronounced bond polarity when the bond is between dissimilar atoms. The surrounding environment can also induce polarity in an otherwise nonpolar covalent bond. The water molecule, H_2O, is the quintessential example of this effect. The oxygen atom of the water molecule forms a covalent bond to two separate hydrogen atoms, thus completely filling the valence shell of all three atoms (the valence shell of an oxygen atom contains six electrons but is capable of containing a maximum of eight). However, because the oxygen atom is very electronegative, meaning that it has a strong tendency to attract electrons, and the hydrogen atoms are very electropositive, the electrons in the covalent bonds between them are pulled toward the oxygen atom, leaving the hydrogen atoms with a positive partial charge and causing the positive end of the O–H bond to be attracted to the negative end of an O–H bond on another molecule. This forms what is known as a hydrogen bond, which is an example of dipole-dipole attraction. A dipole is an object in which the positive and negative charges have separated, such as a molecule that is more positively charged at one end and more negatively charged at the other.

Another example of a hydrogen bond is the organic compound fluoroethane, C_2H_5F. In this compound, the two carbon atoms are bonded to each other, with one also bonded to three hydrogen atoms and the other bonded to two hydrogen atoms and one fluorine atom. The highly electronegative fluorine atom induces a dipole in the covalent bond it forms with the carbon atom, which in turn increases the polarity of the covalent bonds between that carbon atom and the two hydrogen atoms. Not only does the entire fluoroethane molecule have an overall dipole due to the presence of the fluorine atom but each covalent bond within the molecule also has a stronger dipole character in comparison to those of the corresponding covalent bonds in the parent molecule, ethane (C_2H_6).

Covalent sigma bonds are formed directly between two atoms. A second type of covalent bond, the pi bond, forms when orbitals on adjacent atoms do not overlap directly but do so in a side-by-side manner instead. The overlap of *p* orbitals on adjacent carbon atoms, for example, forms very stable pi bonds alongside the sigma bond between the two carbon atoms. The higher-energy *d* orbitals can also take part in pi bonding, but this is not as commonly observed due to the energy differences between *p* and *d* orbitals.

Another type of covalent bond is the coordination bond, also known as a dative bond or a dipolar bond. A coordination bond is formed when one atom provides both of the electrons in a covalent bond. When ions or molecules bond to a central metal atom in this manner, the compound formed is known as a coordination complex, and the ions and molecules surrounding the central atom are called ligands. If the compound has a net electrical charge, it is a complex ion. Like all other types of bonds, coordination bonds range in strength from very weak to very strong.

Ionic or Covalent?

A great many compounds exist in which all atoms are bonded covalently but readily dissociate into ions. Organic acids and bases are typical of this type of compound, as interaction with water has the ability to dissociate the molecule by extracting a hydrogen ion (H^+), leaving the remainder as a negatively charged ion. The ease with which this occurs varies greatly from compound to compound and is highly dependent on the compound's molecular structure. The transmission of charge between adjacent atoms, known as the inductive effect, can greatly enhance the ability of the compound to convert the covalent bond of a particular hydrogen atom to an ionic bond. Acetic acid (CH_3COOH) is a weak acid that dissolves primarily as the whole molecule, with only about one in ten thousand molecules dissociating into H^+ and an acetate ion ($C_2H_3O_2^-$). By contrast, chloroacetic acid ($ClCH_2COOH$), though still considered a weak acid, dissociates one hundred times more readily than acetic acid.

Induction also allows the formation of chemical bonds to atoms that are normally nonreactive. Xenon is one of the so-called inert gases, which means that the valence electron shells of xenon atoms are already completely filled. However, when xenon reacts with the extremely electronegative element fluorine, a stable crystalline compound identified as xenon hexafluoride (XeF_6) can be produced, as can other compounds of xenon and fluorine. The high electronegativity of the fluorine atoms and the relatively large size of the xenon atom allow the fluorine atoms to draw valence electrons from the xenon atom into bond formation. The formation of such compounds, like the dissociation of organic acids and bases, blurs the line between ionic and covalent bonds and reinforces the fact that chemical bonds exist on a continuum between covalent and ionic.

Reactions of metal atoms with relatively electron-rich organic compounds form compounds through coordination bonds. Like all other types of bonds, coordination bonds range in strength from very weak to very strong. Coordination bonding does not necessarily involve the transfer of electrons. In many cases, coordination compounds form just by the overlap of orbitals of appropriate energy configuration. A well-known example of such a compound is bis(benzene) chromium, $Cr(C_6H_6)_2$. This compound is formed in the gas phase by heating a sample of chromium metal in an atmosphere of benzene vapor. Bis(benzene) chromium deposits on the inside of the reaction vessel as a dark brown-black solid. The compound forms as the *p* orbitals of the benzene molecules coordinate to the vacant valence orbitals of the chromium atom. Coordination bonds are highly reversible and are an extremely important aspect of many catalytic processes.

Richard M. Renneboog, MSc

BIBLIOGRAPHY

Berg, Jeremy M., John L. Tymoczko, Gregory J. Gatto, and Lubert Strye. *Biochemistry*. 8th ed. New York: W. H. Freeman, 2015. Print.

Haynes, W. H., PhD, David R. Lide, PhD, and Thomas J. Bruno, PhD, eds. *CRC Handbook of Chemistry and Physics,* 96th Edition. New York: CRC/Taylor & Francis Group, 2015. Print.

Miessler, Gary L., Paul J. Fischer, and Donald A. Tarr. *Inorganic Chemistry*. 5th ed. Upper Saddle River: Prentice Hall, 2013. Print.

Morrison, Robert Thornton, and Robert Neilson. Boyd. *Organic Chemistry*. 7th ed. Englewood Cliffs, N.J.: Prentice Hall, 2003. Print.

Silbey, Robert J., Robert A. Alberty, and Moungi G. Bawendi. *Physical Chemistry*. 5th ed. Hoboken: Wiley, 2012. Print.

Zumdahl, Steven S., and Susan Zumdahl. *Chemistry*. 9th ed. Belmont: Brooks Coles, 2013. Print.

CHEMICAL BUFFERS

FIELDS OF STUDY

Inorganic Chemistry; Geochemistry; Metallurgy

SUMMARY

The nature and function of buffer solutions are presented, and their importance in biological systems is discussed. Buffer solutions resist changes in pH as a result of the equilibrium between the conjugate pair specific to a particular solution.

PRINCIPAL TERMS

- **conjugate acid:** the material formed from a base when it accepts a proton (H^+) from an acid, thus gaining a unit of positive charge.
- **conjugate base:** the material formed from an acid when it donates a proton (H^+) to a base, thus losing a unit of positive charge.
- **dissociation constant:** a characteristic value representing the extent to which a compound dissociates into component ions in a certain solvent and under specific conditions.
- **neutralization:** a chemical reaction between an acid and a base that results in the formation of a salt, usually accompanied by water.
- **pH:** a numerical value that represents the acidity or basicity of a solution, with 0 being the most acidic, 14 being the most basic, and 7 being neutral.

The Nature of Chemical Buffers

Chemical buffers are solutions that resist changes in pH. A buffer solution will generally maintain a constant pH when other materials are added, as long as the quantities of other materials do not overload the ability of the buffer solution to adjust to the change in the system components. The ability of a buffer solution to compensate for the addition of acids or bases, and thus maintain a fairly constant concentration of hydronium ions (H_3O^+), derives from the equilibrium reaction between the particular weak acid or base and the corresponding salt used to prepare the buffer. The key feature of the system is a weak acid, one that does not dissociate well into its component ions when dissolved in water. Acetic acid is a good example. Despite its acrid odor and acidic character, acetic acid is covalently bonded (as are essentially all other carboxylic acids). A weak acid need not necessarily be a carboxylic acid but absolutely must be one that does not dissociate completely into ions when dissolved in water. When this condition is met, a dissociation equilibrium is established, according to the general equation

$$H_2O + HA \rightleftharpoons H_3O^+ + A^-$$

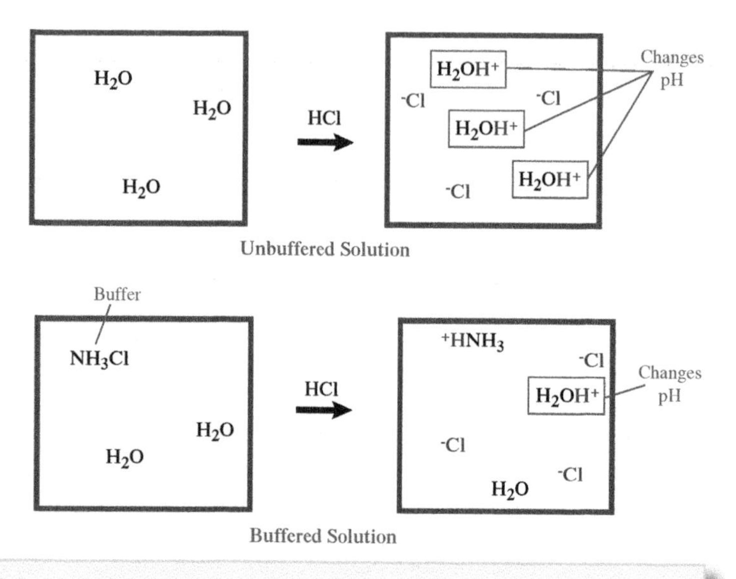

Weak conjugate acid and base pairs can resist a change in pH of the solution. The buffer solution can accept or release protons to maintain an equilibrium.

The equilibrium reaction has two essential characteristics: constant amounts of all reactants and products in a specific characteristic and a constant ratio, defined as the "equilibrium constant." For weak acids, this is the acid dissociation constant, K_a. By adjusting the relative amounts of each component in the system to correspond to the equilibrium's constant value, an equilibrium system automatically adjusts to any effect that perturbs the state of the system. In a buffer system, the introduction of some additional H^+ requires it to combine with an appropriate amount of A^- both to regenerate HA and to maintain the equilibrium constant value. By the same token, the introduction of additional alkalinity, or basicity, means that some of the HA will undergo neutralization. Regardless of the resiliency of the buffer system, however, the introduction of too much acid or base (or of a much stronger acid or base) will overpower the ability of the system to reestablish its fundamental equilibrium.

For a buffer (or any equilibrium) system to behave in this way, the thermodynamic state of the system must be at an energy minimum when it is at equilibrium. The tendency to regain the minimum-energy state is the driving force for the autoadjustment of an equilibrium system, such as a buffer solution.

Preparation of Chemical Buffer Solutions

Simply stated, a buffer solution is prepared by combining a weak acid and a salt of the weak acid in the same solution. An acetate buffer, for example, would be prepared using acetic acid and sodium acetate (or a similar acetate salt). An alkaline buffer might be similarly prepared from a solution of ammonia in water and a corresponding salt, such as ammonium chloride. The obvious characteristic of the buffer solution is that it contains either a weak acid and its conjugate base or a weak base and its conjugate acid, forming a conjugate pair.

This simple description belies the care with which a buffer solution must be prepared. To achieve a specific pH value for the solution, careful calculations are required to determine the correct amounts of a particular conjugate pair

SAMPLE PROBLEM

Use the Henderson-Hasselbalch equation

$$pH = pK_a + \log\left(\frac{[A-]}{[HA]}\right)$$

to determine the pH of a solution of 0.06M acetic acid (HOAc) that is also 0.05M in sodium acetate (NaOAc), given the acid dissociation constant for acetic acid is 1.75×10^{-5}.

Answer:

The acid dissociation constant is calculated from the equilibrium reaction

$$HOAc \rightleftharpoons H^+ + OAc^-$$

as

$$K_a = [H^+]\frac{[OAc-]}{[HOAc]} = 1.75 \times 10^{-5}$$

Therefore, the amount of acetate ion from dissociation of acetic acid in solution is about 0.001M, and the total acetate concentration is (0.05 + 0.001) = 0.051M. The concentration of acetic acid must be (0.06 − 0.001) = 0.059M.

Therefore, the ratio of OAc^- to HOAc in the buffer solution is

$$\left(\frac{0.051}{0.059}\right) = 0.864$$

Convert these values to the pK_a and log values:

$$pK_a = -\log K_a = 4.76$$

$$\log(0.864) = -0.064$$

Substituting these values into the Henderson-Hasselbalch equation yields

$$pH = 4.76 - 0.064 = 4.696$$

Therefore, this buffer solution has a pH of $4.696 \approx 4.7$.

to use. Since buffer systems are used most commonly in biochemical studies, the conjugate pair must be selected according to the nature of the bioreaction system being studied. In biochemical systems, such as the cell cytosol, pH is strictly controlled by materials that are present in the system. With respect to the validity of the results, it would be inappropriate to carry out a biochemical reaction process in a buffer system that is not applicable to cell biochemistry and then apply the observed results to a biochemical system.

Another consideration when discussing chemical buffers is the mode of reaction by which the conjugate pair undergoes neutralization and dissociation reactions. A monoprotic weak acid, such as acetic acid, donates a single proton (H^+) from each molecule in solution and functions in an entirely different manner than does a diprotic or polyprotic acid, such as citric acid, oxalic acid, or phosphoric acid. The mode of reaction also determines both the magnitude and the range of pH in which the buffer is effective. A simple comparison of the relevant equilibrium reactions make this easy to see. Oxalic acid, a diprotic acid, has two neutralization equilibriums that occur simultaneously, as follows:

$$HOOC\text{–}COOH \rightleftharpoons HOOC\text{–}COO^- + H^+$$

$$HOOC\text{–}COO^- \rightleftharpoons {}^-OOC\text{–}COO^-$$

Phosphoric acid, a common component of biochemical systems and a triprotic acid, has three neutralization equilibriums to consider:

$$H_3PO_4 \rightleftharpoons H^+ + H_2PO_4^-$$

$$H_2PO_4^- \rightleftharpoons H^+ + HPO_4^{2-}$$

$$HPO_4^{2-} \rightleftharpoons H^+ + PO_4^{3-}$$

Each separate neutralization equilibrium occurs at a different pH, but all are simultaneously involved in the overall equilibrium condition of the system. The calculations required to achieve a desired pH using such conjugate pairs must take this into account.

THE HENDERSON-HASSELBALCH EQUATION

The $[H_3O^+]$ of a conjugate pair buffer solution can be calculated proportionately as follows:

$$[H_3O^+] = K_a\left(\frac{[acid]}{[conjugate\ base]}\right)$$

Mathematically, this is equivalent to the following logarithmic form:

$$pH=pK_a+\log\left(\frac{[\text{conjugate base}]}{[\text{acid}]}\right)$$

This is known as the Henderson-Hasselbalch equation. For most applications, it works quite well for the calculation of pH at each individual step in an overall equilibrium system. Successive calculations for systems with more than one equilibrium to consider are used to calculate the overall pH for such systems. Knowing the concentrations of the acid and conjugate-base materials allows for calculation of the required masses of the two components, and when properly mixed, the system can come to equilibrium at the desired pH.

BUFFERS IN BIOLOGICAL SYSTEMS

Buffer systems are important in both living systems and the laboratory study of processes that affect living systems. Enzyme-mediated processes are especially sensitive to changes in pH. The fluid components of a cell structure such as the cytosol are very strictly controlled, and complex buffer systems involve the interaction of several kinds of buffers. The various functional groups of amino acids are typically rendered nonfunctional outside of a narrow pH range. The process taking place under the control of the enzyme may result in the release of protons into the reaction medium. The effect that these would have is nullified by the surrounding buffer solution. Structural features such as disulfide bridges, necessary for the enzyme to maintain its proper structure, are sensitive to hydrolysis as well.

Richard M. Renneboog, MSc

BIBLIOGRAPHY

Berg, Jeremy M., John L. Tymoczko, Gregory J. Gatto, and Lubert Strye. *Biochemistry*. 8th ed. New York: W. H. Freeman, 2015. Print.

Lehninger, Albert L. *Biochemistry: The Molecular Basis of Cell Structure and Function*. 2nd ed. New York: Worth, 1975. Print.

Myers, Richard. *The Basics of Chemistry*. Westport: Greenwood, 2003. Print.

Reece, Jane B., et al. *Campbell Biology*. 10th ed. San Francisco: Cummings, 2013. Print.

Skoog, Douglas Arvid, Donald M. West, and F. James. Holler. *Skoog and West's Fundamentals of Analytical Chemistry*. 9th ed. Belmont (Calif.): Cengage Learning, 2014. Print.

CHEMICAL ENERGY (NON-FOSSIL FUELS)

FIELDS OF STUDY

Inorganic Chemistry; Geochemistry; Metallurgy

SUMMARY

The principles underlying bond energy are explained, as well as how bond energy can be used to supply the activation energy required for a chemical reaction. The energy contained in chemical bonds is necessary for chemical reactions to take place.

PRINCIPAL TERMS

- **bond energy:** the amount of energy necessary to break the chemical bonds in a given molecule, measured in kilojoules per mole.
- **combustion:** a reaction between a fuel and an oxidizing agent that results in new chemical compounds and the release of heat; most often takes place between organic material and molecular oxygen, in which case the products include carbon dioxide and water.
- **enthalpy:** the total heat content within a thermodynamic system, defined as internal energy plus the product of pressure and volume; also, the change in heat content associated with a chemical process.
- **potential energy:** the energy contained in an object due to its position, composition, or arrangement that is capable of being translated into kinetic energy or the performance of work; for example, the energy contained in the bonds between atoms, which can be used to fuel a chemical reaction.

THE NATURE OF CHEMICAL ENERGY

Chemical compounds are formed by the interaction of electrons in the outermost valence shells of atoms. In molecules, the atomic orbitals combine and hybridize to form molecular orbitals. According to the

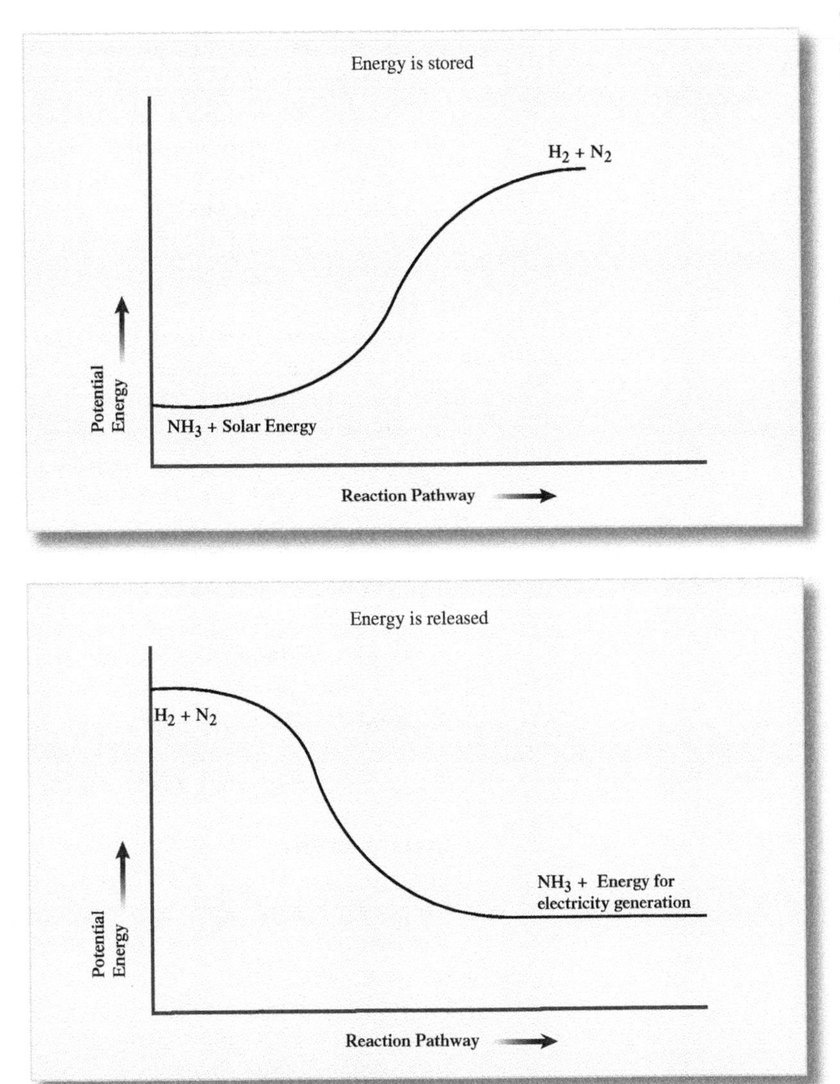

modern theory of atomic structure, the electrons in atomic and molecular orbitals must correspond to very specific energies. The geometrical structures of chemical compounds have ideal values, representing the minimum energy configuration of a particular molecule; for example, the ideal angle between the four bonds of a carbon atom in a tetrahedral geometry is 109.5 degrees. However, chemistry is a real process rather than an ideal one. A number of factors, including electron-electron repulsion and orbital interactions, force bond angles to deviate from their ideal values, contributing to bond-angle strain that forces the bonds out of their minimum energy configuration and into a higher bond energy configuration. Other factors, such as pi-bond formation, add to the stability of the interaction between two chemically bonded atoms, increasing the potential energy contained in the bonds of a molecule.

Atoms in a molecule do not exist in isolation from each other, and conditions that affect one bond in a molecule affect all of the bonds in the molecule. During chemical reactions, some bonds are broken and others are formed. When a bond is broken as part of a reaction, energy is released; when a bond is formed, energy is captured and held within the bond. The enthalpy change of a reaction is the difference between the energy released by bonds being broken and the energy captured by bonds being formed.

ACTIVATION ENERGY IN REACTIONS

Spontaneous reactions begin as soon as the reactants come into contact with each other, although the rate of a spontaneous reaction can be so slow as to be unnoticeable. All other reactions have energy requirements that must be met before a reaction can occur. The activation energy of a reaction is the minimum energy that must be supplied to the reactants so the reaction can proceed. This energy may be needed to break an initial bond, or it may be needed to bring the reactants into the proper conformation for reaction. In an exothermic reaction, in which the reactants contain more energy than the resulting products, the energy that is released by the reaction is typically more than sufficient to provide the activation energy of subsequent reactions.

Governing all chemical reactions are the laws of conservation of energy and conservation of mass. Accordingly, all individual energy transactions that take place during a chemical reaction must combine to produce the overall energy change in the reaction. Just as there can be no extra or missing mass

SAMPLE PROBLEM

What is the total enthalpy change (ΔH) from the conversion of acetylene (C_2H_2, also called ethyne) to carbon dioxide (CO_2) and water (H_2O)? Use the bond energies provided below.

$$C\equiv C: 835\frac{kJ}{mol}$$

$$C-H: 411\frac{kJ}{mol}$$

$$O=O: 498.4\frac{kJ}{mol}$$

$$C=O: 745\frac{kJ}{mol}$$

$$O-H: 463\frac{kJ}{mol}$$

Answer:

Write the balanced chemical equation for the reaction:

$$C_2H_2 + 3O_2 \rightarrow 2CO_2 + H_2O$$

In the reaction, two C–H bonds, one C≡C bond, and three O=O bonds are broken, while four C=O and two O–H bonds are formed. Calculate the energy required to break the C–H, C≡C, and O=O bonds:

$$2\ C-H\ \text{bonds: } 2\times 411\frac{kJ}{mol} = 822\frac{kJ}{mol}$$

$$1\ C\equiv C\ \text{bond: } 1\times 835\frac{kJ}{mol} = 835\frac{kJ}{mol}$$

$$3\ O=O\ \text{bonds: } 3\times 498.4\frac{kJ}{mol} = 1495.2\frac{kJ}{mol}$$

Calculate the energy required to form the C=O and O–H bonds:

$$2\ C=O\ \text{bonds: } 2\times 745\frac{kJ}{mol} = 1490\frac{kJ}{mol}$$

$$2\ O-H\ \text{bonds: } 2\times 463\frac{kJ}{mol} = 926\frac{kJ}{mol}$$

Calculate the change in enthalpy:

$$\Delta H = \left(1490\frac{kJ}{mol} + 926\frac{kJ}{mol}\right) - \left(822\frac{kJ}{mol} + 835\frac{kJ}{mol} + 1495.2\frac{kJ}{mol}\right) = -736.2\frac{kJ}{mol}$$

The change in enthalpy is −736.2 kJ/mol. The negative sign indicates that the reaction is exothermic, meaning that 736.2 kJ/mol is the amount of energy released when the process is completed.

at the end of a reaction, there can be no extra or missing energy.

CHEMICAL ENERGY TRANSACTIONS

Breaking and making bonds is the currency of chemical transformations, especially in biological systems. In respiration—and all other biochemical processes involving ATP, NAD, and similar compounds—energy is recovered and stored by the enzyme-mediated formation of a triphosphate group. In this process, an inorganic phosphate group, PO_4^{3-}, is joined to an existing diphosphate group. The new bond is considered a high-energy bond; when a process requires an input of energy, an enzyme causes this bond to be cleaved, and the energy it releases is used to drive the subsequent reaction.

Great amounts of solar energy are stored in the formation of glucose during photosynthesis. This glucose is used both to fuel cellular processes and to form structural materials such as cellulose. The energy is released through chemical reactions with molecular oxygen, via either respiration within the living cell or combustion. In all cases, the energy stored in the various bonds of the glucose molecules is made available to other processes and eventually released into the environment.

Richard M. Renneboog, MSc

BIBLIOGRAPHY

Berg, Jeremy M., John L. Tymoczko, Gregory J. Gatto, and Lubert Strye. *Biochemistry.* 8th ed. New York: W. H. Freeman, 2015. Print.

Daniels, Farrington, and Robert A. Alberty. *Physical Chemistry.* 3rd ed. New York: Wiley, 1966. Print.

Douglas, Bodie E., Darl H. McDaniel, and John J. Alexander. *Concepts and Models of Inorganic Chemistry.* 3rd ed. New York: Wiley, 1994. Print.

Lehninger, Albert L. *Biochemistry. The Molecular Basis of Cell Structure and Function.* New York: Worth, 1975. Print.

Haynes, W. H., PhD, David R. Lide, PhD, and Thomas J. Bruno, PhD, eds. *CRC Handbook of Chemistry and Physics,* 96th Edition. New York: CRC/Taylor & Francis Group, 2015. Print.

Myers, Richard. *The Basics of Chemistry.* Westport: Greenwood, 2003. Print.

CHEMISORPTION

FIELDS OF STUDY

Chemical Engineering; Biochemistry; Physical Chemistry

SUMMARY

Chemisorption is a process in which actual chemical bonds are formed between a substrate material and an adsorbed material. The process is essential to the function of catalysts including enzymes. Chemisorption allows the modification of material surfaces such as those of gold nanoparticles to provide unique materials with unique properties.

PRINCIPAL TERMS

- **activated complex:** one of the various intermediate molecular structures that exist during a chemical reaction while the original chemical bonds are being broken and new bonds are being formed.
- **activation energy:** the amount of energy a system requires for the formation of an activated complex from which a reaction can proceed.
- **Arrhenius equation:** a mathematical function that relates the rate of a reaction to the energy required to initiate the reaction and the absolute temperature at which it is carried out.
- **bond energy:** the amount of energy necessary to break the chemical bonds in a given molecule, measured in kilojoules per mole.
- **catalyst:** a chemical species that initiates or speeds up a chemical reaction but is not itself consumed in the reaction.
- **chemical bond:** a link between two atoms formed by the interaction of their valence electrons.
- **chemical kinetics:** the branch of chemistry that studies the various factors that affect rates of chemical reactions.
- **enzyme:** a protein molecule that acts as a catalyst in biochemical reactions.
- **equilibrium:** the state that exists when the forward activity of a process is exactly equal to the reverse activity of that process.
- **polymerization:** a process in which small molecules bond together in a chain reaction to form a polymer, which is a much larger molecule composed of repeating structural units.

SORPTION

The general term sorption includes the processes of absorption, adsorption, desorption and chemisorption. Absorption is a physical process in which a fluid suffuses into the spaces between particles or fibers due to physical properties such as the surface tension and viscosity of the fluid, as when water or some other spilled liquid is drawn into the cellulose fiber structure of a paper towel. There is also a driving influence due to weak interactions between the absorbent medium particles and the atoms or molecules of the fluid that is being absorbed, such as dipole-dipole interactions and van der Waals forces.

Adsorption is a process whereby atoms or molecules of a material adhere to the surface of another material without formation of a chemical bond. The driving forces in adsorption are primarily dipole-dipole and van der Waals force attractions, rather than the physical properties that drive absorption. They are thus, in a sense, complimentary processes (perhaps indicated by the complimentary appearance of the letters b in absorption and the d in adsorption). Adsorption, as a pseudochemical process, is more useful in application. The basic principle of chromatography, a method for separating chemical compounds, depends on the equilibrium between the compounds adsorbed onto the particles of a chemically unreactive solid medium and the compounds dissolved in a surrounding fluid medium that flows through the system. In recrystallization technique, powdered charcoal is often used to remove colored impurities from a solution of a compound that will subsequently be allowed to crystallize from the filtered solution in a pure state. This technique relies on the adsorption of the electronically polar colored impurities by the particles of carbon. It is not unusual for this method to change a solution from a dark red color to being completely clear and colorless. In the home, activated charcoal filters are typically used to remove unpleasant odors from the air. They do this in the same manner, by adsorbing the molecules responsible for the odors onto the surface of the carbon particles in the filter. The more finely divided are the

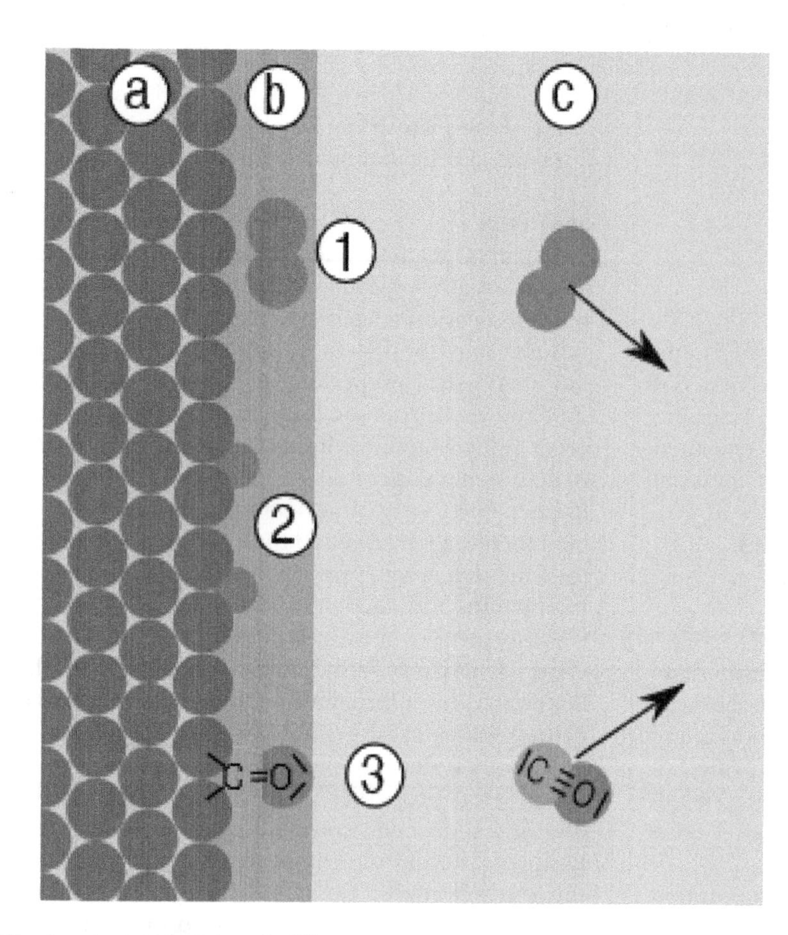

Chemisorption by Roland.chem via Wikimedia Commons

particles, the greater is the surface area available for adsorption and the more effective is the process.

Chemisorption is at first glance similar to adsorption in that the process occurs first at the surface of the substrate material. Unlike adsorption, however, chemisorption involves the formation of actual chemical bonds between the two materials. The result is the formation of an activated complex or even a distinct new molecule, both of which may be extremely short-lived or long-lasting. The formation of this intermediate structure can significantly reduce the activation energy of an overall chemical reaction as defined by the following Arrhenius equation:

$$k = Ae^{-Ea/RT}$$

The value of the constant k in the Arrhenius equation is important in chemical kinetics as the overall rate constant of the particular reaction described by the equation.

Desorption typically refers to the reverse process of absorption and adsorption, often as an equilibrium condition. It can therefore also be used to describe the reverse process of chemisorption in an equilibrium system, but it is much more typically the case that chemisorption results in the functioning of an overall chemical reaction involving a catalyst.

Catalysis

A catalyst is nominally a compound or material that takes part in a chemical reaction but is not consumed in that reaction. Typically, a catalyst acts to reduce the activation energy of a specific reaction, allowing the reaction, and its reverse reaction, to occur at a faster rate. The manner in which a catalyst takes part in a chemical reaction is much more complex than this simple definition would suggest. A simple catalyst may be just a single atom of a metal such as chromium. Catalysts can have even more complex structures, ultimately as complex as the protein structure of an enzyme. Catalysts are central to many chemical engineering applications and processes such as polymerization, the reaction process in which individual small molecules form chemical bonds to each other to form very much larger molecules. As the principal catalysts of biochemistry, enzymes such as those that carry out or moderate the biopolymerization reactions that produce everything from other proteins to DNA are themselves polymers. Chemical engineering processes have traditionally relied solely on inorganic and organometallic compounds to act as catalysts, but the application of enzyme catalysis to chemical engineering processes is an area of increasing importance as "green chemistry" methodologies have become more in demand.

Chemisorption in Catalysis

The process of chemisorption is the essential feature of catalysis, whether the catalyst is of inorganic or biochemical origin. In all cases, the catalyst molecule forms an actual chemical bond to the material undergoing reaction. In the polymerization of ethylene, for example, the most valuable catalysts are the Ziegler-Natta catalysts, a complex of titanium and aluminum atoms. The mechanism of polymerization with these catalysts is believed to involve the formation of a co-ordination bond between the electrons in the C=C double bond of ethylene and an appropriate vacant atomic or molecular orbital in the catalyst complex. The formation of this bond effectively reduces the bond energy of the C=C bond, hence reducing the activation energy of the subsequent reaction to form a new C–C bond. The coordination of the ethylene bond secures the activated ethylene molecule in the correct orientation to form the new C–C bond, which further reduces the activation energy of the reaction. When no further ethylene is available for polymerization, the reaction stops, leaving the catalyst in its original form. In an enzyme-mediated reaction, the lock-and-key analogy is often used to describe the mechanism of the reaction. In this model a portion of the molecule undergoing reaction coordinates to a very specific structure within the enzyme molecule. This is due to the relative position of key functional groups within the three-dimensional structure of the enzyme molecule. The coordination is often made through dipole-dipole interactions and hydrogen bonds, but in other interactions the connection can be the formation of an ester or amide linkage, a disulfide linkage, or other appropriate chemical bond. This coordination is sufficient to alter the conformation of the enzyme-substrate structure to facilitate a modification of the substrate molecule. Once the chemical modification of the substrate is completed it is typically released from the active site of the enzyme. Since all biochemical processes are mediated by enzymes, it is easy to see the importance of the process of chemisorption. It should be remembered that a catalyst facilitates both the forward and the reverse modes of the same reaction, and only the removal of the desired product from the system can prevent the reversion of the product to the starting material of the reaction. An enzyme, for example, can catalyze the oxidation of ethanol to acetaldehyde, but it also catalyzes the reduction of acetaldehyde to ethanol if both compounds are present. (In the metabolism of ethanol, the acetaldehyde that is formed is subsequently quickly removed by further reaction via a different enzyme.)

SAMPLE PROBLEM

Calculate the activation energy for a reaction with an observed rate constant of 49.8 X 105 sec-1 at a temperature of 45°C (318.15K). The value of R = 8.3143JK-1mol-1

Answer:

Use the Arrhenius equation

$$k = Ae^{-Ea/RT}$$

$$k_{obs} = k/A$$

$$\ln(k_{obs}) = (-E_a/RT)$$

$$\ln(49.8 \text{ X } 10^5) = -E_a/RT$$

$$RT\ln(49.8 \text{ X } 10^5) = -E_a$$

$$E_a = -RT\ln(49.8 \text{ X } 10^5)$$

$$E_a = -(8.3143JK^{-1}mol^{-1})(318.15K)(15.42)$$

$$= -40789 \text{ Jmol}^{-1}$$

APPLIED CHEMISORPTION

One of the more important applications of chemisorption is in the formation of self-assembled monolayers, a valuable process in the developing field of nanotechnology. Surface modification of nanoparticles, for example, provides a new and important methodology for the delivery of pharmaceutical compounds to specific sites within the body where medicinal treatment is desired, such as inoperable tumors. Thiols are the sulfur analogs of alcohols, containing the -SH functional group in place of the -OH functional group. Gold nanoparticles are small structures of gold atoms less than 100 nanometers in size, and have been the subject of serious study for their unique properties. When a thiol compound is put into a solution containing gold nanoparticles it is chemisorbed onto the surface of the gold nanoparticles, accompanied by the release of hydrogen gas. The result is the formation of gold nanoparticles encapsulated in a dense coating of whatever molecular structure was attached to the sulfur atom of the -SH functional group. This affords an essentially inert structure that can deliver an active pharmaceutical compound that would normally be decomposed or metabolized before it would be useful in its free form. A similar approach can be used to generate gold nanoparticle-based materials that have specific properties. One such property is the ability to capture light and then emit the captured energy as a specific wavelength as a diagnostic or analytic methodology. Another is the ability to switch between two states, potentially as a means of binary data storage. Yet another is protection of the gold nanoparticles themselves in an otherwise hostile environment.

Another application of chemisorption is the removal of specific components of a gas mixture. Chemisorption of the target component by particles of an appropriate material in a fluid process permits continuous removal of the target component in a safe form and regeneration of the chemisorbant.

The Importance of Chemisorption

Since chemisorption is an essential process in all enzyme-mediated biochemical reactions, the importance of the process in that regard is obvious. Beyond biochemical reactions, however, chemisorption is a central process in many chemical engineering applications, particularly the synthesis of polymers. Catalysts function through chemisorption and the development of reaction specific catalysts as well as catalysts that are more generally applicable is an important area of research and development. This is especially true in medical diagnostic applications for the rapid detection of specific biochemical markers related to specific disease conditions. The use of chemisorption to modify surfaces such as nanoparticles in order to provide specific property enhancements is also an area of very rapid growth in a number of different fields.

Richard M. Renneboog M.Sc.

Bibliography

Boreskov, G.K. *Heterogenous Catalysis* New York, NY: Nova Science Publishers, 2003. Print.

Davison, Sydney George, and K.W. Sulston, *Green-Function Theory of Chemisorption.* New York: Springer, 2006. Print.

Hagen, Jens *Industrial Catalysis. A Practical Approach.* 3rd ed. Weinheim, GER: Wiley-VCH Verlag, 2015. Print.

Hudson, John B. *Surface Science. An Introduction* New York, NY: John Wiley & Sons, 1998. Print.

Kolasinski, Kurt W. *Surface Science. Foundations of Catalysis and Nanoscience* 3rd ed. New York, NY: John Wiley & Sons, 2012. Print.

Ross, Julian R.H. *Heterogeneous Catalysis Fundamentals and Applications* Boston, MA: Elsevier. Print.

Somorjai, Gabor A., and Yimin Li. *Introduction to Surface Chemistry and Catalysis.* 2nd ed. New York: Wiley, 2010. Print.

CRYSTAL GROWTH

FIELDS OF STUDY

Crystallography; Organic Chemistry; Inorganic Chemistry

SUMMARY

Materials in the gas and liquid states are able to adhere to solid surfaces without forming chemical bonds. This physical adsorption is due primarily to van der Waals force and electrostatic interaction. The property is used extensively in chemistry for the separation and purification of compounds.

PRINCIPAL TERMS

- **crystal lattice:** the regular geometric arrangement of atoms and molecules in the internal structure of a crystal.
- **crystallography:** the study of the properties and structures of crystals.
- **derivative:** a compound that is obtained by subjecting a similar parent compound to one or more

chemical reactions that target certain functional groups, leaving the basic molecular structure unaltered.

- **equilibrium:** the state that exists when the forward activity of a process is exactly equal to the reverse activity of that process.
- **functional group:** a specific group of atoms with a characteristic structure and corresponding chemical behavior within a molecule.
- **hydrogen bond:** a weak type of chemical bond formed by the attraction of a hydrogen atom to an electronegative atom—an atom with a strong tendency to attract electrons—in the same or another molecule.
- **polar:** describes a molecule or functional group in which there is a difference in the distribution of electronic charge, causing one part of the molecule or group to be relatively electrically positive and another part to be relatively electrically negative.
- **precipitant:** a substance that, when added to a solution, causes a component of the solution to become solid and separate from the liquid.
- **solubility:** the ability of a particular substance, or solute, to dissolve in a particular solvent at a given temperature and pressure.
- **supernatant:** the liquid or fluid that remains after a substance is precipitated from a solution.

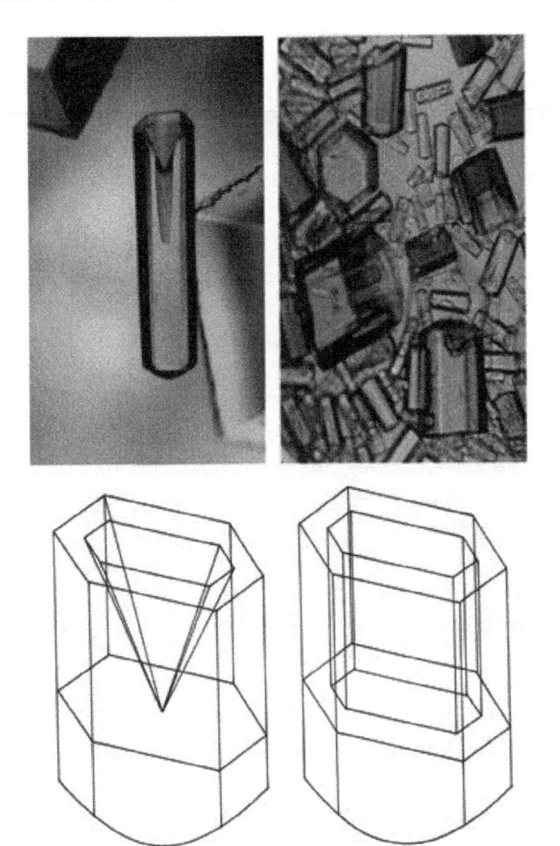

Crystal growth by Alexander McPherson & Lawrence James DeLucas via Wikimedia Commons

Solubility and Crystal Formation

One common method of purifying a compound consists of recrystallizing the material from a suitable solvent. The procedure for recrystallization is to identify a solvent or solvent system in which the solubility of the compound is high when the solvent is hot, but very low when the solvent is cold. Ideally, the compound is then dissolved in the least amount of the hot solvent, and the resulting saturated solution is treated to remove impurities by decolorization and filtration. The solution is then allowed to cool slowly so that the solubility of the solute decreases at a rate that allows the molecules of solute to congregate into a regular crystal form and precipitate from the supernatant liquid. The crystals can then be harvested from the supernatant liquid, or mother liquor, rinsed lightly with very cold fresh solvent, and dried. In some cases it is necessary to add a precipitant to the solution to initiate the formation of crystals. The formation of regular crystals always involves an equilibrium state between the dissolved and crystallized compound. The general rule, like dissolves like, plays well in crystallization processes, as polar materials dissolve well in polar solvents and non-polar materials dissolve well in non-polar solvents. A variety of forces influence the formation of crystals. Ionic compounds generally form highly regular, hard crystals due to electrostatic charge attraction (e.g. sodium chloride). Highly polar compounds can also form hard, regular crystals often due to the formation of hydrogen bonds (e.g. water, sugar) between molecules in the crystal lattice. Non-polar compounds tend to form relatively weak crystals held together by little more than the van der Waals force between molecules.

Crystallography

Study of the regular structure of crystals is the science of crystallography. Initially, the science consisted of the study and classification of the external shapes of crystals. Such observation led Louis Pasteur

(1822–1895) to discover the mirror-image relationship, or enantiomorphism, of tartrate crystals that reflects their isomeric molecular structures. The internal structure of crystals became accessible with the development of X-ray technology and the observation that Bragg scattering produced a regular pattern of reflected X-rays that was dependent upon the ordering of atoms within the crystal lattice.

Crystal growth can be likened to assembling bricks into a construction. Placed in regular rows and layers, the bricks produce a regular structural array. As a compound slowly crystallizes from solution, its molecules assemble into the regular structural array of a crystal lattice. Impurities and inclusions in a crystal have the same effect as oddly-shaped bricks among the regular bricks of a construction. They disrupt the regularity of the array and provide weak points at which the crystal can split apart. Regular crystals normally split, or cleave, along planes defined by the regular arrangement of molecules within the crystal lattice. The growth and recovery of crystals as pure and perfectly shaped as possible is a requirement for good crystallographic analysis.

APPLIED CRYSTAL GROWTH

Many materials have a characteristic functional group in their molecular structure, but are liquids or low-melting solids at room temperature. Such compounds are often reacted with specific reagents to produce a nicely crystalline solid derivative compound with sharp melting point. Ketones and aldehydes, for example, react quickly with 2,4-dinitrophenylhydrazine to produce the corresponding 2,4-diphenylhydraxone (2,4-DNP) derivative that is typically highly colored and is readily purified by recrystallization to exhibit a sharp, distinct, characteristic melting point. New compounds isolated from natural sources or produced synthetically are derivatized to provide characteristic reference data and facilitate handling as crystalline compounds.

Crystals can also be grown from the molten material. This is the method used to produce large single crystals of silicon that are then cut into wafers and etched for the manufacture of computer chips and integrated circuits.

Richard M. Renneboog, M.Sc.

SAMPLE PROBLEM

Phthalic acid has the following solubilities:

water:	0.54 g at 14°,
	18 g at 99°
alcohol	11.71 g at 18°
diethyl ether	0.69 g at 15°
chloroform	insoluble

Which solvent would be the best choice to use to purify by recrystallization 100 g of phthalic acid that has been contaminated?

Answer:

The solubility data shows that phthalic acid dissolves well in hot water but only poorly in cold water. The boiling point of diethyl ether is not much higher than the stated solubility temperature and so would be a poor choice. Phthalic acid dissolves too well in alcohol at room temperature. Therefore water would be the best solvent to use for this recrystallization.

BIBLIOGRAPHY

Bhat, H.L. *Introduction to Crystal Growth. Principles and Practice.* Boca Raton, FL: CRC Press, 2015. Print.

Hall, Judy. The *Crystal Bible. A Definitive Guide to Crystals.* Cincinnati, OH: Walking Stick Press, 2003. Print.

Leubner, Ingo H. *Precision Crystallization. Theory and Practice of Controlling Crystal Size.* Boca Raton, FL: CRC Press, 2010. Print.

Nishinauga, Tatau, ed. *Handbook of Crystal Growth. Fundamentals: Thermodynamics, Kinetics, and Transport and Stability* 2nd ed. Waltham, MA: Elsevier, 2015. Print.

Rudolph, Peter. ed. *Bulk Crystal Growth. Basic Techniques, and Growth Mechanisms and Dynamics.* Waltham, MA: Elsevier, 2015. Print.

Sunagawa, Ichiro. *Crystals. Growth, Morphology and Perfection.* New York, NY: Cambridge University Press, 2005. Print.

D

DEBYE-HÜCKLE THEORY

FIELDS OF STUDY

Physical Chemistry

SUMMARY

The behavior of electrolytes in solution varies from ideal behavior in a manner that is dependent on concentration. The property of chemical activity is defined to account for the observed variance. The relationships are defined by the Debye-Hückel equation.

PRINCIPAL TERMS

- **chemical kinetics:** the branch of chemistry that studies the various factors that affect rates of chemical reactions.
- **concentration:** the amount of a specific component present in a given volume of a mixture.
- **diffusion:** the process by which different particles, such as atoms and molecules, gradually become intermingled due to random motion caused by thermal energy.
- **dissociation constant:** a characteristic value representing the extent to which a compound dissociates into component ions in a certain solvent and under specific conditions.
- **electrolyte:** a material that ionizes in an appropriate solvent to produce an electrically conductive solution.
- **enthalpy:** the total heat content within a thermodynamic system, defined as internal energy plus the product of pressure and volume; also, the change in heat content associated with a chemical process.
- **entropy:** a property of the energy of chemical systems that is related to the total number of degrees of freedom within the system due to the number of components in that system and the number of particles of each component.
- **equilibrium:** the state that exists when the forward activity of a process is exactly equal to the reverse activity of that process.
- **solubility:** the ability of a particular substance, or solute, to dissolve in a particular solvent at a given temperature and pressure.
- **solute:** any material that is dissolved in a liquid or fluid medium, usually water.
- **solution:** an intimate combination of two or more different materials forming a single physical phase, most typically liquid but may also be gaseous as a mixture of gases or solid as an alloy.

ELECTROLYTES AND IDEAL BEHAVIOR

Electrolytes are materials that dissociate into positive and negative ions when they dissolve in an appropriate solvent. Strong electrolytes are characterized by complete separation into dissolved ions and do not have a defined dissociation constant, though they may have limited solubility. A simple example is a solution of sodium chloride in water. The salt dissolves as the positive sodium ions and the negative chloride ions become separated from each other by the surrounding water molecules. Each ion is enclosed within an environment of polar, but electrically neutral, water molecules that serves to keep the ions solvated and separated from each other. Compounds that dissociate incompletely are termed weak electrolytes. An example of a weak electrolyte is acetic acid, which dissociates only by a small percentage into acetate ions and hydrogen ions leaving the majority of the material as dissolved whole acetic acid molecules. The concentrations of ionic species from weak electrolytes are determined by the equilibrium constant of the particular compound. In an ideal electrolytic solution the positive and negative ions are distributed uniformly throughout the solution and do not interact with each other. The solvent molecules in an ideal electrolytic solution also do not interact with the ions in solution. Real electrolytic solutions, however, vary from this ideal behavior in a manner that depends on the concentration of the ionic species in the solution.

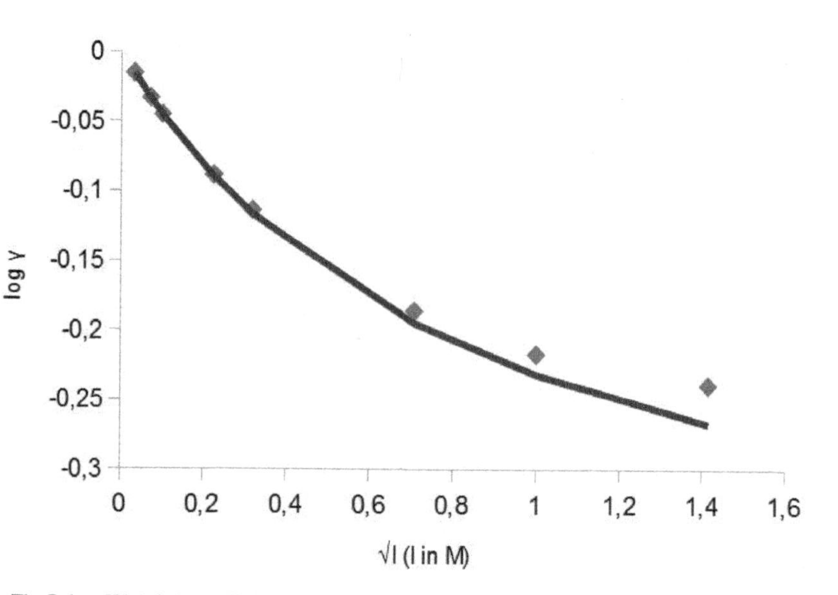

The Debye–Hückel plot by Typhoner via Wikimedia Commons

REAL ELECTROLYTIC SOLUTIONS

In an electrolytic solution the ions produced by the solute carry opposite and often unequal electrical charge, and so are attracted to each other. Similarly, ions carrying the same charge repel each other. Both forces vary according to the distance between the ions, and have the effect of making the distribution of ions non-uniform. Positive ions tend to be surrounded by negative ions and vice versa. Over the entire solution the charges are evenly distributed, but on the microscale the charge distribution is more like collections of charged "blobs" that move about by the process of diffusion. The thermodynamic properties of the system such as enthalpy and entropy that are calculated from the molar concentrations of the electrolytes are found to differ from observed properties. To rectify this, a correction factor called the activity coefficient is used. When multiplied by the molar concentration, the activity coefficient yields the active mass of the particular electrolytes and produces the correct results for the thermodynamic calculations. This can be thought of as the amount of electrolyte that is responsible for the observed thermodynamic values, rather than the total amount of electrolyte. Unfortunately, the complex nature of actual solutions exceeds the ability of Debye-Hückel theory to produce correct values effectively at concentrations greater than 10^{-3} moles per liter, and factors such as the structural symmetry of the ionic species, species carrying multiple charges, and solvation differences tend to cause calculated results to become increasingly varied. Debye-Hückel theory is sometimes described as over-simplified.

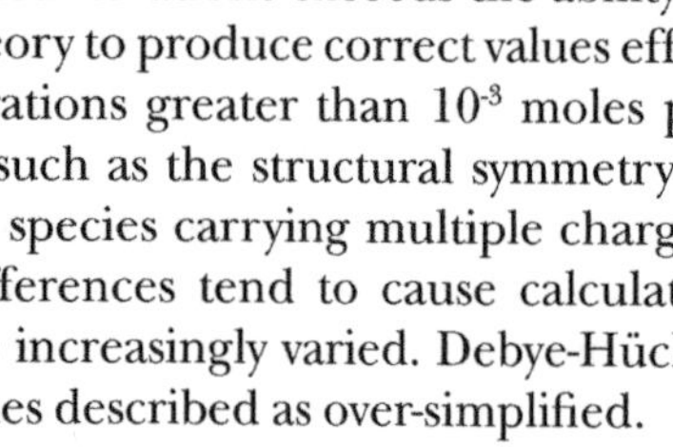

ACTIVITY COEFFICIENTS

Because electrolyte solutions must contain both positive and negative ions it is not possible to measure the activity coefficient of a particular ion in solution directly. Debye-Hückel calculations make use of the mean activity coefficient, $\gamma_{\pm}$, instead. This is defined as the square root of the product of the positive and negative ion activity coefficients and is determined experimentally as its logarithmic value using calculations based on ionic strength, the Maxwell-Boltzmann and Poisson-Boltzmann distributions, and factors describing the limit of closest approach to each other of ions in solution.

IMPORTANCE OF DEBYE-HÜCKEL THEORY

Debye-Hückel theory has been of significant value in understanding the behavior of electrolyte solutions, and especially of those solutions of very low concentrations and ionic strength. The simplicity of the theory is most closely in agreement with simple electrolytes like sodium chloride but becomes decreasingly reliable as solution complexity and ion concentrations increase. It is nevertheless a very good approximation for understanding the chemical kinetics and reaction rates of complex electrolyte solutions such as those developed for use in electromotive cells and the cytoplasm found in living cells.

Richard M. Renneboog, M.Sc.

BIBLIOGRAPHY

Atkins, Peter, and Julio de Paula. *Atkins' Physical Chemistry.* 10th ed. New York: Oxford University Press, 2014. Print.

Brey, Wallace S. *Physical Chemistry and Its Biological Applications.* New York, NY: Academic Press, 1978. Print.

Eu, Byung Chan, and Mazen Al-Ghoul. *Chemical Thermodynamics: With Examples for Non-equilibrium Processes.* Hackensack, NJ: World Scientific, 2010. Print.

Kaufman, Myron. *Principles of Thermodynamics.* New York, NY: Marcel Dekker Inc., 2002. Print.

Rieger, Philip H. *Electrochemistry* 2nd ed. New York, NY: Chapman & Hall, 1994. Print.

Wright, Margaret Robson. *An Introduction to Aqueous Electrolyte Solutions.* Hoboken: Wiley, 2007. Print.

Zemaitis, Joseph F., Diane M. Clark, Marshall Rafal, and Noel C. Scrivner. *Handbook of Aqueous Electrolyte Thermodynamics.* New York: Wiley-Interscience, 2010. Print.

DECOMPOSITION REACTIONS

FIELDS OF STUDY

Organic Chemistry; Chemical Engineering; Environmental Chemistry

SUMMARY

Decomposition reactions are defined, and their importance in various chemical processes is elaborated. Decomposition reactions have broad value, both as a means of destroying unwanted materials and as a useful step in producing desired materials. Through their distinctive products, they also provide a means of identifying the source of materials in forensic investigations.

PRINCIPAL TERMS

- **activation energy:** the amount of energy a system requires for the formation of an activated complex from which a reaction can proceed.
- **combination reaction:** a chemical reaction in which two or more reactants combine to form a single product.
- **product:** a chemical species that is formed as a result of a chemical reaction.
- **reactant:** a chemical species that takes part in a chemical reaction.
- **reaction mechanism:** the sequence of electron and orbital interactions that occurs during a chemical reaction as chemical bonds are broken, made, and rearranged.

CHEMICAL SYNTHESIS IN REVERSE

A decomposition reaction can be defined as the breaking apart of a compound or substance into smaller and less complex components. This breakdown must obey the same rules of chemical behavior as the formation of a compound or substance from its lesser components. In essence, decomposition reactions are the exact opposite of synthesis reactions. When a compound or material is synthesized, the component materials, or reactants, interact with each other in specific ways that are determined by the basic rules of chemical behavior. Decomposition of the material often occurs by reversing the direction of the formation reaction, resulting in products that are identical to the original reactants.

Decomposition reactions can be used to produce a material that is difficult to prepare by normal chemical methods. While it is possible in some cases to prepare a material that decomposes by a different mechanism to produce the desired material, devising and controlling such an application demands a thorough understanding of the electronic principles of reaction mechanisms. In other cases, decomposition reactions make it possible to easily prepare something that cannot be reproduced by synthetic means, such as the caramelization that occurs in cooking when the sugars and other carbohydrates begin to undergo thermal decomposition.

THE OCCURRENCE OF DECOMPOSITION REACTIONS

The term "decomposition" has somewhat different meanings depending on the context in which it is used. In chemistry, it refers to a specific type of chemical reaction, while in forensic analysis and ecological science, for example, materials are described as being in a state of decomposition when the original structure of the material has broken down and is no longer in its living or original state. Thus, in the broadest sense, decomposition can be defined as a process in which materials are broken down to their component parts and elements.

In the natural environment, decomposition is normally caused by chemical reactions such as oxidation (a reaction with oxygen in the atmosphere) or by the activity of bacteria and other living things that actively consume the material. In the context of chemistry, decomposition takes place when a chemical compound splits apart into component pieces that essentially retain the structures they had in the parent molecule. Decomposition reactions can thus be represented by the general symbolic equation:

$$AB \rightarrow A + B$$

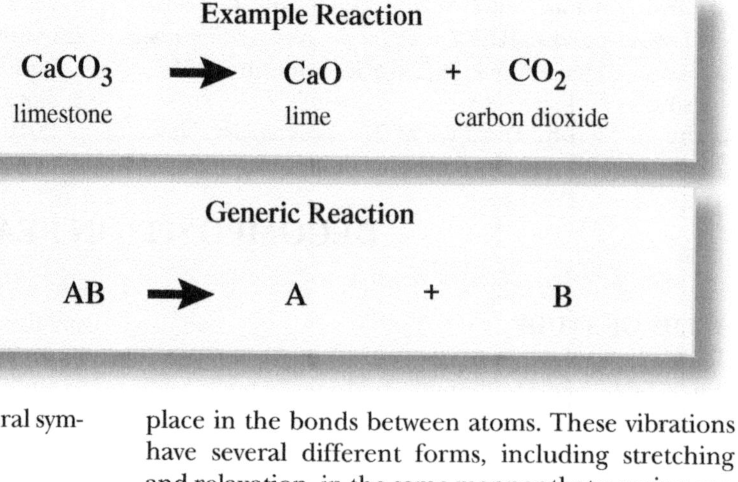

In this representation, the compound AB has two component structures, A and B. The molecule AB decomposes when it splits apart to liberate both A and B.

Decomposition reactions occur in solid, liquid, and gas phases and can pose a problem for purified laboratory chemicals, which often become discolored over time. The essential characteristic of such reactions is that the material undergoing decomposition obeys the general equation given above. It is important to note that the parent material can split into more than two parts, though this is a rare situation. It is far more commonly the case that a product component, once formed, quickly degrades further.

DECOMPOSITION REACTIONS AND ACTIVATION ENERGY

All chemical compounds will undergo decomposition reactions when given sufficient activation energy. In the vast majority of cases, the activation energy for the process is provided by either thermal excitation or photolysis. The mechanisms by which these affect the material are essentially the same: electrons in specific chemical bonds become excited to a higher energy level in a molecular orbital that does not support bonding. When this happens, the parent molecule is free to separate at that location. Thermal excitation acts to stimulate the vibrations that are always taking place in the bonds between atoms. These vibrations have several different forms, including stretching and relaxation, in the same manner that a spring can be stretched and relaxed. The energy of this mode of vibration can become great enough to overpower the strength of the chemical bond and separate the components of the molecule.

The ability of a material to undergo a decomposition reaction is strongly affected by the stability of the product material. One example of this is the equilibrium between carbon dioxide, water, and carbonic acid. One molecule of carbonic acid, H_2CO_3, forms when one molecule of carbon dioxide (CO_2) and one molecule of water (H_2O) undergo a combination reaction. This is the method by which aerobic cells eliminate the carbon dioxide produced during glycolysis, which is the anaerobic metabolism of glucose: an enzyme catalyzes both the formation and the decomposition of carbonic acid, allowing carbon dioxide gas, a by-product of the process, to be expelled. In carbonated beverages, however, the simple chemical equilibrium:

$$H_2O + CO_2 \rightleftharpoons H_2CO_3$$

is established between water and dissolved carbon dioxide. When the cap is removed and the pressure inside the can or bottle is released, numerous bubbles of carbon dioxide appear as the carbonic acid decomposes back to the more stable molecules of water and carbon dioxide. It is easy to demonstrate that the warmer the liquid is when the pressure is released, the more rapid the decomposition. Who has not seen a warm bottle of soda spew its contents violently when opened?

Another example of chemical decomposition due to thermal excitation is the caramelization of white sugar, also known as "sucrose." This is an example of pyrolysis, which is a form of thermal decomposition, or thermolysis, that takes place at high temperatures in the absence of oxygen. The decomposition of sucrose occurs at temperatures around 160 degrees Celsius and higher, when heat drives out the components of water from the carbohydrate structure of the sugar.

SAMPLE PROBLEM

An amateur chef placed 28 grams of sucrose ($C_{12}H_{22}O_{11}$) in a pan to prepare a caramel. Unfortunately, the chef forgot about it and returned to find nothing but carbon remaining in the pan. How much carbon was there? Use the balanced-reaction equation AB → A + B as the basis of your calculation.

Answer:

The molecular formula of sucrose is $C_{12}H_{22}O_{11}$. Complete thermal decomposition eliminates eleven molecules of water per molecule of sucrose, leaving twelve atoms of carbon. The balanced reaction equation is

$$C_{12}H_{22}O_{11} \rightarrow 12C + 11H_2O$$

Determine the molecular weight of sucrose, using the atomic mass for each element as given in the periodic table:

$$(12 \text{ x } 12.01) + (22 \text{ x } 1.01) + (11 \text{ x } 16) = (144.12 + 22.22 + 176) = 342.34$$

The molecular weight of sucrose is 342.34 atomic mass units (u), or grams per mole. Calculate the number of moles in 28 grams of sucrose:

$$28 \text{ g} \div 342.34 \frac{\text{g}}{\text{mol}} = 0.082 \text{ mol}$$

Calculate how many moles of carbon are present in 0.082 mole of sucrose:

$$0.082 \text{ mol x } 12 = 0.984 \text{ mol}$$

Convert this amount to grams, keeping in mind that the atomic mass of carbon is 12.01 g/mol:

$$12.01 \frac{\text{g}}{\text{mol}} \times 0.984 \text{ mol} = 11.818 \text{ g}$$

Therefore, the chef's error produced 11.818 grams of carbon instead of nearly 28 grams of caramel.

DECOMPOSITION IN THE ATMOSPHERE

One alternative to thermolysis is photodecomposition, or photolysis, which occurs when electrons in a chemical bond absorb a photon of light energy and enter an excited state. In this excited state, the strength of the chemical bond is greatly weakened, if not completely eliminated. The normal vibrations of the weakened bond are then sufficient to separate the two components of the molecule. When this occurs, the products are unstable radical entities, which are essentially normal molecules that have all of their electrons but are missing an atom, leaving one free electron that would normally be involved in a chemical bond but for which such a bond does not exist.

This is the typical process that takes place in the atmosphere, and it is especially important in regard to the chemistry of atmospheric pollution. A common example of photolysis is when a molecule of chlorine gas, Cl_2, separates into two neutral chlorine atoms. This phenomenon is also an example of homolytic cleavage, which is when a chemical bond breaks and the resulting products each retain one of the two electrons that formed the bond. Photolysis of a molecule such as chloromethane, CH_3Cl, results in the separation of the neutral chlorine atom Cl• from the neutral methyl radical CH_3• (the dot is used in standard notation to indicate the free electron). The corresponding reaction equations are as follows:

$$Cl_2 \rightleftharpoons 2Cl\bullet$$

$$CH_3Cl \rightleftharpoons CH_3\bullet + Cl\bullet$$

These are shown as equilibria because the process can be easily reversed by simply reforming the bond, and indeed this is primarily what occurs, unless there is some other chemical species available to react with one of the radicals when it is formed. In the atmosphere, molecular oxygen (O_2) and ozone (O_3), as well as several other gaseous materials, are available to enter into a chain reaction that is either a normal process of atmospheric chemistry or one that produces various atmospheric pollutants.

The ozone layer that surrounds Earth has been seriously affected by the release of various halocarbons known as "chlorofluorocarbons," or CFCs, which are similar to chloromethane and were widely used as refrigerants for several decades. The ozone in Earth's stratosphere interacts with incoming ultraviolet radiation from the sun, preventing much of it from reaching the surface of the planet, where it would pose a danger to exposed life forms. This ozone is regenerated from molecular oxygen through a series of radical chain reactions in which decomposition reactions are an essential feature. However, when CFCs reach the planet's upper atmosphere, that same ultraviolet radiation causes a decomposition reaction that frees a radical chlorine atom from the compound, which in turn interacts with existing ozone molecules in such a way that it turns ozone into molecular oxygen, thus depleting the protective ozone layer between Earth and the sun.

Richard M. Renneboog, MSc

BIBLIOGRAPHY

Begon, Michael, Colin R. Townsend, and John L. Harper. *Ecology: From Individuals to Ecosystems.* 4th ed. Malden: Blackwell, 2006. Print.

Berg, Jeremy M., John L. Tymoczko, Gregory J. Gatto, and Lubert Strye. *Biochemistry.* 8th ed. New York: W. H. Freeman, 2015. Print.

Lafferty, Peter, and Julian Rowe, eds. *The Hutchinson Dictionary of Science.* 2nd ed. Oxford: Helicon, 1998. Print.

Morrison, Robert Thornton, and Robert Neilson. Boyd. *Organic Chemistry.* 7th ed. Englewood Cliffs, N.J.: Prentice Hall, 2003. Print.

Silbey, Robert J., Robert A. Alberty, and Moungi G. Bawendi. *Physical Chemistry.* 5th ed. Hoboken: Wiley, 2012. Print.

Zumdahl, Steven S., and Susan Zumdahl. *Chemistry.* 9th ed. Belmont: Brooks Coles, 2013. Print.

DEHYDRATION

FIELDS OF STUDY

Organic Chemistry

SUMMARY

Dehydration is defined and its importance as a chemical processes is described. Dehydration reactions produce the C=C double-bond functionality in molecules and so provide a synthetic pathway for both the preparation and chemical elaboration of more complex molecules and the commercial preparation of numerous compounds used in the production of polymers and other large molecules.

PRINCIPAL TERMS

- **alcohol:** an organic compound in which a hydroxyl is the primary functional group and is bonded to a saturated carbon atom.
- **alkene:** any organic compound that includes two carbon atoms connected by a double bond.
- **decomposition reaction:** a chemical reaction in which a single reactant breaks apart to create several products with smaller molecular structures.
- **double bond:** a type of chemical bond in which two adjacent atoms are connected by four bonding electrons rather than two.
- **hydroxyl group:** a primary functional group consisting of an oxygen atom covalently bonded to a single hydrogen atom.

UNDERSTANDING DEHYDRATION

Dehydration means the loss of water in all contexts. In any organic (carbon-based) molecule that has both a hydroxyl (–OH) group and a hydrogen atom (–H) on adjacent carbon atoms in the molecular structure, the possibility exists for these two substituents to be eliminated from the molecule. Nominally, the loss of the –OH and –H groups from a molecule produces a molecule of water (H–OH, or H_2O), hence the term "dehydration." While the vast majority of dehydration reactions do actually produce water as one of the products, the water molecules produced are not composed of the same –H and –OH groups that were eliminated from the parent molecule. There are two principal elimination reaction mechanisms by which dehydration occurs, depending on the conditions under which the process occurs.

Elimination Reactions

In a dehydration reaction, the components of a molecule of water are eliminated from some parent molecular structure; the process is generally referred to as a decomposition reaction. The manner in which this takes place depends on the actual molecular structure of the parent molecule and the reaction conditions. The two elimination reaction mechanisms are termed E_1, meaning "elimination unimolecular," and E_2, meaning "elimination bimolecular." In E_1 reactions, electronic properties of the parent molecular structure are the significant determining factors of both the rate of the reaction and the structure of the product(s). In E_2 reactions, the three-dimensional stereochemistry, or spatial arrangement of atoms and bonds, of the parent molecule must meet very specific requirements in order for the elimination reaction to proceed. No matter the mechanism by which a dehydration reaction proceeds, the loss of the hydroxyl group and adjacent hydrogen substituent results in the formation of an alkene characterized by the presence of the very reactive carbon-carbon double bond. The presence of other substituents in the parent molecular structure can promote the dehydration by reducing the energy barriers to loss of the substituents or by stabilizing the product double bond, making it more energetically favorable. On the other hand, the presence of other substituents can also make it unfavorable, if not impossible, for the elimination to proceed. The physical size and restricted mobility of substituent groups situated next to an active site in a molecular structure present steric hindrance (interference to the motion of atoms and other chemical species) to activity that would normally occur at that site. Steric hindrance thus slows the rate at which reactions occur and can prevent them from occurring altogether.

Elimination Unimolecular Process

The mechanism of an E_1 reaction depends on the formation of an intermediate chemical species called a "carbonium ion" or "carbocation." The oxygen atom of the –OH functional group has two "lone pairs" of electrons in valence orbitals, exposed to the surrounding environment. An electrophile (an electron-poor chemical species, such as a proton, $FeCl_3$, BCl_3, or other Lewis acid, that reacts preferentially with atoms with a high electron density or negative charge) can form a bond with one of the lone pairs. Simultaneously, the two electrons in the C–O bond become the second lone pair of the O atom. This allows the –OH group to leave the molecule as a neutral molecule of H_2O, and leaves the C atom with a positive charge because of the loss of the electron it donated to the C–O bond. The orbital arrangement on the carbon atom changes from the tetrahedral *sp3* to the trigonal *sp2* arrangement with an

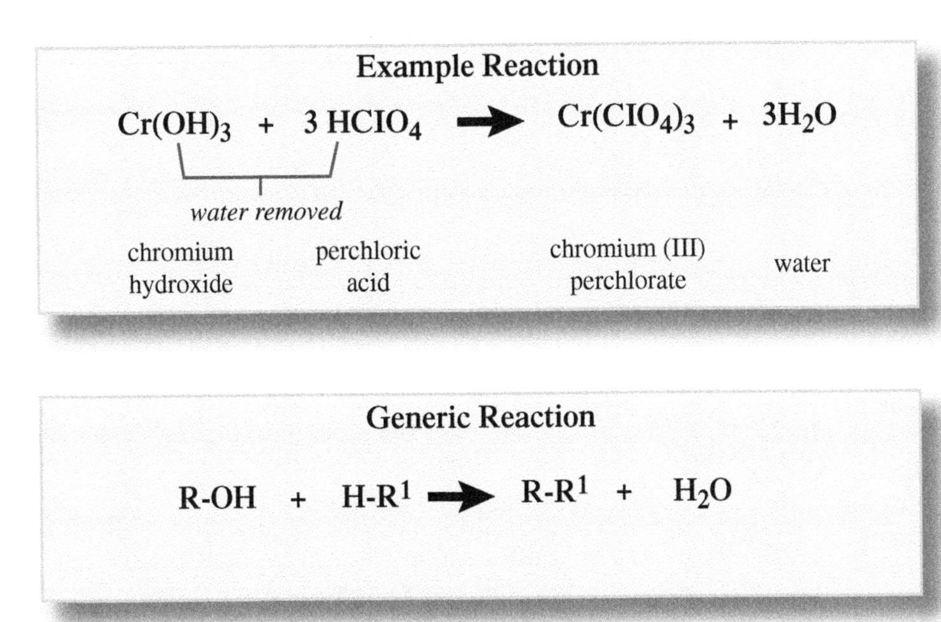

unoccupied *p* orbital. The resulting structure is the carbonium ion intermediate. The carbonium ion can do any of three things once it has been formed: rearrange, add a nucleophile (an electron-rich chemical species, such as a hydroxide ion, amine, phosphine, or other Lewis base), or lose a proton. The dehydration process is completed when the carbonium ion releases the H atom that was adjacent to the original –OH group and forms the C=C double bond. The orbital arrangement of this C atom also changes from *sp3* to sp2, and the two electrons from the C–H bond enter the new *p* orbital, forming a pi bond between the two C atoms and leaving the H atom as H^+.

The process is described as being acid-catalyzed, since acid facilitates but is not consumed in the process. It should be readily apparent, however, that the proton released from the carbonium ion is not the same proton that was originally bonded to the –OH group, and this feature is what allowed the mechanism to be elucidated according to the products of reactions with materials that could serve as tracers. It should also be noted that the water molecule formed by the original protonation of the –OH group can then be added back to the carbonium ion as a nucleophile. Addition of nucleophilic water molecules to either side of the carbonium ion in equal proportions racemizes optically active alcohols in the presence of an acid.

The carbonium ion can also rearrange before losing the proton. This process always produces an energetically more stable, or lower energy, molecular structure for both the carbonium ion and for the C=C double bond in the product. Generally, the more highly substituted the C atoms of the carbonium ion center and the C=C bond are, the more stable they are. Accordingly, a primary carbonium ion formed on the first carbon atom of a chain has only one substituent group and will rearrange by shifting an adjacent H atom, alkyl, or other group from an adjacent carbon atom to form a secondary carbonium ion with two substituent groups. Similarly, a secondary carbonium ion will rearrange to form a tertiary carbonium ion with three substituent groups. The various modes of reaction are in competition, and it is important to carefully consider the molecular structure of a starting alcohol when dehydration is being used to install a C=C bond in a molecule.

ELIMINATION BIMOLECULAR PROCESS

The E_2 mechanism is described as concerted, since it depends on the interaction of the substrate molecule with a base or nucleophile that extracts the proton and would leave the –OH group cleaved off

SAMPLE PROBLEM

Ethanol can be dehydrated to form ethylene when treated with sulfuric acid. The reaction proceeds in the following manner:

$$H_3C{-}CH_2{-}OH + H^+ \rightarrow H_3C{-}CH_2{-}OH_2^+$$

$$H_3C{-}CH_2{-}OH_2^+ \rightarrow H_3C{-}CH_2^+ + H_2O$$

$$H_3C{-}CH_2^+ \rightarrow H^+ + H_2C{=}CH_2$$

What alkene products can be formed when isobutanol, $HO{-}CH_2{-}C(CH_3)_2$, is treated under the same conditions? What is the rate expression for each reaction? (Remember that a carbonium ion can rearrange, add a nucleophile, and lose a proton.)

Answer:

The initial step of the process is protonation of the –OH group, as

$$HO{-}CH_2{-}CH(CH_3)_2 + H^+ \rightarrow H_2O^+{-}CH_2{-}CH(CH_3)_3$$

This is followed by loss of a water molecule to form the primary carbonium:

$$H_2O^+{-}C{-}CH(CH_3)_2 \rightarrow H_2O + {}^+CH_2{-}CH(CH_3)_2$$

The primary carbonium ion can then lose a proton from the second carbon atom to yield isobutylene:

$${}^+CH_2{-}CH(CH_3)_2 \rightarrow H_2C{=}C(CH_3)_2$$

The primary carbonium ion can also shift a proton from the second carbon atom to the first to produce a much more stable tertiary carbonium ion:

continued on next page

SAMPLE PROBLEM CONTINUED

$$^{+}CH_2–CH(CH_3)_2 \rightarrow H_3C–CH_2–CH^{+}–CH_3$$

Loss of a proton from the end carbon atom produces 1-butene, while loss of a proton from the second carbon atom in the chain produces 2-butene:

$$H_3C–CH_2–CH^{+}–CH_3 \rightarrow H_3C–CH_2–CH{=}CH_2$$

and

$$H_3C–CH_2\text{-}CH^{+}\text{-}CH_3 \rightarrow H_3C–CH{=}CH–CH_3$$

The reaction occurs by the E_1 mechanism, and its rate is solely dependent on the rate of initial formation of the carbonium ion, which is determined by the concentration of isobutanol. The kinetic expression for the rate of formation of all of the alkene products will have the first-order kinetics form

$$\text{rate} = k[\text{isobutanol}]$$

but the value of k, the rate constant, will be different for each alkene product.

as OH^- while the C=C bond is formed. However, since OH^- forms a significantly stronger C–O bond than practically all other nucleophilic groups, those groups are incapable of displacing OH–, and dehydration of alcohols to form C=C bonds is essentially never utilized. The mechanism is most commonly utilized in dehydrohalogenation reactions in which the C=C bond is produced by elimination of HX (where X = Cl, Br, I, etc.) from a parent molecular structure. Dehydration of an alcohol to form a C=C bond by this mechanism generally requires that the alcohol be modified in such a way that the O atom of the -OH group becomes part of a substituent group that can be easily displaced. It is not unusual for an alcohol to first be reacted with a material such as toluene sulfonyl chloride (commonly called "tosyl chloride") to replace the H of the -OH group with the tosyl group. This substitution converts the -OH group to the -OTs (tosylate) group, which can be more readily displaced in an E_2 reaction than the -OH group. The E_2 mechanism does not result in rearrangement of the parent molecular structure as would occur with a carbonium ion intermediate. Instead, the H atom to be extracted and the substituent group that will leave the structure as the C=C bond forms must be in the "anti-periplanar" arrangement, on opposite sides of the C–C bond and in the same plane. In this arrangement, the molecular orbitals involved in bond breaking and bond making align with each other, and the transformation is facilitated.

KINETIC ORDER OF DEHYDRATION

The E_1 and E_2 designations also indicate the kinetic order of the reaction. The rate of the E_1 mechanism is controlled by the rate of carbonium formation, since subsequent rearrangement, deprotonation, and addition of a nucleophile occur much more rapidly. The rate of an E_1 reaction can therefore be mathematically described in terms of the concentration of the substrate alcohol molecule. The rate of the E_2 mechanism, on the other hand, is determined by the interaction of the substrate molecule with an appropriate base. This interaction is dependent on the concentrations of both the substrate molecule and the base species, so that the rate of the bimolecular reaction can only be described mathematically in terms of both concentrations and not just by one or the other.

Richard M. Renneboog, MSc

BIBLIOGRAPHY

Carey, Francis A., and Richard J. Sundberg. *Advanced Organic Chemistry: Part A—Structure and Mechanisms.* 5th ed. New York: Springer, 2007. Print.

Clugson, Michael, and Rosalind Flemming. *Advanced Chemistry.* New York: Oxford UP, 2000. Print.

Jacobs, Adam. *Understanding Organic Reaction Mechanisms.* Cambridge: Cambridge UP, 1997. Print.

Morrison, Robert Thornton, and Robert Neilson. Boyd. *Organic Chemistry.* 7th ed. Englewood Cliffs, N.J.: Prentice Hall, 2003. Print.

Smith, Michael B., and Jerry March. *March's Advanced Organic Chemistry: Reactions, Mechanisms, and Structure.* 7th ed. Hoboken: Wiley, 2013. Print.

DIFFUSION

FIELDS OF STUDY

Molecular Biology; Physical Chemistry; Chemical Engineering

SUMMARY

The process of diffusion is defined, and its underlying principles are discussed. Diffusion is presented in various contexts, including fundamental characteristics of living cells and the enrichment of uranium. In all applications and occurrences, the basic principles of the diffusion process are the same.

PRINCIPAL TERMS

- **Brownian motion:** the continuous, random motion of particles in a fluid medium, caused by impacts with the molecules that make up the medium.
- **concentration gradient:** the gradual change in the concentration of solutes in a solution across a specific distance.
- **equilibrium:** the state that exists when the forward activity of a process is exactly equal to the reverse activity of that process.
- **osmosis:** the passage of solvent molecules through a semipermeable membrane from a region of low solute concentration to one of higher concentration; also the primary mechanism by which water moves through cell walls.
- **semipermeable membrane:** a membrane that allows the passage of a material, such as water or another solvent, from one side to the other while preventing the passage of other materials, such as dissolved salts or another solute.

VISUALIZING DIFFUSION

The process of diffusion can be easily modeled using an array of marbles of two different colors and a box. Place the marbles in the box in a single layer, with all the marbles of one color on one side and the marbles of the second color on the other. When the box of marbles is vibrated, marbles of one color will become dispersed throughout those of the other color. When the rate and amplitude of vibration are low, the interspersion of the marbles will be slow, but as either the rate or the amplitude is increased, the marbles will intersperse much more rapidly. At the atomic and molecular levels, diffusion functions in essentially the same manner.

DIFFUSION OF ATOMS AND MOLECULES

Atoms and molecules are constantly in motion, even at the absolute zero of temperature. For the most part atoms in the solid state oscillate around their equilibrium positions but the presence of lattice vacancies and other defects allows a certain amount of hopping. In the liquid state atomic movements are larger and somewhat more difficult to describe. In the gaseous state one can treat atomic motion as largely in straight lines with occasional collisions. In al thee phases of matter the average translational kinetic energy at absolute temperature T is equal to $3RT/N_A$ where R is the gas constant, ant M the molecular mass in kilograms. As a result, the mean distance a molecule or particle will travel in a time t obeys $\langle r^2 \rangle = Dt$, where D is the diffusion coefficient.

In 1905, Albert Einstein showed how the diffusion coefficient governing Brownian motion of particles suspended in a liquid or gas could be related to the bulk transport properties of the material. Many historians of science believe it was this paper that convinced the majority of scientists that the atomic description of matter was valid in all phases: solid, liquid and gas.

APPLICATIONS OF DIFFUSION

The most important application of diffusion occurs naturally in the cell membranes of living systems. The plasma membranes of all animal cells are composed of a lipid bilayer, primarily containing phospholipids. They are hydrophilic (attracted to water) on the phosphate end of the molecule, while the long hydrocarbon chains of the fatty acid are very hydrophobic (repelled by water) in nature. The hydrocarbon chains essentially dissolve in each other, forming a dense hydrocarbon layer sandwiched between two hydrophilic layers that can interact with proteins, ions, and other polar molecules. Sugars, proteins, and other essential compounds enter the cell by passing through the cell membrane, while

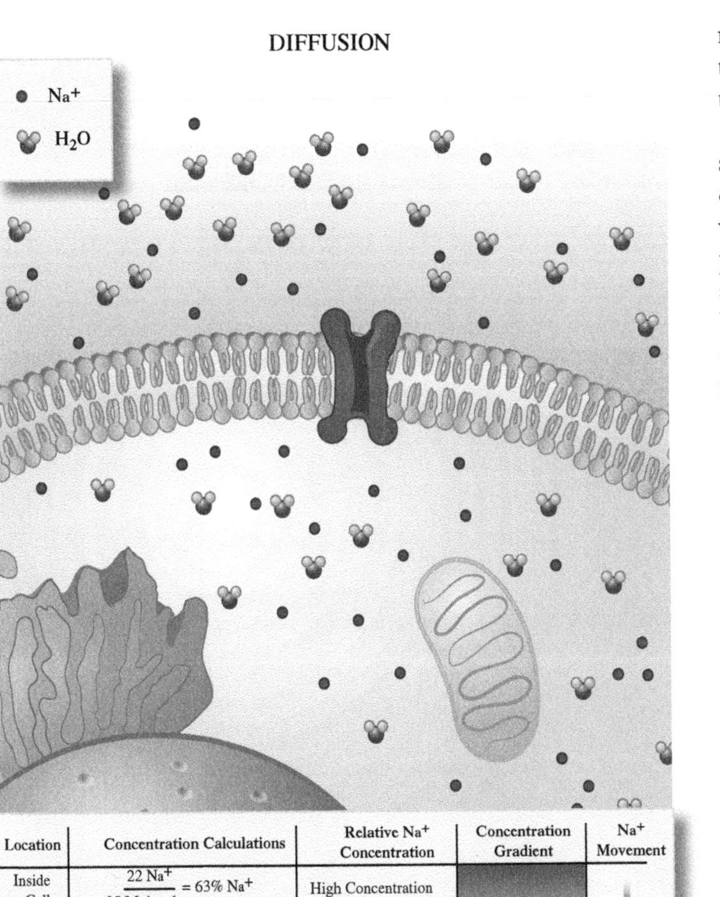

Location	Concentration Calculations	Relative Na^+ Concentration	Concentration Gradient	Na^+ Movement
Inside Cell	$\frac{22\ Na^+}{35\ \text{Molecules}} = 63\%\ Na^+$	High Concentration		↓
Outside Cell	$\frac{16\ Na^+}{37\ \text{Molecules}} = 43\%\ Na^+$	Low Concentration		

by-products of metabolism and respiration must also pass through the membrane in order to be transported away and eliminated.

Molecules pass through cell membranes by either active or passive transport. In active transport, various proteins bind to ions and other molecules that otherwise would not be able to diffuse through the membrane due to their high polarity and the hydrophobic nature; these proteins, known as carrier proteins, form a channel in the membrane through which the polar materials can pass. The difference in composition between the fluid within the cell and the interstitial fluid outside of it forms a concentration gradient. Active transport processes move materials against a concentration gradient, from a region of lower concentration to one of higher concentration. This enables the transfer of signaling compounds between neurons in the nervous system, the passage of metabolic by-products and water through the nephrons of the kidneys, and similar actions in many other such systems.

In strictly chemical systems, diffusion is somewhat more straightforward. The process involves a semipermeable membrane, which is a porous membrane in which the pores are large enough to allow the passage of solvent molecules, typically water, but not large enough to allow the passage of dissolved materials, such as ions. Water molecules that pass through the pores are not actively carried through the membrane; rather, they are driven by the molecular motions of diffusion in an overall process called osmosis. Normally, osmosis works in the same direction as the concentration gradient, from higher to lower concentration, and will continue until equilibrium is attained, at which point the concentrations of the two solutions on either side of the semipermeable membrane are equal and the concentration gradient has disappeared. In the process of reverse osmosis, pressure is applied to the solution of higher concentration to overcome the osmotic pressure of the membrane and drive solvent molecules through the membrane against the concentration gradient. Reverse osmosis has numerous applications, including obtaining fresh potable water from salt-laden seawater and concentrating solutions without the application of heat that would alter or destroy compounds of value.

Diffusion of Nuclear Isotopes

Since the kinetic energy of a molecule depends on its mass, diffusion can be used in the separation of isotopes. In the uranium enrichment process, the volatile compound uranium hexafluoride (UF_6) passes through multiple diffusion stages. At each stage, molecules containing the isotopes uranium-234 and uranium-235 pass through slightly more quickly than those containing the heavier isotope uranium-238. After a sufficient number of repetitions of the diffusion process, the concentration of uranium-235 increases relative to that of uranium-238, rendering the uranium more suitable for use as a fuel source for nuclear reactors. It is also the method

SAMPLE PROBLEM

A reverse osmosis apparatus has been set up to extract freshwater from a saline solution. The concentration of sodium chloride (NaCl) on one side of the semipermeable membrane is 4.7 grams per liter, while the water on the other side has no sodium chloride at all. The membrane is 1 millimeter thick. Calculate the concentration gradient of sodium chloride across the membrane.

Answer:

The concentration gradient is described by the relationship $\frac{\Delta c}{\Delta x}$, where Δc is the change in concentration and Δx is the distance over which the concentration has changed.

First, calculate the concentration of NaCl in the saline solution in terms of moles per liter, using the atomic masses of sodium and chlorine:

$$23\frac{\text{g}}{\text{mol}} + 35.5\frac{\text{g}}{\text{mol}} = 58.5\frac{\text{g}}{\text{mol}}$$

Convert 4.7 grams of NaCl into moles:

$$\frac{4.7\text{ g}}{58.5\text{ g/mol}} = 4.7\text{ g} \times \frac{1\text{ mol}}{58.5\text{ g}} = 0.08\text{ mol}$$

The concentration of NaCl in the saline solution is 0.08 moles per liter, or 0.08 M. Subtract the concentration of NaCl in the recovered freshwater (0 M) from the concentration in the saline solution to determine Δc:

$$0.08\text{ M} - 0\text{ M} = 0.08\text{ M}$$

The distance across the membrane is 1 millimeter, or 0.001 meters. Calculate the concentration gradient across the membrane:

$$\frac{\Delta c}{\Delta x} = \frac{0.08\text{ M}}{0.001\text{ m}} = 80\ \frac{\text{M}}{\text{m}}$$

The concentration gradient of NaCl is 80 moles per liter per meter. This means that the sodium chloride concentration would change by 80 moles per liter if the membrane were 1 meter thick.

of choice for generating the more highly enriched weapons-grade uranium.

Richard M. Renneboog, MSc

BIBLIOGRAPHY

Askeland, Donald R., Wendelin J. Wright, D. K. Bhattacharya, and Raj P. Chhabra. *The Science and Engineering of Materials.* Boston: Cengage Learning, 2016. Print.

Bailey, James E., and David F. Ollis. *Biochemical Engineering Fundamentals.* 2nd ed. New York: McGraw, 1988. Print.

Berg, Jeremy M., John L. Tymoczko, Gregory J. Gatto, and Lubert Strye. *Biochemistry.* 8th ed. New York: W. H. Freeman, 2015. Print.

Lodish, Harvey, et al. *Molecular Cell Biology.* 7th ed. New York: Freeman, 2013. Print.

Pelczar, Michael J., Jr., E. C. S. Chan, and Noel R. Krieg. *Microbiology: Concepts and Applications.* New York: McGraw, 1993. Print.

Silbey, Robert J., Robert A. Alberty, and Moungi G. Bawendi. *Physical Chemistry.* 5th ed. Hoboken: Wiley, 2012. Print.

DIGESTION

FIELDS OF STUDY

Biochemistry; Physical Chemistry; Forensic Chemistry

SUMMARY

The processes of digestion are defined and described in their various contexts. Digestion generally indicates that a system is undergoing chemical changes over time with the input of both thermal energy and physical agitation to maintain a constant elevated temperature.

PRINCIPAL TERMS

- **catabolic reaction:** a metabolic reaction in cells that breaks down large molecules into smaller ones, resulting in a release of energy.
- **decomposition reaction:** a chemical reaction in which a single reactant breaks apart to create several products with smaller molecular structures.
- **enzyme:** a protein molecule that acts as a catalyst in biochemical reactions.
- **hydrochloric acid:** a corrosive aqueous solution of hydrogen chloride, present in the common digestive fluid of the stomach.

VISUALIZING DIGESTION

Digestion refers to several similar processes that occur in different practical contexts. In all cases, starting materials are broken down chemically into smaller molecules derived from the component pieces of a larger molecular structure. Digestion can be demonstrated by grinding up various food items and placing the materials in a jar with a solution of hydrochloric acid, a key component of the gastric acid found in the stomach. After some time, the mixture will have become a more or less homogenous liquid. This closely approximates the process of digestion as it occurs in the human stomach.

FORMS OF DIGESTION

Digestion can take a number of forms. Chemists carrying out preparations in laboratories often set a reaction mixture to digest over a heat source such as a steam bath, essentially slow cooking it to promote a desired reaction. This method might be used when a reaction produces a product in high yield that isomerizes slowly into the desired target molecule at an elevated temperature. (Isomerization is when a molecule transforms into another molecule containing the same atoms in a different configuration.) The process by which this isomerization is carried out is a form of digestion.

Physiological digestion is the most familiar form of the process, as it is the process by which living creatures break down foods into their basic chemical components for use in metabolic processes. Digestion can actually begin with the preparation of food, as techniques such as heating cause various physical and chemical changes. For example, heating vegetables produces steam as the liquids within plant cells evaporate, and the resulting pressure ruptures the cell walls, making the vegetables tender. Freezing has the same effect, but the action is caused by ice crystals forming and puncturing the cell walls.

Once food is taken into the mouth, the physiological process of digestion begins. Saliva contains the enzyme amylase, a protein that promotes the breakdown of starches into their component sugar molecules. The action is fairly rapid, and the release of simple sugars in the process explains why certain foods become sweeter as they are chewed. Amylase, like all other enzymes, is a protein molecule whose three-dimensional shape forms an active site into which only certain molecular structures can fit. When that happens, the amount of energy necessary to cause changes to the chemical bonds within the substrate (the molecule acted upon by the enzyme) is significantly reduced, and specific reactions are promoted. In digestion, these are always decomposition reactions that break large food molecules into smaller components such as sugars, amino acids, and fatty acids. Accordingly, digestion is a catabolic reaction process, by which energy is obtained from the chemical breakdown of various compounds.

The process continues when the masticated (chewed) food is swallowed and enters the stomach. There, gastric acid and various enzymes are secreted through the stomach lining while, at the same time, muscles surrounding the stomach rhythmically

DIGESTION

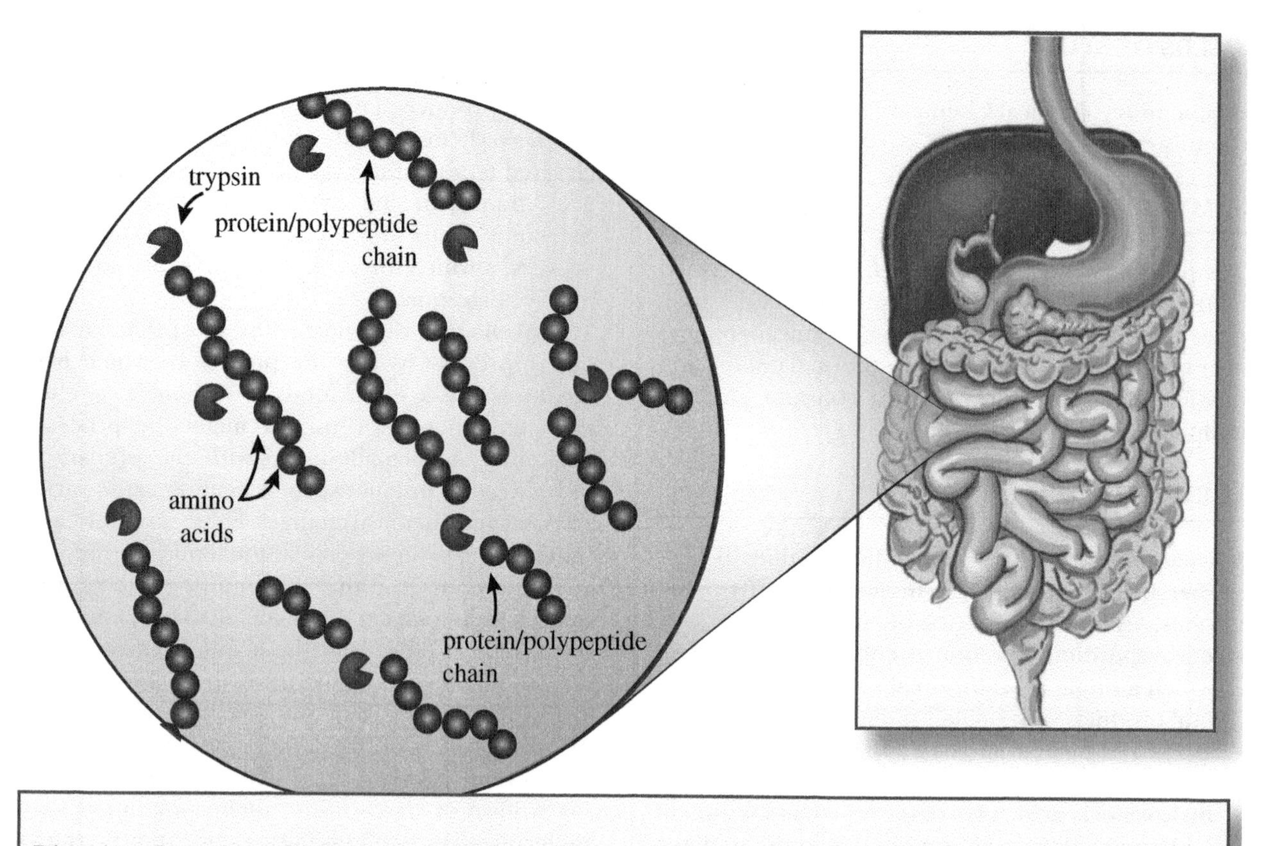

Digestion of protein molecules in the small intestine occurs when trypsin binds to the carbon of a carboxyl group on one amino acid and nitrogen of the next amino acid in a polypeptide chain. Trypsin breaks the polypeptide bond, producing smaller polypeptide chains.

squeeze the stomach and its contents. This churns the mixture in the stomach so that the interaction of the food materials with the digestive materials is maximized. The gastric acid in the stomach acts to break down the peptide bonds in proteins and the ester bonds in fats to release amino acids and long-chain fatty acids, respectively.

Enzyme Activity

Enzyme activity in digestion, as in all other aspects of metabolism, is strictly regulated by the biochemical environment in which it functions. Enzymes specialize in carrying out specific actions. In the enzymatic hydrolysis (chemical breakdown via the addition of water) of proteins and other polypeptides, for example, the enzymes known as trypsin and thrombin function to cleave amide bonds between individual amino acid residues in the molecule. Amino acids are essentially carbon chains with a carboxylic acid group (–COOH) on one end and an amine group ($–NH_2$) on the other. Trypsin will cleave the amide bond on the carboxyl side of arginine (Arg) and lysine (Lys) residues, but thrombin will cleave only the amide bond between arginine and glycine (Gly) residues when they occur in particular amino acid sequences.

SAMPLE PROBLEM

Given a polypeptide segment with the amino acid sequence Arg-Thr-Lys-Arg-Gly-Lys-Gly-Gly-Ser-Lys-Ser, what products will be formed when the material is treated with the enzyme thrombin? With trypsin? With trypsin and then thrombin? Recall that trypsin cleaves the peptide bond on the carboxyl side of arginine and lysine residues, while thrombin cleaves only the peptide bond between arginine and glycine residues. (By convention, polypeptide sequences are written so that the amide group of each amino acid is on its left side and the carboxylic acid group is on the right.)

Answer:

The reaction of the amino acid sequence with thrombin will cleave the peptide bond between arginine and glycine to yield the following fragments:

Arg-Thr-Lys-Arg

Gly-Lys-Gly-Gly-Ser-Lys-Ser

The reaction of the original sequence with trypsin will cleave the peptide bonds between arginine and threonine (Thr), lysine and arginine, arginine and glycine, lysine and glycine, and lysine and serine (Ser) to yield the following fragments:

Arg

Thr-Lys

Arg

Gly-Lys

Gly-Gly-Ser-Lys

Ser

A reaction first with trypsin and then with thrombin will produce the same result, since there is no longer an Arg-Gly linkage in any of the fragments and no further cleavage can take place.

Learning from Digestion

In addition to helping human beings take in the nutrients necessary for survival, the process of physiological digestion can reveal a wealth of information to those who study it. Digestion is a particularly useful source of information in forensic investigations. When a person dies, the physiological process of digestion effectively stops. By examining the stomach contents to determine the extent of digestion that has occurred, a forensic scientist can estimate how much time has elapsed since the person last had something to eat or drink. This information, as well as further analysis of the partially digested materials, can provide valuable clues to the activities of the deceased person prior to his or her demise and allow investigators to isolate the cause of death.

Richard M. Renneboog, MSc

Bibliography

Bailey, James E., and David F. Ollis. *Biochemical Engineering Fundamentals.* 2nd ed. New York: McGraw, 1988. Print.

Berg, Jeremy M., John L. Tymoczko, Gregory J. Gatto, and Lubert Strye. *Biochemistry.* 8th ed. New York: W. H. Freeman, 2015. Print.

Lehninger, Albert L. *Biochemistry: The Molecular Basis of Cell Structure and Function.* 2nd ed. New York: Worth, 1975. Print.

Lodish, Harvey, et al. *Molecular Cell Biology.* 7th ed. New York: Freeman, 2013. Print.

Reece, Jane B., et al. *Campbell Biology.* 10th ed. San Francisco: Cummings, 2013. Print.

DIOLS

FIELDS OF STUDY

Organic Chemistry

SUMMARY

The characteristic properties and reactions of diols are discussed. Diols are a subset of carbohydrate compounds. They have a variety of commercial uses and are also useful in synthesis reactions as starting materials and protecting groups.

PRINCIPAL TERMS

- **functional group:** a specific group of atoms with a characteristic structure and corresponding chemical behavior within a molecule.
- **geminal:** describes the relationship between two functional groups, usually similar or identical, bonded to the same (typically carbon) atom in the same molecule.
- **hydroxyl group:** a primary functional group consisting of an oxygen atom covalently bonded to a single hydrogen atom.
- **polyol:** an organic compound containing multiple hydroxyl groups.
- **vicinal:** describes the relationship between two functional groups bonded to adjacent (typically carbon) atoms in the same molecule.

THE NATURE OF THE DIOLS

The term "diol" indicates that a molecule possesses two hydroxyl groups (–OH) as a primary identifying feature, while "polyol" typically refers to those with three or more. The hydroxyl groups in a diol may be positioned on any of the carbon atoms in the molecule, but they have special relevance when they are on adjacent carbon atoms, as in a vicinal diol (also known as a "glycol"), or on the same carbon atom, as in a geminal diol. The presence of two hydroxyl groups in close proximity brings two electron-rich oxygen atoms near to each other. Because this places strain on the molecule, geminal diols tend to be very unstable and usually decompose to a stable ketone structure by eliminating one of the hydroxyl groups, which bonds with a hydrogen cation (H^+) to form a molecule of water. (A ketone is an organic compound in which the carbon atom of a carbonyl group, C=O, is bonded to two other carbon atoms.) Vicinal diols, on the other hand, tend to be very stable due to the formation of intramolecular hydrogen bonds between the two hydroxyl groups in a five-membered ring formation. Compounds in which the two hydroxyl groups are separated by one carbon atom are called "1,3 diols" and are even more stable, due to both the additional distance between the two oxygen atoms and the formation of intramolecular hydrogen bonds in a six-membered ring formation:

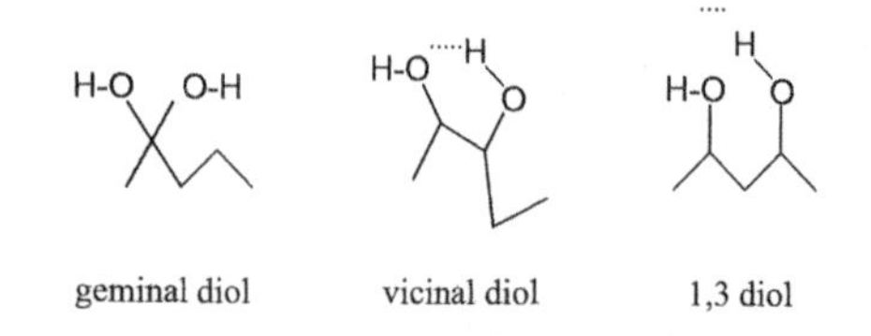

Due to their two hydroxyl groups and their ability to form hydrogen bonds, the smaller diols tend to be soluble in water and other polar solvents. The intramolecular hydrogen bonding presents little or no hindrance to dissolution by water molecules. Solubility does decrease as the size of the alkyl group (C_nH_{2n+1}) increases, however, due to the hydrophobic nature of hydrocarbons.

Diols typically have low melting points. The simplest diol, called ethylene glycol ($C_2H_6O_2$), is commonly used to change the freezing and boiling points of coolant solutions. A mixture of about equal parts of water and ethylene glycol is the standard liquid coolant for most internal combustion engines in cold climates, since the freezing point of such a mixture is about −36°C (−32.8°F) and its boiling point is significantly higher than the boiling point of water alone. Propylene glycol ($C_3H_8O_2$) is also often used in low-temperature applications.

Diols are a sort of intermediate between hydrophilic and hydrophobic compounds. They are often blended into different formulations to improve the solubility of certain materials in aqueous solutions.

NOMENCLATURE OF DIOLS

Diols are named according to the parent hydrocarbon from which they are formed. According to the

International Union of Pure and Applied Chemistry (IUPAC) conventions for systematically naming organic compounds, the basic compound is first identified by the longest carbon chain in the molecular structure. Typically this is an alkane, or saturated hydrocarbon, but there is no stricture against the parent structure being identified as an alkene (unsaturated hydrocarbon with one or more carbon-carbon double bonds, C=C) or an alkyne (unsaturated hydrocarbon with one or more carbon-carbon triple bonds, C≡C). In the case of an alkene or alkyne, the location of the double or triple bond or bonds is indicated after the prefix describing the number of carbon atoms in the chain. For diols, the parent chain must include both alcohol functional groups. Next, the relevant functional groups and their positions in the carbon chain are identified, such that an alcohol group is given the lowest number possible, and the highest-priority functional group determines the final suffix. All other functional groups are listed alphabetically in the formal name of the compound, although certain groups, such as alkyls, are only indicated by prefixes and thus must either come first or follow another prefix. Following these rules, a five-carbon chain with two hydroxyl groups on the first and second carbon atoms, a methyl group ($-CH_3$) on the fourth carbon atom, and a C=C bond between the first and second carbon atoms would be named 4-methylpent-4-en-1,2-diol. The *4-methyl-* prefix describes the location of the methyl group, *pent-* indicates that there are five carbon atoms in the chain, *4-en(e)* describes the location of the C=C bond, and *1,2-diol* describes the locations of the hydroxyl groups.

Some discretion must be exercised in naming diols so as not to confuse naming conventions. The term "glycol" is used exclusively in reference to vicinal diols and always accompanies an alkene name, as in ethylene glycol. Thus, the name hept-3-ene glycol applies to the compound properly named heptan-3,4-diol and not to a seven-carbon alkene with a C=C bond between the third and fourth carbon atoms and two hydroxyl groups elsewhere in the molecular structure. The best practice, at least until one is entirely familiar with naming conventions, is to name diols according to the IUPAC systematic conventions and avoid using the glycol naming method.

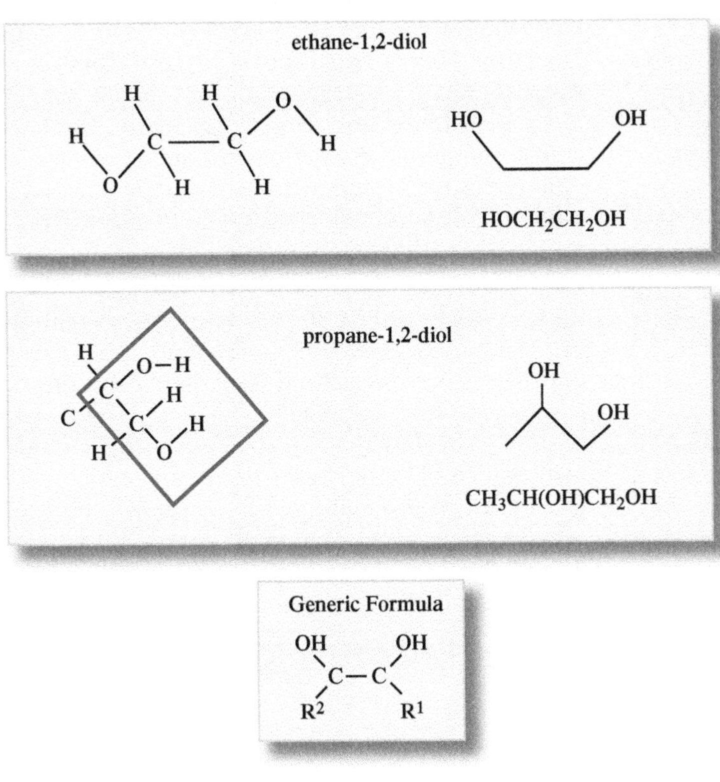

FORMATION OF DIOLS

Ethylene glycol can prepared commercially from ethanol. Dehydration of the ethanol yields ethylene (or ethene) gas, which can undergo a catalyzed reaction with oxygen gas to produce the simple epoxide compound ethylene oxide. (An epoxide is a compound containing a three-membered ring structure made of one oxygen and two carbon atoms.) Hydration of the ethylene oxide then produces ethylene glycol. This is a standard method of producing diols in synthesis.

A compound with a C=C bond can be converted to the corresponding epoxide via reaction with a reagent such as perbenzoic acid or performic acid. These are acids with an extra oxygen atom between the carbonyl group (C=O) and the hydroxyl group. Subsequent hydrolysis of the epoxide produces the corresponding diol. Vicinal diols can similarly be produced by mild oxidation of an alkene C=C bond with potassium permanganate ($KMnO_4$) in alkaline solution. A C=C bond activated by an aryl group, or functional group derived from an aromatic ring, as in the compound styrene (or phenylethylene), can be converted into a vicinal diol by reaction with hydrogen peroxide (H_2O_2) and formic acid (CH_2O_2, also written as HCO_2H). All three mechanisms are shown here:

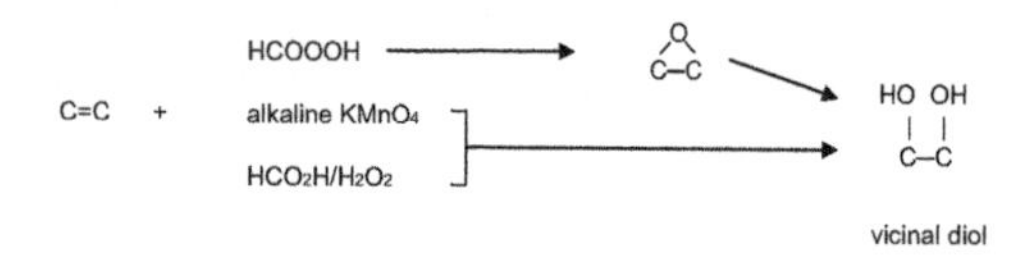

Geminal diols, which dehydrate very readily, are formed by hydration of the carbonyl group in a ketone or aldehyde. They are essentially never isolated because they dehydrate so readily and revert back to the carbonyl group:

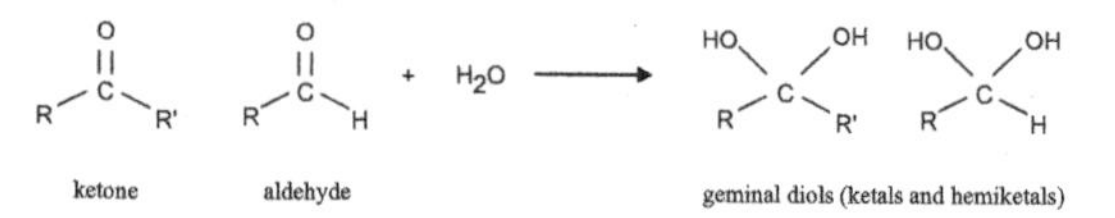

Diols other than vicinal and geminal diols are generally prepared by reducing ketones or aldehydes or by hydrolyzing or substituting appropriate functional groups already present in the molecular structure.

REACTIONS OF DIOLS

Diols are essentially a subset of the carbohydrate class of compounds, and reactions involving diols are generally applicable to the study of carbohydrates. In reactions, diols generally behave either as functional groups to be protected against other reactions or as protecting groups themselves. In a complex synthesis, it is often necessary to protect a carbonyl group against a reaction under conditions designed to alter another part of the molecule. Reducing agents being used to convert a nitrile (organic compound containing a carbon-nitrogen triple bond, or C≡N) into a primary amine, for example, will also reduce the carbonyl group to an alcohol. If the carbonyl group of a ketone or aldehyde reacts with ethylene glycol to form a ketal or acetal first, as shown here, a stable structure is produced that does not react with the reducing agent. After the desired reaction has been carried out, the carbonyl group can be recovered relatively easily by hydrolysis of the ketal or acetal:

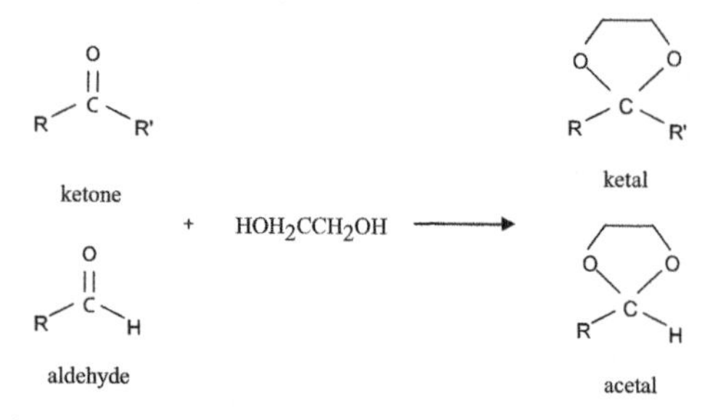

This type of reaction can also be used to produce various substituted heterocyclic ring structures in larger molecules, as well as 1,3-dioxanes. A dioxane is a six-membered ring with two oxygen atoms in its ring structure.

Carbohydrates and closely related compounds contain numerous diol structures, and reaction conditions that affect one hydroxyl group in a carbohydrate molecule will affect all of the hydroxyl groups in the molecule. Thus, in order to alter one part of a carbohydrate molecule without affecting another, it is necessary to protect the hydroxyl groups that are to remain throughout the reaction process. This requires the use of specialized reagents to act as protective groups under the reaction conditions. One such reagent is the dimethyl acetal of 4-methoxybenzaldehyde, which effectively uses the diol to replace the two methyl groups in the acetal. A number of related aryl aldehydes can be used in the same way. Diols and other polyols are also useful in organometallic chemistry as ligands.

Richard M. Renneboog, MSc

SAMPLE PROBLEM

Given the chemical name 2-methylcyclohexan-1,3-diol, draw the skeletal structure of the molecule.

Answer:

The name indicates that the molecule is a six-membered cyclic hydrocarbon, or cyclohexane. There are two hydroxyl groups located on the first and third carbon atoms in the structure and a methyl group on the second. The structure is therefore

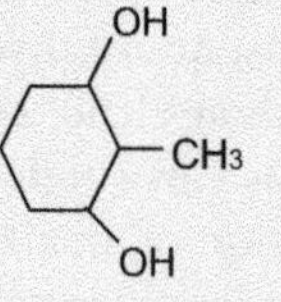

Bibliography

Acheson, R.M. *An Introduction to the Chemistry of Heterocyclic Compounds.* 3rd ed. New York: Wiley, 2008. Print.

Hendrickson, James B., Donald J. Cram, and George S. Hammond *Organic Chemistry.* 3rd ed. New York: McGraw, 1970. Print.

Miessler, Gary L., Paul J. Fischer, and Donald A. Tarr. *Inorganic Chemistry.* 5th ed. Upper Saddle River: Prentice Hall, 2013. Print.

Morrison, Robert Thornton, and Robert Neilson. Boyd. *Organic Chemistry.* 7th ed. Englewood Cliffs, N.J.: Prentice Hall, 2003. Print.

Wuts, Peter G. M., and Theodora W. Greene. *Greene's Protective Groups in Organic Synthesis.* 4th ed. Hoboken: Wiley-Interscience, 2006. Print.

DISPLACEMENT REACTIONS

FIELDS OF STUDY

Organic Chemistry; Inorganic Chemistry; Analytical Chemistry

SUMMARY

Displacement reactions are defined, and their importance in preparative and analytical chemistry is described. Displacement reactions include many types of substitution reactions, such as ionic substitutions, covalent substitutions, and oxidation-reduction (redox) reactions.

PRINCIPAL TERMS

- **anion:** any chemical species bearing a net negative electrical charge, which causes it to be drawn toward the positive pole, or anode, of an electrochemical cell.
- **cation:** any chemical species bearing a net positive electrical charge, which causes it to be drawn toward the negative pole, or cathode, of an electrochemical cell.
- **double displacement:** a substitution reaction in which the atoms of two elements exchange places in their respective compounds.
- **product:** a chemical species that is formed as a result of a chemical reaction.
- **reactant:** a chemical species that takes part in a chemical reaction.
- **single displacement:** a substitution reaction in which atoms of one element replace atoms of another element in a compound.

A Molecular Exchange

The simplest way to recognize a displacement reaction is to look for elements in the molecular formulas of the reactants that have essentially traded places in the products. This is the basic feature of a displacement reaction and is typically readily apparent upon examining a chemical reaction equation. The terms "displacement" and "replacement" are entirely interchangeable in describing such reactions. Another term for replacement is "substitution," though this is often used to refer specifically to the replacement of one functional group with another in an organic compound.

One particularly important type of displacement reaction occurs when one element undergoes oxidation by giving up electrons and another element undergoes reduction by accepting those electrons. This type of replacement reaction is known as a "redox reaction" (from *red*uction and *ox*idation).

The Electrochemical Series

Each element has a characteristic ability to accept or give up electrons, known as its "standard electrode potential." Elements are ranked by this potential in a list called the "electrochemical series," which places elements in descending order of their ability to be oxidized, that is, give up electrons to form a positively charged ion; conversely, they are in ascending order of their ability to be reduced, that is, accept electrons to form a negatively charged ion. In displacement reactions, atoms of an element that is higher in the electrochemical series will replace atoms of an element that is lower in the series and thus has less ability to give up electrons. When two different elements in the electrochemical series are connected to each other through a conductive medium such as an

DISPLACEMENT REACTIONS

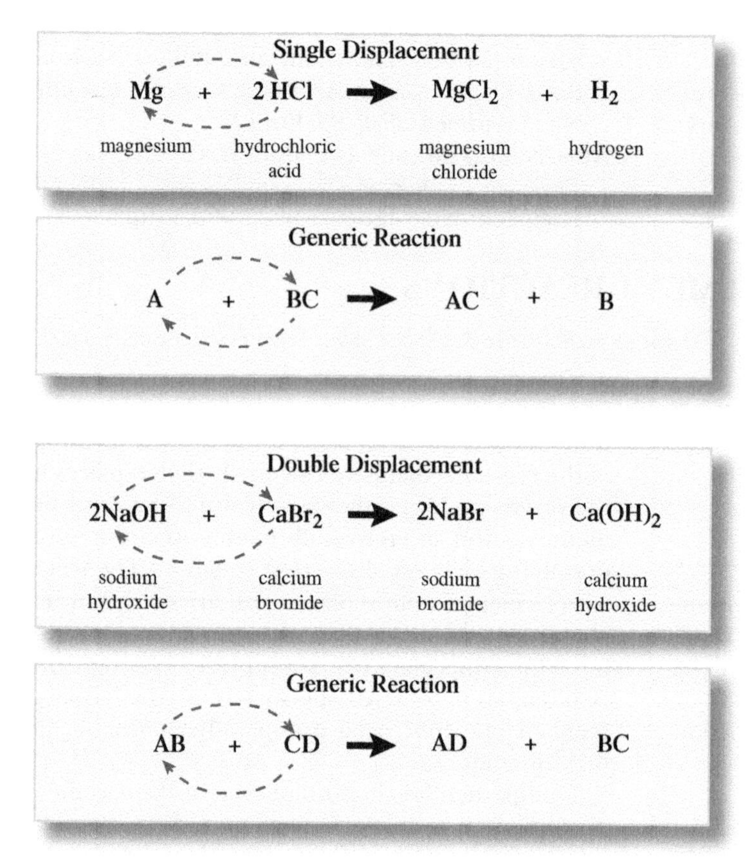

electrolytic solution (a solution containing dissolved ions), electrons will spontaneously flow from the higher-ranked material to the material that is lower in the series and thus has a greater ability to accept electrons. The material from which the electrons flow is called the cathode, and the material that the electrons flow toward is termed the anode.

This process extends to ions in solution, causing negatively charged ions, or anions, to flow toward the anode, while the positively charged ions, or cations, are drawn toward the cathode. The movement of electrons through a conductor is an electrical current and occurs only when there is an electromotive force or voltage differential between the anode and the cathode. Thus, the electrochemical series is the principle behind all batteries and electrochemical cells, as well as the transmission of electrical signals between cells and the extraction of energy from adenosine triphosphate (ATP) in respiration and glycolysis in biochemical systems. All of these processes depend on redox reactions.

SINGLE- AND DOUBLE-DISPLACEMENT REACTIONS

Single-displacement reactions often involve the transformation of a neutral element into cations while cations of the other element are reduced to a neutral form. A simple example is the reaction that occurs when neutral zinc metal is placed into a solution of copper sulfate. The zinc atoms each release two electrons to form Zn^{2+} ions, while the Cu^{2+} ions in the solution each accept two electrons to form neutral copper atoms. Since the neutral copper atoms are not soluble in water, they precipitate out of the solution as copper metal, while the zinc ions are dissolved. The intense blue color of the copper ions in solution gradually disappears as they are replaced by the colorless zinc ions. If the materials are present in the correct proportions, a dark blue solution of copper sulfate with metallic zinc is transformed into a clear, colorless solution of zinc sulfate with metallic copper. The overall reaction equation for the process is as follows:

$$CuSO_4 + Zn \rightarrow ZnSO_4 + Cu$$

The equivalent ionic form is written this way:

$$Cu^{2+} + Zn \rightarrow Cu + Zn^{2+}$$

The replacement is not necessarily a one-to-one replacement in all cases, but it is always in accordance with the number of electrons being transferred.

In a double-displacement reaction, the elements replace each other in their respective compound formulas. For example, in the reaction between chromium sulfate and potassium hydroxide, the products are chromium hydroxide and potassium sulfate, according to the reaction equation:

$$Cr_2(SO_4)_3 + 6KOH \rightarrow 2Cr(OH)_3 + 3K_2SO_4$$

In this equation, the chromium ions and potassium ions replace each other in their respective compounds. Another way of describing the overall result

SAMPLE PROBLEM

The double-displacement reaction between halide ions in solution and silver nitrate can be used to determine the amount of silver that is present in a solution. When a solution of sodium chloride is added to a 25-milliliter sample of a silver-containing solution obtained by stripping the silver emulsion from old photographic film, a mass of 360 milligrams of silver chloride is obtained after recovery and drying. How much silver is contained in 750 milliliters of the solution?

Answer:

The reaction taking place is

$$AgNO_3 + NaCl \rightarrow AgCl + NaNO_3$$

The molecular mass of silver chloride, AgCl, is obtained by adding the atomic masses of silver and chlorine. Atomic mass is measured in unified atomic mass units (u), which are equal to grams per mole.

$$107.87 \text{ g/mol} + 35.45 \text{ g/mol} = 143.32 \text{ g/mol}$$

Convert the mass of the AgCl recovered from milligrams to grams:

$$360 \text{ mg} \div 1000 = 0.36 \text{ g}$$

Divide the mass of the AgCl recovered by the molecular mass of AgCl:

$$0.36 \text{ g} \div 143.32 \text{ g/mol} = 0.0025 \text{ mol}$$

Since the displacement reaction is in a one-to-one mole ratio, there is 0.0025 mole of silver in 25 milliliters of silver chloride. Determine how many times greater 750 is than 25, multiply that by 0.0025, and convert the result back to grams, using the atomic mass of silver:

$$750 \div 25 \text{ ml} = 30$$

$$30 \times 0.0025 \text{ mol} = 0.075 \text{ mol}$$

$$0.075 \text{ mol} \times 107.87 \text{ g/mol} = 8.09 \text{ g}$$

There are 8.09 grams of silver in 750 milliliters of the solution.

would be to say that the sulfate and hydroxide ions have switched partners.

In organometallic compounds, ligands are ions or molecules that are coordinated in a geometric arrangement around central metal atoms. They may be ionic species, or they may be neutral molecules with appropriate electron pairs that can interact with vacant atomic orbitals on the metal atom. Ligands can be replaced by other species that have a better ability to coordinate to the central metal atoms. An example of this occurs when four of the six molecules of H_2O in the complex ion $[Cu(H_2O)_6]^{2+}$ are replaced by neutral ammonia molecules to form the ion $[Cu(NH_3)_4(H_2O)_2]^{2+}$, according to the reaction equation:

$$[Cu(H_2O)_6]^{2+} + 4NH_3 \rightarrow [Cu(NH_3)_4(H_2O)_2]^{2+} + 4H_2O$$

Double-displacement reactions are not normally part of organic reaction mechanisms in any but a stoichiometric sense, which requires all atoms that were present at the start of a reaction to be present after the reaction has occurred. Typically, only the desired product of an organic reaction mechanism is of value.

Substitution in Organic Compounds

In organic compounds, one common form of substitution is nucleophilic substitution, which can proceed by one of two mechanisms. In the S_N1 mechanism, an intermediate carbonium ion is formed by the loss of an appropriate anion from the parent compound. Addition of a nucleophilic species to the carbonium ion generates a new compound in which the departing anion has been replaced. In the S_N2 mechanism, the replacement operation is more direct. A nucleophilic species forms a bond to a carbon atom from one side, while the bond to a less potent nucleophile on the opposite side is weakened and lost. The net result is that the original atom or molecule attached to the carbon atom is replaced by a new atom or molecule. A simple example is the formation of pentanol from the reaction of 1-bromopentane and a hydroxide ion, according to the reaction equation:

$$Br{-}CH_2CH_2CH_2CH_2CH_3 + OH^- \rightarrow HO{-}CH_2CH_2CH_2CH_2CH_3 + Br^-$$

Displacement reactions and substitution reactions are extremely useful in both the preparation of specific compounds and the analysis of materials. Both require understanding the specific reactions that take place and the product that is produced from different reactants under specific reaction conditions.

Richard M. Renneboog, MSc

BIBLIOGRAPHY

Berg, Jeremy M., John L. Tymoczko, Gregory J. Gatto, and Lubert Strye. *Biochemistry.* 8th ed. New York: W. H. Freeman, 2015. Print.

Douglas, Bodie E., Darl H. McDaniel, and John J. Alexander. *Concepts and Models of Inorganic Chemistry.* 3rd ed. New York: Wiley, 1994. Print.

Lew, Kristi. *Chemical Reactions.* New York: Infobase, 2008. Print.

Miessler, Gary L., Paul J. Fischer, and Donald A. Tarr. *Inorganic Chemistry.* 5th ed. Upper Saddle River: Prentice Hall, 2013. Print.

Morrison, Robert Thornton, and Robert Neilson. Boyd. *Organic Chemistry.* 7th ed. Englewood Cliffs, N.J.: Prentice Hall, 2003. Print.

Zumdahl, Steven S., and Susan Zumdahl. *Chemistry.* 9th ed. Belmont: Brooks Coles, 2013. Print.

DNA COMPUTING

FIELDS OF STUDY

Genetics; Molecular Biology

SUMMARY

DNA computing involves performing computations using biological molecules as opposed to traditional digital computer circuitry. It combines ideas from many different disciplines, such as biology, chemistry, and computer science to find solutions to problems by using the genetic molecular material.

PRINCIPAL TERMS

- **algorithm:** a process or set of rules to be followed in calculations
- **DNA base pairing:** two strands of DNA are held together in a double helix shape by weak chemical bonds between base pairs; these base pairs always form the same way, with adenine always forming a base pair with thymine and cytosine always forming a base pair with guanine, following DNA base pair rules.
- **DNA fingerprinting:** a test to identify and evaluate the genetic material in one's cells performed by extracting DNA from a small sample of cells, then amplifying it; called fingerprinting because it is very unlikely that two individuals would have the same genetic material, much as it is unlikely that two individuals have the same fingerprints.
- **electrophoresis:** a lab technique used to identify, quantify, and purify fragments of DNA; samples are placed in agarose gel subjected to an electric field, causing the negatively charged nucleic acids to move toward a positive electrode. Shorter fragments travel furthest, leaving longer fragments behind, resulting in a separation according to length.
- **encoding:** in a DNA sequence, a gene that is responsible for producing a substance or behavior
- **Hamiltonian path:** a path that visits each vertex of a directed graph exactly once without backtracking.
- **ligase:** an enzyme that causes linkage between two molecules of DNA or another substance
- **nucleotide:** the basic structural component of DNA and RNA, consisting of a ribose (in RNA) or deoxyribose (in DNA) sugar molecule bonded to a phosphate group and one of five nucleobases: cytosine, adenine, guanine, thymine (DNA only), or uracil (RNA only).
- **oligomer:** a molecular complex consisting of a few monomers, usually less than 30, as opposed to a polymer which has, in theory, an unlimited number of monomers strung together.
- **polymerase:** any of a variety of enzymes that help synthesize DNA or RNA by using an existing strand as a template.
- **reptation:** the creeping motion that describes how DNA fragments move through agarose gel when they are being separated according to length.
- **thermal cycler:** the instrument that is used to conduct the polymerase chain reaction using

precise temperature control and rapid temperature changes.

As far back at 1959, Richard Feynman presented the idea that individual molecules could be used to perform computations. But it wasn't until a groundbreaking experiment was published in 1994 that Leonard Adleman showed that computations could be carried out at the molecular level. His experiment used the tools of molecular biology to solve an instance of the directed Hamiltonian Path problem. In this experiment, the allowed paths from one vertex of a graph to another were encoded by DNA oligomers, and the computation was performed with standard molecular biology techniques.

Unfortunately, biological processes may involve error, so DNA computing is not completely foolproof, though DNA's ability to only combine according to DNA base pairing rules and the presence of enzymes that repair DNA errors helps reduce errors. DNA is also fragile and sensitive to temperature and UV radiation, requiring special pure reagents that are kept sterile at precise temperatures. Sorting through unwanted results takes time, effort, and resources. But the promise of DNA computing is exciting: As Adleman wrote in his original paper, "the potential of molecular computation is impressive."

"Gel electrophoresis apparatus" by Jeffrey M. Vinocur via Wikimedia Commons

Using DNA in a Computation

If one thinks of a computing as executing an algorithm, then computing is simply a list of step-by-step instructions that takes an input and processes that input according to a certain set of rules to arrive at a result. In a traditional computing situation, binary codes, a series of 0s and 1s, is used to perform the action. In DNA computing, the genetic alphabet, composed of the four nucleotides adenine (A), guanine (G), cytosine (C), and thymine (T), is used to perform the action. This is possible because DNA, in short sequences, may be synthesized as desired.

Adleman's Experiment

Adleman used DNA computing to solve a Hamiltonian Path problem, which involves finding a route through a network that must start and end at a specific point and visit each vertex only once. This is sometimes called "the traveling salesman problem" because it is based on the idea that a salesman would like to use the shortest route to travel through a certain set of towns without backtracking but ensuring a visit to each one. This can be a relatively simple problem when the number of towns is limited but becomes more and more difficult as more and more towns are added. Eventually, if enough towns are added, even with a computer, the problem generates too many possible paths to solve it easily.

Rather simplistically, Adleman's experiment followed this sequence:

- He set up the experiment using "towns" labeled 1 through 7 to represent the vertexes and "one-way roads" to represent the path that would have to be followed to start and end at the same "town" and visit each "town" only once on the "one-way roads."

- Using DNA base pairing rules and a ligase agent, he designed short strands of DNA, or oligomers, to represent the towns and roads, sticking the towns and roads together to represent possible routes
- Using polymerase chain reaction (PCR), he amplified only those DNA sequences that started and ended at the correct towns.
- Using a thermal cycler and the principles of gel electrophoresis, he then sorted the strands of DNA by length, so that only the strands that were the correct length remained.
- Then he used affinity purification and DNA fingerprinting to make sure that each town was represented in the sequences that might contain the answer.
- He was left with strands of DNA that might contain the answer to the problem which were then sequenced to find the solution.
- For a more detailed explanation of Adleman's experiment, please view his original paper. There was nothing special about the problem itself, and it actually took Adleman seven days to solve the problem when it would have taken an average person a few minutes to find the solution with the limited number of "towns" used. This experiment was so significant because Adleman performed small-scale computations only using DNA material, opening up the possibility that biochemical reactions could be programmed and showing that DNA could be a useful computational tool. This led to the field of DNA computing, where scientists began to ask, "What else can we do with DNA?" This field is very interdisciplinary, with biologist, chemists, computer scientists, and mathematicians working together to solve problems that are not necessarily biological.

Applications

DNA computing allows geneticists to do many things with strands of DNA, opening up new venues for DNA manipulation. Some examples follow:

- Separating long and short DNA sequences to isolate the sequence that is to be studied
- Chopping strands that contain a certain sequence out of the DNA molecule to study whether that particular sequence causes a certain type of disease
- Building new types of biochemical systems that can possibly communicate, sense surroundings, or even act on decisions
- Using DNA to control or manipulate other molecules
- Engineering tissues, for example, to generate new skin tissue for burn victims
- Constructing molecules, such as enzymes and RNA sequences, to use in biomolecular engineering
- Using DNA as a high-density storage medium for data because it has four bases as opposed to just two (0s and 1s)
- Encoding genes to behave in a particular way
- Solving problems in game strategy or encryption breaking that were previously considered to be impossible, or at least extremely difficult.

Marianne M. Madsen, MS

Bibliography

Adleman, Leonard M. *Science*, New Series, Vol. 266, No. 5187 (Nov. 11, 1994), 1021-1024.

Amos, Martyn. *Theoretical and Experimental DNA Computation.* Natural Computing Series. Springer (2010)

Hamid, Arabnia R. and Quoc Nam Tran. *Emerging Trends in Computational Biology, Bioinformatics, and Systems Biology: Algorithms and Software Tools.* Emerging Trends in Computer Science and Applied Computing. Morgan Kaufmann (2015).

Ignatova, Zoya, and Israel Martinez-Perez. *DNA Computing Models.* Springer (2010).

Murata, Satoshi, and Satoshi Kobayashi (eds). *DNA Computing and Molecular Programming: 20th International Conference, Kyoto, Japan, September 22-26, 2014.* Springer (2014).

Paun, Gheorghe, Grzegorz Rozenber, and Arto Salmaa. *DNA Computer: New Computing Paradigms.* Texts in Theoretical Computer Science. Springer (2006).

Shasha, Dennis E., and Cathy Lazere. *Natural Computing: DNA, Quantum Bits, and the Future of Smart Machines.* W.W. Norton & Company (2010).

Wang, Zhaocai, Jian Tan, Dongmei Huang, Yingchao Ren, and Zuwen Ji. "A Biological Algorithm to Solve the Assignment Problem Based on DNA Molecules Computation." *Applied Mathematics and Computation.* Volume 244 (2014), pp. 183-190.

DNA/RNA SYNTHESIS

FIELDS OF STUDY

Biochemistry; Genetics; Molecular Biology

SUMMARY

The basic process of DNA and RNA synthesis is described, and its importance in living biochemical systems is discussed. Also described are modern advances such as artificial methods for the synthesis of DNA and RNA and their applications.

PRINCIPAL TERMS

- **complementary strand:** one of the two strands of nucleotides that make up a DNA molecule, with each nucleotide in one strand corresponding to the position of its complementary nucleotide (cytosine for guanine, adenine for thymine, and vice versa) in the other.
- **deoxyribonucleic acid (DNA):** a large molecule formed by two complementary strands of nucleotides that encodes the genetic information of all living organisms.
- **gene expression:** the process by which RNA copies genes, which are specific segments of the DNA molecule, and uses the information to synthesize either proteins or other types of RNA.
- **nucleotide:** the basic structural component of DNA and RNA, consisting of a ribose (in RNA) or deoxyribose (in DNA) sugar molecule bonded to a phosphate group and one of five nucleobases: cytosine, adenine, guanine, thymine (DNA only), or uracil (RNA only).
- **polymerase chain reaction:** a laboratory method in which a very small amount of DNA can be replicated thousands or even millions of times, using free nucleotides and an enzyme called DNA polymerase.
- **ribonucleic acid (RNA):** a category of large molecules, typically consisting of a single strand of nucleotides, that perform various functions in cells, including the transcription of DNA molecules and the transfer of specific genetic information for protein synthesis.

Understanding DNA and RNA Synthesis

It is tempting to oversimplify the formation of deoxyribonucleic acid (DNA) by likening it to zipping up a zipper, but that is perhaps the easiest way to visualize the process. Indeed, DNA synthesis is diagrammed in virtually all biochemistry texts as such. The analogy is even further simplified by associating the "teeth" of the zipper with the purine and pyrimidine nucleotides from which the molecular structures of DNA and ribonucleic acid (RNA) are formed.

A nucleotide is formed when a purine or pyrimidine base and a phosphate group are chemically bonded to a sugar molecule. Only five different purine and pyrimidine bases are utilized in constructing DNA or RNA nucleotides. In DNA nucleotides, these are the bases adenine, cytosine, guanine, and thymine, while RNA uses the base uracil instead of thymine in its nucleotides. The different bases are indicated by the first letter of their names: A, C, T, G, and U. There are different mnemonic devices for recalling the complementarity of the different bases. One is that the curved letters C and G go together, as do the pointed letters A and T. Another is an easily remembered phrase such as "Cary Grant Ate Tacos." In fact, any number of such devices can be used to suit an individual's personal preference.

The second major difference between DNA and RNA is the nature of the sugars with which the nucleotides are constructed. In RNA nucleotides, the sugar molecule is ribose, a five-carbon simple sugar related to fructose. Sugar molecules are carbohydrates, indicating that each carbon atom in the molecule is chemically bonded to both an H atom and an –OH group. These are the components of the water molecule, so the term indicates that each carbon (*carbo-*) atom in the molecule is hydrated (*-hydrate*). In DNA nucleotides, however, the sugar is deoxyribose. The name indicates that the molecule lacks an oxygen atom that is part of the ribose sugar molecular structure. This difference in the structure of the sugar portion of the nucleotide is subtle, but of essential importance because it alone determines whether the molecule, and its role in the biochemical process of life, is DNA or RNA. In both DNA and RNA, the sugar molecules are in the form of a five-member ring structure made up of one oxygen atom and four of the five carbon atoms.

The third component of DNA/RNA nucleotides is the phosphate group, PO_4^{3-}, or Pi, generally referred to as "inorganic phosphate" when in that form and "phosphate" when bonded to another biomolecule, such as adenosine in adenosine triphosphate (ATP). The bonds between an oxygen atom in the phosphate group and other molecules are stable, even though they are deemed "high energy" bonds. In respiration and glycolysis (the decomposition of glucose), the bond between the third and second phosphate group in the triphosphate component is utilized both to store energy by its formation and to release energy when that bond is cleaved.

Both DNA and RNA are large molecules. Their respective molecular structures consist of a long, biopolymer chain of alternating sugar molecules and phosphate groups. In both the DNA and RNA structures, each sugar component has a purine or pyrimidine base molecule bonded to the carbon atom on one side of the ring oxygen atom, and the phosphate group bonded to a carbon atom on the other side of the ring oxygen atom. It is at this point that the difference between ribose and deoxyribose sugar becomes vitally important. The RNA molecule consists of a single strand made up of adenine, cytosine, guanine, and uracil nucleotides. The DNA molecule, however, consists of two complementary strands of adenine, cytosine, guanine, and thymine nucleotides. These match up with each other as the base portions of the nucleotides in one strand and connect to the corresponding nucleotides in the other strand. The end result is that a DNA molecule is considerably bigger than an RNA molecule, and it has the form of a double helix as the two component strands coil around each other. The RNA molecule consists of a single nucleotide strand that assumes different shapes according to its role in transcription and gene expression.

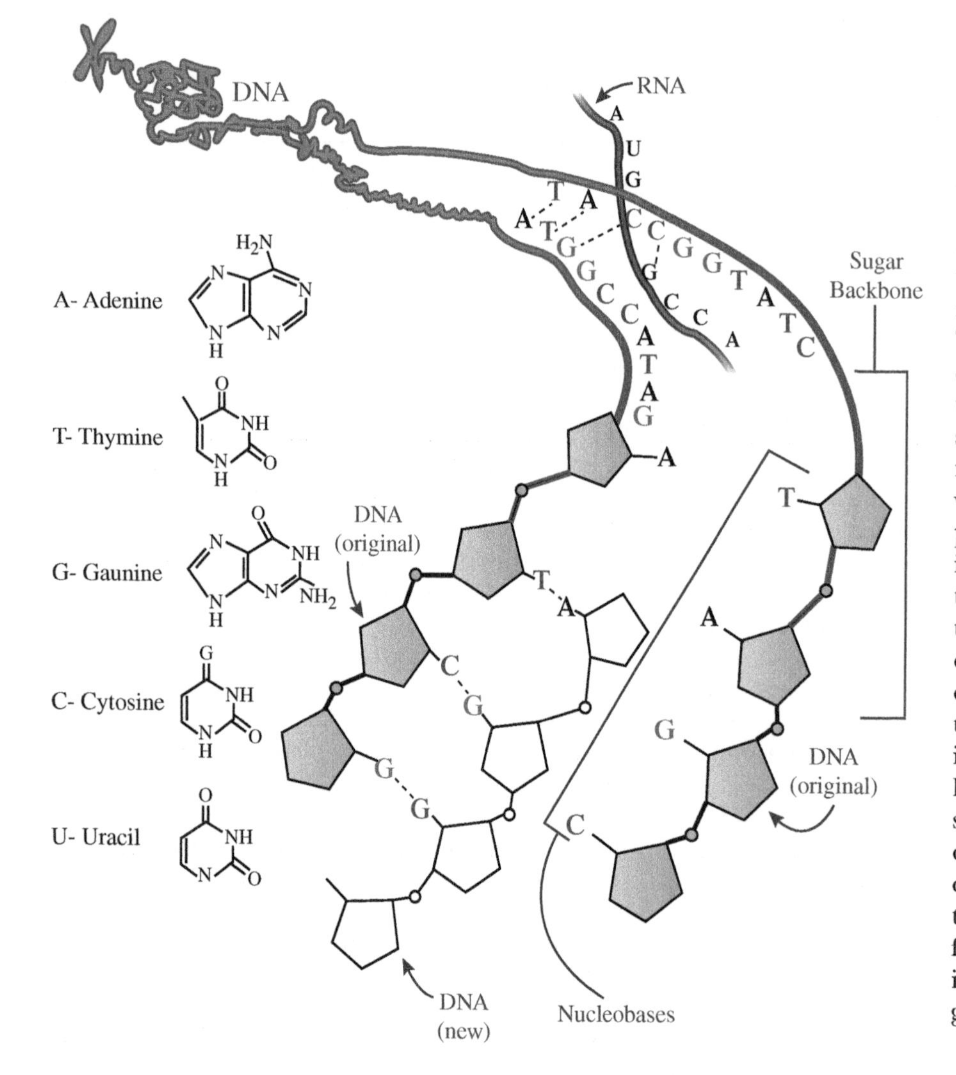

Discovery and Analysis of DNA and RNA Molecules

DNA was isolated from cell nuclei as early as 1869, but the fact that it bears genetic information was not known until 1943, when Oswald Avery, Colin MacLeod, and Maclyn McCarty demonstrated that introducing DNA from a virulent strain of the *pneumococcus* bacterium into a nonvirulent strain could produce the virulent strain. In 1953, James D. Watson and Francis Crick, with significant assistance from their laboratory manager Maurice Wilkins and using the analytical and theoretical work of Rosalind Franklin, first published the discovery that the DNA molecule has a double helix form, an image now so well recognized. Since then, methods and techniques for the manipulation and analysis of DNA have advanced, permitting biochemists and geneticists to obtain a better understanding of the role of genes and chromosomes in the DNA molecule. In 2001, the journal *Nature* published the first complete analysis of the human genome, which demonstrated, among other things, that all humans alive today are descended from a very few human populations that originated in Africa in the distant past. As the genome was deciphered over time, an understanding of the mechanisms by which DNA and RNA are synthesized was also obtained.

Formation of DNA and RNA in Living Cells: Four Basic Rules

Both DNA and RNA are produced by copying a preexisting DNA strand according to the base pairings of adenine to thymine and cytosine to guanine. In order to carry out this process, the DNA molecule must first "unzip," allowing nucleotide fragments to form a complementary RNA strand in which uracil nucleotides replace thymine nucleotides. From this complementary RNA strand, a duplicate of the original DNA strand is assembled from other fragments. This process can be thought of as making a mold from one half of the DNA molecule and then using it to cast a copy of the original.

Second, both RNA and DNA strands grow in one direction only. The phosphate group of each nucleotide is situated at the 5′ position of the sugar molecule, the number indicating a specific location in the molecular structure according to the conventions for naming organic molecules. At the 3′ position of each sugar molecule, there is a free hydroxyl (–OH) substituent that can form the phosphate ester bond with another nucleotide. Formation of a DNA or an RNA strand always proceeds from the 5′-position in one nucleotide to the 3′-position in the next nucleotide as nucleotides are added in sequence.

Third, both DNA and RNA are synthesized by very specific enzymes in polymerase chain reactions. New strands of RNA are produced only by RNA polymerases, and new DNA strands are produced only by DNA polymerases in DNA/RNA polymerase reactions. Through the process of transcription, a new RNA strand is produced as RNA polymerases transcribe the nucleotide pattern of the parent DNA strand. RNA polymerase enzymes are able to initiate the formation of a new strand by coordinating to an appropriate site on a duplex strand of DNA (the double helix form of the molecule), where they temporarily separate the two strands of the DNA molecule and begin the process of assembling a new RNA strand from the corresponding nucleotides. DNA polymerases are not able to initiate the formation of a new strand directly. Instead, the process requires the formation of a primer, a DNA or an RNA segment that is bound to the parent DNA strand and is acting as the template for the new DNA strand. Both RNA and DNA polymerases comprise several different proteins, each of which carries out a specific function or transformation.

Fourth, synthesis of a new duplex DNA strand proceeds only from a particular formation known as a "growing fork." Specific enzymes function to open the duplex strand, allowing other enzymes to assemble the matching complementary strands from the appropriate nucleotides. As the new strands are formed, still other enzymes function to rejoin the strands as the growing fork progresses along the length of the template duplex DNA molecule. An important aspect of duplex DNA replication is that, because the strands only grow in one direction, the directions of growth on the two branches of the growing fork are opposite to each other. The new strand on the "leading" branch of the fork grows continuously, nucleotide by nucleotide. The new strand on the "trailing" branch of the fork is assembled instead in bits and pieces from various nucleotide segments.

SAMPLE PROBLEM

In 2005, a fossilized leg bone of *Tyrannosaurus rex* was found to contain viable tissue. Suppose a fragment of a DNA strand were to be recovered from such material and found to have the nucleotide order AGTTCGCGGAAC-TATTCG. What is the nucleotide order of the complementary strand in duplex DNA? What is its RNA complement?

Answer:

Recall the mnemonic device "Cary Grant Ate Tacos" (or whatever mnemonic you wish to use to signify that C/G and A/T are complementary nucleotides):

The order of the "found" DNA fragment is

AGTTCGCGGAACTATTCG

Place the complementary nucleotide below each one in the series, as

AGTTCGCGGAACTATTCG

TCAAGCGCCTTGATAAGC

This is the complementary sequence that would exist in a duplex DNA molecule.

To generate the RNA complement, recall that RNA uses uracil (U) instead of thymine (T). The RNA complement is then easily defined by substituting U for T in the DNA complement, yielding

UCAAGCGCCUUGAUAAGC

(Note that listing the full sequence of a human DNA molecule in this form would require a book of approximately one million closely printed pages. It is rather unlikely that *T. rex* could be cloned from the above fragment.)

AMPLIFYING DNA FOR ANALYSIS

Among the many techniques and methods that have developed for the manipulation of DNA samples—and that are especially important for the science of DNA "fingerprinting"—probably the most important is DNA amplification. By this method, an extremely minute sample of DNA, as might be obtained from just a few hair follicles found at a crime scene, for example, undergoes repeated replications so that enough of the DNA is present to produce a clear fragmentation pattern that is the "fingerprint" of that particular DNA. This methodology has been used to convict criminals who might otherwise have gone free, as well as to free individuals from prison who had been wrongfully convicted of crimes they did not commit.

Richard M. Renneboog, MSc

BIBLIOGRAPHY

Berg, Jeremy M., John L. Tymoczko, Gregory J. Gatto, and Lubert Strye. *Biochemistry.* 8th ed. New York: W. H. Freeman, 2015. Print.

"The Human Genome." *Nature* 409.6822 (2001): 813–958. Print.

Lafferty, Peter, and Julian Rowe, eds. *The Hutchinson Dictionary of Science.* 2nd ed. Oxford: Helicon, 1998. Print.

Lehninger, Albert L. *Biochemistry: The Molecular Basis of Cell Structure and Function.* 2nd ed. New York: Worth, 1975. Print.

Lodish, Harvey, et al. *Molecular Cell Biology.* 7th ed. New York: Freeman, 2013. Print.

Pelczar, Michael J., Jr., E. C. S. Chan, and Noel R. Krieg. *Microbiology: Concepts and Applications.* New York: McGraw, 1993. Print.

Thro, Ellen. *Genetic Engineering: Shaping the Material of Life.* New York: Facts On File, 1995. Print.

DNA/RNA TRANSCRIPTION

FIELDS OF STUDY

Biochemistry; Genetics; Molecular Biology

SUMMARY

The process of DNA/RNA transcription is defined, and its importance in the biochemistry of living systems is discussed. The process of transcription is essential in living systems for the synthesis of proteins in eukaryotic cells, as well as a variety of other functions that require the extraction of genetic information from DNA.

PRINCIPAL TERMS

- **complementary strand:** one of the two strands of nucleotides that make up a DNA molecule, with each nucleotide in one strand corresponding to the position of its complementary nucleotide (cytosine for guanine, adenine for thymine, and vice versa) in the other.
- **enzyme:** a protein molecule that acts as a catalyst in biochemical reactions.
- **gene expression:** the process by which RNA copies genes, which are specific segments of the DNA molecule, and uses the information to synthesize either proteins or other types of RNA.
- **RNA polymerase:** the enzyme responsible for initiating gene transcription in order to assemble and replicate strands of RNA.
- **transcription factor:** a protein that binds to DNA in order to initiate, regulate, or block gene transcription.

The Structure of DNA versus RNA

The molecular structure of DNA is often likened to a zipper, with the two halves of the molecule matching up in a way that resembles the two halves of a zipper fitting together. However, the actual structure is much more complicated. A DNA molecule contains two complementary strands made up of specific combinations of nucleotides. Each nucleotide in a strand of DNA is composed of a molecule of the sugar deoxyribose bonded to a phosphate group and a nitrogenous base. Only four bases are used in DNA molecules: adenine, thymine, guanine, and cytosine.

Structurally, a DNA strand consists of a very long chain of alternating deoxyribose-and-phosphate units, forming what can be considered the backbone of the molecule, with the various bases appended to the deoxyribose. The complementary strand of a molecule of duplex DNA has the same basic structure but with a different sequence of bases. In duplex DNA, the bases form specific pairs, with adenine complementary to thymine and guanine complementary to cytosine.

The structure of the RNA molecule is very similar to that of a single DNA strand. There are two essential differences between them, however: the sugar component of the RNA molecule is the sugar ribose, not deoxyribose, and the base uracil is used in place of thymine. The difference in the sugar component is what allows DNA to form the familiar double-helix structure, which RNA cannot do.

The DNA molecule carries the genetic information that defines the identity of biological organisms. The sequence of nucleotides in each DNA strand specifies the order in which amino acids are to be assembled into proteins, the basic components of all known life. Each cell in an organism must have a DNA molecule in its nucleus—or, in the case of a prokaryote, its intracellular fluid—in order to produce the proteins and other compounds that are essential to its existence. The mechanism by which the instructions for protein assembly are translated and put into action is the DNA transcription process, which is the fundamental first step in the process of gene expression. Transcription can be described in simplified terms as RNA making a mold of the nucleotide sequence found in the parent DNA molecule and using it to synthesize new proteins.

The Transcription Process

Transcription is the copying of the nucleotide pattern in a strand of DNA by an enzyme called RNA polymerase. An enzyme is a protein that carries out a specific chemical function, which is determined by the relative locations of various atoms and functional groups within its three-dimensional structure. The chemical names of virtually all enzymes have the suffix *-ase*, as in lipase, transcriptase, and polymerase.

Others that were first named as proteins rather than enzymes have names that end with *-in*, such as trypsin and pepsin. The various enzymes that participate in the transcription process, other than RNA polymerase, are called transcription factors.

With very few exceptions, the genetic information of the duplex DNA strand is transcribed from only one of the two strands. Due to the complementary nature of the two strands, the nucleotide sequence in the new RNA strand is identical to the DNA strand that was not transcribed, save for the substitution of uracil for thymine, and both are the reverse of the DNA strand that served as the template. Because of this, the nontemplate strand is alternately called the "coding strand" or the "sense strand," while the template strand from which the RNA is assembled is called the "noncoding strand" or the "antisense strand."

Transcription from DNA to RNA involves a number of types of RNA. Messenger RNA (mRNA) copies genetic information from a DNA molecule to replicate the nucleotide sequence in the coding strand. Transfer RNA (tRNA) carries the amino acids specified by the nucleotide sequence to the growing end of a polypeptide chain. Ribosomal RNA (rRNA) forms the ribosome, which is where polypeptide assembly takes place.

In a eukaryote, transcription begins when an RNA polymerase attaches to an appropriate location on the duplex DNA strand and helicase enzymes temporarily separate the two strands of the DNA molecule at that location. The RNA polymerase begins to assemble nucleotides and attach them to the noncoding strand in the appropriate sequence, building up a hybrid RNA-DNA duplex strand. Due to the structural differences between the ribose and deoxyribose sugars and the complementary pairing of adenine with uracil instead of thymine, this hybrid duplex strand is not stable, so when assembly is complete, the RNA strand separates as mRNA. The mRNA strand then moves to the ribosomes formed by rRNA, where tRNA units carrying different amino acids are matched to the mRNA strand in the order specified. The amino acids are joined to one another with peptide bonds, forming the primary structure of the particular protein that has been encoded. The overall process of synthesizing proteins from genetic information contained in DNA is called "translation," with transcription being the initial step in the overall process.

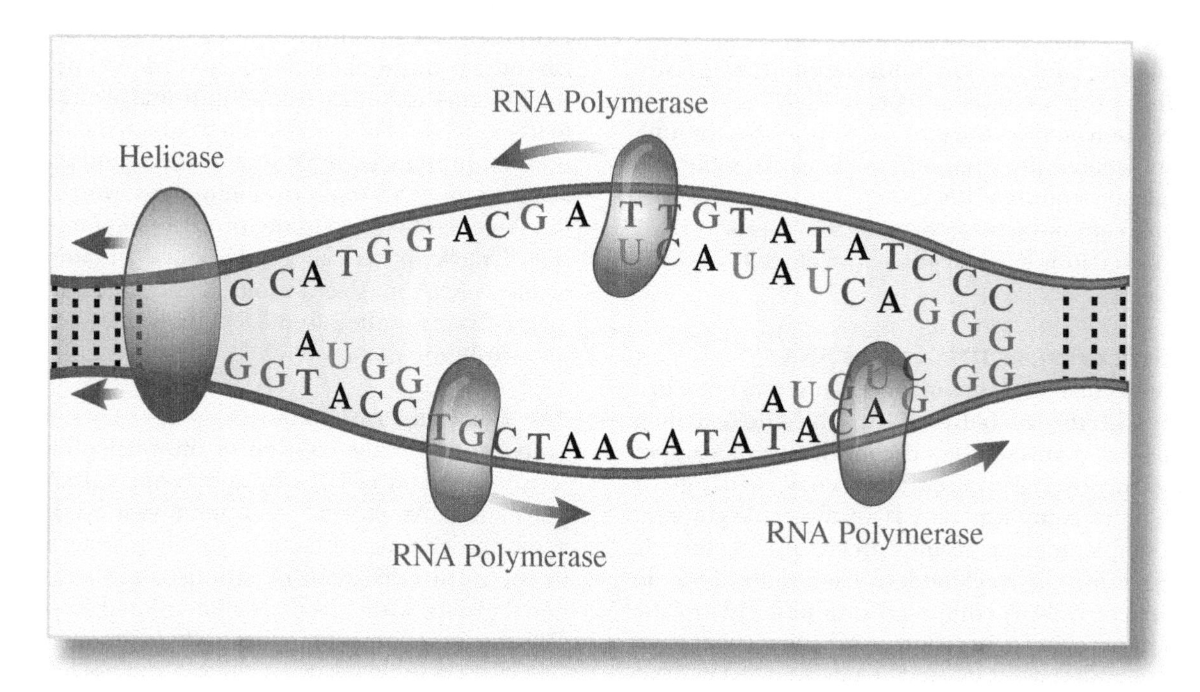

Because DNA and RNA both use only four nucleotides each to specify structure—adenine, cytosine, guanine, and either thymine or uracil—and proteins are synthesized from twenty different amino acids, individual amino acids are specified by unique three-nucleotide sequences called "codons." Since the four nucleotides can produce sixty-four distinct combinations, most amino acids correspond to more than one codon, and some codons serve other purposes, such as initiating protein formation or signaling a stopping point. The codons between a start codon and a stop codon constitute what is called a "reading frame" for a specific nucleotide sequence. It is possible for multiple reading frames to overlap and the same sequence to code for different amino acids, depending on where in the sequence transcription begins.

UNRAVELING THE GENETIC CODE

When biochemists recognized the role of mRNA in transcription, it became possible to investigate the structure of DNA in detail. By assembling synthetic mRNA molecules from just one type of nucleotide base, such as uracil—thus forming polyuridylic acid, or poly(U)—and examining the polypeptides that result, it can be determined which nucleotide sequences code for certain amino acids in protein synthesis. Using this technique, which they developed in 1961, biochemists Marshall Nirenberg and J. Heinrich Matthaei discovered that the codon UUU produces the amino acid phenylalanine—the first time an individual codon was linked to a specific amino acid. Similar experiments with synthetic poly(A) (polyadenylic acid) and poly(C) (polycytidylic acid) determined that the codons AAA and CCC code for lysine and proline, respectively. Poly(G) (polyguanylic acid) was found to form an unusable stacked structure that did not translate into protein synthesis. Synthetic codons of mixed nucleotide units also revealed which codons function as "start" and "stop" signals in protein synthesis and which ones code for the same amino acids.

The sequence of nucleotides in the structure of a DNA molecule is known as a "genome." If one were to transcribe the entire human genome as a sequence of nucleotides, using A, C, T, and G, the result would fill approximately one million densely typed pages. In February 2001, the science journal *Nature* published its report of the first complete analysis of the human genome. Further study of the genome has revealed, among other things, that all humans in the world today are descended from a mere handful of populations that originated in Africa; that the DNA of humans and chimpanzees only differs by approximately 2 percent; and that many modern humans, particularly those of Asian and European descent, carry Neanderthal genes—in some cases as much as 4 percent of their DNA.

SAMPLE PROBLEM

Some researchers think that if they can recover viable DNA from mammoths that were preserved in permafrost, they may be able to clone such a creature. Suppose a sample of mammoth DNA yields an intact DNA strand in which the template strand has the following nucleotide sequence:

AAGTGCACCTGGTATATCCAGTGTCAT

What sequence of nucleotides would be found in the mRNA produced from this DNA fragment?

Answer:

In the transcription of DNA to RNA, thymine nucleotides (T) in DNA coordinate with adenine nucleotides (A) in RNA, and adenine nucleotides in DNA coordinate with uracil (U) nucleotides in RNA. Cytosine (C) coordinates with guanine (G), and guanine with cytosine, in both cases. Therefore, the complementary mRNA sequence would be

UUCACGUGGACCAUAUAGGUCACAGUA

If you are adventurous, use a table of amino acid codons to determine what amino acid sequence a protein made from this mRNA would have.

Richard M. Renneboog, MSc

BIBLIOGRAPHY

Berg, Jeremy M., John L. Tymoczko, Gregory J. Gatto, and Lubert Strye. *Biochemistry*. 8th ed. New York: W. H. Freeman, 2015. Print.

"The Human Genome." *Nature* 409.6822 (2001): 813–985. Print.

Lodish, Harvey, et al. *Molecular Cell Biology*. 7th ed. New York: Freeman, 2013. Print.

Pelczar, Michael J., Jr., E. C. S. Chan, and Noel R. Krieg. *Microbiology: Concepts and Applications.* New York: McGraw, 1993. Print.

Reece, Jane B., et al. *Campbell Biology.* 10th ed. San Francisco: Cummings, 2013. Print.

DUCTILITY

FIELDS OF STUDY

Metallurgy

SUMMARY

The property of ductility is discussed, and its nature as an intensive property of metals is described. Ductility of metals results directly from the nature of their electronic structure and the physical size of their atoms.

PRINCIPAL TERMS

- **deformation:** any permanent change in the shape of an object as a result of the application of force or a change in temperature.
- **fracture:** a dislocation in the internal structure of an object that causes it to break into two or more pieces.
- **intensive properties:** the properties of a substance that do not depend on the amount of the substance present, such as density, hardness, and melting and boiling point.
- **malleability:** the ability of a solid material to be deformed by the application of compressive (pushing) force, such as hammering, without breaking or fracturing.
- **metallic bond:** a type of chemical bond formed by the sharing of delocalized electrons between a number of metal atoms.
- **plasticity:** the ability of a material to undergo deformation without breaking.
- **tensile stress:** a force that acts to pull or stretch a material.

THE NATURE OF DUCTILITY

Ductility is an extremely important property of certain materials, especially metals. The ductility of a metal permits it to undergo deformation via tensile stress (pulling or stretching) without fracture. A common application of this property is in the production of wires and filaments. In this process, a metal bar is subjected to tensile stress as it is pulled through a die that is slightly smaller in diameter than the bar. This extends the length of the bar as it reduces its diameter. The process is repeated with progressively smaller dies until the original rod has been extended so many times that its diameter is reduced to a mere fraction of its original size.

Ductility is usually exploited by applying tensile stress over a slightly extended period of time. A related property, malleability, refers to the ability of a material to undergo compressive force, such as occurs in stamping and hammering, without fracturing. Malleable metals are generally shaped through the instantaneous application of compressive force.

DUCTILITY VS. MALLEABILITY

Both ductility and malleability are aspects of plasticity, which is a temperature-related property of many solid materials, not just of metals. The plasticity of a material increases as temperature increases, until the material reaches its melting point, at which point the material becomes fluid rather than plastic and will not retain its shape as a solid. Ironworkers and steelworkers typically heat metal until it glows with a bright orange color. This increases the plasticity of the metal and makes it easier to shape by the application of force. Plasticity is considered to be an intensive property because neither the ductility nor the malleability of a material depends on how much of it is present.

Nonmetal materials that exhibit plasticity include certain polymers, which were termed "plastics" for that very reason. Polymers that become softer when heated are called "thermoplastic" to indicate that their plasticity increases with heating. Like metals, thermoplastics also become fluid at a certain temperature. Generally, plastics are ductile but not malleable: they can be stretched and bent fairly easily and maintain their structural integrity, but they will fracture into numerous pieces when force is applied too quickly. A temperature range known as the glass-transition

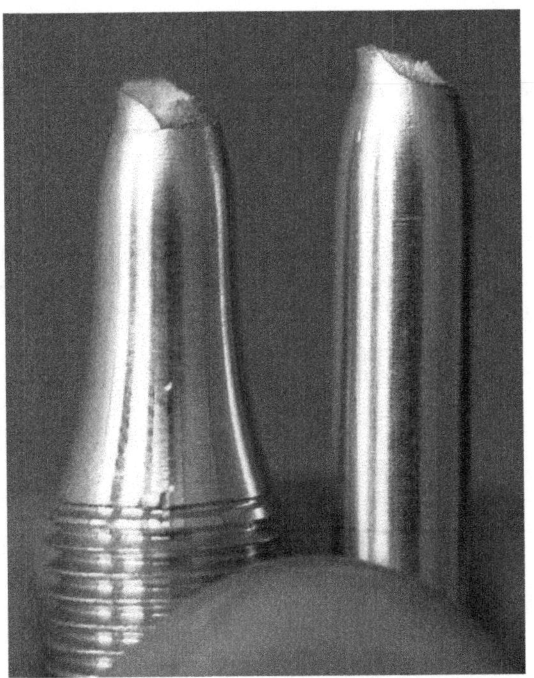

Tensile test of an AlMgSi alloy. This is a ductile fracture type, as seen by the local necking and the cup and cone fracture surfaces. By Sigmund via Wikimedia Commons

temperature marks the dividing line between glass-like behavior and malleability. Metals have a similar temperature range for malleability. For example, it has been demonstrated that the metal used to construct the RMS *Titanic* failed on impact with an iceberg because exposure to the frigid water of the northern Atlantic Ocean lowered its temperature to the point where it lost its malleability.

Ductility and the Electronic Structure of Metals

According to the modern theory of atomic structure, electrons in atoms occupy specific regions of space about the nucleus called orbitals. While the outer electron orbitals of different atoms in a molecule frequently overlap to some degree, combining to form shared molecular orbitals so that the valence electrons of each atom are no longer solely confined to their "parent" atom, this effect is much more pronounced in metals, especially pure metals.

In a metal sample, the orbitals of the individual atoms are able to effectively overlap each other, allowing the valence electrons to be shared by all atoms in the sample and move relatively freely throughout the material. It is a requirement of modern atomic theory that when a number of atomic orbitals overlap, or combine, an equal number of hybrid or molecular orbitals are produced; in a way, this makes the entire body of the metal sample analogous to a single metal "molecule" across which the valence electrons can range. This phenomenon is known as metallic bonding and is characterized by a strong attractive force between the positively charged atomic cores and the negatively charged "sea" of delocalized valence electrons surrounding them. The electrostatic repulsion between the electrons causes them to distribute more or less uniformly throughout the metal so that each atomic core is surrounded by electrons in all directions, resulting in a generalized nondirectional force holding the atoms together. This is an extremely simplified model of metallic bonding, but it serves as a useful introduction to the concept.

The nature of metallic bonding contributes to the characteristic ductility and malleability of metals. The nondirectional attraction and the lack of any strong localized bonds to break make metallic bonds more resilient to applied force than covalent bonds, while the sharing of valence electrons allows the atoms to

SAMPLE PROBLEM

Given the following list of metals and materials, classify them as being ductile or non-ductile:

- iron
- copper
- water
- water ice
- gold
- sodium
- diamond

Answer:

Iron and copper are both ductile and are commonly formed into wire of various sizes. Water is a highly mobile fluid, so it is not ductile. Water ice may be more or less ductile depending on its temperature; in nature, it exists at temperatures fairly close to its melting point and thus is more ductile, as evidenced by the movement of glaciers. Gold is highly ductile; a single ounce of gold can be drawn into a fine wire several kilometers in length. Sodium is highly ductile, and at room temperature it is soft enough to be easily cut with a knife. Diamond, because of its extremely rigid crystal structure, is not ductile.

be packed more closely together, enabling them to slide against each other more easily to produce deformation instead of a fracture. This is especially true when all atoms are of the same size, which is why pure metals are typically more ductile and malleable than alloys.

Richard M. Renneboog, MSc

BIBLIOGRAPHY

Askeland, Donald R., Wendelin J. Wright, D. K. Bhattacharya, and Raj P. Chhabra. *The Science and Engineering of Materials.* Boston: Cengage Learning, 2016. Print.

Douglas, Bodie E., Darl H. McDaniel, and John J. Alexander. *Concepts and Models of Inorganic Chemistry.* 3rd ed. New York: Wiley, 1994. Print.

Fenichell, Stephen. *Plastic: The Making of a Synthetic Century.* New York: Harper, 2009. Print.

Jones, Mark Martin. *Chemistry and Society.* 5th ed. Philadelphia: Saunders College Pub., 1987. Print.

Miessler, Gary L., Paul J. Fischer, and Donald A. Tarr. *Inorganic Chemistry.* 5th ed. Upper Saddle River: Prentice Hall, 2013. Print.

E

ELECTRONS

FIELDS OF STUDY

Inorganic Chemistry; Geochemistry; Metallurgy

SUMMARY

The basic structure of atoms and the role of electrons in that structure are defined, and the development of the modern theory of atomic structure is explained. Atomic structure is fundamental to all fields of chemistry, and especially to fields that rely on the intrinsic properties of individual atoms.

PRINCIPAL TERMS

- **electronegative:** describes an atom that tends to accept and retain electrons to form a negatively charged ion.
- **fundamental particle:** one of the smaller, indivisible particles that make up a larger, composite particle; commonly used to refer to electrons, protons, and neutrons, although these are themselves composed of various actual fundamental particles, such as quarks, leptons, and certain types of bosons.
- **ion:** an atom, molecule, or neutral radical that has either lost or gained electrons and is therefore electrically charged.
- **orbital:** a specific region of space about the nucleus of an atom in which electrons of a given energy level are most likely to be found.
- **valence shell:** the outermost energy level occupied by electrons in an atom.

The Identification of Electrons

Various models of atomic structure arose through the nineteenth century, the most useful being the planetary model and the plum pudding model. In the first, atoms were envisioned to be composed of a nucleus with electrical charges whirling about it like the planets around the sun. In the latter model, developed by British physicist J. J. Thomson (1856–1940), the atom was envisioned to be a round blob with bits of electrical charge embedded throughout. In conducting research on the nature of cathode rays (beams observed when electricity was run through a glass vacuum tube with a cathode, or terminal, at the end) in 1897, Thomson) determined that the behavior of the cathode rays suited the mathematics of streams of charged particles rather than electromagnetic radiation. Logically, they had come from within the metal that made up the cathode, yet their loss had no significant effect on the mass of the cathode. To these particles he assigned a negative charge. Alongside the emission of cathode rays was the emission of so-called canal rays. When German physicist Wilhelm Wien (1864–1928) examined these in a similar way in 1898, he found that they also suited streams of charged particles rather than electromagnetic radiation. However, the canal ray particles were much more massive and possessed the opposite electrical charge from cathode rays, so he therefore assigned them a positive electrical charge. This verified that atoms had an internal structure, but it did not resolve the issue of which model was correct. Further research by British physicists Ernest Rutherford (1871–1937) and James Chadwick (1891–1974) eventually provided the definitive evidence of fundamental particles for the planetary model that has been refined into the modern theory of atomic structure.

Subatomic Sizes

According to the modern theory, atoms are composed of an extremely small, dense nucleus made up of positively charged protons and neutral neutrons, surrounded by a very large, diffuse cloud of electrons. The magnitude of the negative charge on an electron is exactly equal to the positive charge of a proton, but the masses of the two particles are very different. Measurement of the charge-to-mass ratio of electrons demonstrates that the electron has a mass of about 9.1×10^{-31} kilograms, or about 9.1×10^{-28}

ELECTRONS

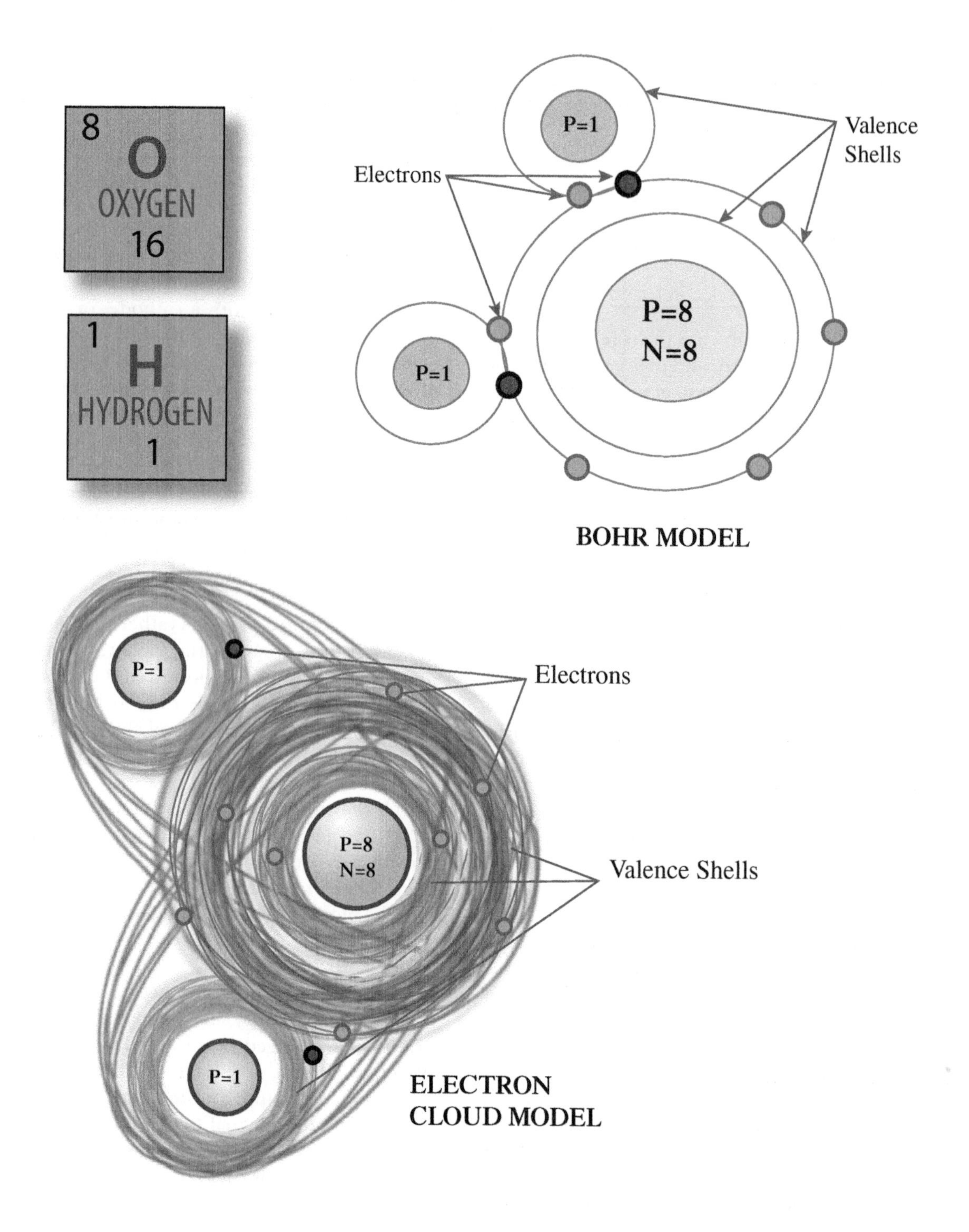
8
O
OXYGEN
16
1
H
HYDROGEN
1
P=1
Valence
Shells
Electrons
P=8
N=8
P=1
BOHR MODEL
P=1
Electrons
P=8
N=8
Valence Shells
P=1
ELECTRON
CLOUD MODEL

SAMPLE PROBLEM

Calculate the number of electrons in the chloride and calcium ions in the compound calcium chloride, $CaCl_2$. What is the electrical charge on the two ions? How many electrons are in an atom of carbon-12? How many electrons are in an atom of carbon-14?

Answer:

The atomic number of chlorine is 17. There are seventeen protons in the nucleus of a chlorine atom and therefore seventeen electrons about the nucleus. The chloride ion, Cl^-, has a single extra negative charge and has therefore added one additional electron. The chloride ion therefore has eighteen electrons, and each has a −1 charge.

The atomic number of calcium is 20. A neutral calcium atom therefore has twenty protons in the nucleus and twenty electrons. The charge on the calcium ion in $CaCl_2$ is balanced by two negative charges. There are therefore two positive charges on the calcium ion, indicating that the calcium atom has given up two of its electrons, leaving eighteen electrons in the ion, which has a +2 charge.

The atomic number of carbon-12 is 6, indicating that there are six protons and six electrons in the neutral atom. Carbon-14 is an isotope of carbon and therefore has the same atomic number, the same number of protons, and the same number of electrons as an atom of carbon-12.

grams. The proton, however, is almost two thousand times more massive than the electron, and the nucleus of any atom makes up about 99.98 percent of the atom's total mass.

Electrons in Atoms

Quantum mechanics, supported by experimental evidence, defines the energies that electrons are allowed to possess in an atom. These energies define spaces about the nucleus called "electron shells," each of which may contain only a certain maximum number of electrons. The outermost or highest energy electron shell in an atom is termed the valence shell, while the electrons within that shell are known as "valence electrons." Atoms are at a minimum energy state when the outermost electron shell contains its full complement of electrons, whether as a neutral atom or as an ion. This drives atoms to give up electrons from an incomplete valence shell to an electronegative atom and to accept electrons from a more electropositive atom in order to have a completely filled outermost electron shell. Within each shell are orbitals, specific regions that can each be occupied by no more than two electrons.

Electrons in Chemical Reactions

The electrons located within the valence shell of an atom determine how that atom interacts with others and thus the chemical behaviors of each element. When two atoms come into contact with each other, they may form a covalent bond by sharing valence electrons. In other cases, one atom may give up valence electrons, which then occupy the valence shell of the other atom, thus causing the previously neutral atoms to become a positive ion and a negative ion, respectively. The transfer of valence electrons is responsible for chemical reactions such as oxidation and reduction.

Richard M. Renneboog, MSc

Bibliography

Gribbin, John. *Science: A History, 1543–2001.* London: Lane, 2002. Print.

Johnson, Rebecca L. *Atomic Structure.* Minneapolis: Lerner, 2008. Print.

Miessler, Gary L., Paul J. Fischer, and Donald A. Tarr. *Inorganic Chemistry.* 5th ed. Upper Saddle River: Prentice Hall, 2013. Print.

Morrison, Robert Thornton, and Robert Neilson. Boyd. *Organic Chemistry.* 7th ed. Englewood Cliffs, N.J.: Prentice Hall, 2003. Print.

Wehr, M. Russell, James A. Richards Jr., and Thomas W. Adair III. *Physics of the Atom.* 4th ed. Reading: Addison-Wesley, 1984. Print.

Winter, Mark J. *The Orbitron: A Gallery of Atomic Orbitals and Molecular Orbitals.* University of Sheffield, n.d. Web.

ENDOCYTOSIS AND EXOCYTOSIS

FIELDS OF STUDY

Biochemistry; Genetics; Molecular Biology

SUMMARY

Endocytosis and exocytosis are described as complementary processes that move materials into and out of cell cytoplasm through the cell membrane. Different types of endocytosis are described, and some of their applications are discussed.

PRINCIPAL TERMS

- **cell membrane:** a biological membrane that forms a semipermeable barrier separating the interior of a cell from the exterior.
- **endosome:** an intracellular compartment that sorts and transports material taken into a cell via endocytosis.
- **phagocytosis:** a type of endocytosis in which solid particles are taken into the cytoplasm of a cell through the cell membrane.
- **pinocytosis:** a type of endocytosis in which extracellular fluid and any substances it may contain are taken into the cytoplasm of a cell through the cell membrane.
- **vesicle:** a small bubble within a cell, surrounded by a double layer of lipids.

COMPLEMENTARY CELLULAR PROCESSES

The two processes of endocytosis and exocytosis are effectively mirror images of each other. In endocytosis, materials are brought into a cell from outside, while in exocytosis, materials are transported out of a cell. These processes are unlike active transport, which functions by opening existing channels in the membrane to allow specific materials to pass through. A model of endocytosis can be seen when two drops of a liquid such as water or mercury come together and combine into a single drop. The outer surfaces of the drops combine to form a slightly larger surface, while the interiors combine and are enclosed within the surface of the new drop. Similarly, a model of exocytosis can be seen when a large drop breaks apart into two or more smaller drops.

ENDOCYTOSIS AND EXOCYTOSIS IN ACTION

Endocytosis and exocytosis are the basic processes by which substances such as enzymes, insulin, carbohydrates, and a variety of other materials are moved about within and without various cells. They are complementary actions in many cases, as endocytosis is used to bring materials into a cell for specific operations to be performed, while exocytosis is used to eliminate the products and by-products of those operations. They are also the processes by which entities such as viruses invade and infect cells.

Endocytosis begins when a material is brought to the cell membrane. A trigger such as a hormone initiates invagination, which is the folding of the membrane to form a hollow pocket around the material. The membrane then closes around the invagination, forming a vesicle that contains the material. The vesicle can thus be thought of as a bubble of the external material within the cytoplasm of the cell. The wall of the vesicle consists of the same material as the cell membrane. A protective coating of the protein clathrin typically forms around the vesicle to allow it to exist safely within the cytoplasm.

When the vesicle arrives at its destination within the cell, such as the Golgi complex or an endosome, it joins to the membrane of that site and essentially grows an opening through the membrane in order to eject its contents. Endosomes are compartments within the cell that serve various functions, including sorting materials introduced via endocytosis and forming pathways for the transport of those materials

SAMPLE PROBLEM

Briefly explain why endocytosis always requires fluidity in the structure of the cell membrane.

Answer:

In the process of endocytosis, the cell membrane must have fluidity so that it can invaginate to form a pocket. When the pocket has been formed, the cell membrane must be able to close up around the pocket and heal over to re-form the cell membrane and create a stable vesicle. This requires the cell membrane to have a fluid character that allows it to "flow" back together and form an unbroken layer.

ENDOCYTOSIS AND EXOCYTOSIS

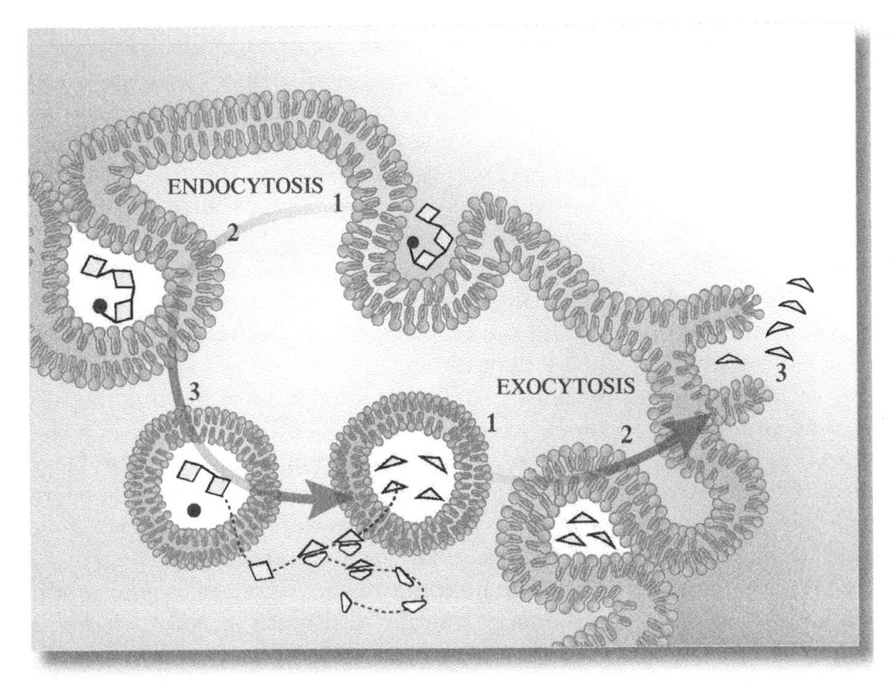

within the cell. Endosomes are first formed when vesicles lose their protein coatings and start to merge; the resulting structure is called an "early endosome." Over time, early endosomes mature into late endosomes, then eventually either mature further to form lysosomes or fuse with existing lysosomes.

There are three basic types of endocytosis. In phagocytosis, cells such as white blood cells extrude themselves about solids such as cellular debris and foreign objects to bring them within the cell cytoplasm, where they can be digested and converted into waste products that can be eliminated. In pinocytosis, cells encapsulate and bring in a bulk quantity of extracellular fluid. In receptor-mediated endocytosis, only specific materials, such as proteins or virus particles (virions), are brought into a cell.

The process of exocytosis within the cell, in which materials are encapsulated and transported through the cell membrane, functions in a similar manner, but in reverse. Mature virions, for example, are eliminated from within infected cells by a receptor-mediated process, while other materials are carried out of the cell and deposited into the extracellular fluid by encapsulation and extrusion at the cell membrane.

ENDOCYTOSIS AND EXOCYTOSIS IN BIOLOGICAL SYSTEMS

Single-celled organisms especially depend on these twin processes for the ingestion of food and the elimination of waste. The amoeba and the paramecium are perhaps the most familiar of such organisms. On encountering a suitable particle of a foodstuff, the amoeba seems to flow around the particle to enclose and engulf it within itself. Once the particle is inside the organism, digestive processes go to work, decomposing it into usable molecules, such as simple sugars, fats, and amino acids. As the various chemical transformations take place, unusable materials are accumulated. These become encapsulated for elimination, and the amoeba then seems to flow away from the waste material. In actuality, the particle was taken into the amoeba's cell interior by endocytosis, and the waste materials were eliminated by exocytosis.

Richard M. Renneboog, MSc

BIBLIOGRAPHY

Berg, Jeremy M., John L. Tymoczko, Gregory J. Gatto, and Lubert Strye. *Biochemistry*. 8th ed. New York: W. H. Freeman, 2015. Print.

Lehninger, Albert L. *Biochemistry: The Molecular Basis of Cell Structure and Function*. 2nd ed. New York: Worth, 1975. Print.

Lodish, Harvey, et al. *Molecular Cell Biology*. 7th ed. New York: Freeman, 2013. Print.

Pelczar, Michael J., Jr., E. C. S. Chan, and Noel R. Krieg. *Microbiology: Concepts and Applications*. New York: McGraw, 1993. Print.

Reece, Jane B., et al. *Campbell Biology*. 10th ed. San Francisco: Cummings, 2013. Print.

Roberts, Michael, Michael Jonathan Reiss, and Grace Monger. *Advanced Biology*. Cheltenham: Nelson, 2000. Print.

ENDOERGIC AND EXOERGIC REACTIONS

FIELDS OF STUDY

Biochemistry; Genetics; Molecular Biology

SUMMARY

The processes of endoergic and exoergic reactions are described, and their importance in biochemical transformations is elaborated. Exoergic reactions release free energy and can occur spontaneously, while endoergic reactions require energy from an outside source in order to proceed. Through the combined action of endoergic and exoergic reactions, many of the essential processes of biochemical systems such as photosynthesis, glycolysis, and respiration are carried out.

PRINCIPAL TERMS

- **endergonic:** synonym for endoergic; describes a reaction process that requires the input of energy in the form of work in order to proceed.
- **equilibrium:** the state that exists when the forward activity of a process is exactly equal to the reverse activity of that process.
- **exergonic:** synonym for exoergic; describes a reaction process that can occur spontaneously and releases energy in the form of work.
- **Gibbs free energy:** the energy in a thermodynamic system that is available to do work.
- **spontaneous reaction:** a chemical reaction that occurs without the input of energy from an outside source.

ENERGY TRANSFERS IN CHEMICAL REACTIONS

All chemical reactions involve the transfer of energy, which can occur in the form of either heat or work. Chemical systems contain energy in the bonds between their atoms. The electrons that form these atomic bonds possess an energy level that is determined by various factors, including the nature of the bond, the extent to which the bond angles are deformed from their ideal angles, and the force of repulsion between the electrons in the bond. During a chemical reaction, the level of energy stored within a chemical system changes as bonds are formed or broken: the formation of a bond releases energy, some of which escapes from the system into its surroundings, while the breaking of a bond consumes energy. The overall change in the level of energy caused by a chemical reaction is called the enthalpy change of the reaction (ΔH). The enthalpy change is determined by subtracting the enthalpy, or total energy content, of the reactants from that of the products. If the enthalpy of the products of a reaction is lower than the enthalpy of the reactants, there is a negative change in enthalpy, meaning that the reaction is exothermic. If the enthalpy change of a reaction is positive, the reaction is considered to be endothermic. Endothermic reactions draw heat energy from their surroundings in order to be carried out, while exothermic reactions release heat energy.

Another important consideration in determining the energy transfer of a reaction is the change in entropy (ΔS) of the system. Entropy is a measure of the degree to which energy has been dispersed throughout the atoms and molecules of a system; high entropy indicates a wide dispersal of energy, causing the components of the system to move in a more disordered manner than in a system with lower entropy. As the enthalpy of a system increases, the entropy of the system also increases, causing the system to become less stable. The entropy change of a reaction is influenced by the temperature at which the reaction occurs; at lower temperatures, the change in entropy is greater than at higher temperatures. Furthermore, as the entropy of a system increases, the entropy of the system's surroundings decreases, as energy flows from the surroundings into the system.

The entropy change of a reaction indicates whether the reaction is spontaneous or non-spontaneous. For a reaction to occur spontaneously, the total entropy change of both the system and its surroundings must be positive, representing a net increase in entropy. This in turn determines whether the reaction is exoergic or endoergic (or exergonic or endergonic; both sets of terms mean the same thing). An exoergic reaction is one that releases free energy, which is energy that is available to do work in a system, while an endoergic reaction requires an input of free energy in order to proceed. In other words, in an exoergic reaction, the change in free energy is negative, indicating that the system expended energy it already possessed in order to do work; in an endoergic reaction, the change in free energy is positive, indicating that energy was supplied to the system from an outside source. Thus, a spontaneous reaction is one that is exoergic, because it can occur at any time without an external energy supply. Spontaneous reactions can occur in one direction only, as reversing one requires an input of energy.

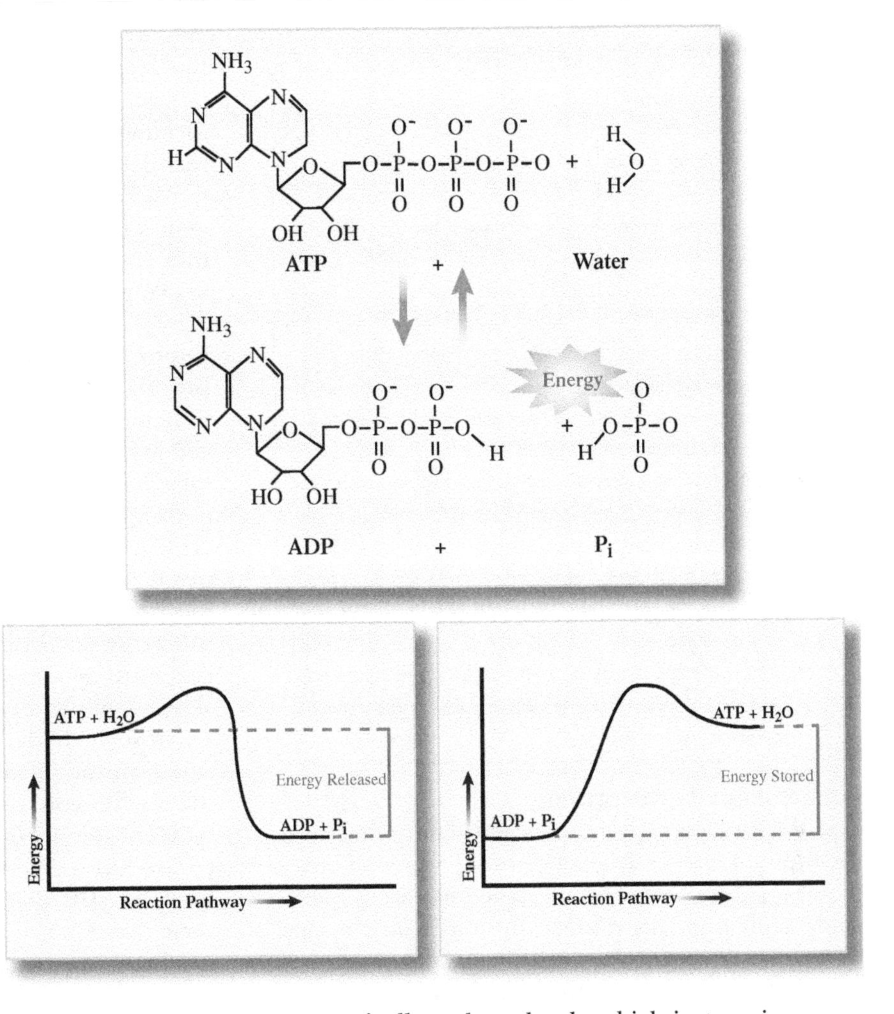

ENTROPY, ENTHALPY, AND GIBBS FREE ENERGY

In thermodynamics, the basic characteristics of chemical reaction systems are described by three factors: entropy, enthalpy, and Gibbs free energy. Gibbs free energy is a measure of free energy as determined by the change of entropy within a system. (Free energy is essentially the total energy of a system minus the entropy, which represents energy that cannot be used to do work.) The three concepts are related by the following expression:

$$\Delta G = \Delta H - T\Delta S$$

where G is Gibbs free energy, H is enthalpy, T is temperature, S is entropy, and the delta symbol (Δ) represents the overall change in the associated factor. From this equation, it can be seen that the change in the Gibbs free energy is equal to the change in enthalpy minus the change in entropy, the effect of which is more significant at higher temperatures.

The relationship can be demonstrated using gas contained in a cylinder that is closed by a piston. Heating the gas increases its thermal energy and thus its enthalpy, causing the gas molecules to move more energetically and randomly, which in turn increases the entropy. As a result, the gas is able to exert more pressure against the piston using the extra energy, but it cannot use the full amount because some of the energy is being used to maintain the increased motion and randomness of the individual gas particles. The portion of the energy that is available to do work to move the piston is the free energy.

Gibbs free energy is named for American scientist Josiah Willard Gibbs (1839–1903), who first formulated the concept of the free energy, as well as several other fundamental principles of chemical thermodynamics.

SAMPLE PROBLEM

Determine whether the following reactions are exoergic (spontaneous) or endoergic (non-spontaneous):

- muscle contraction
- photosynthesis
- the standard formation of acetylene gas ($\Delta G = 209.2$ kJ/mol)
- the standard formation of carbon dioxide gas ($\Delta G = -394.4$ kJ/mol)

Answer:

- Muscle contraction requires the input of energy from a signaling compound to trigger cellular transport functions that consume energy. Muscle contractions are therefore endoergic and non-spontaneous.
- Photosynthesis requires the input of energy from sunlight in order to proceed. Photosynthesis is therefore endoergic and non-spontaneous.
- The Gibbs free energy of formation for acetylene is positive in value. Therefore, the reaction is endoergic and non-spontaneous.
- The Gibbs free energy of formation for carbon dioxide is negative in value. Therefore, the reaction is exoergic and spontaneous.

Free Energy in Reactions

Endoergic reactions are chemical reactions that need the input of energy from an external source in order to take place. Endothermic reactions are superficially similar in this respect; the difference is that endoergic reactions do not occur unless energy is supplied, while endothermic reactions proceed normally, extracting energy from the surroundings in the process. Exoergic reactions, on the other hand, do not need any external input of energy and are therefore spontaneous. In multicomponent chemical reaction systems, the free energy released by an exoergic reaction is available to be used by non-spontaneous endoergic reactions. The Gibbs free energy accordingly plays a significant role in establishing equilibrium between reactants and products, as the energy released by the reactants in order to form the products can then be used by the products to reverse the reaction and re-form the reactants. An unfortunate side effect of this is that when other reaction mechanisms that might lead to undesired products are possible, the energy can be used by those as well. As the system tends toward equilibrium, the free energy in the system tends toward zero. At equilibrium, the free energy of the system is at its minimum value and does not increase unless the equilibrium state of the system is perturbed.

Catalysts function in chemical reactions by forming an activated complex with the reactants that decreases the activation energy of specific reactions. The most effective catalysts are those that ultimately reduce the required activation energy to zero, thus converting the overall reaction from endoergic to exoergic.

Endoergic and Exoergic Reactions in Biochemical Systems

Biological systems make great use of the combined action of exoergic and endoergic reactions. In cellular respiration, for example, glucose ($C_6H_{12}O_6$) is decomposed into carbon dioxide (CO_2) and water (H_2O), accompanied by the release of energy. The products of respiration contain less free energy than was stored in the reactants. The law of conservation of energy states that the energy released by the process must be stored or manifested in another form. While some is manifested as heat, which maintains a constant body temperature, the remainder is stored in the formation of chemical bonds through the conversion of adenosine diphosphate (ADP) to adenosine triphosphate (ATP). The energy is recovered in the reverse process, which converts ATP back to ADP, and the energy released by this reverse reaction is used to drive many other endoergic biochemical reactions.

Richard M. Renneboog, MSc

Bibliography

Abbott, David. *The Biographical Dictionary of Scientists.* New York: P. Bedrick, 1984. Print.

Crowe, Jonathan, and Tony Bradshaw. *Chemistry for the Biosciences: The Essential Concepts.* 2nd ed. Oxford: Oxford UP, 2010. Print.

Lafferty, Peter, and Julian Rowe, eds. *The Hutchinson Dictionary of Science.* 2nd ed. Oxford: Helicon, 1998. Print.

Lehninger, Albert L. *Biochemistry: The Molecular Basis of Cell Structure and Function.* 2nd ed. New York: Worth, 1975. Print.

Lodish, Harvey, et al. *Molecular Cell Biology*. 7th ed. New York: Freeman, 2013. Print.

Reece, Jane B., et al. *Campbell Biology*. 10th ed. San Francisco: Cummings, 2013. Print.

EPITAXY

FIELDS OF STUDY

Crystallography; Chemical Engineering; Inorganic Chemistry

SUMMARY

Epitaxy defines the growth of layers of one material on a substrate of a different material. The methodology is used to produce unique materials and compounds of high purity. Epitaxial growth is achieved in liquid and gas phase media in synthetic methods and in nature to produce crystalline materials.

PRINCIPAL TERMS

- **cohesion:** the tendency for like molecules of a substance to stick together due to their shape and electronic structure.
- **crystal:** a solid consisting of atoms, molecules, or ions arranged in a regular, periodic pattern in all directions, often resulting in a similarly regular macroscopic appearance.
- **crystal lattice:** the regular geometric arrangement of atoms and molecules in the internal structure of a crystal.
- **crystallography:** the study of the properties and structures of crystals.
- **diffusion:** the process by which different particles, such as atoms and molecules, gradually become intermingled due to random motion caused by thermal energy.
- **dipole:** the separation of positive and negative charges within a single molecule due to electron density being relatively high in one part of the molecule and relatively low in another.
- **dipole-dipole:** describes a type of interaction, either attraction or repulsion, between two molecular dipoles, which are molecules that are polarized due to the greater concentration of electrons in one area.
- **equilibrium:** the state that exists when the forward activity of a process is exactly equal to the reverse activity of that process.
- **hydrogen bond:** a weak type of chemical bond formed by the attraction of a hydrogen atom to an electronegative atom—an atom with a strong tendency to attract electrons—in the same or another molecule.
- **phase transition:** the change of matter from one state to another, such as from solid to liquid or liquid to gas, due to the transfer of thermal energy.
- **sublimation:** the phase transformation from solid to gas without passing through an intermediate liquid phase.
- **van der Waals force:** weak forces of attraction between atoms due to proximity that do not affect the electronic structure of those atoms or result in chemical bond formation.

Crystal Growth

In the science of crystallography, the growth of a crystal arises from the organization of atoms or molecules into an organized and orderly array or crystal lattice. This most commonly involves a phase transition from liquid to solid or from gas to solid. In some cases, a change in the structure of a crystal lattice can be effected by a change of pressure and temperature, as is the case with the many different solid phases found in water ice, in which water molecules interact via the dipole of the hydrogen bond. In other substances, atoms and molecules interact via dipole-dipole interactions and by van der Waals forces. Another phase transition occurs from the solid phase to the gas phase when a material undergoes sublimation, but it is the reverse of this process, as a material reverts from the gas phase directly to the solid phase.

Crystal growth in liquid media occurs by two different mechanisms. In one mechanism, a material that has been dissolved in a liquid solvent can precipitate from the solution as the dissolved molecules or ions combine to form a regular solid crystal. This is the methodology typically used in laboratories to purify a material. In the other mechanism, the liquid

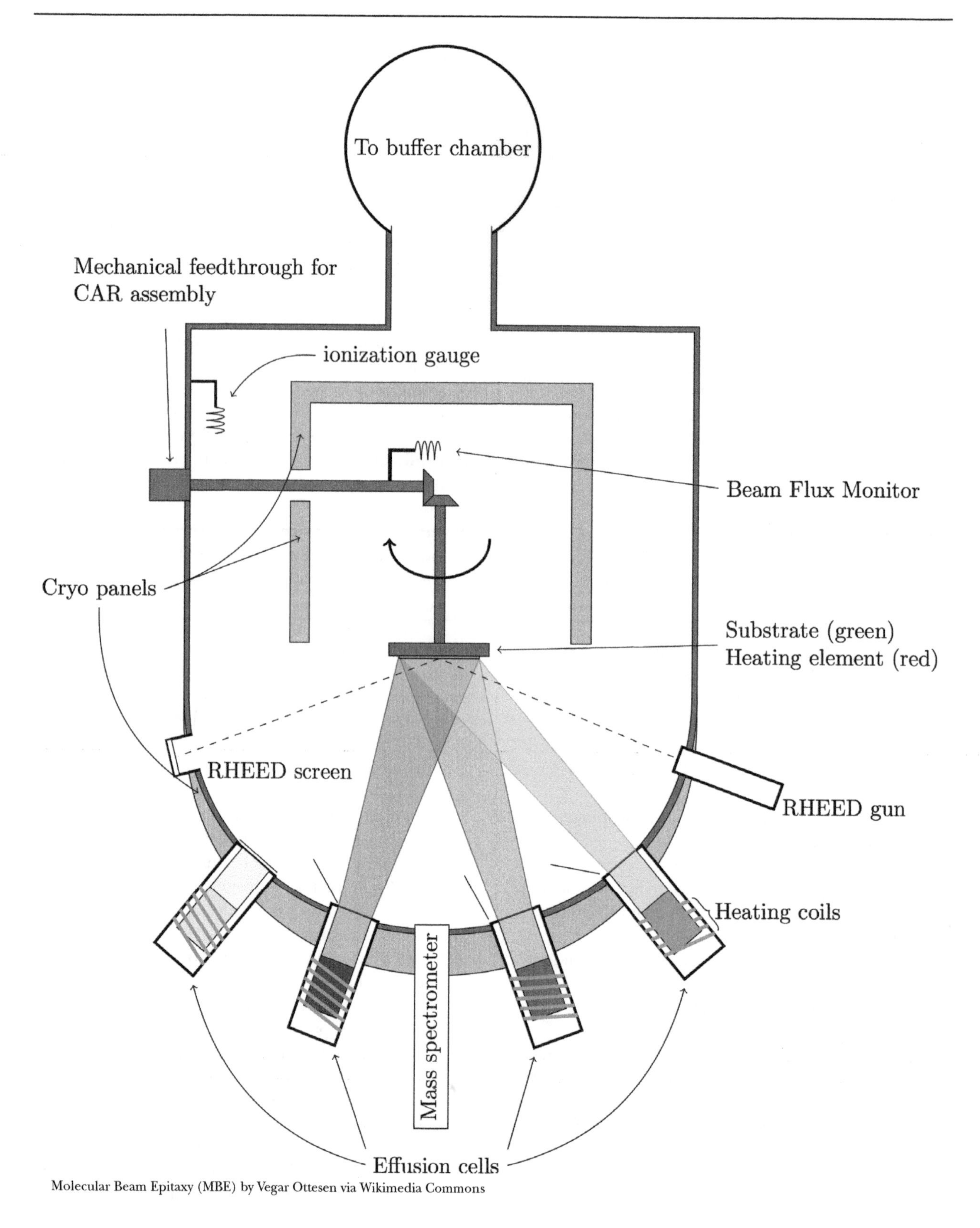

Molecular Beam Epitaxy (MBE) by Vegar Ottesen via Wikimedia Commons

phase is the molten material itself, and crystal formation takes place as the molten material is allowed to cool. This is the methodology used to produce large single crystals of silicon for the production of computer chips and for obtaining materials of very high purity in the zone refining process. Crystal growth as the reverse of sublimation occurs by an entirely different process called chemical vapor deposition, and the process is used to produce exotic materials such as dibenzene chromium and ferrocene, high purity materials, and products such as carbon nanotubes and diamond coatings. In all cases, the structural integrity of the crystal lattice is determined by the cohesion of the component atoms or molecules, and the rate of formation of successive layers in a crystal structure is greatly affected by the rate of diffusion of the component atoms and molecules in the fluid phase.

EPITAXIAL GROWTH

All three methods of crystal growth can be used to produce additional layers of the same or a different material to the surface of an existing crystal lattice. The formation of this secondary external layer, by whatever method used, is termed epitaxial growth. This methodology provides a means of producing composite materials with specific properties. A composite material is one composed of two or more different materials to produce a structure that has properties as though it is a single material. Epitaxial growth occurs in nature in the animal, vegetable and mineral kingdoms. The growth of new wood around the scar of a broken tree branch is a form of epitaxial growth, as is the development of callus on skin. The most important form of epitaxial growth, however, occurs in mineral materials, both naturally and artificially. In naturally occurring crystal growth, changes in the mineral content of the surrounding environment may result in the addition of epitaxial layers as a crystal grows. Large beryl crystals, for example, are seldom a single color, but often change color from one end to the other reflecting the changes of the mineral content of the medium in which the crystal developed. The size and purity of such crystals depends on controlling the equilibrium between the solid and fluid phases. Other mineral crystals may have identical content throughout, but exhibit an irregular surface geometry due to epitaxial growths of new material on the existing surface of the underlying crystal. Even more radical examples can be found in which an entirely different mineral builds an epitaxial crystal structure of its own on the surface of another mineral, such as rutile that has formed on hematite.

CONTROLLED EPITAXY

Under very controlled conditions, epitaxial growth of a crystal structure can be induced and used to produce a material with very specific and predictable properties, as is the case with silicon crystals grown for the manufacture of computer chips and integrated circuits. In this process very pure silicon is melted in a furnace and a specific amount of a doping material such as phosphorus or germanium is added to the liquid silicon to produce a homogeneous solution. A seed crystal is introduced to the surface of the liquid and rotated as it is slowly drawn away from the molten material. The molten material that has adhered to the seed crystal solidifies into the regular crystal structure of the doped silicon alloy in a process of continuous epitaxial growth. When the process is completed, a long cylindrical single crystal of the doped silicon material is formed and is then cut into the wafers that are to become the substrate for the production of silicon computer chips.

Chemical vapor deposition methods are used to produce either pure materials like carbon nanotubes or to form barrier layers on existing materials. One of the more generally useful methods that have been developed for industrial applications forms a layer

SAMPLE PROBLEM

Calculate the growth rate in micrometers per minute of an epitaxial layer of silicon that has attained a thickness of 127 nanometers after 6 minutes.

Answer:

Convert nanometers (10^{-9} m) to micrometers (10^{-6} m) using the relation 1nm = 1μm/1000nm:

$$127\text{nm} = (127\text{nm}/1000\text{nm}/\mu\text{m}) = 0.127\mu\text{m}$$

Calculate R, the rate per minute, by dividing the total growth by the elapsed time:

$$R = 0.127\mu\text{m}/6\text{min} = 0.0212\mu\text{m.min}^{-1}$$

of actual diamond crystal on an appropriate metal surface, usually the cutting surface of a steel tool. Diamond bits for various cutting tools have become quite commonplace as a result of the development of the methodology for forming epitaxial diamond layers on metal surfaces.

THE IMPORTANCE OF EPITAXY

Epitaxy comes in many forms, some of which are beneficial and some of which are not. In chemical and physical applications, epitaxy provides a useful means of producing materials and tools that have considerable value and enhanced utility. While epitaxial growth methods are a valuable means of enhancing the capabilities of many products, it is also the only way many things can be produced. Silicon computer chips and diamond-coated cutting tool bits are prime examples. The use of epitaxial methods for the production of specialty materials, especially in regard to semiconductor applications, is likely to be the only viable means by which those materials can be produced.

Richard M. Renneboog M.Sc.

BIBLIOGRAPHY

Bhat, H.L. *Introduction to Crystal Growth. Principles and Practice.* Boca Raton, FL: CRC Press, 2015. Print.

Capper, Peter, and Michael Mauk, eds. *Liquid Phase Epitaxy of Electronic, Optical and Optoelectronic Materials.* Hoboken, NJ: Wiley, 2007. Print.

Herman, Marian A., Wolfgang Richter and Helmut Sitter. *Epitaxy. Physical Principles and Technical Implementation.* New York, NY: Springer-Verlag, 2004. Print.

Kuech, Thomas F., ed. *Handbook of Crystal Growth III Thin Films and Epitaxy: Basic Techniques, and Materials, Processes and Technology.* 2nd ed. Waltham, MA: Elsevier, 2015. Print.

Orton, John. *Molecular Beam Epitaxy. A Short History.* New York, NY: Oxford UP, 2015. Print.

Pohl, Udo W. *Epitaxy of Semiconductors: Introduction to Physical Principles.* New York, NY: Springer-Verlag, 2013. Print.

Stringfellow, Gerald B. *Organometallic Vapor-Phase Epitaxy: Theory and Practice.* Boston, MA: Academic Press, 1989. Print.

ESTERS

FIELDS OF STUDY

Organic Chemistry; Biochemistry

SUMMARY

The characteristic properties and reactions of esters are discussed. Esters are essential compounds that are very useful in organic synthesis. Esterification is an essential process in biochemistry.

PRINCIPAL TERMS

- **alkoxyl:** a functional group, also called an alkoxy group, consisting of an alkyl group bonded to an oxygen atom.
- **carbonyl group:** a functional group consisting of a carbon atom double bonded to an oxygen atom.
- **ether:** an organic compound consisting of an oxygen atom bonded to two alkyl or aryl groups.
- **functional group:** a specific group of atoms with a characteristic structure and corresponding chemical behavior within a molecule.
- **hydroxyl group:** a primary functional group consisting of an oxygen atom covalently bonded to a single hydrogen atom.

THE NATURE OF THE ESTERS

Esters are the product of a condensation reaction between a carboxylic acid (–RCOOH) and an alcohol (–ROH). They are characterized by the presence of the ester functional group, which consists of a carbonyl group (C=O) bonded via the carbon atom to an alkyl group (derived from a saturated hydrocarbon) and an alkoxyl, or alkoxy group (an alkyl group double-bonded to an oxygen atom). Alternatively, the alkyl group may instead be an aryl group (derived from an aromatic ring), and the alkoxy group may be an aryloxy group. The general formula is often written as RCOOR' or ArCOOAr', with the R– or Ar– indicating an alkyl group substituent and an aryl group substituent, respectively:

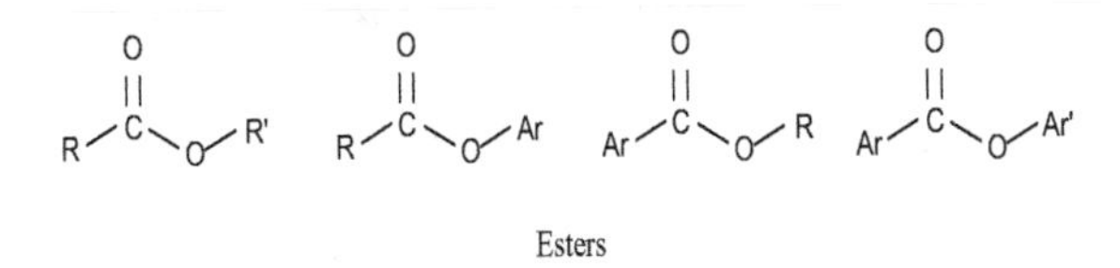

Esters

The resulting structure of the ester represents the replacement of the acid's hydroxyl group (-OH) with the alkoxy or aryloxy group from the alcohol. Reactions between alcohols and acid derivatives such as acid chlorides and anhydrides can also produce esters. In the case of an acid chloride, the alkoxy or aryloxy group replaces the chlorine atom; in an acid anhydride, it replaces the central oxygen atom and one of the acyl groups (a carboxylic acid derivative, -RCO).

Esters feature a C-O-C linkage that is similar to the C-O-C linkage in an ether; however, in an ester, one of the carbon atoms is part of the carbonyl group, causing its chemical behavior to differ vastly from that of an ether. Most common esters are formed from carboxylic acids, but they can also be formed from potentially any other organic acid and many inorganic acids. Esters are generally useful in synthesis procedures to extend the carbon skeleton of another molecule, to stabilize an alcohol or acid functional group, and to add substituents to a molecular structure.

In principle, any carboxylic acid has a corresponding ester that can be derived from it, and indeed hundreds of different esters are available for a variety of purposes. Many naturally occurring esters are responsible for the scent and taste of different foods. They are generally nontoxic liquids with a characteristically fruity smell and lower melting and boiling points than their component acid and alcohol moieties. Esters are also generally noncorrosive, and the ester functional group does not take part in hydrogen bonding.

The most common ester is probably ethyl acetate ($C_4H_8O_2$), a natural component of bananas that is largely responsible for their characteristic scent. It is produced in large quantities for use as a relatively nontoxic solvent notable for its ease of handling. Its hydrolysis products are ethanol (C_2H_6O) and acetic acid ($C_2H_4O_2$), both of which are well tolerated by human metabolism.

Nomenclature of Esters

Esters are named according to the parent acid and alcohol moieties from which they are formed, in the same way that salts of acids are named. The salt formed from sodium hydroxide and benzoic acid, for example, is sodium benzoate; similarly, the ester formed from ethanol and benzoic acid is ethyl benzoate. The name consists of the name of the alkyl or aryl group (ethyl), which comes from the alcohol, followed by the anion of the acid (benzoate). As might be expected, the systematic names of complex esters may become quite complicated, depending on the systematic names of the component acid and alcohol moieties. It is therefore a common practice to use an

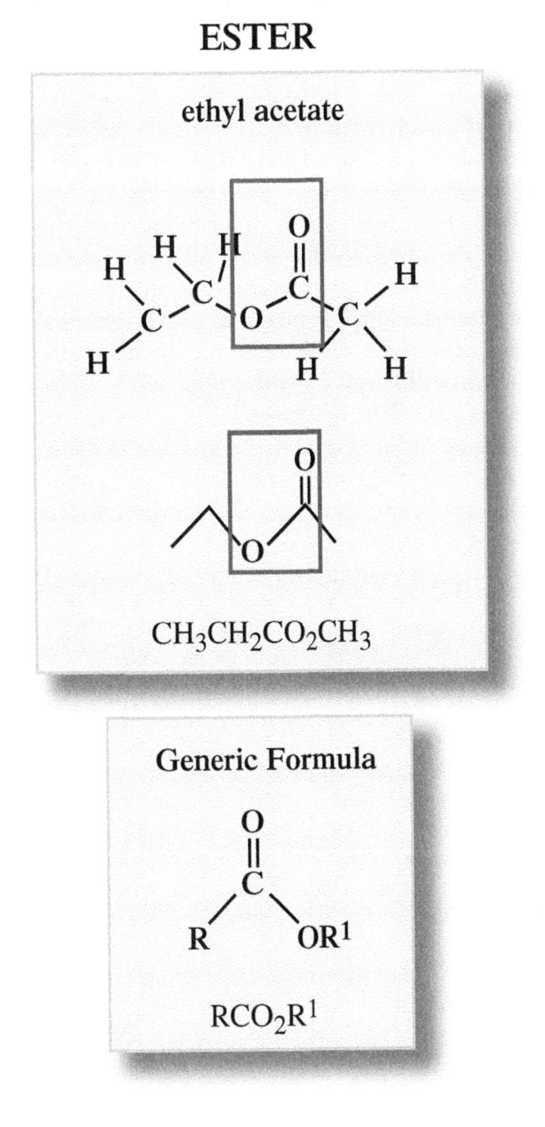

informal name for an acid when possible, and in some cases, a particular ester may simply be referred to as a certain oil or essence. Fortunately for naming purposes, the alcohol moiety of the vast majority of esters is derived from a simple alcohol, such as methanol, ethanol, propanol, or butanol.

Formation of Esters

Simple esters can be formed by dehydration reactions between a carboxylic acid and an alcohol in the presence of an acid catalyst. The reaction proceeds by a stepwise mechanism in which a proton, in the form of the hydrogen cation (H^+), bonds to the carbonyl oxygen atom of the acid, drawn by the two lone pairs of electrons and the pi (ϖ) bond. The pi electrons are attracted to the oxygen atom, leaving the carbonyl carbon atom carrying a positive charge. The lone pair electrons in the oxygen atom of the alcohol's hydroxyl group then add to the positively charged carbon atom, and the hydroxyl group releases its hydrogen atom as a proton, which regenerates the acid catalyst. The intermediate that is formed by this step is unstable and dehydrates readily, eliminating two hydrogen atoms and an oxygen atom to regenerate the carbonyl group. Primary alcohols react to form esters faster than secondary and tertiary alcohols, and tertiary alcohols react the slowest.

Esters can also be produced by reacting an acid chloride or an acid anhydride with an alcohol. Acid chlorides are polarized, with an enhanced positive character on the carbonyl carbon atom. This site acquires the hydroxyl group from the alcohol, and the resulting molecule, or adduct, eliminates hydrogen chloride (HCl) to form the ester linkage. Acid anhydrides react in much the same manner, but the adduct in this case eliminates a molecule of the acid while forming the ester. Given that acid anhydrides are not readily formed from anything other than simple acids, this method of ester formation is not widely used, though it does have the advantage of working with aryl alcohols, known as "phenols," as well as with alkyl alcohols.

The third main method of producing an ester is called "transesterification." In essence, in this process, an existing ester is induced to exchange its alcohol group for another alcohol group, thus transforming from one ester into another. As a simple example, a methyl ester may be dissolved in a large excess of ethanol with a trace of acid catalyst. An equilibrium is set up in which the methyl ester acquires a proton (H^+) and then forms an intermediate adduct with one of the plentiful ethanol molecules (which are nucleophiles, or electron-rich species). This adduct can break down either by eliminating the ethanol molecule to re-form the original methyl ester or by eliminating the methanol to form the new ethyl ester. In the latter case, the methanol is effectively removed from the exchange equilibrium and is dissolved away by the excess ethanol solvent, leaving the ethyl ester in place of the original methyl ester. This method is most useful when the product ester is much more stable than the original ester under the reaction conditions, allowing it to be recovered cleanly after the reaction has been carried out.

Any alcohol group, and any number of alcohol groups in the same molecule, can be esterified. In biochemistry, this vital operation is mediated by enzyme catalysts and produces long-chain fatty-acid esters of glycerol, which are the structural component of the phospholipid bilayer that forms the cell membranes in animals. Enzymes also mediate the esterification of cholesterol by bile acids to facilitate its elimination from the system.

Cyclic Esters

Cyclic esters, or lactones, are formed exclusively by internal esterification of hydroxyacids, which are carboxylic acids with a hydroxyl group at an appropriate position in their molecular structures: the

SAMPLE PROBLEM

Given the chemical name 2,2-dimethylpropanol isobutyrate, draw the skeletal structure of the molecule.

Answer:

The name indicates that the material is an ester of isobutyric acid. Isobutyric acid is derived from butane, a four-carbon chain. The alcohol portion of the ester is 2,2-dimethylpropanol, which has two methyl groups ($-CH_3$), both on the second carbon of its three-carbon chain. The molecule therefore has the following structure:

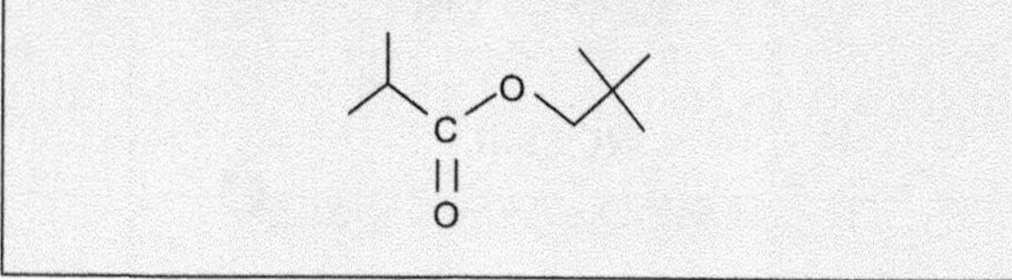

alpha (α) position, on the carbon atom adjacent to the carbon atom of the carboxyl group; beta (β), two carbon atoms away from the carboxyl group; gamma (γ), or three carbon atoms away; or delta (δ), four carbon atoms away. Carboxylic acids in which the hydroxyl group is in the gamma position undergo ring closure by esterification very easily to produce γ-butyrolactone derivatives. Those with the hydroxyl group in the delta position form δ-valerolactones, since pentanoic acid is also known as "valeric acid." β-Propiolactone, the four-membered lactone ring that forms from acids with a hydroxyl group in the beta position, experiences greater strain and does not form as readily:

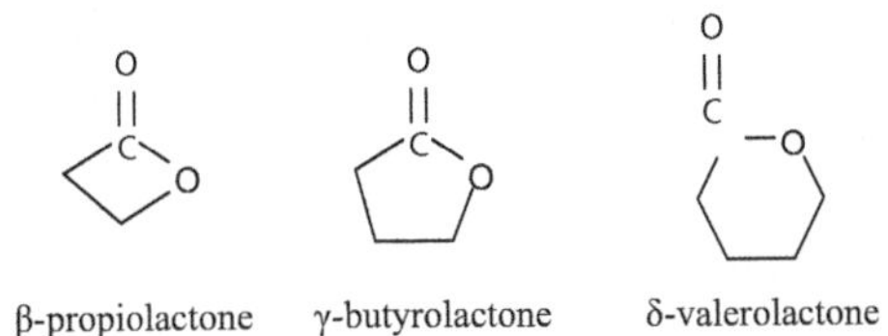

Lactones of greater ring sizes form to the extent that ring strain will allow. As with most cyclization reactions, the larger the ring structure that will form, the more like an acyclic hydrocarbon the structure becomes in terms of strain and the easier it is for ring closure to take place. The names of cyclic esters are derived from their basic lactone structures.

REACTIONS OF ESTERS

Esters are reactive compounds that can be used to synthesize other esters via transesterification. They undergo acid hydrolysis to break the relevant chemical bonds, which allows them to be used as protective groups in reactions that would otherwise affect the carboxyl functional group. Because the carbon atom of an ester's carbonyl group is a common location for the addition of a nucleophile, esters are prone to react with ammonia (NH_3) to produce the corresponding amide. The process is very similar to transesterification, except that the alcohol group of the ester is displaced by an amino group ($-NH_2$) instead of by the alkoxy group of a different alcohol.

Esters can be reduced to primary alcohols either by catalytic hydrogenation, using molecular hydrogen (H_2) and a catalyst, or by chemical reduction, using a reducing agent, such as lithium aluminum hydride ($LiAlH_4$). Catalytic hydrogenation typically cleaves the ester and releases the alcohol group as the carboxylic acid group is reduced to its corresponding alcohol.

The electron-withdrawing effect of the oxygen atom in a carbonyl group imparts an acidic character to any hydrogen atoms on carbon atoms in the alpha position of the molecule. This is an extremely useful feature in organic synthesis reactions and is the basis for a number of reaction types, including aldol condensation, Perkin condensation, Michael addition, Robinson annulation, and malonic ester synthesis. In a malonic ester synthesis reaction, which involves the addition of an alkyl group to an ester of malonic acid ($C_3H_4O_4$), a strong base is used to remove one of the H^+ protons attached to the carbon in the alpha position and form a stabilized anion. The anion is then able to bond as a nucleophile to an appropriate reaction site, such as a primary alkyl halide, forming a new carbon-carbon bond. Typically, one of the two malonic ester groups decomposes and eliminates carbon dioxide. The net result is that the molecule acting as the substrate is enlarged by $-CH_2COOH$. For example, the anion formed from diethyl malonate ($C_7H_{11}O_4^-$) is able to undergo a substitution reaction with isobutyl bromide to yield diethyl 2-isobutylmalonate, which has two carboxylic acid groups. Following hydrolysis and heating, carbon dioxide is eliminated from the molecule to yield 5-methylhexanoic acid. In the diagram here, "Et" represents an ethyl group ($-C_2H_5$):

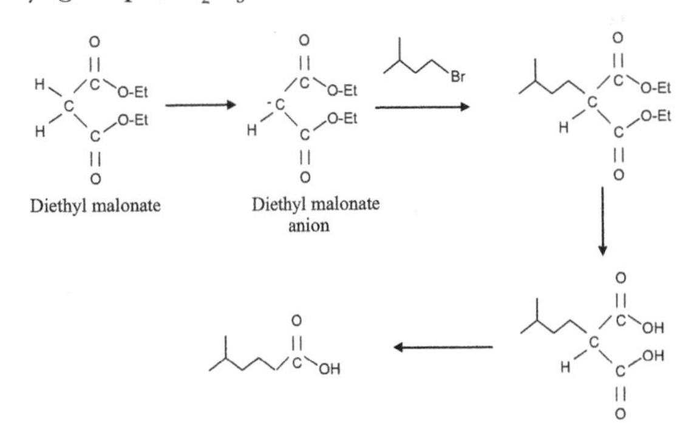

Richard M. Renneboog, MSc

BIBLIOGRAPHY

Berg, Jeremy M., John L. Tymoczko, Gregory J. Gatto, and Lubert Strye. *Biochemistry.* 8th ed. New York: W. H. Freeman, 2015. Print.

Hendrickson, James B., Donald J. Cram, and George S. Hammond. *Organic Chemistry.* 3rd ed. New York: McGraw, 1970. Print.

Jones, Mark Martin. *Chemistry and Society.* 5th ed. Philadelphia: Saunders College Pub., 1987. Print.

Lodish, Harvey, et al. *Molecular Cell Biology.* 7th ed. New York: Freeman, 2013. Print.

Mann, J., et al. *Natural Products: Their Chemistry and Biological Significance.* New York: Wiley, 1994. Print.

Morrison, Robert Thornton, and Robert Neilson. Boyd. *Organic Chemistry.* 7th ed. Englewood Cliffs, N.J.: Prentice Hall, 2003. Print.

O'Neil, Maryadele J., ed. *The Merck Index: An Encyclopedia of Chemicals, Drugs, and Biologicals.* 15th ed. Cambridge: RSC, 2013. Print.

Wuts, Peter G. M., and Theodora W. Greene. *Greene's Protective Groups in Organic Synthesis.* 4th ed. Hoboken: Wiley-Interscience, 2007. Print.

ETHERS

FIELDS OF STUDY

Organic Chemistry

SUMMARY

The characteristic properties and reactions of ethers are discussed. Ethers are useful as reaction media in organic synthesis. They are generally unreactive compounds due to the stability of the ether linkage (C–O–C), but the lone pairs of electrons on their oxygen atoms are capable of coordinating to metal ions as ligands.

PRINCIPAL TERMS

- **alkyl group:** a functional group consisting of a hydrocarbon chain, usually with the general formula C_nH_{2n+1}.
- **aryl group:** a functional group derived from an aromatic ring.
- **base:** a compound that can relinquish one or more hydroxide ions (Brønsted-Lowry acid-base theory) or that possesses lone pairs of electrons that can interact with electron-poor materials (Lewis acid-base theory).
- **functional group:** a specific group of atoms with a characteristic structure and corresponding chemical behavior within a molecule.
- **polyether:** an organic compound characterized by the presence of multiple ether linkages (C–O–C) in its molecular structure.

THE NATURE OF ETHERS

Ethers are characterized by the presence of the ether functional group, which is an oxygen atom bonded to two alkyl groups or aryl groups (or one of each). The general structure is often written in short form as R–O–R', R–O–Ar, or Ar–O–Ar', with R– indicating an alkyl group and Ar– indicating an aryl group.

The molecular geometry of the ether linkage (C–O–C) is similar to that of water, with a bond angle of 109.5 degrees about the oxygen atom (compared to 104.5 degrees in the water molecule). The ether linkage therefore gives the molecule a small electric dipole, which has only a minor effect on physical properties such as melting and boiling points relative to hydrocarbons of similar molecular weight. However, it does allow ether molecules to undergo hydrogen bonding with water to some extent, and they dissolve in water almost as well as alcohols do. Accordingly, ethers have an affinity for water, and it is necessary to subject an ether such as diethyl ether or tetrahydrofuran (THF) to a "drying" protocol in order to remove any traces of water prior to using it as a solvent or reagent.

Polyethers contain more than one ether linkage in their molecular structures. They are useful as solvents and in a variety of hydraulic applications for which analogous hydrocarbons are unsuitable. The stability of the ether functional group provides a similar structural integrity to polyether molecules. A polyether compound can be a mobile liquid, a hard wax, or something in between. Polyethers also have much lower melting points than hydrocarbons of similar weight. Polypropylene glycol [(HO–$(C_3H_6O)_n$H], a major component of brake fluid, is a good example.

The most common ether is probably diethyl ether, once used extensively as a general anesthetic in surgery. It is also used as an ignition booster for diesel engines in cold climates. In laboratory work, diethyl ether is a solvent of choice for Grignard and many other reactions due to its ease of use, inert behavior, and low boiling point.

NOMENCLATURE OF ETHERS

Ethers are named systematically according to the parent hydrocarbon structures from which they are formed. The carbon atoms in the parent hydrocarbon are numbered, either from left or right, depending on which direction gives the ether group the lowest position number possible. The position of the ether group's oxygen atom is indicated by the position number followed by "-oxy"; certain small alkyl or aryl groups are commonly combined with "-oxy" as prefixes, such as "methoxy-," "ethoxy-," "propoxy-," "butoxy-," and "phenoxy-," among others. For example, the ether formed from ethanol and propane would be formally named 1-ethoxypropane, indicating that the substituent group, or side chain, consists of an ethoxy group (CH_3CH_2O–) and is attached to the first atom of the parent structure, propane. Informally, the compound would just be called ethyl propyl ether, indicating that one of the alkyls attached to the oxygen atom is an ethyl group (–C_2H_5, the ethoxy group minus the oxygen) and the other is a propyl group (–C_3H_7). In informal naming, the side chains are listed in alphanumerical order.

Cyclic ethers also follow the systematic name convention. An eight-membered ring in which an ether linkage is located at two of the eight positions in the ring might be named 1,5-dioxacyclooctane, with the prefix "oxa-" indicating that the oxygen atoms are replacing carbon atoms that would otherwise be in cyclooctane. The "oxa-" prefix method is typically used to name ethers in which both substituent groups are complex:

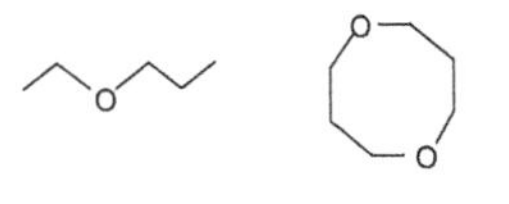

1-ethoxypropane 1,5-dioxacyclooctane

Cyclic polyethers are given the special name and designation "crown ethers." Their overall structure is reminiscent of a crown, and their several oxygen atoms allow them to function as a single ligand to effectively encapsulate metal ions. The crown ethers are identified by a name indicating the size of the ring structure and the number of oxygen atoms it contains. The compound 18-crown-6, for example, is an eighteen-membered ring structure containing six oxygen atoms. The molecule is sufficiently flexible for the six oxygen atoms to coordinate to six positions available around a metal ion:

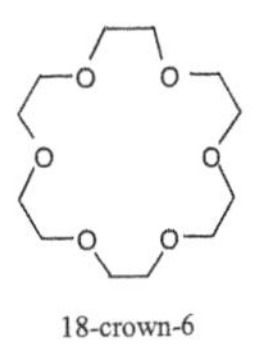

18-crown-6

FORMATION OF ETHERS

Although an ether is ostensibly the product of a condensation reaction between two alcohols, ethers are never formed this way in practice. It is much easier for the alcohols to undergo intermolecular dehydration to form alkenes—organic compounds that include at least one carbon-carbon double bond (C=C)—than to separate the hydroxyl functional group (–OH). Instead, ethers are formed primarily by the Williamson ether synthesis reaction, in which the alkoxide (–OR) of an alcohol is used as a nucleophile, or electron donor, to displace a halide ion from a corresponding alkyl halide. This method is capable of producing asymmetrical as well as symmetrical ethers. It is more generally applicable than the name might suggest, since a great many

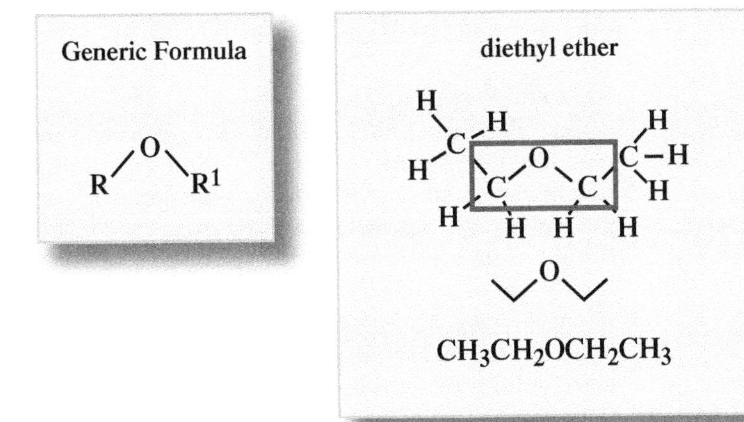

different alkyl and aryl alkoxides can be used as the nucleophilic reagent.

Similarly, numerous substituent groups can be displaced by the nucleophile in bimolecular substitution reactions. One of the most commonly used is the tosylate group, which is the anion of *p*-toluenesulfonyl chloride. The corresponding acid chloride, tosyl chloride, readily reacts with the alcohol functional group to form the tosylate ester. The tosylate ion is a much better leaving group than the hydroxide ion (OH^-) and is therefore much more easily displaced by an alkoxide ion (RO^-) to produce the ether product. Phenols, or aryl alcohols, are more acidic than alkyl alcohols because the electrons in their parent benzene rings are delocalized, moving freely throughout the molecule. Accordingly, phenoxide ions are easily produced by reaction with a hydroxide ion or similar basic compound. Reaction between alkoxide or phenoxide ions and a tosylate ester readily produces the corresponding ether compounds. Other sulfonate esters, such as methyl sulfate, are also useful in producing ethers.

SAMPLE PROBLEM

Given the chemical names benzyl t-butyl ether and 1,4,7,10-tetraoxocyclododecane (or 12-crown-4), draw the structures of the molecules.

Answer:

The first name indicates that the material is an ether with a benzyl group ($C_6H_5CH_2$–) on one side of the oxygen atom and a *t*-butyl group [$(CH_3)_3C$–] on the other. The molecule therefore has the following structure:

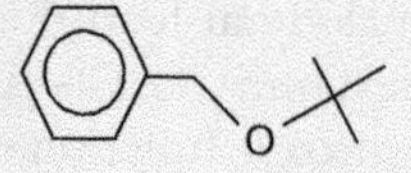

The second name indicates that the compound is a twelve-membered ("dodec-") ring structure ("cyclo-") with four oxygen atoms ("tetraoxo-") instead of methylene bridges (–CH_2–) in the ring, at the first, fourth, seventh, and tenth positions. The molecule therefore has the following structure:

The alkoxymercuration-demercuration reaction can also produce ethers. In this reaction, an alkene substrate molecule and an alcohol react with mercury(II) trifluoroacetate [$Hg(CF_3COO)_2$] to form a complex intermediate. The alkoxyl portion (–OR) of the alcohol adds to one end of the carbon-carbon double bond in the alkene. The other end of the double bond attaches to the mercury ion, displacing one of the trifluoroacetate groups. This is then reduced with sodium borohydride ($NaBH_4$) to eliminate the mercury moiety and form the ether.

Mixed ethers of silicon-based compounds are easily produced from the corresponding silyl chloride (a compound containing one or more silicon-chlorine bonds) and a hydroxyl group, and any number of hydroxyl groups in the same alcohol molecule can be converted into ethers in this way. Such mixed ethers are normally used as protecting groups for the hydroxyl group. Once the silyl ether has served its function, it is easily hydrolyzed back to the alcohol. Silyl ethers are often used to immobilize a carbonyl group (C=O) in a compound in order to direct reactivity to another part of the molecule. This is most useful when there is more than one carbonyl group in a compound but only one of them is meant to undergo a reaction.

A special class of cyclic ethers consists of three-membered rings containing one oxygen atom. These compounds, known as epoxides, are highly reactive because the acute bond angles in the three-membered ring physically strain the system. This additional strain significantly lowers the activation energy and increases the enthalpy released by bond opening. Epoxides are typically formed by reacting an alkene carbon-carbon double bond with a peroxide, though other methods are also used.

Reactions of Ethers

Except for the epoxides, ethers are generally unreactive under many reaction conditions. Though ethers have a certain affinity for moisture, they are generally easy to prepare in anhydrous form. They do not react with strong bases and so are useful as the solvent medium for reactions in which extremely strong bases, such as *n*-butyllithium, are to be used. Similarly, ethers are not prone to react with strongly electrophilic (electron-accepting) reagents, though the oxygen atom of the ether linkage can coordinate to form complexes with metal atoms. This makes ethers useful in Grignard and similar reactions, as their presence tends to stabilize the complex. Diethyl ether and THF are the most commonly used solvents for these types of reactions.

Though ethers are stable toward electrophilic reagents, strong acids, such as hydrogen bromide (HBr) or hydrogen iodide (HI), do cleave them. Because of the two lone pairs of electrons that it carries, the oxygen atom of the ether linkage is a region of high electron density. As a result, the oxygen atom is easily protonated (given a proton in the form of the hydrogen cation, H^+) by acid. But cleavage of the protonated ether requires the formation of a much less stable carbonium ion, in which the carbon atom carries a positive charge. Thus, a relatively high activation energy is required for cleavage to occur, and ether-cleavage reactions typically must be carried out at high temperatures. The reaction produces an alcohol and an alkyl halide.

Ethers can autoxidize, meaning that they can be slowly oxidized by the presence of oxygen in the air. Such reactions produce volatile peroxides and hydroperoxides, so ethers must be carefully handled to avoid contact with air and "dried" to eliminate any peroxides that may form.

The ability to coordinate to metal atoms is an important feature of ethers, especially crown ethers. Metal ions coordinate ligands according to the geometry of their atomic orbitals. A cyclic polyether in which the oxygen atoms are positioned at the proper distances can coordinate to the correct orientations as just a single ligand. This ability has proved useful in different applications, including the development of pharmaceutical compounds.

Epoxides, unlike other ether compounds, are susceptible to ring-opening reactions with either bases or acids. A base, or nucleophile, can initiate ring opening via a bimolecular substitution that effectively uses the oxygen atom of the epoxide ring as a leaving group. Because the oxygen atom is still bonded to one carbon atom, it is capable of bringing the rest of the molecule into a subsequent reaction. Alternatively, an acid, or electrophile, can initiate ring opening via a unimolecular substitution. Protonation of the oxygen atom, for example, can result in formation of a carbonium ion when the ring opens, and the molecule can be carried into subsequent reactions. Epoxy adhesives use this feature as a means of polymerizing the material to produce a strong physical bond.

Richard M. Renneboog, MSc

Bibliography

Acheson, R.M. *An Introduction to the Chemistry of Heterocyclic Compounds.* 3rd ed. New York: Wiley, 2008. Print.

Hendrickson, James B., Donald J. Cram, and George S. Hammond. *Organic Chemistry.* 3rd ed. New York: McGraw, 1970. Print.

Haynes, W. H., PhD, David R. Lide, PhD, and Thomas J. Bruno, PhD, eds. *CRC Handbook of Chemistry and Physics,* 96th Edition. New York: CRC/Taylor & Francis Group, 2015. Print.

Miessler, Gary L., Paul J. Fischer, and Donald A. Tarr. *Inorganic Chemistry.* 5th ed. Upper Saddle River: Prentice Hall, 2013. Print.

Morrison, Robert Thornton, and Robert Neilson. Boyd. *Organic Chemistry.* 7th ed. Englewood Cliffs, N.J.: Prentice Hall, 2003. Print.

O'Neil, Maryadele J., ed. *The Merck Index: An Encyclopedia of Chemicals, Drugs, and Biologicals.* 15th ed. Cambridge: RSC, 2013. Print.

Wuts, Peter G. M., and Theodora W. Greene. *Greene's Protective Groups in Organic Synthesis.* 4th ed. Hoboken: Wiley-Interscience, 2007. Print.

EXOTHERMIC AND ENDOTHERMIC REACTIONS

FIELDS OF STUDY

Organic Chemistry; Inorganic Chemistry; Physical Chemistry

SUMMARY

The processes of exothermic and endothermic reactions are described, and their importance in chemical transformations is elaborated. Reactions that release heat energy into the environment are considered to be exothermic, while those reactions that absorb heat energy from the environment are endothermic.

PRINCIPAL TERMS

- **enthalpy:** the total heat content within a thermodynamic system, defined as internal energy plus the product of pressure and volume; also, the change in heat content associated with a chemical process.
- **heat energy:** the kinetic energy of the random motion of a system's component particles that contributes to the temperature of that system.
- **thermodynamics:** the branch of physics that deals with the relationships between energy, heat, and work within a physical system.

ENERGY TRANSFERS IN CHEMICAL REACTIONS

All chemical reactions involve the transfer of energy, which can occur in the form of either heat or work. Chemical systems contain energy in the bonds between their atoms. The electrons that form these atomic bonds possess an energy level that is determined by various factors, including the nature of the bond, the extent to which the bond angles are deformed from their ideal angles, and the force of repulsion between the electrons in the bond. During a chemical reaction, the level of energy stored within a chemical system changes as bonds are formed or broken: the formation of a bond releases energy, some of which escapes from the system into its surroundings, while the breaking of a bond consumes energy. The overall change in the level of energy caused by a chemical reaction is called the enthalpy change of the reaction (ΔH). The enthalpy change is determined by subtracting the enthalpy, or total energy content, of the reactants from that of the products. If the enthalpy of the products of a reaction is lower than the enthalpy of the reactants, there is a negative change in enthalpy, meaning that the reaction is exothermic. If the enthalpy change of a reaction is positive, the reaction is considered to be endothermic.

Another important consideration in determining the energy transfer of a reaction is the change in entropy (ΔS) of the system. Entropy is a measure of the degree to which energy has been dispersed throughout the atoms and molecules of a system; high entropy indicates a wide dispersal of energy, causing the components of the system to move in a more disordered manner than in a system with lower entropy. When energy is transferred to a system during an endothermic reaction, the entropy of the system increases. The change in entropy is influenced by the temperature at which a reaction occurs; at lower temperatures, the change in entropy is greater than at higher temperatures.

Chemical reaction processes that release heat energy are termed exothermic, while those that absorb heat energy from the environment are termed endothermic. These reactions bear some similarity to exoergic and endoergic (or exergonic and endergonic) reactions. However, the terms "exoergic" and "endoergic" refer to the release and the absorption, respectively, of free energy—that is, energy that is available to do work within a system—during a chemical reaction, while the terms "exothermic" and "endothermic" refer to the release and absorption of heat energy—that is, energy in the form of heat. In an exoergic reaction, the change in free energy is negative, indicating a spontaneous reaction. In an endoergic reaction, the change in free energy is positive, indicating that in order to do work, energy must be provided by a source outside of the reaction system.

ENTHALPY, ENTROPY, AND GIBBS FREE ENERGY

In thermodynamics, the basic characteristics of chemical reaction systems are described by three factors: entropy, enthalpy, and Gibbs free energy. Gibbs free energy is a measure of free energy as determined by the change of entropy within a system. (Free energy is essentially the total energy of a system minus the entropy, which represents energy that cannot be used to do work.) The three concepts are related by the expression:

$$\Delta G = \Delta H - T\Delta S$$

where *G* is Gibbs free energy, *H* is enthalpy, *T* is temperature, *S* is entropy, and the delta symbol (Δ) represents the overall change in the associated factor. From this equation, it can be seen that the change in the Gibbs free energy is equal to the change in enthalpy minus the change in entropy, the effect of which is more significant at higher temperatures. The relationship can be exemplified by the combustion reaction of octane (C_8H_{18}), according to the equation:

$$2C_8H_{18} + 25O_2 \rightarrow 16CO_2 + 18H_2O$$

On the simplest level, it can be seen that the reaction begins with twenty-seven molecules of reactants and ends with thirty-four molecules of products. Entropy increases as the system becomes more disordered. At the same time, a total of seventy-five bonds (thirty-six carbon-hydrogen bonds, fourteen carbon-carbon bonds, and twenty-five oxygen-oxygen bonds) are broken in the reactants, and sixty-eight bonds (thirty-two carbon-oxygen double bonds and thirty-six hydrogen-oxygen bonds) are formed in the products. About 23,063 kilojoules per mole (kJ/mol) of heat energy are absorbed to break the bonds of the octane molecules, and 35,892 kJ/mol are released when forming the bonds in the carbon dioxide and water molecules. In other words, the reaction releases approximately 12,829 kJ/mol more heat energy than it absorbs. The reaction is therefore exothermic, as is apparent by the physical heat that is given off by burning octane (a major component of gasoline). However, it is not possible to tell just from the value of the enthalpy change whether a reaction is spontaneous or not. In this particular example, even though the reaction releases a great deal of heat energy, it nevertheless requires the input of activation energy in order to proceed and is therefore non-spontaneous.

ENDOERGIC AND ENDOTHERMIC REACTIONS

Endoergic reactions are chemical reactions that need the input of energy from an external source in order to take place. Endothermic reactions are superficially similar in this respect. The difference is that endoergic reaction processes do not occur unless energy is supplied, while endothermic reactions proceed normally, extracting energy from the surroundings in the process. The term "endoergic" can be applied to both chemical reaction systems and physical processes, while "endothermic" is used only in reference to specific chemical reactions. (The same is true for "exoergic" and "exothermic," respectively.) Accordingly, while an endoergic reaction can be coupled with an exoergic process so that the latter can provide the energy needed for the former to occur, the reaction is initiated by the free energy released by the exoergic process, not the enthalpy or heat energy of the system or its surroundings. Endothermic reactions cannot be coupled to exothermic reactions in the same way.

The relationship between enthalpy and entropy determines whether a reaction is spontaneous or non-spontaneous by affecting the value of Δ*G*. For a reaction to occur spontaneously, Δ*G* must be negative. In other words, in a spontaneous reaction, the products must have less Gibbs free energy than the reactants. In reactions

EXOTHERMIC AND ENDOTHERMIC REACTIONS

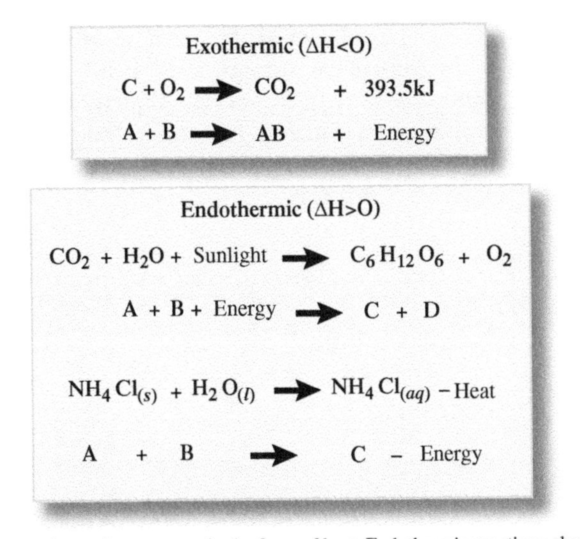

Exothermic reactions release energy in the form of heat. Endothermic reactions absorb energy in the form of heat. Absorption of heat may be written in two ways (shown above).

SAMPLE PROBLEM

Determine whether the following reactions are endothermic or exothermic:

- $2H_2 + O_2 \rightarrow 2H_2O; \Delta H = -241.8$ kJ/mol
- $6CO_2 + 6H_2O + \text{sunlight} \rightarrow C_6H_{12}O_6; \Delta H = 1016$ kJ/mol
- $2Na(s) + Cl_2(g) \rightarrow 2NaCl(s); \Delta H = -145.6$ kJ/mol

Answer:

The value of the enthalpy change (ΔH) is negative for an exothermic reaction and positive for an endothermic reaction.

- The value of ΔH is negative. Therefore, the combustion of hydrogen is exothermic.
- The value of ΔH is positive. Therefore, the formation of glucose by photosynthesis is endothermic.
- The value of ΔH is negative. Therefore, the reaction of sodium metal with chlorine gas to produce solid sodium chloride is exothermic.

where the entropy change (ΔS) is positive and the enthalpy change (ΔH) is negative, the process is spontaneous. When the entropy change is negative and the enthalpy change is positive, the reaction is nonspontaneous and requires the input of energy from an external source.

EXOTHERMIC AND ENDOTHERMIC REACTIONS IN NATURE

All biological systems are formed and supported by endothermic processes such as photosynthesis and exothermic process such as cellular respiration. Organic materials such as carbon-based biopolymers, including celluloses and lignins, and geopolymers, including coal, undergo rapid combustion reactions that are self-sustaining due to the large amounts of heat that they generate. Decomposition reaction processes, such as microbial digestion during composting and the decomposition of biopolymers such as proteins in dead bodies, liberate similar quantities of heat, but at slower rates. All of these examples represent the release of solar energy that was originally captured during the formation of the materials. The first law of thermodynamics, which states that energy cannot be created or destroyed, requires that the total amount of energy released be equal to the total amount of energy that was initially captured.

Richard M. Renneboog, MSc

BIBLIOGRAPHY

Crowe, Jonathan, and Tony Bradshaw. *Chemistry for the Biosciences: The Essential Concepts.* 2nd ed. Oxford: Oxford UP, 2010. Print.

Douglas, Bodie E., Darl H. McDaniel, and John J. Alexander. *Concepts and Models of Inorganic Chemistry.* 3rd ed. New York: Wiley, 1994. Print.

Hendrickson, James B., Donald J. Cram, and George S. Hammond. *Organic Chemistry.* 3rd ed. New York: McGraw, 1970. Print.

Lafferty, Peter, and Julian Rowe, eds. *The Hutchinson Dictionary of Science.* 2nd ed. Oxford: Helicon, 1998. Print.

Morrison, Robert Thornton, and Robert Neilson. Boyd. *Organic Chemistry.* 7th ed. Englewood Cliffs, N.J.: Prentice Hall, 2003. Print.

Reece, Jane B., et al. *Campbell Biology.* 10th ed. San Francisco: Cummings, 2013. Print.

Silbey, Robert J., Robert A. Alberty, and Moungi G. Bawendi. *Physical Chemistry.* 5th ed. Hoboken: Wiley, 2012. Print.

F

FUNCTIONAL GROUPS

FIELDS OF STUDY

Organic Chemistry

SUMMARY

Functional groups are combinations of atoms that influence the reactivity of organic molecules. Although the number of different organic molecules is immense, the number of different functional groups is quite small and their behavior is consistent from compound to compound.

PRINCIPAL TERMS

- **derivative:** a compound that is obtained by subjecting a similar parent compound to one or more chemical reactions that target certain functional groups, leaving the basic molecular structure unaltered.
- **IUPAC:** the International Union of Pure and Applied Chemistry, an organization that establishes international standards and practices for chemistry.
- **moiety:** a specific portion of a molecular structure.
- **nomenclature:** a system of specific names or terms and the rules for devising or applying them; in chemistry, refers mainly to the system of names for chemical compounds as established by the International Union of Pure and Applied Chemistry (IUPAC).
- **organic chemistry:** the study of the chemical identities, behaviors, and reactions of carbon-based compounds and materials.

The Nature of Functional Groups

In organic chemistry, the term "functional group" refers to a specific combination of atoms within a molecular structure that governs the chemical behavior of the molecule. Functional groups exhibit characteristic chemical behavior regardless of the molecule they are found in. Accordingly, although the number of different organic molecules is nearly infinite, the chemical properties and reactions of any organic compound are highly predictable, based on the typical chemical behavior of its component functional groups. This is not a hard and fast rule, as the overall chemical behavior of any particular compound depends on several other factors as well, such as the three-dimensional shape of the molecule and the electronic effects exerted by other groups within the molecule. The environment in which a compound exists, its solvent interactions, and the physical state of the compound are other factors that affect its specific chemical behaviors.

The simplest organic compounds, or carbon-containing molecules, are pure hydrocarbons consisting of only hydrogen and carbon atoms. When all of the carbon atoms in a hydrocarbon compound are bonded to the maximum possible number of atoms, forming four single bonds each, the hydrocarbon compound is said to be saturated. Saturated hydrocarbons, or alkanes, are effectively unreactive compounds under most conditions, though alkanes can be highly reactive in combustion, halogenation, and dehydrogenation reactions.

Hydrocarbon derivatives are hydrocarbon compounds that have additional functional groups in the place of one or more hydrogen atoms. The presence of a functional group, no matter how simple, in a hydrocarbon molecule alters the chemical properties of the molecule in specific ways, according to the nature of the functional group. Through the addition of functional groups, it is possible to convert the simplest hydrocarbon, methane (CH_4), into even the largest and most complex organic molecules imaginable.

Nomenclature of Functional Groups

The nomenclature of an organic compound is based on the characteristic functional group it contains. Because organic molecules can contain any

FUNCTIONAL GROUP

Common Functional Groups

Structure	Name	Example
$-\overset{\vert}{\underset{\vert}{C}}-$	alkane	$CH_3CH_2CH_3$ propane
$>C=C<$	alkene	$CH_3CH\ CH_2$ propene
$-C\equiv C-$	alkyne	CH_3CCH propyne
–F, –Cl, –Br, –I	alkylhalide	CH_3Br methylbromide
$-OH$	alcohol	CH_3CH_2OH ethanol
$-O-$	ether	$CH_3\ OCH_3$ dimethyl ether
$-NH_2$	amine	$CH_3\ NH_2$ methylamine
$-\overset{O}{\overset{\Vert}{C}}-H$	aldehyde	$CH_3\ CHO$ acetaldehyde
$-\overset{O}{\overset{\Vert}{C}}-$	ketone	$CH_3\ COCH_3$ acetone
$-\overset{O}{\overset{\Vert}{C}}-Cl$	acyl chloride	$CH_3\ COCl$ acteyl chloride
$-\overset{O}{\overset{\Vert}{C}}-OH$	carboxylic acid	$CH_3\ CO_2H$ acetic acid
$-\overset{O}{\overset{\Vert}{C}}-O-$	ester	$CH_3\ CO_2CH_3$ methyl acetate
$-\overset{O}{\overset{\Vert}{C}}-NH_2$	amide	$CH_3\ NH_2$ acetamide

Functional groups give the characteristic properties to the compound; they determine the behavior of the compound.

number of functional groups, determining the best systematic name for a compound can be difficult. The International Union of Pure and Applied Chemistry (IUPAC) maintains a set of standardized rules for the nomenclature of chemical compounds, based on the type and position of their functional groups, in order to ensure that the systematic name of a compound accurately reflects its unique molecular structure. Within that system of nomenclature, the essential functional group of a molecular structure is indicated by the suffix of the compound name. For example, the names of all ketones end in "-one," those of all aldehydes end in "-al," those of all alcohols end in "-ol," and so on. The list is quite extensive and is updated as required.

ESSENTIAL FUNCTIONAL GROUPS

Active Hydrocarbons

Hydrocarbons come in two basic types: saturated and unsaturated. The electronic structure of the carbon atom permits it to hybridize its 2*s* and 2*p* atomic orbitals into either three equivalent sp^2 orbitals or four equivalent sp^3 orbitals, which allows it to form two, three, or four bonds of equal lengths. Carbon atoms that have four single bonds to other atoms have saturated their ability to form bonds, and hydrocarbons in which all carbon atoms are saturated are generally the most unreactive. Organic compounds that contain at least one double (C=C) or triple (C≡C) carbon-carbon bond are considered to be unsaturated. These double and triple bonds are reactive sites that define the alkene and alkyne series of hydrocarbon compounds, respectively. The reactive sites of unsaturated hydrocarbons most commonly undergo addition reactions.

Halogen-Containing Compounds

Although alkanes are generally unreactive, they can undergo substitution reactions, primarily through the replacement of a hydrogen atom with a halogen atom (group 17). Removal of a hydrogen atom from an alkane creates an alkyl functional group, which has the general molecular formula C_nH_{2n+1}; in addition, an alkane in which more than one hydrogen atom has been replaced is also considered to be an alkyl group. An alkane that has been substituted with one or more halogen atoms is known as an alkyl halide or a haloalkane. Alkenes and alkynes often undergo addition reactions whereby a halogen atom joins the molecular structure of an unsaturated hydrocarbon at the C=C or C≡C bond.

Oxygen-Containing Compounds

The addition of oxygen to a hydrocarbon molecule forms the basic structure of a number of different classes of compounds. Alcohols are organic compounds than contain the hydroxyl (–OH) functional group bonded to a saturated carbon atom. Phenols feature the hydroxyl group bonded to an aromatic hydrocarbon, which is a hydrocarbon whose carbon atoms form a ring with alternating double and single bonds, distributed in such a way that all bonds are of equal length and strength.

A functional group derived from an aromatic hydrocarbon is known as an aryl group. An alkoxyl, or alkoxy group, is an oxygen atom bonded to an alkyl or aryl group. Another class of oxygen-containing organic compounds is the ethers, characterized by the presence of an ether group, which is similar to an alkoxy group but has a second alkyl or aryl group bonded to the oxygen atom.

Perhaps the most important oxygen-containing compounds are those in which the oxygen atom shares a double bond with a carbon atom, forming a carbonyl group (C=O). The carbonyl group is the fundamental feature of aldehydes, ketones, and all of the derivatives of the carboxylic acids, including acid anhydrides, acid chlorides, esters, amides, and their variations. A carboxylic acid is defined by the presence of a carboxyl group (–COOH), which is a carbonyl group that is attached via the carbon atom to a hydroxyl group. Similarly, an acyl group consists of a carbonyl whose carbon atom is bonded to an alkyl or aryl group, while in an ester group, the carbonyl carbon is bonded to both an alkoxy group and an alkyl or aryl group.

Nitrogen-Containing Compounds

Nitrogen atoms can be incorporated into hydrocarbon molecules as primary, secondary, or tertiary amines. Amines are derivatives of ammonia (NH_3) in which one or more nitrogen-hydrogen bonds have been replaced by nitrogen-carbon bonds; the primary, secondary, and tertiary labels indicate whether one, two, or three of the hydrogen atoms have been replaced. The nitrogen atom can also form double bonds (C=N) and triple bonds (C≡N) with carbon, forming the basis for the imines and the nitriles, respectively.

Sulfur-Containing Compounds

Sulfur forms several different types of hydrocarbon-based compounds. Compounds containing a sulfur atom between carbon and hydrogen (C–S–H) are named thiols. Compounds with a sulfur atom between two carbon atoms (C–S–C) are called sulfides, while those with two sulfur atoms between two carbon atoms (C–S–S–C) are named disulfides. The sulfur atoms in the sulfide and disulfide compounds can also be bonded to a number of oxygen atoms, giving rise to sulfoxides, sulfones, thiosulfoxides, thiosulfones, sulfinates, sulfonates, thiosulfinates, and thiosulfonates.

Phosphorus-Containing Compounds

Phosphorus is not as versatile as sulfur and does not form as many different compounds with hydrocarbon moieties. The most common of the phosphorus-containing compounds are the phosphines, in which three carbon atoms are bonded to one phosphorus atom.

Reactions of Functional Groups

Chemists can deduce the presence of a functional group within a compound based on the compound's reaction when exposed to certain reagents and the electronegativity of the atoms within the molecule. The carbonyl group, for example, is slightly polarized by the difference in electronegativity between the oxygen and carbon atoms. Accordingly, a nucleophilic atom introduced to the carbonyl group will generally join the carbon atom. The same feature also makes hydrogen atoms on carbon atoms adjacent to the carbonyl group somewhat acidic, so that a strong base can remove them to form the corresponding anion. Because functional groups undergo the same types of reactions regardless of the compound they are in, they are an important tool for distinguishing similar compounds from one another.

Richard M. Renneboog, MSc

SAMPLE PROBLEM

Identify the functional groups in the following compounds. Note that in organic chemistry diagrams, each unlabeled atom is assumed to be a carbon atom, and each carbon atom is assumed to be bonded to the maximum possible number of hydrogen atoms unless otherwise noted.

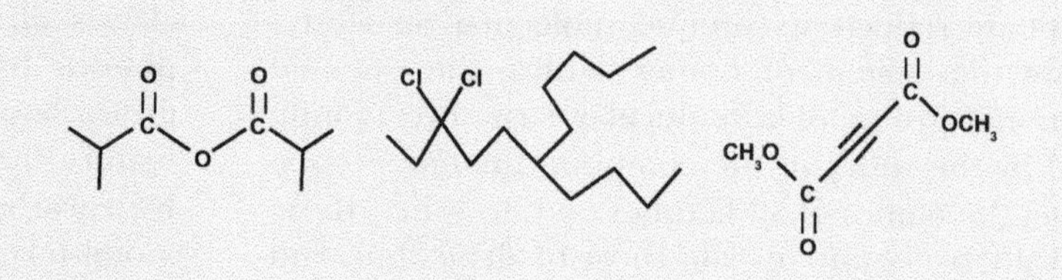

Answer:

The first compound contains two carbonyl groups, as indicated by the double bond between carbon and oxygen. The presence an alkyl group on the carbon atom of each carbonyl group means that both carbonyls are in fact part of acyl groups. The compound is an acid anhydride, which is defined as an organic compound containing two acyl groups bound to the same oxygen atom.

The second compound is a saturated hydrocarbon, or alkane, in which two hydrogen atoms have been replaced with chlorine atoms, creating an alkyl group. Chlorine is a halogen, making the compound an alkyl halide.

The third compound contains an alkyne, as indicated by the carbon-carbon triple bond, and two ester groups, each one consisting of a carbonyl group bonded to the alkoxy group OCH_3, known as methoxy. The presence of the two ester groups makes the compound a diester.

Bibliography

Acheson, R.M. *An Introduction to the Chemistry of Heterocyclic Compounds.* 3rd ed. New York: Wiley, 2008. Print

Berg, Jeremy M., John L. Tymoczko, Gregory J. Gatto, and Lubert Strye. *Biochemistry.* 8th ed. New York: W. H. Freeman, 2015. Print.

Hendrickson, James B., Donald J. Cram, and George S. Hammond. *Organic Chemistry.* 3rd ed. New York: McGraw, 1970. Print.

Jones, Mark Martin. *Chemistry and Society.* 5th ed. Philadelphia: Saunders College Pub., 1987. Print.

Lehninger, Albert L. *Biochemistry: The Molecular Basis of Cell Structure and Function.* 2nd ed. New York: Worth, 1975. Print..

Haynes, W. H., PhD, *David R. Lide, PhD, and Thomas J. Bru*no, PhD, eds. CRC Handbook of Chemistry and Physics, 96th Edition. New York: CRC/Taylor & Francis Group, 2015. Print.

Lodish, Harvey, et al. *Molecular Cell Biology.* 7th ed. New York: Freeman, 2013. Print.

Morrison, Robert Thornton, and Robert Neilson. Boyd. *Organic Chemistry.* 7th ed. Englewood Cliffs, N.J.: Prentice Hall, 2003. Print.

O'Neil, Maryadele J., ed. *The Merck Index: An Encyclopedia of Chemicals, Drugs, and Biologicals.* 15th ed. Cambridge: RSC, 2013. Print.

Wuts, Peter G. M., and Theodora W. Greene. *Greene's Protective Groups in Organic Synthesis.* 4th ed. Hoboken: Wiley-Interscience, 2007. Print.

H

HALF-LIFE

FIELDS OF STUDY

Nuclear Chemistry; Physical Chemistry

SUMMARY

The half-life describes the condition in an exponential rate process at which one-half of any given amount of material is consumed in the process. The concept is particularly useful in determining the age of a radioactive material.

PRINCIPAL TERMS

- **carbon dating:** a method of dating that uses the proportion of radioactive carbon-14 atoms remaining in an organic material to determine the amount of time that has elapsed since it was part of a living organism.
- **exponential decay:** a process of decomposition in which the amount of non-decayed material decreases at a rate proportional to the current amount of material present rather than the original amount.
- **isotope:** an atom of a specific element that contains the usual number of protons in its nucleus but a different number of neutrons.
- **logarithm:** the exponent, or power, to which a specific base number must be raised to produce a given value; commonly abbreviated "log."
- **radioactivity:** the emission of subatomic particles due to the spontaneous decay of an unstable atomic nucleus, the process ending with the formation of a stable atomic nucleus of lower mass.

Half-Life Defined

The term "half-life" refers to the length of time required for one-half of any given amount of a material to be consumed in a process that proceeds according to an exponential rate expression. It is applicable to any and all such processes but is most often used in reference to the process of exponential decay that occurs in radioactive isotopes.

Radioactivity exists when an atomic nucleus contains an unstable combination of protons and neutrons. Because each proton bears a positive electrical charge, a strong force of repulsion exists between protons when more than one is present in a nucleus. The neutron, which is electrically neutral and only very slightly more massive than a proton, seems to have an internal structure that effectively places some negative charge that exists within it next to the adjacent protons, allowing it to function as a sort of nuclear glue to hold the protons together and stabilize the nucleus. There are numbers of protons that cannot be adequately balanced by the presence of any whole-number combination of neutrons. Atomic nuclei with this imbalance, whether due to too many or too few neutrons, are unstable and therefore radioactive.

Radioactive decay is said to be exponential because it obeys the exponential rate law:

$$A_t = A_0 e - kt$$

in which A_0 is the amount of material originally present, A_t is the amount of material present after a specific amount of elapsed time (t), and k is the specific rate constant for the process. The mathematical constant e is known as the natural base, an indeterminate irrational number that has an approximate value of 2.718281828 and is a common feature of the natural world. The exponent $-kt$ is the logarithm of the value A_t/A_0 to the base e, meaning that is the power to which e must be raised in order to equal A_t/A_0. Many may be more familiar with the base-10 use of logarithms, which functions the same way. For example, the logarithm of 100 to base 10 is 2, since $10^2 = 100$. To differentiate the two, the logarithm of base e is commonly called the "natural logarithm," abbreviated ln or $\log_e$.

According to the above rate expression, the half-life is the time (t) that has elapsed when A_t is exactly one-half of the value of A_0, or:

$$A_t = \frac{1}{2}A_0$$

When the expression is rearranged and substituted into the rate expression, it becomes:

$$\frac{A_t}{A_0} = \frac{1}{2} = e^{-kt}$$

The logarithms of both sides of the equation must also be equal. Therefore:

$$\ln\frac{1}{2} = \ln(e^{-kt})$$

$$-0.693 = -kt$$

$$\frac{0.693}{k} = t$$

The half-life of the process ($t_{1/2}$) is thus given by the equation:

$$t_{1/2} = \frac{0.693}{k}$$

In physical terms, this means that when the amount of time equal to the half-life of the process has passed, regardless of the value of k, there will be half the amount of non-decayed material that there was at the beginning of the half-life period. When a second half-life's worth of time has passed, just half of that half—one-quarter of the original amount—will remain. After a third half-life, just one-eighth of the original amount will remain, and so on. The fraction remaining after n half-lives can therefore be expressed as follows:

$$A_n = A_0\left(\frac{1}{2}\right)^n$$

It should be noted that no matter how many half-lives pass, there will always be some amount of the original material left, until only one atom or molecule of the material remains. When that atom or molecule is consumed, the process ends. In radioactive decay, the process ends when a stable nucleus forms.

History of Radioactivity

Though radioactive materials have existed naturally since the beginning of the universe, formed by various nuclear processes, people did not know about them until it became possible to detect their effects. French physicist Henri Becquerel (1852–1908) first recognized radioactive emanations from uranium ore at the end of the nineteenth century. This led to the isolation and identification of radium and other radioactive elements, notably by husband-and-wife physicists Pierre Curie (1859–1906) and Marie Curie (1867–1934).

Unfortunately for many researchers in this area, the adverse health effects of radiation were unknown at the time, and most researchers lived considerably shorter lives than they would have otherwise. General ignorance of the effects of radiation exposure persisted well into the twentieth century. For example, luminous-dial wristwatches, on which the numbers glowed in the dark due to radioactive radium or promethium in the paint, were popular well into the 1960s; tritium, a radioactive isotope of hydrogen, was

HALF-LIFE

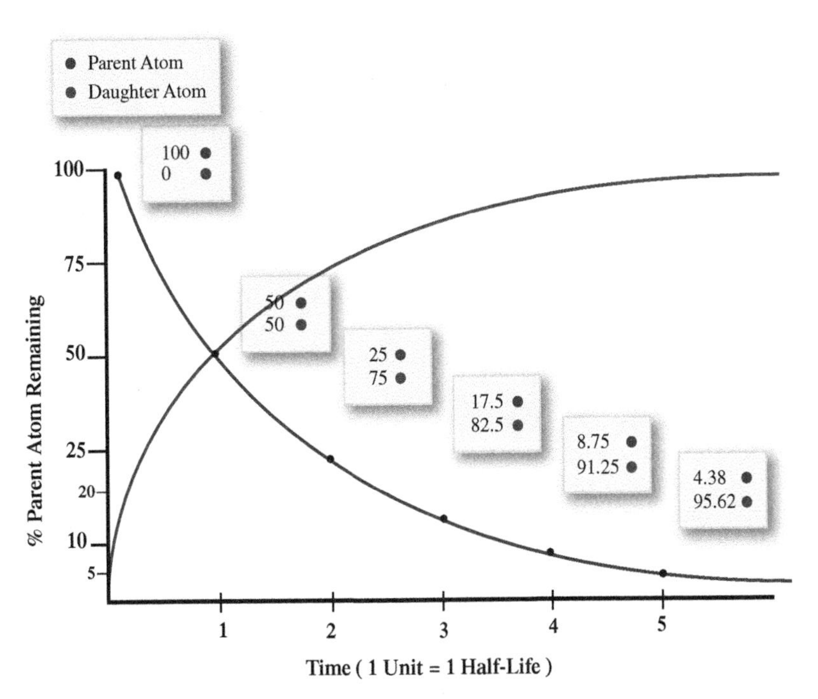

SAMPLE PROBLEM

Radioactive iodine-131 decays to stable xenon-131. After eight days, exactly one-half of a two-gram sample of iodine-131 has disappeared. What mass of the original iodine-131 will remain after an additional twenty-four days have passed?

Answer:

The half-life of iodine-131 is eight days, as indicated by the amount of time that passed before one-half of the original sample decayed into as xenon gas. The total elapsed time is eight days plus twenty-four days, or thirty-two days. Determine how many half-lives will elapse during this period.

$$\frac{32\,\text{days}}{8\,\text{days / half-life}} = 4 \text{ half-lives}$$

The amount remaining after this time is given by the equation

$$A_n = A_0\left(\frac{1}{2}\right)^n$$

where $n = 4$ and $A_0 = 2.0$ grams. Therefore,

$$A_4 = (2.0)\left(\frac{1}{2}\right)^4$$

$$A_4 = 2 \times \frac{1}{16} = \frac{2}{16}$$

$$A_4 = 0.125$$

After thirty-two days have elapsed, 0.125 grams of iodine-131 will remain.

used until the mid-1990s. The manufacture of such products was stopped when it was recognized that long-term exposure to even such small amounts of nuclear radiation was unhealthy.

As understanding of nuclear composition and decay processes grew with the development of the modern atomic theory, high-energy physics researchers sought to apply the knowledge to the identification of previously unknown transuranium elements, including those in the actinide series.

Measuring Age by Half-Life

Using the half-life relationship described above and the rates of decay of various isotopes of different elements, one can determine the age of various materials by measuring the ratio of starting materials of a nuclear-decay process to products. Carbon dating is perhaps the best-known application of this technique.

The radioactive isotope carbon-14 is produced by the near-constant collision of cosmic-ray particles with the nuclei of carbon atoms in atmospheric carbon dioxide. The amount of carbon-14 present in the atmosphere is effectively constant, though there have been some small variations over time, as well as more substantial variations due to nuclear testing and the burning of fossil fuels. When a plant incorporates carbon dioxide into glucose during photosynthesis, or when an animal consumes plant matter that has done so, the proportion of carbon-14 to carbon-12 in the living plant or animal's system also remains constant. However, when the plant or animal dies, incorporation of carbon-14 ceases, and the carbon-14 present in the organic material begins to decrease in proportion. By comparing the ratio of carbon-14 to carbon-12 in a sample with the corresponding ratio in a living system, one can calculate the amount of radioactive decay that has taken place and, from that, the number of carbon-14 half-lives that have passed since the material ceased to incorporate new carbon-14—that is, since it was alive.

Richard M. Renneboog, MSc

Bibliography

Douglas, Bodie E., Darl H. McDaniel, and John J. Alexander. *Concepts and Models of Inorganic Chemistry.* 3rd ed. New York: Wiley, 1994. Print.

Jones, Mark Martin. *Chemistry and Society.* 5th ed. Philadelphia: Saunders College Pub., 1987. Print.

Kean, Sam. *The Disappearing Spoon: And Other True Tales of Madness, Love, and the History of the World from the Periodic Table of the Elements.* New York: Little, Brown, 2010. Print.

Laidler, Keith J. *Chemical Kinetics.* 3rd ed. New York: Harper, 1987. Print.

Miessler, Gary L., Paul J. Fischer, and Donald A. Tarr. *Inorganic Chemistry.* 5th ed. Upper Saddle River: Prentice Hall, 2013. Print.

Silbey, Robert J., Robert A. Alberty, and Moungi G. Bawendi. *Physical Chemistry.* 5th ed. Hoboken: Wiley, 2012. Print.

Skoog, Douglas Arvid, Donald M. West, and F. James. Holler. *Skoog and West's Fundamentals of Analytical Chemistry*. 9th ed. Belmont (Calif.): Cengage Learning, 2014. Print.

Wehr, M. Russell, James A. Richards Jr., and Thomas W. Adair III. *Physics of the Atom*. 4th ed. Reading: Addison-Wesley, 1984. Print

HALOGENS

FIELDS OF STUDY

Inorganic Chemistry; Organic Chemistry; Geochemistry

SUMMARY

The basic structure and properties of the halogen elements are described. Halogens are strongly electronegative elements that form singly charged monatomic anions. They readily form both ionic and covalent compounds with other elements.

PRINCIPAL TERMS

- **halide:** a binary compound consisting of a halogen element (fluorine, chlorine, bromine, iodine, or astatine) bonded to a non-halogen element or organic group; alternatively, an anion of a halogen element.
- **oxidation state:** a number that indicates the degree to which an atom or ion in a chemical compound has been oxidized or reduced.
- **spectroscopy:** an analytical method of studying atoms and molecules based on the absorption or emission of electromagnetic radiation by their electrons.
- **valence electron:** an electron that occupies the outermost or valence shell of an atom and participates in chemical processes such as bond formation and ionization.

THE NATURE OF THE HALOGENS

The halogens consist of the various elements in group 17 of the periodic table. Their atoms have their valence electrons in an outermost *p* orbital and are one electron short of having that orbital filled. This arrangement of electronic structure is responsible for both the high electronegativity of the halogen atoms and the atom's ability to form ionic salts and numerous coordination bonds with neutral organic molecules. The higher halogens are known to form compounds in which they take on different oxidation states. Several of the halogens, such as fluorine, chlorine, bromine, and iodine, are readily recognized and well-known stable elements. Astatine, the remaining halogen element, is a short-lived radioactive element that can be synthesized in high-energy particle accelerators. Astatine-210, the isotope with the longest-known life, has a half-life of just 8.3 hours. There is thought to be a total of only about twenty-five grams of astatine, which is formed by radioactive decay of heavier elements, in Earth's entire crust at any time.

ATOMIC STRUCTURE AND REACTIVITY OF THE HALOGENS

The halogens have been extensively studied using electron spectroscopy to learn about the energy levels of electrons in their respective atoms. The electron distribution of the halogens is characterized by the presence of seven electrons in the outermost, or valence, shell of each one. Within the valence shell, the electrons are distributed between a single *s* orbital, containing two electrons, and three *p* orbitals, containing the remaining five electrons. A completely filled valence shell arrangement is attained by accepting a single electron from another atom, by forming either the corresponding halide anion or a covalent bond, to fill out the electron pair in the third *p* orbital. Atoms are at a minimum energy state when the electron shells are completely filled, with no extra electrons in the outermost orbitals. The stability that this imparts is the driving force behind the high electronegativity of the halogen elements, although the effect becomes weaker as the size of the halogen atom increases. Thus, the effect is strongest in fluorine and weakest in iodine and astatine. The difference in diameter as the halogen atoms increase in size is also an important consideration in the nature of the compounds that they form with other elements. Fluorine compounds tend to be

HALOGENS

extremely stable and unreactive, while the chlorides, bromides, and iodides readily undergo various reactions such as nucleophilic substitution.

Halogen Compounds

All of the halogen elements exist as covalently bonded diatomic compounds: F_2, Cl_2, Br_2, I_2, and At_2. At room temperature, fluorine and chlorine are normally gases, bromine is a liquid, and iodine is a solid. The color of the elemental forms deepens through the series: fluorine is colorless, chlorine is greenish-yellow, bromine is dark red, and iodine is a deep reddish-purple. Fluorine is so highly electronegative that it is able to form compounds with even the most unreactive of materials, even the high inert gases such as xenon (for example, XeF_6 and similar compounds). With organic compounds, fluorine is so reactive (and fluorine compounds so stable) that it will completely destroy an organic compound through fluorination reactions that separate C–C bonds. Perfluorinated organic compounds such as the polymeric material polytetrafluoroethylene (PTFE, or Teflon) are essentially inert to chemical reactions, but they will decompose if heated too strongly. In organic chemistry, fluorinated substituents such as the trifluoromethyl group ($-CF_3$) exert a powerful electron-withdrawing influence on adjacent bonds and are commonly used to activate a specific portion of a molecule to undergo a reaction.

All inorganic and organic compounds that contain halogen atoms or monatomic halogen anions are called halides. All except fluorine are able to assume other oxidation states, especially as oxoanions in the perhalo acid series.

Occurrence of Halogens

Fluorine tends to form very stable minerals, such as fluorite (calcium fluoride, CaF_2), that are poorly soluble and have high melting points. Thermal decomposition of such minerals in Earth's mantle layer occasionally releases quantities of gaseous fluorine, F_2, in volcanic eruptions. The high reactivity of

SAMPLE PROBLEM

Write out the electron distribution for fluorine and iodine, and identify the valence electrons. (Use an electron distribution chart if needed.)

1s
2s 2p
3s 3p 3d
4s 4p 4d 4f
5s 5p 5d 5f 5g
6s 6p 6d 6f 6g 6h
7s 7p 7d 7f 7g 7h

Answer:

The fluorine atom has just nine electrons, distributed as

$$1s^2 2s^2 2p^5$$

The valence electrons are those in the $2p$ orbital, which becomes filled by accepting one electron to achieve $2p^6$ electron occupation.

The iodine atom has fifty-three electrons, distributed as

$$1s^2 2s^2 2p^6 3s^2 2p^6 4s^2 3d^{10} 4p^6 5s^2 4d^{10} 5p^5$$

As in fluorine, the valence electrons are the five electrons located in the $5p$ orbitals, which go from being incompletely filled as $5p^5$ to having the full complement of electrons as $5p^6$.

fluorine, however, quickly dissipates the gas through reactions with other materials. Chlorine, bromine, and iodine are more typically found in ionic compounds. Vast amounts of chlorine are dissolved in seawater, as the chloride ions of the common salt sodium chloride. Quantities of bromine and iodine are also dissolved ions in seawater. These are normally recovered in elemental form by electrolysis of molten sodium chloride or highly concentrated sodium chloride brine. Bromine also occurs naturally in seawater and in certain subterranean brines. Both bromine and iodine are recovered typically through displacement from iodide compounds by chlorine. Being both common and reactive, chlorine is valuable in both inorganic and organic chemical applications, and various chlorinated solvents, such as di-, tri- and tetrachloromethane, are common. Although effective, chlorinated solvents and other organic compounds have been identified as carcinogens, and their common use has been continually reduced. Chlorine, bromine, and iodine compounds are valuable substituents in organic synthesis reactions for their ability to undergo well-defined substitution and elimination reactions.

Richard M. Renneboog, MSc

BIBLIOGRAPHY

Douglas, Bodie E., Darl H. McDaniel, and John J. Alexander. *Concepts and Models of Inorganic Chemistry.* 3rd ed. New York: Wiley, 1994. Print.

Gribbin, John. *Science: A History, 1543–2001.* London: Lane, 2002. Print.

Haynes, W. H., PhD, David R. Lide, PhD, and Thomas J. Bruno, PhD, eds. *CRC Handbook of Chemistry and Physics,* 96th Edition. New York: CRC/Taylor & Francis Group, 2015. Print.

Johnson, Rebecca L. *Atomic Structure.* Minneapolis: Lerner, 2008. Print.

Kean, Sam. *The Disappearing Spoon: And Other True Tales of Madness, Love, and the History of the World from the Periodic Table of the Elements.* New York: Little, Brown, 2010. Print.

Mackay, K. M., R. A. Mackay, and W. Henderson. *Introduction to Modern Inorganic Chemistry.* 6th ed. Cheltenham: Nelson, 2002. Print.

Miessler, Gary L., Paul J. Fischer, and Donald A. Tarr. *Inorganic Chemistry.* 5th ed. Upper Saddle River: Prentice Hall, 2013. Print.

Morrison, Robert Thornton, and Robert Neilson. Boyd. *Organic Chemistry.* 7th ed. Englewood Cliffs, N.J.: Prentice Hall, 2003. Print.

HYDROLYSIS

FIELDS OF STUDY

Organic Chemistry; Inorganic Chemistry; Biochemistry

SUMMARY

The process of hydrolysis is defined and its importance in all chemistry-related fields is elaborated. Hydrolysis is a fundamental and necessary process for living systems, and it is perhaps the single most common chemical reaction process known.

PRINCIPAL TERMS

- **ester:** a class of compounds characterized by a carbonyl group bonded to an alkoxy group, formed by the condensation of an acid and an alcohol.
- **hydrogen ion:** a hydrogen atom that has lost its one electron, represented by the symbol H^+.
- **hydroxyl group:** a primary functional group consisting of an oxygen atom covalently bonded to a single hydrogen atom.
- **monomer:** a molecule capable of bonding to other molecules to form a polymer.
- **polymer:** a large molecule formed by the concatenation of many individual smaller molecules, known as monomers.

Understanding Hydrolysis

Hydrolysis means "cleavage by water" (literally, "dissolution" or "disintegration" by water), and in any hydrolysis reaction, a functional group is cleaved into two separate components by the addition of H to one component and –OH, or a hydroxyl group, to the other component. Overall, the result is the addition of a molecule of HO–H, or H_2O (water), across the principal bond of a functional group. A carboxylic ester, for example, is an acid derivative having the functional group characterized by a carbonyl group (C=O) in which the C atom is bonded to another carbon atom of an alkyl group and to the oxygen atom of an alkoxy group. The addition of HO–H across the bond between the carbonyl carbon atom and the alkoxy oxygen atom produces a carboxylic acid and an alcohol.

Hydrolysis occurs with many types of compounds, including esters, amides, acid anhydrides, ethers, alkyl halides, nitriles, phosphates, carbohydrates, acetals, epoxides, and lactones. The term applies generally to the process rather than to any specific types of compounds. The common feature is that the molecule or part of a molecule undergoing hydrolysis is split apart with the overall addition of a molecule of water to the two separated components. A great many compounds have a cyclic structure in which a functional group on one part of the molecule has formed a bond to a different functional group on another part of the same molecule. In such cases, hydrolysis of the connection separates the two functional groups from each other, but it does not produce two separate molecules.

The same principle applies to both inorganic compounds and organic compounds. Metal halides,

SAMPLE PROBLEM

Ordinary refined white sugar is a naturally occurring compound in most plants; it has the chemical name sucrose. Structural analysis shows it to be a compound formed by a molecule of glucose, a monomer, bonded to a molecule of the sugar fructose. Many foods contain the ingredient "glucose/fructose," which is a one-to-one mixture of the two free sugars produced by the hydrolysis of sucrose. What is the chemical equation for the hydrolysis of sucrose to form glucose/fructose?

Answer:

The chemical formula of sucrose is $C_{12}H_{22}O_{11}$. The chemical formula of glucose is $C_6H_{12}O_6$. The chemical formula of fructose is $C_6H_{12}O_6$.

The difference in the number of atoms between sucrose and glucose/fructose is one O and two H atoms, or one H_2O.

The equation for the hydrolysis of sucrose to glucose/fructose is therefore:

$$C_{12}H_{22}O_{11} + H_2O \rightarrow C_6H_{12}O_6 + C_6H_{12}O_6$$

This reaction, as written, is not informative with regard to the identities and three-dimensional structures of the molecules. Organic chemists normally use symbolic drawings of chemical structures rather than molecular formulas to represent such reactions.

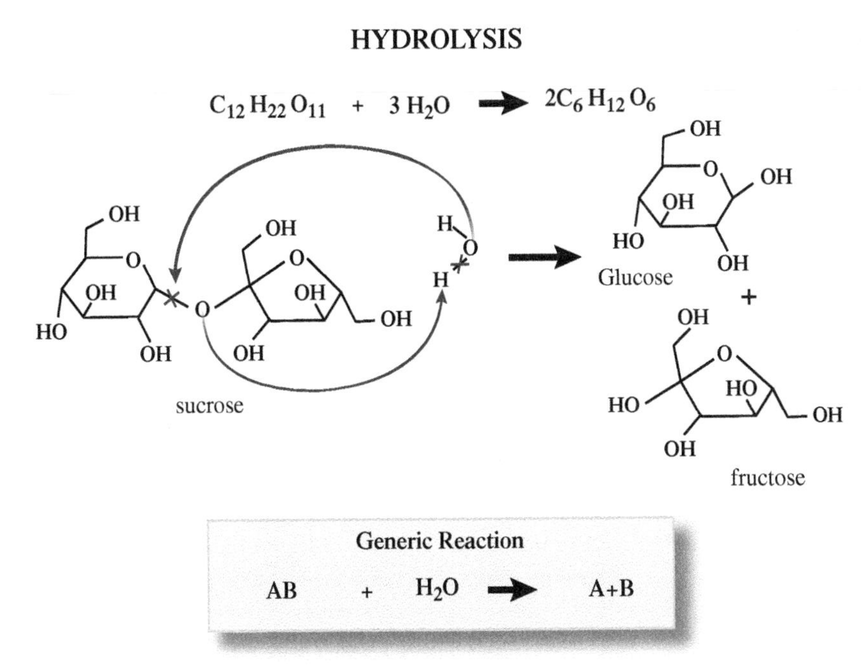

for example, undergo a hydrolysis reaction with water molecules to form the corresponding metal hydroxides and hydrogen halides. The inorganic compound phosphoric acid also forms compounds such as esters that are analogous to their carboxylic acid counterparts. These play a significant role in biochemistry in the formation of DNA, RNA, and the many phosphate ester polymers, such as phosphatidylcholine and adenosine triphosphate (ATP), and the phospholipids that comprise cell membranes.

In chemical synthesis, it is a valuable and common practice to use a protecting group through a series of chemical reactions when there is more than one reactive site in a molecule. The protecting group is bonded to a functional group such as a carbonyl group to prevent it from undergoing undesired reactions. When the desired reactions are completed, the protecting group is removed, typically by a hydrolysis reaction.

Hydrolysis in Respiration

A key feature of the respiration process is the transfer of hydrolysis to extract energy from ATP. When energy is required, the anhydride bond between the second and third units of the triphosphate component of ATP is hydrolyzed (that is, undergoes hydrolysis) to produce adenosine diphosphate (ADP) and inorganic phosphate (PO_4^{3-}), often symbolized in biochemical notation as P_i. This bond is referred to as a "high-energy" bond, and its hydrolysis and reformation is the means by which energy is extracted and stored in biochemical systems.

Richard M. Renneboog, MSc

Bibliography

Douglas, Bodie E., Darl H. McDaniel, and John J. Alexander. *Concepts and Models of Inorganic Chemistry.* 3rd. ed. New York: Wiley, 1994. Print.

Lodish, Harvey, et al. *Molecular Cell Biology.* 7th ed. New York: Freeman, 2013. Print.

Miessler, Gary L., Paul J. Fischer, and Donald A. Tarr. *Inorganic Chemistry.* 5th ed. Upper Saddle River: Prentice Hall, 2013. Print.

Morrison, Robert Thornton, and Robert Neilson. Boyd. *Organic Chemistry.* 7th ed. Englewood Cliffs, N.J.: Prentice Hall, 2003. Print.

Reece, Jane B., et al. *Campbell Biology.* 10th ed. San Francisco: Cummings, 2013. Print.

Stryer, Lubert, Jeremy M. Berg, and John L. Tymoczko. *Biochemistry.* 7th ed. New York: Freeman, 2012. Print.

HYDROPHILIC AND HYDROPHOBIC

FIELDS OF STUDY

Biochemistry; Chemical Engineering; Physical Chemistry

SUMMARY

The characteristic properties of hydrophilic and hydrophobic materials are discussed. Hydrophilic materials readily interact with liquid water, while hydrophobic materials do not. These properties are important in biochemical systems and chemical engineering processes.

PRINCIPAL TERMS

- **cohesion:** the tendency for like molecules of a substance to stick together due to their shape and electronic structure.
- **hydrogen bond:** a weak type of chemical bond formed by the attraction of a hydrogen atom to an electronegative atom—an atom with a strong tendency to attract electrons—in the same or another molecule.
- **polarity:** a characteristic of a molecule or functional group in which there is a difference in the distribution of electronic charge, causing one part of the molecule or group to be relatively electrically positive and another part to be relatively electrically negative.
- **repulsion:** an oppositional force that pushes two entities apart, such as the electrostatic repulsion between particles of like electrical charge.
- **solubility:** the ability of a particular substance, or solute, to dissolve in a particular solvent at a given temperature and pressure.

The Nature of Hydrophilic and Hydrophobic Materials

The terms "hydrophilic" and "hydrophobic" refer specifically to the interaction of a particular material with liquid water. The two words literally mean "water loving" and "water fearing," respectively, and they aptly describe the behavior of the corresponding materials: hydrophilic materials readily interact with water, while hydrophobic materials do not.

To understand this, one must understand the structure and electronic properties of the water molecule. A water molecule consists of two hydrogen atoms covalently bonded to a single oxygen atom. Each hydrogen atom contributes its single electron to the formation of the covalent bond. The electronegativity of the oxygen atom tends to keep the electron density of the oxygen-hydrogen bond locked between those two atoms, leaving the other side of each hydrogen atom bare of electron density and effectively exposing the positive charge of the hydrogen nuclei. At the same time, the oxygen has two lone pairs of electrons essentially bulging out from the other side of the molecule, creating a high electron density and negative charge. There is therefore a high degree of charge separation between the exposed nuclei of the hydrogen atoms and the lone electron pairs of the oxygen atom, giving the molecule as a whole a pronounced electrical polarity and creating what is known as a "dipole moment."

This dipole moment is the source of water's rather unique physical properties. The polarity of the molecules effectively enables them to stick together, positive end to negative end, like little magnets. The resulting cohesion is due to the formation of hydrogen bonds between water molecules. This gives water very high melting and boiling points in comparison to other molecules of similar mass, as well as a great ability to dissolve other materials.

Hydrophilic materials are able to interact with the structural and electronic properties of water to different degrees. They may undergo solvation (that is, dissolve), ionic dissociation, or surface wetting (that is, adhere to the water molecules, thus becoming wet). Hydrophilic materials also tend to have polar molecular structures or similar electronic properties. Hydrophobic materials, on the other hand, are generally nonpolar covalent materials that do not dissolve in water and do not exhibit surface wetting.

Solvent Partitioning and Phase-Transfer Catalysis

Compounds have different solubilities in different solvents. Ionic compounds dissolve well in water but not in organic solvents such as diethyl ether or dichloromethane, while covalently bonded compounds, such as organic molecules, generally do not dissolve well in water. For any particular compound, solubility is a matter of degree, and manipulating the form of the molecule greatly affects its solubility. For example, a neutral amine will dissolve well in acidic water, but if the amine solution is neutralized or made basic by the addition of a stronger base, the amine will precipitate out of the solution as an undissolved solid. Organic acids behave in a similar manner with respect to stronger acids. By adjusting the acidity or basicity of a solution containing different components, it is a simple exercise to isolate the different compounds from each other with a second solvent that is immiscible (that is, does not mix) with the first. If one of the solvents is water or some other hydrophilic substance, the other must be a hydrophobic solvent that will not become dissolved in the existing solution. When the second solvent is added and mixed together with the solution, a material will migrate from one solution into the other, according to its solubility in each one. The process is called solvent partitioning.

A related process, phase-transfer catalysis, uses two immiscible solvents such as water and the organic compound benzene (C_6H_6). Each solution in the process is considered a "phase"—that is, matter with distinct properties that touches but does not mix with other types of matter—and the surface between them is known as the "phase boundary." This technique is used when two reactant materials will not dissolve in the same solvent. Such reactants are typically an ionic compound and a nonionic compound and may even be in different states of matter. In the procedure, each reactant is dissolved in the appropriate solvent, and the two solutions are stirred together vigorously. A third compound, the phase-transfer catalyst, is added and binds reversibly to one of the reactants to form an addition product, or adduct, that will dissolve well in the other solvent. The adduct transfers from one liquid phase into the other, where the two reactant species can contact each other and undergo the desired reaction. As in solvent partitioning, one solvent must be hydrophobic and the other must be hydrophilic.

HYDROPHILIC AND HYDROPHOBIC

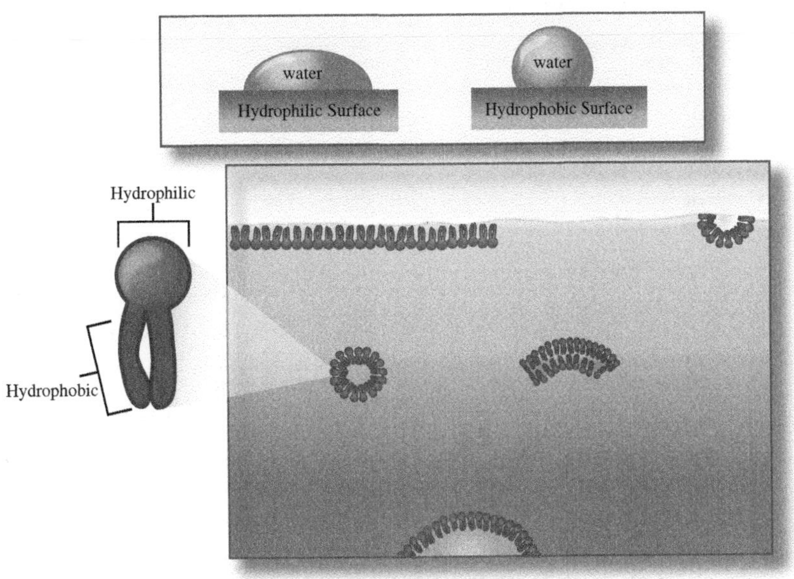

Water with hydrophilic and hydrophobic molecules.

Biological Systems

Hydrophilic and hydrophobic properties are essential in biological systems and are the functional basis of such fundamentally important aspects as cell structure. Every animal cell, and the various organelles that it contains, is enclosed within a cell membrane composed of phospholipid molecules. One common type of phospholipid consists of a glycerol molecule backbone to which are attached two long-chain fatty acids and a phosphate ion (PO_4^{3-}) bonded to an alcohol group. The phosphate ion is bonded to one of the hydroxyl (–OH) substituents of the glycerol molecule, while the carboxyl groups (–COOH) of the fatty acids are bonded to the other two hydroxyl substituents of the glycerol. Overall, a phospholipid molecule has the general structure:

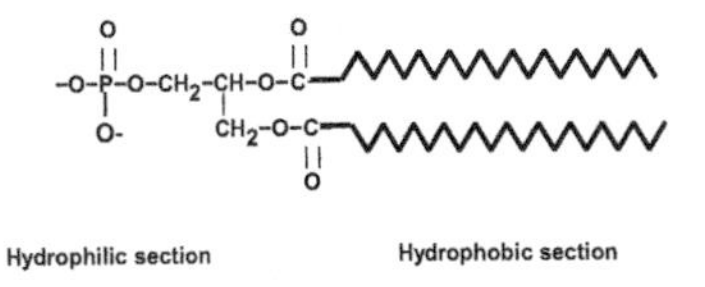

A phospholipid molecule thus has two distinct regions. The phosphate end of the molecule is strongly hydrophilic, while the long-chain fatty acids, which are nonpolar and cannot form hydrogen bonds, are

SAMPLE PROBLEM

Examine the structures of the following compounds and determine whether the compound is hydrophilic or hydrophobic:

- benzene
- ethylamine
- benzoic acid
- *N*-ethylbenzamide
- stearic acid

Answer:

- Benzene (C_6H_6) is a nonpolar hydrocarbon and therefore hydrophobic.
- Ethylamine ($H_3C{-}CH_2{-}NH_2$) has a lone pair of electrons on the nitrogen atom and polarized N–H single bonds. It is therefore hydrophilic.
- Benzoic acid ($C_6H_5{-}COOH$) is a carboxylic acid derived from benzene, with a polar C=O double bond and an O–H single bond in the carboxyl functional group. The benzene ring in the molecule is hydrophobic, but the carboxyl functional group is hydrophilic.
- *N*-Ethylbenzamide is formed from benzoic acid and ethylamine. The ethyl group and the benzene group in the molecule are hydrophobic, and only the C=O double bond in the carbonyl group and the nitrogen atom in the amide have any polarity. Overall, the compound is more hydrophobic than hydrophilic.
- Stearic acid ($H_3C{-}(CH_2)_{16}{-}COOH$) is a long-chain fatty acid. The carboxylic acid functional group (–COOH) is hydrophilic, but the long-chain alkyl substituent is hydrophobic. The compound has both hydrophobic and hydrophilic characteristics.

decidedly hydrophobic and are often referred to as "greasy." The fatty-acid section of the phospholipid molecule is actually much larger than the phosphate portion (the diagram is not to scale). There is a simple rule of solubility that "like dissolves like," meaning that polar solvents such as water generally dissolve polar materials, and nonpolar solvents such as hydrocarbons generally dissolve nonpolar materials. The two different regions of the phospholipid molecule therefore do not interact well with each other but are instead attracted to the corresponding sections of other molecules.

In large quantity in an aqueous environment, the long-chain fatty-acid portions of the phospholipid molecules are forced to aggregate and effectively blend into each other by the surrounding water molecules. The fatty acids cannot dissolve into each other completely, however, due to the repulsion between the phosphate groups to which they are attached. As a result, the molecules automatically form a sandwich-like structure known as a "bilayer." The two outer surfaces of the bilayer consist of the hydrophilic phosphate portions of the phospholipid molecules, often called the "heads." Between them is a thick, greasy hydrophobic layer of intertwined long-chain hydrocarbon groups, often called the "tails." This bilayer fully encloses the interior contents of animal cells, forming an integral membrane that can permit the passage of materials from one side to the other by various mechanisms. Being hydrophilic, the inner and outer surfaces can interact freely with the water-based fluids on either side of the membrane.

Richard M. Renneboog, MSc

Bibliography

Askeland, Donald R., Wendelin J. Wright, D. K. Bhattacharya, and Raj P. Chhabra. *The Science and Engineering of Materials.* Boston: Cengage Learning, 2016. Print.

Berg, Jeremy M., John L. Tymoczko, Gregory J. Gatto, and Lubert Strye. *Biochemistry.* 8th ed. New York: W. H. Freeman, 2015. Print.

Fenichell, Stephen. *Plastic: The Making of a Synthetic Century.* New York: Harper, 2009. Print.

Jones, Mark Martin. *Chemistry and Society.* 5th ed. Philadelphia: Saunders College Pub., 1987. Print.

Lehninger, Albert L. *Biochemistry: The Molecular Basis of Cell Structure and Function.* 2nd ed. New York: Worth, 1975. Print.

Lodish, Harvey, et al. *Molecular Cell Biology.* 7th ed. New York: Freeman, 2013. Print.

Morrison, Robert Thornton, and Robert Neilson. Boyd. *Organic Chemistry.* 7th ed. Englewood Cliffs, N.J.: Prentice Hall, 2003. Print.

Myers, Richard. *The Basics of Chemistry.* Westport: Greenwood, 2003. Print.

Reece, Jane B., et al. *Campbell Biology.* 10th ed. San Francisco: Cummings, 2013. Print.

I

INTERSTELLAR MOLECULES

FIELDS OF STUDY

Spectroscopy; Nuclear Chemistry

SUMMARY

The formation of elements through nuclear fusion reactions within stars is discussed as the source of all natural elements. Atoms and molecules that form as a result are found in the interstellar medium. Known from spectroscopic analysis throughout the past century, only recently has the cause of diffuse interstellar bands been identified as the complex form of carbon called buckminsterfullerene, C_{60}.

PRINCIPAL TERMS

- **alpha particle:** a particle consisting of two protons and two neutrons bound together, identical to a helium nucleus; typically produced in the process of alpha decay.
- **circumstellar space:** the region of space about any particular star, typically the space immediately affected by the activity of that star.
- **deuterium:** an isotope of hydrogen that contains one neutron and one proton; occurs naturally in about 1 in 6,500 hydrogen atoms.
- **element:** a form of matter consisting only of atoms of the same atomic number.
- **ion:** an atom, molecule, or neutral radical that has either lost or gained electrons and is therefore electrically charged.
- **matter:** any substance that occupies physical space and is composed of atoms or the particles that make up atoms.
- **nuclear fusion:** the process in which atomic nuclei combine to form a single nucleus of greater mass; normally refers to the fusion of hydrogen nuclei but is a generally applicable term.
- **polyatomic:** describes a molecule or ion consisting of multiple atoms that are chemically bonded to each other.

Interstellar Molecules and the Stellar Furnace

The phrase "the vacuum of space" suggests that the space between and around stars is entirely vacant of matter yet nothing could be further from the truth. Interstellar space and circumstellar space contain large quantities of matter in a very diffuse state. The source of all matter is the nuclear fusion reactions that take place within stars. An active star such the Sun is composed of a massive quantity of hydrogen atoms. The Sun itself comprises almost 99% of the mass of the Solar system. Within it, nuclear fusion reactions are driven under the intense force of the Sun's gravity. In this reaction, deuterium nuclei are driven to combine to form alpha particles. The conservation of matter and energy requires that these also have the corresponding number of electrons, and so are described as helium atoms. Some beta particles are emitted into space as ions, but within the star's interior they can also undergo further nuclear fusion reactions and produce more massive atoms. Fusion of helium and hydrogen nuclei, for example, would produce lithium atoms, and the fusion of two lithium atoms would produce a carbon atom. In fact, the nuclear fusion reactions occurring within stars produce all of the 92 naturally occurring elements and their isotopes. Other elements higher in the periodic table may also be produced, but are normally so unstable as to decompose very quickly. Given that fusion reactions within stars produce many different elements and eject some of them into space, it should therefore not be at all surprising that many different kinds of molecules and ions are detectable in interstellar and circumstellar space.

DETECTION OF INTERSTELLAR AND CIRCUMSTELLAR MOLECULES

Interstellar molecules are detected by their characteristic spectroscopic properties. Atoms and molecules absorb and emit energy at very specific wavelengths and in similarly specific patterns. By focusing on a specific star or other structure in space, and scanning the spectral properties found in that location it is possible to identify the absorption or emission patterns and wavelengths associated with a particular atom, molecule or ion. This process is carried out through radio telescopes on Earth, and by analysis of the high-resolution observations obtained from space-borne telescopes such as the Hubble and Chandra telescopes. When an atom or molecule absorbs or emits energy, it is due to the movement of electrons between different energy levels within the atom or molecule. Accordingly, the energy absorbed or emitted is specific to the difference between those energy levels. No other wavelengths of energy are absorbed or emitted. By 2009, some 150 different molecular and ionic species had been identified, but this number has increased dramatically since then. The type and complexity of compounds that have been discovered has also greatly increased from simple diatomic species like hydrogen (H_2) and the methylidyne radical (CH) to various polyaromatic hydrocarbons and buckminsterfullerene (C_{60}). Astrophysicists have recognized the existence of complex molecules in interstellar and circumstellar space over the past century due to observations of diffuse interstellar bands that could nevertheless not be explained. The identification of buckminsterfullerene as the source of the diffuse interstellar bands not only solved the mystery of their source, but also indicated that such large molecules are actually quite common in interstellar and circumstellar space. Lists of the interstellar molecules and ions that have been identified can be readily found online.

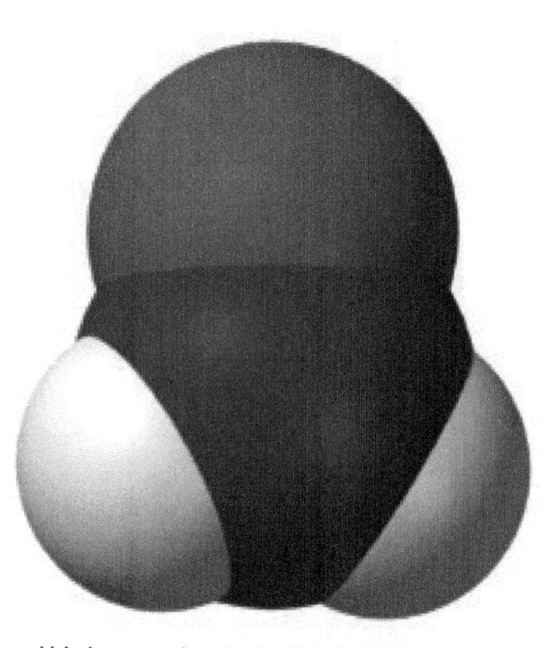

Formaldehyde, an organic molecule widely distributed in the interstellar medium. Public Domain via Wikimedia Commons

SAMPLE PROBLEM

Devise a nuclear fusion scheme for the generation of oxygen and carbon (atomic number 6, atomic mass 12) (atomic number 8, atomic mass 16) from Helium, accounting for the relative abundance of there elements.

Answer:

Fusion of two He nuclei can produce the beryllium (Be) nucleus, as

$$He + He \rightarrow Be$$

Fusion of two Be nuclei can produce the oxygen (O) nucleus, as

$Be + Be \rightarrow O$, while fusion of a single Be nucleus with a He can produce carbon

$$Be + He \rightarrow C$$

NUCLEAR SYNTHESIS

The relative abundance of heavier elements in respect to hydrogen is a good reflection of the probability of the nuclear fusion reactions for their formation within the interior of a star. In a star like our Sun, beginning with two hydrogen atoms, which have only one proton and one electron apiece, nuclear weak force must first produce a heavy hydrogen nucleus by combining one proton and one electron. The heavy hydrogen nucleus then collides with a normal hydrogen to form a nucleus of ^3He, two such nuclei can then collide to form a normal ^4He nucleus and release two protons. In the case of the sun, once it has burned much of its core hydrogen to form He, its core will contract, become hotter, and start fusing 3He to form carbon and 4He to form oxygen. From this point on, the possible combinations of nuclei to produce different elements, and eventually molecular species, increase exponentially. At the same time, however, the number of different nuclei available to take part in nuclear fusion reactions remains extremely small in proportion to the hydrogen that is present. That very abundance of hydrogen directs which elements are formed preferentially simply by competition with the nuclei present in lesser quantities.

The Importance of Interstellar Molecule Research

The greatest value in studying and understanding interstellar and circumstellar molecules lies in the development of methodologies that enable their identification. Spectroscopic analysis is a fundamental practice of terrestrial science, and the refinement of the methods used to identify molecules hundreds of light-years distant in space find valuable application on Earth in different fields. Molecular analysis in medicine identifies chemical imbalances and indicators of disease, facilitating treatment that can be life-saving. In addition, ongoing research into the mechanism of conditions and diseases that have eluded curative treatment can be investigated at ever finer scales in order to identify the biochemical weakness of the disease-causing agent that leads to a cure. Communications technology also benefits from research on interstellar molecules as the ever finer resolution required also demands the development of ever more robust and efficient means of sending and receiving information to accommodate the massive amounts of data that are generated by the research. In geosystems research, analysis of the materials that exist in circumstellar and interstellar space leads to better understanding of the manner in which a star functions and the effects that Solar activity can have on satellites and terrestrial systems such as the national power grid system. These are generally affected by such events as flares, bursts and sunspots, which also tend to be the primary sources of matter in circumstellar and interstellar space.

Richard M. Renneboog, MSc

Bibliography

Dumé, Belle. "Bountiful buckyballs resolve interstellar mystery" *physicsworld.* Web. 20 Aug. 2015. <http://physicsworld.com/cws/article/news/2015/jul/16/bountiful-buckyballs-resolve-interstellar-mystery>

Hudgins, Douglas M. "Interstellar Polycyclic Aromatic Compounds and Astrophysics." *The Astrophysics & Astrochemistry Laboratory.* NASA Ames Research Center, n.d. Web.

Herbst, Eric and van Dishoeck, Ewine F. "Complex Organic Interstellar Molecules" *The Annual Review of Astronomy and Astrophysics* 2009. 47:427-480. Retrieved from <http://www.annualreviews.org>

Yamada, Koichi M.T. and Winnewissner, Gisbert, eds. *Interstellar Molecules. Their Laboratory and Interstellar Habitat* New York, NY: Springer, 2011. Print.

Kwok, Sun. *Physics and Chemistry of the Interstellar Medium* Herndon, VA: University Science Books, 2007. Print.

Tielens, A.G.G.M.. *The Physics and Chemistry of the Interstellar Medium* New York, NY: Cambridge University Press, 2005. Print.

Verschuur, Gerrit. *The Invisible Universe: The Story of Radio Astronomy.* New York, NY: Springer, 2015. Print.

ION IMPLANTATION

FIELDS OF STUDY

Crystallography; Metallurgy; Chemical Engineering

SUMMARY

Ion implantation, builds a doped region in a crystal by actually injecting ions into the crystal lattice of the substrate material.

PRINCIPAL TERMS

- **alloy:** a mixture of a metal and at least one other element, often another metal; also known as a solid solution.
- **alpha particle:** a particle consisting of two protons and two neutrons bound together, identical to a helium nucleus; typically produced in the process of alpha decay.
- **amorphous:** a physical form that has no defined or regular crystal structure.
- **composition:** the identities and relative proportions of different elements or components in a compound, mixture, or other material.
- **concentration gradient:** the gradual change in the concentration of solutes in a solution across a specific distance.
- **conductivity:** the ability of a material to transfer heat (thermal conductivity) or electricity (electrical conductivity) from one point to another.
- **crystal lattice:** the regular geometric arrangement of atoms and molecules in the internal structure of a crystal.
- **dopant:** an element or material that is included in the crystal lattice of another material in small quantities in order to enhance or alter the electronic properties of that material.
- **ion:** an atom, molecule, or neutral radical that has either lost or gained electrons and is therefore electrically charged.
- **monatomic:** describes a molecule or ion consisting of only one atom.
- **repulsion:** an oppositional force that pushes two entities apart, such as the electrostatic repulsion between particles of like electrical charge.
- **valence electron:** an electron that occupies the outermost or valence shell of an atom and participates in chemical processes such as bond formation and ionization.

Ion Implantation and Epitaxy

The two processes of epitaxy and ion implantation are complimentary processes in many ways and can be used to obtain similar results in regard to the materials and characteristics that they produce. In epitaxy a layer of material, typically crystalline in nature, builds onto the surface of an existing material, expanding the dimensions of the structure outward. Ion implantation, on the other hand, builds the material inward from the surface instead of outward by actually injecting ions into the crystal lattice of the substrate material. Epitaxial growth can be used to produce semiconductor materials with specific properties by adding a layer of doped material to the outside of the substrate. Ion implantation is used to add the doping element directly into the substrate. Epitaxy can be used to add a layer of material such as diamond to the surface of an object to enhance properties such as conductivity. Ion implantation can be used to enhance the surface properties of the material itself by altering its composition and effectively becoming at least in part an alloy.

Methods of Ion Implantation

The principles of ion implantation are common to all applications, and are relatively independent of the type of ions that are to be implanted. Implantation can be carried out using either positive or negative ions, and ions can be monatomic ions or compound ions, although the procedure is most amenable to the implantation of monatomic ions and simple ions such as alpha particles. Typically, a focused beam of ions is generated in a plasma and accelerated towards the target material. The ion beam is scanned over the surface of the substrate material until the desired degree of implantation has been achieved. For objects such as silicon wafers for the production of computer chips, the process can be highly automated, and equipment for this is commercially available from some laboratory instrumentation manufacturers. The process is accordingly very well controlled,

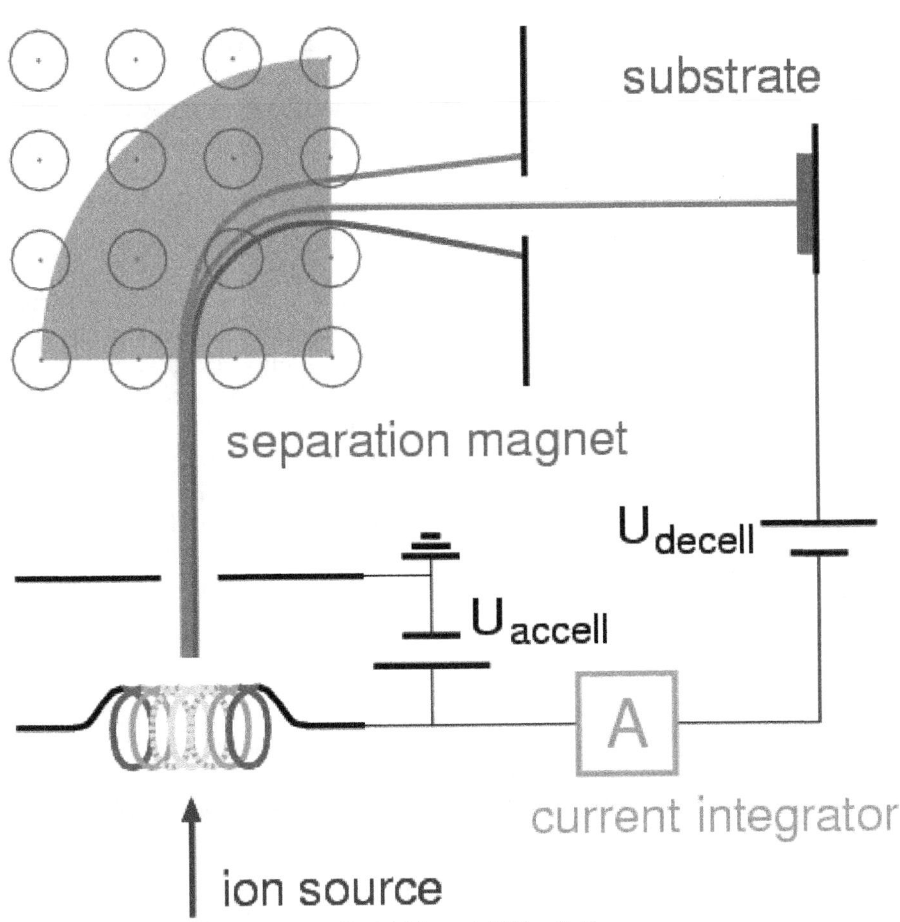

Ion implantion setup with mass separator. Daniel Schwen via Wikimedia Commons

allowing implantation of very low amounts of a desired dopant ion, on the order of just 10^{11} per square centimeter. Recall that one mole of a monatomic material contains 6.023 X 10^{23} atoms, and it becomes easier to see that this dose of a dopant corresponds to about one picogram of additional matter per square centimeter. The accelerated ions impact the surface of the target material with enough force to be driven to some depth within the material, and do not merely accumulate on the surface. This is controlled by the energy with which the ions are accelerated from the plasma ion source. The result is effectively an alloy, at least to the depth of penetration of the ions. Accordingly, the dose of dopant in the material can be properly described in terms of a concentration gradient. The kinetic energy of the accelerated ions must be sufficient to overcome interaction with the valence electrons of the substrate material. However, if the kinetic energy of the accelerated ions is great enough to overcome the force of repulsion between the nuclei in the substrate, it is possible that nuclear interactions can take place. It should be remembered that the same principle by which ions are accelerated in an ion implantation process is used to accelerate ion positively-charged atomic nuclei in high-energy studies of nuclear fusion. In ion implantation, the energies are controlled so that the ions are merely inserted into the crystal lattice of the substrate material to produce enhanced material properties. If the implantation dose is too high, however, the crystal lattice of the substrate will be seriously compromised, resulting in the formation of an amorphous layer rather than a coherent crystal lattice.

DEVICES

The structure of an ion implantation device is complex and consists of several distinct stages. The initial stage is the plasma generator. In this device the material that is to be the ion source is heated to form a plasma consisting of gaseous ions. This can be achieved most efficiently by an electrical spark of an appropriate energy to an electrode of the desired material. Commonly used ion source materials for implantation into silicon wafers, for example, include boron, phosphorus and arsenic. Formation of the plasma state is non-specific, however, and produces a variety of ions of the implantation material. Ion implantation itself must be well controlled if the desired material properties are to be obtained, so it is important that only the desired ions are selected from the plasma. The second stage of an implantation device is, accordingly, extraction of the ions from the plasma source. This introduces the ions into the third stage, which is the separation of the ions according to their charge-to-mass ratio in a magnetic field. This is the same principle by which mass

SAMPLE PROBLEM

Calculate the dose for the implantation of 3.2 picograms of arsenic over an area of 1.3 square centimeters.

Answer:

Begin with the atomic weight of arsenic and determine the number atoms this corresponds to using Avogadro's constant (6.02259×10^{23} atoms per mole):

atomic weight of arsenic = $74.9216\ g.mol^{-1}$

3.2 picogram = 3.2×10^{-12} g or ($3.2 \times 10^{-12}/74.9216$ =) 4.2711×10^{-14} mole

dose = (# of moles X # of atoms per mole) / area

= ($4.2711 \times 10^{-14} \times 6.02259 \times 10^{23}$) / 1.3 cm^2

= 1.9787×10^{10} atoms per centimeter

spectrometers operate. The magnetic field strength is adjusted or controlled so that only the ions having the correct mass-to-charge ratio are directed into the fourth stage of the implantation device, which is the ion accelerator column. This is essentially a tube in which an applied electromagnetic field accelerates the ions to the energy needed for implantation to take place. When the accelerated ions exit this stage they immediately strike the target material that is to be implanted. While this structure has been commercialized into stand-alone units for the production of silicon wafers for the production of computer chips, it is the general principal by which all methods of ion implantation operate.

The Importance of the Ion Implantation

While ion implantation technology has potential value for material modification in many fields, both for the enhancement of existing materials and the development of entirely new alloys that are fine tuned to specific applications, by far the most important and common use of the technology is in the precision production of doped silicon wafers for the production of computer chips. The first semiconductor transistor was demonstrated in 1947, and since that time these central components of digital electronics have become progressively smaller in accord with Moore's Law. As the limit of this law is approached and new technologies are sought to replace the current technology, fine control of the materials and methods is ever more important, and ion implantation technology has the potential to allow customization of semiconductor materials at the atomic scale.

Richard M. Renneboog M.Sc.

Bibliography

Bréchignac, Catherine, Philippe Houdya, and Marcel Lahmani, eds. *Nanomaterials and Nanochemistry.* New York, NY: Springer-Verlag, 2007. Print

Doering, Robert and Yoshio Nishi, eds. *Handbook of Semiconductor Manufacturing Technology.* 2nd ed. Boca Raton, FL: CRC Press, 2008. Print.

Knystautas, Émile, ed. *Engineering Thin Films and Nanostructures With Ion Beams.* Boca Raton, FL: Taylor & Francis/CRC Press, 2005. Print.

Nastasi, M., and Mayer, J.W. *Ion Implantation and Synthesis of Materials.* New York, NY: Springer-Verlag, 2006. Print.

Rimini, Emanuele. *Ion Implantation: Basics to Device Fabrication.* Boston, MA: Kluwer Academic Publishing, 1995. Print.

Widmann, D., H. Mader, and H. Friedrich. *Technology of Integrated Circuits.* New York, NY: Springer-Verlag, 2000. Print.

IONS

FIELDS OF STUDY

Inorganic Chemistry; Organic Chemistry; Physical Chemistry

SUMMARY

The basic structure of ions is defined, and the modern theory of atomic structure is elaborated. Ionic materials are not only ubiquitous in nature but also essential to the biochemical processes that support life.

PRINCIPAL TERMS

- **anion:** any chemical species bearing a net negative electrical charge, which causes it to be drawn toward the positive pole, or anode, of an electrochemical cell.
- **cation:** any chemical species bearing a net positive electrical charge, which causes it to be drawn toward the negative pole, or cathode, of an electrochemical cell.
- **dissociation:** the separation of a compound into simpler components.
- **electrovalent bond:** an alternate term for an ionic bond, which is a type of chemical bond formed by mutual attraction between two ions of opposite charges.
- **ionization:** the process by which an atom or molecule loses or gains one or more electrons to acquire a net positive or negative electrical charge.

The Nature of Ions

Ions are simply atoms or chemically bonded groups of atoms that have a net electrical charge. Ions carrying a positive charge do not have the full complement of electrons necessary to balance the positive charge of their nuclei, while ions carrying a negative charge have more electrons than they have protons in their nuclei. Positive and negative ions are termed cations and anions, respectively. Oppositely charged ions typically interact to form crystalline compounds through electrovalent bonds, or ionic bonds. Many such compounds are highly soluble in water and other highly polar solvents; dissociation of these compounds occurs when they are dissolved by the solvent molecules, causing them to split into their component ions. The compound can typically be re-formed by evaporating the solvent.

The strength of the electrovalent bonds in an ionic compound, and thus its ability to dissolve in polar solvents, is highly dependent on the charge density of the ions. "Charge density" refers to the net electrical charge on the ion relative to its physical size. Generally, the higher the charge density of the two opposing ions, the less soluble the material. Other factors, such as the structure of the material's crystal lattice, also affect its solubility. The ionic compound sodium chloride (NaCl), for example, is highly soluble in water, while sodium fluoride (NaF) is much less so due to the smaller size and higher charge density of the fluoride ion compared to the chloride ion. Magnesium chloride ($MgCl_2$) will dissolve 54.6 grams in 100 milliliters of cold water, but magnesium fluoride (MgF_2) is barely soluble and dissolves only slightly fewer than 8 milligrams in the same amount of cold water. The more soluble an ionic compound is, the more easily the solvent molecules can surround each ion in solution. "Solvation energy" is the

IONS

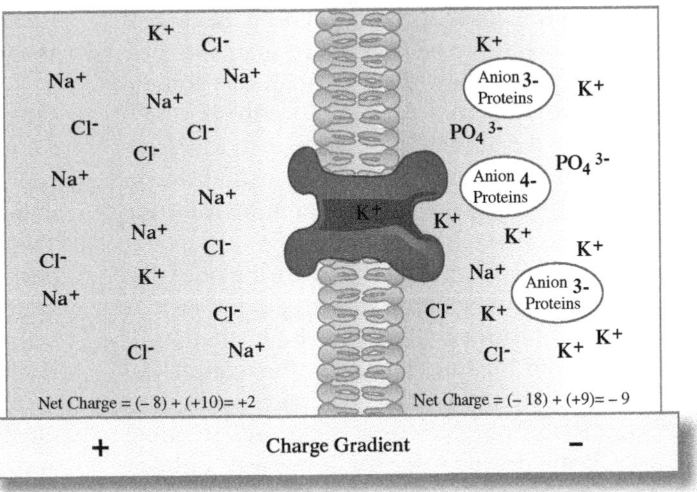

Ions transport across a cell membrane and reduce a concentration gradient.

term used to describe the difference in energy between dissolved ions and the undissociated ions in the compound.

Atomic Structure and the Formation of Ions

All ions are formed by the loss or addition of electrons in the outermost valence shell of an atom. The modern theory of atomic structure describes atoms as being composed of a small, dense nucleus containing a fixed number of protons and a variable number of neutrons. The number of protons in the nucleus defines the identity of the atom as an element. The neutrons that are present act as a sort of nuclear glue that counteracts the force of electrostatic repulsion between the positively charged protons. Together, the protons and neutrons make up almost all of the mass of an atom.

Surrounding the nucleus is a large, diffuse cloud of negatively charged electrons. This cloud is approximately one hundred thousand times larger in diameter than the atomic nucleus and contains the same number of electrons as there are protons in the nucleus, rendering the atom electrically neutral. According to quantum mechanical theory, these electrons may possess only very specific energies within the electron cloud. The limits on these energy levels are determined by the structure of the nucleus, and each allowed energy level defines a shell of electrons about the nucleus and the number of electrons each shell can contain. Within each shell are one or more subshells, designated either *s*, *p*, *d*, or *f* and containing either one, three, five, or seven mathematically defined regions of space called orbitals, each of which is allowed to contain no more than two electrons at any time. Thus, the *s* subshell can contain up to two electrons, the *p* subshell up to six, the *d* subshell up to ten, and the *f* subshell up to fourteen. The outermost shell is called the valence shell, and in most atoms, this shell does not contain the full complement of electrons that is allowed.

The electrons in the valence shell are the ones that are involved in interactions between atoms, such as chemical bond formation, and in the formation of ions. Atoms and ions are in a state of maximum stability when their outermost electron shell contains its full complement of electrons allowed, and attaining this state is the driving force for atoms with unfilled valence shells to either donate electrons to or accept electrons from other atoms. This ionization process has an energy cost, however, and the amount of energy required to remove an electron from its stable location in an atomic or molecular orbital is called ionization energy. Each electron in a particular orbital has its own associated ionization energy, which increases dramatically as the shells get closer to the nucleus.

SAMPLE PROBLEM

A carbon atom combines with three oxygen atoms to form an ion with the molecular formula CO_3. Draw a Lewis dot structure for this compound, showing how the atoms are arranged, which electrons are shared to form bonds, and which electrons remain unshared. Identify whether the molecule will accept or donate electrons to form an ion, and predict the charge on the ion.

Answer:

A carbon atom has four valence electrons, and an oxygen atom has six. Each oxygen atom can accept two electrons to fill out its valence shell. Thus, one oxygen atom will accept two electrons from the carbon atom, leaving two electrons to be apportioned to the remaining two oxygen atoms, as seen below:

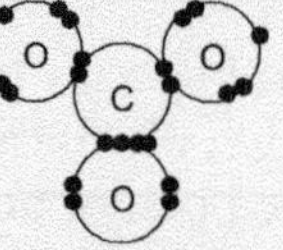

This Lewis structure shows that two of the oxygen atoms will have unfilled valence shells and will need to accept one more electron each to complete their electron octets. The resulting ion will therefore have two negative charges, forming a carbonate ion, CO_3^{2-}.

Occurrence of Ions

Atoms are described as "electronegative" if they accept electrons to form anions and as "electropositive" if they give up their valence electrons to form cations. The electronegativity of atoms increases across the periodic table, with hydrogen and the alkali metals being the least electronegative and the halogens being the most electronegative. At the extreme right-hand side of the table are the noble gases, also called inert gases because they have very stable, completely filled valence shells and are thus very resistant to ionization.

Ions are readily formed as the result of reduction-oxidation (redox) reactions, in which one reactant gives up electrons and another reactant gains those electrons. For example, the reaction between chlorine gas (Cl_2) and sodium metal (Na) produces sodium chloride (NaCl), composed of sodium ions (Na^+) and chloride ions (Cl^-). Each sodium atom has ejected its lone valence shell electron, leaving a filled outermost electron shell. Each chlorine atom has accepted one electron into its valence shell, adding to the seven that are normally there to make a full complement of eight electrons. In the reaction, a great deal of heat and light energy are given off, indicating the difference in stability between the reactants and the product.

Multiple Possibilities

Many elements, particularly the transition metals, are able to form more than one ion. As the atoms of the elements increase in size, overlapping electron-shell energies result in different orbitals being of the same or very similar energy. Orbital interactions between different atoms also become easier, allowing the formation of different types of bonds, such as coordination bonds. These factors provide atoms with a number of ways to achieve filled outermost electron shells, usually by releasing different numbers of valence shell electrons. Copper, for example, can lose either one or two electrons to form either the Cu^+ or the Cu^{2+} ion, while iron atoms typically eject either two or three electrons, forming either the Fe^{2+} or the Fe^{3+} ion.

Richard M. Renneboog, MSc

Bibliography

Douglas, Bodie E., Darl H. McDaniel, and John J. Alexander. *Concepts and Models of Inorganic Chemistry.* 3rd ed. New York: Wiley, 1994. Print.

Hendrickson, James B., Donald J. Cram, and George S. Hammond. *Organic Chemistry.* 3rd ed. New York: McGraw, 1973. Print.

Johnson, Rebecca L. *Atomic Structure.* Minneapolis: Lerner, 2008. Print.

Lehninger, Albert L. *Biochemistry: The Molecular Basis of Cell Structure and Function.* 2nd ed. New York: Worth, 1975. Print.

Mackay, K. M., R. A. Mackay, and W. Henderson. *Introduction to Modern Inorganic Chemistry.* 6th ed. Cheltenham: Nelson, 2002. Print.

Miessler, Gary L., Paul J. Fischer, and Donald A. Tarr. *Inorganic Chemistry.* 5th ed. Upper Saddle River: Prentice Hall, 2013. Print.

Myers, Richard. *The Basics of Chemistry.* Westport: Greenwood, 2003. Print.

Winter, Mark J. *The Orbitron: A Gallery of Atomic Orbitals and Molecular Orbitals.* University of Sheffield, n.d. Web.

ISOTOPES

FIELDS OF STUDY

Nuclear Chemistry; Geochemistry

SUMMARY

The basic characteristics of isotopes are defined in the context of modern atomic theory. Isotopes are characterized by different numbers of neutrons in the nuclei of atoms with the same number of protons. All radioactive isotopes decay exponentially, which can be used to determine the age of materials.

PRINCIPAL TERMS

- **atomic mass:** the total mass of the protons, neutrons, and electrons in an individual atom.
- **carbon dating:** a method of dating that uses the proportion of radioactive carbon-14 atoms remaining in organic material to determine how much time has elapsed since it was part of a living organism.
- **deuterium:** an isotope of hydrogen that contains one neutron and one proton; occurs naturally in about 1 in 6,500 hydrogen atoms.
- **neutron:** a fundamental subatomic particle in the atomic nucleus that is electrically neutral and about equal in mass to the mass of one proton.
- **protium:** the essential form of hydrogen, containing one proton and no neutron; the most common form of matter in the known universe.
- **radioisotope:** any radioactive isotope of an element that undergoes spontaneous nuclear fission until a stable, nonradioactive isotope is formed.

- **tritium:** an unstable radioisotope of hydrogen that contains one proton and two neutrons.

The Nature of Isotopes

An element is any material that consists of one and only one type of atom, identified by the number of protons that are in the nucleus of each atom. This definition does not depend in any way on the number of neutrons that are in the nucleus; indeed, for every element, there are atoms with the same number of protons but different numbers of neutrons. Such atoms are called isotopes. The essential carbon atom, for example, is carbon-12 (^{12}C), which has six protons and six neutrons in the nucleus. Another kind of carbon atom occurs naturally: carbon-14 (^{14}C), in which the nucleus contains eight neutrons, instead of just six, along with the six protons. Due to the extra neutrons, the atomic mass of carbon-14 is 14 atomic mass units (u) instead of 12. The extra neutrons also render the carbon nucleus unstable, causing it to undergo spontaneous fission. In other words, the carbon-14 atom is radioactive. Radioactive isotopes are generally referred to as radioisotopes. The presence of different isotopes that occur naturally in the elements is responsible for the fractional values of atomic masses in the periodic table. These are the atomic weights, or weighted averages of the different atomic masses, according to the naturally occurring proportions of each isotope.

ISOTOPES

1 H hydrogen 1

	Average	H-1	H-2	H-3
Protons	1	1	1	1
Neutrons	0	0	1	2
Electrons	1	1	1	1
Atomic Mass	1.01	1	2	3
% in Nature		99.9884%	0.0115%	$4x10^{-15}$%
Name	Hydrogen	Protium	Deuterium	Tritium

6 C carbon 12

	Average	C-12	C-13	C-14
Protons	6	6	6	6
Neutrons	6	6	7	8
Electrons	6	6	6	6
Atomic Mass	12.0107	12	13	14
% in Nature		98.93%	1.07%	$<1x10^{-12}$%
Name	Carbon	Carbon-12	Carbon-13	Carbon-14

Atomic Structure of the Isotopes

Atoms are defined as having a very small, dense nucleus composed of a fixed number of protons and a variable number of neutrons. These constitute essentially all of the mass of the atom. Each proton bears a single positive charge, and electrostatic repulsion between like charges would drive the protons apart if not for the presence of the neutrons, which bind the protons together in the nucleus and make it stable. Some combinations of neutrons and protons are not stable, however, and such nuclei can undergo spontaneous nuclear fission, with half-lives ranging from millionths of a second (e.g., polonium-212) to several billion years (e.g., uranium-238).

The simplest example of this phenomenon is hydrogen, which is known in three different isotopes, all of which exhibit the normal chemistry associated with hydrogen. The simplest and most abundant, protium, consists of one proton and one electron, with no neutrons in the nucleus. Deuterium, the isotope used to prepare heavy water (D_2O), also occurs naturally in about one of every 6,500 hydrogen atoms. Deuterium has one proton and one neutron in its nucleus. The third isotope is tritium, which is synthesized by nuclear reactions, either in the upper atmosphere or in reactors. Tritium has one proton and two neutrons in its nucleus. This unstable arrangement breaks down spontaneously, with one neutron becoming a proton and a beta particle, or high-energy electron, that is then ejected. The result is a stable helium atom.

HALF-LIVES OF ISOTOPES

All radioactive isotopes decay according to the same exponential rate law. The half-life of an isotope is defined as the time required for half of an amount of the material to decay or be consumed in a process. Mathematically, this is stated as follows:

$$T_{1/2} = \frac{\ln 2}{k} = \frac{0.693}{k}$$

where k is the disintegration constant of a particular isotope, determined by counting the number of disintegrations that occur in a certain period of time. Carbon dating is perhaps the best-known application of this principle. The isotope carbon-14 is incorporated into living tissues according to its natural abundance; this process ceases when life ceases, and the amount of carbon-14 present in an organic material then decreases at a known rate. Comparing the amount of carbon-14 that remains with the amount that would have been present in the living material determines the number of half-lives that have passed since the tissue was alive, thereby giving its age. A similar method uses the proportions of uranium-238 and the specific isotope of lead that it produces to determine the age of rocks.

ANALYTICAL APPLICATIONS

Isotopes are also used as tracers in various analytical methods. Incorporating a radioactive element into the molecular structure of a bioactive compound allows the analyst to deduce the pathways by which materials are synthesized or metabolized biologically. A radioactive element can also be incorporated into a pharmaceutical compound in order to deliver radiation treatment to specific diseased tissues, as opposed to general radiation.

Richard M. Renneboog, MSc

SAMPLE PROBLEM

Lead occurs naturally as the four isotopes below. Find the number of neutrons in each isotope and calculate the atomic weight of lead (as would be listed in the periodic table).

Isotope	Natural Abundance
Pb-204	1%
Pb-206	24%
Pb-207	23%
Pb-208	52%

Answer:

Every atom of lead contains only eighty-two protons. To find the number of neutrons in each isotope, subtract the number of protons (82) from the number of the isotope (204, 206, 207, or 208). Therefore, lead-204 has 122 neutrons, lead-206 has 124 neutrons, lead-207 has 125 neutrons, and lead-208 has 126 neutrons.

The atomic weight of naturally occurring lead is the weighted or proportional average of the masses of the four different isotopes, calculated as follows:

$$\text{atomic weight} = (0.01 \times 204) + (0.24 \times 206) + (0.23 \times 207) + (0.52 \times 208)$$

$$\text{atomic weight} = (2.04) + (49.44) + (47.61) + (108.16)$$

$$\text{atomic weight} = 207.25$$

The atomic weight given in periodic tables is 207.19, which is derived using more accurate percentage values. The difference between it and the amount calculated here arises from rounding error.

BIBLIOGRAPHY

Gribbin, John. *Science: A History, 1543–2001.* London: Lane, 2002. Print.

Haynes, W. H., PhD, David R. Lide, PhD, and Thomas J. Bruno, PhD, eds. *CRC Handbook of Chemistry and Physics,* 96th Edition. New York: CRC/Taylor & Francis Group, 2015. Print.

Johnson, Rebecca L. *Atomic Structure.* Minneapolis: Lerner, 2008. Print.

Kean, Sam. *The Disappearing Spoon: And Other True Tales of Madness, Love, and the History of the World from the Periodic Table of the Elements.* New York: Little, Brown, 2010. Print.

Mackay, K. M., R. A. Mackay, and W.Henderson. *Introduction to Modern Inorganic Chemistry.* 6th ed. Cheltenham: Nelson, 2002. Print.

Myers, Richard. *The Basics of Chemistry.* Westport: Greenwood, 2003. Print.

Wehr, M. Russell, James A. Richards Jr., and Thomas W. Adair III. *Physics of the Atom.* 4th ed. Reading: Addison-Wesley, 1984. Print.

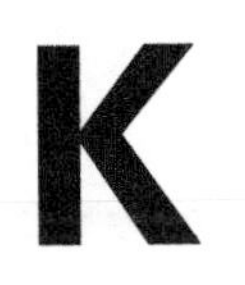

KETONES

FIELDS OF STUDY

Organic Chemistry

SUMMARY

The characteristic properties and reactions of ketones are discussed. Ketones are very useful compounds for the relatively easy formation of complex organic molecular structures from very simple starting materials.

PRINCIPAL TERMS

- **carbonyl group:** a functional group consisting of a carbon atom double bonded to an oxygen atom.
- **cyclic:** describes a compound whose molecular structure forms a closed ring.
- **double bond:** a type of chemical bond in which two adjacent atoms are connected by four bonding electrons rather than two.
- **functional group:** a specific group of atoms with a characteristic structure and corresponding chemical behavior within a molecule.
- **polar:** describes a molecule or functional group in which there is a difference in the distribution of electronic charge, causing one part of the molecule or group to be relatively electrically positive and another part to be relatively electrically negative.

The Nature of Ketones

Ketones are organic molecules that have a carbonyl group (C=O) as their principal functional group. A ketone is often represented by an R– or Ar– attached to the carbon atom of the carbonyl group, where R– indicates an alkyl group (derived from a saturated hydrocarbon) and Ar– indicates an aryl group (derived from an aromatic ring). They are reactive compounds due to the polar nature of the carbonyl group. The oxygen atom of the carbonyl group is more electronegative than the carbon atom and so tends to draw the electron density in the C=O double bond toward it, reducing the electron density around the carbon atom and causing it to become more susceptible to reactions with nucleophiles (electron-rich species). Meanwhile, the oxygen atom becomes susceptible to reactions with electrophiles (electron acceptors) and especially to gaining a proton in an acid medium.

The carbonyl group shares several structural features with the carbon-carbon double bond (C=C). As both oxygen and carbon are second-period elements, with valence electrons in the $2p$ orbitals, they are both able to hybridize their s and p atomic orbitals into sets of energetically equivalent sp, sp^2, and sp^3 orbitals. The sp^3 hybrid orbitals are typical of saturated compounds. An sp^3-hybridized atom can have four pairs of electrons filling the four sp^3 orbitals, either as bonding pairs or as nonbonding lone pairs of electrons. When this condition is met, the ability of the orbitals to contain electrons is "saturated," and no more electrons can be held in those orbitals.

When a carbon, nitrogen, or oxygen atom only has enough electrons to fill three orbitals, the hybridization involves only two of the three p orbitals, forming three sp^2 hybrid orbitals. The three orbitals are oriented at 120 degrees to each other in a plane with the nucleus of the atom at its center. The third p orbital remains a p orbital and is perpendicular to the plane of the sp^2 orbitals. Direct overlap of sp^2 orbitals on two adjacent atoms forms a covalent chemical bond called a sigma (σ) bond, while side-by-side overlap of the adjacent p orbitals forms a pi (ϖ) bond parallel to the sigma bond. Together, the sigma bond and the pi bond form the double bond. In the carbonyl group, both the carbon and the oxygen are sp^2 hybridized, giving the group a trigonal planar geometry overall. Every carbonyl-containing compound, including ketones, has a region within its molecular structure where the angles between bonds are wider, forming a site that is more susceptible to reaction. This form, coupled with the polarity of the carbonyl group, greatly enhances the reactivity of the compounds.

In principle, the number of possible ketone compounds is infinite, beginning with the simplest alkyl ketones, dimethyl ketone (or acetone) and cyclopropanone. The simplest aryl ketone is diphenyl ketone (or benzophenone). Cyclic ketones are less physically strained than cyclic alkenes (organic compounds containing a C=C double bond) because the oxygen atom of the carbonyl is directed outward from the ring rather than incorporated into it. Accordingly, cyclic ketones can effectively be of any ring size.

Though quite reactive, ketones tend also to be rather less toxic than many other types of compounds. They make excellent general-purpose solvents, and precautions against unwanted solvent effects are necessary when handling bulk quantities of the materials. The polarity of the carbonyl group gives ketones good solubility in water, especially the smaller ketone molecules. Acetone, for example, is miscible in all proportions with water, meaning that it will mix without reacting. The carbonyl functional group also makes ketones significant components of many biochemical cycles.

Nomenclature of Ketones

Ketones are named systematically according to their parent hydrocarbon structure, as outlined by the International Union of Pure and Applied Chemistry (IUPAC) rules for the nomenclature of organic compounds. The longest unbranched chain of carbon-carbon bonds is identified, the position of the ketone carbonyl group is stated, and the name is given the suffix "-one." The carbon atoms in the parent chain must be counted; the count may start from either right or left, depending on which direction gives the ketone group the lowest position number possible.

By historical convention, familiar names are commonly formed by identifying the functional groups on either side of the carbonyl, followed by the word "ketone." For example, the three-carbon chain of propane in which the central carbon atom is part of a carbonyl group is systematically named "propan-2-one" (or sometimes "2-propanone") but known familiarly as "acetone" and less familiarly as "dimethyl ketone." The four-carbon chain of butane with a carbonyl group on the second carbon atom is systematically named "butan-2-one" but known familiarly as the common solvent "methyl ethyl ketone," or MEK.

For cyclic ketones, the systematic naming method is the same. The parent cyclic structure is identified, and the carbonyl carbon atom is typically designated as the first carbon atom in the cyclic structure. Accordingly, a five-membered cyclic ketone with a methyl group bonded to the carbon atoms on either side of the carbonyl group would be systematically named 2,5-dimethylcyclopentanone. There is no equivalent method of familiar naming for cyclic ketones.

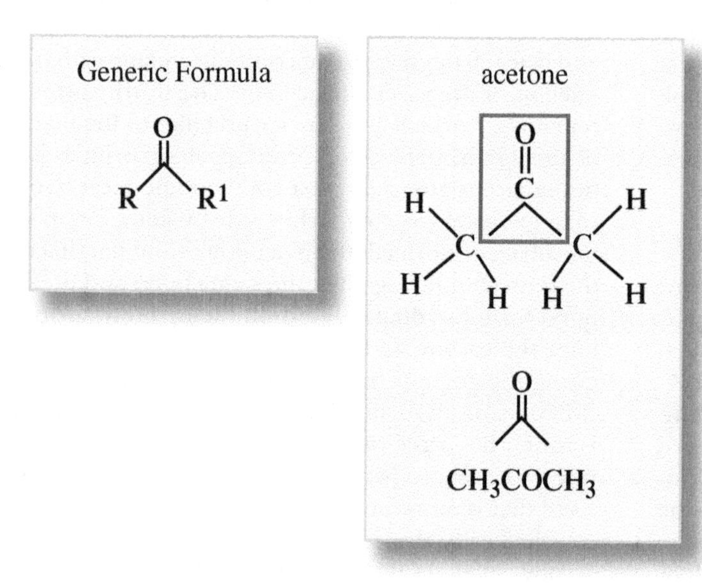

An alkyl or aryl group bonded to a carbonyl group is also the acyl radical moiety of carboxylic acids (R–COOH) and their derivatives. Many common ketone names reflect this. The acyl radical of acetic acid, for example, is the acetyl group, and that of benzoic acid is the benzoyl group. Generally, the acyl radical name of the corresponding carboxylic acid is used to identify the ketone portions of a molecule when they are not part of the basic molecular skeleton.

Formation of Ketones

Ketones are prepared directly by three main methods. The first and most common method is the oxidation of secondary alcohols (hydrocarbons of the general form R–CHOH–R'). Reaction of the secondary alcohol with an oxidizing agent, such as chromium trioxide (CrO_3) or potassium dichromate ($K_2Cr_2O_7$), eliminates a molecule of hydrogen (H_2) from the alcohol to produce the C=O double bond of the carbonyl group. For example, treatment of cyclohexanol with chromium trioxide in acetic acid yields cyclohexanone. The method is generally amenable to the use of protective groups to prevent oxidation of other susceptible functional groups in the molecule.

The second principal method of forming ketones is the Friedel-Crafts acylation reaction, in which an acid chloride reacts with an appropriate aryl compound using a Lewis acid, such as aluminum trichloride ($AlCl_3$), as a catalyst. The reaction adds the acyl group directly to the aryl ring, displacing the hydrogen atom that was originally present. This method is most useful for producing aromatic ketone compounds.

The third method, useful for producing more complex ketones, is a coupling reaction involving an acid chloride and the lithium salt of an organocopper compound. An alkyl or aryl halide (of the form RX or ArX, where X is a halogen) is first reacted with lithium metal (Li) to produce the corresponding organolithium compound. This is then reacted with a cuprous halide (CuX) to form a dialkyl or diaryl copper compound (R_2CuLi). Reaction of this material with an acid chloride produces the ketone.

Neither the Friedel-Crafts method nor the organocopper method generates the ketone carbonyl function directly; rather, both work to augment a preexisting carbonyl group. Other synthetic methods also incorporate an existing carbonyl group into a reaction process to produce a ketone product. One such technique is the acetoacetic ester method. In ethyl acetoacetate or a similar ester, the hydrogen atoms on the carbon atom between the two carbonyl groups are somewhat acidic and can be extracted by a strong base. The resulting anion is then added onto another compound as a nucleophile, donating an electron and forming a new carbon-carbon single bond (C–C). The ester carboxylate group is then hydrolyzed, or cleaved by the addition of a water molecule, producing the parent carboxylic acid, from which the components of carbon dioxide (CO_2) are removed to give the ketone product.

SAMPLE PROBLEM

Given the chemical names isobutyrophenone and 1-(2,4,6-trimethylphenyl)butan-2-one, draw the skeletal structures of the molecule.

Answer:

The "-phenone" suffix in the first compound's name indicates a ketone with a phenyl group (benzene substituent) on one side of the carbonyl group and, based on the prefix "isobutyro-," an isobutyl group on the other. The molecule therefore has the following structure:

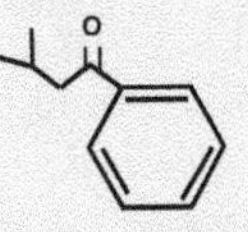

The root of the second name, "butan-," indicates the four-carbon chain of butane, while "2-one" indicates a carbonyl group on the second carbon. The first carbon atom bonds to a phenyl group with three methyl groups, located in the second, fourth, and sixth positions, with the bond to the butane structure being position one. The molecule therefore has the following structure:

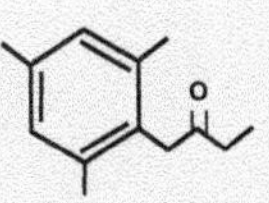

Reactions of Ketones

Ketones are reactive compounds that are widely used to form different types of compounds. The ease with which the ketone carbonyl group undergoes nucleophilic addition reactions facilitates the conversion of ketones into product compounds. In introductory organic chemistry courses, for example, the student's knowledge and practical laboratory skills are developed through classification tests of unknown compounds, among other procedures. The goal of these tests is to identify various compounds using simple reactions. To identify a ketone or an aldehyde (R-COH), one of the simplest tests is to combine a small amount of the suspected ketone or aldehyde with the compound 2,4-dinitrophenylhydrazine (DNPH). Either type of compound will immediately form the highly colored crystalline material 2,4-dinitrophenylhydrazone. The reaction is almost instantaneous and very distinct, proceeding as illustrated here:

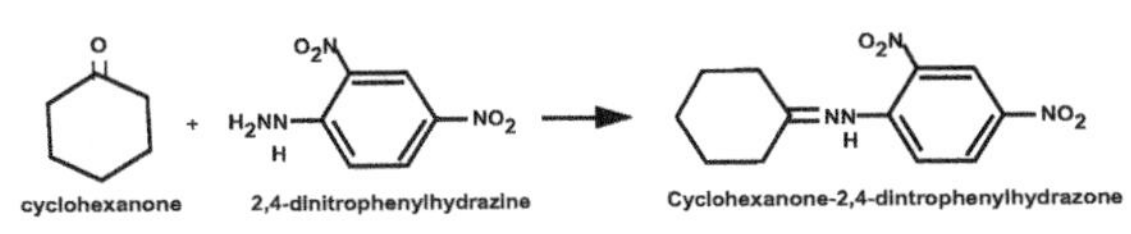

A variety of reducing agents can reduce ketones, formed by oxidizing secondary alcohols, back to secondary alcohols and even, in some cases, to methylene bridges ($-CH_2-$). The secondary alcohols produced in this way may be end products or intermediate stages in an overall synthetic process. Ketones can also react with 1,2- or 1,3-diols to produce the corresponding acetal (or ketal) in order to protect the carbonyl group against undesired reactions.

Richard M. Renneboog, MSc

BIBLIOGRAPHY

Berg, Jeremy M., John L. Tymoczko, Gregory J. Gatto, and Lubert Strye. *Biochemistry.* 8th ed. New York: W. H. Freeman, 2015. Print.

Hendrickson, James B., Donald J. Cram, and George S. Hammond. *Organic Chemistry.* 3rd ed. New York: McGraw, 1970. Print.

Jones, Mark Martin. *Chemistry and Society.* 5th ed. Philadelphia: Saunders College Pub., 1987. Print.

Lehninger, Albert L. *Biochemistry: The Molecular Basis of Cell Structure and Function.* 2nd ed. New York: Worth, 1975. Print.

Mann, J., et al. *Natural Products: Their Chemistry and Biological Significance.* New York: Wiley, 1994. Print.

Morrison, Robert Thornton, and Robert Neilson. Boyd. *Organic Chemistry.* 7th ed. Englewood Cliffs, N.J.: Prentice Hall, 2003. Print.

Wuts, Peter G. M., and Theodora W. Greene. *Greene's Protective Groups in Organic Synthesis.* 4th ed. Hoboken: Wiley-Interscience, 2007. Print.

L

LANTHANIDES

FIELDS OF STUDY

Inorganic Chemistry

SUMMARY

The basic structure of atoms is defined, and the development of the modern theory of atomic structure is elaborated. Atomic structure is fundamental to all fields of chemistry, and especially to fields that rely on the intrinsic properties of individual atoms.

PRINCIPAL TERMS

- **f-block:** the portion of the periodic table containing the elements whose valence electrons are in their *f* orbitals, namely the lanthanides and the actinides.
- **rare earth elements:** a group of chemical elements in the periodic table that includes the fifteen lanthanide elements (lanthanum, cerium, praseodymium, neodymium, promethium, samarium, europium, gadolinium, terbium, dysprosium, holmium, erbium, thulium, ytterbium, and lutetium) as well as scandium and yttrium.
- **trivalent:** describes an atom that has the ability to exist in three different oxidation states and to form compounds accordingly.

THE NATURE OF THE LANTHANIDES

The lanthanides are a group of metallic elements in the periodic table with the atomic numbers 57 through 71. They are in the f-block of the periodic table, though it is a matter of some debate whether one of the lanthanides, either lanthanum or lutetium, should be considered a d-block element instead. Elements in the f-block have valence electrons in their *f* orbitals, while some of their inner atomic orbitals may remain incompletely filled. This arrangement of electrons is responsible for the multivalent nature of lanthanide ions, which allows them to form compounds in which they take on different oxidation states.

Although they are relatively abundant for the most part, lanthanides tend not to occur in concentrated deposits; that is why they were historically called "rare earth metals" or "rare earth elements," from the old usage of the term "earth" for metal oxide minerals. In addition to the lanthanides, the transition metals scandium and yttrium are typically categorized as rare earth elements as well, as they are chemically similar to lanthanides and often occur alongside them.

ATOMIC STRUCTURE OF THE LANTHANIDES

The *f* orbitals first become relevant in atomic structure in the elements of the fourth period of the periodic table. Lanthanum is the first fourth-period element to be able to have electrons in its *f* orbital, due to the ordering of the energy levels of atomic orbitals. Electrons usually fill atomic orbitals in the order 1*s*, 2*s*, 2*p*, 3*s*, 3*p*, 4*s*, 3*d*, 4*p*, 5*s*, 4*d*, 5*p*, 6*s*, 4*f*, 5*d*, 6*p*, 7*s*, rather than in strict numerical order. However, the allowed energy levels of the orbitals become closer and closer together as the distance from the nucleus increases, permitting electrons to shift from orbital to orbital in response to environmental influences, such as the types and numbers of ions or molecules that surround the particular atom in a compound. As a result, the distribution of electrons in these higher orbitals may differ slightly from the expected configuration.

The lanthanides are known to be strong reducing agents, meaning that they are able to readily take part in reduction-oxidation (redox) reactions due to the ease with which they can donate three valence electrons to form cations with a 3+ charge. This is facilitated by the low difference in the energy levels of the outer electron orbitals.

LANTHANIDE COMPOUNDS

The radii of the various lanthanide cations with 3+ charges are very similar, which allows them to form

LANTHANIDES

60 Nd neodymium 144

1 H hydrogen 1																	2 He helium 4
3 Li lithium 7	4 Be beryllium 9											5 B boron 11	6 C carbon 12	7 N nitrogen 14	8 O oxygen 16	9 F fluorine 19	10 Ne neon 20
11 Na sodium 28	12 Mg magnesium 29											13 Al aluminium 27	14 Si silicon 28	15 P phosphorous 31	16 S sulfur 32	17 Cl chlorine 35	18 Ar argon 40
19 K potassium 39	20 Ca calcium 40	21 Sc scandium 45	22 Ti titanium 48	23 Vi vanadium 51	24 Cr chromium 52	25 Mn manganese 55	26 Fe iron 56	27 Co cobalt 59	28 Ni nickel 59	29 Cu copper 64	30 Zn zinc 65	31 Ga gallium 70	32 Ge germanium 73	33 As arsenic 75	34 Se selenium 79	35 Br bromine 80	36 Kr krypton 84
37 Rb rubidium 85	38 Sr strontium 88	39 Y yttrium 89	40 Zr zirconium 91	41 I niobium 93	42 Mo molybdenum 96	43 Tc technetium 98	44 Ru ruthenium 101	45 Rh rhodium 103	46 Pd palladium 106	47 Ag silver 108	48 Cd cadmium 112	49 In indium 115	50 Sn tin 119	51 Sb antimony 122	52 Te tellerium 128	53 I iodine 127	54 Xe xenon 131
55 Cs caesium 133	56 Ba barium 137	*	72 Hf hafnium 178	73 Ta tantalum 181	74 W tungsten 184	75 Re rhenium 186	76 Os osmium 190	77 Ir iridium 192	78 Pt platinum 195	79 Au gold 197	80 Hg mercury 201	81 Tl thallium 204	82 Pb lead 207	83 Bi bismuth 209	84 Po polonium 209	85 At astatine 210	86 Rn radon 222
87 Fa francium 223	88 Ra radium 226	**	104 Rf rutherfordium 267	105 Db dubnium 268	106 Sg seaborgium 271	107 Bh bohrium 272	108 Hs hassium 270	109 Mt meitnerium 276	110 Ds darmstadtium 281	111 Rg roentgenium 280	112 Cn copernicum 285	113 Uut ununtrium 284	114 Fl flerovium 289	115 Uup ununpentium 288	116 Lv livermorium 293	117 Uus ununseptium 294	118 Uuo ununoctium 289

57 * La lanthanum 139	58 Ce cerium 140	59 Pr praseodymium 141	60 Nd neodymium 144	61 Pm promethium 145	62 Sm samarium 150	63 Eu europium 152	64 Gd gadolinium 157	65 Tb terbium 159	66 Dy dysprosium 163	67 Ho holmium 165	68 Er erbium 167	69 Tm thulium 168	70 Yb ytterbium 173	71 Lu lutetium 175
89 ** Ac actinium 227	90 Th thorium 232	91 Pa protactinium 231	92 U uranium 238	93 Np neptunium 237	94 Pu plutonium 244	95 Am americium 243	96 Cm curium 247	97 Bk berkelium 247	98 Cf californium 251	99 Es einsteinium 252	100 Fm fermium 257	101 Md mendelevium 258	102 No nobelium 259	103 Lr lawrencium 284

very similar compounds. Typically these are all ionic compounds with 3+ cations, though some compounds of europium and ytterbium 2+ cations are known as well. It is theorized that the trivalent ions form readily to achieve an electron distribution corresponding to that of a noble gas, with the outermost valence orbitals vacant. The divalent ions that form are thought to reflect an electron distribution in which all of the valence orbitals are half filled, each containing a single electron. Cerium and terbium are also known to produce cations bearing a 4+ charge.

The lanthanide metal oxides have the general formula Ln_2O_3, similar to that of aluminum oxide and iron oxide, and a high negative enthalpy of formation, meaning their formation releases a great deal of heat. This makes the lanthanide metals useful in thermite reactions in which the metal very actively reduces other compounds to elemental form, just as aluminum metal reduces iron oxide to elemental iron in the thermite reaction. A mixture of cerium and lanthanum, with minor quantities of iron and other lanthanide metals, is so reductive as to spontaneously ignite upon contact with air when finely divided. Lanthanide metals are used primarily to lend specific properties to alloys of other metals.

Occurrence of Lanthanides

Lanthanides almost always occur together in nature as compounds of the 3+ ions. They are normally found in nature as ores and minerals in which they exist primarily as oxides, although sulfides, silicates, carbonates, sulfates, and other inorganic compounds are also possible. One of the most common sources is a mineral called monazite, which occurs in at least four different variations, each one containing several different lanthanides. There are various methods

SAMPLE PROBLEM

Determine the electron distribution of gadolinium (atomic number 64), and identify its valence electrons. Use the following electron distribution chart if necessary.

```
1s
2s 2p
3s 3p 3d
4s 4p 4d 4f
5s 5p 5d 5f 5g
6s 6p 6d 6f 6g 6h
7s 7p 7d 7f 7g 7h
```

Answer:

Gadolinium contains sixty-four electrons, theoretically distributed as

$$1s^2 2s^2 2p^6 3s^2 3p^6 4s^2 3d^{10} 4p^6 5s^2 4d^{10} 5p^6 6s^2 4f^8$$

However, reference books list the actual distribution as . . . $4f^7 5d^1$, based on observation of the energy levels of electrons revealed by spectroscopic studies of the lanthanide elements. This is a result of the $4f$ and $5d$ energy levels being so close together.

Gadolinium has ten valence electrons, in the $4f$, $5d$, and $6s$ orbitals. Atoms form ions in order to attain stable electron configurations, which are represented by the electron distributions of the noble gases (helium, neon, argon, krypton, xenon, and radon). The electron distributions of all elements apart from hydrogen can also be conveyed by putting the symbol of the noble gas immediately preceding the element in brackets, followed by the extra electrons not contained in the noble gas. For example, the last noble gas before gadolinium in the periodic table is xenon (Xe), which has an electron configuration of

$$1s^2 2s^2 2p^6 3s^2 3p^6 4s^2 3d^{10} 4p^6 5s^2 4d^{10} 5p^6$$

so the electron configuration of gadolinium can be written as $[Xe]6s^2 4f^7 5d^1$. For elements in the f-block, the valence electrons are the ones outside the noble gas configuration, as these are the electrons the gadolinium atom would have to lose to achieve a stable electron configuration.

for extracting lanthanides from monazite, including heating monazite sand with sulfuric acid, which produces a lanthanide sulfate solution, and combining monazite sand with a heated sodium hydroxide solution, which separates the phosphate content of the mineral from the lanthanides. The lanthanides are never found as "native" metals, only as mineral compounds.

Richard M. Renneboog, MSc

Bibliography

Douglas, Bodie E., Darl H. McDaniel, and John J. Alexander. *Concepts and Models of Inorganic Chemistry.* 3rd ed. New York: Wiley, 1994. Print.

Gribbin, John. *Science: A History, 1543–2001.* London: Penguin, 2002. Print.

Haynes, W. H., PhD, David R. Lide, PhD, and Thomas J. Bruno, PhD, eds. *CRC Handbook of Chemistry and Physics,* 96th Edition. New York: CRC/Taylor & Francis Group, 2015. Print.

Johnson, Rebecca L. *Atomic Structure.* Minneapolis: Lerner, 2008. Print.

Kean, Sam. *The Disappearing Spoon: And Other True Tales of Madness, Love, and the History of the World from the Periodic Table of the Elements.* New York: Little, Brown, 2010. Print.

Mackay, K. M., R. A. Mackay, and W. Henderson. *Introduction to Modern Inorganic Chemistry.* 6th ed. Cheltenham: Nelson, 2002. Print.

Winter, Mark J. *The Orbitron: A Gallery of Atomic Orbitals and Molecular Orbitals.* University of Sheffield, n.d. Web.

LEWIS STRUCTURE AND DIAGRAM

FIELDS OF STUDY

Physical Chemistry; Inorganic Chemistry; Organic Chemistry

SUMMARY

The visual representation of electron distribution known as Lewis structures and the Lewis dot diagram are discussed. The method represents the valence electrons in atoms and illustrates their combination in shared pairs to form bonds between atoms. Lewis structures are a simple means of relating molecular structure to valence.

PRINCIPAL TERMS

- **covalent bond:** a type of chemical bond in which electrons are shared between two adjacent atoms.
- **electron:** a fundamental subatomic particle with a single negative electrical charge, found in a large, diffuse cloud around the nucleus.
- **lone pair:** two valence electrons that share an orbital and are not involved in the formation of a chemical bond; also called a nonbonding pair.
- **multiple bond:** a bond formed by two atoms sharing two or more electron pairs; includes double bonds and triple bonds.
- **shared pair:** the two electrons shared between two atoms in a normal covalent bond.

GILBERT N. LEWIS AND VALENCE THEORY

In 1916, American chemist Gilbert N. Lewis (1875–1946) published "The Atom and the Molecule," his first paper on the role of valence in chemical bonding. Electrons had been identified as charged subatomic particles just nineteen years earlier, and protons were about to be discovered the following year, while the existence of neutrons as the third subatomic particle would not be demonstrated until 1932. The modern theory of atomic structure was therefore in its very early stages and had not yet been defined by quantum mechanics, although the basic underlying mathematical principles had been developed. Any well-defined concept of atomic orbitals and the wave-particle duality of electrons (that is, their tendency to behave as both particles and waves) was still very controversial and very much open to discussion. These are all concepts now taken for granted by chemists and physicists the world over, but in Lewis's time, they were very new ideas. The theoretical aspects did not yet have the support of experimental evidence. The electronic measurement devices to which he had access were crude by modern standards, predating the invention of the transistor by more than thirty years. Accordingly, explanations of chemical behavior based on electron shells and orbitals were subject to intense questioning by traditional chemists.

In this environment, Lewis proposed that atoms with an atomic mass greater than that of helium have inner shells of electrons with the same distribution as in the noble gas preceding them in the periodic table (as the six noble gases all have their outer electron shells filled and are therefore chemically inert). A sodium atom, for example, would have all of its electrons but one in the same configuration as the electrons in a neon atom; the extra electron would lie outside of these inner shells of electrons, in the outermost, or valence, shell. The natural corollary of this hypothesis was that the valence electrons of an atom could be easily given up to form a positively charged ion, or cation, with all of its electron shells filled in the same way as those of the preceding noble-gas element. Conversely, electrons could be added from other atoms to form a negatively charged ion, or anion, with all of its electron shells filled in the same way as those of the next noble-gas element in the periodic table.

LEWIS DOT STRUCTURES

According to the modern theory of atomic structure, electrons are arranged in shells about the nucleus of an atom. The first shell can hold up to two electrons, the second shell can hold a total of eight, the third shell a total of eighteen, and the fourth shell a total of thirty-two; while subsequent shells could theoretically hold more, in practice, the fifth, sixth, and seventh electron shells also hold a maximum of thirty-two electrons. The outermost shell of an atom, however, only ever has a maximum of eight electrons (at least in theory—transition metals can have more), beyond which the next shell starts filling. For example,

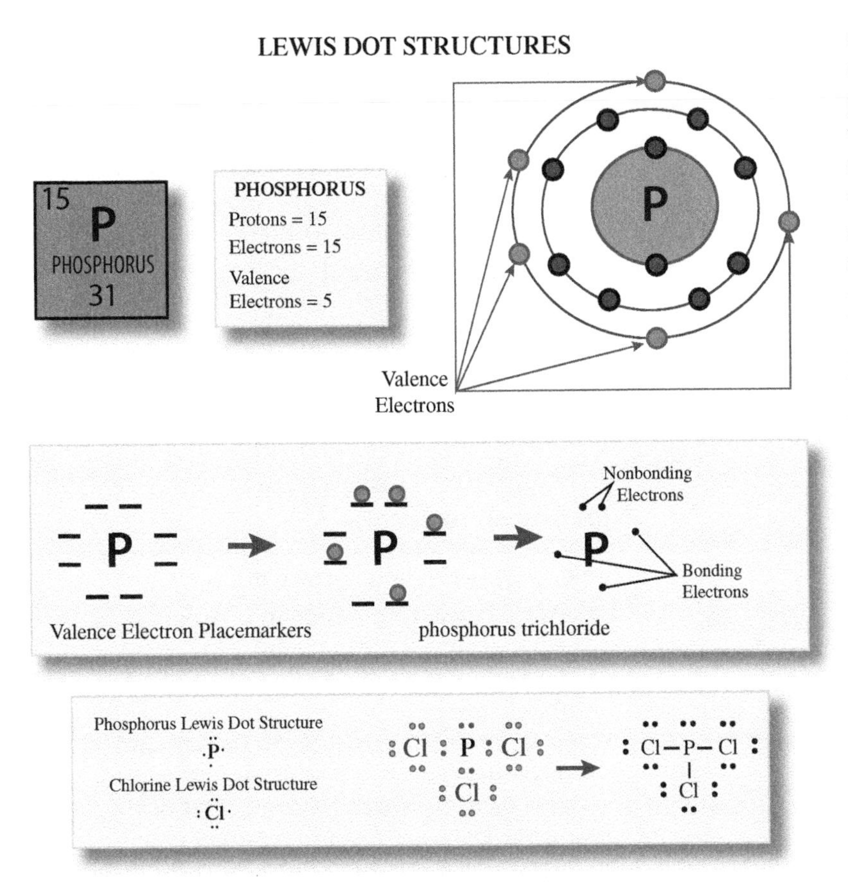

A Lewis Dot Structure shows the valence electrons (electrons in the outermost shell, capable of bonding) surrounding the atom, represented by the element's symbol. Valence electrons are spaced evenly on the four sides of the symbol. Paired electrons are nonbonding, single electrons are used to show covalent bonds. Bonding electron pairs are drawn as dashes instead of dots.

a calcium atom has twenty electrons, but it has two electrons in the first shell, eight in the second, eight in the third, and two in the fourth, rather than ten in the third and none in the fourth. The tendency of atoms to combine or form ions in such a way that they have eight electrons in their valence shells is known as the octet rule.

In a Lewis dot diagram of an atom, the chemical symbol of the element is shown, representing the nucleus and the inner electron shells, and is surrounded by up to eight dots, generally in pairs, representing the electrons in the atom's outermost shell. For example, using calcium again, the Lewis diagram of this element consists of the element symbol, "Ca," with two dots next to it, representing the two electrons in the fourth shell.

In a Lewis structure of a molecule, the chemical symbols of the constituent elements are shown connected by lines in place of dots, each one representing the shared pair of electrons that forms a covalent bond between the atoms; a single line represents a single bond, while double and triple lines represent multiple bonds. (Ionic bonds are represented differently; an ionic compound is depicted as adjacent but separate ions.) Any electrons not involved in chemical bonds, be they unpaired electrons or lone pairs, are still shown as dots next to their respective elements. There are variations on this system, some of which retain the circles representing the atomic orbitals, others of which use the dots in place of lines to show the bonds between atoms in a compound; however, the system described here is the most widely used, as it permits a comprehensible two-dimensional representation of the molecule and the bond system it contains.

The Cubical Atom

Lewis's 1916 paper introduced several important ideas: the sharing of electrons to form a covalent bond, the transfer of electrons from one atom to another to form an ionic bond, the octet rule, and of course his dot diagram. He also proposed that, counter to the planetary model of the atom introduced by Niels Bohr (1885–1962) in 1913, atoms were in fact cubical in shape, with valence electrons positioned at some or all of the cube's eight corners. Two cubical atoms could form a single covalent bond by sharing a single edge so that they had two corners in common, while a double covalent bond was formed by the atoms sharing a full face, giving them four corners in common. Because this model could not account for triple bonds—two cubes cannot share more than four corners at one time—Lewis suggested that in some cases, the electrons of an atom would rearrange themselves from a cubical to a tetrahedral shape (a three-sided pyramid) with two electrons at each corner, allowing two atoms to share six electrons by sharing a single face.

SAMPLE PROBLEM

Draw the Lewis dot diagram for silicon. Then draw a Lewis diagram showing how the atoms are bonded in the acetic acid molecule ($C_2H_4O_2$).

Answer:

Silicon is in group 14 of the periodic table, the same group as carbon, and has fourteen electrons, four of which are in the valence shell (the $n = 3$ shell). Its Lewis dot diagram is the symbol Si surrounded by four electrons:

```
  •
• Si •
  •
```

Acetic acid has two carbon atoms, four hydrogen atoms, and two oxygen atoms. Three of the hydrogen atoms are bonded to one of the carbon atoms, which in turn is bonded to the second carbon atom. The second carbon atom is double bonded to one of the oxygen atoms and single bonded to the second oxygen atom. The second oxygen atom is bonded to the fourth hydrogen atom. Both oxygen atoms have two lone pairs of electrons that do not form bonds. The Lewis dot diagrams for this is as follows:

```
    H  ..
    |  O:
    | //
  H-C-C
    |
    H-:O-H
       ..
```

While the cubical model of the atom was consistent with Lewis's valence theory, it never found widespread acceptance, although Irving Langmuir (1881–1957) built on Lewis's ideas to further refine valence theory and propose his own model of atomic structure. Lewis's atomic model, like all other contemporary models, was eventually disproved and replaced by the quantum mechanical model, which was superficially similar to Bohr's model but incorporated the idea of wave-particle duality and defined electron orbitals as the areas of the atom with the greatest probability of containing a given electron, rather than defined paths for electrons to follow around the nucleus.

Richard M. Renneboog, MSc

BIBLIOGRAPHY

Abbott, David. *The Biographical Dictionary of Scientists.* New York: P. Bedrick, 1984. Print.

Askeland, Donald R., Wendelin J. Wright, D. K. Bhattacharya, and Raj P. Chhabra. *The Science and Engineering of Materials.* Boston: Cengage Learning, 2016. Print.

Douglas, Bodie E., Darl H. McDaniel, and John J. Alexander.Concepts an*d Models of Inorganic Chemistry. 3rd ed.* New York: Wiley, 1994. Print.

Jones, Mark Martin. *Chemistry and Society.* 5th ed. Philadelphia: Saunders College Pub., 1987. Print.

Mackay, K. M., R. A. Mackay, and W. Henderson. *Introduction to Modern Inorganic Chemistry.* 6th ed. Cheltenham: Nelson, 2002. Print.

Myers, Richard. *The Basics of Chemistry.* Westport: Greenwood, 2003. Print.

LITMUS TEST

FIELDS OF STUDY

Analytical Chemistry; Inorganic Chemistry; Organic Chemistry

SUMMARY

Litmus is a complex mixture of compounds isolated from various lichens. The material is sensitive to changes in pH, turning red in acidic solutions and blue in alkaline solutions. The intensity and shade of the color correlate to the pH of the solution.

PRINCIPAL TERMS

- **acid:** a compound that can relinquish one or more hydrogen ions (Brønsted-Lowry acid-base theory) or that possesses vacant atomic orbitals to interact with electron-rich materials (Lewis acid-base theory).
- **alkaline:** describes a material that tends to increase the concentration of hydroxide ions in an aqueous solution, as well as conditions produced by the presence of bases.
- **base:** a compound that can relinquish one or more hydroxide ions (Brønsted-Lowry acid-base

theory) or that possesses lone pairs of electrons that can interact with electron-poor materials (Lewis acid-base theory).

- **neutral:** describes a chemical solution that has a pH of 7 and thus is neither acidic nor basic.
- **pH indicator:** a compound that changes color according to the pH of a solution, or a device that measures the pH of a solution electronically.
- **water soluble:** describes a compound that can be dissolved by water.

THE THEORY BEHIND THE LITMUS TEST

Water-soluble materials, particularly acids and bases, alter the natural amounts of hydrogen ions present in a solution. Pure water undergoes an equilibrium dissociation reaction called autoprotolysis, in which a molecule of water splits into a hydrogen ion (H^+) and a hydroxide ion (OH^-), according to the equation:

$$2H_2O \rightleftharpoons H^+ \cdot H_2O + OH^-$$

Since equal amounts of positively charged hydronium ions (H_3O^+) and negatively charged hydroxide ions are produced, water is neutral, having neither a net negative nor a net positive charge. Thus, it is neither acidic nor alkaline. When a compound is dissolved in water, it may disrupt this equilibrium so that the hydronium ions and the hydroxide ions are no longer in balance. This does not mean that there will be a net negative or positive electrical charge, however, since the material that was dissolved will always have the appropriate number of opposite charges, so the solution will always be electrically neutral. The important feature is that the hydroxide ions and the hydronium ions specifically will not be in balance. For example, dissolving sodium hydroxide (NaOH) in water adds extra hydroxide ions to the solution, but their negative charge is balanced by the positive charges carried by the sodium ions. However, because there are more hydroxide ions than hydronium ions, the autoprotolysis equilibrium cannot be maintained, so a solution of sodium hydroxide in water is alkaline rather than neutral. In the same sense, dissolving an acidic compound, such as sulfuric acid (H_2SO_4), in water introduces an excess of hydronium ions, creating a solution that is acidic in nature.

THE NATURE OF LITMUS

Many aqueous solutions of acid are colorless, exactly like plain, pure water. A method of indicating which solution is acidic, which is alkaline, and which is neutral is required to tell them apart. This is the purpose of the litmus test.

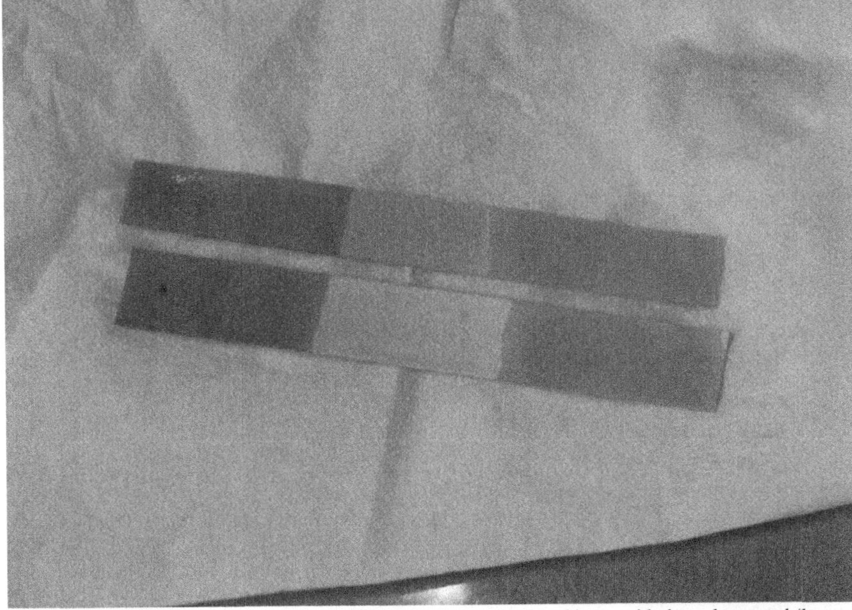

Blue (top) and red (bottom) litmus paper. When litmus paper is placed in an acid, the red stays red (bottom right) and the blue turns red (top right). When litmus paper is placed in a base, the blue stays blue (top left) and the red turns blue (bottom left). By Chemicalinterest, via Wikimedia Commons

The material known as "litmus" is a combination of several similar compounds isolated from a number of species of lichens. The compounds are sensitive to the acidic or alkaline character of aqueous solutions. They are also water soluble and so can be added to or mixed directly with other solutions in water. Litmus reacts with hydronium and hydroxide ions differently, and as the molecular structures of the litmus compounds change, the materials change color accordingly. In acidic solutions, litmus becomes red, while in alkaline solutions, it becomes blue. The intensity of the color relates directly to the concentration of hydronium and hydroxide ions in the solution. In this way, litmus can be used as a pH indicator.

Litmus is now used only as a quick means of approximating the pH of a solution. Most pH measurements are obtained using electronic instruments that are more precise and do not add contaminants that can alter the pH of the solution.

LITMUS AND THE PH SCALE

It is generally inconvenient to use numerical values for hydronium and hydroxide ion concentrations. In neutral water, the concentration of each is just 10^{-7} molar (M), or moles per liter. Such a numerical value is necessary in precise calculations, but for general purposes, it is much more convenient to describe acidic and alkaline solutions in terms of their respective pH values. The pH of a solution is defined as:

$$pH = -\log[H_3O^+]$$

where $[H_3O^+]$ is the molar concentration of hydronium ions in the solution. Accordingly, the pH of pure water is 7. The pOH, which is defined as:

$$pOH = -\log[OH^-]$$

is therefore also 7, and the equilibrium constant of water autoprotolysis is $10^{-7} \times 10^{-7}$, or 10^{-14}, which has a logarithmic value of 14. Because the constant is an equilibrium value, the pH and the pOH of a solution must always add up to 14, which is an extremely useful rule when carrying out numerical calculations involving pH. As the concentration of hydronium ions increases from the neutral value of 10^{-7} M, the pH must decrease in value. Thus, acidic solutions have a pH value of less than 7, and alkaline solutions have a pH greater than 7.

SAMPLE PROBLEM

Litmus pH indicator is added to samples of three different solutions. Solution A produces a pale orange color, B a rich red color, and C a medium blue color. Rank the solutions in order of increasing acidity.

Answer:

Litmus turns blue in alkaline solutions and red in acidic solutions. The blue solution must therefore be alkaline and the red solution acidic. The pale orange solution is also acidic, but much less so than the red solution. Therefore, in order of increasing acidity, the solutions can be ranked as C (blue), A (pale orange), and B (red).

THE COLORS OF LITMUS

The interaction between litmus and hydronium ions is also an equilibrium process. In solid powder form, litmus appears blue. The color-producing functional groups, or chromophores, of the litmus compounds absorb light such that only blue light is reflected. When litmus becomes protonated—gains a proton—in an acidic solution, as here:

$$\text{litmus} + H^+ \rightleftharpoons \text{litmus}H^+$$

the bonds of the chromophores change so that light in the blue region of the spectrum is absorbed, allowing light in the red region to pass through. The more hydronium ions are present, the more the equilibrium between litmus and protonated litmus shifts toward the protonated form, and so the intensity of the red color increases with increasing hydronium ion concentration. But when litmus interacts with hydroxide, it gives up a proton, acting as an acid:

$$\text{litmus} + OH^- \rightleftharpoons \text{litmus}^- + H_2O$$

When this happens, the chromophores absorb light from the red region of the spectrum, permitting light from the blue region of the spectrum to pass through. Again, the amount of hydroxide ions present shifts the equilibrium of the litmus-hydroxide reaction, causing the intensity of the blue color to increase with increasing hydroxide ion concentration.

Richard M. Renneboog, MSc

BIBLIOGRAPHY

Jones, Mark Martin. *Chemistry and Society.* 5th ed. Philadelphia: Saunders College Pub., 1987. Print.

Mackay, K. M., R. A. Mackay, and W. Henderson. *Introduction to Modern Inorganic Chemistry.* 6th ed. Cheltenham: Nelson, 2002. Print.

Myers, Richard. *The Basics of Chemistry.* Westport: Greenwood, 2003. Print.

Skoog, Douglas Arvid, Donald M. West, and F. James. Holler. *Skoog and West's Fundamentals of Analytical Chemistry.* 9th ed. Belmont (Calif.): Cengage Learning, 2014. Print.

Wink, Donald J., Sharon Fetzer-Gislason, and Sheila McNicholas. *The Practice of Chemistry.* New York: Freeman, 2004. Print.

M

MALLEABILITY

FIELDS OF STUDY

Metallurgy

SUMMARY

The property of malleability is discussed, and its nature as an intensive property of metals is described. Malleability of metals results directly from the nature of their electronic structure and the physical size of their atoms.

PRINCIPAL TERMS

- **compressive stress:** a force that acts to push on or compress a material.
- **deformation:** any permanent change in the shape of an object as a result of the application of force or a change in temperature.
- **ductility:** the ability of a solid material to be deformed by the application of tensile (pulling) force, such as bending, without breaking or fracturing.
- **fracture:** a dislocation in the internal structure of an object that causes it to break into two or more pieces.
- **intensive properties:** the properties of a substance that do not depend on the amount of the substance present, such as density, hardness, and melting and boiling point.
- **metallic bond:** a type of chemical bond formed by the sharing of delocalized electrons between a number of metal atoms.
- **plasticity:** the ability of a material to undergo deformation without breaking.

THE NATURE OF MALLEABILITY

Malleability is an extremely important property of certain materials, especially metals. The malleability of a metal permits it to undergo deformation via compressive stress without fracture. A prime example of this property is the forging or shaping of a metal object by hammering or stamping in a press. In this process, a metal blank is subjected to compressive stress as it is pounded by a hammer or in a die that produces a pattern such as those on coins. This alters the physical shape of the blank as well as its dimensions. The process is repeated until the original blank has been converted into the desired object. Stamping, pressing, and forging are all methods by which a malleable metal may be shaped by the application of force; the difference is the nature of the force that is applied.

Malleability is usually exploited by the instantaneous application of compressive force, such as occurs in stamping and hammering. A related property, ductility, refers to the ability of a material to undergo tensile stress (pulling or stretching) without fracturing. Tensile stress is generally applied to ductile metals over an extended period of time.

MALLEABILITY VS. DUCTILITY

Both malleability and ductility are aspects of plasticity, which is a temperature-related property of many materials, not just of metals. The plasticity of a material increases as temperature increases, until the material reaches its melting point, at which point the material becomes fluid rather than plastic and will not retain its shape as a solid. Ironworkers and steelworkers typically heat metal until it glows with a bright orange color. Plasticity is considered to be an intensive property because neither the malleability nor the ductility of a material depends on how much of it is present.

Nonmetal materials that exhibit plasticity include certain polymers, which were termed "plastics" for that very reason. Polymers that become softer when heated are called "thermoplastic" to indicate that their plasticity increases with heating. Like metals, thermoplastics also become fluid at a certain temperature. Generally, plastics are ductile but not malleable: they can be stretched and bent fairly easily and maintain their structural integrity, but they will fracture into numerous pieces when force is applied too quickly. A temperature range known as the glass-transition temperature marks the dividing line between glass-like behavior and malleability. Metals have a similar temperature range for malleability. For example, it has been demonstrated that the metal used to construct the RMS *Titanic* failed on impact with an iceberg because exposure to the frigid water of the northern Atlantic Ocean lowered its temperature to the point where it lost its malleability.

Corrugated galvanized iron rolling machine. This metal sheet can be pressed into a specific form without fracturing due to its malleability. By Peripitus via Wikimedia Commons

MALLEABILITY AND THE ELECTRONIC STRUCTURE OF METALS

According to the modern theory of atomic structure, electrons in atoms occupy specific regions of space about the nucleus called orbitals. While the outer electron orbitals of different atoms in a molecule frequently overlap to some degree, combining to form shared molecular orbitals so that the valence electrons of each atom are no longer solely confined to their "parent" atom, this effect is much more pronounced in metals, especially pure metals.

In a metal sample, the orbitals of the individual atoms are able to effectively overlap each other, allowing the valence electrons to be shared by all atoms in the sample and move relatively freely throughout the material. It is a requirement of modern atomic theory that when a number of atomic orbitals overlap, or combine, an equal number of hybrid or molecular orbitals are produced; in a way, this makes the entire body of the metal sample analogous to a single metal "molecule" across which the valence electrons can range. This phenomenon is known as metallic bonding and is characterized by a strong attractive force between the positively charged atomic cores and the negatively charged "sea" of delocalized valence electrons surrounding them. The electrostatic repulsion between the electrons causes them to distribute more or less uniformly throughout the metal so that each atomic core is surrounded by electrons in all directions, resulting in a generalized nondirectional force holding the atoms together. This is an extremely simplified model of metallic bonding, but it serves as a useful introduction to the concept.

The nature of metallic bonding contributes to the characteristic malleability and ductility of metals. The nondirectional attraction and the lack of any strong localized bonds to break make metallic bonds more resilient to applied force than covalent bonds, while the sharing of valence electrons allows the atoms to be packed more closely together, enabling them to slide against each other more easily to produce deformation instead of a fracture. This is especially true when all atoms are of the same size, which is why pure metals are typically more malleable and ductile than alloys.

Richard M. Renneboog, MSc

SAMPLE PROBLEM

Given the following list of metals and materials, classify them as being malleable or non-malleable:

- iron
- copper
- water
- water ice
- gold
- sodium
- diamond

Answer:

Iron and copper are both malleable and are commonly shaped by hammering or stamping, though copper is much more malleable than iron at any temperature. Water is a highly mobile fluid, so it is not malleable. Water ice is somewhat malleable and will absorb some impact force without fracturing. Gold is highly malleable; an ounce of gold can be hammered thinner than fine paper without fracturing. Sodium is highly malleable, and at room temperature it is soft enough to be easily cut with a knife. Diamond, because of its extremely rigid crystal structure, is not malleable.

Bibliography

Askeland, Donald R., Wendelin J. Wright, D. K. Bhattacharya, and Raj P. Chhabra. *The Science and Engineering of Materials.* Boston: Cengage Learning, 2016. Print.

Douglas, Bodie E., Darl H. McDaniel, and John J. Alexander. *Concepts and Models of Inorganic Chemistry.* 3rd ed. New York: Wiley, 1994. Print.

Fenichell, Stephen. *Plastic: The Making of a Synthetic Century.* New York: Harper, 2009. Print.

Jones, Mark Martin. *Chemistry and Society.* 5th ed. Philadelphia: Saunders College Pub., 1987. Print.

Miessler, Gary L., Paul J. Fischer, and Donald A. Tarr. *Inorganic Chemistry.* 5th ed. Upper Saddle River: Prentice Hall, 2013. Print.

MAN-MADE ELEMENTS

FIELDS OF STUDY

Nuclear Chemistry; Inorganic Chemistry

SUMMARY

The occurrence and properties of man-made elements are discussed. Man-made elements are the transuranium elements in the actinide series and beyond. Uranium is the heaviest naturally occurring actinide element and is commonly used as a starting material in nuclear synthesis reactions. All man-made elements are radioactive.

PRINCIPAL TERMS

- **half-life:** the length of time required for one-half of a given amount of material to decompose or be consumed through a continuous decay process.
- **plutonium:** atomic number 94, an extremely toxic and dangerously radioactive element of the actinide group that has several known isotopes.
- **radioactivity:** the emission of subatomic particles due to the spontaneous decay of an unstable atomic nucleus, the process ending with the formation of a stable atomic nucleus of lower mass.
- **synthetic:** produced by artificial means or manipulation rather than by naturally occurring processes.
- **transuranium elements:** the elements in the periodic table that have an atomic number greater

than that of uranium (92), all of which are unstable and radioactive.

THE NATURE OF MAN-MADE ELEMENTS

Only the first ninety-eight elements of the periodic table, hydrogen through californium, are known to occur naturally, yet the periodic table lists well over one hundred elements. All of the transuranium elements, which are all elements with an atomic number greater than that of uranium (92), are radioactive, and all were first discovered by being artificially created in a laboratory. Each isotope of these elements has a specific half-life, some of which are very short, measured in mere fractions of a second. Others, such as plutonium, have half-lives of several thousand years. All transuranium elements were originally thought to be entirely synthetic, or man-made, but atomic numbers 93 (neptunium) through 98 (californium) were later discovered to occur in nature in trace amounts.

One consistent feature of man-made elements is that they are very unstable and have short half-lives. The longest-lived known isotope of a purely synthetic element is einsteinium-252, which has a half-life of 471.7 days. Thus, even if some of these elements existed naturally in the distant past, they would have long since decayed into lower elements. Radioactive isotopes, or radioisotopes, decay in a series of nuclear-fission events in which their nuclei lose mass by spontaneously emitting particles in the form of ionizing radiation, ultimately transforming the isotopes into stable elements with less massive nuclei.

Accelerating charged particles to high kinetic energies can initiate the opposite type of reaction, nuclear fusion, in which highly energetic atomic nuclei are made to collide and fuse with a uranium or transuranium nucleus, adding new protons and neutrons to produce a more massive nucleus. The addition of a carbon-12 nucleus to a curium-246

MAN-MADE ELEMENTS

99 Es einsteinium 252

1	2	3	4	5	6	7	8	9	10	11	12	13	14	15	16	17	18
1 H hydrogen 1																	2 He helium 4
3 Li lithium 7	4 Be beryllium 9											5 B boron 11	6 C carbon 12	7 N nitrogen 14	8 O oxygen 16	9 F fluorine 19	10 Ne neon 20
11 Na sodium 28	12 Mg magnesium 29											13 Al aluminium 27	14 Si silicon 28	15 P phosphorus 31	16 S sulfur 32	17 Cl chlorine 35	18 Ar argon 40
19 K potassium 39	20 Ca calcium 40	21 Sc scandium 45	22 Ti titanium 48	23 Vi vanadium 51	24 Cr chromium 52	25 Mn manganese 55	26 Fe iron 56	27 Co cobalt 59	28 Ni nickel 59	29 Cu copper 64	30 Zn zinc 65	31 Ga gallium 70	32 Ge germanium 73	33 As arsenic 75	34 Se selenium 79	35 Br bromine 80	36 Kr krypton 84
37 Rb rubidium 85	38 Sr strontium 88	39 Y yttrium 89	40 Zr zirconium 91	41 I niobium 93	42 Mo molybdenum 96	43 Tc technetium 98	44 Ru ruthenium 101	45 Rh rhodium 103	46 Pd palladium 106	47 Ag silver 108	48 Cd cadmium 112	49 In indium 115	50 Sn tin 119	51 Sb antimony 122	52 Te tellurium 128	53 I iodine 127	54 Xe xenon 131
55 Cs caesium 133	56 Ba barium 137	*	72 Hf hafnium 178	73 Ta tantalum 181	74 W tungsten 184	75 Re rhenium 186	76 Os osmium 190	77 Ir iridium 192	78 Pt platinum 195	79 Au gold 197	80 Hg mercury 201	81 Tl thallium 204	82 Pb lead 207	83 Bi bismuth 209	84 Po polonium 209	85 At astatine 210	86 Rn radon 222
87 Fa francium 223	88 Ra radium 226	**	104 Rf rutherfordium 267	105 Db dubnium 268	106 Sg seaborgium 271	107 Bh bohrium 272	108 Hs hassium 270	109 Mt meitnerium 276	110 Ds darmstadtium 281	111 Rg roentgenium 280	112 Cn copernicium 285	113 Uut ununtrium 284	114 Fl flerovium 289	115 Uup ununpentium 288	116 Lv livermorium 293	117 Uus ununseptium 294	118 Uuo ununoctium 289

57 * La lanthanum 139	58 Ce cerium 140	59 Pr praseodymium 141	60 Nd neodymium 144	61 Pm promethium 145	62 Sm samarium 150	63 Eu europium 152	64 Gd gadolinium 157	65 Tb terbium 159	66 Dy dysprosium 163	67 Ho holmium 165	68 Er erbium 167	69 Tm thulium 168	70 Yb ytterbium 173	71 Lu lutetium 175
89 ** Ac actinium 227	90 Th thorium 232	91 Pa protactinium 231	92 U uranium 238	93 Np neptunium 237	94 Pu plutonium 244	95 Am americium 243	96 Cm curium 247	97 Bk berkelium 247	98 Cf californium 251	99 Es einsteinium 252	100 Fm fermium 257	101 Md mendelevium 258	102 No nobelium 259	103 Lr lawrencium 284

nucleus, for example, produces the man-made isotope nobelium-254. A similar method is neutron capture, in which the original nucleus is collided with individual neutrons rather than whole nuclei. Such reactions may require a number of years of continual bombardment to produce barely enough material to examine, and some of these elements have been identified from electronic traces produced by no more than a few individual atoms.

Electronic Structures

The man-made elements that have been studied most extensively are those in the actinide series, which all have valence electrons in their *f* orbitals and appear to exhibit the properties expected of elements with such electron configurations. The *f* orbitals can hold up to fourteen electrons, and a number of oxidation states are possible for these elements, although the purely synthetic actinides—einsteinium, fermium, mendelevium, nobelium, and lawrencium—are less variable in their oxidation states than the others. Atoms appear to be most stable when the outermost electron shell on the atom corresponds to that of the nearest noble gas. Thus, elements toward the left-hand side of the periodic table are prone to give up electrons, while elements toward the right-hand side of the periodic table are prone to accept extra electrons into their valence shells. An intermediate stability level is attained when the outermost electron shell is half filled. The man-made actinides most commonly take on a +3 oxidation state—that is, they lose three electrons to become a positively charged ion, or cation—although nobelium is most stable in a +2 oxidation state, and einsteinium, fermium, and mendelevium are occasionally stable in the +2 state as well.

SAMPLE PROBLEM

In a particle accelerator, atoms of uranium-238 are bombarded with highly accelerated boron-10 nuclei. If all of the particles in the boron-10 nucleus are absorbed by the uranium-238 nucleus, what isotope of which element might be created by this interaction?

Answer:

The number of protons in an atom's nucleus is equal to its atomic number, and the number in the name of an isotope is equal to its atomic mass, measured in atomic mass units (u). The atomic number of boron is 5, meaning that the boron-10 nucleus contains five protons and five neutrons (because 10 u – 5 protons = 5 neutrons). The atomic number of uranium is 92, indicating that the nucleus contains ninety-two protons. The nuclear reaction would add five neutrons and five protons to the uranium nucleus, increasing its mass by ten atomic mass units and increasing the number of protons to ninety-seven. Thus, the element would be berkelium, atomic number 97, and its atomic mass would be 238 u+ 10 u, or 248 u, making the name of the isotope berkelium-248.

Applications

Most of the elements heavier than plutonium do not exist in sufficient quantities to have any practical applications. Uranium, generally the fuel of choice for nuclear reactors, is also the starting material for the production of plutonium—perhaps the most virulently poisonous material known to exist, as well as being dangerously radioactive and the fuel of choice for nuclear warheads. While miniscule amounts of plutonium are found in nature, typically as decay products in concentrated uranium deposits, most of the plutonium that exists was created in nuclear reactors by bombarding uranium-238 with neutrons or deuterons (the nuclei of deuterium, or heavy hydrogen). Plutonium is a high-output fuel for nuclear reactors, but it has a half-life of nearly twenty-five thousand years, making storage of depleted plutonium fuel rods a serious safety problem in terms of both containing it and preventing it from being repurposed for destructive uses.

Richard M. Renneboog, MSc

Bibliography

Gribbin, John. *Science: A History, 1543–2001.* London: Lane, 2002. Print.

Haynes, W. H., PhD, David R. Lide, PhD, and Thomas J. Bruno, PhD, eds. *CRC Handbook of Chemistry and Physics,* 96th Edition. New York: CRC/Taylor & Francis Group, 2015. Print.

Johnson, Rebecca L. *Atomic Structure.* Minneapolis: Lerner, 2008. Print.

Kean, Sam. *The Disappearing Spoon: And Other True Tales of Madness, Love, and the History of the World from the Periodic Table of the Elements.* New York: Little, Brown, 2010. Print.

Mackay, K. M., R. A. Mackay, and W. Henderson. *Introduction to Modern Inorganic Chemistry.* 6th ed. Cheltenham: Nelson, 2002. Print.
Wehr, M. Russell, James A. Richards Jr., and Thomas W. Adair III. *Physics of the Atom.* 4th ed. Reading: Addison-Wesley, 1984. Print.
Winter, Mark J. *The Orbitron: A Gallery of Atomic Orbitals and Molecular Orbitals.* University of Sheffield, n.d. Web.

MELTING POINT

FIELDS OF STUDY

Physical Chemistry; Metallurgy; Analytical Chemistry

SUMMARY

The physical property of the melting point is discussed. Melting is the result of the energetic motions of atoms and molecules exceeding the ability of intermolecular attractive forces to maintain a rigid formation, allowing the component atoms or molecules to move freely about each other while maintaining contact.

PRINCIPAL TERMS

- **freezing point:** the temperature at which a liquid undergoes a phase change to become a solid.
- **intensive properties:** the properties of a substance that do not depend on the amount of the substance present, such as density, hardness, and melting and boiling point.
- **liquid:** a state of matter in which material is fluid, has definite volume but indefinite shape, and maintains a relatively constant density.
- **phase transition:** the change of matter from one state to another, such as from solid to liquid or liquid to gas, due to the transfer of thermal energy.
- **solid:** a state of matter in which material is non-fluid, has definite volume and shape, and maintains a near-constant density.

MELTING AND FREEZING EXPLAINED

All matter is composed of atoms, which can combine with other atoms to form molecules of compounds. Each element consists of just a single kind of atom, while compounds consist of atoms of two or more different elements. On Earth, matter typically exists at any one time in one of three physical phases: solid, liquid, or gas.

The phase in which matter exists is temperature dependent. When a material is in the solid phase, its atoms or molecules are held so rigidly in place that the material cannot alter its shape to conform to the shape of its container, as a liquid or gas can; the intermolecular attraction between the component particles prevents them from moving freely about relative to each other. As the temperature is raised, however, the atoms or molecules vibrate more and more energetically until, at a certain energy level, the motion is sufficient to overcome the intermolecular forces, allowing the particles to move relative to each other. At that point the material goes through a phase transition as it changes from solid to liquid. The melting point of a material is an intensive property and is

MELTING POINT

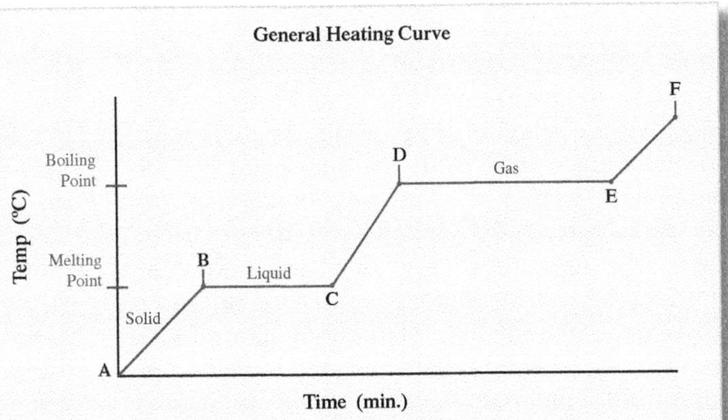

A general heating curve includes the following features. Segment A-B: Substance in a solid state; Point B: Melting point/freezing point. Segment C-D: Substance in a liquid state; Point D: Boiling point. Segment E-F: Substance in a gaseous state.

characteristic of the material; this means a sample of a pure compound weighing just a few milligrams melts at the same temperature as thousands of kilograms of the same material—though it would take a much longer time for the larger mass of material to achieve that temperature, as greater mass requires the input of more heat energy to reach the same temperature.

The opposite mechanism takes place as the temperature of the material is reduced and heat energy is removed from the material. Atomic or molecular motion decreases to the point where it is no longer sufficient to defeat the intermolecular forces that act to bind the particles in place, and the material passes from the liquid phase to the solid phase. The temperature at which this occurs is the freezing point of the material. In theory, the law of conservation of energy requires the melting point and the freezing point of a pure element or compound to be equal, although in practice there are certain exceptions.

Impurities and the Melting Point

The particles that make up certain solids, especially solid ionic compounds, are arranged in a regular, rigidly structured array called a crystal lattice. Such solids are said to be crystalline, as opposed to amorphous, which describes solids with irregular particle structures.

The presence of a foreign material in a quantity of a crystalline compound will disrupt the regularity of the crystal structure, and any impurity can alter the melting behavior of the compound. An impurity is typically an undesired component that is present only in very small proportions. The presence of any amount of foreign material in an otherwise pure compound has the effect of lowering the melting point of the mixture relative to the melting point of each component when pure. This phenomenon is the basis of the mixed melting-point test, which can be used to determine the identity of an unknown compound. In this test, the melting point of the unknown compound is determined, a known compound with a similar melting point is obtained, and the two compounds are combined. If the melting point stays the same, they are both the same compound; if the melting point drops, then they are not the same, and further tests are necessary to determine the identity of the unknown compound.

The amount by which the melting point of a mixture is lowered is directly related to the proportions of the materials in the mixture, which gives rise to an analytical procedure called freezing-point depression analysis. In a freezing-point depression analysis, a precisely weighed sample of a compound is mixed with a certain mass of another material used as a standard. The standard material must be of very high purity and have a very sharp melting point—that is, a melting point that exists within a very narrow range of temperatures. The amount by which the freezing point of the mixture is lowered, or depressed, relative to the melting point of the standard material can then be used to calculate the molecular mass of the unknown material from the mass of the sample.

Mixtures of different compounds often have broad melting points, meaning that one compound in the mixture melts sooner than another, causing the overall melting process to take place over a broad temperature range. In some cases, however, specific

SAMPLE PROBLEM

An unknown white crystalline material was found to have a melting point of 121°C. Five samples of the material were mixed with one additional compound each: benzoic acid, benzamide, naphthalene, urea, and fructose. When the melting points of the five mixture samples were obtained, one sample had a sharp melting point of 120–122°C, while the other four melted over several degrees at different temperatures. Identify the unknown material, using a suitable reference to find the melting points of the five test compounds.

Answer:

The melting points of the test materials are as follows:

- benzoic acid: 121°C
- benzamide: 130°C
- naphthalene: 80°C
- urea: 132°C
- fructose: 104°C

In a mixed melting-point determination, if the unknown material is different from the test compound, mixing the two together will both lower the melting point of the test compound and broaden the range of temperatures at which it melts. However, if the unknown material and the test compound are the same, the melting point of the mixture will be unchanged and remain within a narrow range of temperatures. The benzoic acid is the only compound whose melting point falls within the sharp melting-point range of 120–122°C. Thus, the unknown compound must also be benzoic acid.

proportions of compounds form what is termed a "eutectic" mixture, characterized by a sharp melting point. For example, an alloy (a mixture of a metal and some other element) is said to be eutectic when it is formed from a precise proportion of components that gives it a narrowly defined melting point; different proportions of the same components would cause it to have a much broader melting point, at which point it would no longer be eutectic.

The Effects of Molecular Structure

A crystal lattice is an orderly arrangement that can be compared to the arrangement of bricks in a wall; all of the component pieces have the same size, shape, and electronic properties, including polarity (the distribution of electric charge). These electronic properties are the source of the intermolecular attraction that holds the material together in the solid phase. Typically, the more polar the molecules of a compound, the more tightly they combine in the solid phase and the higher the melting point of the solid. In metals, the strength of the atomic interactions that maintain the material's solid phase is due to the metallic bonds between the atoms, which in effect make the entire mass of the metal behave as a single molecule; as a result, metals tend to have very high melting points. Conversely, molecules that are nonpolar or of very low polarity also have lower melting points.

Richard M. Renneboog, MSc

Bibliography

Askeland, Donald R., Wendelin J. Wright, D. K. Bhattacharya, and Raj P. Chhabra. *The Science and Engineering of Materials.* Boston: Cengage Learning, 2016. Print.

Fenichell, Stephen. *Plastic: The Making of a Synthetic Century.* New York: Harper, 2009. Print.

Haynes, W. H., PhD, David R. Lide, PhD, and Thomas J. Bruno, PhD, eds. *CRC Handbook of Chemistry and Physics,* 96th Edition. New York: CRC/Taylor & Francis Group, 2015. Print.

Holden, Alan, and Phylis Morrison. *Crystals and Crystal Growing.* New York: Doubleday, 1960. Print.

Jones, Mark Martin. *Chemistry and Society.* 5th ed. Philadelphia: Saunders College Pub., 1987. Print.

Kean, Sam. *The Disappearing Spoon: And Other True Tales of Madness, Love, and the History of the World from the Periodic Table of the Elements.* New York: Little, Brown, 2010. Print.

Lewis, Richard J., and Gessner G. Hawley. *Hawley's Condensed Chemical Dictionary.* 15th ed. New York: Wiley, 2007. Print.

Myers, Richard. *The Basics of Chemistry.* Westport: Greenwood, 2003. Print.

Silbey, Robert J., Robert A. Alberty, and Moungi G. Bawendi. *Physical Chemistry.* 5th ed. Hoboken: Wiley, 2012. Print.

Skoog, Douglas Arvid, Donald M. West, and F. James. Holler. *Skoog and West's Fundamentals of Analytical Chemistry.* 9th ed. Belmont (Calif.): Cengage Learning, 2014. Print.

METALLOIDS

FIELDS OF STUDY

Inorganic Chemistry; Geochemistry; Metallurgy

SUMMARY

The basic properties of the metalloid elements are discussed as they relate to the electronic structure of their atoms. The metalloids are also an essential component of semiconductors, and their use in modern digital electronics based on the transistor is their principal application.

PRINCIPAL TERMS

- **alloy:** a mixture of a metal and at least one other element, often another metal; also known as a solid solution.
- **amphoteric:** describes a compound with the ability to act as either an acid or a base, depending on its environment and the other materials present.
- **conductivity:** the ability of a material to transfer heat (thermal conductivity) or electricity (electrical conductivity) from one point to another.
- **ductility:** the ability of a solid material to be deformed by the application of tensile (pulling)

force, such as bending, without breaking or fracturing.

- **malleability:** the ability of a solid material to be deformed by the application of compressive (pushing) force, such as hammering, without breaking or fracturing.

THE NATURE OF THE METALLOIDS

The metalloids consist of a small group of elements that include germanium (Ge), arsenic (As), antimony (Sb), tellurium (Te), boron (B), and silicon (Si). These elements are termed "metalloids" because they have some metallike properties, such as luster, but not sufficiently so to be classified as metals. For example, the metalloids do not conduct electricity well at room temperature, though they can be made to become conductive at higher temperatures or with small amounts of other elements in the lattices of their crystalline structure. The metalloids have intermediate electronegativity, so they are also neither as electronegative as the nonmetal group 16 and 17 elements (the chalcogen and halogen groups) nor as electropositive as the transition metals. Accordingly, the metalloids are characterized by intermediate ionization that typically falls between that of metals and nonmetals. The metalloids are not inert elements but quite reactive under the right conditions, depending on the properties of other elements in the reaction. Numerous compounds of metalloid elements are known, though the metalloids are most often used in their elemental form as components of alloys, in which they enhance the ductility and malleability of the particular combination of materials.

ELECTRONIC STRUCTURE OF THE METALLOIDS

The electrical conductivity of metalloids can be described in terms of molecular orbitals, which form from the combination of and interactions between

METALLOIDS

14 Si silicon 28

1 H hydrogen 1																	2 He helium 4
3 Li lithium 7	4 Be beryllium 9											5 B boron 11	6 C carbon 12	7 N nitrogen 14	8 O oxygen 16	9 F fluorine 19	10 Ne neon 20
11 Na sodium 28	12 Mg magnesium 29											13 Al aluminium 27	14 Si silicon 28	15 P phosphorous 31	16 S sulfur 32	17 Cl chlorine 35	18 Ar argon 40
19 K potassium 39	20 Ca calcium 40	21 Sc scandium 45	22 Ti titanium 48	23 Vi vanadium 51	24 Cr chromium 52	25 Mn manganese 55	26 Fe iron 56	27 Co cobalt 59	28 Ni nickel 59	29 Cu copper 64	30 Zn zinc 65	31 Ga gallium 70	32 Ge germanium 73	33 As arsenic 75	34 Se selenium 79	35 Br bromine 80	36 Kr krypton 84
37 Rb rubidium 85	38 Sr strontium 88	39 Y yttrium 89	40 Zr zirconium 91	41 I niobium 93	42 Mo molybdenum 96	43 Tc technetium 98	44 Ru ruthenium 101	45 Rh rhodium 103	46 Pd palladium 106	47 Ag silver 108	48 Cd cadmium 112	49 In indium 115	50 Sn tin 119	51 Sb antimony 122	52 Te tellerium 128	53 I iodine 127	54 Xe xenon 131
55 Cs caesium 133	56 Ba barium 137	*	72 Hf hafnium 178	73 Ta tantalum 181	74 W tungsten 184	75 Re rhenium 186	76 Os osmium 190	77 Ir iridium 192	78 Pt platinum 195	79 Au gold 197	80 Hg mercury 201	81 Tl thallium 204	82 Pb lead 207	83 Bi bismuth 209	84 Po polonium 209	85 At astatine 210	86 Rn radon 222
87 Fa francium 223	88 Ra radium 226	**	104 Rf rutherfordium 267	105 Db dubnium 268	106 Sg seaborgium 271	107 Bh bohrium 272	108 Hs hassium 270	109 Mt meitnerium 276	110 Ds darmstadtium 281	111 Rg roentgenium 280	112 Cn copernicium 285	113 Uut ununtrium 284	114 Fl flerovium 289	115 Uup ununpentium 288	116 Lv livermorium 293	117 Uus ununseptium 294	118 Uuo ununoctium 289

57 * La lanthanum 139	58 Ce cerium 140	59 Pr praseodymium 141	60 Nd neodymium 144	61 Pm promethium 145	62 Sm samarium 150	63 Eu europium 152	64 Gd gadolinium 157	65 Tb terbium 159	66 Dy dysprosium 163	67 Ho holmium 165	68 Er erbium 167	69 Tm thulium 168	70 Yb ytterbium 173	71 Lu lutetium 175
89 ** Ac actinium 227	90 Th thorium 232	91 Pa protactinium 231	92 U uranium 238	93 Np neptunium 237	94 Pu plutonium 244	95 Am americium 243	96 Cm curium 247	97 Bk berkelium 247	98 Cf californium 251	99 Es einsteinium 252	100 Fm fermium 257	101 Md mendelevium 258	102 No nobelium 259	103 Lr lawrencium 284

the atomic orbitals of bonded atoms. In a simple bond between two atoms, two molecular orbitals of different energy levels are formed. As more atoms are bonded together in a molecule, an increasing number of atomic orbitals become available to form molecular orbitals, and the resulting energy levels become more closely spaced, ultimately forming a continuum of unoccupied energy levels called a "band." The large number of atoms in a macroscopic quantity of metal form an equally huge number of molecular orbitals, creating a tightly spaced band of orbitals that allows electrons to move more easily between the various energy levels. In metals, the presence of empty molecular orbitals (called the "conduction band") that are close in energy to filled molecular orbitals (the valence band) enables the electrons to become highly mobile and delocalized from the atom, which facilitates the flow of a current through a material. In a metalloid, there is a larger energy gap between the valence band and the conduction band than in metals, which prevents electrons from flowing easily through the conduction band. However, some electrons naturally acquire enough energy to "jump" into the conduction band. Accordingly, the electrical conductivity of the metalloids increases when temperature is elevated, as more electrons are able to acquire the energy needed to move into the conduction band. Likewise, the conductivity of the metalloids decreases relative to metals when the temperature decreases.

SAMPLE PROBLEM

Determine the electron distribution in germanium and antimony, and identify the valence electrons. (Use the electron distribution chart if needed.)

1s
2s 2p
3s 3p 3d
4s 4p 4d 4f
5s 5p 5d 5f 5g
6s 6p 6d 6f 6g 6h
7s 7p 7d 7f 7g 7h

Answer:

Germanium has thirty-two electrons, and antimony has fifty-one electrons. The electron distribution of germanium is therefore

$$1s^2 2s^2 2p^6 3s^2 3p^6 4s^2 3d^{10} 4p^2$$

and the electron distribution of antimony is

$$1s^2 2s^2 2p^6 3s^2 3p^6 4s^2 3d^{10} 4p^6 5s^2 4d^{10} 5p^3$$

The valence electrons are the 4*p* and 5*p* electrons, respectively.

Metalloid Compounds

All of the metalloids form compounds according to the number of valence electrons they have, but their bonding tends to be more covalent than ionic in character. Germanium, for example, forms compounds with the halogens and oxygen by sharing its four valence electrons rather than by giving them up to form ions. This permits the recovery of the metalloids in extremely pure form from the minerals and inorganic sources, such as coal ash, in which they naturally occur. Germanium ore is first reacted with hydrochloric acid (HCl), producing the volatile liquid germanium tetrachloride ($GeCl_4$) as an intermediate. Germanium tetrachloride can be hydrolyzed to form germanium dioxide (GeO_2), which is then reduced to elemental germanium using hydrogen gas (H_2). Zone refining, or melting, of the material yields germanium with just one part of impurity in 10^{10}, or 99.9999999 percent purity. Of the metalloid elements, germanium poses no health risks, but antimony can form the toxic compound stibine (SbH_3) by reaction with strong acids. All compounds of arsenic and tellurium are highly toxic.

Applications of the Metalloids

The semiconducting properties of the metalloids, particularly germanium and silicon, have enabled the development of solid-state electronics, such as semiconductor diodes, integrated circuits, light-emitting diodes (LEDs), and liquid-crystal displays (LCDs). Semiconductors are essential components of transistors, and the production of computer chips is the primary use of germanium and silicon. The complex process begins with the growth of large single crystals of silicon or germanium from the molten state. In order to improve the conductivity of these semiconducting elements, which both have four valence electrons, a small amount of another element, called a "dopant," is introduced into the silicon or germanium lattice. Dopants for silicon and germanium are typically group 3 or group 5 elements (which have three and five valence electrons, respectively), thereby increasing the number of charges that can move through the lattice and greatly enhancing the material's conductivity. The doped semiconducting crystal is then sliced into wafers and subjected to numerous etching processes that assemble minute transistor structures on its surface. These are the heart of all digital devices.

Richard M. Renneboog, MSc

BIBLIOGRAPHY

Haynes, W. H., PhD, David R. Lide, PhD, and Thomas J. Bruno, PhD, eds. *CRC Handbook of Chemistry and Physics*, 96th Edition. New York: CRC/Taylor & Francis Group, 2015. Print.

Kean, Sam. *The Disappearing Spoon: And Other True Tales of Madness, Love, and the History of the World from the Periodic Table of the Elements.* New York: Little, Brown, 2010. Print.

Mackay, K. M., R. A. Mackay, and W. Henderson. *Introduction to Modern Inorganic Chemistry.* 6th ed. Cheltenham: Nelson, 2002. Print.

Miessler, Gary L., Paul J. Fischer, and Donald A. Tarr. *Inorganic Chemistry.* 5th ed. Upper Saddle River: Prentice Hall, 2013. Print.

O'Neil, Maryadele J., ed. *The Merck Index: An Encyclopedia of Chemicals, Drugs, and Biologicals.* 15th ed. Cambridge: RSC, 2013. Print.

Riordan, Michael, and Lillian Hoddeson. *Crystal Fire: The Birth of the Information Age.* New York: Norton, 1997. Print.

METALS

FIELDS OF STUDY

Inorganic Chemistry; Metallurgy; Biochemistry

SUMMARY

The basic properties of metals are discussed as they relate to the electronic properties of metal atoms. Metals constitute 80 percent of the elements in the periodic table and are also important cofactors in biochemistry.

PRINCIPAL TERMS

- **alloy:** a mixture of a metal and at least one other element, often another metal; also known as a solid solution.
- **conductivity:** the ability of a material to transfer heat (thermal conductivity) or electricity (electrical conductivity) from one point to another.
- **ductility:** the ability of a solid material to be deformed by the application of tensile (pulling) force, such as bending, without breaking or fracturing.
- **malleability:** the ability of a solid material to be deformed by the application of compressive (pushing) force, such as hammering, without breaking or fracturing.
- **oxide:** a compound formed by the reaction of any element with oxygen, such as carbon dioxide, carbon monoxide, iron oxide, or diphosphorus pentoxide.

THE NATURE OF METALS

All but twenty-one of the more than one hundred known elements are metals. Metals are characterized by having valence electrons that can be given up to form corresponding metal cations. This arrangement of electrons is responsible for both the multivalent nature of many metal ions and their ability to form numerous compounds by coordinating to neutral organic molecules. Many metals are able to form compounds in which they must take on different oxidation states. Several of the transition metals, such as titanium, chromium, iron, nickel, copper, zinc, silver, gold, platinum, and mercury, are readily recognized and well known. Others, such as osmium and iridium, are scarce and thus much less familiar.

In their familiar elemental forms, the metals have various physical properties of conductivity, ductility, and malleability that make them useful in different applications. Most metals will form alloys with each other, though some, such as aluminum and lead, will not. Various small proportional quantities of metalloid and other elements can be included in the composition of an alloy to enhance specific physical characteristics or impart new ones. The most common compounds of metals are the oxides, products of redox (reduction-oxidation) reactions between elemental metals and atmospheric oxygen.

METALS

79 Au gold 197

1 H hydrogen 1																	2 He helium 4
3 Li lithium 7	4 Be beryllium 9											5 B boron 11	6 C carbon 12	7 N nitrogen 14	8 O oxygen 16	9 F fluorine 19	10 Ne neon 20
11 Na sodium 28	12 Mg magnesium 29											13 Al aluminium 27	14 Si silicon 28	15 P phosphorus 31	16 S sulfur 32	17 Cl chlorine 35	18 Ar argon 40
19 K potassium 39	20 Ca calcium 40	21 Sc scandium 45	22 Ti titanium 48	23 Vi vanadium 51	24 Cr chromium 52	25 Mn manganese 55	26 Fe iron 56	27 Co cobalt 59	28 Ni nickel 59	29 Cu copper 64	30 Zn zinc 65	31 Ga gallium 70	32 Ge germanium 73	33 As arsenic 75	34 Se selenium 79	35 Br bromine 80	36 Kr krypton 84
37 Rb rubidium 85	38 Sr strontium 88	39 Y yttrium 89	40 Zr zirconium 91	41 I niobium 93	42 Mo molybdenum 96	43 Tc technetium 98	44 Ru ruthenium 101	45 Rh rhodium 103	46 Pd palladium 106	47 Ag silver 108	48 Cd cadmium 112	49 In indium 115	50 Sn tin 119	51 Sb antimony 122	52 Te tellerium 128	53 I iodine 127	54 Xe xenon 131
55 Cs caesium 133	56 Ba barium 137	*	72 Hf hafnium 178	73 Ta tantalum 181	74 W tungsten 184	75 Re rhenium 186	76 Os osmium 190	77 Ir iridium 192	78 Pt platinum 195	79 Au gold 197	80 Hg mercury 201	81 Tl thallium 204	82 Pb lead 207	83 Bi bismuth 209	84 Po polonium 209	85 At astatine 210	86 Rn radon 222
87 Fa francium 223	88 Ra radium 226	**	104 Rf rutherfordium 267	105 Db dubnium 268	106 Sg seaborgium 271	107 Bh bohrium 272	108 Hs hassium 270	109 Mt meitnerium 276	110 Ds darmstadtium 281	111 Rg roentgenium 280	112 Cn copernicum 285	113 Uut ununtrium 284	114 Fl flerovium 289	115 Uup ununpentium 288	116 Lv livermorium 293	117 Uus ununseptium 294	118 Uuo ununoctium 289

57 * La lanthanum 139	58 Ce cerium 140	59 Pr praseodymium 141	60 Nd neodymium 144	61 Pm promethium 145	62 Sm samarium 150	63 Eu europium 152	64 Gd gadolinium 157	65 Tb terbium 159	66 Dy dysprosium 163	67 Ho holmium 165	68 Er erbium 167	69 Tm thulium 168	70 Yb ytterbium 173	71 Lu lutetium 175
89 ** Ac actinium 227	90 Th thorium 232	91 Pa protactinium 231	92 U uranium 238	93 Np neptunium 237	94 Pu plutonium 244	95 Am americium 243	96 Cm curium 247	97 Bk berkelium 247	98 Cf californium 251	99 Es einsteinium 252	100 Fm fermium 257	101 Md mendelevium 258	102 No nobelium 259	103 Lr lawrendium 284

Electronic Structure of Metals

All metals have a certain number of unpaired electrons in their outermost, or valence, electron shell. Atoms have a higher stability when the outermost shell has the same electron distribution as the nearest noble-gas element; attaining this more stable state drives the metal atoms to relinquish some or all of their valence electrons to form positively charged ions. The alkali metals each have just a single electron in an *s* orbital that is easily relinquished to produce the M^+ ion. Alkaline earth metals have two electrons in the *s* orbital and readily form the corresponding M^{2+} ion. In the transition metals, the sequential order of electron shells begins to change; the name "transition metal" derives from the fact that the electrons of these elements are able to transition between orbitals of different electron shells that have similar energy restrictions.

In compounds, the orbitals of the individual atoms behave as atomic orbitals. In their elemental forms, however, metals are essentially very large molecules. In a pure elemental metal, the atomic orbitals of each atom of the metal combine to form the corresponding number of molecular orbitals, all of which are very close in energy, so that the electrons of one particular atom are no longer confined to orbiting that specific atom. The movement of electrons between the molecular orbitals of the metal enables electrical conductivity when a potential difference exists between two points in the metal.

Metal atoms and ions form a large number of ionic and coordination compounds that are useful in their own right and can also be used as intermediate stages in various processes. Uranium metal, for example, is enriched with the isotope uranium-235 by using the small mass difference between it and uranium-238 to separate the two isotopes when they occur in the gaseous compound uranium hexafluoride, UF_6. Pure nickel is obtained through the purification and decomposition of the volatile compound nickel tetracarbonyl, $Ni(CO)_4$.

SAMPLE PROBLEM

Given the following metal compounds, identify the oxidation state of each metal atom. Look up the ionic charge of the other species in each compound if necessary, and keep in mind that in a single atom, the oxidation state is equal to the ionic charge.

$Ni(CO)_4$ $Fe(OH)_3$ $CuNO_3$ Al_2O_3 Au-ClO_4 $[Cr(CN)_6]^{3-}$

Answer:

$Ni(CO)_4$: Carbon monoxide (CO) is a neutral compound with no electrical charge. Therefore, the nickel (Ni) atom must also have no electrical charge, so its oxidation state is 0.

$Fe(OH)_3$: Hydroxide ions (OH) each bear a single negative charge, so three of them have a cumulative charge of 3−. Therefore, the iron (Fe) ion has a charge of 3+ and is in the +3 oxidation state.

$CuNO_3$: The nitrate ion (NO_3^-) bears a single negative charge. Therefore, the copper (Cu) ion has a single positive charge and is in the +1 oxidation state.

Al_2O_3: Oxygen (O) ions in oxides each bear two negative charges. There are three oxygen atoms, for a total of six negative charges. Therefore, the two aluminum (Al) ions must each have a charge of 3+, meaning each one is in the +3 oxidation state.

$AuClO_4$: The perchlorate ion (ClO_4^-) bears a single negative charge. Therefore, the gold (Au) ion has a single positive charge and is in the +1 oxidation state.

$[Cr(CN)_6]^{3-}$: The cyanide (CN) ions each bear a single negative charge. There are six of them, for a total charge of 6−, and the complex ion has a net negative charge of 3−. Therefore, the chromium (Cr) ion must be in the +3 oxidation state.

Many metal ions are vitally important factors in biochemistry, and almost all are present in normal biochemistry to some extent. Sodium, potassium, and calcium are the most common, but many others, such as iron, magnesium, and chromium, are essential biochemical elements as well. Other metal ions that can take part in biochemical reactions in place of their nonionic counterparts are severely poisonous to those processes. Thus, heavy metals such as mercury, lead, manganese, and cadmium pose severe health hazards in ionic form but are typically less harmful as neutral atoms. Other metals, such as the lanthanides, are too uncommon to be of concern, while the actinides are all inimical to biochemical systems due to their innate radioactivity. The actinide metal plutonium and all of its compounds are considered the most severely toxic materials known.

OCCURRENCE OF METALS

The metal elements typically occur as ores composed of various oxides and salts. Chemical treatment or thermal refinement is the usual means of recovering the metal in its elemental form. Some of the least reactive metals can be found as "native" metals requiring no further purification. Gold and platinum especially, as well as copper, silver, and mercury, can be found in metallic form in their respective ores rather than as compounds. The metals all occur in varying abundances on Earth, but some, such as iridium, are far more common in asteroids.

Richard M. Renneboog, MSc

BIBLIOGRAPHY

Agricola, Georgius. *De re metallica.* Trans. Herbert Clark Hoover and Lou Henry Hoover. New York: Dover, 1950. Print.

Berg, Jeremy M., John L. Tymoczko, Gregory J. Gatto, and Lubert Strye. *Biochemistry.* 8th ed. New York: W. H. Freeman, 2015. Print.

Douglas, Bodie E., Darl H. McDaniel, and John J. Alexander. *Concepts and Models of Inorganic Chemistry.* 3rd ed. New York: Wiley, 1994. Print.

Gribbin, John. *Science: A History, 1543–2001.* London: Lane, 2002. Print.

Haynes, W. H., PhD, David R. Lide, PhD, and Thomas J. Bruno, PhD, eds. *CRC Handbook of Chemistry and Physics,* 96th Edition. New York: CRC/Taylor & Francis Group, 2015. Print.

Johnson, Rebecca L. *Atomic Structure.* Minneapolis: Lerner, 2008. Print.

Kean, Sam. *The Disappearing Spoon: And Other True Tales of Madness, Love, and the History of the World from the Periodic Table of the Elements.* New York: Little, Brown, 2010. Print.

Lehninger, Albert L. *Biochemistry: The Molecular Basis of Cell Structure and Function.* 2nd ed. New York: Worth, 1975. Print.

Mackay, K. M., R. A. Mackay, and W. Henderson. *Introduction to Modern Inorganic Chemistry*. 6th ed. Cheltenham: Nelson, 2002. Print.

Miessler, Gary L., Paul J. Fischer, and Donald A. Tarr. *Inorganic Chemistry*. 5th ed. Upper Saddle River: Prentice Hall, 2013. Print.

MOLARITY AND MOLALITY

FIELDS OF STUDY

Physical Chemistry

SUMMARY

The characteristics of molarity and molality are discussed, and the methods of calculating both the molarity and the molality of a solution are described. Molarity is the number of moles of solute per volume of a solution, and molality is the number of moles of solute per mass of the solvent.

PRINCIPAL TERMS

- **Avogadro's number (NA):** $6.02214129 \times 10^{23}$, often rounded to 6.022×10^{23}; the number of particles (atoms or molecules) that constitute one mole of any element or compound, the mass of which in grams is numerically equal to the atomic or molecular weight of the material.
- **concentration:** the amount of a specific material present in a given volume of a mixture.
- **density:** the amount of a material contained within a particular space, usually expressed as mass per unit volume.
- **mole:** the amount of any pure substance that contains as many elementary units (approximately 6.022×10^{23}) as there are atoms in twelve grams of the isotope carbon-12.
- **solute:** any material that is dissolved in a liquid or fluid medium, usually water.
- **solvent:** any fluid, most commonly water, that dissolves other materials.

THE CONCEPT OF THE MOLE

The mole is a fundamental tool of calculation in chemistry. It directly relates mass quantities of atoms or molecules to the mass of the individual atom or molecule. Based on experimental observations of the behavior of gases, Italian scientist Amadeo Avogadro (1776–1856) theorized that equal volumes of gases, at the same temperature and pressure, contain the same number of "particles" (a viable theory of atomic structure did not exist at that time). The key to this conclusion was that the mass of a specific volume of any gas, under identical conditions of temperature and pressure, was found to be directly related to the mass of the particles that composed that gas, or what is known now as the molecular mass. From the point of view of the modern theory of atomic structure, this is easy to understand. Because atoms interact only at the level of their outermost, or valence, electrons, only whole atoms are involved in compounds and chemical transformations. Accordingly, an individual atom has a specific atomic mass that does not change. A molecule composed of any number of atoms therefore has a molecular mass that is the sum of the mass of its component atoms.

The mole was developed as a unit of measure to describe atoms and molecules in amounts large enough to work with in the laboratory. Originally, scientists sought to determine the number of hydrogen atoms in one gram of hydrogen, which has an atomic mass of 1. This number, approximately 6.022×10^{23}, was eventually named Avogadro's number (now also called the Avogadro constant) and corresponds to one mole. The mass in grams of one mole of any element or compound—its molar mass—is numerically the same as its atomic or molecular mass; thus, oxygen has an atomic mass of 16, and one mole of oxygen has a mass of sixteen grams. Ways of calculating Avogadro's number have changed over time, and the modern standard is the number of atoms in twelve grams of the pure isotope carbon-12.

MOLES IN SOLUTION

The use of weights and measures in science, especially in the practice of chemistry, requires the scientist to know what quantities of material are being used. For solid materials, this is not a great problem, as such materials are easily weighed. But for materials in solution, the quantities cannot be so easily measured, and it may be necessary to know how much of a specific material is present in a specific amount of the solution. This amount is called the concentration of the solution.

A solution is prepared by dissolving a given amount of a material, or solute, in an amount of a fluid, or solvent. Many characteristic behaviors of solutions, and of materials in solutions, are found to be related to the amount of solute per unit volume of the solution; this is called the molarity of the solution. Other properties are found to be related to the amount of solute per unit mass of the solvent, which is called the molality of the solution.

Molarity versus Molality

The important distinction between molarity and molality is that molarity is defined as the number of moles of a solute per liter of solution (called molars, abbreviated M), while molality is defined as the number of moles of a solute per kilogram of solvent (originally called molal, now given in units of moles per kilogram, abbreviated m). To comprehend this difference, it is helpful to visualize the preparation of two solutions, each containing one mole of a particular solute. To prepare the one-molar (1 M) solution, one mole of the solute is added to a quantity of solvent significantly less than one liter in volume and dissolved. Next, additional solvent is added to increase the volume of the solution to one liter. This produces a solution that contains one mole of solute per liter of solution. To prepare a solution with a molality of one mole per kilogram (1 m), the mole of solute is added directly to an amount of solvent weighing one kilogram and dissolved. This produces a solution that contains one mole of solute per kilogram of solvent.

For low concentrations of solute in an aqueous solution (that is, a solution in which water is the solvent), there may be little difference between the values of molarity and the molality, and the properties of an aqueous solution with a molarity of one mole per liter will be only very slightly different from those of the same solution with a molality of one mole per kilogram. As concentrations increase, however, the difference between molarity and molality increases as well. A one-mole-per-kilogram solution of acetone in water, for example, will have a volume of 1.0733 liters and a molarity of 0.932 molar, but a four-mole-per-kilogram solution of acetone in water will have a volume of 1.293 liters and a corresponding molar concentration of just 3.094 molars.

Calculating Molarity and Molality

The calculation of molarity uses the general formula:

$$[x]_M = \frac{\text{moles of } x}{\text{liters of solution}} = \text{molarity}$$

For example, 0.25 gram of sodium hydroxide (NaOH) is dissolved in 20 milliliters of water, and the final volume is brought up to 25 milliliters by the addition of the necessary amount of water. The molecular weight of sodium hydroxide is 40 grams per mole, so the number of moles of sodium hydroxide is 0.25 divided by 40, or 0.00625 mole. The concentration

SAMPLE PROBLEM

At 20°C, the density of benzene is $0.88\frac{\text{g}}{\text{mL}}$. Calculate the molarity and the molality of 0.11 g of benzoic acid dissolved in 6 mL of benzene at that temperature. Assume a negligible change in volume when the solid benzoic acid is dissolved.

Answer:

The chemical formula of benzoic acid is $C_7H_6O_2$. Calculate its molar mass:

$$7(12)\frac{\text{g}}{\text{mol}} + 6(1)\frac{\text{g}}{\text{mol}} + 2(16)\frac{\text{g}}{\text{mol}} = 122\frac{\text{g}}{\text{mol}}$$

Calculate the number of moles of benzoic acid in 0.11 g:

$$\frac{122\text{g}}{1\text{mol}} = \frac{0.11\text{g}}{x\text{ mol}}$$

$$(x\text{ mol})(122\text{ g}) = (0.11\text{ g})(1\text{ mol})$$

$$x\text{ mol} = \frac{(0.11\text{g})(1\text{ mol})}{122\text{g}} = 0.0009\text{ mol}$$

There is 0.0009 mol, or 0.9 mmol, in 0.11 g. Thus, the molarity of the solution is

$$\frac{0.0009\text{ mol}}{0.006\text{ L}} = \frac{0.9\text{ mmol}}{6\text{ mL}} = 0.15\text{ M}$$

Calculate the mass of 6 mL of benzene:

$$6\text{ mL} \times 0.88\frac{\text{g}}{\text{mL}} = 5.28\text{ g}$$

The molality of the solution is therefore

$$\frac{0.0009\text{ mol}}{0.00528\text{ kg}} = \frac{0.9\text{ mmol}}{5.28\text{ g}} = 0.17\text{ m}$$

of the solution is therefore 0.00625 mole divided by 0.025 liter, which is 0.25 mole per liter, or 0.25 molar. It is common when dealing with small quantities such as these to state the moles as "millimoles" (mmol) and keep the volume in milliliters, thus eliminating the need to convert to moles and liters. In the above example, 0.00625 mole of sodium hydroxide is the same as 6.25 millimoles. The calculation using this value and the volume in milliliters also yields a molar concentration of 0.25 molar (6.25 divided by 25).

The calculation of molality uses this general formula:

$$[x]_m = \frac{\text{moles of } x}{\text{kilograms of solvent}} = \text{molality}$$

For example, if the same 0.25 gram of sodium hydroxide, or 0.00625 mole, were added directly to 25 milliliters of water and dissolved, the molality of the solution would be 0.00625 mole divided by 0.025 kilogram, or 0.25 moles per kilogram. The same use of millimoles is appropriate for the calculation of molality; 6.25 millimoles divided by 25 grams also equals 0.25 moles per kilogram. The mass of the solvent is determined from its density, which is dependent on both the temperature and the pressure. For aqueous solutions, except when very precise measurements are required, a density of 1 gram per milliliter is sufficient. For other solvents, however, the density can vary considerably with temperature, and the mass of the solvent must be adjusted accordingly.

Solid and Gaseous Solutions

The fluid component of a solution is most commonly a liquid, but can also be a gas or even a solid. Metal alloys, for example, are also known as solid solutions, having solidified after being prepared in the liquid state from molten components. Solutions of gases, commonly though incorrectly called gas mixtures, are very well known, the atmosphere of the planet being one such combination. Both solid and gaseous solutions have the same component relationships as liquid solutions.

Richard M. Renneboog, MSc

Bibliography

Askeland, Donald R., Wendelin J. Wright, D. K. Bhattacharya, and Raj P. Chhabra. *The Science and Engineering of Materials.* Boston: Cengage Learning, 2016. Print.

Jones, Mark Martin. *Chemistry and Society.* 5th ed. Philadelphia: Saunders College Pub., 1987. Print.

Laidler, Keith J. *Chemical Kinetics.* 3rd ed. New York: Harper, 1987. Print.

Myers, Richard. *The Basics of Chemistry.* Westport: Greenwood, 2003. Print.

Silbey, Robert J., Robert A. Alberty, and Moungi G. Bawendi. *Physical Chemistry.* 5th ed. Hoboken: Wiley, 2012. Print.

Skoog, Douglas Arvid, Donald M. West, and F. James. Holler. *Skoog and West's Fundamentals of Analytical Chemistry.* 9th ed. Belmont (Calif.): Cengage Learning, 2014. Print.

MOLECULAR FORMULA

FIELDS OF STUDY

Organic Chemistry

SUMMARY

The basic principles of molecular structures and how they are used to communicate chemical information are presented. A chemical formula identifies and numbers each type of atom in a molecule, but it cannot differentiate between isomers. A molecular formula provides some information about the structure of the molecule. A structural formula describes the spatial relationship of the atoms in a molecule and differentiates between isomeric molecules.

PRINCIPAL TERMS

- **isomeric:** describes chemical species that have the same molecular formula but different molecular structures.
- **Lewis structure:** a simplified representation of the bonds and unbonded electron pairs associated with the atoms of a molecule.

- **molecular formula:** a chemical formula that indicates how many atoms of each element are present in one molecule of a substance.
- **semi-structural formula:** a chemical formula that does not show the complete molecular structure of a compound but gives sufficient information to differentiate it from its isomers.
- **stereochemistry:** the relative spatial arrangement of atoms and bonds in a molecular structure.

COMMUNICATING CHEMICAL INFORMATION

Through the formation of chemical bonds, atoms combine to form molecules. The identity of any particular molecule is determined by the identity and number of the individual atoms from which it is composed, as well as by their positions relative to each other in the molecule (the molecular structure of the compound). The chemical formula of a compound identifies the kinds and number of atoms that comprise that particular molecule. For example, the simple alcohol methanol consists of one carbon atom, four hydrogen atoms, and one oxygen atom. It has the chemical formula CH_4O. The rules of chemical bond formation are such that this is the only compound that can have this chemical formula. Ethanol, next in the alcohol series, is composed of two carbon atoms, six hydrogen atoms, and one oxygen atom. These atoms can be arranged and bonded in different ways. Thus, more than one compound can have the chemical formula C_2H_6O.

The chemical formula alone is incapable of differentiating between these isomeric forms. To overcome this problem, chemists sometimes use a molecular formula or semi-structural formula to identify whether the formula C_2H_6O refers to dimethyl ether or to ethanol. The former is written as $H_3C–O–CH_3$ and the latter as $C_2H_5–OH$, $CH_3CH_2–OH$, or Et–OH (in which Et indicates the ethyl group, C_2H_5). For many compounds, this amount of differentiation is sufficient. For more complex compounds, however, knowing the three-dimensional structure of the molecule is vital, but the number of possible isomers of the same molecular formula mitigates against any such simplistic representation. To overcome this problem, chemists resort to drawing "stick figure" structural formulas that communicate the chemical and structural information about a molecule. These structural formulas chart the electron movements in bond formation and the sequential organization of substituents about the asymmetric centers in optical isomers. Some basic rules have standardized the practice.

A chemical formula is specific to the atoms that make up a corresponding molecular structure. The chemical formula NaCl, for example, identifies sodium chloride (and only sodium chloride). Thus, each chemical formula comprises a specific "word" in a chemical equation and communicates to the

STRUCTURAL FORMULAS

Chemical Name	Molecular Formula	Semi-Structural Formula	Structural Formula	Simplified Structural Formula	3D Structural Formula
butane	C_4H_{10}	C_4H_{10}	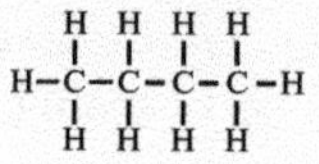 H H H H / H–C–C–C–C–H / H H H H		
acetic acid	$C_2H_4O_2$	CH_3COOH	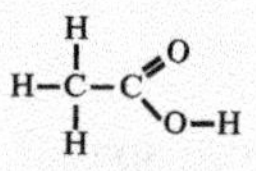 H / H–C–C=O / H O–H	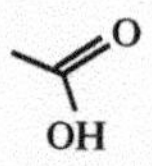 O / OH	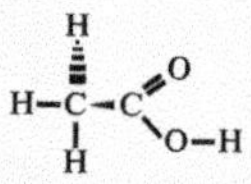 H / H–C–C=O / H O–H
glucose	$C_6H_{12}O_6$	$(CHOH)_4OCHCH_2OH$	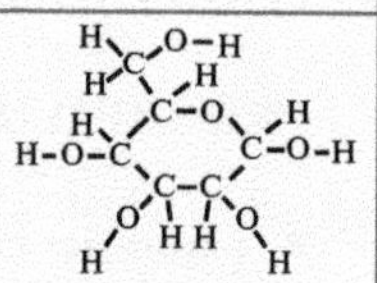 H O–H / H H / H C–O H / H–O–C C–O–H / C–C / O H H O / H H	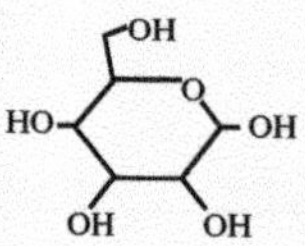 OH / O / HO OH / OH OH	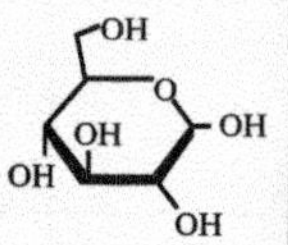 OH / O / OH OH / OH / OH
galactose	$C_6H_{12}O_6$	$(CHOH)_4OCHCH_2OH$	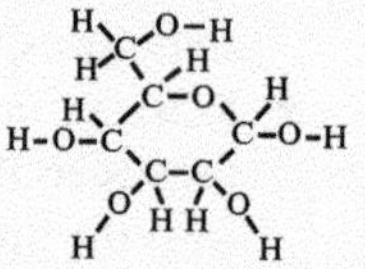 H O–H / H H / H C–O H / H–O–C C–O–H / C–C / O H H O / H H	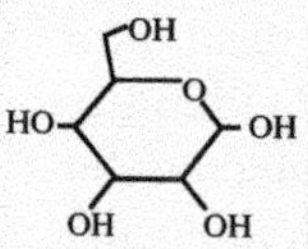 OH / O / HO OH / OH OH	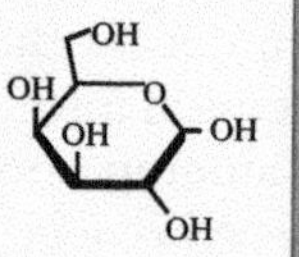 OH / OH O / OH OH / OH

chemist that only that specific chemical entity is meant. In the following chemical equation:

$$C_{12}H_{22}O_{11} + 12O_2 \rightarrow 12CO_2 + 11H_2O$$

the chemist is informed explicitly that exactly one molecule of $C_{12}H_{22}O_{11}$ reacts with exactly twelve molecules of oxygen to produce exactly twelve molecules of carbon dioxide and eleven molecules of water. But which of the many possible compounds having that chemical formula is the $C_{12}H_{22}O_{11}$ that is referred to by this equation? A structural molecular formula would present this information quite clearly to chemists able to read the structural formula.

Nomenclature

The first rule of structural molecular formulas is that they must correspond to the exact name of the compound according to the International Union of Pure and Applied Chemistry (IUPAC) rules of standard nomenclature. If this rule is not followed, the compound being illustrated will have an entirely different chemical identity. The effect is the same as using the wrong words in an oral or written description.

The second rule is that the drawing must show only the bonds between carbon atoms in organic compounds and neither the carbon atoms themselves nor any hydrogen atoms that are bonded to them. The drawing must also include the proper bond angles for the structure, types, and numbers of ions or molecules that surround the particular atom in a compound. Each end of a stick, and each vertex where two sticks connect, is understood to represent a carbon atom with the proper number of hydrogen atoms. Two lines between points indicate a double bond, and three lines a triple bond.

The third rule is that any atoms in the structure other than carbon or hydrogen are represented by their element symbol: O for oxygen, N for nitrogen, Si for silicon, and so on. Thus, cyclopentane, diethylamine, 2-pentanone, and diphenylmethane can be represented respectively as:

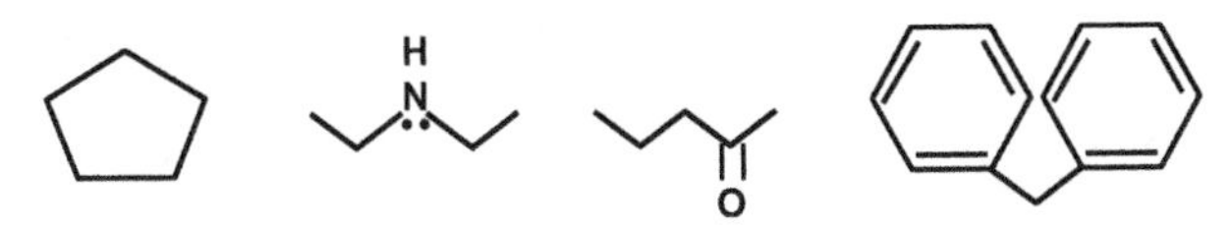

(Note that the circle inside the hexagons of diphenylmethane indicates the alternating single and double bonds of the benzene ring.) Such structural formulas illustrate the identity of each compound.

Isomers of Organic Compounds

Many different compounds can be represented by the exact same chemical formula. A Lewis structure can do much to identify the particular compound, but it is an inefficient, two-dimensional means of representing a three-dimensional molecule. Only the simplest isomeric relationships, such as ethanol and dimethyl ether, can be differentiated by a Lewis structure. There are many types of isomers, however, and a structural formula is the single best means of identifying them in a chemical equation.

All isomeric forms have one thing in common: they are represented by the same chemical formula, no matter how different the actual compounds are. Structural isomers exist when a basic structure has the same substituents but in different positions on the base structure. Two examples of this include 2-methylpentanol and 3-methylpentanol (both $C_6H_{12}O$) and 1,2-dimethylbenzene, 1,3-dimethylbenzene, and 1,4-dimethylbenzene (all three C_8H_{10}). These are represented clearly as:

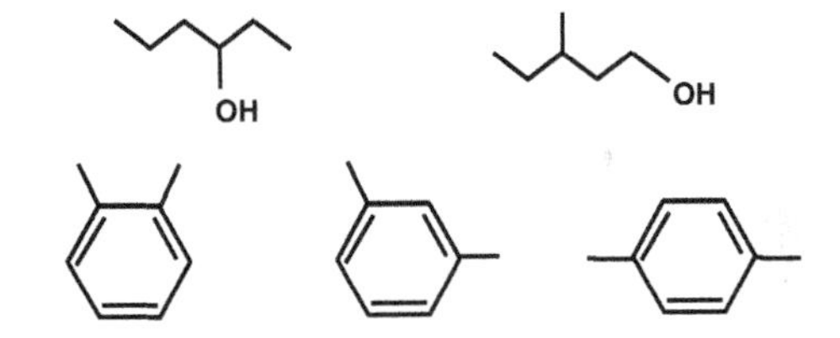

The four bonds to a carbon atom (like those in the pentanol derivatives pictured above) are arranged tetrahedrally. When all four substituents on the four bonds are different, they can be arranged equally in two different ways to form two different isomers. The four substituents are "enantiomers," meaning that the parts are mirror images of each other. A structural molecular formula differentiates between these forms as well, but visualizing the relationship takes some practice.

To visualize an enantiomeric structure, the substituents must be arranged in a sequential order according to certain rules of priority. A C–H bond is given the lowest priority and is accorded the position pointing directly away from the viewer. Substituent priorities are then assigned according to degree of oxidation. In the 2-methylpentanol example, the prioritized order of the substituents is $-CH_2OH$, then

$-C_3H_7$, and finally $-CH_3$. In one enantiomer, these follow a clockwise sequence, while in the other, their sequence is counterclockwise. These are illustrated in a structural molecular formula by drawing two substituents in their proper relative orientation and using a solid wedge-shaped line to indicate the substituent pointing toward the viewer and a dashed wedge-shaped line for the substituent pointing away from the viewer, as:

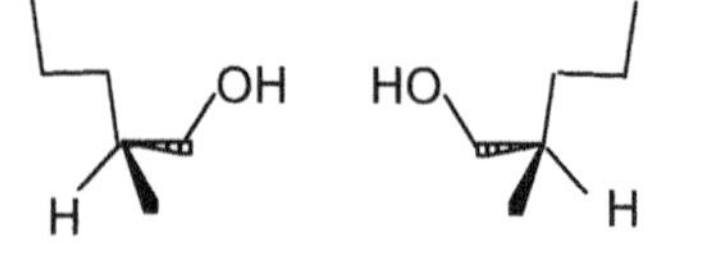

Careful examination of the two structures will reveal that they cannot be inverted, flipped, or manipulated in any way that makes them identical to each other. The stereochemistry about the "asymmetric center" is clearly represented in this simple way.

This simple representation becomes extremely important for more complex molecules in which the structure effectively extends in three directions, a common feature of biochemically active compounds. A simple tricyclic compound such as tricyclo[3.2.1.0^{2,6}]octane is difficult enough to name, let alone keep track of through a word-based identifier. A structural molecular formula, however, makes keeping track of the structure a simple task when the molecule in question can be readily illustrated as:

The presence of double bonds and ring structures present the possibility of geometric isomers, due to the restricted motion about the bonds that make up those sites in the molecule. In a simple substituted alkene (an unsaturated hydrocarbon with one or more C=C bonds), substituents can be situated either on the same side of the C=C bond relative to each other or on opposite sides of the C=C bond, as are the two methyl groups in the following:

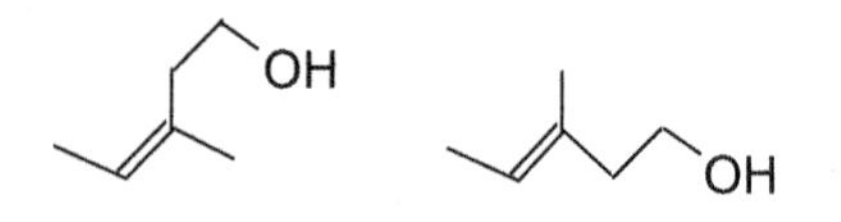

SAMPLE PROBLEM

Given the compound 1-methyl-3-isopropyl-6-phenylcyclooct-2-ene, determine the chemical formula, a semi-molecular formula, and the structural molecular formula that correspond exactly to the name. Are any isomeric forms possible?

Answer:

To determine the chemical formula, tabulate the different atoms in the substituents and the base molecular structure, then add them together.

methyl group	$-CH_3$	1C, 3H
isopropyl group	$-C_3H_7$	3C, 7H
phenyl group	$-C_6H_5$	6C, 5H
cyclooctene	C_8H_{14}	8C, 14H
subtract 1H for each substituent	-3H	

Therefore, the chemical formula is $C_{18}H_{26}$.

A semi-molecular formula may be devised by using the abbreviations Me for methyl, i-Pr for isopropyl, and Ph or φ (the Greek letter *phi*) instead of the full words. Identifying the structure is problematic for this compound, however.

The structural molecular formula can be drawn as

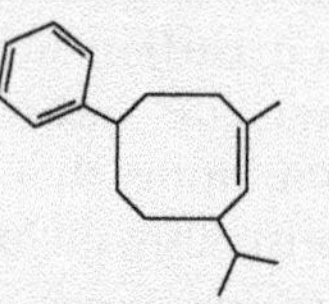

Several isomers are possible for this structure, since there are four different substituent groups bonded to the third and sixth carbon atoms in the ring. These can be arranged as enantiomers. The phenyl and isopropyl groups can also be on the same or opposite side of the cyclooctene ring, and another structure can be drawn in which the C=C bond has C–C bonds on opposite sides instead of on the same side, as shown. These can all be identified precisely by the proper structural representation.

A ring structure has the same composition, as two substituents can be on either the same side or opposite sides of the ring.

Using Structural Molecular Formulas in Reaction Equations

A structural molecular formula presents information about the identity, structure, and behavior of a specific compound in a clear, understandable manner that, were it written in words, would often require several pages. Molecular formulas are especially valuable in presenting a proposed mechanism that shows the movement of electrons as bonds are reorganized during a chemical reaction. It is not always necessary to use a structural molecular formula, as when stereochemical considerations and other isomers are not important or when the compound is unambiguous based on the chemical formula. A chemist who can read the "hieroglyphics" of structural molecular formulas will know precisely what is being presented in a reaction equation.

Richard M. Renneboog, MSc

BIBLIOGRAPHY

Haynes, W. H., PhD, David R. Lide, PhD, and Thomas J. Bruno, PhD, eds. *CRC Handbook of Chemistry and Physics,* 96th Edition. New York: CRC/Taylor & Francis Group, 2015. Print.

Hendrickson, James B., Donald J. Cram, and George S. Hammond. *Organic Chemistry.* 3rd ed. New York: McGraw, 1973. Print.

Morrison, Robert Thornton, and Robert Neilson. Boyd. *Organic Chemistry.* 7th ed. Englewood Cliffs, N.J.: Prentice Hall, 2003. Print.

Myers, Richard. *The Basics of Chemistry.* Westport: Greenwood, 2003. Print.

Stevens, T. S., and W. E. Watts. *Selected Molecular Rearrangements.* New York: Van Nostrand, 1973. Print.

Winter, Mark J. *The Orbitron: A Gallery of Atomic Orbitals and Molecular Orbitals.* University of Sheffield, n.d. Web.

MOLECULAR ORBITAL THEORY

FIELDS OF STUDY

Physical Chemistry

SUMMARY

Molecular orbital theory is a theoretical model that applies the principles of quantum mechanics to describe the behavior of electrons in molecules. Molecular orbital theory posits that the interaction of atoms to form molecules results in the formation of electron orbitals that span the entire molecule rather than being localized to individual atoms.

PRINCIPAL TERMS

- **alkene:** any organic compound that includes two carbon atoms connected by a double bond.
- **aromatic hydrocarbon:** a hydrocarbon in which the carbon atoms form a ring with alternating double and single bonds, distributed in such a way that all bonds are of equal length and strength; also called an arene.
- **aromaticity:** a characteristic of certain ring-shaped molecules in which alternating double and single bonds are distributed in such a way that all bonds are of equal length and strength, giving the molecule greater stability than would otherwise be expected.
- **carbonyl group:** a functional group consisting of a carbon atom double bonded to an oxygen atom.
- **conjugation:** the overlap of *p* orbitals between three or more successive carbon atoms in a molecular structure, creating a system of alternating double and single bonds through which electrons can move freely.
- **cyclic delocalization:** a property of some ring molecules, such as benzene, in which the overlap of orbitals between the atoms that make up the ring allows their electrons to move freely about the molecule.
- **hybrid structure:** a representation of molecular structure that averages a number of possible molecular structures that are equivalent in terms of the arrangement of orbitals and electrons within them.
- **multiple bond:** a bond formed by two atoms sharing two or more electron pairs; includes double bonds and triple bonds.
- **orbital:** a specific region of space about the nucleus of an atom in which electrons of a given energy level are most likely to be found.
- **resonance:** a method of graphically representing a molecule with both single and multiple covalent bonds whose valence electrons are not associated with one particular atom or bond; two or more diagrams, or resonance structures, depict each possible arrangement of single and multiple bonds,

while the true structure of the molecule is somewhere between the different resonance structures, an intermediate form known as the resonance hybrid.

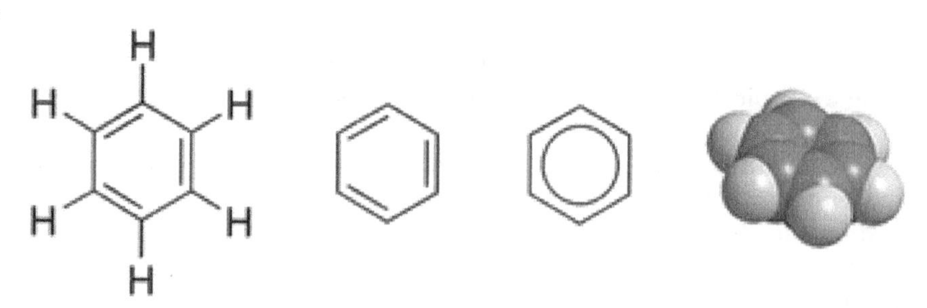

Structure of benzene. Arrowsmaster via Wikimedia Commons via Wikimedia Commons

Electrons and Valence Bond Theory

A number of theories have been developed historically to describe the structure of molecules in terms of the formation of bonds between atoms. The behavior of electrons in atoms is observed to obey very strict mathematical relationships with regard to the absorption and emission of electromagnetic energy. The realization that light exhibits two types of behavior, known as wave-particle duality, was historically a pivotal point in the development of the modern theory of atomic structure and by extension, molecular geometry. This behavior and the observation that atoms and molecules absorb or emit light of specific wavelengths cannot be readily described by classical physics. Early models of atomic structure such as those of Nils Bohr and Ernest Rutherford were based on the behavior of electrons in atoms as though they were small planets in orbits about the nucleus of the atom. These early classical models could not reconcile the observed behaviors of electrons satisfactorily, however. Albert Einstein's mathematical explanation of the photoelectric effect in 1905 lead to the resolution of the dual nature of light as both an electromagnetic wave having no mass and a physical particle having mass. This enabled the development of quantum mechanics as a means of describing the behavior of electrons in atoms. According to the mathematics of the quantum mechanical model of atomic structure, the energies of electrons in atoms are determined by the particular atomic nucleus and the electrons surrounding a nucleus can exist stably only in regions that correspond to a specific energy level. Electrons in atoms absorb or emit light only in discrete units called quanta, hence the name quantum theory. A quantum is the unit particle of energy that corresponds to a specific wavelength. Under the restriction that the speed of light, *c*, is a universal constant the wavelength of light is directly related to its specific energy. For individual atoms, quantum mechanics theory describes the regions corresponding to the specific allowed energies by the probability of finding an electron of specific energy about the nucleus. These regions are determined by the quantum numbers associated with the atom. As a result of the mathematical relationships involved, the regions, called orbitals, correspond to specific geometries centered on the atomic nucleus. The lowest atomic orbitals in any quantum level, termed the *s* orbitals, are spherical in shape. The next highest atomic orbitals are termed the three *p* orbitals, and each has a figure-eight shape. The *p* orbitals are oriented at 90 degrees relative to each other. Next are the five *d* orbitals, followed by the seven *f* orbitals. The geometry of the *d* and *f* orbitals is much more complex due to their multi-lobed shapes. Electrons fill the orbitals of an atom in pairs from the lowest energy to the highest, with the outermost or highest energy being termed the valence orbitals and valence electrons. In valence bond theory, all bonds between atoms are formed as pairs of electrons from orbitals on different atoms are exchanged. That is to say that the two atoms each contribute one electron that they both then share as an electron pair in a valence orbital. In some cases, particularly carbon, the number of single electrons differs from the number of available valence orbitals, but the orbitals are sufficiently alike in energy that they are able to undergo hybridization and combine to form an equal number of atomic orbitals of equal energy and geometry about the nucleus of the atom.

How Molecular Orbital Theory Differs

Molecular orbital theory extends valence bond theory by positing that atomic orbitals involved in bonds between atoms combine to produce orbitals that are not localized between the two atoms. Instead, the orbitals that are formed extend across several nuclear centers. As with valence bond theory the number of molecular orbitals produced must be equal to the number of atomic orbitals that are combined. For example, the formation of a sigma (σ) bond between the *s* orbital of two hydrogen atoms forms two molecular orbitals that encompass the nucleus of both hydrogen atoms. One of those orbitals is termed the bonding orbital and is of lower energy that either of the original *s* orbitals. The two electrons that are shared in the chemical bond between the two atoms occupy this orbital. The other molecular orbital that is formed is termed the antibonding orbital and is of higher energy than either of the original *s* orbitals. In principle, an electron promoted from the bonding to the antibonding orbital by the absorption of the appropriate amount of energy, the bond is weakened and may or may not remain intact. The promotion of both electrons to the energy level of the antibonding orbital eliminates the bond and separates the two atoms. This works the same way no matter which atomic orbitals are involved in the bond formation.

SAMPLE PROBLEM

Benzene, C_6H_6, has 6 π electrons and obeys the 4n + 2 rule. Determine if the cyclic compounds C_4H_4, C_8H_8 and the cyclic ion $C_7H_7^+$ also obey this rule.

Answer:

Benzene has three π bonds, for a total of 6 π electrons. For n = 1, (4n + 2) = 6, which corresponds to an aromatic compound.

Cyclobutadiene (C_4H_4) has two π bonds with 4 π electrons. No integer value of n produces a result of 4, so the compound can not be aromatic.

Cyclooctatetraene (C_8H_8) has four π bonds with 8 π electrons. No integer value of n produces a result of 8, so the compound cannot be aromatic.

The tropylium ion ($C_7H_7^+$) has three π bonds with 6 π electrons, just like benzene, and is therefore aromatic.

Note that the term aromatic in this context does not refer to the odor of the compound, but to its electronic configuration and properties.

HYBRID ORBITALS, MOLECULAR ORBITALS AND DELOCALIZATION

In both valence bond theory and molecular orbital theory atomic orbitals of similar quantum characteristic on the same atom can combine to reform as a group of equivalent hybrid atomic orbitals with the corresponding geometric arrangement about the nucleus. The *s* and one, two or three of the *p* orbitals can hybridize to produce two *sp*, three sp^2 or four sp^3 hybrid atomic orbitals. The remaining *p* orbitals retain their original geometry. The formation of a bond utilizing any of these hybrid orbitals also generates a bonding and an antibonding orbital. This is rather straightforward. The system becomes more complex and far-reaching when the non-hybridized *p* orbitals are considered. In any system in which two adjacent carbon atoms are hybridized to either *sp* or sp^2 configuration those two atoms also have *p* orbitals that are not involved in the σ bond. Their proximity to each other allows them to overlap in a sideways manner in the space they occupy about their respective carbon nuclei and form a multiple bond. The formation of a σ bond is mathematically defined according to wave mechanics as having the minimum energy at the maximum end-to-end overlap of the atomic orbitals involved in the σ bond. The sideways overlap of the non-hybridized *p* orbitals does not fulfill this mathematical description. However, because the wave functions describing the *p* orbitals have the same quantum characteristics they also have a point of minimum energy and maximum overlap. This equivalence allows the movement of electrons freely between them. That is, the two *p* orbitals can share their single electrons as an electron pair and form a secondary bond called a pi (ϖ) bond. In this arrangement, a second bonding and antibonding orbital pair spanning both carbon atoms is produced. The simplest example of this bonding arrangement is the lowest member of the alkene hydrocarbons, ethylene ($H_2C{=}{=}CH_2$). The arrangement is not restricted to carbon atoms, but is commonly found in compounds of period 2 elements, especially those containing C==N bonds and the C==O carbonyl group. Because the *p* orbitals of the system cannot be distinguished from each other in their quantum characteristics, they are mathematically equivalent and the electrons within them cannot be localized to either atom. They are therefore deemed to have free movement between atoms within the ϖ orbitals, and the overall process is known as delocalization.

Conjugation, Resonance and Aromaticity

Molecular orbital theory and valence bond theory become very different with the addition of a third partially hybridized atom or more to the system. The molecular orbitals produced as a result now span three or more adjacent atoms, with three or more bonding and antibonding orbital pairs for the ϖ orbital overlap in addition to the bonding and antibonding orbital pairs associated with the bonds. Two double bonds spanning four adjacent atoms are said to be in conjugation with each other, or conjugated, and electron movement can occur across the entire system. A special circumstance arises in such arrangements of alternating multiple and single bonds that are readily seen in molecular structure diagrams, in which the positions of the bonds in the molecular structure can be shifted without changing the electronic configuration of the molecule. The classic example of this resonance is the cyclic compound benzene, C_6H_6, in which three C==C alternate with three C—C bonds. The positions of all six bonds can be shifted around the ring structure of the molecule in two hybrid structures. The molecular orbitals of the system span all six atoms of the structure, and the electrons within those orbitals move freely about the system through cyclic delocalization. The benzene structure exhibits a special electronic property called aromaticity as a result. It, and all other aromatic hydrocarbons, obeys the Hückel "4n + 2" rule, which defines the number of electrons in the ϖ molecular orbitals of the compound.

The Importance of Molecular Orbital Theory

It is important to remember that molecular orbital theory, like valence bond theory, is a mathematics-based model that predicts the properties of compounds in a way that allows them to be better understood. Prediction in this sense does not mean that some property is foretold and then discovered. Instead, it means that the mathematics of the theory agree with observed and measured electronic properties. This is the function of any theoretical model. Molecular orbital theory provides a means of more precisely knowing how compounds are structured and of understanding the processes by which they interact. This in itself is the very essence of chemistry and leads to the identification of new compounds with new uses in medicine, materials science, and other fields. Though the number of elements is small, it is not an exaggeration to state that the number of compounds based on just carbon is infinite. The carbon compounds buckminsterfullerene (C_{60}) and graphene, and related structures called carbon nanotubes hold promise for a revolution in digital electronics, and understanding how those materials behave electronically is vital to their use.

Richard Renneboog, M.S.

Bibliography

Barrett, Jack. *Structure and Bonding* Cambridge, UK: The Royal Society of Chemistry, 2001. Print.

Carey, Francis A., and Richard J. Sundberg. *Advanced Organic Chemistry: Part A—Structure and Mechanisms.* 5th ed. New York: Springer, 2007. Print.

Fleming, Ian. *Molecular Orbitals and Organic Chemical Reactions.* Hoboken: Wiley, 2010. Print.

Jorgensen, William L., and Lionel Salem. *The Organic Chemist's Book of Orbitals.* New York: Academic, 1973. Print.

Rauk, Arvi. *Orbital Interaction Theory of Organic Chemistry.* 2nd ed. Hoboken, NJ: John Wiley & Sons, 2001. Print.

Suggs, J. Wm. *Organic Chemistry.* Hauppauge, NY: Barron's Educational Series, 2002. Print.

Winter, Arthur. *Organic Chemistry 1 for Dummies.* Hoboken: Wiley, 2014. Print.

MONOMERS AND POLYMERS

FIELDS OF STUDY

Organic Chemistry; Biochemistry

SUMMARY

The basic relationship of monomers and polymers is defined, and the importance of polymeric materials in both living and nonliving systems is elaborated. Polymers are formed by simple molecules bonding together to create large and often complex molecules that have specific properties and serve specific functions.

PRINCIPAL TERMS

- **bonding:** the formation of a link between two atoms as a result of the interaction of their valence electrons.
- **dehydration reaction:** a chemical reaction in which hydrogen and oxygen atoms are removed from the reactants and combine to form water.
- **molecule:** the basic unit of a compound, composed of two or more atoms connected by chemical bonds.
- **polymerization:** a process in which small molecules bond together in a chain reaction to form a polymer, which is a much larger molecule composed of repeating structural units.
- **synthesis reaction:** a chemical reaction in which two or more reactants combine to form a single, more complex product; also, a reaction executed with the goal of creating a specific product or products.

THE NATURE OF MONOMERS AND POLYMERS

All polymerization reactions are synthesis reactions in which a product molecule is formed by the combination and chemical bonding of reactant molecules. The molecular structure of a polymer consists of repeating monomer units, monomers being the reactant molecules specific to a polymerization reaction. For example, the monomer of polyethylene is the simple ethylene molecule, C_2H_4. The monomer of cellulose is the sugar glucose, $C_6H_{11}O_6$, formed through the process of photosynthesis. The monomer of a protein molecule is the set of twenty amino acids from which it is produced, according to the genetic code in the DNA molecule—which is itself a complex polymer made up of repeating units of four nucleotides.

FORMATION OF POLYMERS

Polymers can be formed from very simple monomers in a straightforward manner, by bonding the monomer molecules together in a head-to-tail fashion. The properties of the polymer are determined by the number and type of substituent groups, or side chains, on the monomer molecules. Combinations of different monomers can also be used. The "linear" polymers produced in this way tend to be resilient materials that are thermoplastic, meaning they can be softened by heating.

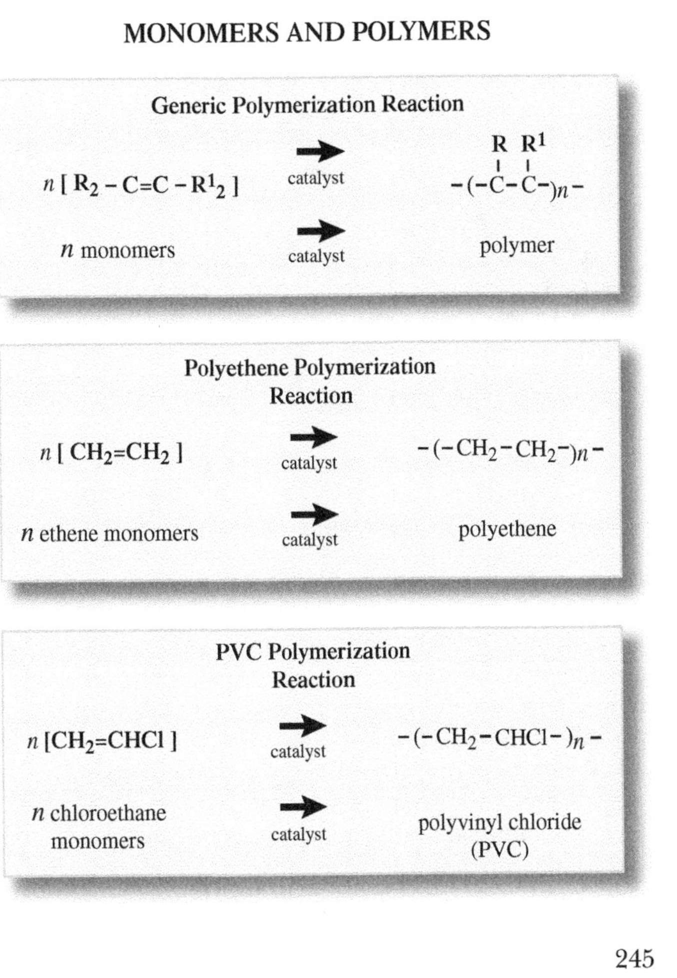

SAMPLE PROBLEM

Given the following monomeric compounds, write general formulas for the structure of the polymers they would each produce. What kind of chemical bonding takes place in the polymerization reaction?

- ethylene (C_2H_4)
- styrene (C_6H_5–CH=CH_2)
- glycine-asparagine (Gly–Asp)

Note that glycine and asparagine are two amino acids, and all amino acids contain both a carboxyl group (–COOH) and an amide group (–NH_2).

Answer:

Ethylene forms the polymeric material polyethylene through covalent bonds between the carbon atoms of two separate molecules. Additional ethylene units are added to the chain by the successive formation of covalent bonds between carbon atoms on the corresponding ethylene molecules. The general equation for the formation of a polyethylene molecule from an indeterminate number of ethylene monomers is

$$nH_2C{=}CH_2 \rightarrow -(CH_2CH_2)_n-$$

Styrene forms polystyrene in the same manner, without involving the phenyl substituent (–C_6H_5). The general equation for this is

$$nC_6H_5{-}CH{=}CH_2 \rightarrow -(C_6H_5{-}CH{-}CH_2)_n-$$

The third monomer is a dipeptide consisting of the amino acids glycine and asparagine. A dehydration reaction created the amide bond linking the carboxyl functional group of glycine to the amide group of asparagine. The formation of the polypeptide created from this material has the general equation

$$n(\text{Gly–Asp}) \rightarrow -(\text{Gly–Asp})_n-$$

indicating a chain of alternating glycine and asparagine molecules that begins with a glycine group and ends with an asparagine group.

Cross-linking in the polymerization process occurs when there is more than one active site in the monomer molecules, allowing each one to take part in different polymerization chain reactions. Cross-linked polymers are generally hard materials that, when heated, will decompose before ever reaching the temperature at which they would melt (the "glass transition temperature"). Cross-linking can be so extensive and complete that, in principle, the contents of an entire cross-linkable resin could become a single, extremely large molecule.

The bonds between monomer molecules in a polymer can be formed by many different reactions. Dehydration reactions create bonds by eliminating the components of water from two different monomer molecules to form ester or amide linkages, as in the peptide bond between two amino acids. Chemical bonds between monomers with carbon-carbon double bonds (C=C bonds) can be formed by the addition of a catalyst molecule with an unpaired electron, which breaks apart the double bond in one of the monomers so the catalyst can bond to one of the carbon atoms, thus leaving an unpaired electron on the other carbon atom to repeat the process with the next monomer. In fact, all types of bond-forming reactions can be applied to the synthesis of polymers from monomers.

In biological systems, polymerization is mediated by enzymes. Photosynthesis converts carbon dioxide, water, and solar energy into the molecule α-d-glucose, which is then polymerized to produce a broad range of compounds, from simple disaccharides (two-sugar units) to large cellulose molecules. Amino acids are another major monomeric component of biological systems, used to synthesize the polymeric structure of all proteins and enzymes. Technically, a protein is not a polymer but a "copolymer," produced through polymerization of a number of different monomers (the amino acids). The synthesis of proteins is enzyme mediated and involves both DNA and RNA. In a similar manner, DNA and RNA are produced by the copolymerization of a combination of four different nucleotides with inorganic phosphate groups. The pattern for all of the many thousands of proteins required by living systems is contained in the genetic code, which is the sequence of nucleotides that make up the DNA molecule.

Advanced Composite Materials

A special application of monomers and polymers is the production of advanced composite materials. Typically these materials are prepared from a structured array of mineral (such as aluminum, glass, basalt, or carbon) or polymeric (such as nylon or Kevlar polyamide) fibers bound in a matrix of thermosetting resin, which is a resin that becomes permanently set when heated. These are the materials of choice for a wide variety of applications, especially for the manufacture of aircraft and other structures requiring the combination of extreme strength and low weight. Formation of advanced composite materials requires advanced training, though the principles are fairly simple. These materials do not exist in nature, and they would not exist without monomers and polymers.

Richard M. Renneboog, MSc

Bibliography

Berg, Jeremy M., John L. Tymoczko, Gregory J. Gatto, and Lubert Strye. *Biochemistry.* 8th ed. New York: W. H. Freeman, 2015. Print.

Campbell, Neil A. *Biology.* Menlo Park: Benjamin, 1987. Print.

Fenichell, Stephen. *Plastic: The Making of a Synthetic Century.* New York: Harper, 2009. Print.

Jones, Mark Martin. *Chemistry and Society.* 5th ed. Philadelphia: Saunders College Pub., 1987. Print.

Lehninger, Albert L. *Biochemistry: The Molecular Basis of Cell Structure and Function.* 2nd ed. New York: Worth, 1975. Print.

Lodish, Harvey, et al. *Molecular Cell Biology.* 7th ed. New York: Freeman, 2013. Print.

Morrison, Robert Thornton, and Robert Neilson. Boyd. *Organic Chemistry.* 7th ed. Englewood Cliffs, N.J.: Prentice Hall, 2003. Print.

Myers, Richard. *The Basics of Chemistry.* Westport: Greenwood, 2003. Print.

MULTIPLE VALENCES

FIELDS OF STUDY

Inorganic Chemistry; Organic Chemistry

SUMMARY

Valence is central to the formation and study of various compounds. Multivalent atoms form a number of bonds to other atoms, depending on both the number of valence electrons they possess and the vacant orbitals in their valence shells.

PRINCIPAL TERMS

- **electron shell:** a region surrounding the nucleus of an atom that contains one or more orbitals capable of holding a specific maximum number of electrons.
- **orbital:** a specific region of space about the nucleus of an atom in which electrons of a given energy level are most likely to be found.
- **oxidation state:** a number that indicates the degree to which an atom or ion in a chemical compound has been oxidized or reduced.
- **transition metals:** the elements of the periodic table that have valence electrons in their *d* orbitals.
- **VSEPR theory:** the valence-shell electron-pair repulsion theory, which states that electron pairs in the valence shell of an atom will arrange themselves so that the electrostatic repulsion between them is minimized.

The Behavior of Electrons

In chemistry, "valence" refers to the number of bonds that a given atom can form with other atoms, which is determined by the number of electrons that atom has in its outermost electron shell. Each electron shell contains between one and four subshells (a fifth subshell is theoretically possible but has yet to be observed), which in turn contain a specific number of orbitals. There are four known orbital shapes, designated *s*, *p*, *d*, and *f*. These shapes appear in subshells in set numbers, so that a subshell may contain one *s* orbital, three *p* orbitals, five *d* orbitals, or seven *f* orbitals, each occupied by a maximum of two electrons. For simplicity's sake, the term "orbital" is often applied to the subshell as a whole, so that a subshell containing three *p* orbitals, each with two electrons, may also be referred to as a single *p* orbital containing six electrons.

Electrons fill the orbitals in order from the lowest energy level to the highest, usually progressing from one electron shell to the next only when a shell has all of the electrons it is allowed to have. Each shell is designated by its principal quantum number (n), so that the first electron shell is called the $n = 1$ shell,

MULTIPLE VALENCES

ELEMENT	ION	LATIN NAME	IUPAC NAMES
Fe	Fe^{2+}	ferrous	iron II
	Fe^{3+}	ferric	iron III
Cu	Cu^{+}	cuprous	copper I
	Cu^{2+}	cupric	copper II
Au	Au^{+}	aurous	gold I
	Au^{3+}	auric	gold III
Sn	Sn^{2+}	stannous	tin II
	Sn^{4+}	stannic	tin IV
Pb	Pb^{2+}	plumbous	lead II
	Pb^{4+}	plumbic	lead IV
Hg	Hg^{+}	mercurous	mercury I
	Hg^{2+}	mercuric	mercury II
Cr	Cr^{2+}	chromous	chromium II
	Cr^{3+}	chromic	chromium III
	Cr^{6+}		chromium VI
Mn	Mn^{2+}	manganous	manganese II
	Mn^{3+}	manganic	manganese III
	Mn^{4+}		manganese IV
Co	Co^{2+}	cobaltous	cobalt II
	Co^{3+}	cobaltic	cobalt III
Ni	Ni^{2+}	nickelous	nickel II
	Ni^{3+}	nickelic	nickel III
Ag	Ag^{+}	argentous	silver I
	Ag^{2+}	argentic*	silver II*

*Silver is not always recognized as a multivalent element.
The 2+ valence, latin name: argentic, and IUPAC name: silver II, is often unobserved.

the second is the $n = 2$ shell, and so on. Subshells are labeled 1*s*, 2*s*, 2*p*, et cetera. Due to variations in the energy levels of the subshells, once the *d* orbitals first appear in the $n = 3$ shell, electrons stop filling the orbitals in a strictly numerical order, and the *s* orbital of the next shell is filled before the *d* orbitals of the current shell. The *f* orbitals, which first appear in the $n = 4$ shell, add an additional layer of complexity.

Electrons are typically distributed individually rather than in pairs, so that each orbital in a subshell contains a single unpaired electron before the first orbital gains its second electron. In the case of most atoms, when all of an atom's electrons have filled the available electron shells, the outermost electron shell will contain fewer than the maximum number. This outermost electron shell is called the valence shell, and it is where all normal chemistry takes place, through the interactions of each atom's valence-shell electrons and orbitals.

A central feature of valence-shell electron configuration is what is known as the octet rule. The elements in group 18 of the periodic table (the farthest-right column), known as the noble-gas elements, each have full *s* and *p* orbitals in their outermost electron shells, for a total of eight valence electrons. (The one exception to this is helium, which contains only a single *s* orbital and thus two electrons total.) This noble-gas configuration seems to be the most stable energy state an atom can have, and other atoms either give up or gain electrons to form cations or anions (positively or negatively charged ions), respectively, in order to achieve the same electron configuration as the nearest noble-gas element.

Whether an atom forms an anion or a cation depends on its number of valence electrons. For example, the potassium atom (atomic number 19) readily gives up its single valence electron to form a potassium cation (K^+) with the same electron configuration as the noble-gas element argon (atomic number 18), while the chlorine atom (atomic number 17), which has seven valence electrons, readily takes an eighth electron into its valence shell to form a chloride anion (Cl^-), which also has the same electron configuration as argon.

Valence and Oxidation State

Valence-shell electrons and orbitals determine the chemical behavior of atoms in a variety of ways. In many compounds, the valence of an atom corresponds exactly to the number of electrons it has either gained or lost. However, in others, the situation is not as clear. One way of keeping track of valence electrons in a compound is to assign an oxidation state to each atom. Essentially an accounting method for electrons, the oxidation state is just the number of electrons that an atom has gained or lost. This system treats all atoms as though they actually lose or gain electrons when forming a compound, regardless of whether the electrons are simply shared between atoms, as in covalent bonds, rather than transferred from one to another.

There are basic rules for assigning the oxidation state. Hydrogen is typically assigned the oxidation state of +1, and oxygen is typically −2, while an atom bonded to an identical atom is assigned an oxidation state of 0. Ions are always assigned the oxidation state that corresponds to their electrical charge, and the sum of all oxidation states within a single neutral molecule must be zero. For example, in the compound potassium dichromate ($K_2Cr_2O_7$), each potassium ion would be assigned an oxidation state of +1, for a total oxidation number of +2; each oxygen atom would have an oxidation state of −2, for a total oxidation number of −14; and, to balance the charges, each of the two chromium atoms must have an oxidation state of +6, for a total oxidation number of +12.

While valence and oxidation are related and often convey the same information—that is, the number of electrons "lost" or "gained" by an atom in a compound is usually equal to the number of its chemical bonds—the oxidation state is used more often than the valence number, as it tends to be less ambiguous. One advantage of oxidation state is that valence refers to the number of chemical bonds an atom can form and thus is always a positive number, while the oxidation number is positive or negative depending on whether electrons have been lost or gained (or donated or accepted in order to form a covalent bond).

Oxidation States of Transition Metals

An element with multiple valences is one that can take on more than one oxidation state in order to form different numbers of bonds within a compound. While any element can potentially have multiple valences, according to the number of electrons present in the valence shell, this is rarely observed in elements that are either one or two valence electrons away from a noble-gas configuration; such atoms almost exclusively form compounds in which they have only one or two bonds, respectively. The assigning of different oxidation states to nonmetal atoms, such as carbon and nitrogen, in their various compounds is more a formality than it is an indication of multiple valences.

Rather, the property of multiple valences is the province of the transition metals, including the lanthanides and the actinides (often called "inner transition metals"). These are elements that have incomplete *d* or *f* orbitals, which seems to be the determining factor in whether an element can exhibit multiple valence behavior. Because the *d* and *f* orbitals of each electron shell are filled in a slightly different order from the *s* and *p* orbitals, a transition metal can have an incomplete inner electron shell; in first-row transition metals, for example, the $n = 4$ shell is the outermost electron shell, but the 4*s* orbital is filled before the 3*d* orbitals, meaning the 3*d* orbitals may be incomplete when the 4*s* orbital is not. The electrons in this incomplete inner shell can also act as valence electrons when needed, giving the atom the ability to lose more electrons than the number of electrons in its outermost shell would suggest. Transition metals, like ordinary metals, form cations rather than anions, as they lose electrons much more easily than they could gain them.

Because transition metals may have different valences in different compounds, the best way to

SAMPLE PROBLEM

Given the valences of tin as 4, carbon as 4, and oxygen as 2, determine the chemical formula of a neutral ionic compound these three elements can form.

Answer:

Tin is a metal, so it forms a cation rather than an anion. Its valence is 4, giving it an oxidation number of +4. Because it is forming an ionic compound, the tin cation (Sn^{4+}) must be balanced by an anion with four negative charges. Oxygen usually takes an oxidation state of −2, which is consistent with a valence of 2, and when two oxygen atoms combine with a carbon atom, they create the neutral compound carbon dioxide (CO_2), suggesting that carbon's valence of 4 corresponds to an oxidation state of +4. While carbon dioxide will not form a compound with tin, three atoms of oxygen combined with one atom of carbon forms a carbonate ion with two negative charges (CO_3^{2-}). Two carbonate ions can balance the +4 oxidation state of tin, producing tin(IV) carbonate, which has the formula $Sn(CO_3)_2$.

What is the valence of the iron atoms in Fe_3O_4?

Answer:

The oxidation state of the oxygen atom is −2. There are four oxygen atoms in Fe_3O_4, for a total oxidation number of −8, to be balanced by the positive oxidation states of the three iron atoms. If all iron atoms were in the same oxidation state, they would have to have an oxidation number of $\frac{8}{3}$, or $2\frac{2}{3}$. Such a state is not possible, since there cannot be partial electrons in an atom. However, if two of the iron atoms are in the +3 oxidation state and the third is in the +2 oxidation state, then all of the charges balance in whole-number proportions. The compound Fe_3O_4 is therefore a combination of iron(II) oxide (FeO) and iron(III) oxide (Fe_2O_3), containing one iron atom with a valence of 2 and two iron atoms with a valence of 3.

determine the valence of such an element in a particular compound is to use the oxidation-state accounting method described in the potassium dichromate ($K_2Cr_2O_7$) example above. Of the three elements in the compound—potassium, chromium, and oxygen—only chromium is a transition metal; the other two have relatively consistent oxidation states. Thus, the oxidation state of the transition metal can be determined by balancing the known oxidation states of the other elements. In this compound, the oxidation state of chromium is +6; accordingly, each chromium atom is only bonded to four oxygen atoms, but two of those chromium-oxygen bonds are double bonds, making six bonds total and giving a valence of 6.

In chemical notation, if an element has multiple valences, the oxidation number of the element in a particular compound is displayed either as a superscript or in parentheses, written in roman numerals to distinguish it from the charge number. For example, the +6 oxidation state of the chromium atoms in potassium dichromate could be represented as Cr^{VI} or Cr(VI). This oxidation state can also be conveyed in the name of the compound by putting the roman numerals in parentheses immediately after the name of the element, so that an alternative name for this compound would be potassium dichromate(VI).

Typically, however, this notation is only used if the same name could be applied to two or more compounds formed with the same element in different oxidation states. Thus, because the name "iron chloride" could refer to either the compound $FeCl_2$, in which the iron has an oxidation state of +2, or the compound $FeCl_3$, in which it has an oxidation state of +3, the two compounds are differentiated by the names iron(II) chloride and iron(III) chloride, respectively. In addition, atoms of the same element in different oxidation states can be present in the same compound, as in iron(II,III) sulfide (Fe_3S_4), where the −2 oxidation states of the four sulfur atoms (for a total of −8) are balanced by one iron(II) atom (+2) and two iron(III) atoms (+6). The formula of such a compound is often given in the format $FeS \cdot Fe_2S_3$.

VSEPR Theory

Valence-shell electron-pair repulsion theory, or VSEPR theory for short, is related directly to the quantum-mechanical description of atomic orbitals in the modern theory of atomic structure. The electrons in the valence shell of an atom all carry the same electrical charge and therefore exert a force of repulsion on each other, the magnitude of which is proportional to their proximity to each other. Atoms in compounds have contributed electrons to form bonds to other atoms. Each atom typically contributes one electron per bond. The carbon atom, for example, has four electrons in its valence shell and so can contribute an electron to the formation of four bonds.

Because atoms are spherical, the forces of repulsion are minimized when these bonds are radially distributed about the nucleus. The four electrons in the carbon atom are distributed so that two are in the 2*s* orbital and the third and fourth are in two of the three 2*p* orbitals. The *s* orbital is spherical and the *p* orbitals are at right angles to each other, making the spatial distribution of the charge on their electrons asymmetrical. By combining the *s* and *p* orbitals into four "hybrid" atomic orbitals oriented toward the apexes of a tetrahedron, the repulsive force between the electron pairs in the four bonds is minimized. The same principle applies to all other compounds. VSEPR theory essentially states that the chemical bonds within compounds are oriented in ways that minimize the repulsive forces between their electron pairs, including any lone (nonbonding) pairs of electrons.

Richard M. Renneboog, MSc

Bibliography

Douglas, Bodie E., Darl H. McDaniel, and John J. Alexander. *Concepts and Models of Inorganic Chemistry.* 3rd ed. New York: Wiley, 1994. Print.

Haynes, W. H., PhD, David R. Lide, PhD, and Thomas J. Bruno, PhD, eds. *CRC Handbook of Chemistry and Physics,* 96th Edition. New York: CRC/Taylor & Francis Group, 2015. Print.

Hendrickson, James B., Donald J. Cram, and George S. Hammond. *Organic Chemistry.* 3rd ed. New York: McGraw, 1970. Print.

Mackay, K. M., R. A. Mackay, and W. Henderson. *Introduction to Modern Inorganic Chemistry.* 6th ed. Cheltenham: Nelson, 2002. Print.

Miessler, Gary L., Paul J. Fischer, and Donald A. Tarr. *Inorganic Chemistry.* 5th ed. Upper Saddle River: Prentice Hall, 2013. Print.

Morrison, Robert Thornton, and Robert Neilson. Boyd. *Organic Chemistry.* 7th ed. Englewood Cliffs, N.J.: Prentice Hall, 2003. Print.

Myers, Richard. *The Basics of Chemistry.* Westport: Greenwood, 2003. Print.

Powell, P., and P. L. Timms. *The Chemistry of the Non-Metals.* London: Chapman, 1974. Print.

N

NAMING ORGANIC MOLECULES

FIELDS OF STUDY

Organic Chemistry

SUMMARY

The method of naming organic molecules, as standardized by the International Union of Pure and Applied Chemistry (IUPAC), is described. The method is based on parent hydrocarbon molecular structures and the functional groups contained in their derivative compounds.

PRINCIPAL TERMS

- **functional group:** a specific group of atoms with a characteristic structure and corresponding chemical behavior within a molecule.
- **IUPAC:** the International Union of Pure and Applied Chemistry, an organization that establishes international standards and practices for chemistry.
- **nomenclature:** a system of specific names or terms and the rules for devising or applying them; in chemistry, refers mainly to the system of names for chemical compounds as established by the International Union of Pure and Applied Chemistry (IUPAC).
- **parent chain:** the longest continuous hydrocarbon chain in an organic compound, used as the basis for its unique name identification.
- **side chain:** a group of atoms that branches off from the main chain, or backbone, of an organic molecule.

Understanding the Naming of Organic Molecules

The principal structural component of an organic molecule is a framework of carbon atoms. The various different organic molecules number well into the millions, and identifying each one with a given familiar name would be difficult and inefficient. As such, a system of standardized nomenclature based on chemical formulas has been developed instead.

Chemical and molecular formulas can be thought of as the words of the language of chemistry, spelled out with the symbols of the elements of the periodic table, which serve as the letters of the chemical alphabet. For organic compounds, the chemical alphabet is very limited, consisting almost entirely of the symbols C (carbon), H (hydrogen), O (oxygen), N (nitrogen), F (fluorine), Cl (chlorine), Br (bromine), I (iodine), S (sulfur), and P (phosphorus). Other element symbols may at times appear in chemical formulas, but such occurrences are relatively rare. Yet despite being generally limited to these few elements, organic chemistry encompasses what amounts to an infinite series of molecules of increasing complexity, ranging from those with just a single carbon atom to giant molecules containing millions of carbon atoms. Each series of organic molecules increases in complexity just one carbon atom at a time, however, which enabled the development of an equally systematic method of naming organic molecules.

Historically, molecules were given whatever name the person who discovered the molecule deemed appropriate at the time, and many names were derived from such sources as the substance from which the material was isolated, the alchemical name of the material, and even the name of the discoverer's love interest. Since relatively few molecules were known, it was not overly difficult to keep track of chemical identities. Indeed, some of the best-known common names survived into the twenty-first century: acetone and acetic acid, for example, are most commonly referred to by those names instead of the proper names 2-propanone and ethanoic acid.

As more and more organic molecules were discovered, however, the system quickly became chaotic and confusing. In order to standardize the language of chemistry and end this confusion, the International

NAMING ORGANIC MOLECULES

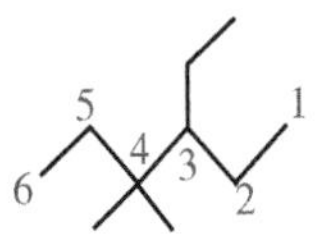

3 ethyl-4, 4 dimethylhexane

Br 3 2 1 4 5

3 bromo, 1 ethylcyclopentane

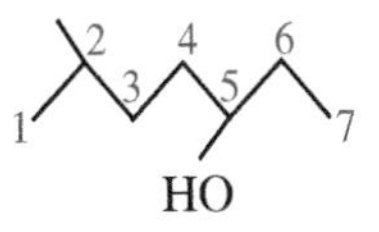

2 methyl-5-heptanol

2 4 1 3

1 butene

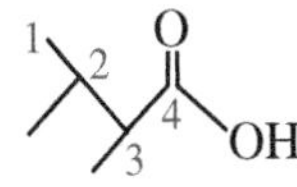

2, 3-dimethylpentanoic acid

2 4 1 3

2 butene

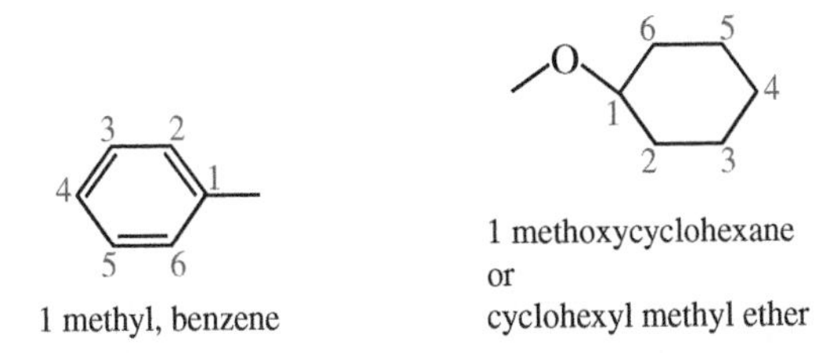

1 methyl, benzene

1 methoxycyclohexane
or
cyclohexyl methyl ether

Union of Pure and Applied Chemistry (IUPAC) developed a systematic method of nomenclature for organic compounds, based on the structures of the simplest class of organic compounds, a hydrocarbon series called alkanes. In many cases, particularly for very complex molecular structures, the common name of the compound was adopted as the official base name for all of its related compounds within the systematic method of nomenclature.

NOMENCLATURE AND HYDROCARBONS

The hydrocarbons, as the name suggests, consist of only hydrogen atoms and carbon atoms. The series begins with the simplest hydrocarbon, which is methane (CH_4), and progresses through C_2H_6 (ethane), C_3H_8 (propane), and so on. Beginning with C_4H_{10}, the same chemical formula can represent different molecular structures that contain exactly the same numbers and types of atoms. These are known as isomers, and without an unequivocal means of identifying the exact molecular structure in a systematic name, nomenclature would become as chaotic as the old common-name methods. In the IUPAC system of nomenclature, the name of a compound is based on the longest single chain of carbon atoms in its structure, known as the parent chain. C_4H_{10} has two isomers, *n*-butane and 2-methylpropane (commonly known as isobutane). *n*-Butane contains of a chain of four carbon atoms, while the longest chain of carbon atoms in 2-methylpropane consists of just three carbon atoms, with the fourth bonded to the middle carbon atom as a side chain. The two structures can be represented through simple diagrams, often called skeletal formulas:

n-butane, C_4H_{10} 2-methylpropane, C_4H_{10}

Each straight line segment begins and ends with a carbon atom, and each carbon atom is assumed to be bonded to (4 – *x*) hydrogen atoms, where *x* is the number of other carbon atoms the atom in question is bonded to.

As the number of carbon atoms increases, the number of possible structures for the same chemical formula also increases. The basic method of the IUPAC system is able to accommodate this.

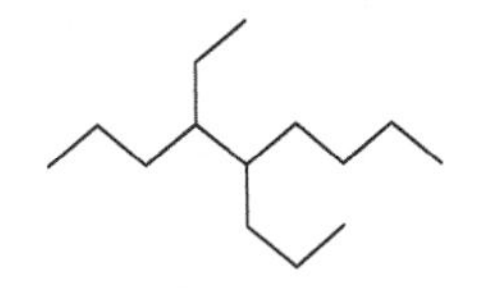

To determine the nomenclature of the above molecular structure, first identify the longest unbroken linear chain of carbon atoms. This is easiest to see when the structure is drawn out as above. In this case, it is the horizontal nine-carbon chain.

Second, identify the various side chains that are attached to the parent chain. Here there is a two-carbon ethyl group and a three-carbon propyl group, the names of which derive from the two-carbon hydrocarbon ethane and the three-carbon hydrocarbon propane, respectively.

Third, number the carbon atoms in the parent chain. Start the numbering on whichever end would result in the side chains being attached to the carbon atoms with the lowest numbers. For this compound, the carbon atoms in the parent chain should be numbered from left to right:

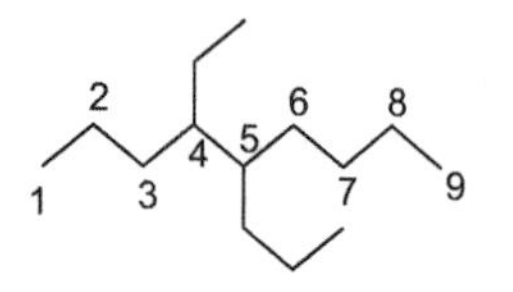

This places the ethyl and propyl side chains on C-4 and C-5. Numbering the chain from right to left would place them on C-5 and C-6, which are higher numbers and therefore not allowed.

The parent chain, being nine carbons in length, is named nonane, from the Latin for "nine." The full IUPAC name of the compound specifies the name of the parent chain and the location of each side chain on the parent chain. In this case, the basic name of the compound is 4-ethyl-5-propylnonane.

There are other considerations when establishing the full, unique name of the compound. For example, the carbon atoms at C-4 and C-5 in this compound are defined as asymmetric centers, since they are each attached to four different groups of atoms; that is, even though C-4 is attached to three carbon atoms and just one hydrogen atom, each of those three carbon atoms is part of a group containing different numbers of carbon and hydrogen atoms, called a substituent group, and the same is true for C-5. Every asymmetric center exists in two isomeric forms, and the full name would also specify the order of the substituents about each asymmetric center.

The hydrocarbon series of linear compounds also includes the alkenes (characterized by the presence of double carbon, or C=C, bonds) and the alkynes (characterized by the presence of triple carbon, or C≡C, bonds). There are also corresponding series of hydrocarbons with cyclic structures rather than linear structures. The naming process for each series employs the same basic approach as described above and specifies the locations and orientations of the structural features as well as the substituents, as shown in the following examples:

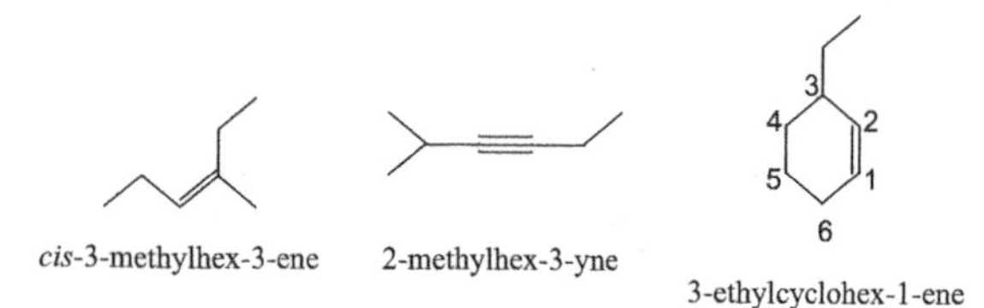

cis-3-methylhex-3-ene 2-methylhex-3-yne 3-ethylcyclohex-1-ene

HETEROATOMS AND FUNCTIONAL GROUPS

Heteroatoms are atoms of any element other than carbon—typically oxygen, nitrogen, sulfur, or phosphorus—that are included in the basic structure of an organic molecule. The presence of heteroatoms in a structure complicates the naming system somewhat by introducing several common names as the base structures of various classes of organic compounds. Heteroatoms are also present in certain functional groups. The alkanes are very stable compounds and do not react with many typical reagents, such as acids, bases, reducing agents, and oxidizers. The presence of a functional group in an organic molecule provides a reactive site where such reagents can function and bring about a reaction. Typical functional groups include alcohols, ethers, carboxylic acids, amines, thiols or mercaptans, ketones, aldehydes, and several others. The common feature of all functional groups is that the specific atoms that compose any functional group impart the same behavior in chemical reactions regardless of the remaining structure of the molecule. The technique for generating a unique IUPAC-standard name for such compounds is the same as for hydrocarbons, in that the name must identify the location and type of the functional group. The following structures are examples of such naming:

SAMPLE PROBLEM

Determine the IUPAC name for the following compound:

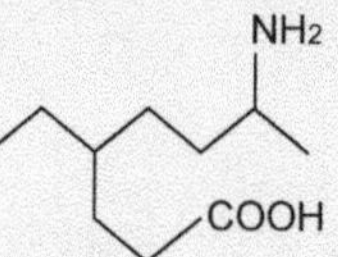

Answer:

The compound is characterized by the presence of the –COOH functional group as the terminator of a carbon atom chain and can therefore be named as a carboxylic acid.

Determine the number of carbon atoms in the longest chain, beginning from the C atom of the –COOH group. This is an eight-carbon chain with no C=C bonds or heteroatoms, so the compound can be given the base name of octanoic acid.

Identify the substituent groups on the parent chain, counting the C atom of the –COOH group as the first atom in the chain. There is an ethyl (two-carbon) group on C-4 and an amino group on C-7. Since the –COOH group must occupy the lowest numbered position in the chain, according to the IUPAC naming convention, the compound can therefore be named 4-ethyl-7-aminooctanoic acid.

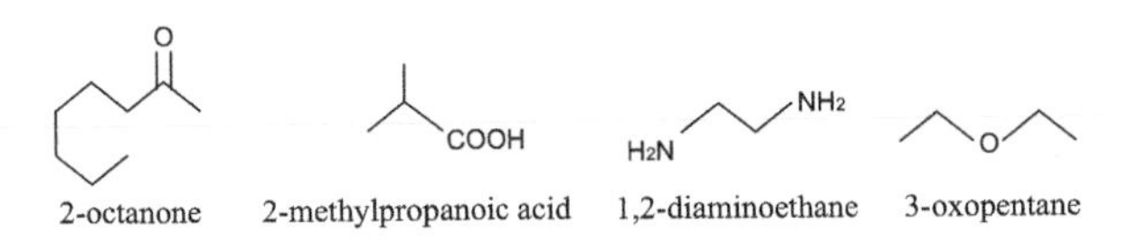

2-octanone 2-methylpropanoic acid 1,2-diaminoethane 3-oxopentane

NAMING ORGANIC COMPOUNDS IN BIOCHEMISTRY

The IUPAC naming convention is appropriate for many small compounds in biochemical systems. However, the complexity of most biochemical compounds makes such a naming approach wholly inadequate. It would be exceedingly difficult to attempt to ascribe an IUPAC standard name to the deoxyribonucleic acid (DNA) molecule, for example. Even very small proteins are impossible to name by the IUPAC conventions. They are typically referred to by a biology-based reference to their role, as in the cases of aconitase and RNA polymerase.

Richard M. Renneboog, MSc

BIBLIOGRAPHY

Favre, Henry, and Warren H. Powell. *Nomenclature of Organic Chemistry*. Cambridge: Royal Soc. of Chemistry, 2014. Print.

Haynes, W. H., PhD, David R. Lide, PhD, and Thomas J. Bruno, PhD, eds. *CRC Handbook of Chemistry and Physics*, 96th Edition. New York: CRC/Taylor & Francis Group, 2015. Print.

Lodish, Harvey, et al. *Molecular Cell Biology*. 7th ed. New York: Freeman, 2013. Print.

Morrison, Robert Thornton, and Robert Neilson. Boyd. *Organic Chemistry*. 7th ed. Englewood Cliffs, N.J.: Prentice Hall, 2003. Print.

Myers, Richard. *The Basics of Chemistry*. Westport: Greenwood, 2003. Print.

NEUTRONS

FIELDS OF STUDY

Nuclear Chemistry

SUMMARY

The discovery and basic properties of neutrons are described. Neutrons are found in all but the simplest hydrogen atom. They stabilize the protons against the force of electrostatic repulsion. They are electrically neutral but seem to have an internal structure that includes electrical charge.

PRINCIPAL TERMS

- **atomic mass:** the total mass of the protons, neutrons, and electrons in an individual atom.
- **deuterium:** an isotope of hydrogen that contains one neutron and one proton; occurs naturally in about 1 in 6,500 hydrogen atoms.
- **fundamental particle:** one of the smaller, indivisible particles that make up a larger, composite particle; commonly used to refer to electrons, protons, and neutrons, although these are themselves composed of various actual fundamental particles, such as quarks, leptons, and certain types of bosons.
- **isotope:** an atom of a specific element that contains the usual number of protons in its nucleus but a different number of neutrons.
- **nucleus:** the central core of an atom, consisting of specific numbers of protons and neutrons and accounting for at least 99.98 percent of the atomic mass.

MODELS OF THE ATOM

Various models of atomic structure were proposed throughout the nineteenth and twentieth centuries, the most useful being the "planetary" model and the "plum pudding" model. In the first, an atom was envisioned to be composed of a nucleus with electrical charges whirling about it like the planets around the sun. In the latter model, the atom was envisioned to be a round, positively charged blob with dots of negative electrical charge embedded throughout. In 1897, while conducting research on the nature of cathode rays in cathode ray tubes, J. J. Thomson (1856–1940) discovered that the cathode rays behaved more like streams of charged particles than like electromagnetic radiation. He called the particles "corpuscles" at first, though they soon became known as electrons. Based on this discovery, Thomson proposed the plum-pudding model of the atom in 1904.

NEUTRONS

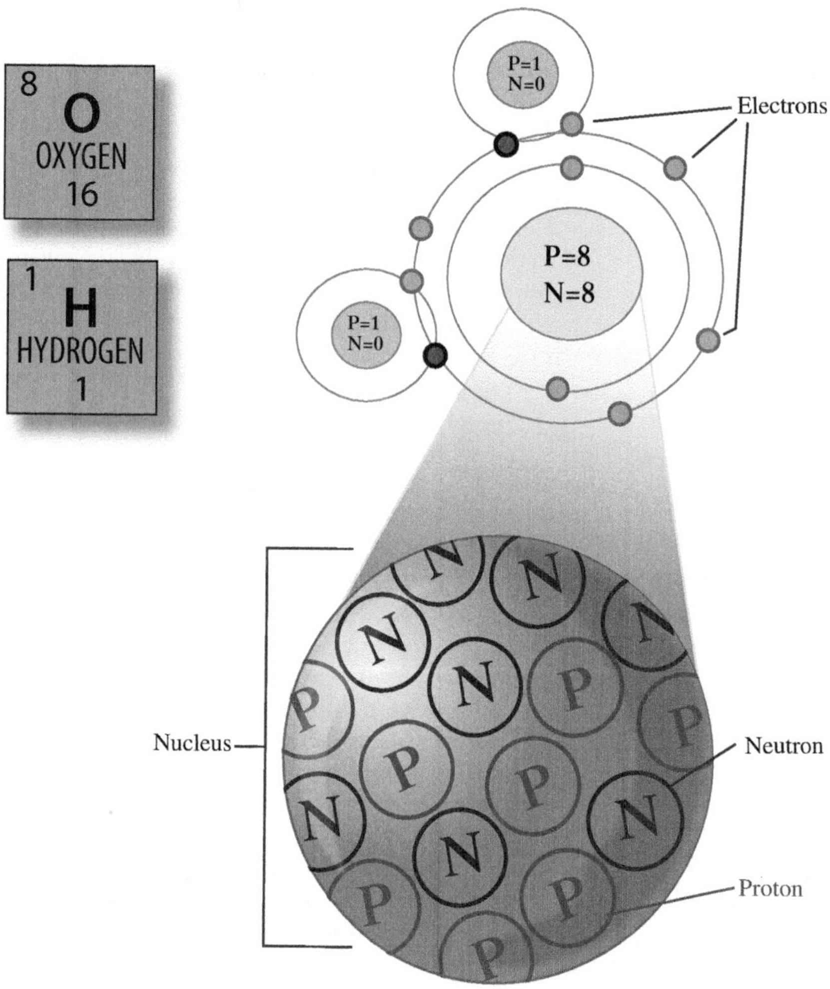

Thomson's discoveries verified that atoms had an internal structure consisting of fundamental particles, but the existence of subatomic particles other than electrons had yet to be proved. In 1909, Ernest Rutherford (1871–1937), Hans Geiger (1882–1945), and Ernest Marsden (1889–1970) performed what is known as the gold-foil experiment, which produced more conclusive results. By directing a beam of (alpha) particles at a very thin target of gold foil, they found that while most of the particles passed directly through the foil as though it were not there, some of the particles were deflected at various angles, as though they had bounced off something very small and very dense inside the metal. These observations led Rutherford to propose the planetary model of atomic structure, with the majority of the atom's mass concentrated in a small, dense nucleus. He continued experimenting with alpha particles, and in 1917 he discovered that the collision of an alpha particle with the nucleus of a nitrogen atom released a positively charged particle identical to a hydrogen nucleus—in other words, a proton, as the most common isotope of hydrogen contains no neutrons. Rutherford also posited the existence of an as-yet-undiscovered neutral particle, though he believed it to be composed of one electron and one proton, which was later disproved. The existence of the neutron was finally experimentally verified by James Chadwick (1891–1974) in 1932.

PROPERTIES AND FUNCTION OF NEUTRONS

The atomic number of an element is equal to the number of protons in its nucleus, as it is the number of protons that defines an atom's identity as an element. In contrast, the number of neutrons in an atom's nucleus can vary without changing its elemental identity. Rather, atoms containing the same number of protons but different numbers of neutrons are said to be different isotopes of the same element. The simplest example of this is deuterium, an isotope of hydrogen that contains one proton and one neutron in its nucleus (the most common isotope of hydrogen has one proton and no neutron). Deuterium can also be called hydrogen-2, where the "2" is the total number of protons and neutrons in the nucleus and is approximately equal to the atomic mass of the isotope.

The positively charged protons would readily fly apart from each other due to electrostatic repulsion between like charges if not for the presence of neutrons, which counteract this repulsion. Studies in high-energy particle physics suggest that the neutron has an internal structure that effectively places a negative charge within the neutron next to the positive charge of the proton, as shown here:

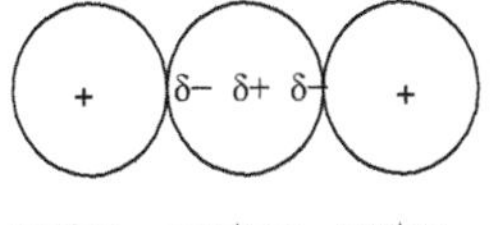

In this way, the neutron acts as a sort of glue for protons. Neutrons interact strongly with nuclei of other atoms but produce only weak ionization from collisions with electrons. Free neutrons have a half-life of about twelve minutes and decompose into a proton, an electron, and a neutrino, which is an electrically neutral physical particle having no detectable mass.

The Discovery of the Neutron

James Chadwick had studied under Rutherford and later worked alongside him. He, too, was convinced of the existence of the neutron, and throughout the 1920s he conducted several experiments designed to prove it. However, it was not until the end of the decade that new developments paved the way for Chadwick to make his discovery.

In 1930, Walther Bothe (1891–1957) and his student Herbert Becker reported on an experiment in which they had bombarded beryllium and other light elements, including boron and lithium, with alpha particles. The bombardment produced an unidentified radiation that easily penetrated other materials, which they believed to consist of γ (gamma) rays. Irène Joliot-Curie (1897–1956) and her husband, Frédéric Joliot-Curie (1900–1958), investigated this new radiation further by directing it at a sheet of paraffin, a type of wax composed of hydrogen and carbon atoms. They reported in early 1932 that when the paraffin was struck by the unidentified radiation, it emitted protons with high kinetic energy.

In an elastic collision between two particles, the kinetic energy of one particle is transferred completely to the other, and the velocity of the rebounding particles is directly related to their mass. Chadwick realized that the Joliot-Curies' discovery was consistent with the idea that the unidentified radiation was composed not of gamma rays but of neutrally charged particles and that, if this were true, the mass of such a particle had to be similar to that of the proton to account for the transfer of energy. In January 1932, the same month that the Joliot-Curies published their observations, Chadwick immediately began to replicate their experiment with some modifications. He theorized, and his research confirmed, that when the nucleus of an atom of beryllium-9 was struck by an alpha particle, the product would not be an atom of carbon-13 accompanied by the emission of a gamma ray, but rather an atom of carbon-12 accompanied by the ejection of a neutron—the source of the unidentified radiation.

SAMPLE PROBLEM

Given the following isotopes, determine the number of neutrons in each:

- iridium-186
- calcium-40
- carbon-12
- oxygen-18
- gold-197

Answer:

The atomic number of iridium (Ir) is 77, indicating that it has 77 protons in its nucleus. Calculate the number of neutrons:

$$186 - 77 = 109$$

Therefore, iridium-186 has 109 neutrons in its nucleus.

Following the same calculation, calcium (Ca) has 20 protons in its nucleus, so calcium-40 must have 20 neutrons; carbon (C) has 6 protons in its nucleus, so carbon-12 must have 6 neutrons; oxygen (O) has 8 protons in its nucleus, so oxygen-18 must have 10 neutrons; and gold (Au) has 79 protons in its nucleus, so gold-197 must have 118 neutrons.

Richard M. Renneboog, MSc

Bibliography

Gribbin, John. *Science: A History, 1543–2001.* London: Lane, 2002. Print.

Johnson, Rebecca L. *Atomic Structure.* Minneapolis: Lerner, 2008. Print.

Kean, Sam. *The Disappearing Spoon: And Other True Tales of Madness, Love, and the History of the World from the Periodic Table of the Elements.* New York: Little, Brown, 2010. Print.

Myers, Richard. *The Basics of Chemistry.* Westport: Greenwood, 2003. Print.

Wehr, M. Russell, James A. Richards Jr., and Thomas W. Adair III. *Physics of the Atom.* 4th ed. Reading: Addison-Wesley, 1984. Print.

Winter, Mark J. *The Orbitron: A Gallery of Atomic Orbitals and Molecular Orbitals.* University of Sheffield, n.d. Web.

NITRILES

FIELDS OF STUDY

Organic Chemistry

SUMMARY

The characteristic properties and reactions of nitriles are discussed. Nitriles are linear molecules or substituent groups. They are very useful for the relatively easy formation of complex organic molecular structures from very simple starting materials.

PRINCIPAL TERMS

- **cyanocarbon:** an organic compound containing multiple cyano groups, which consist of a carbon atom triple bonded to a nitrogen atom.
- **dipole:** the separation of positive and negative charges within a single molecule due to electron density being relatively high in one part of the molecule and relatively low in another.
- **functional group:** a specific group of atoms with a characteristic structure and corresponding chemical behavior within a molecule.
- **triple bond:** a type of chemical bond in which two adjacent atoms are connected by six bonding electrons rather than two.

The Nature of Nitriles

A nitrile is an organic compound with at least one cyano functional group (C≡N) in its molecular structure. It is essentially a "disguised" carboxylic acid (of the general form RCOOH, where R– indicates a hydrocarbon derivative), since splitting a nitrile with water, a process known as hydrolysis, yields first the amide $RCONH_2$ and then the carboxylic acid. Most common nitriles are formed by adding the negatively charged cyanide ion (CN^-) in an appropriate substitution reaction. Substituting cyanide for a halide, for example, is an easy way to lengthen a target molecule by one carbon atom. The cyanide ion itself and the related compound cyanogen (N≡C–C≡N) are often referred to as a "pseudohalide" and "pseudohalogen," respectively, indicating that the cyanide ion and cyanogen exhibit chemical behaviors similar to those of the halides and halogens. Cyanogen also qualifies as a cyanocarbon, which is a nitrile containing more than one cyano group.

The characteristic feature of a nitrile is the cyano group's carbon-nitrogen triple bond. The triple bond, like a double bond, is formed by the side-to-side overlap of *p* orbitals on adjacent atoms. Both the carbon atom and the nitrogen atom have their valence electrons in the 2*s* and 2*p* atomic orbitals, and both are able to hybridize their *s* and *p* atomic orbitals into sets of energetically equivalent *sp*, sp^2, or sp^3 orbitals. The sp^3 hybrid orbitals are typical of saturated compounds. The term arises because an sp^3-hybridized atom can have up to four pairs of electrons filling the four sp^3 orbitals, at which point the

NITRILE

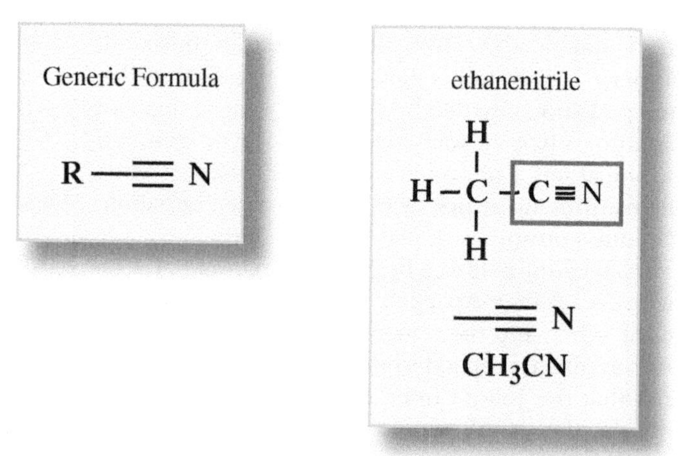

ability of the orbitals to contain electrons is said to be saturated, meaning that no more electrons can be acquired. When a carbon or nitrogen atom only has enough electrons to fill two orbitals, however, the hybridization involves only the *s* orbital and one of the three *p* orbitals, forming two *sp* hybrid orbitals. The two orbitals are oriented at 180 degrees to each other, in a line with the nucleus of the atom at its center. The other two *p* orbitals remain unaltered and are perpendicular to each other about the axis of the *sp* orbitals. The end-to-end overlap of the carbon and nitrogen *sp* orbitals forms a sigma (σ) bond, or primary covalent chemical bond, between the two atoms, while the side-by-side overlap of the adjacent *p* orbitals forms two secondary pi (ϖ) bonds parallel to the sigma bond, leaving one *sp* orbital remaining to form a fourth bond with another atom.

Overall, the nitrile system is linear and extends outward, away from the rest of the molecular structure, forming a site that is highly susceptible to reaction. The carbon-nitrogen triple bond creates an electric dipole because the nitrogen atom is more electronegative than the carbon atom. This, coupled with the electron-rich nature of the triple bond, makes the nitrile functional group a valuable reactive site.

In principle, the series of nitrile compounds is infinite, but it is also restricted to linear compounds. Any addition to the carbon-nitrogen bond destroys its identity as a cyano group. Accordingly, there are no cyclic nitriles in the same sense that there are cyclic hydrocarbons. Rather, a cyano group can only exist in a cyclic compound as a substituent, or side chain.

The most common organic nitrile is acetonitrile (CH_3CN). The reactivity of nitriles is such that they are good solvents in many applications, and acetonitrile is the least toxic of the nitriles, only posing a lethal risk to humans at concentrations of about 2,700 parts per million (ppm) in air and being relatively safe when handled properly. Butyronitrile (C_3H_7CN), in comparison, can be lethal at concentrations of about 250 ppm.

Nomenclature of Nitriles

Nitriles, being similar to carboxylic acids, are commonly named according to the parent acid from which they are formed, as in the case of acetonitrile, the nitrile form of acetic acid. To name the compound systematically, following the rules of the International Union of Pure and Applied Chemistry (IUPAC), the parent hydrocarbon structure is first identified and named as its nitrile. For example, acetonitrile is systemically named "ethanenitrile," indicating that it has a nitrile group attached to an ethane parent molecule. A ring compound in which the cyano group is the principal functional group is called a carbonitrile. When it is not the principal functional group in a compound, the cyano group is named as a substituent group, along with its numerical position in the structure, and listed alphabetically with the other functional groups before the suffix identifying parent structure.

Typically, the carbon atom containing the cyano group is assigned the lowest position number in the molecule, as the group is usually found at the end of a hydrocarbon chain, in which case the number may be omitted from the name. For example, a cyclohexene molecule (a six-membered ring containing one carbon-carbon double bond) with a cyano functional group on the third carbon atom in the ring, counting clockwise from the first carbon in the double bond, would be informally named 3-cyanocyclohexene. To name the same molecule systematically, the cyano carbon atom is assigned the position number 1 and the other carbon atoms are numbered counterclockwise from it, so the double bond begins on the second carbon atom. The molecule would thus have the IUPAC name cyclohex-2-ene-1-carbonitrile, but it could also be called cyclohex-2-enecarbonitrile.

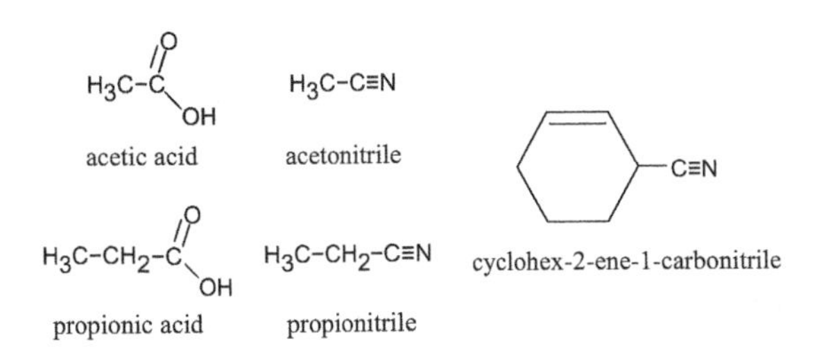

Reactions of Nitriles

Nitriles are reactive compounds that are widely used to form primary amines, carboxylic acids, and their corresponding derivatives. Carboxylic acids are formed by complete hydrolysis of the nitrile in acid. Under mildly acidic conditions, the nitrile becomes protonated (gains a proton, in the form of the hydrogen cation H^+) and carries the positive charge at its carbon atom, which changes the carbon-nitrogen triple bond to a double bond. The resulting cation is known as the iminium ion (an imine is a group or compound containing a carbon-nitrogen double bond), has the form $R_2C=N^+R_2$, and is electronically similar to a protonated carbonyl group ($C=O^+$). The positively charged carbon atom becomes the point of attachment for a molecule of water, which then loses one of its hydrogen atoms as another proton. This leaves the molecule as a hydroxy imine, with the hydroxyl group (–OH) bonded to the carbon atom of the imine structure. This is a very unstable arrangement and is prone to intramolecular rearrangement. The hydrogen atom of the hydroxyl group transfers to the nitrogen atom as the electrons from the carbon-nitrogen double bond reorganize to form a carbonyl group (C=O) and an amino group ($-NH_2$). The resulting product is the corresponding amide.

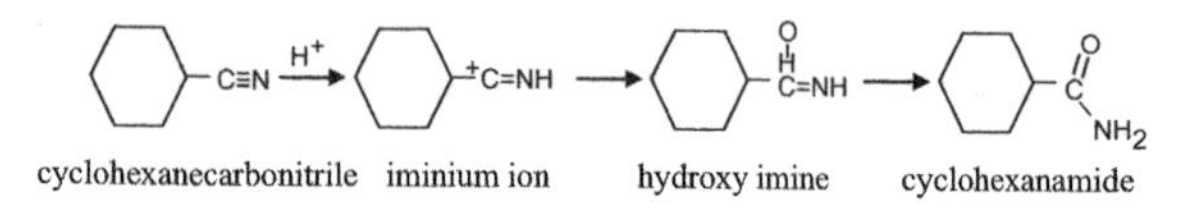

More vigorous hydrolysis, using a stronger acid medium, hydrolyzes the amide completely to form the carboxylic acid and ammonia (NH_3).

A good reducing agent, such as lithium aluminum hydride ($LiAlH_4$), can reduce a nitrile directly to the corresponding primary amine. This method is a very useful way of augmenting an existing structure with $-CH_2-NH_2$. Milder reducing agents, such as boranes, can reduce the nitrile to a corresponding imine. Compared to the iminium ion, the neutral imine is much less susceptible to reaction with water and other nucleophiles (electron-rich species), and since the reduction is carried out in an unreactive anhydrous solvent such as diethyl ether or tetrohydrofuran, it is quite stable. The imine itself can subsequently be used in different reactions to produce other compounds, either as desired final products or as intermediate compounds in a larger synthetic protocol.

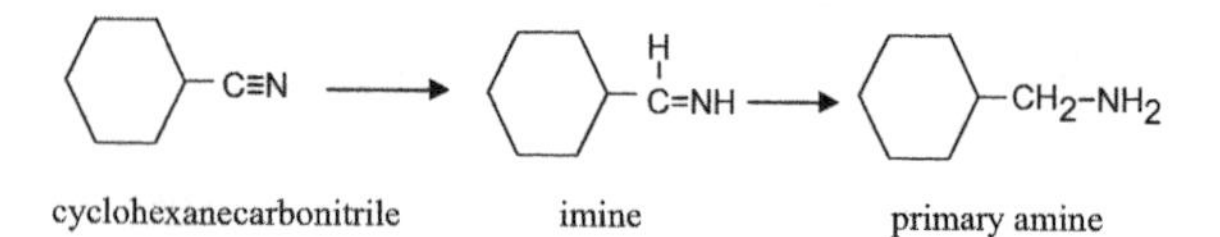

Nitriles in which the cyano group is attached directly to a carbon-carbon double bond undergo polymerization reactions very easily. The most well known of these compounds is acrylonitrile, which is commonly used in the plastics industry. In an acrylonitrile molecule, the cyano group is connected by a single bond to a methine group (=CH–), which in turn is carbon-carbon double bonded to a methylene group ($=CH_2$).The nitrogen atom of the cyano group is sufficiently electron withdrawing that it polarizes the carbon-carbon double bond, making it highly susceptible to polymerization reactions in which the double bond becomes a single bond and numerous acrylonitrile molecules link to each other in a head-to-tail fashion, as here:

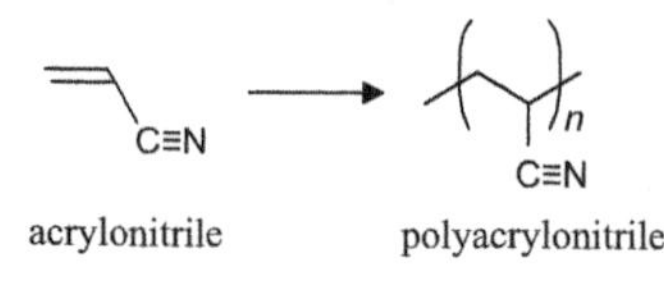

FORMATION OF NITRILES

Nitriles are prepared by various methods. One of the most common ways to produce a nitrile is to dehydrate an amide. For example, acetonitrile, the simplest nitrile, is prepared industrially by the dehydration of acetamide (CH_3CONH_2) with thionyl chloride ($SOCl_2$) or phosphorus pentoxide (P_2O_5), according to the reaction:

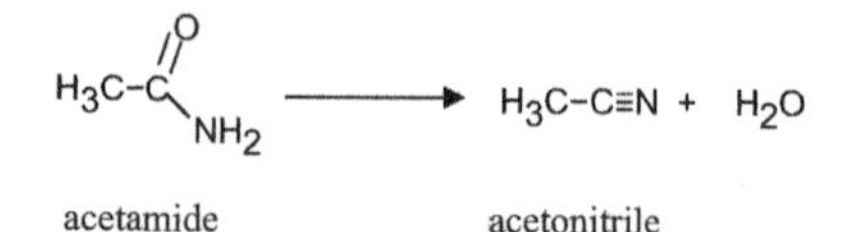

SAMPLE PROBLEM

Given the chemical name 2-isobutyl-3-phenylbenzonitrile, draw the skeletal structure of the molecule.

Answer:

The parent name, benzonitrile, indicates that the material is a benzene ring containing a cyano group. Because there is no number indicating the placement of the cyano group, it is assumed to be attached to carbon atom number one The principal benzene ring has an isobutyl group at the second carbon atom and a second benzene ring (indicated by the prefix "phenyl-") bonded to the third carbon atom. The structure is therefore as follows:

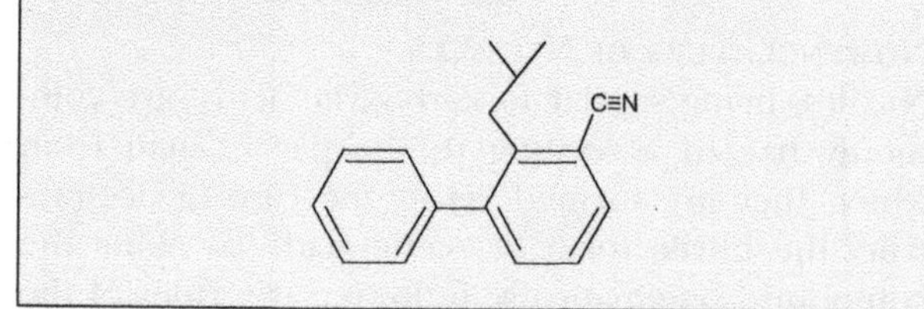

More complex nitriles are synthesized by using a preexisting nitrile group as a cyanide ion. The cyanide ion, as nucleophile, is prone to react with the carbonyl group in acid solution. The components of hydrogen cyanide (HCN) add to the carbonyl group of an aldehyde (RCOH) or a ketone (RCOR') to produce the corresponding hydroxynitrile, or cyanohydrin. Hydroxynitriles are characterized by having a hydroxyl group and a cyano group bonded to the same carbon atom in the hydrocarbon structure. Formation of the hydroxynitrile adds one carbon atom to the basic structure of the molecule and sets the stage for other reactions. Halogenation would then add a halide functional site, dehydration would form an alkene (an unsaturated hydrocarbon with at least one carbon-carbon double bond), reduction with a catalyst would produce a primary amine (RNH_2), and hydrolysis would form the carboxylic acid. A third way to form a nitrile is through a nucleophilic substitution reaction with an alkyl halide, which replaces the halide with the cyano group.

The synthesis of ring-based aryl nitriles is somewhat more difficult and is normally achieved by the formation of a diazonium salt ($R-N^+\equiv N$). In this process, an aryl amine such as phenylamine is converted to the corresponding diazonium salt by reaction with sodium nitrite ($NaNO_2$) and an acid, such as hydrochloric acid (HCl), at low temperature. The diazonium salt intermediate, being unstable and prone to explode, is never isolated in practice but rather treated with copper cyanide (CuCN) to convert it to the aryl nitrile.

Richard M. Renneboog, MSc

BIBLIOGRAPHY

Acheson, R.M. *An Introduction to the Chemistry of Heterocyclic Compounds.* 3rd ed. New York: Wiley, 2008. Print.

Haynes, W. H., PhD, David R. Lide, PhD, and Thomas J. Bruno, PhD, eds. *CRC Handbook of Chemistry and Physics,* 96th Edition. New York: CRC/Taylor & Francis Group, 2015. Print.

Hendrickson, James B., Donald J. Cram, and George S. Hammond. *Organic Chemistry.* 3rd ed. New York: McGraw, 1970. Print.

Jones, Mark Martin. *Chemistry and Society.* 5th ed. Philadelphia: Saunders College Pub., 1987. Print.

Morrison, Robert Thornton, and Robert Neilson. Boyd. *Organic Chemistry.* 7th ed. Englewood Cliffs, N.J.: Prentice Hall, 2003. Print.

Wuts, Peter G. M., and Theodora W. Greene. *Greene's Protective Groups in Organic Synthesis.* 4th ed. Hoboken: Wiley-Interscience, 2007. Print.

NOBLE GAS COMPOUNDS

FIELDS OF STUDY

Inorganic Chemistry; Physical Chemistry; Crystallography

SUMMARY

The noble gases were traditionally believed to be entirely unreactive due to their valence electron arrangement. In 1962 Neil Bartlett synthesized the first known compound of an inert gas element, disproving the long-standing belief. Many more noble gas compounds have since been synthesized and studied.

PRINCIPAL TERMS

- **combination reaction:** a chemical reaction in which two or more reactants combine to form a single product.
- **covalent bond:** a type of chemical bond in which electrons are shared between two adjacent atoms.
- **dipole:** the separation of positive and negative charges within a single molecule due to electron density being relatively high in one part of the molecule and relatively low in another.
- **electronegative:** describes an atom that tends to accept and retain electrons to form a negatively charged ion.
- **electrovalent bond:** an alternate term for an ionic bond, which is a type of chemical bond formed by mutual attraction between two ions of opposite charges.

- **entropy:** a property of the energy of chemical systems that is related to the total number of degrees of freedom within the system due to the number of components in that system and the number of particles of each component.
- **ionization energy:** the amount of energy required to remove an electron from an atom in a gaseous state.
- **oxidizing agent:** any atom, ion, or molecule that accepts one or more electrons from another atom, ion, or molecule in a reduction-oxidation (redox) reaction and is thus reduced in the process.
- **valence electron:** an electron that occupies the outermost or valence shell of an atom and participates in chemical processes such as bond formation and ionization.

ELECTRONS IN ATOMS

Electrons in atoms obey very strict mathematical relationships described by quantum mechanics. Within these limitations electrons are restricted to distinct regions about the nucleus of an atom. No more than two electrons can occupy any one of these regions, called orbitals, at any one time. Electrons fill the orbitals in pairs from the lowest energy orbitals to the highest. The highest energy or outermost group of orbitals contains the valence electrons of the atom. The number of electrons in the valence shell of an atom increases by one across each period of the periodic table. The noble gases are the terminal case of this progression since their valence orbitals contain all of the electrons that are allowed for that particular quantum level. These are the elements helium, neon, argon, xenon and radon. Elements containing a small number more of electrons in the next quantum level tend to give up those outermost electrons. Sodium atoms, for example, have one more electron than neon, in the next quantum level, and readily give it up to form a positively-charged sodium ion. Fluorine has one less electron than

INERT GASES

10 Ne neon 20

1 H hydrogen 1																	2 He helium 4
3 Li lithium 7	4 Be beryllium 9											5 B boron 11	6 C carbon 12	7 N nitrogen 14	8 O oxygen 16	9 F fluorine 19	10 Ne neon 20
11 Na sodium 28	12 Mg magnesium 29											13 Al aluminium 27	14 Si silicon 28	15 P phosphorus 31	16 S sulfur 32	17 Cl chlorine 35	18 Ar argon 40
19 K potassium 39	20 Ca calcium 40	21 Sc scandium 45	22 Ti titanium 48	23 Vi vanadium 51	24 Cr chromium 52	25 Mn manganese 55	26 Fe iron 56	27 Co cobalt 59	28 Ni nickel 59	29 Cu copper 64	30 Zn zinc 65	31 Ga gallium 70	32 Ge germanium 73	33 As arsenic 75	34 Se selenium 79	35 Br bromine 80	36 Kr krypton 84
37 Rb rubidium 85	38 Sr strontium 88	39 Y yttrium 89	40 Zr zirconium 91	41 I niobium 93	42 Mo molybdenum 96	43 Tc technetium 98	44 Ru ruthenium 101	45 Rh rhodium 103	46 Pd palladium 106	47 Ag silver 108	48 Cd cadmium 112	49 In indium 115	50 Sn tin 119	51 Sb antimony 122	52 Te tellerium 128	53 I iodine 127	54 Xe xenon 131
55 Cs caesium 133	56 Ba barium 137	*	72 Hf hafnium 178	73 Ta tantalum 181	74 W tungsten 184	75 Re rhenium 186	76 Os osmium 190	77 Ir iridium 192	78 Pt platinum 195	79 Au gold 197	80 Hg mercury 201	81 Tl thallium 204	82 Pb lead 207	83 Bi bismuth 209	84 Po polonium 209	85 At astatine 210	86 Rn radon 222
87 Fa francium 223	88 Ra radium 226	**	104 Rf rutherfordium 267	105 Db dubnium 268	106 Sg seaborgium 271	107 Bh bohrium 272	108 Hs hassium 270	109 Mt meitnerium 276	110 Ds darmstadtium 281	111 Rg roentgenium 280	112 Cn copernicium 285	113 Uut ununtrium 284	114 Fl flerovium 289	115 Uup ununpentium 288	116 Lv livermorium 293	117 Uus ununseptium 294	118 Uuo ununoctium 289

57 * La lanthanum 139	58 Ce cerium 140	59 Pr praseodymium 141	60 Nd neodymium 144	61 Pm promethium 145	62 Sm samarium 150	63 Eu europium 152	64 Gd gadolinium 157	65 Tb terbium 159	66 Dy dysprosium 163	67 Ho holmium 165	68 Er erbium 167	69 Tm thulium 168	70 Yb ytterbium 173	71 Lu lutetium 175
89 ** Ac actinium 227	90 Th thorium 232	91 Pa protactinium 231	92 U uranium 238	93 Np neptunium 237	94 Pu plutonium 244	95 Am americium 243	96 Cm curium 247	97 Bk berkelium 247	98 Cf californium 251	99 Es einsteinium 252	100 Fm fermium 257	101 Md mendelevium 258	102 No nobelium 259	103 Lr lawrencium 284

neon, and has a very strong ability to extract an electron from another atom to form a negatively-charged fluoride ion. The noble gases, having the exactly correct number of electrons in their valence shell, have no impetus to either donate or accept electrons to form ions or chemical bonds. Since their discovery, this lack of chemical reactivity earned them the name inert gases. The stability of this arrangement of electrons is a consequence of quantum mechanics, but, as a rule of thumb, atoms having a different number of electrons will gain or lose electrons readily to achieve the same arrangement, and the ions so formed are themselves more stable than their non-ionized parent atoms.

Structure of XeF 4, one of the first noble gas compounds to be discovered. Public Domain, via Wikimedia Commons

Atomic Diameter and Polarization

As the number of electrons in an atom increases, the size of the atomic orbitals and the diameter of the atoms also increase. The electrons occupy a volume around the nucleus that is approximately 100,000 times larger in diameter. Within this volume, the lowest energy electrons are held in orbitals that are the closest to the nucleus. The valence electrons are the greatest distance from the nucleus and so experience the least influence from the nucleus. This diffuse cloud of electrons can become polarized by the presence of an attractive or repulsive force such as the electronegativity of another atom, and if the polarization is sufficient the valence electron can be drawn into the formation of an electrovalent bond. In some cases a covalent bond can be formed as part of a more complex ionic structure. The smaller the diameter of the atom, the more strongly is the attraction of the nucleus felt by the valence electrons, and the higher is the ionization energy of the atom. This in turn demands the action of a more powerful oxidizing agent to extract a valence electron from the atom. In the noble gases, this effect is compounded by the extra stability given by the filled valence shell orbitals. Accordingly, it requires 1335 kJ per mole more energy to ionize helium than is required to ionize radon (2372 $kJ.mol^{-1}$ vs. 1037 $kJ.mol^{-1}$).

The Noble Gases

The traditional view of the Group VIIIa elements as inert gases was based on their observed lack of chemical reactivity. Some chemists, notably Walther Kossel (1888 – 1956) and Linus Pauling (1901 – 1994), felt that compounds of xenon might be possible. It took one of chemistry's "happy accidents" to prove this true. In 1961, Neil Bartlett, working at the University of British Columbia, in Canada, discovered an unexpected compound while working with platinum hexafluoride (PtF_6), an extremely strong oxidizing agent. The new material was dark red in color, and on analysis proved to be a compound of molecular oxygen as O_2^+ and PtF_6^- due to the extraction of an electron from the oxygen molecule by the strongly electronegative fluorine atoms on the platinum compound. Bartlett realized that the ionization potential of molecular oxygen and xenon were very close in value and theorized that a similar reaction between Xe and PtF_6 should occur. In a simple apparatus, he allowed the two gases to come into contact with each other and witnessed the immediate formation of a yellow-gold colored material that he identified as $Xe^+PtF_6^-$ the first known compound of an inert gas. The terminology changed slowly as more compounds of the so-called inert gases were subsequently discovered, and they are now called the noble gases. Strictly speaking, the noble gases are not exactly rare, as a volume by volume comparison demonstrates. In a volume of 10^6 liters (visualize a cube that is 10 meters, or about 33 feet, on a side) there are approximately 5.24 liters of helium, 18.3 liters of neon, 9.34 liters of argon, 1.14 liters of krypton and 0.087 liters of xenon at standard temperature and pressure.

To date no compounds of helium, the second most abundant element in the universe after hydrogen, are known. The next element in the noble gas series, neon, is still very small in size, and the valence electrons bound too strongly by the attractive force of the nucleus to become part of a chemical bond. Argon is to date the smallest noble gas atom to be observed in a chemical compound, and that only at a very low temperature.

Xenon, Krypton, and Radon Compounds

The first true noble gas compound was synthesized via a straightforward combination reaction. Subsequent research has resulted in the synthesis of more than one hundred related compounds. The structure of that original compound was thought to be an ionic structure $Xe^+PtF_6^-$ but on closer examination is believed to be the compound formed by the process:

$$Xe + 2PtF_6 \rightarrow [XeF]^+[PtF_6]^- + PtF_5 \rightarrow [XeF]^+[Pt_2F_{11}]^-$$

in which the dipole induced in the valence shell of the xenon atom by the fluorine atom facilitates the extraction of an electron by the hexafluoroplatinum species. Gas-phase reaction between xenon and fluorine gas was also proved to be an expedient means of producing xenon compounds, and the reaction is sufficiently facile that XeF_2 can be prepared by simply mixing the two gases in a nickel-lined tube at little more than atmospheric pressure. The compounds XeF_4 and XeF_6 require higher pressures and temperatures, but proceed as simply. Combined products based on the elements ruthenium, rhodium, silicon, phosphorus and antimony are also known having structures analogous to that of $XePtF_6$. The silicon, antimony and phosphorus compounds are formed by reaction of xenon and fluorine gases in the presence of the non-metal fluoride, as for example:

$$2\,Xe_{(g)} + F_{2(g)} + SiF_{4(g)} \rightarrow XeSiF_{6(s)}$$

Due to the large number of gas-phase molecules involved in the production of a solid-phase product, the reaction system undergoes a very large decrease in entropy. The oxygen-based compounds $XeOF_4$, $XeOF_3$ and $XeOF_2$ as well as the very strongly oxidizing XeO_3 are well known. A xenic acid series is also known, consisting of $XeO_3 - H_2XeO_4 - H_2XeO_6$, all of which are extremely strong oxidizing agents. The analogous krypton compounds do not form, except for KrF_2 and KrF_4, though it was expected that they should form at least as easily as the xenon compounds. Radon, being dangerously radioactive in all of its isotopes, has not been well studied in regard to the formation of compounds, although it is expected that it will exhibit similar chemistry.

SAMPLE PROBLEM

The equation for the formation of $XeSiF_6$ has an extra Xe in the reactants that does not appear on the product side. What might this indicate in regard to the products?

Answer:

The product may actually include the Xe_2^+ ion rather than Xe^+. The neutral Xe atom is not strictly a component of the $XeSiF_6$ molecule and therefore is not included in the molecular formula.

The Importance of Noble Gas Compounds

The lack of chemical reactivity makes the noble gases themselves useful in regard to the conduct of normal chemical reactions and physical processes that are sensitive to the presence of moisture or oxygen. Helium is often utilized as the carrier gas in gas-phase separation techniques such as gas chromatography. Argon is commonly used to provide an inert atmosphere for chemical reactions, and for welding operations. Their lack of reactivity makes them ideal for these purposes. The high oxidizing power of xenon compounds makes them useful in chemical synthesis reactions requiring that ability. Xenon fluorides are also effective agents for fluorinating aromatic ring structures that cannot be formed by other methods, and have been used in the synthesis of 5-fluorouracil, an effective anticancer drug. That radon is expected to exhibit a similar ability to form compounds suggests that chemical scrubbing might be used to remove the radioactive element as an atmospheric contaminant in homes and mining operations.

Richard M. Renneboog M.Sc.

BIBLIOGRAPHY

Bartlett, Neil. *The Oxidation of Oxygen and Related Chemistry: Selected Papers of Neil Bartlett.* Singapore: World Scientific, 2001. Print.

Halka, Monika, and Brian Nordstrom. *Halogens and Noble Gases.* New York, NY: Facts on File Inc., 2010. Print.

Sears, William M. Jr., *Helium: The Disappearing Element.* New York, NY: Springer, 2015. Print.

Mackay, K. M., R. A. Mackay, and W.Henderson. *Introduction to Modern Inorganic Chemistry.* 6th ed. Cheltenham: Nelson, 2002. Print.

House, James E., and Kathleen A. House. *Descriptive Inorganic Chemistry.* Waltham, MA: Academic Press, 2016. Print.

Greenwood, N.N. and Earnshaw, A. *Chemistry of the Elements* 2nd ed. Burlington, MA: Elsevier, Butterworth-Heinemann, 2005. Print.

NONMETALS

FIELDS OF STUDY

Organic Chemistry; Inorganic Chemistry; Geochemistry; Metallurgy

SUMMARY

The basic properties of the nonmetals and their chemical behavior are described. The nonmetal elements make up all biochemicals and can form a near-infinite variety of compounds due to their properties of electronic structure and atomic size.

PRINCIPAL TERMS

- **allotrope:** one of two or more principal physical forms in which a single pure element occurs, due to differences in chemical bonding or the structural arrangement of the atoms; for example, diamond and graphite are two allotropes of carbon.
- **anion:** any chemical species bearing a net negative electrical charge, which causes it to be drawn toward the positive pole, or anode, of an electrochemical cell.
- **diatomic:** describes a molecule or ion consisting of two atoms that are chemically bonded to each other.
- **monatomic:** describes a molecule or ion consisting of only one atom.
- **polyatomic:** describes a molecule or ion consisting of multiple atoms that are chemically bonded to each other.

THE NATURE OF THE NONMETALS

Seventeen elements in the periodic table are generally considered to be nonmetals. These include four of the five halogens (astatine is more often classified as a metalloid) as well as the six noble gases. The remaining seven elements—hydrogen (H), carbon (C), nitrogen (N), oxygen (O), phosphorus (P), sulfur (S), and selenium (Se)—make up what is known as the nonmetals group. The term "nonmetals" can refer either to these seven elements specifically or to all seventeen nonmetallic and non-metalloid elements.

Nonmetals are elements that do not have metallic character and do not exhibit the behavior of metals. They are not malleable or ductile, are poor conductors of heat and electricity, and are characterized by their high electronegativity. The noble gases (group 18)—helium (He), neon (Ne), argon (Ar), krypton (Kr), xenon (Xe), and radon (Rn)—are noted for their almost complete lack of chemical activity and exist as monatomic gases. Xenon and radon are known to form compounds with the extremely electronegative element fluorine. The halogens (group 17)—fluorine (F), chlorine (Cl), bromine (Br), iodine (I), and astatine (At)—are noted for their very high reactivity. The halogens exist in elemental form as diatomic molecules, either as a gas (fluorine and chlorine), a liquid (bromine), or a solid (iodine and possibly astatine). The halogen group is the only group on the periodic table to contain elements that exist in all three states of matter at standard temperature and pressure.

Seven nonmetals exist as diatomic molecules in their standard state: hydrogen, nitrogen, oxygen, fluorine, chlorine, bromine, and iodine. Sulfur, in the same group as oxygen (group 16), exists in elemental form as a bright yellow crystalline solid consisting of a polyatomic molecule with the formula S_8. Elemental sulfur can be found in a number of

NONMETALS

6 C carbon

1	2	3	4	5	6	7	8	9	10	11	12	13	14	15	16	17	18
1 H hydrogen 1																	2 He helium 4
3 Li lithium 7	4 Be beryllium 9											5 B boron 11	6 C carbon 12	7 N nitrogen 14	8 O oxygen 16	9 F fluorine 19	10 Ne neon 20
11 Na sodium 28	12 Mg magnesium 29											13 Al aluminium 27	14 Si silicon 28	15 P phosphorus 31	16 S sulfur 32	17 Cl chlorine 35	18 Ar argon 40
19 K potassium 39	20 Ca calcium 40	21 Sc scandium 45	22 Ti titanium 48	23 Vi vanadium 51	24 Cr chromium 52	25 Mn manganese 55	26 Fe iron 56	27 Co cobalt 59	28 Ni nickel 59	29 Cu copper 64	30 Zn zinc 65	31 Ga gallium 70	32 Ge germanium 73	33 As arsenic 75	34 Se selenium 79	35 Br bromine 80	36 Kr krypton 84
37 Rb rubidium 85	38 Sr strontium 88	39 Y yttrium 89	40 Zr zirconium 91	41 I niobium 93	42 Mo molybdenum 96	43 Tc technetium 98	44 Ru ruthenium 101	45 Rh rhodium 103	46 Pd palladium 106	47 Ag silver 108	48 Cd cadmium 112	49 In indium 115	50 Sn tin 119	51 Sb antimony 122	52 Te tellerium 128	53 I iodine 127	54 Xe xenon 131
55 Cs caesium 133	56 Ba barium 137	*	72 Hf hafnium 178	73 Ta tantalum 181	74 W tungsten 184	75 Re rhenium 186	76 Os osmium 190	77 Ir iridium 192	78 Pt platinum 195	79 Au gold 197	80 Hg mercury 201	81 Ti thallium 204	82 Pb lead 207	83 Bi bismuth 209	84 Po polonium 209	85 At astatine 210	86 Rn radon 222
87 Fa frandum 223	88 Ra radium 226	**	104 Rf rutherfordium 267	105 Db dubnium 268	106 Sg seaborgium 271	107 Bh bohrium 272	108 Hs hassium 270	109 Mt meitnerium 276	110 Ds darmstadtium 281	111 Rg roentgenium 280	112 Cn copernicum 285	113 Uut ununtrium 284	114 Fl flerovium 289	115 Uup ununpentium 288	116 Lv livermorium 293	117 Uus ununseptium 294	118 Uuo ununoctium 289

57 * La lanthanum 139	58 Ce cerium 140	59 Pr praseodymium 141	60 Nd neodymium 144	61 Pm promethium 145	62 Sm samarium 150	63 Eu europium 152	64 Gd gadolinium 157	65 Tb terbium 159	66 Dy dysprosium 163	67 Ho holmium 165	68 Er erbium 167	69 Tm thulium 168	70 Yb ytterbium 173	71 Lu lutetium 175
89 ** Ac actinium 227	90 Th thorium 232	91 Pa protactinium 231	92 U uranium 238	93 Np neptunium 237	94 Pu plutonium 244	95 Am americium 243	96 Cm curium 247	97 Bk berkelium 247	98 Cf californium 251	99 Es einsteinium 252	100 Fm fermium 257	101 Md mendelevium 258	102 No nobelium 259	103 Lr lawrendum 284

different allotropes, depending on the manner in which the S_8 molecules form the crystal lattice of the solid. Most nonmetal elements, except the noble gases, normally form anions by accepting electrons from other atoms that are electropositive in character. Other nonmetal elements—particularly carbon, nitrogen, and phosphorus—typically do not form anions but are most likely to participate in covalent bonds due to the prohibitive energy levels associated with either gaining or losing enough electrons to form a complete octet of electrons in the valence shell.

CHNOPS

Life on earth depends on the characteristic chemical behavior of the nonmetals, particularly carbon, hydrogen, nitrogen, oxygen, phosphorus, and sulfur (CHNOPS). Carbon has a total of four electrons in its outermost electron shell, or valence shell—that is, half of a complete octet. Because of this, carbon molecules typically share their four valence electrons through covalent bonds, thereby forming an octet electron configuration, rather than losing or gaining molecules to form ionic bonds. Carbon's valence electrons are distributed as $2s^22p^2$, meaning that there are two electrons in the 2*s* orbital and one each in two of the three 2*p* orbitals. The *s* and *p* orbitals, belonging to the same electron shell, are sufficiently close in energy that they are able to combine to form four equivalent hybrid sp^3 orbitals. Furthermore, carbon's intermediate electronegativity makes a carbon atom more likely to share electrons than to gain or lose them. This unique combination of characteristics enables carbon atoms to form strong, stable covalent bonds with other carbon atoms, facilitating the formation of molecules that contain long chains or rings of carbon. Carbon is also capable of forming strong bonds with other nonmetals, such as hydrogen, oxygen, sulfur, and the halogens. These large, complex

SAMPLE PROBLEM

What is the electron distribution in carbon and silicon? Use the chart below to fill in the electrons in the appropriate order, and keep in mind that *s* orbitals hold a maximum of two electrons, while *p* orbitals hold a maximum of six electrons (two each in three individual orbitals).

1s
2s 2p
3s 3p 3d
4s 4p 4d 4f
5s 5p 5d 5f 5g
6s 6p 6d 6f 6g 6h
7s 7p 7d 7f 7g 7h

Answer:

The atomic number of carbon is 6, meaning it has six protons and thus, in a neutral atom, six electrons. There are two electrons in the 1*s* atomic orbital, two electrons in the 2*s* orbital, and two in the 2*p* orbitals. The electron distribution is therefore $1s^2 2s^2 2p^2$.

The atomic number of silicon is 14. There are two electrons in the 1*s* orbital, two electrons in the 2*s* orbital, six electrons in the 2*p* orbitals, two electrons in the 3*s* orbital, and two electrons in the 3*p* orbitals. The electron distribution is therefore $1s^2 2s^2 2p^6 3s^2 3p^2$.

carbon-based molecules are the structural basis of the many compounds that make up living cells.

Nitrogen and oxygen have similar electronic properties to carbon, though with one and two more electrons, respectively. These extra valence electrons increase the electronegativity of nitrogen and oxygen relative to carbon, but they do not significantly affect the size of the atoms. Both nitrogen and oxygen, like carbon, have their valence electrons in the 2*s* orbital and the three 2*p* orbitals, and they can easily form covalent bonds with carbon atoms. The elements phosphorus and sulfur are also necessary to the molecules of biochemistry, though they have their valence electrons in the 3*s* and 3*p* orbitals. These five elements, along with hydrogen, form the basic chemical components of life itself and are the constituents of essentially all biochemical compounds. Minor quantities of metals (such as iron, sodium, and magnesium) and other nonmetal elements (such as iodine and chlorine) are significant cofactors in biochemical processes, but the molecular structures with which they interact consist primarily of the CHNOPS elements.

Nonmetal Compounds

The nonmetals, especially carbon, can form a nearly infinite variety of chemical compounds. Carbon atoms can catenate without limit, often bonding with hydrogen to form a series of linear saturated hydrocarbons, or alkanes. Ring structures, double bonds, and triple bonds between carbon atoms produce an even greater variety of corresponding organic compounds, including alkenes (which contain at least two carbon atoms connected by a double bond) and alkynes (two carbon atoms connected by a triple bond). The presence of oxygen, nitrogen, and sulfur significantly increases the number of possible organic compounds. Carbon's unusual versatility in forming bonds enables the formation of isomeric compounds, which are compounds that share the same molecular formula but have different molecular structures due to the various possible combinations of the molecule's substituent groups. Though several million organic compounds of carbon, hydrogen, oxygen, nitrogen, and sulfur are known, many millions more have yet to be identified and studied. Oxygen and sulfur also commonly form oxides and sulfides with most other elements.

The other nonmetal elements have markedly different chemical behaviors. The noble gases are also called inert gases because they typically do not form compounds under standard conditions. The exceptions are xenon and radon, which form compounds when combined with elemental fluorine (F_2). The extremely high electronegativity and charge density of the fluorine atom is able to induce sufficient polarity within the xenon atom to create a bond. The halogens, as mentioned, are highly electronegative and readily form halide anions with a charge of 1–. Halogen elements are also able to form covalent bonds with carbon atoms to create halocarbon compounds that are commonly found in nature. Organic compounds isolated from marine organisms often include a chlorine or other halogen atom in the place where a hydrogen atom would normally be found in a similar compound from a terrestrial organism.

Occurrence of Nonmetals

Although there are over five times more metal than nonmetal elements in the periodic table, nonmetal compounds are more readily found in nature, making up nearly 99 percent of the earth's atmosphere (particularly N_2, O_2, and CO_2) and more than 97 percent of a human body (especially oxygen, carbon, hydrogen, nitrogen, and phosphorus) by mass. The noble gases are found in the gaseous state in nature. A large quantity of each noble-gas element is also believed to exist in clathrate compounds, which are compounds that trap atoms or molecules within a crystal structure. All halides, with the exception of many fluoride compounds, exist as either ionic solids or ions in solution. Seawater, for example, contains sodium and chloride ions that form sodium chloride (NaCl).

Oxygen is found in huge quantities as a component of various minerals, especially silicates. Oxygen also forms carbon dioxide (CO_2) in the atmosphere. Carbon dioxide can react with water (H_2O) to form carbonic acid (H_2CO_3), which can then be trapped as the carbonate salts of various metallic elements, such as calcium carbonate ($CaCO_3$). Plants also trap oxygen by converting carbon dioxide into glucose ($C_6H_{12}O_6$) in photosynthesis. In respiration, atmospheric oxygen (O_2) is used to convert the glucose back into carbon dioxide. Oxygen is the most abundant element by mass in the earth's crust and oceans and, after nitrogen, the second-most abundant element in its atmosphere.

Richard M. Renneboog, MSc

BIBLIOGRAPHY

Acheson, R.M. *An Introduction to the Chemistry of Heterocyclic Compounds.* 3rd ed. New York: Wiley, 2008. Print.

Berg, Jeremy M., John L. Tymoczko, Gregory J. Gatto, and Lubert Strye. *Biochemistry.* 8th ed. New York: W. H. Freeman, 2015. Print.

Gribbin, John. *Science: A History, 1543–2001.* London: Lane, 2002. Print.

Hendrickson, James B., Donald J. Cram, and George S. Hammond. *Organic Chemistry.* 3rd ed. New York: McGraw, 1970. Print.

Johnson, Rebecca L. *Atomic Structure.* Minneapolis: Lerner, 2008. Print.

Kean, Sam. *The Disappearing Spoon: And Other True Tales of Madness, Love, and the History of the World from the Periodic Table of the Elements.* New York: Little, Brown, 2010. Print.

Mackay, K. M., R. A. Mackay, and W. Henderson. *Introduction to Modern Inorganic Chemistry.* 6th ed. Cheltenham: Nelson, 2002. Print.

Morrison, Robert Thornton, and Robert Neilson Boyd. *Organic Chemistry.* 7th ed. Englewood Cliffs, N.J.: Prentice Hall, 2003. Print.

Powell, P., and P. L. Timms. *The Chemistry of the Non-Metals.* London: Chapman, 1974. Print.

Winter, Mark J. *The Orbitron: A Gallery of Atomic Orbitals and Molecular Orbitals.* University of Sheffield, n.d. Web.

NUCLEOSYNTHESIS

FIELDS OF STUDY

Nuclear Chemistry; Spectroscopy

SUMMARY

The formation of elements through nuclear fusion reactions within stars is discussed as the source of all natural elements. Atoms and certain molecular forms that result are identified in stars by spectroscopy and are also found in the interstellar medium. Analysis of isotopic abundances provides evidence of the nuclear reactions that have occurred.

PRINCIPAL TERMS

- **alpha particle:** a particle consisting of two protons and two neutrons bound together, identical to a helium nucleus; typically produced in the process of alpha decay.
- **atomic mass:** the total mass of the protons, neutrons, and electrons in an individual atom.
- **beta particle:** an electron produced in the process of beta decay, in which a neutron of an unstable atom decomposes into a proton, electron and neutrino, and with the electron emitted from the nucleus at high speed.
- **deuterium**: an isotope of hydrogen that contains one neutron and one proton; occurs naturally in about 1 in 6,500 hydrogen atoms.

- **element:** a form of matter consisting only of atoms of the same atomic number.
- **gamma ray:** ionizing electromagnetic radiation of higher energy and shorter wavelength than x-rays, typically produced by the nucleus of an atom undergoing radioactive decay.
- **half-life:** the length of time required for one-half of a given amount of material to decompose or be consumed through a continuous decay process.
- **matter:** any substance that occupies physical space and is composed of atoms or the particles that make up atoms.
- **neutron:** a fundamental subatomic particle in the atomic nucleus that is electrically neutral and about equal in mass to the mass of one proton.
- **nuclear fusion:** the process in which atomic nuclei combine to form a single nucleus of greater mass; normally refers to the fusion of hydrogen nuclei but is a generally applicable term.
- **proton-proton chain:** the nucleosynthesis process in which protons are added sequentially together to produce helium nuclei.

THE STELLAR FURNACE

Interstellar and circumstellar space contain large quantities of matter in a very diffuse state. Active stars like the Sun, however, are composed of a massive quantity of matter as hydrogen atoms. The Sun itself comprises almost 99% of the mass of the Solar system. The source of all matter is the nuclear fusion reactions that take place within stars. Within them, nuclear fusion reactions are driven under the intense force of their gravity and the energy released by the conversion of a small percentage of their matter into energy. These reactions, termed nucleosynthesis reactions, occur by specific sequences that construct the nuclei of elements of greater atomic mass than hydrogen from the basic starting components of protons and electrons. A nucleosynthesis reaction is entirely different from a synthesis reaction that occurs in "normal" chemistry that does not involve the nucleus of any atoms at all. Normal chemical reactions occur only at the level of the outermost electrons of the atoms that are involved. Historically, theories of how the Sun worked tended to parallel the current technical knowledge of the time. In the 18th century and earlier the dark areas known as sunspots, for example, inspired the idea that the light and heat from the Sun was due to it being a gigantic mass of burning coal. The development of spectroscopic analysis methods in the 19th century revealed that there is some carbon in the Sun, but the composition of the solar matrix consists almost entirely of hydrogen. In the early 20th century, the development of the quantum mechanical model of atomic structure led to the realization of both nuclear fission and nuclear fusion as real processes. Subsequently, astrophysicists have been able to identify the processes occurring

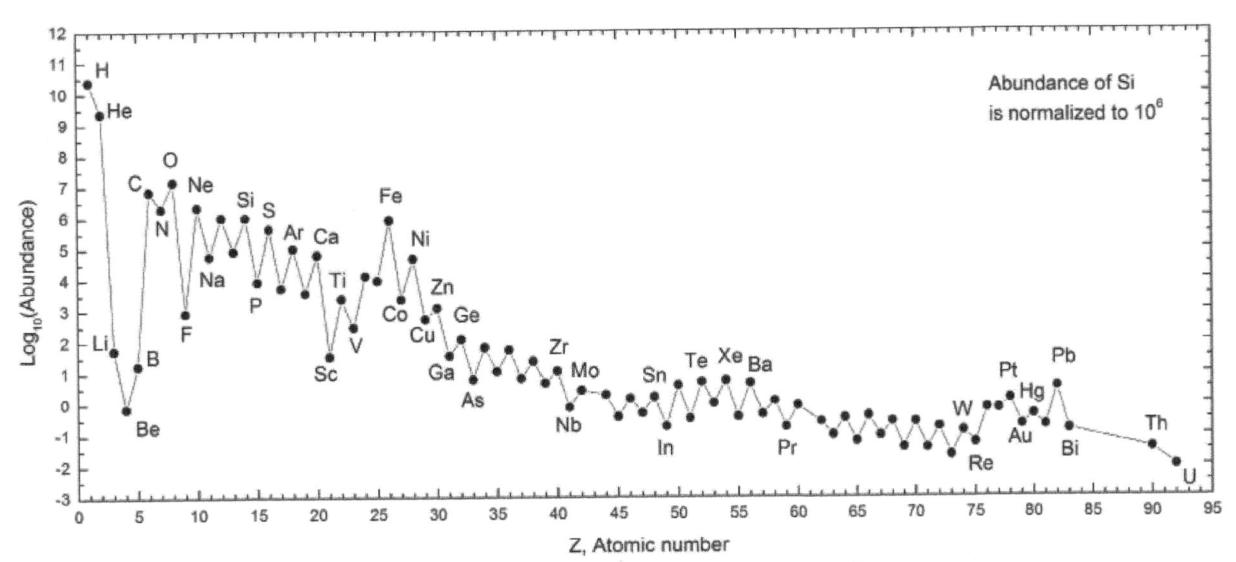

Abundances of the chemical elements in the Solar System. By Orionus at English Wikipedia via Wikimedia Commons

within the Sun and all other stars by which different atomic nuclei are produced.

BASIC NUCLEOSYNTHESIS

Stellar nucleosynthesis reactions begin with protons and electrons. Under the conditions within a star, deuterium nuclei are formed when a proton and an electron are combined to form a neutron, which is then combined with another proton to form a deuterium nucleus. The actual process is not that simple, however, as it also liberates an electron and a massless particle called a neutrino. The overall nuclear reaction for this can be written as

$$2\,{}^{1}H \rightarrow {}^{2}H + {}^{0}e^{-} + {}^{0}\nu$$

in which ^{1}H is a standard hydrogen atom, ^{2}H is an atom of the hydrogen isotope deuterium, e^{-} is the electron, and ν is the neutrino. (Isotopes have the same number of protons and electrons, but a different number of neutrons in their respective atomic structures.) Once formed, two deuterium nuclei could be driven to combine to form helium nuclei, or alpha particles. However, the amount of deuterium nuclei formed is not significant relative to the number of protons that are available, and the proton-proton chain produces helium nuclei by adding protons sequentially, as the reaction series:

$$2\,{}^{1}H \rightarrow {}^{2}H + {}^{0}e^{-} + {}^{0}\nu$$

$${}^{2}H + {}^{1}H \rightarrow {}^{3}He + {}^{0}\gamma$$

$${}^{3}He + {}^{3}He \rightarrow {}^{4}He + 2\,{}^{1}H$$

where γ represents a high-energy photon called a gamma ray. The conservation of matter and energy requires that these also have the corresponding number of electrons, and so are described as helium atoms. In stellar nucleosynthetic processes some 0.71% of the mass in each reaction system is converted into energy, in accord with Einstein's equation $e = mc^2$, as positrons (the antimatter equivalent of electrons) are emitted and immediately collide with electrons in the stellar matrix. This is the source of all of the light and heat that streams outward from the Sun and all other stars in the early stage of their lifetimes. Alpha particles, or helium atoms, within the star's interior can also undergo further nuclear fusion reactions and produce more massive atoms. Fusion of helium and hydrogen nuclei, for example, would produce lithium atoms, and the fusion of two lithium atoms would produce a carbon atom.

SAMPLE PROBLEM

Given that the Sun converts about 4 X 10^{12} grams (4 million tons) of mass into energy each second, use Einstein's equation to calculate the amount of energy produced by the Sun on a 24-hour period.

Answer:

The time period is calculated as (24 hr/day X 60 min/hr X 60 sec/min) = 86,400 seconds.

The mass converted to energy per second of time is 4 X 10^{12} grams, or 4 X 10^{9} kilograms.

The speed of light, c, is approximately 3 X 10^{8} meters per second.

Einstein's equation is $E = mc^2$

Energy produced in 1 second is

$$E = (4 \text{ X } 10^{9}\text{ kg})(3 \text{ X } 10^{8}\text{ m.sec}^{-1})^2$$

$$= (4 \text{ X } 10^{9}\text{ kg})(9 \text{ X } 10^{16}\text{ m}^2\text{.sec}^{-2})$$

$$= 36 \text{ X } 10^{25}\text{ kg.m}^2\text{sec}^{-2}$$

$$= 36 \text{ X } 10^{25}\text{ joules } (3.396 \text{ X } 10^{23}\text{ BTU})$$

Total energy produced in 24 hours is therefore

$$36 \text{ X } 10^{25} \text{ X } 86400 \text{ seconds/day}$$

$$= 3.1104 \text{ X } 10^{31}\text{ joules } (2.934 \text{ X } 10^{28}\text{ BTU})$$

ALTERNATIVE NUCLEOSYNTHESIS CYCLES

The relative abundance of heavier elements in respect to hydrogen is a good reflection of the probability of the nuclear fusion reactions for their formation within the interior of a star. Spectroscopic analyses of starlight also reveal the presence of several other elements in addition to hydrogen and helium, and energy calculations based on the proton-proton chain alone do not agree with observed stellar temperatures. To account for the differences, a sequence termed the CNO cycle utilizes carbon, nitrogen, and oxygen atoms as catalysts in processes that overall produce one helium atom from four hydrogen atoms. A third alternative process is called the triple alpha process, in which three helium nuclei theoretically collide and are combined to form a single carbon-12 atom. Statistically, such a collision is effectively impossible to achieve, but due to the vast number of atoms within a star, it cannot be entirely discounted.

Nuclide Abundances and Terrestrial Nuclide Synthesis

Every single atom of matter is believed to be produced by nucleosynthesis occurring in stars. As stars age, matter is naturally ejected into the surrounding space both continuously and through cataclysmic events. The formation of various nuclides during nucleosynthesis is reflected in the matter that is observed beyond the interiors of stars, on Earth and other planets, meteors and meteorites, comets, and so on. Interaction with high-energy particles can also disrupt the nuclei of stable atoms and produce unstable or radioactive isotopes. Each has its own half-life and the rate at which such elements break down can be used in many ways. One common methodology uses the rate of decay or the relative abundances of various elements to determine the age of the material being examined. Of more immediate value is the artificial production of specific nuclides under controlled conditions for use as medical or analytical markers and effective treatment of disease. Such nucleosynthetic reactions are typically carried out through the use of specially designed nuclear reactors, and the recovered products are harvested and sent to their intended uses.

Richard M. Renneboog, MSc

Bibliography

Boyd, Richard N. *An Introduction to Nuclear Astrophysics.* Chicago, IL: University of Chicago Press, 2008. Print.

Kwok, Sun. *Physics and Chemistry of the Interstellar Medium.* Herndon, VA: University Science Books, 2007. Print.

Ryan, Sean G. and Andrew J. Norton. *Stellar Evolution and Nucleosynthesis.* New York, NY: Cambridge UP, 2010. Print.

Schramm, David N. *The Big Bang and Other Explosions in Nuclear & Particle Astrophysics.* River Edge, NJ: World Scientific Publishing, 1996. Print.

Shaviv, Giora. *The Synthesis of the Elements.* New York, NY: Springer, 2012. Print.

Tielens, A.G.G.M. *The Physics and Chemistry of the Interstellar Medium* New York, NY: Cambridge University Press, 2005. Print.

Verschuur, Gerrit *The Invisible Universe. The Story of Radio Astronomy* New York, NY: Springer, 2015. Print.

O

ORBITALS

FIELDS OF STUDY

Inorganic Chemistry, Organic Chemistry

SUMMARY

Atomic and molecular orbitals are theoretical constructs of the modern theory of atomic structure, based on the mathematical description of quantum mechanics. They are ascribed specific geometries about the atomic nucleus according to the probability that an electron with the appropriate energy level will be in that region of space. Orbitals describe the physical and chemical behavior of electrons in atoms and molecules.

PRINCIPAL TERMS

- **electron configuration:** the order and arrangement of electrons within the orbitals of an atom or molecule.
- **electron shell:** a region surrounding the nucleus of an atom that contains one or more orbitals capable of holding a specific maximum number of electrons.
- **probability density:** in reference to electrons, the probability of finding a particular electron in a given region of space within an atom or molecule; also called electron density.
- **quantum number:** one of four numbers that describe the energy level, orbital shape, orbital orientation, and spin of an electron within an atom.
- **valence:** the maximum number of bonds an atom can form with other atoms based on its electron configuration; also called valency.

The Dual Nature of Electrons

The modern theory of atomic structure, based on quantum mechanics, dates back to the 1920s and is still being developed and refined by researchers. The foundations of the theory were laid in 1897, when J. J. Thomson (1856–1940) identified electrons as charged particles with negligible mass, and were reinforced in 1909, when Ernest Rutherford (1871–1937), Hans Geiger (1882–1945), and Ernest Marsden (1889–1970) identified protons as much heavier particles bearing the opposite charge. Subatomic particles had not been identified as discrete entities before that time, though a number of theories about the structure of the atom had already been proposed. A more complete description of the internal structure of atoms required a third particle that was electrically neutral, termed the "neutron." Because neutrons bear no electrical charge, their direct observation by electrical means was not possible, and it was not until James Chadwick (1891–1974) demonstrated their existence by indirect means in 1932 that the basic theoretical principles of atomic structure were resolved.

A large pool of experimental observations and measurement data relating to the interaction of atoms and molecules with light also existed. This information was crucial to ascribing energy values to electrons in atoms and molecules, with perhaps the most important discovery being the relationship between mass and energy, defined by Albert Einstein (1879–1955) in 1905 as:

$$E = mc^2$$

where E is energy, m is mass, and c is the speed of light. Earlier the same year, Einstein had proposed that electromagnetic energy is transferred to electrons not at a constant rate but in minute, individual units—that is, as particles. These particles, known as "photons," have no mass and exhibit wave-particle duality, meaning that they display properties characteristic of both waves and particles. In 1924, Louis de Broglie (1892–1987) combined Einstein's mass-energy equation with his observations about photons to propose that all particles, including electrons, exhibit the same wave-particle duality. This duality makes the concept of electron orbitals central to the understanding of chemical behavior.

QUANTUM NUMBERS

The quantum-mechanical model of the atom describes the electrons in each atom in terms of quantum numbers. A "quantum" (plural "quanta") is the smallest possible unit of some physical property, such as energy or magnetism. In atoms, each electron is described by four quantum numbers: the principal, orbital, magnetic, and spin quantum numbers, each of which can take only certain specific values.

The principal quantum number (n) defines the energy level of the electron and its likely distance from the nucleus, effectively identifying which electron shell it inhabits. Each electron shell can be thought of as containing one or more subshells. The orbital or azimuthal quantum number (l) determines the electron's angular momentum and therefore the shape of its orbital; it is related to the probability density function, which describes the probability that an electron of a specific energy level will be found in a given location within the atom. The magnetic quantum number (m) relates to the electron's energy within a magnetic field and determines the number and orientation of the orbitals within each subshell. Depending on the orbital number, a subshell may contain one, three, five, or seven orbitals. These first three quantum numbers are all mathematically related and can only take integer values.

The fourth number is the spin quantum number (s), which describes the orientation of the electron's intrinsic angular momentum—in other words, the direction of its spin—and can take one of only two values, +1/2 (up) or −1/2 (down). Because of this, only two electrons can occupy any individual orbital at any one time, and they must spin in opposite directions. This is due to the Pauli exclusion principle, which states that two electrons in the same atom cannot have the same values for all four quantum numbers. Because two electrons in the same orbital must have the same principal, orbital, and magnetic quantum numbers, their spin quantum numbers must be different, and because there are only two possible spin quantum numbers, there cannot be more than two electrons in one orbital.

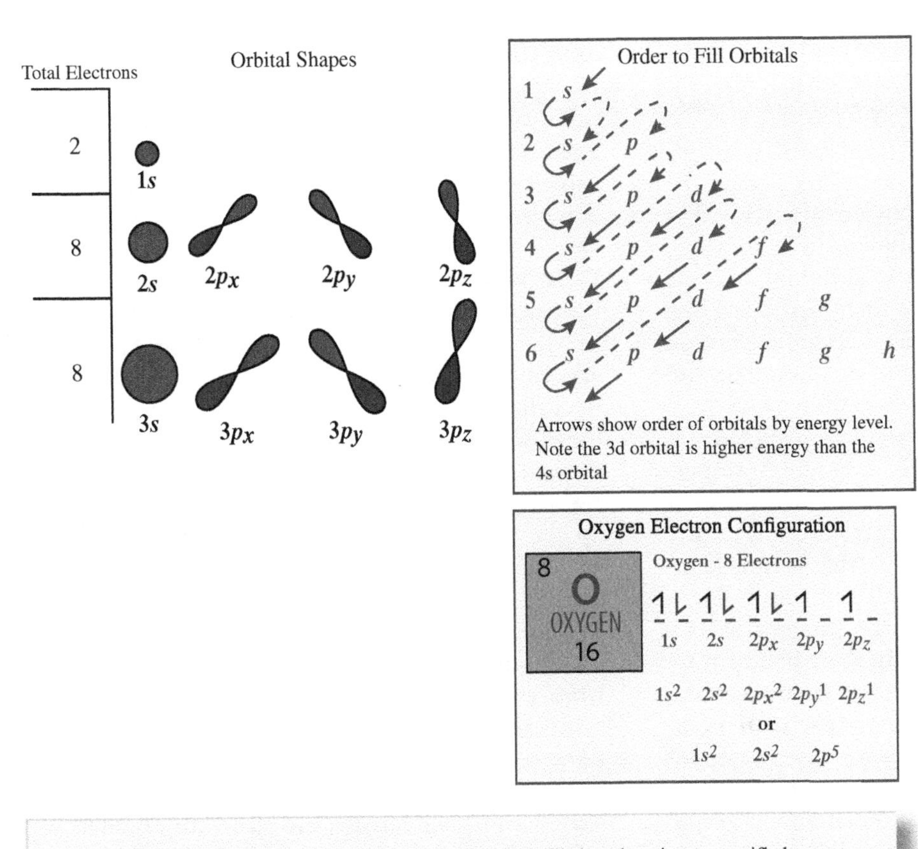

Orbitals represent the volume in which an electron is highly likely to be, given a specified energy state. The general rule is that electrons fill orbitals from top to bottom and left to right, filling all red portions before any blue portions. The first five orbital shapes are visually represented above.

ORBITAL SHAPES

Atomic orbitals are most easily visualized as three-dimensional shapes. Because

SAMPLE PROBLEM

Use the chart below, following the red arrows from top to bottom, to rank the atomic orbitals from 1*s* through 4*p* in order of increasing energy. Then write the electron configuration for the arsenic atom and identify its valence electrons.

1s
2s 2p
3s 3p 3d
4s 4p 4d 4f
5s 5p 5d 5f 5g
6s 6p 6d 6f 6g 6h
7s 7p 7d 7f 7g 7h

Answer:

The order of the atomic orbitals according to their energy levels is 1*s*, 2*s*, 2*p*, 3*s*, 3*p*, 4*s*, 3*d*, 4*p*. The atomic number of arsenic is 33; therefore, it has thirty-three electrons. These are configured as follows:

$$1s^2 2s^2 2p^6 3s^2 3p^6 4s^2 3d^{10} 4p^3$$

The valence electrons are the three electrons in the 4*p* orbitals.

the regions of space around a nucleus are defined by wave functions, there are specific locations in which an electron cannot exist. These locations are the "nodes" of the particular wave function and are best thought of as boundary surfaces, where the electron can be on one side or the other of the boundary surface but not at any point actually on the surface.

There are four known orbital shapes, labeled *s*, *p*, *d*, and *f*, corresponding to $l = 0$, 1, 2, and 3, respectively. *S* orbitals are spherically symmetrical about the nucleus, and the *s* orbital corresponding to the principal quantum number $n = 1$ (that is, the 1*s* orbital) can be thought of as a ball with the nucleus at its center. The 2*s* orbital can be thought of as a ball surrounding another ball, the 3*s* orbital can be thought of as a ball surrounding a ball surrounding a third ball, and so on. The surface of each inner ball is the boundary-surface node of the wave function. However, orbitals are merely mathematical representations of the probability of an electron with specific energy being in that particular space; electrons do not actually follow the paths of their orbitals as planets do in a solar system. The most difficult concept to grasp in visualizing the shapes of the orbitals is that all of them exist in the same space at the same time.

The *s* orbitals are the simplest to visualize, especially for higher principal quantum numbers, as each *s* subshell only contains one orbital. A *p* subshell contains three *p* orbitals, oriented at right angles to each other; they are shaped like figure eights and have the same relationship to each other as the *s* orbitals as the principal quantum number increases. There are five *d* orbitals, four of which are shaped like four-leaf clovers and oriented in the spaces between the *p* orbitals; the fifth *d* orbital has the shape of a figure eight with a "donut" around its middle and is oriented in the same direction and space as one of the *p* orbitals. There are seven *f* orbitals, again shaped like four-leaf clovers, oriented in as many different directions about the nucleus. For the sake of simplicity, although technically a *p* orbital is one of three orbitals within a *p* subshell (and a *d* orbital is one of five, and an *f* orbital is one of seven), each containing only two electrons, the term "orbital" is sometimes used to refer to the entire subshell instead, and it is said that a *p* orbital can contain up to six electrons, a *d* orbital can contain up to ten electrons, and an *f* orbital can contain up to fourteen electrons.

Electrons in Orbitals

The electrons in atoms occupy the different orbitals in pairs within each electron subshell, and each atom accordingly has an electron configuration that reflects this principle. Electrons fill orbitals in order of increasing energy, from the 1*s* orbital outward. While the order of the atomic orbitals according to their respective quantum numbers is 1*s*, 2*s*, 2*p*, 3*s*, 3*p*, 3*d*, 4*s*, 4*p*, 4*d*, 4*f*, and so on, with each electron shell after $n = 4$ containing all four orbital types, the order in which electrons fill these orbitals is slightly different, as some orbitals of different quantum numbers have similar energy requirements. The number of electrons contained in a particular subshell is given as a superscript number following the orbital designation, so that a completely full $n = 2$ electron shell, for example, would be represented as $2s^2 2p^6$, while an $n = 3$ electron shell missing two electrons would be represented as $3s^2 3p^6 3d^8$ (as the *d* subshell can hold up to ten electrons).

When all of the electrons in an atom have occupied their proper orbitals, the outermost electron shell of almost all elements contains fewer electrons than the number needed to fill its orbitals. This outermost shell is the valence shell, and the electrons in it are the valence electrons. The number of valence electrons corresponds to the atom's valence, which is essentially the number of bonds it can form with other atoms.

Hybrid Orbitals

In molecules, the atomic orbitals can combine in different ways to produce hybrid atomic and molecular orbitals. These have their own theoretical shapes, determined by the manner in which they "overlap" mathematically to form chemical bonds. The most common hybrid orbitals are sp, sp^2, and sp^3, formed by the combination of an *s* orbital with one, two, or three *p* orbitals, respectively; the superscript number indicates the proportion of each type of orbital involved. Other possible hybrid orbitals include dsp^3, d^2sp^3, and dsp^2.

Richard M. Renneboog, MSc

Bibliography

Douglas, Bodie E., Darl H. McDaniel, and John J. Alexander. *Concepts and Models of Inorganic Chemistry.* 3rd ed. New York: Wiley, 1994. Print.

Haynes, W. H., PhD, David R. Lide, PhD, and Thomas J. Bruno, PhD, eds. *CRC Handbook of Chemistry and Physics,* 96th Edition. New York: CRC/Taylor & Francis Group, 2015. Print.

Hendrickson, James B., Donald J. Cram, and George S. Hammond. *Organic Chemistry.* 3rd ed. New York: McGraw, 1970. Print.

Jones, Mark Martin. *Chemistry and Society.* 5th ed. Philadelphia: Saunders College Pub., 1987. Print.

Mackay, K. M., R. A. Mackay, and W. Henderson. *Introduction to Modern Inorganic Chemistry.* 6th ed. Cheltenham: Nelson, 2002. Print.

Miessler, Gary L., Paul J. Fischer, and Donald A. Tarr. *Inorganic Chemistry.* 5th ed. Upper Saddle River: Prentice Hall, 2013. Print.

Morrison, Robert Thornton, and Robert Neilson Boyd. *Organic Chemistry.* 7th ed. Englewood Cliffs, N.J.: Prentice Hall, 2003. Print.

Wehr, M. Russell, James A. Richards Jr., and Thomas W. Adair III. *Physics of the Atom.* 4th ed. Reading: Addison-Wesley, 1984. Print.

Winter, Mark J. *The Orbitron: A Gallery of Atomic Orbitals and Molecular Orbitals.* University of Sheffield, n.d. Web.

OSMOSIS

FIELDS OF STUDY

Biochemistry; Genetics; Molecular Biology

SUMMARY

The process of osmosis is defined, and its importance in chemistry-related fields is elaborated. Osmosis is a fundamental and necessary process for living systems, as well as the basis of a valuable technology used to produce potable water and to control water-based fluid systems.

PRINCIPAL TERMS

- **concentration gradient:** the gradual change in the concentration of solutes in a solution across a specific distance.
- **diffusion:** the process by which different particles, such as atoms and molecules, gradually become intermingled due to random motion caused by thermal energy.
- **equilibrium:** the state that exists when the forward activity of a process is exactly equal to the reverse activity of that process.
- **hypertonic:** describes a solution with a greater concentration of solutes than the solution to which it is being compared; in biology, a solution with a greater solute concentration than the cytoplasm of a cell.
- **hypotonic:** describes a solution with a lower concentration of solutes than the solution to which it is being compared; in biology, a solution with a lower solute concentration than the cytoplasm of a cell.
- **isotonic:** describes a solution with the same concentration of solutes as the solution to which it is

being compared; in biology, a solution with the same solute concentration as the cytoplasm of a cell.

- **osmotic pressure:** the pressure that would have to be applied to a solution to prevent the flow of solvent through a semipermeable membrane.
- **reverse osmosis:** the application of pressure to a solution in order to overcome the osmotic pressure of a semipermeable membrane and force water to pass through it in the direction opposite to normal osmotic flow.
- **semipermeable membrane:** a membrane that allows the passage of a material, such as water or another solvent, from one side to the other while preventing the passage of other materials, such as dissolved salts or another solute.
- **solute:** any material that is dissolved in a liquid or fluid medium, usually water.
- **solvent:** any fluid, most commonly water, that dissolves other materials.

Visualizing the Concept of Osmosis

Osmosis is the process by which molecules of a solvent pass through a semipermeable membrane that is separating two solutions with differing concentrations of solute. The solvent moves from the solution with the lower concentration (the hypotonic solution) toward the one with the higher concentration (the hypertonic solution). The process will continue until the concentrations of both solutions are equal and equilibrium is achieved.

A semipermeable membrane is necessary for osmosis to occur. Such a membrane acts as a porous barrier that will allow the passage of solvent molecules but not dissolved materials, such as various mineral salts. Water molecules, though polar in nature, are electrically neutral and very small. When salts such as sodium chloride are dissolved in water, they dissociate into ions, which are both electrically charged and significantly larger in size than the surrounding water molecules. A porous membrane that has pores big enough to allow electrically neutral water molecules to pass through, but not the larger electrically charged dissolved ions, is said to be semipermeable. Other types of dissolved materials, such as various sugars and proteins, are also too large to pass through the pores of the membrane and so are subject to the process of osmosis as well.

The presence of a semipermeable membrane can produce an osmotic system. In an osmotic system, the hypotonic solution is confined to one side of the membrane and the hypertonic solution is contained on the other side. The process of osmosis will occur spontaneously and continue until the two solutions become isotonic, meaning that both solutions have the same concentration of solutes.

Diffusion and Osmotic Pressure

Diffusion can be demonstrated simply by adding a few drops of food coloring to a container of water, being careful not to mix them together, and then letting the water stand undisturbed. At first, the food coloring will remain where it was placed, but over time it will become evenly distributed throughout the water. The water molecules are in constant motion, and as they continually bump into the molecules of food

OSMOSIS

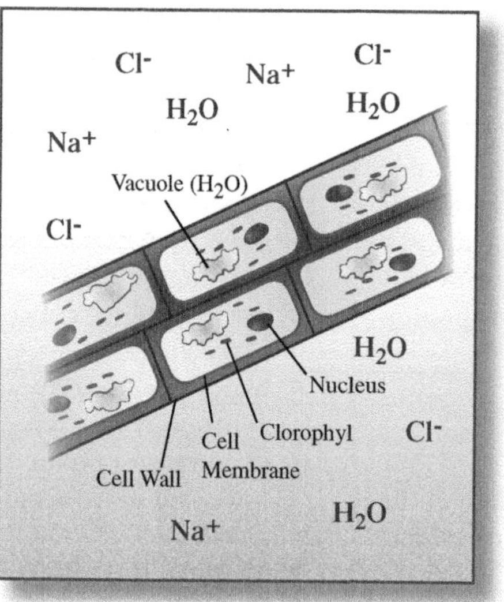

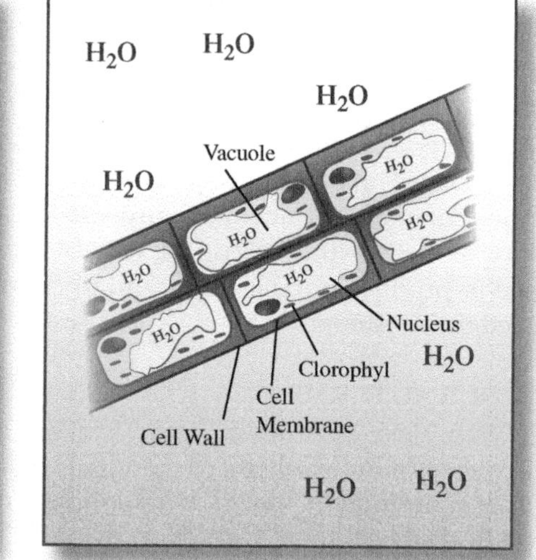

coloring, they eventually spread them throughout so that the two become mixed together. During this process, the distribution of the food coloring in the water follows a concentration gradient, which is the difference in concentration of a solute when the concentration is not constant throughout the solution. Once the food coloring is evenly distributed, the result of this mixing is the same as if the solution had been stirred or agitated, but it requires a much longer time.

Diffusion is the mechanism that drives water molecules through the pores of a semipermeable membrane. Once they are through the membrane, the water molecules interact with the dissolved salts in that solution and remain. The process is reversible, so some water molecules are driven through the membrane in the opposite direction at the same time. However, the difference in concentration ensures that the net flow of water molecules is toward the hypertonic solution until equilibrium is achieved.

Osmosis can be prevented by applying pressure to the hypertonic solution. The amount of pressure that must be applied to stop osmosis is termed the osmotic pressure of the membrane, and it is dependent on both the temperature and the difference in concentration between the two solutions. Osmotic pressure was first described by Jean-Antoine Nollet (1700–1770), also known as Abbé Nollet, in 1748 and first measured directly by Jacobus van't Hoff (1852–1911) in 1877. The osmotic pressure is given the symbol ϖ and, in the case of an ideal solution, is defined by the van't Hoff equation as:

$$\varpi = RT(C_B - C_A)$$

where T is temperature in kelvins and C is the concentration in moles per liter (mol/L), or molars (M). R is the gas constant and can be written as:

$$R = 0.0821 \frac{\text{L atm}}{\text{mol K}}$$

where L is liters, atm is atmosphere (a unit of pressure), mol is moles, and K is kelvins. An ideal solution is a solution in which the molecules of solute and solvent interact with each other in the same way they interact with themselves. If the solution is not ideal, then an osmotic coefficient, , must be included in the equation.

SAMPLE PROBLEM

Use the van't Hoff equation to determine the osmotic pressure (π) between two ideal solutions of sodium chloride, one with a concentration of 0.01 M and one with a concentration of 0.02 M, in water at a temperature of 20°C. Use $R = 0.0821 \frac{\text{L atm}}{\text{mol K}}$.

Answer:

Convert the temperature from degrees Celsius (°C) to kelvins (K):

$$K = °C + 273.15$$

$$K = 20 + 273.15$$

$$K = 293.15$$

The van't Hoff equation is

$$\pi = RT(C_B - C_A)$$

Substitute in the values of $R\ (0.0821 \frac{\text{L atm}}{\text{mol K}})$, T (temperature), C_B (concentration of solution B), and C_A (concentration of solution A) and calculate, paying attention to the units throughout:

$$\pi = RT(C_B - C_A)$$

$$\pi = (0.0821 \frac{\text{L atm}}{\text{mol K}})(293.15\ \text{K})(0.02 \frac{\text{mol}}{\text{L}} - 0.01 \frac{\text{mol}}{\text{L}})$$

$$\pi = 0.24067615\ \text{atm}$$

The osmotic pressure between 0.01 M and 0.02 M solutions of salt in water is 0.24067615 atm, or 24386.51 pascals (Pa).

By applying pressure in excess of the osmotic pressure, the process can be driven in reverse. Reverse osmosis forces water molecules to pass through a semipermeable membrane from the hypertonic solution into the hypotonic solution. It is through this process that salt-free potable water can be produced from salty seawater or other non-potable sources.

A Demonstration of Osmosis

The process of osmosis and osmotic pressure can be readily demonstrated and observed. The essential feature of the demonstration is that two solutions are separated by a semipermeable membrane and so are not able to mix. This can be done by using the membrane as a partition to separate one half of the inside of a beaker from the other half. The solution on one side of the membrane is a solution of salt in water, while on the other side is just plain water. As osmosis takes place, the level of the saltwater solution will increase as the level of the unadulterated water decreases. The rate at which the levels change depends on the area of the membrane that is exposed to both solutions.

If both solutions contain dissolved salt but in different amounts, the same effect will be observed, but it will cease when the concentrations of the two solutions become equal. As water molecules pass through the membrane, the concentration of dissolved salt in the hypertonic solution decreases and the concentration of dissolved salt in the hypotonic solution increases. At the equilibrium point, when the two solutions become isotonic, water molecules pass through the membrane in both directions at the same rate.

Osmosis in Biological Systems

Cell membranes function as semipermeable membranes in living systems, allowing water, oxygen, carbon dioxide, sugars, enzymes, ions, hormones, metabolites, and various other cellular components to pass through as necessary. In order to maintain the proper amount of water in cells and prevent dehydration, living systems use a complex mechanism of osmoregulation that actively brings water into the cells to replace water that is lost through osmosis. Anything that interferes with this mechanism, such as the consumption of alcohol, use of drugs, smoking, or lack of sufficient water in the diet, adversely affects the viability of the system. Hangovers, for example, are partly the result of dehydration of cell fluids as alcohol is metabolized and often persist until the osmotic balance of the cells is restored.

Richard M. Renneboog, MSc

Bibliography

Costanzo, Linda S. *Physiology: Cases and Problems.* 4th ed. Baltimore: Lippincott, 2012. Print.

Kucera, Jane. *Reverse Osmosis: Design, Processes, and Applications for Engineers.* Hoboken: Wiley, 2010. Print.

Lafferty, Peter, and Julian Rowe, eds. *The Hutchinson Dictionary of Science.* 2nd ed. Oxford: Helicon, 1998. Print.

Lodish, Harvey, et al. *Molecular Cell Biology.* 7th ed. New York: Freeman, 2013. Print.

Pelczar, Michael J., Jr., E.C.S. Chan, and Noel R.Krieg. *Microbiology: Concepts and Applications.* New York: McGraw, 1993. Print.

Reece, Jane B., et al. *Campbell Biology.* 10th ed. San Francisco: Cummings, 2013. Print.

P

PERICYCLIC REACTIONS

FIELDS OF STUDY

Organic Chemistry; Physical Chemistry

SUMMARY

The processes of pericyclic reactions are described, and their importance in chemical synthesis is explained. Pericyclic reactions represent a simple means of producing six-membered rings in molecular structures that are not easily produced by other means.

PRINCIPAL TERMS

- **concerted reaction:** a reaction that proceeds from starting materials to end products in a single molecular process rather than via a stepwise mechanism
- **Diels-Alder reaction:** a reaction in which an activated conjugated diene compound reacts with a suitable dienophile to form a six-membered ring structure; named for Otto Diels and Kurt Alder.
- **diene:** an organic compound that contains two carbon-carbon double bonds (C=C) in its molecular structure.
- **dienophile:** an organic compound containing a carbon-carbon double bond (C=C) that reacts preferentially with a suitable diene in a Diels-Alder reaction.

Understanding Pericyclic Reactions

In pericyclic reactions, the reacting compounds reorganize the electrons in the bonds between their atoms to form a new ring structure in the product of the reaction. Such reactions are unusual in that they involve neither any ionic or free radical mechanism nor any nucleophiles (electron donors) or electrophiles (electron acceptors) as reagents, and they are generally not affected by catalysts or different solvents. Perhaps the most startling fact about pericyclic reactions is that they break and make multiple bonds at once in what is termed a "concerted mechanism." Pericyclic reactions are not triggered by another reagent in the reaction mixture; instead, they are frequently triggered by light (in photolysis) or heat (in thermolysis).

Concerted Reactions

Pericyclic reactions are equilibrium reactions in accordance with their nature as concerted reactions. In a concerted reaction, everything can be thought of as happening at once. For example, a compound containing a conjugated diene system based on 1,3-butadiene (C=C–C=C) can undergo electrocyclic ring closure. (A conjugated diene is a diene that has alternating double and single bonds and a *p* orbital on three or more successive carbon atoms, which allows their electrons to move freely through them.) In that process, electrons in the molecular orbitals shift all at once to produce a cyclobutene structure by forming a new bond between the two carbon atoms at the ends of the system, changing the C=C–C=C bond sequence to a C–C=C–C bond sequence. These changes take place all at the same time rather than in a step-by-step manner. This corresponding simultaneous change is typical of all pericyclic reactions.

A conjugated diene system is not an absolute requirement for a concerted pericyclic reaction to occur, although the vast majority of pericyclic reactions involve such a system. What is required is that the nonbonding *p* orbitals of the reacting components be in the correct positions so that electrons can shift through them effectively unhindered to form new bonds. Because pericyclic reactions are equilibrium reactions, the reverse must also be true: that is, the electrons in bonds must be able to shift unhindered through molecular orbitals to regenerate the nonbonding *p* orbitals and bonds of the original structures.

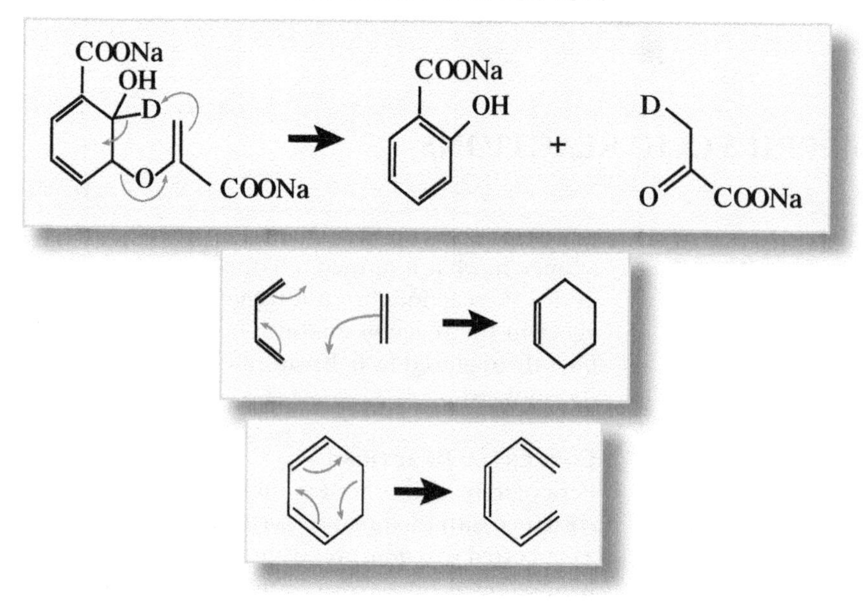

two separate molecules. For this to occur, however, the bonds that would be taking part in the reaction must be able to assume the requisite spatial relationship.

Electrocyclic ring closure is another type of pericyclic reaction. In this type of reaction, the *p* orbitals on the terminal carbon atoms of the butadiene structure behave in exactly the same way as in the Diels-Alder reaction: that is, they hybridize with the *sp2* atomic orbitals to form *sp3* orbitals that are part of a primary covalent, or sigma (σ), bond. In the electrocyclic reaction, the relative position of the two terminal carbon atoms determines whether a reaction takes place. Not surprisingly, electrocyclic ring closures can take place in more extensive conjugated systems as well. A compound in which three conjugated C=C bonds can assume the configuration of a six-membered ring can undergo electrocyclic ring closure. Electrons and orbitals shift to form a sigma bond between the two carbon atoms at either end of the conjugated triene system and a conjugated cyclohexadiene structure. In other molecules that are large enough, even three isolated C=C bonds that can assume the proper *p* orbital configuration can shift electrons and orbitals to form a six-membered ring. However, the more complex the molecular structure, the less likely it is that the appropriate configuration will be achieved.

Although pericyclic reactions do not function by an ionic mechanism, it is possible for certain ions to take part in electrocyclic reactions. The most common of these are termed "1,3-cycloadditions" and normally involve the transformation of a cyclopropyl ring structure and a carbocation in which the carbon atom carries the positive charge, called an "allyl cation." Ring strain in the cyclopropyl system imparts significant additional energy into the C–C bonds due to the difference in their actual bond angles and the ideal bond angles determined by the geometry of their atomic orbitals. There is thus a fairly strong incentive for the cyclopropyl ring to open up in an electrocyclic manner to form an allyl cation

TYPES OF PERICYCLIC REACTIONS

The best-known pericyclic reaction is the Diels-Alder reaction, named for its discoverers, German chemists Otto Diels (1876–1954) and Kurt Alder (1902–58), who reported it in the chemical literature in 1928. In a Diels-Alder reaction, a conjugated diene undergoes a concerted reaction with an activated alkene (unsaturated hydrocarbon with one or more C=C bonds) called a dienophile. The interaction forms two new bonds to connect the two carbon atoms of the dienophile to the two end carbons of the diene system. When this happens, the *p* orbitals of those four carbon atoms hybridize into the *sp2* atomic orbitals to produce *sp3* orbitals. At the same time, the electrons shift to form a pi (ϖ) bond, or secondary covalent bond, with the two remaining *p* orbitals between the two interior carbon atoms of the diene system. In the end, where there was once a conjugated diene molecule and an alkene molecule, there is a single six-membered ring molecule with a C=C bond in its structure. The most basic example of the Diels-Alder reaction is the reaction between 1,3-butadiene and ethylene to form cyclohexene, with the C=C bond between the two carbon atoms that had been the second and third carbon atoms in the 1,3-butadiene molecule. A Diels-Alder reaction can take place within a single relatively large molecular structure in the same way that it would occur between

system (C=C–C$^+$) when the structure of the molecule makes that possible.

Factors Affecting Pericyclic Reactions

Given that pericyclic reactions take place readily under the influence of either light or heat, it should not be surprising that compounds in which such reactions are possible easily become contaminated by the products of those reactions. Since all pericyclic reactions are equilibrium reactions, this sort of contamination is nearly impossible to avoid. In an equilibrium system, both the forward and the reverse reactions take place to attain a steady state in which constant quantities of both reactants and products are present. It is possible to prepare stable systems with the right components for a pericyclic reaction to occur, but the reaction is prevented from occurring by steric hindrance, the physical interference to the motion of atoms and other chemical components resulting from the physical size and restricted mobility of substituent groups, or side chains, within the molecular structure. The presence of relatively large substituent groups in the molecular structure can effectively prevent the orbitals in the conjugated system from attaining the proper trajectory for interaction to take place. The three orbitals about an *sp2* hybridized carbon atom are arranged in a plane, with the *p* orbital perpendicular. For the two terminal *p* orbitals at the ends of the system to interact in an electrocyclic reaction, they must be able to rotate. Similarly, the *p* orbitals involved in a Diels-Alder reaction must be able to overlap. In both cases, the order of substituents about the two carbon atoms becomes rigidly fixed by the ring structure. Any substituent group that either physically gets in the way of orbital alignment or acts to stabilize the *p* orbitals can effectively prevent a pericyclic reaction from taking place.

SAMPLE PROBLEM

When writing out reaction equations involving organic compounds, chemists adhere to a standardized pictorial method that uses curved arrows to indicate how electrons shift about in atomic and molecular orbitals to reorganize existing bonds and form new ones. Arrows with complete arrowheads indicate that two electrons are involved, and arrows with half arrowheads indicate the movement of single electrons. Given the basic components of a Diels-Alder reaction involving 1,3-butadiene and ethylene to form cyclohexene, draw the corresponding equation showing electron movement.

Answer:

Remember that pericyclic reactions such as the Diels-Alder reaction are equilibrium reaction systems, indicated in equations by two half arrows, one pointing forward toward products and the other back toward reactants. The equation is therefore written as

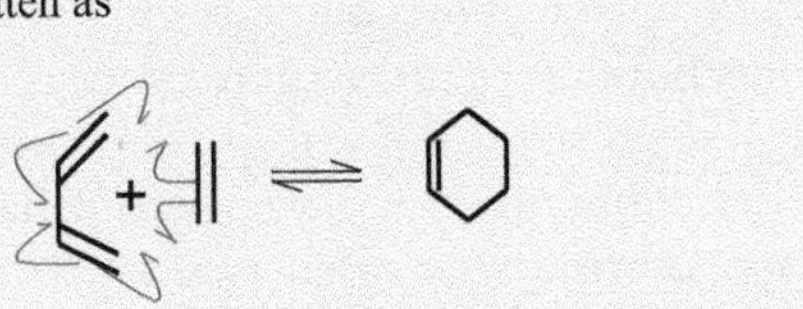

Pericyclic Reactions in Synthesis

The use of pericyclic reactions in the synthesis of specific compounds demands that close attention be paid to the manner in which orbitals must be able to shift in order to come into the proper configuration for bond formation. This will determine the stereochemical (spatial) arrangement of substituents about the carbon atoms involved in the new bonds. The compound cantharidin is a naturally occurring compound produced as a defensive material by certain beetles. The three-dimensional structure of cantharidin immediately suggests that a simple Diels-Alder reaction between the compounds furan and 2,3-dimethylmaleic anhydride should yield the cantharidin structure directly. However, when researchers attempted to carry out this synthesis, the reaction failed, largely due to steric hindrance. Steric hindrance in the structure was sufficient to render the material unstable and drive the reverse reaction to re-form the starting materials. The actual synthesis of cantharidin, first successfully carried out by chemist Gilbert Stork in 1951, thus requires a prolonged series of transformations in order to produce a compound stereochemically identical to natural cantharidin.

As is often the case, equilibrium reactions are useful in synthesis but require some special considerations in order to obtain the best yields of their desired products. This is best achieved when the desired product can isolated in some way be from the reaction mixture as it is formed. In some cases, it is possible to carry out a reaction using a solvent in which the desired product is insoluble and so can be filtered out. In other cases, the product may be sufficiently volatile that it can be isolated by distillation as the reaction proceeds. However, in many cases, the chemist has no choice but to identify the stage in the reaction when the desired product is at its highest concentration, quench the reaction mixture to stop the reaction, and then separate and identify the various components in order to isolate the desired product of the reaction.

Richard M. Renneboog, MSc

BIBLIOGRAPHY

Bachrach, Steven M. *Computational Organic Chemistry.* Hoboken: Wiley, 2007. Print.

Fleming, Ian. *Molecular Orbitals and Organic Chemical Reactions.* Hoboken: Wiley, 2010. Print.

Hendrickson, James B., Donald J. Cram, and George S. Hammond. *Organic Chemistry.* 3rd ed. New York: McGraw, 1970. Print.

Morrison, Robert Thornton, and Robert Neilson Boyd. *Organic Chemistry.* 7th ed. Englewood Cliffs, N.J.: Prentice Hall, 2003. Print.

Sankararaman, Sethuraman. *Pericyclic Reactions.* Weinheim: Wiley, 2005. Print.

Stevens, T. S., and W. E. Watts. *Selected Molecular Rearrangements.* New York: Van Nostrand, 1973. Print.

pH

FIELD OF STUDY

Organic Chemistry, Inorganic Chemistry, Biochemistry

SUMMARY

The acidity and basicity of aqueous solutions are dependent on the balance of hydronium (H_3O^+) and hydroxide (OH^-) ions in the solution. Molar concentrations of both ions are typically low, making calculations unwieldy. The pH scale, based on the logarithmic values of the concentrations, greatly facilitates such calculations.

PRINCIPAL TERMS

- **acid:** a compound that can relinquish one or more hydrogen ions (Brønsted-Lowry acid-base theory) or that possesses vacant atomic orbitals to interact with electron-rich materials (Lewis acid-base theory).
- **alkaline:** describes a material that tends to increase the concentration of hydroxide ions in an aqueous solution, as well as conditions produced by the presence of bases.
- **base:** a compound that can relinquish one or more hydroxide ions (Brønsted-Lowry acid-base theory) or that possesses lone pairs of electrons that can interact with electron-poor materials (Lewis acid-base theory).
- **concentration:** the amount of a specific component present in a given volume of a mixture.
- **logarithm:** the exponent, or power, to which a specific base number must be raised to produce a given value; commonly abbreviated "log."
- **neutral:** describes a chemical solution that has a pH of 7 and thus is neither acidic nor basic.

THE NATURE OF WATER

In order to understand the pH scale, one must first understand the nature and properties of water. Water is the most common known solvent. It is unique in terms of its molecular structure, physical and chemical properties, and interaction with other materials.

All atoms consist of a central nucleus containing positively charged protons and neutral neutrons, surrounded by negatively charged electrons. Electrons can be thought of as existing in specific regions called orbitals, which in turn are contained within layers known as electron shells. In neutral atoms and molecules, the protons and electrons are equal in number; when an atom or molecule has an unequal number of electrons and protons, it is known as an ion.

The water molecule consists of two hydrogen atoms bonded to an oxygen atom. As an element in the second period of group 16 of the periodic table of the elements, the oxygen atom has six electrons in its outermost, or valence, shell. When the oxygen atom bonds to the two hydrogen atoms, it shares one electron from each, giving the oxygen atom a filled valence shell. The one 2*s* and three 2*p* orbitals that contain those eight electrons, arranged in four pairs, hybridize to form four equivalent sp^3 orbitals oriented to the four apexes of a tetrahedron. This orbital arrangement has the effect of minimizing the force of repulsion experienced between the electron pairs. Two of the sp^3 orbitals form the bonds to the hydrogen atoms; the other two are not involved in bonds, and each contains a lone pair of electrons (a pair of electrons not shared with a hydrogen atom). This gives the water molecule a high degree of polarity, as the lone pairs create a region of high negative charge, while the placement of the hydrogen electrons within the oxygen-hydrogen bonds exposes the positive charge of the hydrogen nuclei. These areas of positive and negative charge cause water molecules to act somewhat like magnets and stick to each other, resulting in a cohesion between molecules that gives water an unexpectedly high boiling point relative to its molecular weight.

In liquid water, the concentrations of acidic hydronium ions (H_3O^+) and basic hydroxide ions (OH^-) are exactly equal; as such, liquid water is neither acidic nor alkaline but neutral. Experimental analysis has determined that the concentration of both types of ions in water is 10^{-7} moles per liter, or molars (M). (A mole is the amount of any pure substance that contains as many elementary units—approximately 6.022 × 10^{23}—as there are atoms in twelve grams of the isotope carbon-12.) The equilibrium constant for the reaction is defined as

$$K_{eq} = [H_3O^+][OH^-] = (10^{-7})(10^{-7}) = 10^{-14}$$

where [*x*] is equal to the concentration of *x*. When the concentration of one ion increases, the concentration of the other ion must decrease by a corresponding amount. This principle is the foundation of the pH scale.

pH SCALE

	pH	$[H^+]$	$[OH^-]$	Household Substance
	0	1 X 10^{0}	1 X 10^{-14}	**Battery Acid**
ACIDIC	1	1 X 10^{-1}	1 X 10^{-13}	
	2	1 X 10^{-2}	1 X 10^{-12}	**Lemon Juice Vinegar**
	3	1 X 10^{-3}	1 X 10^{-11}	
	4	1 X 10^{-4}	1 X 10^{-10}	
	5	1 X 10^{-5}	1 X 10^{-9}	**Black Coffee**
	6	1 X 10^{-6}	1 X 10^{-8}	**Saliva**
NEUTRAL	7	1 X 10^{-7}	1 X 10^{-7}	**Pure Water Sea Water**
	8	1 X 10^{-8}	1 X 10^{-6}	**Baking Soda**
	9	1 X 10^{-9}	1 X 10^{-5}	
	10	1 X 10^{-10}	1 X 10^{-4}	**Lime Water**
	11	1 X 10^{-11}	1 X 10^{-3}	
BASIC	12	1 X 10^{-12}	1 X 10^{-2}	**Ammonia**
	13	1 X 10^{-13}	1 X 10^{-1}	**Bleach**
	14	1 X 10^{-14}	1 X 10^{0}	

THE PH SCALE

The term "pH" can be defined as a numerical value that represents the acidity or basicity of a solution, with 0 being the most acidic, 14 being the most basic, and 7 being neutral. The pH of an aqueous solution is determined using this equation:

$$pH = -\log[H_3O^+]$$

The notation $[H^+]$ is often used instead of $[H_3O^+]$, though both forms are correct. The pH scale is a logarithmic scale, meaning that every increase of one unit of pH corresponds to a tenfold change in the concentration of hydronium ions. The use of logarithmic values greatly simplifies calculations of both acid and base concentrations.

The logarithm of a value, or "log" for short, is the exponent to which a base number, in this case 10, must be raised to produce that value. For example, the log of 10^3 is 3, and the log of 10^{-2} is −2. The ease of using logs is due to what are known as the power rules of mathematics. When two numbers are multiplied, their respective exponents, or powers, are added together. For example, 100 multiplied by

1,000 is 100,000. In scientific notation, this is written as follows:

$$10^2 \times 10^3 = 10^5$$

It is easy to see from this that the exponent values 2 and 3 can be added to equal 5. Similarly, to divide two values expressed with exponents, one can subtract one exponent from the other.

In the case of water, the logarithmic expression of the equation for the equilibrium constant of the process is:

$$\log(10^{-14}) = \log[(10^{-7})(10^{-7})] = (-7) + (-7)$$

SAMPLE PROBLEM

The $[H^+]$ of a solution is 1.48×10^{-3} M. What is the pH of the solution?

Answer:

The pH of a solution is defined as $-\log[H^+]$. The logarithm of 1.48×10^{-3}, or 0.00148, is the exponent that 10 must be raised to in order to equal 0.00148. Thus, the equation can be set up as follows:

$$10^{-pH} = 0.00148$$

However, the easiest way to solve this problem is to use the log function of a scientific calculator. This gives the value of $\log(1.48 \times 10^{-3})$ as −2.83 (rounded). Therefore, the pH is 2.83.

The $[OH^-]$ of a solution is 1.48×10^{-3} M. What is the pH of the solution?

Answer:

The pH and the pOH of a solution must always add up to 14. The pOH of the solution is defined as $-\log[OH^-]$. In this case, the $[OH^-]$ is 1.48×10^{-3} M, and $-\log(1.48 \times 10^{-3})$ was determined in the previous problem to equal −2.83, so the pOH is 2.83. Simply subtract this value from 14 to get the pH:

$$14 - 2.83 = 11.17$$

Therefore, the pH is 11.17.

Thus,

$$-\log(10^{-14}) = -\log(10^{-7})(10^{-7}) = -[(-7) + (-7)]$$

In modern chemical notation, the *p* of pH represents the cologarithm, or colog, of the value that follows, which is equal to the logarithm of the reciprocal of the value—or, more simply, the negative logarithm (−log). Therefore,

$$pK_{eq} = 14 = p[H_3O^+] + p[OH^-] = 7 + 7$$

The term "pOH" is sometimes used to refer to the concentration of hydroxide ions in the solution. The pH and the pOH must always add up to the neutral value of 14. According to Le Chatelier's principle, named for the French chemist Henry Louis Le Châtelier (1850–1936), adding any material that changes the concentration of either the hydronium or hydroxide ion causes the other concentration to shift to compensate and restore the equilibrium condition. Thus, adding an acidic material such as hydrogen chloride (HCl) to pure water increases the $[H^3O^+]$ and forces the $[OH^-]$ to decrease by the same amount to maintain the equilibrium. Similarly, adding a base such as sodium hydroxide (NaOH) increases the $[OH^-]$ and causes the $[H_3O^+]$ to decrease by the same amount.

The pH scale is used to indicate an aqueous solution's acidity or basicity, with acidic solutions falling between 0 and 7 on the scale and basic solutions between 7 and 14. Common acidic substances include vinegar and lemon juice, while well-known basic substances include baking soda and ammonia. Pure water is generally considered to have a neutral pH of 7, but this can change when other substances are added to the water, as in the case of the environmental phenomenon known as acid rain.

pH IN BIOLOGICAL SYSTEMS

Maintaining a constant pH is particularly important in biological systems, as changes in pH can have harmful effects on living things. Enzymes, like many other chemical systems, can only function within a specific pH range, outside of which they become degraded and lose their enzyme functionality. Very minor changes in blood pH affect the solubility of many of the compounds that are transported in the blood, as well as the function of receptor sites on any cells that come in contact with it. Materials such as uric acid that are normally dissolved in blood may instead precipitate as solids in the blood vessels and sinovial fluids of the joints, causing a variety of often painful and even life-threatening conditions. Biological systems prevent harmful changes in pH through the use of buffer solutions, which are mixtures of acids and bases that have the ability to absorb limited quantities of additional acids or bases without changing their pH.

Richard M. Renneboog, MSc

BIBLIOGRAPHY

Berg, Jeremy M., John L. Tymoczko, Gregory J. Gatto, and Lubert Strye. *Biochemistry.* 8th ed. New York: W. H. Freeman, 2015. Print.

Jones, Mark Martin. *Chemistry and Society.* 5th ed. Philadelphia: Saunders College Pub., 1987. Print.

Laidler, Keith J. *Chemical Kinetics.* 3rd ed. New York: Harper, 1987. Print.

Lehninger, Albert L. *Biochemistry: The Molecular Basis of Cell Structure and Function.* 2nd ed. New York: Worth, 1975. Print.

Lodish, Harvey, et al. *Molecular Cell Biology.* 7th ed. New York: Freeman, 2013. Print.

Mackay, K. M., R. A. Mackay, and W. Henderson. *Introduction to Modern Inorganic Chemistry.* 6th ed. Cheltenham: Nelson, 2002. Print.

Myers, Richard. *The Basics of Chemistry.* Westport: Greenwood, 2003. Print.

Rang, H. P., et al. *Rang and Dale's Pharmacology.* 7th ed. New York: Elsevier, 2012. Print.

PHENOLS

FIELDS OF STUDY

Organic Chemistry

SUMMARY

The characteristic properties and reactions of phenols are discussed. Phenols are limited in type because of their molecular structure, which also endows them with chemical reactivity. Phenols and phenolic compounds form an important class of naturally occurring compounds as well as industrially important materials.

PRINCIPAL TERMS

- **aromaticity:** a characteristic of certain ring-shaped molecules in which alternating double and single bonds are distributed in such a way that all bonds are of equal length and strength, giving the molecule greater stability than would otherwise be expected.
- **functional group:** a specific group of atoms with a characteristic structure and corresponding chemical behavior within a molecule.
- **hydroxyl group:** a primary functional group consisting of an oxygen atom covalently bonded to a single hydrogen atom.
- **phenyl group:** a cyclic functional group with the formula $-C_6H_5$, similar to a benzene molecule but with one fewer hydrogen atom.
- **unsaturated:** describes an organic compound in which carbon atoms are attached to other atoms via double or triple bonds, preventing the compound from containing the maximum possible number of hydrogen atoms.

THE NATURE OF PHENOLS

Phenol, or hydroxybenzene, is the simplest of the class of compounds known as phenols and related phenolic compounds. Structurally, phenol consists of a ring-shaped benzene molecule in which one of the hydrogen atoms has been replaced by a hydroxyl group (–OH). The phenols are often referred to as aryl alcohols or aromatic alcohols. They are characterized by the presence of a hydroxyl group bonded to an aryl group (a broad class of ring-shaped functional groups that includes phenyl groups), and although the number of atoms that makes up the benzene ring is small, the class of phenols and phenolics is quite large.

The aryl group significantly affects the reactivity and chemical behavior of the hydroxyl group due to the aromaticity of its unsaturated bond system. Some of the electron orbitals of the hydroxyl group's oxygen atom overlap with the orbitals of the carbon atoms in the aryl ring, which has the effect of reducing the energy required to separate the oxygen-hydrogen bond. This renders phenol and similar compounds sufficiently acidic that they can be deprotonated, or lose one or more hydrogen cations (H^+), by reaction with a base such as sodium hydroxide. Alkyl alcohols, in comparison, typically require a much stronger base, such as *n*-butyllithium or sodium hydride, to deprotonate the alcohol hydroxyl functional group.

The hydroxyl group of a phenol allows the oxygen atom to react readily by radical mechanisms as well. In a radical reaction mechanism, the oxygen-hydrogen bond cleaves in such a way that the electrons in the bond are distributed equally, creating two radicals, which are atoms or molecules with unpaired valence electrons. Radicals can undergo a number of different reactions that are not available to ionic species. Two radicals can join together simply by forming a normal covalent bond with their unpaired electrons. The most exotic reaction available to a radical is insertion, in which a radical inserts itself into a bond between two atoms in another molecule. Due to the low energy required to break the aryl oxygen-hydrogen bond, phenols and phenolic compounds produce, or are involved in the production of, a large number of compounds found in nature, including the polymeric material known as lignin. Along with cellulose, lignin is a main structural component of plants. Industrially, lignin decomposition is a source of a number of important chemical products.

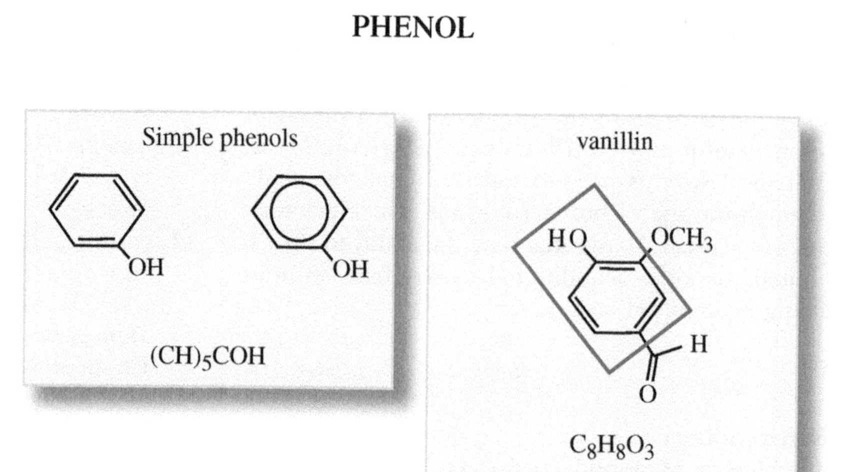

Nomenclature of Phenols

The systematic nomenclature of phenols, as devised by the International Union of Pure and Applied Chemistry (IUPAC), is straightforward. Since the structure of the phenol is based on the six-membered benzene ring, with the carbon atom carrying the hydroxyl group designated as the first atom in the ring, naming any simple phenolic compound is a matter of identifying the various other substituent groups and their positions relative to the hydroxyl group. The parent compound is typically identified in the name as phenol rather than benzene. For example, a benzene ring with a hydroxyl functional group on one carbon atom and an isopropyl alkyl substituent on a carbon atom two positions removed can be named 1-hydroxy-3-isopropylbenzene (or 3-isopropylhydroxybenzene), but it can also be named 3-isopropylphenol. Another common naming convention uses the *ortho-*, *meta-*, and *para-* designations to identify the position of a single substituent relative to the primary ring position. Under this naming convention, 3-isopropylphenol can be named *m*-isopropylphenol, indicating that the isopropyl substituent is in the meta position of the six-membered ring, two carbon atoms away from the hydroxyl functional group. (The ortho position is next to the primary carbon, while the para position is directly opposite it.)

Nomenclature becomes more complicated when there is more than one hydroxyl functional group on the molecule. Because of the six-membered structure of the benzene ring, there can be as many as six hydroxyl functional groups in one molecule. The IUPAC system for naming polyhydroxylated phenols is the simplest method of identifying such a molecule, since each of the basic structures has its own common name. In the IUPAC convention, the substituents are identified first, and then the hydroxyl groups are assigned their corresponding positions such that the substituents occupy the lowest-numbered positions, with one of the hydroxyl groups assigned the position of priority. The full name then consists simply of the substituent groups in alphabetical order with their corresponding position number on the ring. For example, a compound with the structure:

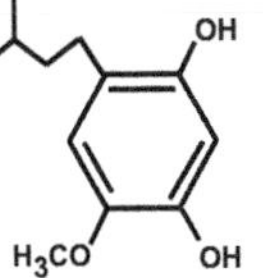

would have the systematic name 1,5-dihydroxy-4-isopentyl-2-methoxybenzene, starting with the hydroxyl group at the bottom of the diagram and numbering in a clockwise direction. With such compounds, the simplest way to decide which carbon atom is in the primary position and in which direction the numbering should proceed is to determine which potential combination of position numbers adds up to the lowest sum. In this example, the position numbers used are 1 (hydroxyl group), 2 (methoxy group, H_3CO–), 4 (isopentyl group, $(CH_3)_2CH–CH_2–CH_2$–), and 5 (second hydroxyl group), which add up to 12. Numbering in the other direction gives the position numbers 1, 3, 4 and 6, which add up to 14. Since 12 is less than 14, that name is the correct one according to the systematic convention. (While the sum would be the same if the carbon with the other hydroxyl group were numbered 1 and the numbering proceeded in a counterclockwise direction, the IUPAC system ranks alkoxy groups, such as methoxy, as higher in priority than alkyl groups, such as isopentyl, meaning that the alkoxy should have the lower number. The priority of various functional groups is a more advanced topic in the nomenclature of organic chemistry.)

It is also possible to name polyphenolic compounds such as this using the common name for the parent structure. In such cases, the parent structure used typically is a matter of choice. All of these names identify the compound unequivocally and are therefore correct in their own right, but only one name is the proper systematic IUPAC name.

REACTIONS OF PHENOLS

Phenols are reactive compounds that undergo all of the reactions typical of alcohols. One important aspect of the reactivity of phenols is the enhanced acidity of the hydroxyl group. This makes salt formation, in the form of the corresponding phenoxide ion, a much easier process. A simple reaction with hydroxide, for example, yields the phenoxide ion salt, whereas much stronger bases are typically required to produce the corresponding salts in other substances. The phenoxide ion reacts with alkyl halides to form ethers in a reaction known as the Williamson ether synthesis. Several methods have been developed to use the ethers formed by the phenoxide ion as protecting groups, which are used to prevent certain portions of molecules from being affected when the surrounding atoms are taking part in a reaction.

As aryl compounds, phenols are also able to undergo all of the reactions typical of those compounds. Friedel-Crafts alkylations and acylations, diazonium salt reactions, and other electrophilic substitution reactions generally proceed faster and with higher yields than when benzene reacts under the same conditions. The hydroxyl group is termed an activating ortho-para directing group because the products of electrophilic substitution reactions result from

SAMPLE PROBLEM

Given the chemical names 4,6-dibromo-2-methylphenol and *p*-methoxythiophenol, draw the skeletal structure of the molecules.

Answer:

The name of the first compound identifies that material as a phenol molecule with the hydroxyl function on the first carbon atom in the benzene ring, bromine atoms on the fourth and sixth carbon atoms, and a methyl group on the second carbon atom. The structure of the molecule is therefore

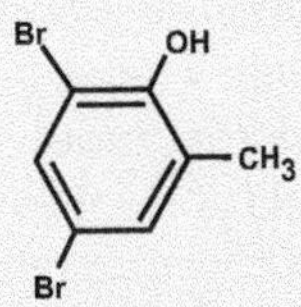

The name of the second compound identifies it as the sulfur analog of phenol ("thio-" means a sulfur atom is present), with a sulfhydryl group (–SH) in place of the hydroxyl group. The methoxy group (H_3CO–) is attached to the carbon atom directly across the ring from the sulfhydryl group, in the para position, as indicated by the prefix *p*-. The structure of the molecule is therefore

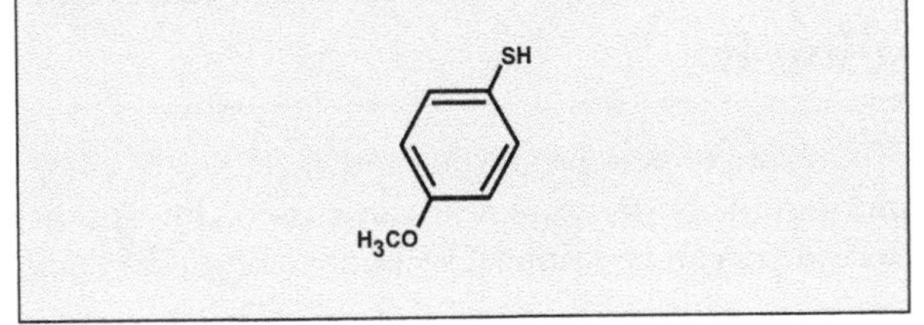

substitution in those positions relative to the hydroxyl group. For example, the reaction of phenol with nitric acid produces predominantly *p*-nitrophenol and *o*-nitrophenol.

PRODUCTION OF PHENOLS

Like many chemical compounds, phenols are produced using a wide variety of methods. Phenol itself is a component of coal tar, which is obtained by the destructive distillation of bituminous coal. The process decomposes the molecular structure of the compounds within the coal into many different smaller molecules. In addition to phenols, the distillation of bituminous coal can produce a number of other aryl and polyaryl compounds, such as naphthalene.

Other phenols are synthesized by various reactions typical of benzene and other aromatic compounds. The majority of phenol produced industrially is made using cumene (isopropylbenzene or 2-phenylpropane, $C_6H_5CH(CH_3)_2$) as the starting material. Reaction of cumene with molecular oxygen (O_2) produces cumene hydroperoxide, the chemical bonds of which are easily cleaved by aqueous acid in a process known as hydrolysis to yield phenol and acetone. Another process, named the Dow process after Canadian-born chemist and industrialist Herbert Dow (1866–1930), reacts chlorobenzene with aqueous sodium hydroxide at a high temperature. Phenol is formed in the reaction as hydroxide displaces chloride by nucleophilic substitution, a process in which one nucleophile, or chemical that tends to react with positively charged or electron-poor species, is substituted for another. A third important industrial method for the production of phenol involves the fusion of sodium benzene sulfonate with an alkali, a high-temperature process in which hydroxide from the molten alkali salt displaces the sulfate ion to yield phenol and sodium sulfate. In laboratory-scale preparations, some phenol compounds can be produced by the hydrolysis of diazonium salts.

Richard M. Renneboog, MSc

BIBLIOGRAPHY

Berg, Jeremy M., John L. Tymoczko, Gregory J. Gatto, and Lubert Strye. *Biochemistry.* 8th ed. New York: W. H. Freeman, 2015. Print.

Fenichell, Stephen. *Plastic: The Making of a Synthetic Century.* New York: Harper, 2009. Print.

Hendrickson, James B., Donald J. Cram, and George S. Hammond. *Organic Chemistry.* 3rd ed. New York: McGraw, 1970. Print.

Herbert, R. B. *The Biosynthesis of Secondary Metabolites.* 2nd ed. London: Chapman & Hall, 1989. Print.

Jones, Mark Martin. *Chemistry and Society.* 5th ed. Philadelphia: Saunders College Pub., 1987. Print.

Mann, J., et al. *Natural Products: Their Chemistry and Biological Significance.* New York: Wiley, 1994. Print.

Morrison, Robert Thornton, and Robert Neilson Boyd. *Organic Chemistry.* 7th ed. Englewood Cliffs, N.J.: Prentice Hall, 2003. Print.

O'Neil, Maryadele J., ed. *The Merck Index: An Encyclopedia of Chemicals, Drugs, and Biologicals.* 15th ed. Cambridge: RSC, 2013. Print.

Robinson, Trevor. *The Organic Constituents of Higher Plants.* 6th ed. North Amherst: Cordus, 1991. Print.

Wuts, Peter G. M., and Theodora W. Greene. *Greene's Protective Groups in Organic Synthesis.* 4th ed. Hoboken: Wiley-Interscience, 2007. Print.

PHOSPHINE

FIELDS OF STUDY

Organic Chemistry, Inorganic Chemistry

SUMMARY

The characteristic properties and reactions of phosphine and its derivatives are discussed. Phosphine derivatives can be primary, secondary, or tertiary and are useful in organic synthesis reactions for the formation of complex molecular structures.

PRINCIPAL TERMS

- **functional group:** a specific group of atoms with a characteristic structure and corresponding chemical behavior within a molecule.

- **organophosphorus compound:** an organic compound containing one or more carbon-phosphorus bonds.
- **phosphane:** a class of compounds with the general formula P_nH_{n+2}; also, an alternative name for the compound phosphine (PH_3).
- **primary:** describes an organic compound in which one of the hydrogen atoms bonded to a central atom is replaced by another atom or group of atoms, called a substituent.
- **secondary:** describes an organic compound in which two of the hydrogen atoms bonded to a central atom are replaced by other atoms or groups of atoms, called substituents.
- **tertiary:** describes an organic compound in which three of the hydrogen atoms bonded to a central atom are replaced by other atoms or groups of atoms, called substituents.

PHOSPHINE

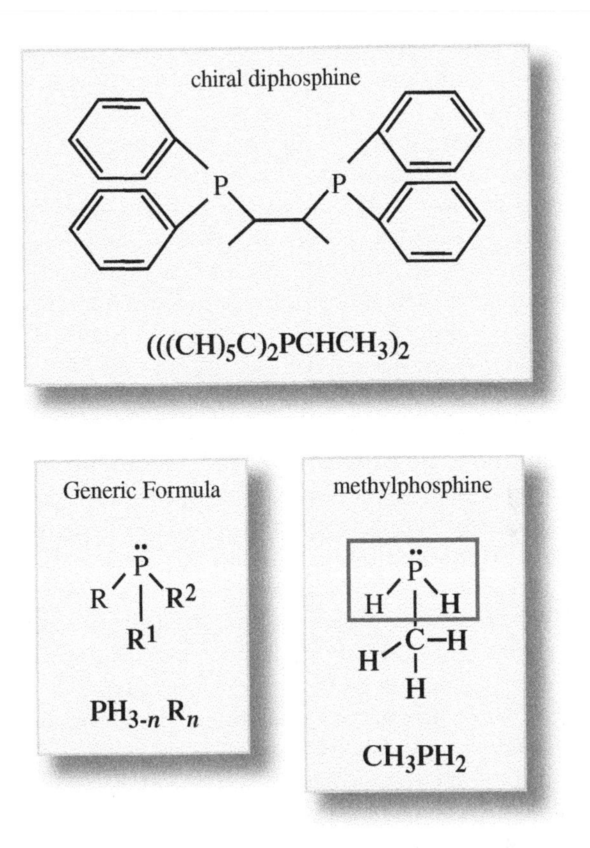

Characteristics and Derivatives of Phosphine

Phosphine is an inorganic compound with the molecular formula PH_3. It belongs to a class of molecules called phosphanes, which, according to the International Union of Pure and Applied Chemistry (IUPAC), are compounds of trivalent phosphorus (that is, phosphorus that can form three single bonds) that have the general formula P_nH_{n+2}. However, "phosphines" is also the name of a class of organophosphorus compounds derived by replacing one or more of phosphine's three hydrogen atoms with a hydrocarbon group. Adding to the confusion is the fact that the official IUPAC name of the parent compound, phosphine, is actually phosphane, as this is more analogous to the IUPAC method for naming alkanes.

The phosphorus atom has five electrons in its valence electron shell, meaning that it can accept three electrons in order to achieve a full complement of eight valence electrons. Because it falls in period 3 of the periodic table of the elements, it has two electrons in its 3*s* orbital and three in its 3*p* orbitals; in order to minimize electron-electron repulsion between the orbitals, the *s* and *p* orbitals can combine, or hybridize, to allow the phosphorus atom to form double or triple carbon-phosphorus bonds. Organophosphorus compounds containing a double carbon-phosphorus bond (C=P) are termed "phosphaalkenes," and those containing a triple carbon-phosphorus bond (C≡P) are termed "phosphaalkynes."

Phosphines can be divided into three categories. The primary phosphines are those in which just one hydrogen atom has been substituted; they are represented generally as $R–PH_2$, where R is either an alkyl functional group (basic hydrocarbon) or an aryl group (ring-shaped hydrocarbon). In secondary phosphines, two hydrogen atoms have been substituted, as represented by the general formula $R_2–PH$. Secondary phosphines also form when the phosphorus atom is substituted for a carbon atom in a ring-shaped compound. The tertiary phosphines have had all three hydrogen atoms substituted, giving the general molecular formula $R_3–P$.

Nomenclature of Phosphines

Because they are not common compounds, the phosphines are named systematically as phosphine derivatives, although the official parent compound name "phosphane" is used instead. The alkyl and aryl groups are named as substituents, or side chains, according to the identity of the parent hydrocarbons from which they were derived, and the appropriate prefixes for those groups are then appended to the "-phosphane" suffix. Thus, a primary phosphine that has been substituted with a methyl group ($-CH_3$) is called methylphosphane, one substituted with an ethyl group ($-C_2H_5$) is ethylphosphane, one with a propyl group ($-C_3H_7$) is propylphosphane, and so on, progressing through the prefixes based on the number of carbon atoms in the substituent: "butyl-" (four), "pentyl-" (five), "hexyl-" (six), et cetera. A secondary or tertiary phosphine substituted with two or three identical functional groups adds a numerical prefix to the substituent prefix, so that a secondary phosphine substituted with two methyl groups is dimethylphosphane, and a tertiary phosphine substituted with three methyl groups is trimethylphosphane. If the substituents are not the all same, the various prefixes are simply listed in alphabetical order. For example, a tertiary phosphine with one benzyl substituent ($-CH_2C_6H_5$) and two phenyl substituents ($-C_6H_5$) is named benzyldiphenylphosphane.

If the hydrocarbons substituted for the hydrogen atoms contain their own substituent groups, determining the name of the phosphine can be more difficult. The first step is to establish the name of each substituent carried by the phosphine. A hydrocarbon structure is named for the number of carbon atoms that make up the longest unbranched chain in the molecular structure, as in the case of the prefixes given above, with the positions of substituent groups identified accordingly. After the base chain has been determined, the carbon atoms are numbered in order so that each substituent has the lowest position number possible in the chain, and these position numbers are given immediately before the name of the corresponding substituent. For example, a seven-carbon chain with two amino groups bonded to the second and third carbon atoms would be named 2,3-diaminoheptane if it were a standalone molecule; in this case, as it is a substituent, it takes the prefix form "2,3-diaminoheptyl-." This entire name is then enclosed in parentheses and prefixed by a position number indicating where on the substituent the bond to the phosphorus atom occurs. If the bond is located on the third carbon atom of the 2,3-diaminoheptyl substituent (the same atom as one of the amino groups), the resulting phosphine is named 3-(2,3-diaminoheptyl)phosphane.

The structures of organic compounds are typically represented by line drawings that depict the angles between the carbon atoms in such a way that every angle vertex and line end represents a single carbon atom. Each carbon atom is assumed to be surrounded by the maximum possible number of hydrogen atoms unless otherwise specified, and all other atoms are represented by their elemental symbol. Such line drawings allow chemists to communicate a large amount of chemical information in a small space. For example, both molecular structures below depict 3-(2,3-diaminoheptyl)phosphane, but the image on the right is much more efficient—and, to a trained eye, readily comprehensible—than the one on the left.

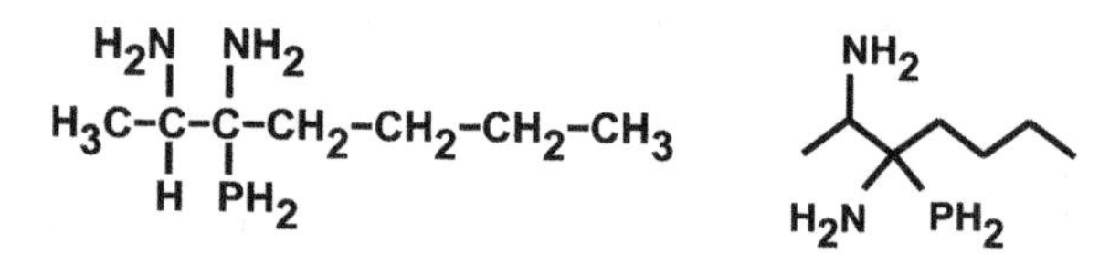

Formation of Phosphine and Its Derivatives

Phosphine is a relatively stable, colorless gas with a freezing temperature of −133.5 degrees Celsius (−208.3 degrees Fahrenheit) and a boiling point of −87.7 degrees Celsius (−125.9 degrees Fahrenheit). Depending on its purity, it may have an acrid or garlic-like odor. It can be prepared via the hydrolysis (breaking of chemical bonds due to the presence of water) of a metal phosphide, such as calcium phosphide (CaP). Hydrolysis of calcium phosphide also produces minor quantities of both diphosphane (P_2H_4) and triphosphane ($H_2P-PH-PH_2$). Another way to produce phosphine is to react elemental phosphorus with activated hydrogen or potassium hydroxide.

Alkyl phosphines can be formed by the reaction of the compound phosphine with an alkyl halide, which is an alkane-derived compound that contains a halogen. A modified Grignard reaction can be used to synthesize aryl phosphines such as triphenylphosphane ($P(C_6H_5)_3$), a tertiary phosphine substituted with three ring-shaped phenyl groups. A Grignard reaction is one in which a Grignard reagent, formed by the combination of an alkyl or aryl halide with magnesium metal, attacks the carbon-oxygen double bond (C=O) of a carbonyl group, turning it into a single bond by adding another atom or group to the carbon atom. One such reagent is bromo(phenyl)magnesium, also called phenylmagnesium bromide (C_6H_5MgBr), which can react with phosphorus trichloride (PCl_3) to produce triphenylphosphane via a similar mechanism, with the target being the phosphorus atom rather than the carbon atom of a carbonyl group.

Reactions of Phosphines

Phosphines are reactive compounds used for the formation of ylides and phosphonium salts. An ylide is a neutral dipolar molecule consisting of a negatively charged ion, or anion—usually a carbanion, which is a polyatomic anion containing a carbon atom with an unshared pair of electrons—bonded to an atom that is neither carbon nor hydrogen. A phosphonium salt is a type of ionic compound in which the positively charged ion, or cation, is phosphonium (PR_4^+, where R is either a hydrogen atom or a substituted hydrocarbon).

Phosphonium salt is typically produced by the reaction of a tertiary phosphine, such as the triphenylphosphane produced by a modified Grignard reaction, with an alkyl halide. It can then be combined with a strong base in an appropriate solvent to produce a phosphonium ylide, which features a carbon-phosphorus double bond (C=P). For example, the reaction of triphenylphosphane with bromomethane (CH_3Br), an alkyl halide, produces the salt methyltriphenylphosphonium bromide, which consists of a bromine anion (Br^-) and the methyltriphenylphosphonium cation ($(C_6H_5)_3P\text{–}CH_3^+$). When this salt is dissociated in solution and a strong base is added, a hydrogen cation (H^+) is removed from the methyl group of the methyltriphenylphosphonium, turning it into a methylene group ($=CH_2$) and forming the neutral compound methylenetriphenylphosphorane ($(C_6H_5)_3P=CH_2$), an ylide. The double bond between the carbon and the phosphorus is indicated by the "-ene" ending of the "methylene-" prefix.

The ability of the phosphorus atom to form these two extra bonds, in addition to the original three of the phosphine molecule, is due to its position in the periodic table. The fact that phosphorus has five valence electrons, leaving it three short of a full complement of eight, suggests that it is trivalent—that is, able to form a maximum of three chemical bonds—and indeed this is often the case. However, because phosphorus is in period 3 of the periodic table, it has accessible *d* orbitals, permitting it to undergo chemical processes that are not available to its period 2 counterpart, the nitrogen atom. (Period 3 elements have their valence electrons in the n = 3 electron shell, which is the first one to contain d orbitals.) One such process is the use of the 3*d* orbitals to form additional bonds, as the phosphorus atom does when forming an ylide, becoming pentavalent (able to form five chemical bonds).

Phosphonium ylides serve as reagents in a type of reaction known as the Wittig reaction. In this reaction, the phosphonium ylide reacts with an aldehyde or ketone, which are organic compounds that contain a carbonyl group, to form an alkene (a compound containing a carbon-carbon double bond) and a phosphine oxide (a phosphine that is double bonded to an oxygen atom via the phosphorus atom).

Reduction of Ylides

Just as phosphines can be used to produce ylides, the reverse is also true. Reaction of the

SAMPLE PROBLEM

Given the chemical name 4-(N-methyl-N-phenylaminocyclohexyl)phosphane, draw the skeletal structure of the molecule.

Answer:

The name indicates that the material is a primary phosphine with a cyclohexane substituent. The cyclohexane ring contains an amino group ($-NH_2$) whose hydrogen atoms have been substituted with a methyl group and a phenyl group; the "N-" prefix represents the chemical symbol for nitrogen, indicating that the specified groups are bonded to the nitrogen atom. Because the location of the amino group is not given, the carbon atom containing it is designated position number 1. The "4-" immediately precedes the substituent in parentheses, indicating that the phosphorus atom is bonded to the cyclohexane ring at its fourth carbon atom—that is, three carbon atoms away from the amino group. Therefore, the compound has the following structure:

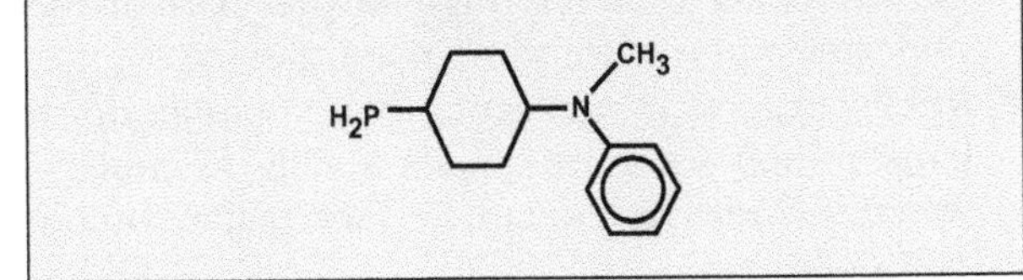

carbon-phosphorus double bond with one molar equivalent of molecular hydrogen (H_2) reduces the ylide to the corresponding phosphine. However, when this type of reduction is too powerful and may affect other functional groups or portions of the ylide molecule, a variety of other reducing agents are available. Reagents such as lithium aluminum hydride and sodium borohydride are more selective about the site of reduction and operate under gentler conditions. Typically, the formation of an ylide and its subsequent reduction to the corresponding phosphine is used in synthetic procedures to increase the selectivity of a reaction that would otherwise produce an unmanageable mixture of products. Often the ylide will undergo a reaction in such a mixture as well, but the ylide product can be recovered cleanly, then reduced to the phosphine and cleaved from the rest of the product.

Richard M. Renneboog, MSc

BIBLIOGRAPHY

Haynes, W. H., PhD, David R. Lide, PhD, and Thomas J. Bruno, PhD, eds. *CRC Handbook of Chemistry and Physics,* 96th Edition. New York: CRC/Taylor & Francis Group, 2015. Print.

Hendrickson, James B., Donald J. Cram, and George S. Hammond. Organic Chemistry. 3rd ed. New York: McGraw, 1970. Print.

Jones, Mark Martin. *Chemistry and Society.* 5th ed. Philadelphia: Saunders College Pub., 1987. Print.

Miessler, Gary L., Paul J. Fischer, and Donald A. Tarr. *Inorganic Chemistry.* 5th ed. Upper Saddle River: Prentice Hall, 2013. Print.

Morrison, Robert Thornton, and Robert Neilson Boyd. *Organic Chemistry.* 7th ed. Englewood Cliffs, N.J.: Prentice Hall, 2003. Print.

Myers, Richard. The Basics of Chemistry. Westport: Greenwood, 2003. Print.

Powell, P., and P. L. Timms. The Chemistry of the Non-Metals. London: Chapman, 1974. Print.

Wuts, Peter G. M., and Theodora W. Greene. *Greene's Protective Groups in Organic Synthesis.* 4th ed. Hoboken: Wiley-Interscience, 2007. Print.

PHYSISORPTION

FIELDS OF STUDY

Physical Chemistry; Chemical Engineering

SUMMARY

Materials in the gas and liquid states are able to adhere to solid surfaces without forming chemical bonds. This physical adsorption is due primarily to van der Waals force and electrostatic interaction. The property is used extensively in chemistry for the separation and purification of compounds.

PRINCIPAL TERMS

- **adhesion:** the tendency for atoms and molecules of a substance to maintain contact with a surface due to cohesion and their electronic structure.
- **aromatic hydrocarbon:** a hydrocarbon in which the carbon atoms form a ring with alternating double and single bonds, distributed in such a way that all bonds are of equal length and strength; also called an arene.
- **aromaticity:** a characteristic of certain ring-shaped molecules in which alternating double and single bonds are distributed in such a way that all bonds are of equal length and strength, giving the molecule greater stability than would otherwise be expected.
- **bonding:** the formation of a link between two atoms as a result of the interaction of their valence electrons.
- **chromophore:** the part of a molecule that absorbs specific wavelengths of electromagnetic radiation, often causing it to appear as a certain color.
- **dipole:** the separation of positive and negative charges within a single molecule due to electron density being relatively high in one part of the molecule and relatively low in another.
- **lipid:** a type of biomolecule that is soluble in organic nonpolar solvents and generally insoluble in water; includes fats, waxes, and the major components of organic oils.

- **lone pair:** two valence electrons that share an orbital and are not involved in the formation of a chemical bond; also called a nonbonding pair.
- **polar:** describes a molecule or functional group in which there is a difference in the distribution of electronic charge, causing one part of the molecule or group to be relatively electrically positive and another part to be relatively electrically negative.
- **van der Waals force:** weak forces of attraction between atoms due to proximity that do not affect the electronic structure of those atoms or result in chemical bond formation.

VAN DER WAALS ATTRACTION

Unlike absorption, physisorption does not depend on surface tension for its function. Physisorption depends instead primarily on the weak force of attraction between atoms that is called the van der Waals force. The van der Waals force can be visualized as an electrostatic interaction between induced dipole moments of the atoms electrostatic attraction between atoms. As such, it is easy to understand that it is a very weak force of attraction, though remarkably effective in many cases. A good example of physisorption in everyday life can be found in the fingerprints that are left on glass. The oils and other lipids that exude from skin adhere to the glass surface so well that the use of a surfactant or detergent is normally required to remove them, yet they have not formed any kind of chemical bond to the glass. They bind to the glass solely by the power of the van der Waals force of attraction between the hydrogen atoms of the lipid molecules and the silicon and oxygen atoms of the glass. The attractive force can be modeled by different mathematical methods. The image charge method describes the forces as though each atom interacts with its mirror image within the surface. Alternatively, the attractive force can be modeled using the quantum mechanical properties and wave functions of an oscillator.

Leaftail gecko climbing glass using van der Waal's force. Lpm via Wikimedia Commons

ELECTRONIC INTERACTIONS

While the van der Waals force of attraction is the primary force maintaining adhesion of atoms and molecules to a surface, there are several other factors that influence physisorption, or physical adsorption. For polar molecules, or molecules having a structure that generates an electronic dipole, non-bonding electronic interactions become more important than the van der Waals force of attraction in physisorption. Atoms such as nitrogen and oxygen contain lone pairs of electrons and so have regions of relatively high electron density or exposed negative charge. Accordingly, they adsorb well to materials that having relatively low electron densities. High electron density is also a feature of the chromophore of various large molecules, a portion of the molecular structure that is composed of segments like aromatic hydrocarbons such as the benzene and cyclopentadiene rings. The aromaticity of such structures is a feature of the distribution of electrons through the secondary ϖ bond system rather than the primary σ bond system

SAMPLE PROBLEM

In thin layer chromatography, the R_f value of a compound is the ratio of the distance it has moved across the plate relative to the distance the solvent has moved across the plate. Calculate the R_f value for three spots that have moved 0.5 cm, 0.9 cm and 2.9 cm relative to the solvent front that has moved a distance of 5.5 cm

Answer:

The R_f value is just the proportion of each distance d_c relative to the distance of the solvent front d_s, or

$$R_f = d_c/d_s$$

Therefore,

$$d_1 = (0.5/5.5) = 0.09$$

$$d2 = (0.9/5.5) = 0.16$$

$$d3 = (2.9/5/5) = 0.53$$

of the molecular structure. The contribution to physisorption by electrostatic or dipole attraction is most significant for such systems, and the quantum mechanical oscillator model more appropriately describes the force of attraction mathematically as the energy levels of the ϖ electrons more closely match those of the atoms and molecules of the surface.

PHYSISORPTION AS A PRACTICAL TECHNIQUE

The practice of chemistry utilizes physisorption for many purposes. One use for physisorption in practice is to remove colored contaminants from a solution so that crystals recovered from the solution will be as pure as possible. This is typically carried out by simply adding an appropriate amount of activated charcoal powder to the solution, and then removing it with the adsorbed impurities by filtration. The most important use, however, is chromatography. This is a technique used to separate the different components of a mixture or a solution by the different extent to which they adhere to a surface under the influence of a flowing medium. In chromatography, each material becomes partitioned between adsorbing on the surface of the stationary medium and being dissolved in the flowing medium. The more strongly a material is adsorbed on the stationary medium, the more time is required for it to pass through the chromatography system in the flowing medium. Thus, different materials pass through the same system from the same starting point in a manner that allows them to separate from each other.

OTHER USES OF PHYSISORPTION

Various solid materials can adsorb materials that are gases or liquids by being exposed to them. The adsorbed material can be recovered at a later time and analyzed to determine the amount of a particular material in the atmosphere or in a particular environment. This information is typically used to monitor the amount of exposure individuals experience to potentially harmful materials in the workplace and elsewhere. New methods for detecting specific compounds using adsorption methodology are constantly being developed.

Richard M. Renneboog, MSc

BIBLIOGRAPHY

Bottani, Eduardo J., and Juan Tascón, M.D., eds. *Adsorption by Carbons.* Jordan Hill, UK: Elsevier, 2008. Print.

Ibach, Harald. *Physics of Surfaces and Interfaces.* New York, NY: Springer, 2006. Print.

Kolasinski, Kurt W. *Surface Science. Foundations of Catalysis and Nanoscience* 3rd ed. New York, NY: John Wiley & Sons, 2012. Print.

Lüth, Hans. *Solid Surfaces, Interfaces and Thin Films* 6th ed. New York, NY: Springer, 2015. Print.

Madou, Marc J. and Morrison, S. Roy. *Chemical Sensing with Solid State Devices.* San Diego, CA: Academic Press, 1989. Print.

Poole, Colin. *Gas Chromatography.* Waltham, MA: Elsevier, 2012. Print.

Wixom, Robert L., and Charles W. Gehrke. *Chromatography: A Science of Discovery.* Hobboken: Wiley, 2010. Print.

POINT DEFECTS AND THEIR EQUILIBRIA

FIELDS OF STUDY

Crystallography, Physical Chemistry

SUMMARY

At any temperature above absolute zero even the most perfect crystal lattice is found to have a number of point defects. Vacancies, interstitials and atoms out of place are typical. The Kröger-Vink notation provides a convenient description of the defects usually encountered in ionic and other crystals, and can be used to express equilibrium relations between defect concentrations. The concentration of intrinsic point defects increases as the temperature is increased and melting begins.

PRINCIPAL TERMS

- **lattice:** The idealized version of the crystal structure. A lattice may consist of several sub lattices, each for a separate species of ion.
- **vacancy:** A lattice point not occupied by the usual atom or ion.
- **interstitial:** An atom placed between normal lattice points.
- **aliovalent impurity:** An impurity atom taking the place of a lattice ion of different charge.
- **Schottky disorder:** the presence of anion and cation vacancies in roughly equal numbers to achieve electrical neutrality. This type of disorder is present in most alkali halides.
- **Frenkel disorder:** The presence of vacancies and interstitials of the same sign. Typical of ionic crystals with anions much larger than cations or vice-versa (sometimes called anti-Frenkel disorder).
- **electron (mobile):** In an ionic crystal or semiconductor, an electron which has been lifted from the valence band and is free to move about the crystal.
- **hole (mobile):** a vacant electronic state in the valence band, behaving as an effective positive charge.

DESCRIBING POINT DEFECTS: THE KRÖGER-VINK NOTATION

Several notations have been developed to describe point defects in crystals. Possibly the most widely employed is the Kröger-Vink notation in which defects are described by the symbol:

$$S_P^C$$

where S is the type of defect, C is its charge with respect to the perfect lattice and P is its crystallographic position. S is writes as V to denote a vacancy, e a free electron and h denoting an electron hole. C denotes the effective charge (x = neutral, ' = negative and . = positive), P is the chemical symbol of the lattice site in the perfect lattice.

The following table provides a few examples for ionic crystals of the NaCl type and MgO type:

	MX(NaCl)	MX(MgO)
Cation vacancy	$V_M^{'}$	$V_M^{''}$
Anion vacancy	V_X	V_x
Interstitial cation	M_i	M_i
Interstitial anion	$X_i^{'}$	$X_i^{''}$
Mobile electron	e'	e'
Mobile electron hole	h	h
Intentional impurities:		
N^+ on cation site	N_M^X	$N_M^{'}$
N^{2+} on a cation site	N_M	N_M^X
Y^- on an anion site	Y_X^X	Y_X
Y^{2-} on an anion site	$Y_X^{'}$	Y_X^X
N+ on an interstitial site	N_i	N_i
Y− on an interstitial site	$Y_i^{'}$	Y_i^-

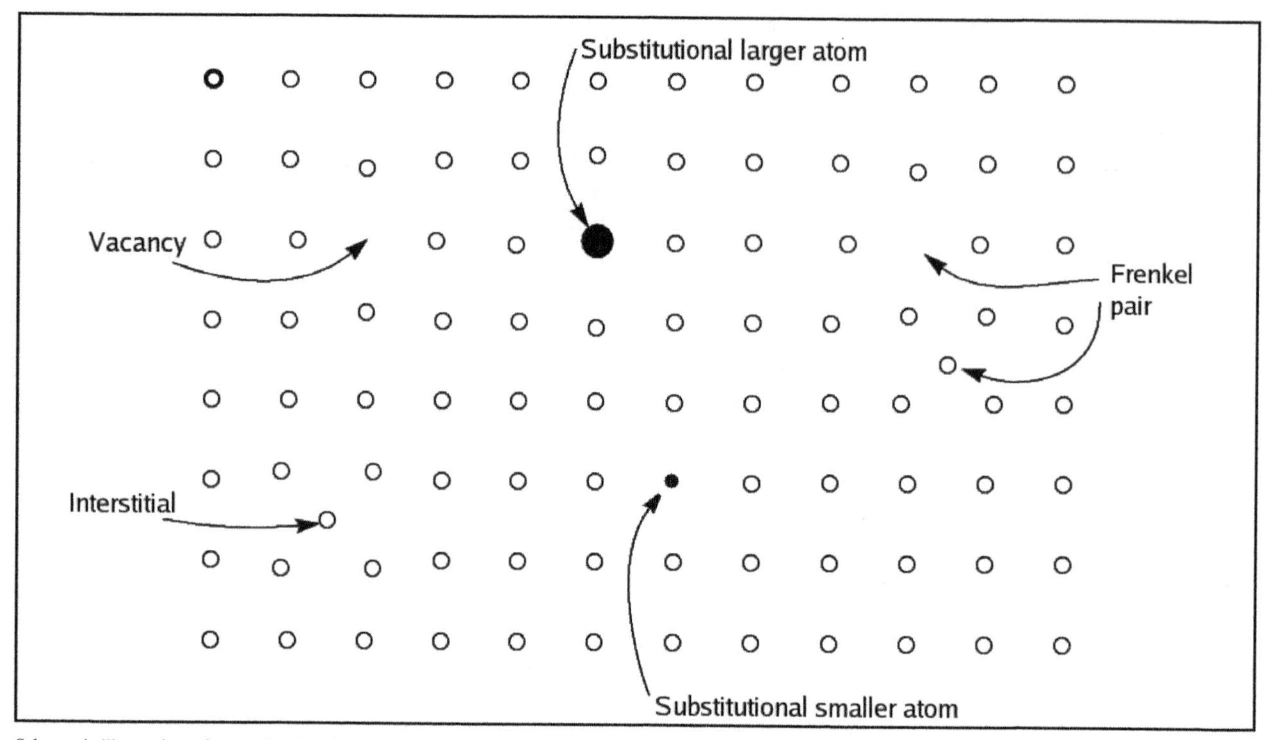

Schematic illustration of some simple point defect types in a monatomic solid. By Knordlun at English Wikipedia via Wikimedia Commons

With a notation such as this the spontaneous appearance of a cation and anion vacancy can be written:

$$0 = V'_M + V_{\dot{X}}$$

where 0 represents the perfect crystal lattice. Note that, physically, this can only be done by moving a cation and an anion to the surface of the crystal. This can be compared to the solution of a sparingly soluble salt in water. Just as in the aqueous case one has a solubility product constant expression:

$$[V'_M][V_{\dot{X}}] = K_{sp}$$

Just as in the case of a sparingly soluble salt one can induce a larger concentration of cation vacancies by adding an indifferent cation of higher valence (an aliovalent cation) say N^{++}.

Electrical charge balance then requires that $[N_M^{\cdot}]$ $+[V_{\downarrow}X^{\uparrow}\cdot]$ $V_{\downarrow}X^{\uparrow}\cdot]=[V_M']$ so the concentration of cation vacancies will go up as the concentration of anion vacancies will go down. (Common ion effect).

A unique set of point defects consists of free electrons trapped an anion vacancies,

$$e_X^x$$

This is the factor responsible for the additive coloration of alkali halides. If one suspends, say, a crystal of sodium chloride above a crucible of molten sodium metal one finds that the crystal expands and acquires a distinct yellow coloration. The only way the crystal can expand is through the introduction of additional cations with the anion sites holding enough electrons for electroneutrality. The trapped electrons are known farbe (or color) centers, or F centers.

SAMPLE PROBLEM

Write the equations governing defect equilibrium in silicon doped with phosphorus.

$[n'][p] = c_{i0}^2$ (The product of electron and hole concentrations = the concentration of either in the absence of impurities)

Answer:

$[N_D][n'] = K[N_D^X]$ where $N_D^0 = N_D^X + N_D^+$

Nonstoichiometry

While Proust's law of definite proportions applies fairly well to a majority of crystalline solids, that is that the atomic ratios found in solids tend to be the ratios of small whole numbers, the law is strictly speaking only approximate. The existence of point defects means that most solid materials will exist within a range of stoichiometry. A crystal of silver bromide grown in an atmosphere of Bromine vapor will have a silver to bromine ratio of 1+x while that grown in contact with metallic silver will have a ratio of 1-y. Here x and y are numbers of the order of 0.01.

Semiconductors and the Emergence of "Dirt Physics"

We live in the computer age, which is to say the age of semiconductors. Were it not for the ability to fabricate structures out of semiconducting material, with carefully controlled feature sizes and high reproducibility, our computing machinery would have to rely on vacuum tubes or mechanical gears, and necessarily be a million or more times larger and subject to frequent failures. The reliability of the modern computer owes much to the ability to produce silicon of extreme purity and then to add precise quantities of dopants in carefully designed regions.

Semiconducting devices are usually based on the element silicon, which can now be zone refined to extreme purity. Early in their development semiconductors were a puzzle to solid-state physicists and chemists. The traditional way to deal with materials which displayed an interesting property was to purify them and measure the properties of the purified material. When applied to the electrical conductivity of silicon, this approach got nowhere. As it was purified, silicon lost its unusual electrical behaviors. It was found then, with the clarity of hindsight, that what made silicon interesting was not the material itself but rather the effect of impurities on it. The study of semiconductor physics was really the study of the impurities or "dirt" to be found there. The field of semiconductor physics thus became known as dirt physics.

The impurities to be found in semiconductors could be classified as electron acceptors or electron donors. In silicon acceptors typically were elements like boron and aluminum, which had one less valence electron than silicon and served as a source of electron holes in the material, while phosphorus or nitrogen served as electron donors. P-type material has an excess of acceptors while n-type material had an excess of donors. Putting p-type in contact with n-type produced a pn junction which could act as a rectifier, putting a thin layer of p type between a small n-type emitter and a bigger n-type collector produced a transistor, a semiconductor device that could amplify signals. Alternatively the device becomes a high-speed switch, ideal for computer applications.

Donald R. Franceschetti, Ph.D.

Bibliography

Kittel, Charles. Introduction to Solid State Physics. 8th ed. Hoboken, NJ: Wiley, 2005. Print.

Kröger, Ferdinand A. *The Chemistry of Imperfect Crystals.* Amsterdam: North-Holland Publ., 1974. Print.

Mott, Nevill F., and Ronald Wilfried. Gurney. *Electronic Processes in Ionic Crystals.* New York, NY: Dover, 1964. Print

POLYMERASE CHAIN REACTION

FIELDS OF STUDY

Molecular Biology; Genetics

SUMMARY

Polymerase chain reaction (PCR is a technology that creates billions of copies of a DNA sequence from a single copy of a small piece of DNA. PCR revolutionized the study of DNA because any sequence could be, in principle, duplicated permitting further study. This method of creating large amounts of DNA from minimal source material won the Nobel Prize in 1993 for its creator, Kary B. Mullis.

PRINCIPAL TERMS

- **DNA base pairing:** two strands of DNA are held together in a double helix shape by weak chemical bonds between base pairs; these base pairs always form the same way, with adenine always forming a base pair with thymine and cytosine always forming a base pair with guanine, following base pair rules.

- **DNA fingerprinting:** a test to identify and evaluate the genetic material in one's cells performed by extracting DNA from a small sample of cells, then making a large number of copies, then braking up the copies with restriction enzymes and finally sorting the copies by electrophoresis. The pattern thus produced is called a fingerprint, because it is very unlikely that two individuals would have the same genetic material, much as it is unlikely that two individuals would have the same fingerprints.
- **electrophoresis:** a lab technique used to identify, quantify, and purify fragments of DNA; samples are placed in agarose gel and electrified, causing the negatively charged nucleic acids to move toward a positive electrode. Shorter fragments travel rapidly, leaving longer fragments behind, resulting in a separation according to length.
- **nucleotide:** the basic structural component of DNA and RNA, consisting of a ribose (in RNA) or deoxyribose (in DNA) sugar molecule bonded to a phosphate group and one of five nucleobases: cytosine, adenine, guanine, thymine (DNA only), or uracil (RNA only).
- **polymerase:** any of a variety of enzymes that help synthesize DNA or RNA by using an existing strand as a template.
- **reptation:** the creeping motion that describes how DNA fragments move through agarose gel when they are being separated according to length.
- **thermal cycler:** the instrument that is used to conduct the polymerase chain reaction using precise temperature control and rapid temperature changes.

G storm thermal recycler for PCR. By Rror (Own work) via Wikimedia Commons

Until PCR was created, replicating DNA was a laborious process that took place in bacteria and took several weeks to complete. Synthesis took place in test tubes that had to be carefully monitored and timed. After each cycle of synthesis, the reaction had to be reheated to cause it to take place again. The reheating destroyed the enzymes used in the reaction, so fresh enzymes had to be added at each step. PCR, sometimes called molecular photocopying, is a fast and inexpensive way to generate multiple copies of DNA to analyze.

The creation of PCR is a great example of scientists building on work by previous scientists to envision and create something new. All the components necessary to create this innovative technology had been described by 1980. Even in Watson and Crick's original paper describing the structure of DNA published in 1953, they postulated that copying genetic material was possible with the type of structure, the double helix, they were proposing. Other major highlights regarding components that needed to be created, isolated, or described before PCR could be created include the following:

- Kornberg identifying the first DNA polymerase
- Khorana creating synthetic DNA nucleotides
- Brock isolating Taq from a hot spring in Yellowstone Park
- Klenow finding a modified version of DNA Polymerase I from E. coli

- Sanger reporting a method for sequencing DNA

Mullis first became interested in DNA synthesis and gene cloning after attending a seminar on the topics. He quickly realized that DNA could be created chemically. At the time, replicating DNA was a laborious, difficult process and prone to many quality issues. Based on the work of many other scientists before him, Mullis's idea involved using large quantities of the four necessary nucleotides along with primers and a second complementary DNA strand to do multiple rounds of replication. He originally thought that the DNA would simply replicate on its own when provided with the necessary materials, but he quickly learned that fresh enzymes and repetitive heat cycles were necessary to provoke the needed reactions. In 1985, *Taq* polymerase was found to be thermostable enough to use in these reactions, and in 1987, a thermal cycler had been invented. These two components eliminated the need for scientists to continually add fresh enzymes and time heating applications when replicating DNA, greatly simplifying the process and shortening the time needed to produce these reactions.

When Mullis first described his vision for PCR technology at a scientific conference, no one understood the value of his creation. The prestigious journal *Science* rejected his paper on this topic, only to later accept it and name PCR and *Taq* polymerase as the Molecule of the Year three years later.

Components Needed to Perform PCR:

- DNA template: The sample of DNA that is to be replicated
- DNA polymerase: The enzyme that will be used to synthesize the complementary strands of DNA. Often *Taq* polymerase is used, but *Pfu* DNA polymerase can be used. *Pfu* DNA polymerase has a higher fidelity rate in copying DNA, but either enzyme is suitable because both are (1) heat resistant and (2) can generate new strands of DNA if they are provided with a DNA template and primers.
- Primers: Short pieces of single-stranded DNA that provide the complementary nucleotides to the DNA template.
- Nucleotides: Single units of cytosine, adenine, guanine, and thymine; the building blocks of the new DNA strands.

How PCR Replicates DNA

To amplify a segment of DNA, the following steps take place. The entire process happens inside a thermocycler, which alters the temperature every few minutes so that the reaction continues to take place as long as is necessary to generate the required copies of the DNA template.

- The DNA sample is heated so that it separates from its double helix structure into its two component pieces of single-stranded DNA, or denatured.
- A polymerase is added. This enzyme builds two new strands of DNA, using the original strand as a template according to the rules of DNA base pairing.
- Each of these strands is then again heated, denatured, and used to replicate two new copies.
- The process repeats over and over until the desired amount of copies has been achieved.
- The process is fully automated and can be completed in a few hours. It can be repeated as many as 30 or 40 times and generate more than one billion exact copies of the original segment. The DNA fragments can be separated according to length by using gel electrophoresis; fragments move through the electrified gel using a motion called reptation. Smaller segments of DNA move more quickly through the gel, leaving larger segments behind and thus separating them according to length.

Practical Applications

The copies of DNA generated by PCR can be used in a variety of ways, such as genotyping, sequencing, cloning, and analyzing genetic expression. Once DNA fingerprinting is performed, the results can be used for several different applications, including the following:

- Determine if two or more people are in the same family using a small sample of DNA from each of them, for example, to determine if a man is the father of a particular child
- Solve crimes from small samples of DNA left at a crime scene by the perpetrator of a crime

- Identify a body from a small amount of DNA; particularly helpful if a body has decayed significantly or if only a small portion of a body is recovered
- Map the human genome
- Diagnose genetic defects in individuals and families
- Detect viruses, such as HIV, in cells
- Study the evolution of humans and track their migration throughout history
- Engineer genetically modified substances

Marianne Moss Madsen, MS

Bibliography

Basu, Chhandak, ed. *PCR Primer Design (Methods in Molecular Biology)*. Humana Press, 2015.

Biasonni, Roberto, and Alessandro Raso, eds. *Quantitative Real-Time PCR: Methods and Protocols (Methods in Molecular Biology)*. Humana Press, 2014.

Bustin, Stephen A., ed. *The PCR Revolution: Basic Technologies and Applications.* Cambridge University Press, 2014.

Lo, Y. M. Dennis, Rossa W. K. Chiu, and K. C. Allen Allen Chan, eds. *Clinical Applications of PCR.* 2nd ed. Totowa, NJ: Humana, 2006. Print.

Moller, Simon. *PCR (The Basics).* Taylor & Francis, 2006.

Nolan, Tania, and Stephen A. Bustin, eds. *PCR Technology: Current Innovations.* CRC Press, 2013.

van Pelt-Verkuil, Elizabeth, Alex van Belkum, and John P. Hays. *Principals and Technical Aspects of PCR Amplification.* Springer, 2010.

PRECIPITATION

FIELDS OF STUDY

Analytical Chemistry; Crystallography; Geochemistry

SUMMARY

The process of precipitation is defined, and its importance in chemistry-related fields is elaborated. Precipitation is a fundamental process in geochemical systems as well as an essential technique in the preparation and purification of materials in synthesis and for crystallographic study, analytic assays, and many other procedures.

PRINCIPAL TERMS

- **precipitant:** a substance that, when added to a solution, causes a component of the solution to become solid and separate from the liquid.
- **sedimentation:** the gradual accumulation of solid particles as they settle out of suspension in a fluid.
- **solute:** any material that is dissolved in a liquid or fluid medium, usually water.
- **solvent:** any fluid, most commonly water, that dissolves other materials.
- **supernatant:** the liquid or fluid that remains after a substance is precipitated from a solution.

Visualizing the Concept of Precipitation

Precipitation is the process by which a substance separates from the supernatant liquid or fluid of a solution, generally in the form of a solid; the separated substance is then called a precipitate. Precipitation can occur in any kind of solution and under any conditions. Minerals and native metals form by precipitating from molten rock or hot, highly pressurized aqueous solutions in the earth's crust. Such processes take place very slowly, over long periods of time, and the results can be quite spectacular. In colder climates, precipitation of a solid from a fluid solution is often seen when weather conditions produce snow. Rain also constitutes precipitation, although in that case the precipitate is a liquid rather than a solid.

Principles of Solutions

Precipitation can only occur in solutions. A solution has only two types of components: the solvent and the solute. Though typically thought of as a liquid system, it is more correct to regard a solution as a fluid system, since gases also form solutions that behave according to the same rules as liquid solutions. In any solution, the solvent is a fluid material that has the ability to dissolve other materials, and the solute is any material that has been dissolved in the solvent. A liquid solvent can dissolve solids, liquids, and gases; the most common solvent, water, readily dissolves salt,

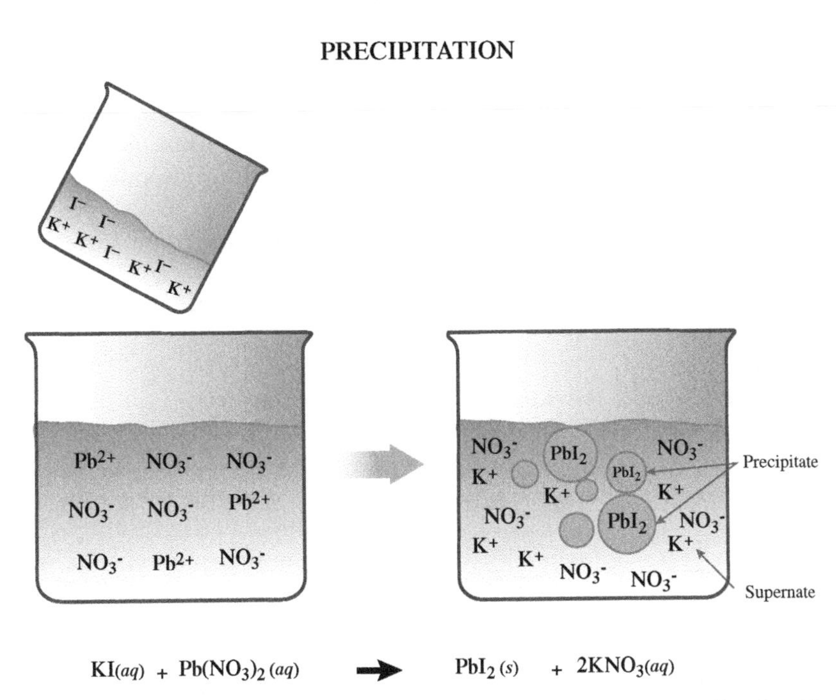

sugar, alcohol, acetone, carbon dioxide, oxygen, and innumerable other materials. However, when the solvent is a gas, it can only dissolve other gases. Air, for example, is a solution of nitrogen and oxygen gases, plus several other minor constituents, and it readily dissolves water vapor but cannot dissolve liquid water.

A solute dissolves in a particular fluid only to the extent that the quantity of fluid is able to dissolve the material. Accordingly, any material can be described by its solubility properties. Two liquids that effectively dissolve each other, such as water and acetone, are described as being miscible. When two substances dissolve each other completely regardless of how much of either one is present, they are said to be miscible in all proportions. One can, for example, pour any quantity of acetone into any quantity of water, and the two will immediately form a solution as a single phase. However, if the solubility of one liquid in the other is limited, two separate phases will form when the solubility limit is reached. For example, *n*-butanol ($HO-CH_2-CH_2-CH_2-CH_3$) dissolves in water at twenty-five degrees Celsius to a maximum of 9.1 milliliters per 100 milliliters of water. Any further amount of *n*-butanol added to such a solution will not dissolve but will sit as a separate layer on the surface of the solution. Gases generally tend to be less soluble in liquids, though their solubility is of extreme importance in physiological systems.

The solubility of solids in liquids is generally limited and is normally expressed as the mass of a solid material that will dissolve in a specific quantity of a liquid. The organic material benzoic acid, for example, will dissolve in water to the extent of 3.4 grams per liter at twenty-five degrees Celsius. At lower temperatures, less benzoic acid will dissolve; at higher temperatures, more will dissolve. For many compounds, particularly ionic compounds, dissolution occurs as the ions are separated from the crystal lattice of the solid phase and surrounded by water molecules. Thus, an ionic material such as sodium chloride (NaCl) dissolves as sodium ions, Na^+(*aq*), and chloride ions, Cl^-(*aq*). The abbreviation "(*aq*)" is added to show that the ions are dissolved in water, an aqueous solution. Similarly, "(*g*)" indicates that the substance is dissolved in a gas solution; "(*l*)," a nonaqueous liquid solution; and "(*s*)," a solid solution.

Solubility Product, Solubility, and Precipitation

Materials that dissolve in solutions as whole molecules rather than ions are described by their solubility in a specific quantity of a particular solvent. Sucrose (refined white sugar), for example, dissolves in water to the maximum extent of 2,000 grams per liter. Materials that dissolve into ions can also be described by the overall solubility of the original material, expressed in grams per liter. The solubility of an ionic compound depends on its solubility product, which is the product of the concentrations of the individual component ions expressed in moles per liter. When the product of the concentrations of component ions in a solution equals or exceeds the solubility product of the compound from which they dissociated, then that compound will begin to precipitate from the solution. For example, in a solution containing potassium ions (K^+) and chloride ions (Cl^-), increasing the concentration of chloride ions to exceed the solubility product of potassium chloride (KCl) will cause crystals of solid potassium chloride to form and precipitate from the solution. In this scenario, the additional chloride ions act as a precipitant. The particles of a precipitate are normally held in suspension in the mixture, eventually settling to the bottom of the container in a process called sedimentation.

Precipitation always occurs when the ability of a certain quantity of solvent to dissolve a material is exceeded. Solubility is temperature dependent, a feature that is used in the laboratory to purify compounds via recrystallization. In this procedure, an impure crystalline material is dissolved in the minimum amount of an appropriate preheated solvent. At this point, the solution may be treated to remove colored impurities. It is then allowed to cool slowly.

SAMPLE PROBLEM

In the reactions below, identify the solvent, precipitant, supernatant, and precipitate. The downward-pointing arrow is used to indicate that a material has precipitated from solution during the reaction.

$$AgNO3(aq) + NH_4Cl(aq) \rightarrow NH_4NO_3(aq) + AgCl(s)\downarrow$$

$$SO_4^{2-}(aq) + Ba^{2+}(aq) \rightarrow BaSO_4(s)\downarrow$$

Answer:

Subscripts and arrows are the chemist's shorthand for such information as solvent, phase, and activity. First, check that all chemical equations are properly balanced.

Recall that (*aq*) indicates that the compound it accompanies is dissolved in water, making an aqueous solution, while (*s*) indicates that the material is present as a solid and not as a dissolved species in solution. Thus, in both equations, the solvent is water.

In the first equation, silver ions and chloride ions combine to form a precipitate of solid silver chloride, $AgCl(s)$. Therefore, the chloride is the precipitant of the silver ions; conversely, it can also be said that the silver ions are the precipitant of the chloride. After the reaction has occurred, an aqueous solution of ammonium nitrate, $NH_4NO_3(aq)$, remains. This is the supernatant.

In the second equation, written in ionic form, sulfate ions and barium ions combine to form solid barium sulfate has precipitated. Thus, the barium ions are the precipitant of the sulfate ions, and the sulfate ions are the precipitant of the barium ions. Because the original solutions are both aqueous, the only known supernatant in this case is water.

As the temperature of the solution decreases, so does the ability of the solvent to dissolve the solute that it contains. When the solute can no longer remain dissolved, it begins to precipitate from the solution, ideally as crystals of high purity. Manipulating the procedure is as much art as it is science, however, and does not always work as expected. Many a student has encountered the "oiling out" effect, in which the desired material simply separates as an impure oil rather than clean crystals.

APPLICATIONS OF PRECIPITATION

The process of precipitation is effectively used in analytical chemistry and synthetic chemistry. To analyze mineral or metal materials, a sample is typically dissolved in acid and then manipulated to cause specific materials to precipitate from the solution. The precipitated compounds can then be separated, dried, and weighed. The relative quantities of these compounds allow the analyst to accurately determine the amounts of specific elements, such as silver or uranium, that were present in the original sample. In synthetic procedures in all branches of chemistry, the desired materials are often isolated and recovered after being precipitated from solution. Modern techniques permit the precipitation and recovery of very small amounts of materials such as DNA.

In chemical engineering, precipitation is an effective way to recover the product of an industrial-scale reaction process, especially when the reaction is carried out in the gas phase. Two gaseous materials may react to form a material that then precipitates out of the gas, automatically isolating itself from the reaction mixture. The reaction of uranium oxide with fluorine gas, for example, produces gaseous uranium hexafluoride, UF_6, which is easily transported to a collection chamber and then readily precipitated as solid UF_6 powder known as yellowcake.

PRECIPITATION IN NATURAL SYSTEMS

Molten rock accumulates in large formations during tectonic activity. As the magma cools, various types of minerals crystallize and may precipitate from the fluid matrix. This is the principal means by which deposits of metal ore form. Mineral-laden aqueous solutions underground also may be maintained at high temperatures and pressures, permitting the slow precipitation of specific compounds that are often highly crystalline in structure. This is the principal means of formation of many valuable materials, including various gemstones. Miners have recovered beryl crystals as large as one meter or more in length that have been produced in this way. One of the most striking examples of mineral precipitation is found in a cave several hundred meters below the Sonoran Desert in Mexico, where gypsum crystals the size of transport trucks have been discovered. The crystals were formed over thousands of years by slow precipitation from the hot, mineral-rich water that filled the caves.

Richard M. Renneboog, MSc

BIBLIOGRAPHY

Haynes, W. H., PhD, David R. Lide, PhD, and Thomas J. Bruno, PhD, eds. *CRC Handbook of Chemistry and Physics,* 96th Edition. New York: CRC/Taylor & Francis Group, 2015. Print.

Holden, Alan, and Phylis Morrison. *Crystals and Crystal Growing.* New York: Doubleday, 1960. Print.

Lafferty, Peter, and Julian Rowe, eds. *The Hutchinson Dictionary of Science.* 2nd ed. Oxford: Helicon, 1998. Print.

Lodish, Harvey, et al. *Molecular Cell Biology.* 7th ed. New York: Freeman, 2013. Print.

O'Neil, Maryadele J., ed. *The Merck Index: An Encyclopedia of Chemicals, Drugs, and Biologicals.* 15th ed. Cambridge: RSC, 2013. Print.

Skoog, Douglas Arvid, Donald M. West, and F. James. Holler. *Skoog and West's Fundamentals of Analytical Chemistry.* 9th ed. Belmont (Calif.): Cengage Learning, 2014. Print.

PROPERTIES OF MATTER: COMPOSITION

FIELDS OF STUDY

Physical Chemistry; Organic Chemistry; Inorganic Chemistry

SUMMARY

The principles of chemical identity and composition are discussed. The atomic number defines the identity of the atom as an element, and the differing amounts and types of elements in a molecule define the identity of a compound. The empirical chemical formula of a compound can be calculated using percent composition.

PRINCIPAL TERMS

- **bonding:** the formation of a link between two atoms as a result of the interaction of their valence electrons.
- **chemical formula:** the combination of symbols and numerical coefficients that specifies the number and identity of the different atoms involved in a chemical reaction or molecular structure.
- **element:** a form of matter consisting only of atoms of the same atomic number.
- **homogeneous mixture:** a physical combination of different materials that has a generally uniform distribution of composition, and therefore properties, throughout its mass; called a solution when liquid and an alloy when solid metal.
- **molecule:** the basic unit of a compound, composed of two or more atoms connected by chemical bonds.

THE NATURE OF MATTER

All matter is composed of atoms, which are in turn composed of electrons, protons, and neutrons. Each atom has a unique chemical identity as one of the chemical elements, determined by the number of protons in its nucleus. Each atom also has a specific mass that is essentially determined by the combined masses of the protons and neutrons that it contains. The elemental identity of an atom is maintained in normal, nonnuclear chemical reactions, since all such reactions take place only at the level of the outermost electrons of the atoms as they undergo chemical bonding and molecular formation. The specific combination of different atoms in a molecule is unique to the identity of the compound in the same way that atoms are unique to the identity of an element. Adding, removing, or even just changing one single atom in a molecule of a specific compound changes the chemical identity of the compound, no matter how large and complex the molecule may be. The different types of atoms and the number of each

PROPERTIES OF MATTER COMPOSITION

COMPOSITION: Chemical Components and Their Proportions	PROPERTIES: Physical and Chemical Qualities and Attributes
Water Components: 2 parts hydrogen, 1 part oxygen	Physical Properties: clear, (when pure) non-conductive, density –1g/cm^3, specific heat capacity – 4.186 J/ (g•K) Chemical Properties: pH neutral, readily dissolves many substances, reacts with acid anhydrides to form acids with K, Na, and Ca to form hydroxides
Ice Components: 2 parts hydrogen, 1 part oxygen	Physical Properties: clear, non-conductive, density –.9179/cm^3, specific heat capacity –2.05J / (g•K) Chemical Properties: HCl solvates readily on ice surfaces
Iron(II) Oxide Components: 2 parts iron, 3 parts oxygen	Physical Properties: reddish-brown, magnetic, density 5.745 g/cm^3 Chemical Properties: insoluble in water, reacts with aluminum

type in the molecule are given in the standard form of the chemical formula.

Matter and Composition

Any number of molecules with as many different chemical identities can be mixed together, even to the extent that no adjacent molecules are of the same compound. Macroscopically, such a homogeneous mixture would have consistent physical properties throughout its mass due to the uniform distribution of its components. This is a matter of scale, depending on how broadly the mass is defined. Concrete, for example, consists of water, cement, and innumerable sand and stone components of different sizes and shapes, yet it has a uniform consistency and distribution of components throughout, making it a homogeneous mixture. At the molecular level, however, the different component molecules each retain their specific, unique chemical identities. This is not affected by the combination of components, unless a reaction can occur between two or more of the components.

The composition of matter, whether as bulk mixtures or discrete molecules, is typically described in terms of the percent of each component. For a molecule, this corresponds to the proportion of each type of atom in the molecule's unique structure. Analysis of a sample of a pure compound can identify the proportion by weight of each individual element present in its molecular structure. This allows calculation of the molecular formula. For example, a sample of a compound may be found to consist of x percent by weight of one type of atom, y percent by weight of a second, and z percent by weight of a third. Since the percentage by weight of each element must add up to 100 percent, the easiest way to solve a problem like this is to assume a starting weight of one hundred grams to simplify the calculations. In one hundred grams, there are therefore x grams of the first type of atom, y grams of the second, and z grams of the third. Having made this assumption, the next step is to use the appropriate atomic weights to determine the number of moles of each element present. Next, use the molar quantities of each element to determine their lowest whole-number ratios. It is helpful to use a table to track the different values. This procedure returns the empirical formula of the

SAMPLE PROBLEM

A compound is analyzed and found to contain only potassium, carbon, and oxygen. It is 47 percent potassium (K) by weight, 14.5 percent carbon (C) by weight, and 38.5 percent oxygen (O) by weight. What is the empirical formula of the compound?

Answer:

Assume a sample size of 100 grams for the unknown compound. Determine the moles of each element, dividing the weight in grams, which is equal to the percent by weight, by the atomic weight of each. Then find the molar ratios by dividing the mole value for each element by the smallest mole value.

Element	*Weight (g)*	*Atomic Wt.*	*Moles*	*Molar Ratio*
K	47	39	$\frac{47}{39}=1.21$	$\frac{1.21}{1.21}=1$
C	14.5	12	$\frac{14.5}{12}=1.21$	$\frac{1.21}{1.21}=1$
O	38.5	16	$\frac{38.5}{16}=2.41$	$\frac{2.41}{1.21}=1.99\approx 2$

The empirical formula of the compound is therefore KCO_2.

compound, giving the proportions, but not the actual numbers, of each element present.

DERIVING THE MOLECULAR FORMULA

The actual molecular formula of a compound will be a whole-number multiple of its empirical formula, but because every multiple will produce exactly the same proportions by weight, it requires more information than this to determine the molecular formula. A second analysis, such as a freezing-point depression study, is needed to determine the number of moles in a specific weight of the compound and return the molecular weight of the compound. Comparison between the molecular and empirical weights easily reveals the molecular formula.

Richard M. Renneboog, MSc

BIBLIOGRAPHY

Askeland, Donald R., Wendelin J. Wright, D. K. Bhattacharya, and Raj P. Chhabra. *The Science and Engineering of Materials.* Boston: Cengage Learning, 2016. Print.

Douglas, Bodie E., Darl H. McDaniel, and John J. Alexander. *Concepts and Models of Inorganic Chemistry.* 3rd ed. New York: Wiley, 1994. Print.

Jones, Mark Martin. *Chemistry and Society.* 5th ed. Philadelphia: Saunders College Pub., 1987. Print.

Kean, Sam. *The Disappearing Spoon: And Other True Tales of Madness, Love, and the History of the World from the Periodic Table of the Elements.* New York: Little, Brown, 2010. Print.

Lafferty, Peter, and Julian Rowe, eds. *The Hutchinson Dictionary of Science.* 2nd ed. Oxford: Helicon, 1998. Print.

Miessler, Gary L., Paul J. Fischer, and Donald A. Tarr. *Inorganic Chemistry.* 5th ed. Upper Saddle River: Prentice Hall, 2013. Print.

Morrison, Robert Thornton, and Robert Neilson Boyd. *Organic Chemistry.* 7th ed. Englewood Cliffs, N.J.: Prentice Hall, 2003. Print.

Myers, Richard. *The Basics of Chemistry.* Westport: Greenwood, 2003. Print.

Skoog, Douglas Arvid, Donald M. West, and F. James. Holler. *Skoog and West's Fundamentals of Analytical Chemistry.* 9th ed. Belmont (Calif.): Cengage Learning, 2014. Print.

PROTEIN SYNTHESIS

FIELDS OF STUDY

Biochemistry; Genetics; Molecular Biology

SUMMARY

The process of protein synthesis is described, and its importance in the biochemistry of living systems is elaborated. Protein synthesis is a vital process for the production of enzymes and other polypeptides, which are in turn responsible for carrying out many functions.

PRINCIPAL TERMS

- **amino acid:** an organic compound that contains both an amine and a carboxylic acid functional group and can, in some cases, combine with other amino acids to form proteins and other polypeptides.
- **anticodon:** a sequence of three nucleotide bases in transfer RNA (tRNA) that bonds to a complementary codon in messenger RNA (mRNA).
- **codon:** a sequence of three nucleotide bases that specifies a particular amino acid or control point in the process of protein synthesis.
- **ribosome complex:** a structure consisting of ribosomal RNA (rRNA) and enzymes that decodes messenger RNA (mRNA) and coordinates the assembly of proteins from amino acids carried by transfer RNA (tRNA).
- **translation:** the overall process of RNA-mediated protein synthesis, entailing transcription of genetic information from the DNA molecule by messenger RNA (mRNA), assembly of the corresponding amino acids by transfer RNA (tRNA), and formation of the protein molecule in ribosomes by ribosomal RNA (rRNA).

Visualizing Protein Synthesis

The process of protein synthesis is similar to any modern manufacturing facility. DNA is analogous to the design department, where all the product specifications are kept. The communications department is the role of mRNA, transmitting the product specifications to the production line. The production line is found in the ribosome, where rRNA operates the protein assembly line. All of the components needed to keep the assembly line running are delivered by tRNA, the just-in-time suppliers.

The actual process of protein synthesis takes place at the molecular level, beginning with the transcription of genetic code from a molecule of DNA. Enzymes act on the DNA molecule first to straighten out the double helix structure of the duplex DNA strand and then to initiate a separation of the two DNA strands. At the separation, mRNA assembles along one of the strands, forming a fragment of RNA that is complementary to the nucleotide sequence of the DNA strand. The nucleotides pair up in the process, the cytosine nucleotides pairing with guanine nucleotides and the adenine nucleotides pairing with thymine nucleotides. However, when the adenine nucleotide is in the DNA strand, the complementary nucleotide in the RNA strand is the uracil nucleotide, instead of its thymine counterpart.

The genetic information contained in DNA is "written" as a series of nucleotides called genes. The genes differ in length, but each gene spans several thousand nucleotides. Specific sequences indicate both the starting and end points of a gene. The overall sequence of nucleotides is transcribed into the mRNA strand by this process of gene expression, and when completed, the RNA strand separates as a molecule of mRNA. Enzymes then act on the DNA molecule again to seal up the location at which it had been separated for transcription and gene expression.

mRNA, tRNA, and rRNA

Three forms of RNA are involved in the translation of the genetic code from DNA to protein synthesis. The mRNA carries a transcribed copy of genetic code from DNA. Once freed from its association with the DNA molecule, mRNA makes its way to the nearest ribosome complex, a structure consisting of about 60 percent ribosomal RNA and 40 percent various enzyme proteins. The function of the ribosome is to direct the assembly of new proteins according to the code carried by the mRNA. At the ribosome, the mRNA coordinates with the rRNA in a manner that reveals the nucleotide sequence as codons, which are then used for recognition by corresponding units of tRNA. A codon is a series of three nucleotides in mRNA that correspond to a complementary anticodon on a tRNA molecule and, thus, to the specific amino acid carried by that tRNA molecule. Because RNA, like DNA, is assembled from only four different nucleotides and proteins are assembled from twenty different amino acids, it is impossible for a single nucleotide to specify a single amino acid. Thus, there must be combinations of nucleotides, with three being the minimum number capable of specifying all twenty essential amino acids. In fact, the system of three-nucleotide codons is capable of specifying more than twenty amino acids, which has resulted in a number of redundancies in the codon-anticodon series. Some amino acids can correspond with up to six different codons, and there are codons that specify the starting and ending points of amino acid sequences in the protein molecules being synthesized.

"Amino acid" is a general term for any organic compound that has both amine and acid functional groups in its molecular structure. Therefore, there are millions of possible amino acid molecules. The genetic code specifying the sequence of amino acids in proteins in essentially all biochemical systems uses only a limited set of twenty essential amino acids. Each one has both a carboxylic acid group (–COOH) and an amine group ($-NH_2$) bonded to the same carbon atom. The third substituent, or side chain bonded to the carbon atom, is different in each of the twenty amino acids, and ranges from a simple hydrogen atom in glycine to more complex cyclic, bicyclic, and electrically charged substituents in others. The substituents in nine of the amino acids are neutral and nonpolar. Six others are also neutral but polar, and the remaining five are electrically charged and thus highly polar. The sequence of the amino acid residues in a protein molecule determines the structure and function of that protein.

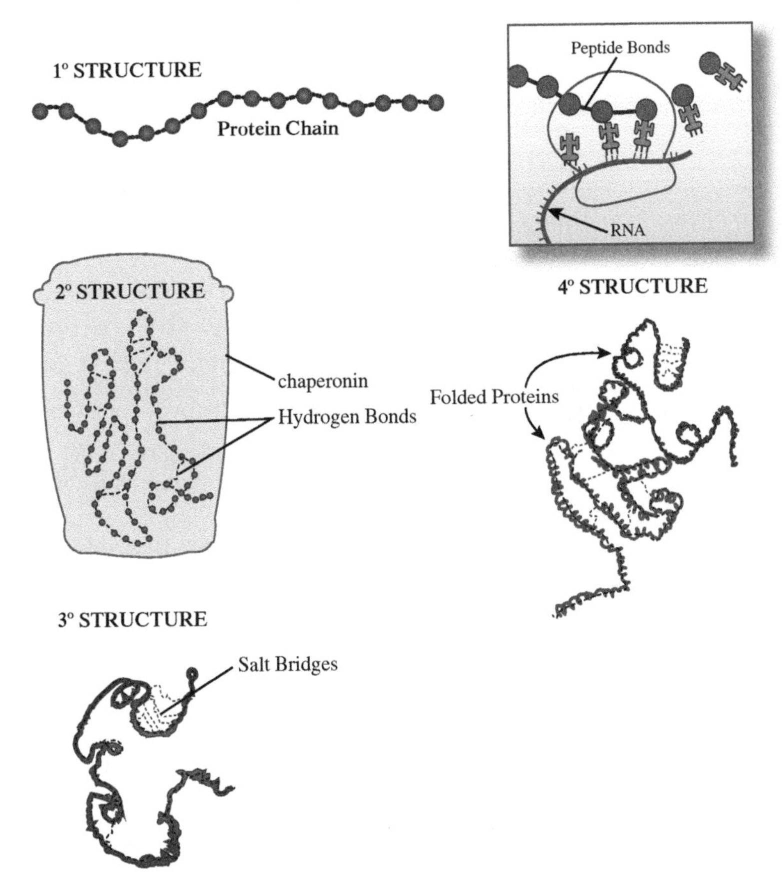

PROTEIN STRUCTURE AND FUNCTION

Protein molecules have four aspects of their structure and function. The first aspect, known as the "primary structure," is the sequence of amino acid residues in the molecule. The shapes of molecules are guided by strict rules with regard to their geometric shape and conformation, corresponding to the geometry of the electronic orbitals in the atoms. Therefore, the bonds between component atoms of a molecule must form specific angles with each other, determined by the geometry of the atomic orbitals. As a result, protein molecules (and effectively all others) are not straight-line structures, but rather twist and turn as bond angles demand. This motion produces the secondary structure of protein molecules, forming various segments into coil and sheetlike conformations. The tertiary structure of protein molecules is produced by the interaction of amino acid substituents at different locations, and especially by the formation of disulfide bridges between methionine residues in different parts of the molecule. A disulfide bridge bonds the two residues together chemically,

SAMPLE PROBLEM

Describe the four levels of structure in proteins with respect to the interactions that determine each aspect.

Answer:

The primary level of structure in a protein is the sequence of amino acids. This is due to the formation of peptide, or amide, bonds between the carboxylic-acid functional group of one amino acid and the neighboring amine-functional group.

The secondary level of structure is the gross shape of the parts of the protein molecule as coils and sheets. These formations arise from the restrictions of the bond angles between the atoms.

The third level of structure results from the interaction of certain functional groups and atoms with other functional groups and atoms in other parts of the molecule. Both the electronic affects of polar and charged side chains and the formation of disulfide linkages play roles in this process.

The fourth level of structure is the overall conformation of a protein complex involving two or more separate protein molecules, which results from both the manner in which the different proteins fill out the space available in the other molecules and the interactions between atoms and functional groups in the different molecules.

locking the protein molecule into a certain conformation. The fourth aspect of protein structure, termed the "quaternary structure," arises when two or more separate protein molecules intertwine and form a single functional entity.

Taken altogether, the four levels of structure determine the protein's function and specificity. Many proteins serve structural purposes, helping to create muscle, cartilage, hair, and fingernails. Others function as enzymes to facilitate and mediate biochemical processes such as digestion, respiration, and metabolism. In enzymes, the overall shape of the protein molecule forms an active site into which only materials having the proper corresponding structure can fit.

Applications of Protein Synthesis

Many medical conditions are caused by improper protein synthesis in individuals. The genetic code in DNA may be lacking the proper trigger for the transcription and synthesis of certain proteins, and as a result, the corresponding protein or enzyme is lacking. Many people suffer from lactose intolerance, for example. Those that do lack the ability to produce the enzyme lactase, which is responsible for the digestion of lactose, the sugar found in milk and other dairy products. The lactose affects such individual's biochemistry in much the same way that a chemical poison would, producing symptoms of discomfort. Avoiding milk and dairy products can diminish the negative effects, but developing a method by which the body naturally produces lactase provides an absolute cure for lactose intolerance.

Richard M. Renneboog, MSc

Bibliography

Berg, Jeremy M., John L. Tymoczko, Gregory J. Gatto, and Lubert Strye. *Biochemistry*. 8th ed. New York: W. H. Freeman, 2015. Print.

Lafferty, Peter, and Julian Rowe, eds. *The Hutchinson Dictionary of Science*. 2nd ed. Oxford: Helicon, 1998. Print.

Lehninger, Albert L. *Biochemistry: The Molecular Basis of Cell Structure and Function*. 2nd ed. New York: Worth, 1975. Print.

Lodish, Harvey, et al. *Molecular Cell Biology*. 7th ed. New York: Freeman, 2013. Print.

Pelczar, Michael J., Jr., E. C. S. Chan, and Noel R. Krieg. *Microbiology: Concepts and Applications*. New York: McGraw, 1993. Print.

Reece, Jane B., et al. *Campbell Biology*. 10th ed. San Francisco: Cummings, 2013. Print.

PROTEINS, ENZYMES, CARBOHYDRATES, LIPIDS, AND NUCLEIC ACIDS

FIELDS OF STUDY

Organic Chemistry

SUMMARY

The basic characteristics of proteins, enzymes, carbohydrates, lipids, and nucleic acids are presented, and their polymeric nature is described. These macromolecules are the building blocks of biological systems and have a variety of biochemical functions.

PRINCIPAL TERMS

- **biochemistry:** the chemistry of living organisms and the processes incidental to and characteristic of life.
- **functional group:** a specific group of atoms with a characteristic structure and corresponding chemical behavior within a molecule.
- **macromolecule:** a very large molecule; most often refers to polymers but can also refer to single molecules with extended, non-polymeric structures.
- **monomer:** a molecule capable of bonding to other molecules to form a polymer.
- **organic compound:** generally, a compound containing one or more carbon atoms, although some carbon-containing compounds are considered inorganic.

THE NATURE OF BIOMOLECULES

Biomolecules are the organic molecules that make up living organisms. Five particular classes of biomolecules are particularly significant in the study of biochemistry: proteins, enzymes, carbohydrates, lipids, and nucleic acids, all of which can be described as macromolecules. A macromolecule is, quite simply, a molecule with a large mass; macromolecules are often polymers, but this is not always the case.

All biochemical macromolecules are organic compounds, their essential molecular structure being composed of carbon atoms. The particular chemical and biochemical behaviors of the macromolecules derive from the various functional groups that are present in their molecular structures.

PROTEINS AND ENZYMES

The genetic code in deoxyribonucleic acid (DNA) carries the blueprint for more than tens of thousands of different polypeptide compounds that combine in polymeric chains to form proteins and enzymes. A polypeptide consists of four or more amino-acid subunits chemically bonded together in a linear head-to-tail fashion by peptide bonds, which are simply amide bonds formed between a carboxyl functional group (–COOH) and an amine, or amino, functional group ($-NH_2$).

A protein is a polypeptide that performs a specific role in biochemical processes, typically either as an enzyme or as part of the many types of tissue. Fibrous proteins provide structural support to tissues such as muscle, hair, and cartilage, while globular proteins transport and store nutrients and can act as catalysts for a number of biochemical reactions necessary to maintaining life. A globular protein that catalyzes or mediates specific biochemical reactions is considered to be an enzyme. Essentially all biochemical processes are enzyme mediated, with each particular enzyme serving to catalyze or facilitate a specific biochemical transformation. Enzymes control the transcription and translation of genetic information by which proteins are formed.

The broad versatility of proteins and enzymes in carrying out biochemical functions is a result of the nearly infinite possible combinations of their component amino acids, which number approximately twenty and can be repeated all but indefinitely. The linear sequence of the amino acids in a protein chain determines the protein's primary structure, and even slight differences in the order of the component amino acids can create an entirely different protein. The secondary structure of a protein is largely determined by hydrogen bonding between the carboxyl and amino groups of the various amino acids, which can cause various segments of the macromolecule to assume either sheetlike or coiled shapes, giving the protein flexibility and strength. The tertiary structure of a protein derives from the three-dimensional shape of the protein molecule, which can be determined by interactions between specific side-chain functional groups. These interactions may determine the shape of the active site of an enzyme or the specific compound or part of a compound that is amenable to enzyme catalysis. The quaternary structure of an enzyme results when two separate enzyme proteins interact without bonding chemically to each other to form a protein complex with specific enzymic activity.

CARBOHYDRATES

Carbohydrates, including sugars, starches, and cellulose, serve as a food source for most organisms and provide structural support to plants. Most carbohydrates have the empirical formula CH_2O, meaning that there are typically two hydrogen atoms for every carbon and oxygen atom. Generally, however, the term "carbohydrate" is used to refer to the sugars, or saccharides, that form the most basic units of carbohydrates.

Simple sugars such as glucose and fructose are called monosaccharides because they cannot be broken down further through hydrolysis, a chemical reaction that splits bonds in the presence of water. Monosaccharides are characterized by the presence of one carbonyl group (C=O) and a hydroxyl group (-OH) on each of the non-carbonyl carbon atoms. They may be either ketoses (polyhydroxy ketones), in which the carbonyl group is attached to two carbons atoms, or aldoses (polyhydroxy aldehydes), in which it is attached to a hydrogen atom. When two monosaccharides are chemically bonded to each other, they form a disaccharide, which has a molecular structure that can be broken down by hydrolysis to form two monosaccharides. The disaccharide sucrose ($C_{12}H_{22}O_{11}$), commonly known as table sugar, is formed from the monosaccharides glucose and fructose (both isomers with the molecular formula $C_6H_{12}O_6$) by the elimination of water (H_2O) through a condensation reaction to form a carbon-oxygen-carbon bond, called a glycosidic bond. Polysaccharides, such as cellulose, glycogen, and starch, feature long chains or rings of monosaccharide units.

Glucose is made in green plants by the process of photosynthesis, in which atmospheric carbon dioxide (CO_2) and water are combined through the heat energy of sunlight. The glucose molecules that are formed link together to form much larger polysaccharides called starches, which often serve as energy stores for living organisms, or celluloses, which form the structural material of plants and trees. When consumed as food, carbohydrates are subjected to hydrolysis that separates the individual monosaccharides from the polymeric chain. In respiration, each glucose molecule is oxidized back into carbon dioxide and water, resulting in the release of energy stored in the glucose's chemical bonds, which can then be used by the organism to fuel other biochemical processes.

PROTEINS/ENZYMES, CARBOHYDRATES, LIPIDS, NUCLEIC ACIDS

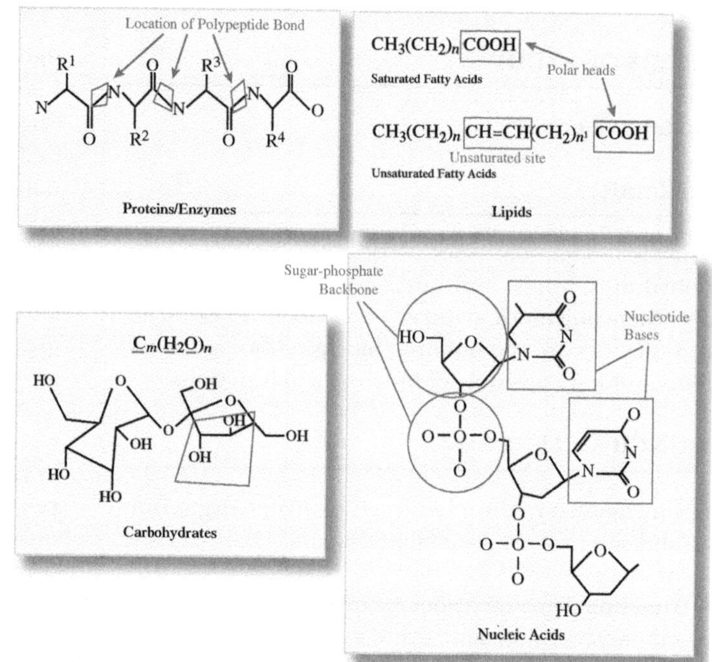

Proteins and enzymes have a characteristic bond formation of a carbon double-bonded to oxygen and single-bonded to the nitrogen of another amino acid. Lipids are composed of hydrocarbon tails and polar heads, often with a carbon double-bonded to an oxygen and hydrogen group. Carbohydrates have a characteristic molecular formula of oxygen and hydrogen in a 1:2 ratio. Nucleic acids have a characteristic sugar-phosphate backbone and nucleotide bases.

LIPIDS

The term "lipids" describes a wide variety of compounds, including fats, phospholipids, waxes, and steroids. It is most often used in reference to fats, which are formed as esters of glycerol and fatty acids. A fatty acid is a long-chain carboxylic acid with the basic formula R-COOH, where R is a hydrocarbon chain of variable length, most often containing between twelve and twenty carbon atoms. The glycerol end of the esters may become bonded to a phosphate group, making the lipid into a phospholipid. The phosphate group enhances the hydrophilic (literally

SAMPLE PROBLEM

Classify the following compounds as protein, nucleic acid, lipid, or carbohydrate:

- fructose
- glyceryl tripalmitate
- amylase
- mRNA

Answer:

- Fructose is a simple sugar with the chemical formula $C_6H_{12}O_6$. The formula shows that there are two hydrogen atoms and one oxygen atom for each carbon atom, so fructose is a carbohydrate. Carbohydrate names typically end in "-ose."
- Glyceryl palmitate is the triester of glycerol and palmitic acid. It is a lipid. Ester names typically end in "-ate."
- Amylase is the starch-cleaving enzyme secreted in saliva. It is a protein. Enzyme names typically end in "-ase."
- mRNA is the abbreviation for messenger ribonucleic acid. It is a nucleic acid.

"water loving," meaning it is attracted to water and other polar substances) character of that end of the molecular chain. The long hydrocarbon chains, on the other hand, are hydrophobic ("water fearing," repelled by water and other polar substances). This leads to the natural formation of a structure called a lipid bilayer. The phospholipid bilayer is a major component of cell membranes, as well as the membranes surrounding many of the organelles and other components within cells.

Nucleic Acids

The nucleic acids DNA and ribonucleic acid (RNA) are complex polymeric molecules constructed by the sequential addition of hundreds of thousands of similar structural units called nucleotides. In DNA, each nucleotide contains a molecule of deoxyribose sugar ($C_5H_{10}O_4$), while in RNA, each nucleotide contains a ribose sugar ($C_5H_{10}O_5$); in both, the sugar is bonded to either a purine or pyrimidine nitrogenous base at one end and an inorganic phosphate group at the other. The structure of both DNA and RNA thus consists of a long chain of sugar molecules alternating with phosphate groups. In the case of DNA, the base on each nucleotide matches up with a complementary base on a nucleotide in a second DNA strand, and the two strands twist together to form the double-helix structure of a DNA molecule. Only four nitrogenous bases are used to construct either DNA or RNA. In DNA, the component bases are adenine, guanine, thymine, and cytosine; in RNA, the base uracil is used instead of thymine, but all the others are the same. The order of nucleotides in the DNA molecule specifies the order of amino acids in all of the proteins and enzymes in an organism.

Polymeric and Non-Polymeric Biomolecules

Polymers are formed by the sequential addition of individual units called monomers, which are small molecules that easily bond together to form a chain. Proteins and enzymes are polymers that are formed from amino acids that link together via peptide bonds. Nucleic acids, including DNA and RNA, are also a form of polymer, constructed from hundreds of thousands of individual units called nucleotides, which themselves are composed of a nucleoside (a five-carbon sugar and a nitrogen-containing base) and a phosphoric acid. Carbohydrates, particularly the starches, celluloses, and glycogen, are formed by the sequential addition of thousands of molecules of monosaccharides such as glucose and fructose.

Lipids are the smallest of the biochemical macromolecules and are insoluble in water. There are various classes of lipids, the most well known of which are the triglycerides, or fats. Triglycerides generally consist of a molecule of glycerol, an alcohol containing three hydroxyl groups, that has been esterified by long-chain carboxylic acids called fatty acids. Because the molecules that make up a lipid are not structured in the form of a repetitive chain, lipids are non-polymeric macromolecules.

Richard M. Renneboog, MSc

Bibliography

Berg, Jeremy M., John L. Tymoczko, Gregory J. Gatto, and Lubert Strye. *Biochemistry.* 8th ed. New York: W. H. Freeman, 2015. Print.

Lehninger, Albert L. *Biochemistry: The Molecular Basis of Cell Structure and Function.* 2nd ed. New York: Worth, 1975. Print.

Lodish, Harvey, et al. *Molecular Cell Biology.* 7th ed. New York: Freeman, 2013. Print.

Morrison, Robert Thornton, and Robert Neilson Boyd. *Organic Chemistry.* 7th ed. Englewood Cliffs, N.J.: Prentice Hall, 2003. Print.

Myers, Richard. *The Basics of Chemistry.* Westport: Greenwood, 2003. Print.

PROTONS

FIELDS OF STUDY

Nuclear Chemistry; Inorganic Chemistry; Organic Chemistry

SUMMARY

The discovery and basic properties of protons are described. Protons are found in all atoms and provide half of the mass for about the first twenty elements. Each proton bears a positive charge that must be countered by the presence of neutrons in the nucleus.

PRINCIPAL TERMS

- **atomic number:** the number of protons in the nucleus of an atom, used to uniquely identify each element.
- **electropositive:** describes an atom that tends to lose electrons to form a positively charged ion.
- **fundamental particle:** one of the smaller, indivisible particles that make up a larger, composite particle; commonly used to refer to electrons, protons, and neutrons, although these are themselves composed of various actual fundamental particles, such as quarks, leptons, and certain types of bosons.
- **nucleus:** the central core of an atom, consisting of specific numbers of protons and neutrons and accounting for at least 99.98 percent of the atomic mass.
- **protium:** the essential form of hydrogen, containing one proton and no neutron; the most common form of matter in the known universe.

The Identification of Protons

Various models of atomic structure were proposed throughout the nineteenth and twentieth centuries, the most useful being the "planetary" model and the "plum pudding" model. In the first, an atom was envisioned to be composed of a nucleus with electrical charges whirling about it like the planets around the sun. In the latter model, the atom was envisioned to be a round, positively charged blob with dots of negative electrical charge embedded throughout. In 1897, while conducting research on the nature of cathode rays in cathode ray tubes, J. J. Thomson (1856–1940) discovered that the cathode

PROTONS

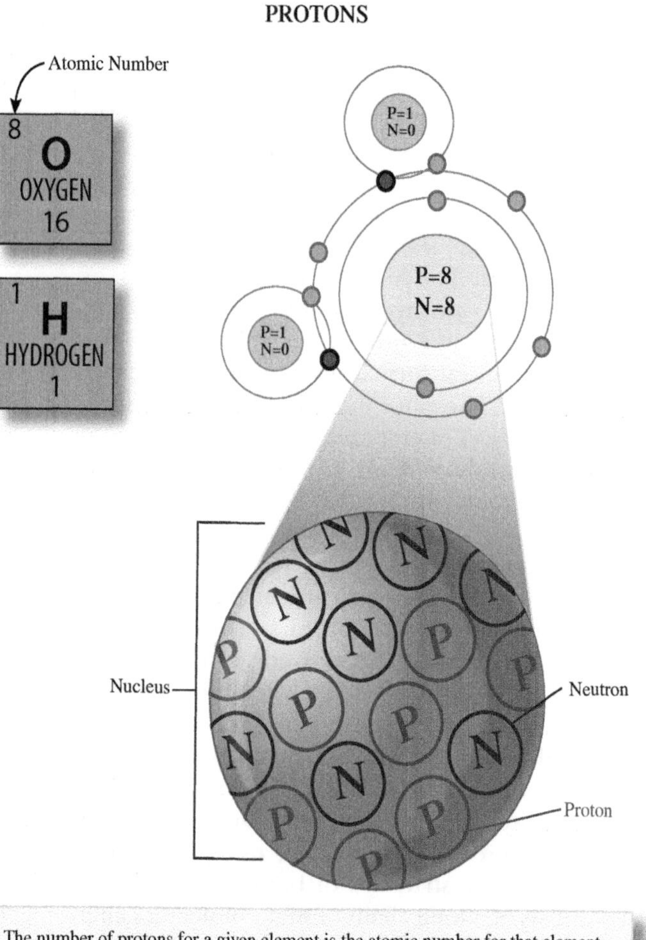

The number of protons for a given element is the atomic number for that element. If there's a different number of protons, it is a different element.

rays behaved more like streams of charged particles than like electromagnetic radiation. He called the particles "corpuscles" at first, though they soon became known as electrons. Based on this discovery, Thomson proposed the plum-pudding model of the atom in 1904.

Thomson's discoveries verified that atoms had an internal structure consisting of fundamental particles, but the existence of subatomic particles other than electrons had yet to be proved. In 1909, Ernest Rutherford (1871–1937), Hans Geiger (1882–1945), and Ernest Marsden (1889–1970) performed what is known as the gold-foil experiment, which produced more conclusive results. By directing a beam of α (alpha) particles at a very thin target of gold foil, they found that while most of the particles passed directly through the foil as though it were not there, some of the particles were deflected at various angles, as though they had bounced off something very small and very dense inside the metal. These observations led Rutherford to propose the planetary model of atomic structure, with the majority of the atom's mass concentrated in a small, dense nucleus. He continued experimenting with alpha particles, and in 1917 he discovered that the collision of an alpha particle with the nucleus of a nitrogen atom released a positively charged particle identical to a hydrogen nucleus—in other words, a proton, as the most common isotope of hydrogen contains no neutrons. This breakthrough is widely credited as the discovery of the proton.

SAMPLE PROBLEM

Calculate the number of protons in 0.6 mole of the inorganic compound gold(I) perchlorate ($AuClO_4$). Note that the roman numeral indicates the oxidation state of the gold, which relates to the number of electrons possessed by the gold atom and has no effect on the number of protons.

Answer:

The atomic number of gold (Au) is 79; therefore, the gold atom contains 79 protons. The atomic number of chlorine (Cl) is 17, so the chlorine atom has 17 protons. The atomic number of oxygen (O) is 8, so the four oxygen atoms have 8 protons each. Calculate the number of protons in one molecule of gold(I) perchlorate:

$$79 + 17 + (4 \times 8) = 128$$

One mole of any pure compound contains exactly $6.02214129 \times 10^{23}$ molecules. This value is known as the Avogadro constant or Avogadro's number, and it is commonly rounded to 6.022×10^{23}. Multiply this number by the number of protons in one molecule of gold(I) perchlorate:

$$128 \times (6.022 \times 10^{23}) = 7.70816 \times 10^{25}$$

Therefore, one mole of gold(I) perchlorate contains 7.70816×10^{25} protons.

Properties and Function of Protons

The elemental identity of every atom is determined by the number of protons in its nucleus, which is also the element's atomic number. With the exception of hydrogen, the numbers of protons and neutrons in the nuclei of the first twenty elements are approximately equal (allowing for some variation in different isotopes), which means that protons account for about half of the mass of the atoms of these first twenty elements. Hydrogen is a special case because its most common form by far is protium, which has one proton and no neutrons in its nucleus. Protium is what is typically referred to when discussing hydrogen; the isotope with one proton and one neutron is called deuterium, and it is significantly less abundant. If a protium atom loses its sole electron, it forms the positively charged ion H^+, which has no electrons or neutrons and thus is essentially a proton.

After element 20, which is calcium, the protons of each element are increasingly outnumbered by the neutrons. For example, zinc (atomic number 30) typically has 35 neutrons and 30 protons, while uranium (92) has 146 neutrons but only 92 protons. Each proton bears a single positive electrical charge, and the protons in a given atom must be countered by an equal number of electrons for that atom to be neutral. The neutrons in the nucleus interact more closely with the protons and stabilize them against the force of electrostatic repulsion that would otherwise drive them apart. Electropositive elements are those that tend to give up one or more valence electrons in order to form positively charged ions that are more stable than their neutral forms. The positive charge on such an ion results from the fact that the protons are no longer balanced by an equal number of electrons.

PROTONS IN CHEMICAL REACTIONS

Protons are most important in normal chemical reactions when those reactions involve the protium atom. Because the hydrogen atom has only a single electron, that electron's involvement in bond formation tends to leave the positive charge of the hydrogen nucleus exposed. If hydrogen forms a compound with an element of high electron density, such as oxygen, nitrogen, or fluorine, this creates a polar molecule, in which the positive charge is concentrated at the hydrogen end of the molecule and the negative charge is concentrated at the other end. The polar molecules are then attracted to each other, with the electron-dense end of one molecule forming a hydrogen bond with the hydrogen atom of another molecule. (A hydrogen bond is not a true chemical bond, but rather a strong electrical attraction between molecules.) This fact is the reason that water, a highly polar molecule, has such high melting and boiling points relative to other compounds of similar molecular weight. Compounds that are able to release the H^+ ion—that is, the proton—from their molecular structure when dissolved in water or another polar solvent are defined as acids. Acids are valuable active components of reaction processes as well as indispensable analytical tools.

Richard M. Renneboog, MSc

BIBLIOGRAPHY

Gribbin, John. *Science: A History, 1543–2001.* London: Lane, 2002. Print.

Johnson, Rebecca L. *Atomic Structure.* Minneapolis: Lerner, 2008. Print.

Kean, Sam. *The Disappearing Spoon: And Other True Tales of Madness, Love, and the History of the World from the Periodic Table of the Elements.* New York: Little, Brown, 2010. Print.

Myers, Richard. *The Basics of Chemistry.* Westport: Greenwood, 2003. Print.

Wehr, M. Russell, James A. Richards Jr., and Thomas W. Adair III. *Physics of the Atom.* 4th ed. Reading: Addison-Wesley, 1984. Print.

Winter, Mark J. *The Orbitron: A Gallery of Atomic Orbitals and Molecular Orbitals.* University of Sheffield, n.d. Web.

R

RADIOACTIVE DECAY

FIELDS OF STUDY

Nuclear Chemistry; Geochemistry; Analytical Chemistry

SUMMARY

The characteristics and principles of radioactive decay are discussed. Radioactive decay is a spontaneous process that occurs in the nuclei of unstable atoms, causing various subatomic particles to be ejected from the nucleus.

PRINCIPAL TERMS

- **alpha particle:** a particle consisting of two protons and two neutrons bound together, identical to a helium nucleus; typically produced in the process of alpha decay.
- **beta particle:** an electron or positron produced in the process of beta decay, in which a neutron of an unstable atom decomposes into a proton, and emitted from the nucleus at high speed.
- **electron:** a fundamental subatomic particle with a single negative electrical charge, found in a large, diffuse cloud around the nucleus.
- **gamma ray:** ionizing electromagnetic radiation of higher energy and shorter wavelength than x-rays, typically produced by the nucleus of an atom undergoing radioactive decay.
- **positron:** the antiparticle of the electron, with identical physical properties but the opposite (positive) electrical charge.

THE NATURE OF RADIOACTIVE DECAY

According to the modern theory of atomic structure, matter is composed of atoms. Each atom has a nucleus composed of a very specific number of protons and neutrons. The identity of each atom as an element is determined by the number of protons in its nucleus, while the total number of nucleons (neutrons and protons) determines the mass of the atom. Atoms that have the same number of protons but different numbers of neutrons in their nuclei are different isotopes of the same element.

Each atom also has the same number of electrons as there are protons in its nucleus. The electrons surround the nucleus in a sort of cloud that is up to hundreds of thousands of times greater in diameter than the nucleus. Because electrons are so much smaller than nucleons and the electron cloud is so diffuse, the vast majority of an atom is just empty space. The normal chemical behavior of all atoms, particularly in regard to bond formation and rearrangement, takes place at the level of the outermost, or valence, electrons.

Each proton carries a single positive charge, while each electron carries a single negative charge, making each atom electrically neutral overall, unless it has either accepted or donated one or more electrons to become an ion. Because the protons in the nucleus are packed together in a dense mass, the repulsive force created by their like electrical charges would drive them apart if not for the neutrons, which act as a sort of nuclear glue to keep the protons bound together.

According to particle physics, protons and neutrons are composed of combinations of even smaller particles called "quarks." (Quarks are one of two classes of elementary fermionic particles, the other being leptons, a category that includes electrons.) Accordingly, the neutron is believed to have an internal structure that positions negative energy at the surface of the particle to counter the positive charge of an adjacent proton.

For the first dozen or so of the elements, a number of neutrons equal to the number of protons is sufficient to maintain the stability of the nucleus. But as the number of protons increases, the number of neutrons required to maintain a stable nucleus becomes even greater. In some nuclei, the number of protons is so great that no amount of neutrons is

RADIOACTIVE DECAY

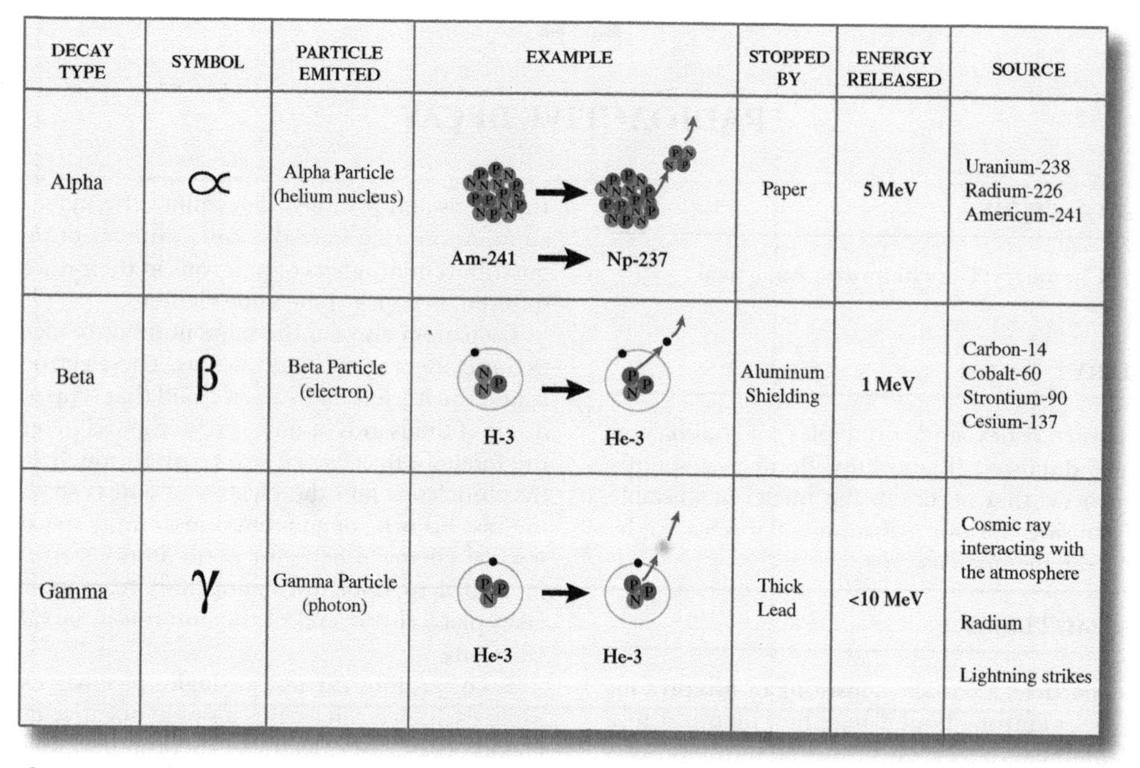

DECAY TYPE	SYMBOL	PARTICLE EMITTED	EXAMPLE	STOPPED BY	ENERGY RELEASED	SOURCE
Alpha	α	Alpha Particle (helium nucleus)	Am-241 → Np-237	Paper	5 MeV	Uranium-238 Radium-226 Americum-241
Beta	β	Beta Particle (electron)	H-3 → He-3	Aluminum Shielding	1 MeV	Carbon-14 Cobalt-60 Strontium-90 Cesium-137
Gamma	γ	Gamma Particle (photon)	He-3 → He-3	Thick Lead	<10 MeV	Cosmic ray interacting with the atmosphere Radium Lightning strikes

enough to properly counteract the repulsive forces between them without creating additional instability. Nuclei with either too few or too many neutrons are unstable and are said to be radioactive, meaning that they are undergoing the process of radioactive decay.

MODES OF RADIOACTIVE DECAY

When an unstable nucleus decays, it does so by emitting various nuclear particles, energy, or both. All such emissions are referred to generally as "radiation." In the process of alpha decay, the unstable nucleus emits an alpha (α) particle, which consists of two protons and two neutrons, making it essentially identical to the nucleus of a helium atom. The alpha particle then acquires the two electrons that the decayed atom no longer requires, becoming a neutral helium atom in the process. Since the decaying nucleus has lost two protons and two neutrons, it has become an atom of an element with an atomic number two less than the original element, and its mass is four atomic mass units (u) lower than before. For example, the emission of an alpha particle from an atom of cobalt-60, which has an atomic number of 27 and a mass of 60 atomic mass units, transmutes that atom into manganese-56, which has an atomic number of 25 and a mass of 56 atomic mass units. (When discussing isotopes such as cobalt-60 and manganese-56, the number that follows the element name identifies the isotope by its atomic mass.)

In beta-decay processes, the nucleus emits a beta (β) particle—which can be either an electron (beta-minus decay, or β^-) or a positron (beta-plus decay, or β^+) and is produced when a neutron transmutes into a proton (β^-) or vice versa (β^+)—along with either an antineutrino (β^-) or a neutrino (β^+). This process also changes the elemental identity of the atom, transmuting it into an element with an atomic number either one higher (β^-) or one lower (β^+). In both types of beta decay, the mass of the atom remains effectively unchanged, as the transformed nucleon remains in the nucleus and the only particles emitted are the electron (or positron) and the antineutrino (or neutrino), which are of negligible mass. For example, the beta-plus decay of an atom of germanium-69 (atomic number 32) transmutes it into an atom of gallium-69 (atomic number 31).

Other, more unusual modes of decay include neutron emission, in which a nucleus containing too many neutrons simply ejects one; proton emission, which is a similar process but with protons; electron capture, in which the nucleus of an atom with excess protons absorbs one of its own inner electrons and uses it to transform a proton into a neutron; and neutron capture, in which an atom collides with a free neutron and absorbs it into its nucleus. Very heavy elements may also undergo spontaneous fission, in which the nucleus splits into two smaller, stable nuclei, accompanied by several free neutrons. All forms of decay may involve the emission of gamma (γ) rays, or high-energy photons, from the excited nucleus as it settles into a lower energy state. Gamma rays represent the difference in energy between the original nucleus and the decay product.

THE RATE OF RADIOACTIVE DECAY

Radioactive decay is a first-order kinetic process, which means that it is dependent only on a single component in the system—in this case, the radioactive element. The rate at which radioactive decay (and other first-order processes) occurs is described mathematically by the equation:

$$A_t = A_0e^{-}kt$$

where A_t is the amount of the element remaining after a certain length of time (t), A_0 is the amount of the element present prior to decay, k is the specific rate constant for the decay process, and t is the length of time that has passed. This relationship is used to define the half-life of a particular decay process, which, coupled with the ratio of starting material to decay product, can be used to determine the amount of time that has passed. It is the foundation of radioactive dating methods such as carbon-14 dating, which is commonly used in archaeology to determine the age of organic remains.

A BRIEF HISTORY OF RADIOACTIVE DECAY

Since ancient times, miners were often plagued by unexplained illnesses. It is now suspected that this was due to the presence of uranium and other naturally radioactive elements in the ores they mined. The phenomenon of radioactivity was first identified in 1896 by Henri Becquerel (1852–1908), who realized that the effect he was observing in uranium salts was different from the x-rays that had been discovered only a year before. Marie Curie (1867–1934) and Pierre Curie (1859–1906), who were students of Becquerel at the time, went on to discover radium and polonium in the same ores that yielded uranium.

Because radioactive emissions are not visible to the human eye, some means of detecting them was needed. The Geiger counter was invented in 1928 by Hans Geiger (1882–1945) and Walther Müller (1905–1979), based on a technique Geiger had developed twenty years earlier. Geiger counters work by detecting the interaction of ionizing particles

SAMPLE PROBLEM

Calculate the number of atoms of cobalt-60 (^{60}Co) remaining from a 2.0-gram sample after exactly 168 hours. Use the first-order equation $A_t = A_0e^{-}kt$ and the decay-rate constant $k = 4.15\times10^{-9}\frac{1}{\text{sec}}$

Answer:

The atomic weight of ^{60}Co is 60 u, which is equivalent to 60 g/mol. Convert the weight of the sample from grams to moles:

$$\frac{60\text{g}}{1\text{ mol}} = \frac{2\text{g}}{n\text{ mol}}$$

$$(60\text{ g})(n\text{ mol}) = 2\text{ g mol}$$

$$n\text{ mol} = \frac{2\text{g mol}}{60\text{g}} = 0.0333\text{ mol}$$

Multiply 0.0333 mol by Avogadro's number (6.022×10^{23}) to determine the number of atoms in the original sample (A_0):

$$A_0 = (0.0333)(6.022 \times 10^{23}) = 2.005326 \times 10^{22}\text{ atoms}$$

Convert the time passed (t) from hours to seconds:

$$t = 168 \times 60 \times 60 = 604{,}800\text{ sec}$$

Plug the values of k, A_0, and t into the first-order equation and solve for A_t:

$$A_t = (2.005326\times10^{22}\text{ atoms})e^{-(4.15\times10^{-9}\frac{1}{\text{sec}})(604800\text{ sec})}$$

$$A_t = (2.005326\times10^{22}\text{ atoms})e^{-0.00251}$$

$$A_t = 2.000299\times10^{22}\text{ atoms}$$

The amount of ^{60}Co remaining after 168 hours is approximately 2.000299×10^{22} atoms. This corresponds to 99.7 percent of the original amount.

and photons with a tube of inert gas. Scintillation counters, developed in 1944 by Samuel Curran (1912–1998), use a type of material called a "scintillator," which emits photons when struck by ionizing radiation. Unlike Geiger counters, scintillation counters can differentiate between alpha, beta, and gamma radiation and determine the energy levels of each. They are also sensitive to a wider range of energies than are Geiger counters. Radiation detection is an important aspect of nuclear power generation and hazardous-waste management, as well as an important diagnostic tool in medical, biochemical, and historical research.

Richard M. Renneboog, MSc

BIBLIOGRAPHY

Askeland, Donald R., Wendelin J. Wright, D. K. Bhattacharya, and Raj P. Chhabra. *The Science and Engineering of Materials.* Boston: Cengage Learning, 2016. Print.

Douglas, Bodie E., Darl H. McDaniel, and John J. Alexander. *Concepts and Models of Inorganic Chemistry.* 3rd ed. New York: Wiley, 1994. Print.

Haynes, W. H., PhD, David R. Lide, PhD, and Thomas J. Bruno, PhD, eds. *CRC Handbook of Chemistry and Physics,* 96th Edition. New York: CRC/Taylor & Francis Group, 2015. Print.

Jones, Mark Martin. *Chemistry and Society.* 5th ed. Philadelphia: Saunders College Pub., 1987. Print.

Mackay, K. M., R. A. Mackay, and W. Henderson. *Introduction to Modern Inorganic Chemistry.* 6th ed. Cheltenham: Nelson, 2002. Print.

Miessler, Gary L., Paul J. Fischer, and Donald A. Tarr. *Inorganic Chemistry.* 5th ed. Upper Saddle River: Prentice Hall, 2013. Print.

Myers, Richard. *The Basics of Chemistry.* Westport: Greenwood, 2003. Print.

Silbey, Robert J., Robert A. Alberty, and Moungi G. Bawendi. *Physical Chemistry.* 5th ed. Hoboken: Wiley, 2012. Print.

Skoog, Douglas Arvid, Donald M. West, and F. James. Holler. *Skoog and West's Fundamentals of Analytical Chemistry.* 9th ed. Belmont (Calif.): Cengage Learning, 2014. Print.

Wehr, M. Russell, James A. Richards Jr., and Thomas W. Adair III. *Physics of the Atom.* 4th ed. Reading: Addison-Wesley, 1984. Print.

RADIOACTIVE ELEMENTS

FIELDS OF STUDY

Inorganic Chemistry; Geochemistry; Metallurgy

SUMMARY

The basic properties of radioactive elements are described in the context of nuclear composition. Radioisotopes are typically described by their particular half-life, since all radioactive isotopes decompose according to the same rate law.

PRINCIPAL TERMS

- **half-life:** the length of time required for one-half of a given amount of material to decompose or be consumed through a continuous decay process.
- **isotope:** an atom of a specific element that contains the usual number of protons in its nucleus but a different number of neutrons.
- **man-made element:** an element or isotope that does not occur naturally but is synthesized in high-energy particle accelerators by bombarding other elements with streams of nuclear particles.
- **radioactive decay:** the loss of particles from the nucleus of an unstable atom in the form of ionizing radiation.
- **unstable:** describes a chemical species with a structure or composition that is prone to spontaneous decomposition.

THE NATURE OF RADIOACTIVE ELEMENTS

Essentially all elements have known radioactive isotopes. Only a small number of naturally occurring isotopes are radioactive, however. The majority of radioactive elements are man-made elements that have been produced and identified by nuclear synthesis reactions in high-energy particle collision experiments. Some of these are valuable tools in diagnostic medicine and other scientific applications, but most are of value only as empirical data

RADIOACTIVE ELEMENTS

92 U uranium 238

1 H hydrogen 1																	2 He helium 4
3 Li lithium 7	4 Be beryllium 9											5 B boron 11	6 C carbon 12	7 N nitrogen 14	8 O oxygen 16	9 F fluorine 19	10 Ne neon 20
11 Na sodium 28	12 Mg magnesium 29											13 Al aluminium 27	14 Si silicon 28	15 P phosphorous 31	16 S sulfur 32	17 Cl chlorine 35	18 Ar argon 40
19 K potassium 39	20 Ca calcium 40	21 Sc scandium 45	22 Ti titanium 48	23 Vi vanadium 51	24 Cr chromium 52	25 Mn manganese 55	26 Fe iron 56	27 Co cobalt 59	28 Ni nickel 59	29 Cu copper 64	30 Zn zinc 65	31 Ga gallium 70	32 Ge germanium 73	33 As arsenic 75	34 Se selenium 79	35 Br bromine 80	36 Kr krypton 84
37 Rb rubidium 85	38 Sr strontium 88	39 Y yttrium 89	40 Zr zirconium 91	41 I niobium 93	42 Mo molybdenum 96	43 Tc technetium 98	44 Ru ruthenium 101	45 Rh rhodium 103	46 Pd palladium 106	47 Ag silver 108	48 Cd cadmium 112	49 In indium 115	50 Sn tin 119	51 Sb antimony 122	52 Te tellerium 128	53 I iodine 127	54 Xe xenon 131
55 Cs caesium 133	56 Ba barium 137	*	72 Hf hafnium 178	73 Ta tantalum 181	74 W tungsten 184	75 Re rhenium 186	76 Os osmium 190	77 Ir iridium 192	78 Pt platinum 195	79 Au gold 197	80 Hg mercury 201	81 Tl thallium 204	82 Pb lead 207	83 Bi bismuth 209	84 Po polonium 209	85 At astatine 210	86 Rn radon 222
87 Fa francium 223	88 Ra radium 226	**	104 Rf rutherfordium 267	105 Db dubnium 268	106 Sg seaborgium 271	107 Bh bohrium 272	108 Hs hassium 270	109 Mt meitnerium 276	110 Ds darmstadtium 281	111 Rg roentgenium 280	112 Cn copernicum 285	113 Uut ununtrium 284	114 Fl flerovium 289	115 Uup ununpentium 288	116 Lv livermorium 293	117 Uus ununseptium 294	118 Uuo ununoctium 289

57 * La lanthanum 139	58 Ce cerium 140	59 Pr praseodymium 141	60 Nd neodymium 144	61 Pm promethium 145	62 Sm samarium 150	63 Eu europium 152	64 Gd gadolinium 157	65 Tb terbium 159	66 Dy dysprosium 163	67 Ho holmium 165	68 Er erbium 167	69 Tm thulium 168	70 Yb ytterbium 173	71 Lu lutetium 175
89 ** Ac actinium 227	90 Th thorium 232	91 Pa protactinium 231	92 U uranium 238	93 Np neptunium 237	94 Pu plutonium 244	95 Am americium 243	96 Cm curium 247	97 Bk berkelium 247	98 Cf californium 251	99 Es einsteinium 252	100 Fm fermium 257	101 Md mendelevium 258	102 No nobelium 259	103 Lr lawrencium 284

for academic study and theoretical models of atomic behavior. Radioactive elements are characterized by the spontaneous radioactive decay of their unstable atomic nuclei. Radioactive decay can emit a number of different kinds of radiation, but the conversion of one radioactive element into an element having a lower atomic number always requires the emission of protons.

Nuclear Structure of Radioactive Elements

All atoms, except the simplest isotope of hydrogen (protium), have both neutrons and protons in their nuclei. For the first twenty elements, the number of both neutrons and protons are almost exactly equal; for the remaining elements, however, the number of neutrons increases rapidly relative to the number of protons. Each proton bears a single positive electrical charge. This charge is stable in hydrogen atoms but not in any other nucleus. The force of electrostatic repulsion that exists between similar electrical charges would drive the protons apart from each other and destroy the nucleus. The presence of neutrons counteracts this effect or, at the very least, provides a stabilizing factor.

Neutrons themselves are electrically neutral, but advanced studies in particle physics has suggested that the neutron has an internal structure that includes a negative charge. It is theorized that the internal structure of the neutron rearranges in such a way as to interpose its internal negative charge between the positively charged protons in a nucleus. Since the supposed internal structure of the neutron would not be allowed to change in composition, there must then be a point at which the one-to-one ratio of neutrons to protons is not able to provide the necessary stabilization. It appears to be at twenty protons that additional stabilization becomes necessary. Thus, all elements higher than calcium have increasingly more neutrons than protons. The stable element with the highest atomic number is uranium, which occurs naturally as uranium-238. Only a small percentage of naturally occurring uranium has a different atomic weight, that which comprises uranium-235. All known elements having an atomic number higher than polonium (84) are unstable and radioactive. The implication is that there is a limit to the amount of nuclear stabilization that neutrons can provide neutrons; at some point the number of neutrons becomes unstable. (In an ironic reversal, the neutrons that function to stabilize protons above atomic number 20 can themselves apparently no longer be stabilized by the protons when there are more than 125 of them.)

Radioactive Decay Processes

Unstable atomic nuclei decay at an exponential rate is described by the following mathematical equation:

$$[A]_t = [A]_0 e^{-kt}$$

where $[A]_0$ is the amount of a specific material present at some starting point in time, $[A]_t$ is the amount of that material remaining after a certain amount of time has passed, t is the amount of time that has passed, and k is the specific rate constant for that material. For radioactive nuclei, the decay rate can be measured directly using a Geiger or scintillation counter. Counting the number of radioactive emissions given off by a specific amount of material over a specific period of time gives a direct measurement of the rate constant for that material. The most generally recognized description of radioactive elements is by the half-life, sometimes called the "lifetime," of that specific isotope. The half-life of the isotope thorium-226, for example, has been determined to be just 30.9 minutes, while the half-life of uranium-235 is 7.1×10^8 years and that of uranium-238 is even longer, at 4.51×10^9 years. The half-life is calculated by defining the ratio of $[A]_t$ as exactly half that of $[A]_0$, or:

$$\frac{[A]_t}{[A]_0} = e^{-kt} = \frac{1}{2}$$

$$\ln(\tfrac{1}{2}) = \ln(e^{-kt}) = -kt\tfrac{1}{2} = \ln(1) - \ln(2) = -\ln(2) = -0.693$$

$$t\frac{1}{2} = \frac{0.693}{k}$$

Radioactive nuclei decay by different mechanisms. The type of decay and the observed products of decay offer some indication that the elementary particles (protons, neutrons, and electrons) are themselves composed of even smaller particles, such as quarks and bosons. For several decades, these were just hypothetical constructs proposed in order to fit the mathematics of quantum mechanics. However, in the late twentieth century and early twenty-first century, high-energy experiments have demonstrated that quarks and bosons have physical reality. Their existence makes understandable the conversion of a neutron into a proton accompanied by the emission of an electron or a positron (the positively charged "antimatter" version of the electron).

Nuclei decompose, or decay, by the emission of different kinds of subnuclear particles and energy. Decay of one nucleus to another of smaller atomic mass, such as the conversion of uranium-238 to thorium-234, occurs with the emission of an alpha (α) particle. The α particle consists of two protons and two neutrons and bears two positive charges as a result. Once emitted, the α particle quickly acquires two electrons to become an atom of helium.

The nucleus can also decay with the emission of a beta (β) particle. A β particle is an electron, but it may also be a positron. These are differentiated as β^- and β^+, respectively. A β particle is typically emitted when a neutron decomposes into a proton, which increases the atomic number of the nucleus but not the atomic mass. This process occurs in the decay of ^{90}thorium-234, and produces ^{91}protactinium-234. A third radioactive emission is the gamma (γ) ray. Gamma rays are high-frequency, or short-wavelength, x-rays. They are emitted from the nucleus rather than from the electron shells surrounding the nucleus.

Radioactive decay processes pose complex problems with regard to the amounts of materials. It is often the case that a short-lived radioisotope decays into an isotope with a much longer half-life. Therefore, a sequential series by which one radioactive isotope decays to a stable nonradioactive isotope of a different element is neither short nor consistent. For example, an isotope having a half-life of just several hours may decay initially to another radioactive isotope having a half-life measured in tens of thousands of years. That isotope, in turn, may decay to yet another radioactive isotope having a shorter or even longer half-life, and so on until a final decay step

Marie Curie was a pioneer in radioactivity research. Working with her husband, she discovered the radio-active elements radium and polonium, and developed techniques for isolating radioactive isotopes. By Nobel Foundation, via public domain, Wikimedia Commons

produces a nonradioactive nucleus. This is similar to considerations of the rate-determining step in a series of chemical reactions.

Occurrence of Radioactive Elements

The list of commonly known radioactive elements is short: uranium, plutonium, carbon-14, and the few that are used in medical treatment and research. A few others occur naturally, including polonium, astatine, radon, thorium, uranium, and protactinium. Uranium, polonium, radium, radon, and similar radioactive isotopes are typically obtained from the ore pitchblende. Uranium can be purified by formation of the volatile compound uranium hexafluoride (UF_6), a solid that sublimes into the gas phase at a temperature of just 56 degrees Celsius (133 degrees Fahrenheit). Thorium is mined commercially as the mineral monazite, which is from 3 to 9 percent thorium oxide (ThO_2) by weight. Metallic thorium, useful as a fuel for nuclear reactors and a number of other commercial applications, can be produced from thorium oxide by reduction with calcium metal in a thermite-type reaction. Radon, the heaviest of the inert gases, is produced by the radioactive decay of radium. Francium, the heaviest of the alkali metals, is also the most reactive. It is estimated that no more than fifteen grams of francium exist in the entire crust of the planet at any one time, and no bulk amount of the element has ever been available in sufficient quantities to carry out even the simplest chemical tests.

Carbon-14 occurs naturally in the atmosphere as a result of the interaction of atmospheric carbon dioxide with cosmic rays. The incoming cosmic ray particles strike normal carbon-12 nuclei and fuse to produce carbon-14. The normal abundance of carbon-14 is low but constant. Thus, living organisms that consume the glucose and other carbohydrates produced from atmospheric carbon dioxide

SAMPLE PROBLEM

Write a balanced nuclear reaction equation for the conversion of uranium-238 to thorium-234. What kind(s) of particles or energy is emitted in the process?

Answer:

The atomic number of uranium is 92, indicating that there are 92 protons in the nucleus. The atomic number of thorium is 90, indicating that there are 90 protons in the nucleus. The atomic mass of uranium-238 corresponds to 146 neutrons in the nucleus. The atomic mass of thorium-234 corresponds to 144 neutrons in the nucleus.

$$\text{Mass difference} = 238 - 234 = 4 \text{ atomic mass unit}$$

$$\text{Protons lost} = 92 - 90 = 2$$

$$\text{Neutrons lost} = 146 - 144 = 2$$

Therefore, the decay of the uranium-238 nucleus to the thorium-234 nucleus represents the loss of two protons and two neutrons. This corresponds to the loss of an α particle, which is the nucleus of a helium atom.

A balanced nuclear reaction equation for the process can be written as

$$^{92}U_{238} \rightarrow {}^{90}Th_{234} + He^{2+} + 2e^{-}$$

incorporate a similarly constant amount of carbon-14 into their tissues. When the organism dies, it ceases to maintain the constant carbon-14 level in its tissues, and the radioactive isotope begins to decrease in quantity. The half-life of carbon-14 is fairly precisely known to be 5,730 years, and the amount remaining in an object of organic origin, relative to the amount that would be present in the living organic source, can be used to determine the amount of time that has passed since the organism ceased to take in carbon-14 from the atmosphere.

Essentially all of the elements higher than uranium in the actinide series are synthetic, having been prepared by high-energy physics experiments. Some are known only by the traces left in detection systems from the formation of a mere handful of atoms. All of the hundreds of known isotopes of the lower elements have also been formed by synthetic nuclear reactions in which stable nuclei have been bombarded by beams of high-energy particles in particle accelerators and nuclear reactors.

Richard M. Renneboog, MSc

Bibliography

Gribbin, John. *Science: A History, 1543–2001.* London: Lane, 2002. Print.

Haynes, W. H., PhD, David R. Lide, PhD, and Thomas J. Bruno, PhD, eds. *CRC Handbook of Chemistry and Physics,* 96th Edition. New York: CRC/Taylor & Francis Group, 2015. Print.

Johnson, Rebecca L. *Atomic Structure.* Minneapolis: Lerner, 2008. Print.

Kean, Sam. *The Disappearing Spoon: And Other True Tales of Madness, Love, and the History of the World from the Periodic Table of the Elements.* New York: Little, Brown, 2010. Print.

Mackay, K. M., R. A. *Mackay, and W.Henderson. Introduction to Modern Inorganic Chemistry.* 6th ed. Cheltenham: Nelson, 2002. Print.

Wehr, M. Russell, James A. Richards Jr., and Thomas W. Adair III. *Physics of the Atom.* 4th ed. Reading: Addison-Wesley, 1984. Print.

Winter, Mark J. The Orbitron: A Gallery of Atomic Orbitals and Molecular Orbitals. University of Sheffield, n.d. Web.

REACTANTS AND PRODUCTS

FIELDS OF STUDY

Organic Chemistry; Inorganic Chemistry; Biochemistry

SUMMARY

The concept of reactants and products in chemical reactions is described, and their importance in all related fields is elaborated. Reactants and products are an obligate pair in that one cannot occur without the other. They are also relative terms defined by the direction in which a chemical reaction proceeds.

PRINCIPAL TERMS

- **chemical formula:** the combination of symbols and numerical coefficients that specifies the number and identity of the different atoms involved in a chemical reaction or molecular structure.
- **chemical reaction:** a process in which the molecules of two or more chemical species interact with each other in a way that causes the electrons in the bonds between atoms to be rearranged, resulting in changes to the chemical identities of the materials.
- **compound:** a chemically unique material whose molecules consist of several atoms of two or more different elements.
- **element:** a form of matter consisting only of atoms of the same atomic number.
- **equilibrium:** the state that exists when the forward activity of a process is exactly equal to the reverse activity of that process.

The Relationship between Reactants and Products

The concept of "up" is inseparable from the concept of "down." In the same way, the concept of reactants is inseparable from that of products. When atoms or molecules are brought together under the proper conditions, the energy relationships that govern how electrons behave in chemical bonds allow new bonds to form and existing bonds to change in specific ways, and a chemical reaction takes place. Bonds may break apart, form anew, or simply rearrange. When this happens, the chemical identities of the reactants change. They cease to be the elements or compounds they were before the interaction and become new compounds as a result of the various changes they have gone through. In other words, they transform from reactants, or starting materials, into products.

A chemical reaction can be indicated by the general equation:

$$A + B \rightarrow C + D$$

This particular representation indicates that two compounds, A and B, undergo a reaction to produce two products, C and D. It should be obvious that many other representations, indicating different numbers and types of materials, are possible. By convention, the starting materials—the reactants—are presented on the left, the products on the right. Since the arrow always points from the reactants to the products, however, it is equally valid to write an equation from right to left. It is also true that virtually all reactions that proceed in one direction can be made to proceed in the opposite direction as well, causing the products of a reaction to undergo the "reverse reaction" and re-form the original reactants. That is, the equation:

$$A + B \rightarrow C + D$$

indicating that A and B react to produce C and D can usually also be written as:

$$C + D \rightarrow A + B$$

indicating that C and D react to produce A and B. What were originally the reactants in the first reaction are the products in the second. The designation of which compounds or elements are reactants and which are products is thus relative, depending on which reaction is desired.

Many reaction systems enter a condition called equilibrium. In such cases, the forward reaction, in which the reactants form the products, occurs at the same time as the reverse reaction, in which the products re-form the reactants. At equilibrium, these two opposing reaction processes take place at exactly the same rate, meaning that A and B are forming C and D at the same time and to the same extent that C and D are re-forming A and B. This condition is indicated in the reaction equation by replacing the one-way arrow with a double half arrow that points in both directions:

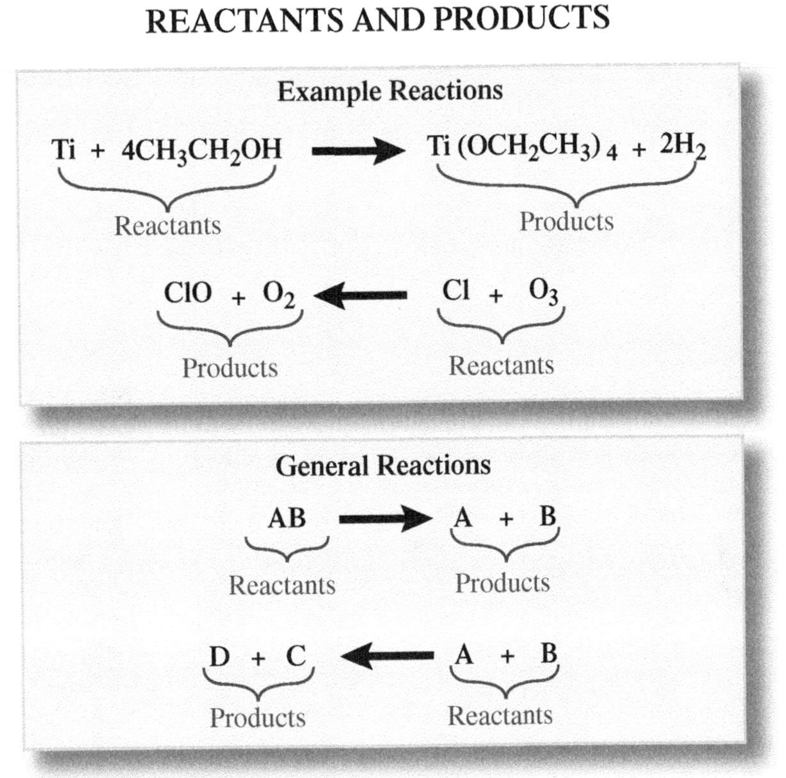

$$A + B \rightleftharpoons C + D$$

Writing Equations with Reactants and Products

In order to provide the precise, usable information that a reaction equation should convey, such as the chemical identities of the reactants and products, a chemical formula that uniquely and unequivocally identifies each reactant and product is required. The chemical formula also uses numerical subscripts and coefficients to identify all the atoms or molecules

SAMPLE PROBLEM

Identify the reactants and the products in these reactions:

$$N_2O_4 \rightleftharpoons 2NO_2$$

$$AlCl_3 + 3H_2O \rightarrow Al(OH)_3 + 3HCl$$

Answer:

The first reaction is an equilibrium in which each reactant molecule of dinitrogen tetroxide decomposes into two product molecules of nitrogen dioxide. In the reverse reaction, two reactant molecules of nitrogen dioxide combine to form one product molecule of dinitrogen tetroxide.

In the second reaction, the reactants are aluminum chloride and three molecules of water, and the products are aluminum hydroxide and three molecules of hydrogen chloride.

that are present and how many of each. For example, the equation:

$$2Al + Fe_2O_3 \rightarrow 2Fe + Al_2O_3$$

clearly states that two atoms of aluminum react with one molecule of iron oxide to produce two atoms of iron and one molecule of aluminum oxide. It also states that each molecule of iron oxide is composed of two atoms of iron and three of oxygen, while each molecule of aluminum oxide is composed of two atoms of aluminum and three of oxygen.

For most organic compounds, the reaction equation is written using a pictorial representation of the three-dimensional molecular structure of each reactant and product rather than just their chemical formulas.

Richard M. Renneboog, MSc

BIBLIOGRAPHY

Berg, Jeremy M., John L. Tymoczko, Gregory J. Gatto, and Lubert Strye. *Biochemistry.* 8th ed. New York: W. H. Freeman, 2015. Print.

Douglas, Bodie E., Darl H. McDaniel, and John J. Alexander. *Concepts and Models of Inorganic Chemistry.* 3rd ed. New York: Wiley, 1994. Print.

Lodish, Harvey, et al. *Molecular Cell Biology.* 7th ed. New York: Freeman, 2013. Print.

Miessler, Gary L., Paul J. Fischer, and Donald A. Tarr. *Inorganic Chemistry.* 5th ed. Upper Saddle River: Prentice Hall, 2013. Print.

Morrison, Robert Thornton, and Robert Neilson Boyd. *Organic Chemistry.* 7th ed. Englewood Cliffs, N.J.: Prentice Hall, 2003. Print.

Reece, Jane B., et al. *Campbell Biology.* 10th ed. San Francisco: Cummings, 2013. Print.

REACTION CALCULATIONS: AVOGADRO'S LAW

FIELDS OF STUDY

Organic Chemistry; Physical Chemistry; Chemical Engineering

SUMMARY

The background and basic principles of Avogadro's law (1811) are elaborated. The principle is fundamental to all branches and applications of chemistry and provides a concrete reference point for stoichiometric calculations, those involving the relative quantities of substances that make up compounds or participate in reactions. Amadeo Avogadro (1776–1856) was an Italian scientist, considered one of the principal founders of the science of physical chemistry.

PRINCIPAL TERMS

- **Avogadro's number (NA):** $6.02214129 \times 10^{23}$, often rounded to 6.022×10^{23}; the number of particles (atoms or molecules) that constitute one mole of any element or compound, the mass of which in grams is numerically equal to the atomic or molecular weight of the material.
- **Boyle's law:** the principle that states that the volume occupied by a gas varies in inverse proportion to the pressure.

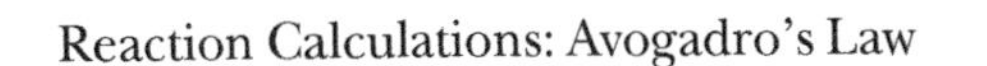

- **Charles's law:** the principle that states that the volume occupied by a gas varies in direct proportion to the temperature.
- **ideal gas law:** the principle that states that the product of the pressure and volume of a gas is directly proportional to the product of the absolute temperature and the number of moles of the gas, expressed by the equation $PV = nRT$.
- **mole:** the amount of any pure substance that contains as many elementary units (approximately 6.022×10^{23}) as there are atoms in twelve grams of the isotope carbon-12.

FORMULATING AVOGADRO'S LAW

Avogadro's law states that, if measured at the same temperature and pressure, equal volumes of gases contain exactly the same number of molecules. It does not differentiate the gases by their mass and says, in fact, that under identical conditions of temperature and pressure, a given volume of a very light gas such as hydrogen (H_2, molecular weight = 2) and a very heavy gas such as uranium hexafluoride (UF_6, molecular weight = 352) contains exactly the same number of molecules of either gas.

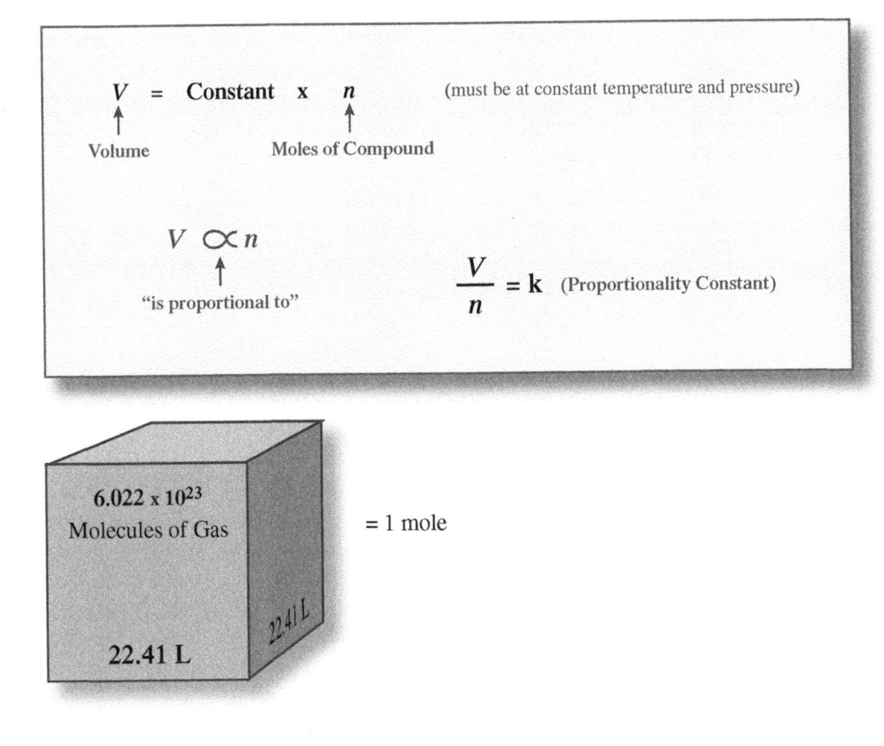

Amadeo Avogadro (1776–1856) based this statement on Joseph Gay-Lussac's (1778–1850) observation in 1802 that when different gases are exposed to the same increase in temperature, they expand by exactly the same amount. From this, Avogadro deduced that the different gases in Gay-Lussac's experiments must contain the same number of particles. Avogadro also introduced the term "molecules" at this time to indicate materials that were composed of a number of atoms. It should be remembered that at the time this work was being carried out, many of the chemical elements had not yet been identified, and there was nothing like the modern theory of atomic structure. Because of this lack of a workable atomic theory and the imprecise nature of analytical methods, the basic composition of materials was neither known nor understood. The term "atom" simply meant the smallest particle of a material, not the elemental unit that is understood in the modern atomic theory. Many scientists believed that atoms of a material were unique to that material. The identification of electrons, protons, and neutrons as the elementary particles of atomic structure would not occur until nearly a century later. Avogadro's work, published in 1811, was essentially ignored for some fifty years before its value began to be recognized in the scientific community.

Eventually, it was realized that Avogadro's hypothesis could provide a rationale for basic chemical behaviors such as the combining volumes of gases. It also provided the basis for the concept of the mole as a unit of measurement.

HINDSIGHT AND MOLES

From the point of view of modern atomic theory, the impact of Avogadro's law on the calculation of molecular masses is entirely logical. Given that atoms, in the modern theory of atomic structure, are composed of a small, dense nucleus of protons and neutrons surrounded by a large, diffuse cloud of electrons and that interactions between atoms involve only the outermost of the electrons, the elemental identity of any atom is not affected in a chemical reaction. No atoms are subdivided into smaller component pieces, and so only whole atoms can be involved in molecular structures and transformations. Therefore, since one particle of H_2 weighs 2 units and one particle of UF_6 weighs 352 units, the number of particles composing 2 grams of H_2 must be exactly equal to the number of particles composing 352 grams of UF_6. This relationship defines the "gram molecular weight" of a material as the weight in grams that is numerically equal to the atomic or molecular weight of the material. For example, if H_2 has an atomic weight of 2 units and O has an atomic weight of 16 units, then the gram molecular weight of H_2O is 18 g. The number of atoms or molecules that comprises the gram molecular weight is termed the "mole" and given the unit abbreviation "mol."

AVOGADRO'S LAW AND THE BEHAVIOR OF GASES

The study of the behavior of gases was the basis for Avogadro's deduction regarding the number of particles in a given volume of different gases. The work of Jacques Charles (1746–1823) in 1787 and of Joseph Gay-Lussac in 1802 demonstrated that gases increased in volume by the same amount when their temperature was increased by the same amount. The principle that the volume of a gas is directly proportional to its absolute temperature is known as Charles' law. Robert Boyle (1627–91) studied the relationship between gas volume and pressure. His work demonstrated that the volume of a gas decreases inversely to the pressure exerted against it. The principle that the volume of a gas is inversely proportional to the pressure is known as Boyle's law. Combining the two laws yields the proportion:

$$V \propto \frac{T}{P}$$

From this proportion describing the behavior of gases, Avogadro reasoned that the number of particles making up the mass of different gases must be the same if the same variation of temperature or pressure produced the same change in the volume of the gas. Furthermore, the proportion must become an equivalency with the number of particles, and the number of particles must be the same from gas to gas for an equal change in volume. A greater or lesser change implies a greater or lesser number of particles. Therefore, the proportionality equation can be rewritten as:

$$V = x\frac{T}{P}$$

where T is temperature in kelvins, V is volume, P is pressure, and x is the factor related to the number of particles present. By manipulating the equation to obtain an expression for x, it can be seen that:

$$x = \frac{PV}{T}$$

By using different values within a consistent set of units, such as atmospheres, liters, and kelvins, and different quantities of gases, it was observed that x is the product of two components. The first is n, the quantity of the gas in moles, and the second is the gas

SAMPLE PROBLEM

Under controlled conditions of standard temperature and pressure (STP; equal to 0°C and 1 atmosphere), a sample of 36 grams of pure water was hydrolyzed by electric current, and the hydrogen and oxygen were collected. Three students who carried out the same experiment at the same time, under identical conditions, collected 20.6 liters, 21.2 liters, and 19.3 liters of O_2 from their respective experiments. Given that a mole of any gas at STP has a volume of approximately 22.4 liters, how many moles of O_2 did each student collect?

Answer:

The molecular weight of H_2O is 18 atomic mass units (u), equal to 18 grams per mole. Therefore, the amount of H_2O that underwent hydrolysis was two moles. According to the hydrolysis equation

$$2H_2O(l) \rightarrow 2H_2(g) + O_2(g)$$

two moles of H_2O can produce only one mole of O_2, which at STP has a volume of 22.4 liters.

According to Avogadro's law, the volume of one gas at a specific temperature and pressure and the volume of a second gas at the same temperature and pressure are in proportion to the number of particles, or moles of particles, that they contain. Therefore, the first student's sample contained $(20.6 \div 22.4) = 0.92$ mole of O_2, the second student's sample contained $(21.2 \div 22.4) = 0.95$ mole of O_2, and the third student's sample contained $(19.3 \div 22.4) = 0.86$ mole of O_2.

constant, R. Thus, the relationship evolves into the ideal gas law:

$$nR = \frac{PV}{T}$$

or

$$PV = nRT$$

Without Avogadro's law—an equal volume of different gases contains an equal number of particles at the same temperature and pressure—it is not possible to make the generalization that leads to the ideal gas law.

Avogadro's Law and the Combining Volumes of Gases

In Avogadro's time, the observation had been made that subjecting water to an electric current produced two different gases. The volume of the gas hydrogen was twice the volume of the other gas, oxygen. Without a valid model of atomic and molecular structure, this could not be understood. Avogadro reasoned, on the basis of his statement that equal volumes of gases contain an equal number of particles, that the atoms of hydrogen and the atoms of oxygen must combine to form "atoms" (what are now known to be molecules) of water in the ratio of 2:1. Conversely, two volumes of hydrogen must combine with one volume of oxygen to produce one volume of water (as water vapor) at the same temperature and pressure. This yields directly the molecular formula for water as H_2O. The respective weights of the two gases also provide the relative weights of the hydrogen and oxygen particles. Avogadro used the relative weights to hypothesize that both hydrogen and oxygen existed as the diatomic compounds "H_2" and "O_2," as he referred to them, for subscripts were not yet in use. Unfortunately, this notion was not given the credit it was due because the major scientists of the day thought that electrostatic theory forbade the combination of two particles of like character, believing that they would repel each other rather than join together.

Avogadro's Law and the Bends

Inexperienced and careless divers who ascend from depth too quickly can experience a painful and potentially deadly affliction known as "the bends," or nitrogen narcosis. As pressure increases, the nitrogen gas in the lungs becomes compressed and its solubility in the blood increases. But as the pressure decreases when the diver rises toward the surface, the dissolved nitrogen leaves the blood and is exchanged in the lungs again as nitrogen gas. When the diver rises too quickly, the same amount of nitrogen gas that was dissolved in the blood at the higher pressure must "un-dissolve" from the blood at the lower pressure, forming bubbles in the blood before the gas can be exchanged in the lungs and exhaled. This causes intense pain and may bring about the sudden death of the diver.

Richard M. Renneboog, MSc

Bibliography

Abbott, David. *The Biographical Dictionary of Scientists.* New York: P. Bedrick, 1984. Print.

Crystal, David, ed. *The Cambridge Biographical Encyclopedia.* 2nd ed. New York: Cambridge UP, 1998. Print.

Gribbin, John. *Science: A History, 1543–2001.* London: Penguin, 2002. Print.

Lafferty, Peter, and Julian Rowe, eds. *The Hutchinson Dictionary of Science.* 2nd ed. Oxford: Helicon, 1998. Print.

Zumdahl, Steven S., and Susan Zumdahl. *Chemistry.* 9th ed. Belmont: Brooks Coles, 2013. Print.

REACTION CALCULATIONS: BALANCING EQUATIONS

FIELDS OF STUDY

Organic Chemistry; Inorganic Chemistry; Physical Chemistry

SUMMARY

The process of balancing chemical equations is essential to the practice of all branches of chemistry. Because chemical reactions take place only at the level of the outer electrons of atoms, it is essential to understand that only whole atoms are involved in any chemical transformation. The law of conservation of mass requires that all atoms that were present before the transformation must also be present after the transformation is complete.

PRINCIPAL TERMS

- **coefficient:** the number preceding a molecular, ionic, or atomic formula in a chemical equation that identifies the exact amount of each component present in the reaction being described.
- **law of conservation of mass:** the principle that states that all atoms present at the beginning of a process must also be present after the completion of the process; matter can be neither created nor destroyed, only changed from one form to another.
- **product:** a chemical species that is formed as a result of a chemical reaction.
- **reactant:** a chemical species that takes part in a chemical reaction.
- **stoichiometry:** the relative quantities of substances that make up compounds or participate in chemical reactions, or the branch of chemistry that deals with these relative quantities.

WHY BALANCE EQUATIONS?

Chemical equations detail the specific numbers and kinds of atoms that are involved in a chemical reaction. Atoms are the fundamental identical particles of the elements and are neither created nor destroyed in a chemical reaction. Since chemical interactions, such as bond formation and ionization, only involve the electrons in the outermost, or valence, shells, the nuclei of the various atoms are not affected in chemical reactions. Thus, only whole atoms can be involved. This is the basic principle of the law of conservation of mass, which states that matter (that is, atoms) can be neither created nor destroyed in a chemical reaction, only changed from one form into another. A properly balanced chemical equation keeps an absolute record of the number and type of each atom in the system, as well as the chemical identities of the forms in which those atoms exist both before and after the reaction has taken place. A fully detailed chemical equation also specifies the physical state of the materials and the environmental conditions under which the reaction took place.

THE MEANING OF CHEMICAL EQUATIONS

If a clear, colorless solution of the ionic compound silver nitrate is added to a likewise clear, colorless solution of sodium chloride, a white solid material forms instantly. This material can be separated from the mixture by filtration, leaving another clear, colorless solution. What has happened in this reaction, and how can it be efficiently described? Clearly, a reaction has changed the materials that were present in the two solutions into entirely different materials that must be composed of the same atoms. The white solid is the ionic compound silver chloride, which does not dissolve in water. Since it is not water soluble, it immediately precipitates from the liquid as a separate solid phase. The clear, colorless solution that remains after the solid is removed must contain the sodium ions from sodium chloride and the nitrate ions from silver nitrate and is, therefore, a solution of sodium nitrate.

In words, the chemical reaction could be described as either "silver ions and chloride ions in solution combine to form insoluble silver chloride" or "silver nitrate in solution reacts with sodium chloride in solution to produce insoluble silver chloride." Other wordings are also possible, as long as they convey the same information. This is very impractical, however, since it demands a lot of reading and is subject to interpretation. In a proper chemical equation, all of the information needed is presented in a short, easy-to-understand symbolic phrase that uses molecular formulas, arrows, coefficients, and subscripts.

Molecular and Structural Formulas

The atom is the smallest individual particle that defines the identity of a chemical element. Similarly, a molecule is the smallest individual particle that identifies the chemical identity of a compound. In both cases, breaking such a particle into its component parts destroys the chemical identity of the material.

The molecular formula of a compound specifies the kind and quantity of atoms of each individual element that constitutes the molecule. For instance, a molecule of the silver nitrate described in the above example is composed of one atom of silver, one atom of nitrogen, and three atoms of oxygen, while silver chloride is composed of one atom of silver and one atom of chlorine. The overall reaction is a double-displacement reaction, in which the sodium atoms and silver atoms trade places from their original compounds, with one silver atom exchanging places with one sodium atom. From this, it is possible to write the corresponding molecular formulas of the various compounds in the reaction mixture as $AgNO_3$ (silver nitrate), $NaNO_3$ (sodium nitrate), NaCl (sodium chloride), and AgCl (silver chloride). Each of these inorganic molecular formulas unequivocally identifies the corresponding material.

Organic compounds are compounds in which the framework of the molecule is constructed of carbon atoms. Hydrocarbons are the basic structures upon which all other organic compounds are made. The simplest of the hydrocarbons are methane (CH_4), ethane (C_2H_6), and propane (C_3H_8). The next one in the series is butane (C_4H_{10}), at which point the usefulness of the molecular formula for identifying organic compounds disappears due to the ability of organic structures to have isomers, which are molecules containing the same numbers and types of atoms but in different arrangements, giving them different chemical identities. For example, the compound 2-methylpropane, or isobutane, is an isomer of butane and has the same molecular formula, C_4H_{10}, but the two compounds are both chemically and physically different. The situation worsens exponentially as the number of carbon atoms increases: there are three isomers with the formula C_5H_{12}, five with C_6H_{14}, nine with C_7H_{16}, and so on. When the number of carbon atoms reaches ten, the number of isomers is over seventy. Which one is meant by the simple molecular formula $C_{10}H_{22}$?

The problem is solved by the use of diagrammatic representations of the molecular structure, a feature that unequivocally defines the identity of the molecule and links it to a specific molecular formula. Using the element symbols for carbon and hydrogen in a structural formula is not efficient, again due to the amount of "reading" involved. Rather than having to read the identity of the molecule from the symbols, chemists use a line-drawing convention that simply indicates the presence and angles of bonds between carbon atoms. A carbon atom is assumed to be at the end of each line and at each angle vertex, and each carbon atom is assumed to have its full complement of hydrogen atoms. Any different bond angles, bond types, and atoms are indicated by their chemical symbol.

REACTION CALCULATIONS BALANCING EQUATIONS

$$2\,C_6H_5COOH + 15\,O_2 \longrightarrow 14\,\not{7}CO_2 + 6\,\not{3}H_2O$$

Element	Before	After	Balanced
C	~~7~~ 14	~~1~~ ~~7~~ 14	√
H	~~6~~ 12	~~2~~ ~~6~~ 12	√
O	~~4~~ ~~6~~ 34	~~3~~ ~~5~~ ~~17~~ ~~31~~ 34	√

A balanced equation will have the same number of atoms for each element on the reactant side (before) and the product side (after) of the reaction equation. When determining the number of atoms present for each element, the subscript is multiplied by the corresponding coefficient. Only coefficients may be changed when balancing an equation, and it may take several changes to balance all elements in the reaction equation.

Stoichiometry

The stoichiometry of a chemical reaction is the proportion of the reactants required for the completion of the reaction according to the properly balanced chemical equation. In the silver nitrate–sodium chloride example, exactly one molecule of $AgNO_3$ reacts with exactly one molecule of NaCl to produce one molecule of $NaNO_3$ and one molecule of AgCl. The balanced reaction can be written in the basic form as:

$$AgNO_3 + NaCl \rightarrow NaNO_3 + AgCl$$

In another example, consider the equation of the reaction between hydrogen (H_2) and oxygen (O_2) that produces water (H_2O). Examination of the molecular formulas reveals that each oxygen atom forms just one molecule of H_2O and requires two hydrogen atoms to do so. Since there are two oxygen atoms in O_2, the reaction must produce two molecules of H_2O, requiring four hydrogen atoms. Therefore, two molecules of H_2 are required to react with each molecule of O_2. The stoichiometric equation can thus be written as the following:

$$2H_2 + O_2 \rightarrow 2H_2O$$

The coefficient 2 that precedes H_2 and H_2O shows how many molecules of each are needed for the correct stoichiometry of the balanced equation.

BALANCED EQUATIONS IN PRACTICE

In the working chemistry lab, balanced equations are essential for calculating the amounts of different compounds to use in a preparation. More importantly, balanced equations tell the chemist how much of the product would be obtained if the reaction were to go 100 percent to completion. The difference between this amount and the amount actually obtained from a reaction is a measure of the effectiveness of a reaction. A chemist can use this information to improve the conditions under which the reaction is carried out. These conditions can then also be written into the balanced equation for other chemists to read. In the silver nitrate–sodium chloride reaction above, the complete reaction equation would be written as:

$$AgNO_3(aq) + NaCl(aq) \rightarrow NaNO_3(aq) + AgCl(s)\downarrow$$

telling anybody who sees it that when silver nitrate dissolved in water is added to sodium chloride dissolved in water, a solution of sodium nitrate dissolved in water is produced, from which silver chloride precipitates out as a solid.

In biochemical reactions, the rules for balancing equations hold as rigidly as they do for inorganic and organic chemical reactions. The complexity of the systems involved in RNA transcription, for example, makes writing a balanced equation for the reactions essentially impossible. Accordingly, biochemists assume a great deal of background knowledge and rely on a lot of abbreviations and acronyms in writing out "balanced" equations such as:

$$ADP + Pi \rightarrow ATP$$

in which adenosine diphosphate and an inorganic phosphate ion combine to produce adenosine triphosphate, a process necessary for cellular metabolism.

Richard M. Renneboog, MSc

SAMPLE PROBLEM

Write the balanced equation for the combustion reaction of ethanol (C_2H_6O) and oxygen (O_2) to produce carbon dioxide (CO_2) and water (H_2O).

Answer:

Because there are two atoms of carbon in C_2H_6O, each molecule of ethanol must produce two molecules of CO_2. Each molecule of CO_2 produced requires one molecule of O_2.

There are six atoms of hydrogen in C_2H_6O, which can produce three molecules of H_2O. This would require three atoms of oxygen, but since there is already one atom of oxygen in C_2H_6O, only two additional atoms of oxygen, or one molecule of O_2, are required to account for the other four atoms of hydrogen.

Taking all of this into account, three molecules of O_2 are required for the complete combustion of each molecule of C_2H_6O. The reaction equation can therefore be written as follows:

$$C_2H_6O + 3O_2 \rightarrow 2CO_2 + 3H_2O$$

BIBLIOGRAPHY

Berg, Jeremy M., John L. Tymoczko, Gregory J. Gatto, and Lubert Stryer. *Biochemistry*. 8th ed. New York: W. H. Freeman, 2015. Print.

Douglas, Bodie E., Darl H. McDaniel, and John J. Alexander. *Concepts and Models of Inorganic Chemistry.* 3rd ed. New York: Wiley, 1994. Print.

Miessler, Gary L., Paul J. Fischer, and Donald A. Tarr. *Inorganic Chemistry.* 5th ed. Upper Saddle River: Prentice Hall, 2013. Print.

Morrison, Robert Thornton, and Robert Neilson Boyd. *Organic Chemistry.* 7th ed. Englewood Cliffs, N.J.: Prentice Hall, 2003. Print.

Reece, Jane B., et al. *Campbell Biology.* 10th ed. San Francisco: Cummings, 2013. Print.

Zumdahl, Steven S., and Susan Zumdahl. *Chemistry.* 9th ed. Belmont: Brooks Coles, 2013. Print.

REACTION CALCULATIONS: CONSERVATION OF MASS

FIELDS OF STUDY

Inorganic Chemistry; Organic Chemistry; Chemical Engineering

SUMMARY

The conservation of mass through chemical and physical transformations is a fundamental principal of the universe. The composition of matter as atoms requires that all atoms present before a chemical reaction are also present after the reaction. Thus, matter can be neither created nor destroyed in a chemical reaction, but only changed from one form to another.

PRINCIPAL TERMS

- **closed system:** a physical or chemical reaction system defined by certain boundary conditions that prevent any components, reactants, or products from entering or exiting the system.
- **mass:** an intrinsic property of matter that determines the extent to which it can be acted on by a force.
- **matter:** any substance that occupies physical space and is composed of atoms or the particles that make up atoms.
- **standard temperature and pressure (STP):** standardized conditions under which experiments are conducted in order to maintain consistency between results; established as 273.15 kelvins and 10^5 pascals by the International Union of Pure and Applied Chemistry (IUPAC), though other organizations use different standards.
- **stoichiometry:** the relative quantities of substances that make up compounds or participate in chemical reactions, or the branch of chemistry that deals with these relative quantities.

THE CONCEPT OF CONSERVATION OF MASS

Matter does not appear and disappear in chemical reactions by magic. All chemical transformations obey the law of conservation of mass (LCM). In its simplest form, the LCM requires that all atoms (and thus all mass) present before a reaction must also be present after it. This is the basic principle behind balancing chemical equations and of particular importance in chemical-engineering processes.

Unlike nuclear reactions that alter the structure of the atomic nucleus, chemical reactions take place only at the level of the outermost, or valence, electrons when atoms interact with one another. The

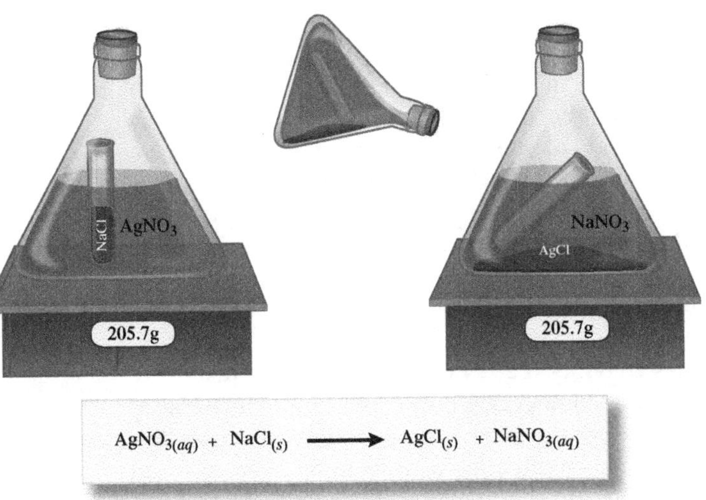

atoms themselves do not change their elemental identity in a chemical reaction. Because of this, only whole atoms can be involved in a chemical reaction; therefore, the total mass of the atoms both before and after the reaction must remain constant. This principle can be readily seen in the process of balancing a chemical equation. If one says, for example, that hydrogen reacts with oxygen to produce water, the meaning of the statement is generally understood. The gas hydrogen reacts with the gas oxygen, and the product is water. However, if the reaction equation were to be written as:

$$H + O \rightarrow H_2O$$

the observer would immediately recognize that something is amiss. Hydrogen gas exists as H_2, and oxygen gas as O_2, so the equation would have to be written as:

$$H_2 + O_2 \rightarrow H_2O$$

While more correct, this equation is still wrong. There are two oxygen atoms in O_2, but only one in H_2O. Where did the second one go? The short answer is that it has not gone anywhere; it just has not been accounted for properly. If the equation is rewritten yet again to include the coefficient "2" before both the H_2 and H_2O terms, the point is clearly made that the stoichiometry of the reaction requires two molecules of H_2 for each molecule of O_2, producing two molecules of H_2O, as:

$$2H_2 + O_2 \rightarrow 2H_2O$$

There are as many H and O atoms in the products as there are in the reactants; thus, there is as much mass after the reaction as there was before. While "mass" and "matter" are not interchangeable terms, there must be as much total matter present in various forms as there was before the reaction occurred. It is only in nuclear reactions that matter, and therefore mass, can be lost through the conversion of matter into energy.

Mass Differences in Chemical Processes

Chemical engineering is the application of chemical reactions for the production of various materials. For a chemical-engineering process to be as economical as possible, all of the matter involved in the process

SAMPLE PROBLEM

What is the total mass of iron oxide (Fe_2O_3) that is produced from the reaction of 50 grams of iron metal and 50 grams of oxygen (O_2) gas? How much iron metal or oxygen remains when the reaction has used up either the oxygen or the iron?

Answer:

As with all chemical reactions, write out the balanced reaction equation as the first step:

$$4Fe + 3O_2 \rightarrow 2Fe_2O_3$$

Next, tabulate the atomic and molecular masses of the reactants and products:

$$\text{Fe atomic weight: } 55.85\frac{g}{mol}$$

$$O_2 \text{ molecular weight: } 32\frac{g}{mol}$$

$$Fe_2O_3 \text{ molecular weight: } 159.7\frac{g}{mol}$$

Tabulate the number of moles of Fe and O_2 in the given quantities:

$$71.9\ g = 0.45\ mol \times 159.7\frac{g}{mol}$$

The reaction equation reveals that four molar units of Fe require three molar units of O_2 stoichiometrically.

Therefore, 0.9 mole of Fe would require

$$0.9\ mol \times \frac{3}{4} = 0.68\ mol\ of\ O_2$$

Since there is more O_2 than required, all of the Fe can be converted to Fe_2O_3.

The stoichiometry of the reaction also dictates that 4 moles of Fe produce 2 moles of Fe_2O_3. Therefore, 0.9 mol of Fe will produce 0.45 mol of Fe_2O_3, or

$$\text{Fe: } 50\ g \div 55.85\frac{g}{mol} = 0.9\ mol$$

$$O_2\text{: } 50\ g \div 32\frac{g}{mol} = 1.56\ mol$$

Since only 0.68 mole of O_2 was consumed to produce the Fe_2O_3, there must be

0.88 mol = 1.56 mol − 0.68 mol, or 28.2 g = 0.88 mol x 32 g/mol, of O_2 remaining.

must be accounted for. The particular chemical reaction applied in the process need not be the one that produces the highest yield. In most cases, a low-yielding reaction that permits easy recovery of the desired product in a readily usable form—the recovery and recycling of unused starting material—is the preferred method. For example, a reaction between two gases that produces either a solid or a liquid under the reaction conditions automatically isolates the product from the reactants while the reaction proceeds, whereas a significant investment in additional methods of manipulation would be required to isolate a gaseous product from the gaseous reaction stream.

Handling waste and excess materials that are not reusable is another aspect of the chemical-engineering process. Due diligence demands that these materials be minimized so as to minimize costs associated with their postprocess disposal. Therefore, knowing how much there is of such materials is integral.

In some processes, it may seem that mass has disappeared, especially when some of the by-products of a reaction process are gases—such as water vapor—that are allowed to escape into the atmosphere. However, Earth is a closed system, bounded by the planet's gravity; therefore, materials that escape into the atmosphere remain within the system and do not simply disappear.

Richard M. Renneboog, MSc

BIBLIOGRAPHY

Douglas, Bodie E., Darl H. McDaniel, and John J. Alexander. *Concepts and Models of Inorganic Chemistry.* 3rd ed. New York: Wiley, 1994. Print.

Lafferty, Peter, and Julian Rowe, eds. *The Hutchinson Dictionary of Science.* 2nd ed. Oxford: Helicon, 1998. Print.

Matson, Michael L., and Alvin W. Orbaek. *Inorganic Chemistry for Dummies.* Hoboken: Wiley, 2013. Print.

Morrison, Robert Thornton, and Robert Neilson Boyd. *Organic Chemistry.* 7th ed. Englewood Cliffs, N.J.: Prentice Hall, 2003. Print.

Smith, Michael B., and Jerry March. *March's Advanced Organic Chemistry: Reactions, Mechanisms, and Structure.* 7th ed. Hoboken: Wiley, 2013. Print.

Weir, M. Russell, James A. Richards Jr., and Thomas W. Adair III. *Physics of the Atom.* 4th ed. Reading: Addison, 1984. Print.

REACTION CALCULATIONS: DEFINITE PROPORTIONS

FIELDS OF STUDY

Organic Chemistry; Inorganic Chemistry

SUMMARY

The principle of definite proportions in chemical compounds is discussed as one of the fundamental principles of chemistry, though it is not a universal law. The principle laid the foundation for stoichiometry and was a breakthrough realization in the history of chemistry. The modern theory of atomic structure provides the framework on which the principle is founded.

PRINCIPAL TERMS

- **composition:** the identities and relative proportions of different elements or components in a compound, mixture, or other material.
- **law of conservation of mass:** the principle that states that all atoms present at the beginning of a process must also be present after the completion of the process; matter can be neither created nor destroyed, only changed from one form to another.
- **law of multiple proportions:** the principle that states that the atoms that comprise any particular compound are present in simple, whole-number ratios.
- **stoichiometry:** the relative quantities of substances that make up compounds or participate in chemical reactions, or the branch of chemistry that deals with these relative quantities.

THE LAW OF DEFINITE PROPORTIONS

The law of definite proportions originated at the end of the eighteenth century, fully one hundred years before the basic components of the modern theory of atomic structure were realized. The principle was first conceived in 1793 by French chemist Joseph-Louis Proust (1754–1826) and published in 1795 in Spanish journals. It was an idea that could not be well supported at the time for many reasons, not least because it was at odds with the statements of other, better-known scientists of the time, particularly Claude-Louis Berthollet (1748–1822).

Proust's hypothesis, which came to be known as the "law of definite proportions," states that compounds always maintain the exact same proportions of atoms in their composition. That is to say, the proportions of different atoms in any particular pure compound are definite and not variable. Proust developed his theory on the basis of careful analysis and observation using inorganic compounds that had a consistent level of purity. At the time, laboratory methods did not include standardized systems of weights and measures, making Proust's observations all the more noteworthy.

SAMPLE PROBLEM

Compare the proportional masses of reactants and products in the thermite reaction when 27 g of aluminum react with 79.85 g of Fe_2O_3 to produce 55.85 g of Fe and 51 g of Al_2O_3. Compare the unit masses of the atoms when two atoms of Al react with one molecule of Fe_2O_3 to produce two atoms of Fe and one molecule of Al_2O_3.

Answer:

Gather the atomic and molecular masses in atomic mass units (u):

Fe: 55.85 u

Al: 27 u

O: 16 u

Fe_2O_3: 159.7 u

Al_2O_3: 102 u

The balanced equation for the thermite reaction is

$$2Al + Fe_2O_3 \rightarrow 2Fe + Al_2O_3$$

Total mass of reactants: (27 g + 79.85 g) = 106.85 g

Total mass of products: (55.85 g + 51 g) = 106.85 g

Total unit masses of reactants: (2 atoms of Al x 27 u) + (1 molecule of Fe_2O_3 x 159.7 u) = 213.7 u

Total unit masses of products: (2 atoms of Fe x 55.85 u) + (1 molecule of Al_2O_3 x 102 u) = 213.7 u

The Modern Atomic Theory and the Law of Definite Proportions

The identification of the subatomic particles—electrons, protons, and neutrons—between 1897 and 1932 provided the foundation for the modern theory of atomic structure. According to this model, atoms are composed of a very small, dense nucleus in which the protons and neutrons, representing almost the entire mass of the atom, are found. Electrons surround the nucleus in a relatively large, diffuse cloud. The electrons occupy different levels according to the specific energies they are allowed to possess. Chemical reactions and bonds between atoms involve only their outermost electrons but not their nuclei. No atoms undergo a change of elemental identity in chemical reactions, and thus only whole atoms are involved in chemical interactions and the formation of molecules. Within this framework, the elemental identity of individual atoms and the chemical identity of individual compounds are unequivocally defined.

A compound's chemical identity is the specific number of atoms of each element in one of its molecules. From the point of view of the modern atomic theory, it is easy to see that molecules having different numbers and types of atoms have different chemical identities and are different compounds. Without the modern theory, however, the hypothesis that compounds are defined by specific numbers of atoms in specific proportions was nothing short of revolutionary.

The Laws of Definite Proportions and Conservation of Mass

The law of definite proportions, though not a universal law, is nevertheless fundamentally related to the law of conservation of mass. The law of conservation of mass has several ramifications. In its most fundamental statement, this law requires that all atoms that were present before a chemical change takes place must also be present after that change has taken place. Accordingly, the total mass of each individual atom must be the same before and after a chemical reaction takes place. For example, if there are six atoms of oxygen in the reactants of the reaction, there must be six atoms of oxygen in the products of the reaction. Likewise, the mass of the six oxygen atoms in the reactants must be the same as the mass of the six oxygen atoms in the products. So, while the elements in the reactants are the same as those in the products, the reaction changes the chemical identity and composition, the identity and relative proportions of different elements, of each reactant. The reactants break down and form different products. These products have different chemical identities and compositions from the reactants.

As an example, consider thermite, a compound used in welding and in bombs and known for the large quantity of heat it gives off when ignited. During the thermite reaction, aluminum metal (Al) reacts with iron oxide (Fe_2O_3) to produce iron metal (Fe) and aluminum oxide (Al_2O_3) according to the following balanced equation:

$$2Al + Fe_2O_3 \rightarrow 2Fe + Al_2O_3$$

The reactants contain two aluminum atoms, two iron atoms, and three oxygen atoms; the products also contain two aluminum atoms, two iron atoms, and three oxygen atoms. The mass of each of the different elements is therefore the same before and after the reaction in accordance with the law of conservation of mass. The chemical identities (Al, Fe_2O_3) and compositions (2Al, Fe_2O_3) of the reactants and those of the products are different, however. Each compound in the reaction is defined by the combination of its atoms in fixed proportions.

THE LAWS OF DEFINITE PROPORTIONS AND MULTIPLE PROPORTIONS

John Dalton (1766–1844) extended the concept of definite proportions by hypothesizing the law of multiple proportions. This law, which was also based on the observation of composition and relative masses, states that when atoms form different compounds the ratios of the masses of the elements in the compounds are small, whole number multiples. Thus, the law of definite proportions states that the elemental composition of molecules in any particular compound is constant, and the law of multiple proportions states that the ratios of the masses of the individual elements in any particular compound are also constant. In light of the modern theory of atomic structure, with its stricture that only whole atoms are involved in chemical reactions and only at the level of their outermost electrons, it can be readily seen that the law of multiple proportions applies at the atomic level as well as at the macroscopic level. For example, elemental analysis of a large sample of a compound such as the sugar glucose would reveal specific quantities of carbon (C), hydrogen (H), and oxygen (O), with a mass ratio of 72:12:96 (72 parts C to 12 parts H to 96 parts O). These are the definite proportions of the compound. By dividing the proportional mass of each element in the compound by the respective atomic mass units (u) of each element (12 u for C, 1 u for H, and 16 u for O), the unit ratios of the different elements are 6:12:6 (6 atoms of C, 12 atoms of H, and 6 atoms of O). The number of atoms is written as a subscript after each elemental symbol in the molecular formula $C_6H_{12}O_6$ and indeed in all molecular formulas. Taken together, the laws of definite proportions and multiple proportions are the foundation of stoichiometry, the specific relative amounts of reactants and products that are involved in a chemical reaction process.

DEFINITE PROPORTIONS IN CHEMICAL ENGINEERING

Chemical engineering is an economic enterprise, and as such, costs and cost savings play a very significant role. In a chemical engineering process, chemical reactions are applied to the production of commercially valuable materials. Often, a low-yield reaction that allows easy separation of the desired product is favored over a high-yield reaction that requires a great deal of isolation and purification of the product. But even the most economical methods demand close attention to mass balance to ensure not only that all possible product is recovered but that all leftover reactants can be recovered for use in future reactions. Engineering designers work diligently to ensure that materials are used efficiently and with a minimum of waste.

Richard M. Renneboog, MSc

Bibliography

Gribbin, John. *Science: A History, 1543–2001.* London: Penguin, 2002. Print.

Lafferty, Peter, and Julian Rowe, eds. *The Hutchinson Dictionary of Science.* 2nd ed. Oxford: Helicon, 1998. Print.

Myers, Richard. *The Basics of Chemistry.* Westport: Greenwood, 2003. Print.

Zumdahl, Steven S., and Susan Zumdahl. *Chemistry.* 9th ed. Belmont: Brooks Coles, 2013. Print.

REACTION CALCULATIONS: MOLECULAR FORMULA

FIELDS OF STUDY

Organic Chemistry; Inorganic Chemistry; Biochemistry

SUMMARY

Molecular formulas are explained and discussed as a central concept of all chemistry-related fields. A molecular formula specifies precisely the elemental identities and absolute numbers of the atoms that make up the molecules of any particular compound. Because of the existence of isomers, however, a molecular formula may not necessarily uniquely identify a compound.

PRINCIPAL TERMS

- **chemical formula:** the combination of symbols and numerical coefficients that specifies the number and identity of the different atoms involved in a chemical reaction or molecular structure.
- **coefficient:** the number preceding a molecular, ionic, or atomic formula in a chemical equation that identifies the amount of each component present in the reaction being described.
- **empirical formula:** a chemical formula that indicates the relative proportion of atoms of each element present in one molecule of a substance, which may not be equal to the total number of atoms present.
- **law of conservation of mass:** the principle that states that all atoms present at the beginning of a process must also be present after the completion of the process; matter can be neither created nor destroyed, only changed from one form to another.
- **structural formula:** a graphical representation of the arrangement of atoms and bonds within a molecule.

What's in a Molecule?

In the modern theory of atomic structure, each element's unique identity is determined by the number of protons contained in the nucleus of each atom. Isotopes of elements are determined by the number of neutrons accompanying the protons in the nucleus. Each atom has a large, diffuse cloud of electrons surrounding its much smaller, denser nucleus; when the number of electrons is equal to the number of protons in the nucleus, the atom has no electrical charge. Chemical reactions occur via interactions between the outermost electrons in these clouds and do not affect the nuclei of any of the atoms. Thus, the elemental identity of each atom is retained throughout any normal chemical reaction.

Interactions between different atoms typically result in the formation of chemical bonds, which are subject to various geometric constraints that determine the relative positions of the atoms in the structure of any resulting molecule. The specific three-dimensional arrangement of atoms and bonds is the compound's molecular structure, which gives it a unique chemical identity. This arrangement can be represented by a diagram called a structural formula.

Each molecule of a compound contains a specific number of each type of atom that makes up the compound. This information is communicated in a general way by the molecular formula, which is a type of chemical formula that shows how many atoms of each element are in a molecule. For example, a molecule consisting of exactly twenty-two atoms of carbon, forty atoms of hydrogen, six atoms of oxygen, one atom of nitrogen, and two atoms of sulfur would have the molecular formula C22H40O6NS2. The letters are the element symbols as they appear in the periodic table of the elements.

REACTION CALCULATIONS MOLECULAR FORMULA

Chemical Name	Molecular Formula	Empirical Formula	Skeletal Formula
pentane	C_5H_{12}	C_5H_{12}	H HH HH H / H C C C C C H / H H H H
pentane	C_5H_{12}	C_5H_{12}	H / H–C–H / H H / H–C—C—C–H / H H / H–C–H / H
ammonia	NH_3	NH_3	H–N̈–H / H
acetic acid	$C_2H_4O_2$	CH_2O	H / H–C–C=O / H O–H
methyl formate	$C_2H_4O_2$	CH_2O	O=C–O–C–H / H H H

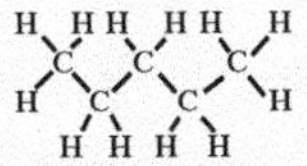

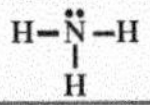

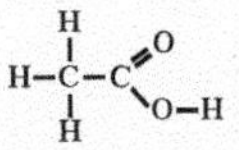

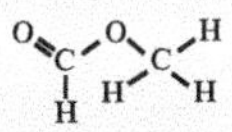

Molecular Formulas and Isomers

Isomers are molecules that have the same molecular formula but different molecular structures. A simple example is the formula C_3H_6, which describes two different compounds, propene and cyclopropane. It is not possible to determine from the molecular formula which of the two isomers is being indicated. This is true of all isomers of any particular molecule and is of extreme importance in organic chemistry, because the number of possible isomers increases drastically as the number of carbon atoms in the molecular formula increases. While only two isomers of C_3H_6 are possible, the molecular formula $C_{10}H_{22}$ describes more than seventy different isomers, and when elements other than carbon and hydrogen are part of the formula, the number of possible isomers is even higher. The chemical formula can differentiate between isomeric structures in some cases; for example, the compounds diethyl ether and 1-butanol have the same molecular formula, $C_4H_{10}O$, but can be readily identified by the chemical formulas H_5C_2-O-C_2H_5 and HO-C_4H_9. This type of formula is called a "semi-structural formula" because it gives some indication of the arrangement of the molecule but is not a true structural formula.

Determining the Molecular Formula

Molecular formulas can be determined by quantitative elemental analysis. The molecular formulas of many compounds were initially determined by subjecting a precisely weighed sample to combustion in pure oxygen and measuring the amounts of the products obtained. Knowing the weight of the sample at the beginning of the procedure is essential to calculating the result. In a combustion analysis of cyclopropane, for example, the products of combustion are simply carbon dioxide and water, according to the equation:

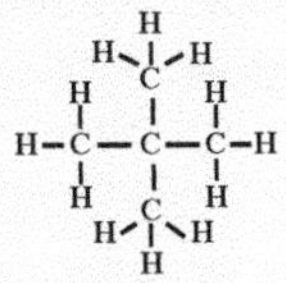

$$2C_3H_6 + 9O_2 \rightarrow 6CO_2 + 6H_2O$$

Since the combustion takes place in a closed system, all of the carbon in the recovered carbon dioxide (CO_2) must have come from the carbon atoms in the sample. The same is true of the hydrogen in the recovered water (H_2O). This is because, as mentioned above, atomic nuclei, and therefore the elemental identities of the atoms, are not affected in a chemical reaction. Therefore, there must be exactly the same number of each type of atom present after the reaction as there was before the reaction. This is the law of conservation of mass.

By comparing the masses of carbon dioxide and water produced, it is possible to relate them on a molar basis. This is accomplished by using the atomic weight of each element—provided in the periodic table in unified atomic mass units (u), which are equivalent to grams per mole—to determine the molecular weight of each product, converting the amount of each product from grams to moles, and then determining the ratio of moles of carbon to moles of hydrogen. This ratio yields the empirical formula CH_2 for cyclopropane, meaning that there are two hydrogen atoms present for every carbon atom. Next, by relating the weights of carbon dioxide and water to the weight of the original sample, the

SAMPLE PROBLEM

After subjecting a sample of 1.44 grams of a hydrocarbon to combustion analysis, the analyst recovered 4.4 grams of carbon dioxide and 2.16 grams of water. What is the empirical formula of the hydrocarbon? What is its molecular formula?

Answer:

The precise reaction equation is not known, so it cannot be used as the basis of calculations. The formulas must therefore be determined from weight ratios.

Use the atomic weights of carbon, hydrogen, and oxygen to determine the molecular weights of the products:

$$\text{atomic weight of C: } 12.011\frac{g}{mol}$$

$$\text{atomic weight of H: } 1.008\frac{g}{mol}$$

$$\text{atomic weight of O: } 15.999\frac{g}{mol}$$

$$\text{molecular weight of } CO_2 = 12.011 + (2 \times 15.999) = 44.009\frac{g}{mol}$$

$$\text{molecular weight of } H_2O = (2 \times 1.008) + 15.999 = 18.015\frac{g}{mol}$$

Determine how many moles of carbon dioxide and water were recovered:

$$CO_2\text{: } 4.4g \div \frac{44.009g}{1mol} = 4.4g \times \frac{1mol}{44.009g} = 0.100 \text{ mol}$$

$$H_2O\text{: } 16g \div \frac{18.015g}{1mol} = 2.16g \times \frac{1mol}{18.015mol} = 0.120mol$$

All of the carbon atoms in the recovered carbon dioxide and the hydrogen atoms in the recovered water came from the original hydrocarbon sample. Therefore, 1.44 grams of the hydrocarbon contain 0.1 mole of carbon and 0.24 mole of hydrogen (because there are two hydrogen atoms in one molecule of water, and thus two moles of hydrogen for every mole of water). The ratio of moles of carbon to moles of hydrogen in the hydrocarbon is therefore 1:2.4.

To obtain the empirical formula, the ratio must be expressed in the smallest whole numbers possible. In this case, multiplying both sides by 5 yields a molar ratio of 5:12. The empirical formula of the hydrocarbon is therefore C_5H_{12}.

Determine the empirical weight (that is, the weight of a theoretical molecule with a molecular formula identical to the empirical formula) of C_5H_{12}:

$$(5 \times 12.011\frac{g}{mol}) + (12 \times 1.008\frac{g}{mol}) = 72.151\frac{g}{mol}$$

Use the empirical formula to generate a balanced equation for the combustion reaction:

$$C_5H_{12} + 8O_2 \rightarrow 5CO_2 + 6H_2O$$

The amounts of carbon dioxide and water that would be produced by 1.44 grams of C_5H_{12} according to this reaction equation correspond with the amounts that were recovered when the combustion reaction was carried out. The molecular formula of the hydrocarbon is therefore also C_5H_{12}.

number of moles of the original material can be determined, and from this the molecular formula. The molecular formula is always a simple multiple of the empirical formula, in accordance with the law of multiple proportions. In this example, the molecular formula C_3H_6 is three times the empirical formula CH_2.

BALANCING CHEMICAL EQUATIONS

Chemical reactions are described by chemical equations like the one in the example above. In all cases, a chemical reaction equation is true if, and only if, it is properly balanced. The coefficient preceding each molecular formula in a chemical equation represents the precise number of those molecules that are involved in the reaction. In the above example, it would be correct but entirely insufficient to say that cyclopropane reacts with oxygen to produce carbon dioxide and water, as this does not convey the quantitative nature of the reaction. To be precise, the properly balanced equation says that exactly two molecules of cyclopropane react with exactly nine molecules of oxygen gas to produce exactly six molecules of carbon dioxide and six molecules of water—no more and no less.

MOLECULAR FORMULAS IN BIOLOGICAL SYSTEMS

Molecular formulas in biological systems range from very simple molecules, such as CO_2 and H_2O, to very complex and extremely large molecules, such as DNA and various proteins. For smaller molecules such as steroid hormones, fatty acids, neurotransmitters, and many others, the typical molecular or structural formulas are useful. For larger molecules such as proteins and enzymes, however, these formulas are not useful at all and indeed may not even be known.

Because of this, instead of formulas, biochemists use assigned names and symbols, such as RNA and DNA for ribonucleic acid and deoxyribonucleic acid, P_i for inorganic phosphate, ATP for adenosine triphosphate, NAD^+ for nicotinamide adenine dinucleotide, and so on. The three-dimensional structure of an enzyme is typically represented by symbols denoting structural features, with the relative locations of specific functional groups identified.

Richard M. Renneboog, MSc

BIBLIOGRAPHY

Berg, Jeremy M., John L. Tymoczko, Gregory J. Gatto, and Lubert Strye. *Biochemistry.* 8th ed. New York: W. H. Freeman, 2015. Print.

Douglas, Bodie E., Darl H. McDaniel, and John J. Alexander. *Concepts and Models of Inorganic Chemistry.* 3rd ed. New York: Wiley, 1994. Print.

Myers, Richard. *The Basics of Chemistry.* Westport: Greenwood, 2003. Print.

Miessler, Gary L., Paul J. Fischer, and Donald A. Tarr. *Inorganic Chemistry.* 5th ed. Upper Saddle River: Prentice Hall, 2013. Print.

Morrison, Robert Thornton, and Robert Neilson Boyd. *Organic Chemistry.* 7th ed. Englewood Cliffs, N.J.: Prentice Hall, 2003. Print.

O'Neil, Maryadele J., ed. *The Merck Index: An Encyclopedia of Chemicals, Drugs, and Biologicals.* 15th ed. Cambridge: RSC, 2013. Print.

Pine, Stanley H. *Organic Chemistry.* 5th ed. New York: McGraw, 1987. Print.

Skoog, Douglas Arvid, Donald M. West, and F. James. Holler. *Skoog and West's Fundamentals of Analytical Chemistry.* 9th ed. Belmont (Calif.): Cengage Learning, 2014. Print.

REACTION CALCULATIONS: MULTIPLE PROPORTIONS

FIELDS OF STUDY

Physical Chemistry; Analytical Chemistry

SUMMARY

The principle of multiple proportions in chemical compounds, one of the fundamental principles of chemistry, is discussed. The development of the concept, which has strong ties to the principle of definite proportions and the law of conservation of mass, was a breakthrough in the field of chemistry.

PRINCIPAL TERMS

- **atomic weight:** the average of the atomic masses of an element's isotopes, weighted according to their proportions in nature; often used interchangeably with "atomic mass."
- **composition:** the identities and relative proportions of different elements or components in a compound, mixture, or other material.
- **law of conservation of mass:** the principle that states that all atoms present at the beginning of a process must also be present after the completion of the process; matter can be neither created nor destroyed, only changed from one form to another.
- **law of definite proportions:** the principle that states that the proportions by mass of the elements that combine to form a specific compound are definite and do not vary.
- **stoichiometry:** the relative quantities of substances that make up compounds or participate in chemical reactions, or the branch of chemistry that deals with these relative quantities.

DEVELOPING THE LAW OF MULTIPLE PROPORTIONS

The modern theory of atomic structure did not yet exist when English chemist John Dalton (1766–1844) hypothesized the law of multiple proportions. While Dalton and other scientists believed in the existence of atoms, their understanding of the concept was greatly informed by the ancient Greek philosophical view that matter was composed of infinitesimally small particles, called "atoms," that were unique to each material. The identification of the subatomic particles that form the basis of atomic structure would not begin until 1897, over half a century after Dalton's death.

Without a workable theoretical framework of atomic structure, Dalton and his contemporaries focused their work on the use of weights and measures as the means of describing chemical properties and interactions. Through careful measurement of combining weights, the proportions by weight at which elements combine, Dalton concluded that when two elements combine to form more than one compound, their proportions are always in simple, whole-number ratios such as 1:1, 1:2, 3:4, and so on. According to the modern theory of atomic structure, this phenomenon is readily understandable in terms of the interactions of single atoms. A single atom can bond to one or two other atoms or in more complex molecules (the makeup of a compound is referred to as its composition). This may seem obvious to twenty-first-century chemists, but in Dalton's time, the structure and behavior of atoms were not understood within any comprehensive theoretical framework.

REACTION CALCULATIONS MULTIPLE PROPORTIONS

Element A (nitrogen)	Element B (oxygen)	Compound	Molecular Formula
7 g	4 g	nitrous oxide	N_2O
7 g	8 g	nitrogen oxide	NO
7 g	8 g	hyponitrite	N_2O_2
7 g	16 g	nitrous dioxide	NO_2
7 g	20 g	dinitrogen pentoxide	N_2O_5
7 g	12 g	dinitrogen trioxide	N_2O_3
7 g	16 g	dinotrogen tetroxide	N_2O_4

Dalton's observation of combining weights and subsequent hypothesis of the law of multiple proportions was one of the first steps on the path to the modern theory of atomic structure. The other principal steppingstone was the law of definite proportions, developed by French chemist Joseph-Louis Proust (1754–1826), which states that the proportions of the weights of different elements in a compound are definite and unchanging.

THE MODERN ATOMIC THEORY AND THE LAW OF MULTIPLE PROPORTIONS

The atomic model in the modern theory of atomic structure describes atoms as composed of a very small, dense nucleus that contains essentially all of the mass of an atom, surrounded by a much larger, diffuse cloud of electrons. All of the positive electrical charge in the atom is located in the protons of the nucleus, which also contains the accompanying neutral particles, neutrons. The electrons in the surrounding cloud are equal in number to the number of protons in the nucleus and contain all of the negative electrical charge in the atom. The mass of the protons, neutrons, and electrons within the atom dictates the atom's total atomic mass; an element's atomic weight is an average of the atomic masses of its isotopes by their frequency in nature.

The regions around the nucleus, called "orbitals," are constrained by the structure of the nucleus to specific geometric arrangements and structures, and the electrons they contain are able to form chemical bonds through their interaction with the electrons within the orbitals of different atoms. The number of electrons required to occupy the outermost electron shell of the atom fully determines the number of bonds that are possible. A carbon atom, for example, can in most cases form a maximum of four bonds to other atoms, while a hydrogen atom can form only one bond. This is the basic principle behind Dalton's observations of combining weights and the law of multiple proportions.

STOICHIOMETRY AND THE LAW OF MULTIPLE PROPORTIONS

The laws of multiple proportions and of definite proportions were fundamental steps in the development of the concept of stoichiometry. A third component in the development of stoichiometry was the law of conservation of mass. According to this law, matter can be neither created nor destroyed in a normal chemical reaction, but only changed from one form to another. Accordingly, every atom that was present

SAMPLE PROBLEM

Nitrogen and oxygen combine to form the two compounds NO and NO_2. Calculate the percent composition of nitrogen in each compound and show how they obey the law of multiple proportions. Use 14 g/mol as the molecular weight of nitrogen and 16 g/mol as the molecular weight of oxygen.

Answer:

Calculate the molecular weights of the two compounds:

$$\text{molecular weight of NO: } 14\frac{g}{mol}+16\frac{g}{mol}=30\frac{g}{mol}$$

$$\text{molecular weight of NO}_2\text{: } 14\frac{g}{mol}+\left(16\frac{g}{mol}\times 2\right)=46\frac{g}{mol}$$

Calculate the percent by weight from nitrogen:

$$\text{percent N by weight in NO: } \frac{14}{30}\times 100=46.7\text{ percent}$$

$$\text{percent N by weight in NO}_2\text{: } \frac{14}{46}\times 100=30.4\text{ percent}$$

Calculate the percent by weight from oxygen:

$$\text{percent O by weight in NO: } 100-46.7=53.3\text{ percent}$$

$$\text{percent O by weight in NO}_2\text{: } 100-30.4=69.6\text{ percent}$$

Calculate the ratio by weight of N to O in each compound:

$$\text{percent N/percent O in NO: } \frac{46.7}{53.3}=0.876$$

$$\text{percent N/percent O in NO}_2\text{: } \frac{30.4}{69.6}=0.437$$

Calculate the numerical proportion of the ratios of N to O in the two compounds:

$$\frac{0.876}{0.437}=2$$

From this, it can be seen that the combining weights of the two elements are in a small whole number ratio, as predicted by the law of multiple proportions.

at the beginning of a reaction must also exist after the reaction has been completed.

The stoichiometry of a reaction is a detailed statement of the exact number of atoms and molecules that are required to take part in a complete reaction. For example, the addition reaction between ethylene, C2H4 (or H2C=CH2), and chlorine, Cl2, to produce 1,2-dichloroethane, C2H4Cl2 (or Cl–CH2–CH2–Cl) occurs strictly according to the reaction equation:

$$\text{H2C=CH2 + Cl2} \rightarrow \text{Cl–CH2–CH2–Cl}$$

Exactly one molecule of ethylene reacts with exactly one molecule of chlorine to produce exactly one molecule of 1,2-dichloroethane. All atoms present at the beginning of the reaction are present at the end, and no additional atoms are present.

The Law of Multiple Proportions in Chemical Engineering

Chemical engineering is an economic enterprise, and as such, costs and cost savings play a very significant role. In a chemical-engineering process, chemical reactions are applied to the production of commercially valuable materials. Often, a low-yield reaction that allows very easy separation of the desired product is favored over a high-yield reaction that requires a great deal of isolation and purification of the product. In any chemical process in which reactants can combine to form more than one distinct compound, the control of process quantities is dictated by the law of multiple proportions. That is, the reaction conditions and reactant quantities must be carefully controlled so that only the desired product is produced and other reactions that consume often expensive starting materials and catalysts are eliminated or at least minimized.

Richard M. Renneboog, MSc

Bibliography

Gregersen, Erik, ed. *The Britannica Guide to the Atom.* New York: Britannica, 2011. Print.

Gribbin, John. Science: *A History, 1543–2001.* London: Penguin, 2002. Print.

Jones, Mark Martin. *Chemistry and Society.* 5th ed. Philadelphia: Saunders College Pub., 1987. Print.

Lafferty, Peter, and Julian Rowe, eds. *The Hutchinson Dictionary of Science.* 2nd ed. Oxford: Helicon, 1998. Print.

Myers, Richard. *The Basics of Chemistry.* Westport: Greenwood, 2003. Print.

Wehr, M. Russell, James A. Richards Jr., and Thomas W. Adair III. *Physics of the Atom.* 4th ed. Reading: Addison-Wesley, 1984. Print.

REACTION CALCULATIONS: PERCENTAGE COMPOSITION

FIELDS OF STUDY

Physical Chemistry; Chemical Engineering; Forensic Chemistry

SUMMARY

The concept of percentage composition is discussed, and the method of calculating percentage composition is presented. Percentage composition by weight is the relative proportions of the weight of a molecule to the individual elements of which it is comprised.

PRINCIPAL TERMS

- **atomic weight:** the average of the atomic masses of an element's isotopes, weighted according to their proportions in nature; often used interchangeably with "atomic mass."
- **chemical formula:** the combination of symbols and numerical coefficients that specify the number and identity of the different atoms involved in a chemical reaction or molecular structure.
- **composition:** the identities and relative proportions of different elements or components in a compound, mixture, or other material.
- **empirical formula:** a chemical formula that indicates the relative proportion of atoms of each element present in one molecule of a substance, which may not be equal to the total number of atoms present.
- **proportion:** the quantity of a unit component relative to the others in the same system.

VISUALIZING PERCENTAGE COMPOSITION

Percentage composition is a very useful means of relating the quantities of individual components in a material to the overall or total quantity of the material. Each component is present in the material in a specific proportion according to the composition of the particular material. The concept is easily demonstrated with a collection of one hundred marbles of different colors: twenty-five red, thirteen blue, twenty-seven green, and thirty-five colorless. The entire collection of one hundred marbles has the composition 25 percent red, 13 percent blue, 27 percent green, and 35 percent colorless. Similarly, the percentage composition of a molecule can be described in this way, according to the elemental identities of the individual atoms that make up that particular molecule.

Percentage composition is always determined on the basis of a physical parameter, and the statement of percentage composition is always accompanied by a qualifying factor, such as "by weight," "by volume,"

REACTION CALCULATIONS
PERCENT COMPOSITION

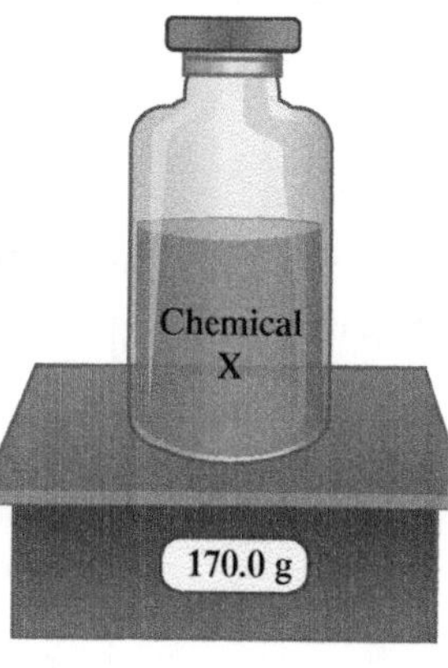

Element	Mass	% Mass
C	60.8	5.07
H	4.1	4.1
O	48.6	3.04
N	56.5	4.04

Empirical Formula	Chemical Name
$C_5H_4N_4O_3$	Uric Acid

SAMPLE PROBLEM

A 2.5-gram sample of a white crystalline substance was subjected to elemental analysis and yielded the equivalent of 1.0528 grams of carbon, 0.1623 grams of hydrogen, and 1.2849 grams of oxygen. The molecular weight of the unknown compound is 342 u. What is the empirical or chemical formula of the compound? Is this sufficient to identify the compound?

Answer:

For the unknown crystalline compound, realize that the total weight of the individual elements recovered may not correspond exactly to the weight of the unknown material used, since there are always some cumulative losses of material in the various steps of any procedure. Round-off errors in calculations also work against exact agreement of the quantities. With that in mind, first calculate the proportion by weight of the recovered materials and the starting weight of the unknown compound:

$$\text{C: } \frac{1.0528\text{ g}}{2.5\text{ g}} \times 100 = 42.11 \text{ percent by weight}$$

$$\text{H: } \frac{0.1623\text{ g}}{2.5\text{ g}} \times 100 = 6.49 \text{ percent by weight}$$

$$\text{O: } 100 - (42.11 + 6.49) = 51.40 \text{ percent by weight}$$

Now assume a 100-gram quantity of the material that would consist of 42.11 grams of carbon, 6.49 grams of hydrogen, and 51.4 grams of oxygen and use the atomic weights of all three elements to calculate the molar quantities of each element in 100 grams:

$$\text{C: } \frac{42.11\text{ g}}{12.01\text{ g/mol}} = 3.51\text{ mol}$$

$$\text{H: } \frac{6.49\text{ g}}{1.01\text{ g/mol}} = 6.43\text{ mol}$$

$$\text{O: } \frac{51.4\text{ g}}{16.00\text{ g/mol}} = 3.21\text{ mol}$$

Divide each molar quantity by the lowest value to get the basic molar ratios:

$$\text{C: } \frac{3.51}{3.21} = 1.09$$

$$\text{H: } \frac{6.43}{3.21} = 2.00$$

$$\text{O: } \frac{3.21}{3.21} = 1.00$$

To obtain the empirical formula, the basic molar ratios are then raised by the lowest factor that produces integer values for all of them. In this case, the lowest multiplying factor is 11, which gives

$$\text{C: } 1.09 \times 11 = 11.99 \approx 12$$

$$\text{H: } 2 \times 11 = 22$$

$$\text{O: } 1 \times 11 = 11$$

This corresponds to the empirical formula $C_{12}H_{22}O_{11}$, which has the empirical weight of 342 u. Because the empirical weight is equal to the measured molecular weight, the molecular formula must also be $C_{12}H_{22}O_{11}$.

Due to the number of possible isomers having this same molecular formula, this by itself is not sufficient to identify the compound. At best, this information can only serve to eliminate all of the vast number of other white crystalline compounds that do not correspond to this molecular formula.

or even "weight to volume." Mixtures of solids, liquid solutions, combinations of different gases, individual molecules, bulk quantities of specific compounds, fruits and vegetables, and in fact all matter can be described by its percentage composition. Astrophysicists describe the entire known universe, for example, as having a composition of over 90 percent by weight of hydrogen. In a chemical compound, the percentage composition is almost always stated in terms of the atomic weight of individual component elements relative to the molecular weight of the compound.

Determining Percentage Composition

The absolute relationship of atomic structure and atomic weight derives from the fact that only whole atoms can be involved in the structure of molecules. This enables the use of weights to determine the percentage composition of any compound and, from this, the empirical formula or chemical formula of the compound. A simple calculation is all that is needed to determine the percentage composition of a compound if the chemical formula is known; one simply multiplies the atomic weight of the particular element by the number of atoms of that element that are present, then divides that by the molecular weight of the compound. For example, the compound arsenic trichloride has the chemical formula $AsCl_3$. The atomic weight of arsenic (As) is 74.9 atomic mass units (u), and the molecular weight of $AsCl_3$ is 181.25 u. The equation

$$\frac{74.9 \text{ u}}{181.25 \text{ u}} \times 100 = 41.32$$

shows that the molecule of $AsCl_3$ is 41.32 percent by weight arsenic and 58.68 percent by weight chlorine (Cl). Note that when determining the percentage composition of a compound, it is only necessary to calculate all but one of the components; since they must all total 100 percent, the percentage of the last component is equal to 100 minus the sum of all the other percentages.

One valuable use of percentage composition is in elemental analysis of an unknown material to determine its chemical formula. Historically, this procedure helped establish the basic principles of chemical reactions that eventually led to the development of a functional atomic theory by early physical chemists. By using various chemical reactions that allow each element to be captured quantitatively, the weights of each element in a known weight of a compound can be determined. The number of moles of each element is then determined using the atomic weights of the respective elements. Relating these molar quantities to each other then gives the ratio of the number of atoms of each element in the chemical formula of the compound. A second type of physical test is often required to determine the number of moles of the compound that were in the original sample. With these data in hand, the chemical formula of the compound can be determined.

Percentage Composition in Forensic Applications

In 1912, the giant ocean liner RMS *Titanic* sank after striking an iceberg in the Atlantic Ocean. The ship had been a marvel of engineering, deemed by some to be "unsinkable." The rapidity with which it sank proved them wrong. When a sample of steel plates from the *Titanic* was subjected to analysis, it was found that the percent composition of the metal was wrong, which resulted in the metal becoming brittle when exposed to the frigid waters of the North Atlantic Ocean.

In 1937, the great hydrogen-filled airship *Hindenburg* ignited and was quickly engulfed in flames, spelling the end of such airships. Subsequent analysis of a sample of the same fabric from which *Hindenburg* had been constructed revealed that it had been coated with the compound iron(II) oxide and very fine particles of aluminum, the components of the highly energetic thermite reaction. It is believed that a spark from static-electricity buildup may have initiated the reaction.

Percentage composition analysis is one of the basic tools of forensic crime analysis, although the wet chemistry methods on which it relies have been replaced in large part by faster, more sensitive instrumental techniques. In some cases, classical elemental analysis and percentage composition determination remain the most effective means of identifying certain materials recovered during an investigation.

Richard M. Renneboog, MSc

Bibliography

Deutch, Yvonne, ed. *Science against Crime.* London: Marshall, 1982. Print.

Myers, Richard. *The Basics of Chemistry.* Westport: Greenwood, 2003. Print.

Miessler, Gary L., Paul J. Fischer, and Donald A. Tarr. *Inorganic Chemistry.* 5th ed. Upper Saddle River: Prentice Hall, 2013. Print.

Morrison, Robert Thornton, and Robert Neilson Boyd. *Organic Chemistry.* 7th ed. Englewood Cliffs, N.J.: Prentice Hall, 2003. Print.

Skoog, Douglas Arvid, Donald M. West, and F. James. Holler. *Skoog and West's Fundamentals of Analytical Chemistry.* 9th ed. Belmont (Calif.): Cengage Learning, 2014. Print.

REACTION CALCULATIONS: STOICHIOMETRY

FIELDS OF STUDY

Physical Chemistry; Chemical Engineering

SUMMARY

The concept of stoichiometry is described, and its importance in all chemistry-related fields is explained. Stoichiometry describes the precise and very specific molar relationships within any particular chemical reaction process. It is based in the law of conservation of mass and was an important factor in the development of the modern theory of atomic structure.

PRINCIPAL TERMS

- **Avogadro's number (NA):** $6.02214129 \times 10^{23}$, often rounded to 6.022×10^{23}; the number of particles (atoms or molecules) that constitute one mole of any element or compound, the mass of which in grams is numerically equal to the atomic or molecular weight of the material.
- **atomic mass unit:** a unit of mass defined as one-twelfth of the mass of one atom of carbon-12 (approximately 1.66×10^{-27} kilograms), equivalent to 1 gram per mole.
- **law of conservation of mass:** the principle that states that all atoms present at the beginning of a process must also be present after the completion of the process; matter can be neither created nor destroyed, only changed from one form to another.
- **law of definite proportions:** the principle that states that the proportions by mass of the elements that combine to form a specific compound are definite and do not vary.
- **mole:** the amount of any pure substance that contains as many elementary units (approximately 6.022×10^{23}) as there are atoms in twelve grams of the isotope carbon-12.

UNDERSTANDING STOICHIOMETRY

Stoichiometry is one of the most fundamental concepts in chemistry. The term can have slightly different meanings based on the field of chemistry in which it is used, but it typically refers to the quantitative relationships—that is, relationships based on specific amounts—between chemical components in a reaction. It can also refer to the study of these relationships.

A simple analogy for the concept of stoichiometry can begin with a standard egg carton in which some of the spaces are marked to be filled with white-shelled eggs and the remainder with brown-shelled eggs. It does not matter how many additional white and brown eggs are available; only the specified numbers of each will fill the carton properly. Similarly, in a chemical reaction, only the specific number of atoms or molecules required are involved in the conversion from reactants to products. The logic of this, which is the foundation of stoichiometry, derives from the modern theory of atomic structure and the law of conservation of mass, which states that matter can be neither created nor destroyed. All matter present at the beginning of a chemical process must remain after the process is complete; likewise, no additional matter can be present at the end of the process.

THE FOUNDATION OF STOICHIOMETRY

The concept of stoichiometry brings together several fundamental principles of chemical behavior. The first of these is the law of definite proportions. Near the beginning of the nineteenth century, French chemist Joseph Proust (1754–1826) proposed that the proportions of different elements in a compound were definite, or constant, and did not change unless the chemical identity of the compound changed and new fixed proportions were achieved. This insight allowed English chemist John Dalton (1766–1844) to demonstrate through his own experimentation that when two elements form more than one compound, they do so in small, whole-number ratios or small multiples, thus indicating that the atoms in a compound are present in whole-number ratios. This observation became known as the law of multiple proportions. The law provided the first clue to the nature of atomic interactions at the level of electrons and was later explained more completely in the modern theory of atomic structure.

In the early nineteenth century, Italian chemist Amadeo Avogadro (1776–1856) studied the combining ratios of different gases and the relationships between temperature, pressure, and volume. Through his research, he determined that equal volumes of any gases under the same conditions of temperature and pressure contain the same number of particles. This discovery, known as Avogadro's law, proved highly influential, and Avogadro was later recognized for his important work when the value known as Avogadro's number or Avogadro's constant was named in his honor. This number, $6.02214129 \times 10^{23}$, is equal to the number of particles (atoms, molecules, or similar units) in one mole of any element or compound. For example, one mole of monatomic oxygen contains $6.02214129 \times 10^{23}$ oxygen atoms, while one mole of water contains $6.02214129 \times 10^{23}$ molecules of H_2O. The value of a mole, and thus of Avogadro's number, is based on the number of atoms in twelve grams of carbon-12, an isotope of carbon with six protons and six neutrons in its nucleus. Because twelve grams of carbon-12 contains $6.02214129 \times 10^{23}$ atoms—equal to Avogadro's number—one mole of any substance is the amount of that substance that is made up of $6.02214129 \times 10^{23}$ atoms, molecules, or other particles or entities.

The concepts of the mole and Avogadro's number are particularly important because they provide a direct relationship between bulk quantities of compounds and the atomic and molecular identities of those compounds. The mass in grams of a mole of an element is numerically equal to that element's atomic mass. For example, one mole of carbon-12 weighs exactly twelve grams. The relationship between a mole of particles and its weight in grams holds true for every atom and molecule. One-twelfth of the mass of one atom of carbon-12, 1.66×10^{-27} kilograms, is known as an atomic mass unit.

Another basic principle of atomic behavior is explained by the modern theory of atomic structure, which states that chemical reactions occur only at the level of an atom's outermost, or valence, electrons. The nucleus of any atom is not affected in a chemical reaction, though its particular combination of protons and neutron determines the distribution of electrons about the nucleus and

REACTION CALCULATIONS STOICHIOMETRY

	REACTANTS	PRODUCTS
Compound Names	silicon dioxide + fluorine	silicon tetrafluoride + oxygen
Molecular Formulas	SiO_2 + F_2	SiF_4 + O_2
Balanced Formulas	SiO_2 + $2F_2$	SiF_4 + O_2
Moles	1 mol SiO_2 + 2 mol F_2	1 mol SiF_4 + 1 mol O_2
Mass	60.1 g SiO_2 + 38 g F_2	66.1 g SiF_4 + 32 g O_2

CONVERSION RATIOS

1 mol SiO_2 : 2 mol F_2 : 1 mol O_2 : 1 mol SiF_4

60.1 g SiO_2 : 38 g F_2 : 32 g O_2 : 66.1 g SiF_4

the behavior of electrons in the formation of bonds with other atoms in molecules. Thus, only whole individual atoms can be involved in chemical reactions. Since no nuclei are changed in a chemical reaction, no atoms change their identity as elements. Atoms can be neither created nor destroyed in a chemical reaction; they merely combine in different ways. Therefore, each atom, and its corresponding mass, that was present at the beginning of a chemical reaction must also be present after the reaction has been completed. This rule is known as the law of conservation of mass.

Finally, quantum mechanics describes the behavior of electrons in atoms and restricts them to specific regions about the nucleus according to the particular energy levels they possess. The regions are called orbitals, and the restrictions to which they are subject dictate that only the valence electrons, which inhabit the outermost orbitals, have the ability to form bonds. This allows the formation of different numbers of bonds between atoms of the same type.

NONSTOICHIOMETRIC COMPOUNDS

There exists a class of discrete compounds that do not seem to adhere to the rules of stoichiometry. Strictly speaking, "stoichiometry" refers to the weight relationships that exist in chemical compounds, but the term is often used to refer to the molar relationships instead. In nonstoichiometric compounds, the molar relationship does not seem to apply, as this relationship relies on ratios based on whole atoms, and nonstoichiometric compounds display a distinct identity based on structural components that seem to be only parts of atoms or molecules.

This occurs most commonly in mixtures of compounds, such as metal alloys or combined salt melts. A mixture with a specific combination of components may display properties characteristic of a unique compound rather than a mixture of compounds, such as a crystalline structure and a melting or boiling point within a very narrow range of temperatures. Analysis of the composition of such materials often reveals a molecular formula in which the individual elements do not appear in a whole-number ratio. Crystalline minerals often have nonstoichiometric formulas; for example, the silver-copper sulfide mineral jalpaite has the formula $Ag_{1.55}Cu_{0.45}S$.

SAMPLE PROBLEM

Write a balanced equation representing the action of sodium hydroxide (NaOH) on aluminum nitrate, which has the formula $Al(NO_3)_3 \cdot 9H_2O$. The dot followed by "$9H_2O$" indicates that the compound is a crystalline hydrate known as aluminum nitrate nonahydrate, meaning that the crystal contains water at a ratio of nine water molecules for each molecule of aluminum nitrate. If 75.03 grams of aluminum nitrate nonahydrate are dissolved in water and reacted with 20.47 grams of sodium hydroxide, what is the mass in grams of the aluminum hydroxide, $Al(OH)_3$, that will be formed?

Answer:

Aluminum nitrate nonahydrate reacts with sodium hydroxide in solution to produce aluminum hydroxide. The initial reaction equation can be written as

$$Al(NO_3)_3 \cdot 9H_2O + NaOH \rightarrow Al(OH)_3 + NaNO_3$$

The properly balanced equation is

$$Al(NO_3)_3 \cdot 9H_2O + 3NaOH \rightarrow Al(OH)_3 + 3NaNO_3 + 9H_2O$$

To determine the mass of the aluminum hydroxide produced, first determine the molar masses of aluminum nitrate nonahydrate and sodium hydroxide:

$$Al(NO_3)_3 \cdot 9H_2O\text{: } 27\frac{g}{mol} + 3(14)\frac{g}{mol} + 18(16)\frac{g}{mol} + 18(1)\frac{g}{mol} = 375\frac{g}{mol}$$

$$NaOH\text{: } 23\frac{g}{mol} + 16\frac{g}{mol} + 1\frac{g}{mol} = 40\frac{g}{mol}$$

Next, calculate the number of moles of each reactant that are present:

$$Al(NO_3)_3 \cdot 9H_2O\text{: } \frac{75.03\,g}{375\,g/mol} = 75.03\ g \times \frac{1\ mol}{375\,g} = 0.200\ mol$$

$$NaOH\text{: } \frac{20.47\,g}{40\,g/mol} = 20.47\ g \times \frac{1\ mol}{40\,g} = 0.512\ mol$$

The balanced reaction equation states that three moles of OH^- are consumed for each mole of $Al(NO_3)_3 \cdot 9H_2O$. Therefore, 0.512 mole of NaOH can produce only $0.512 \div 3 = 0.171$ mole of $Al(OH)_3$. Calculate how much $Al(OH)_3$ this can produce:

$$\text{molar mass of } Al(OH)_3\text{: } 27\frac{g}{mol} + 3(16)\frac{g}{mol} + 3(1)\frac{g}{mol} = 78\frac{g}{mol}$$

$$\text{mass of } Al(OH)_3 \text{ in grams: } 0.171\ mol \times 78\frac{g}{mol} = 13.338\ g$$

A total of 13.338 grams of aluminum hydroxide can be produced by this reaction.

STOICHIOMETRY AND BALANCING CHEMICAL EQUATIONS

Since only whole atoms can be involved in the formation of a chemical compound, the individual atoms join together to form a molecule of the compound. If the number of each kind of atom in a molecule is multiplied by Avogadro's number, the proportional relationship between moles of atoms in a mole of molecules is exactly the same as the proportional relationship between the individual atoms in the individual molecule. For example, one atom of nitrogen can combine with three atoms of hydrogen to form one molecule of ammonia, which has the formula NH_3. By extension, one mole of nitrogen atoms can combine with three moles of hydrogen atoms to form one mole of ammonia.

However, while this is strictly correct, it does not reflect the normal reaction. Normal chemical reactions are written using the normal form of an element or compound. The actual reaction between nitrogen and hydrogen would be written using the N_2 and H_2 molecular forms rather than the monatomic N and H forms. This is where the concept of stoichiometry becomes most important. The strict molar relationships between atoms and the law of conservation of mass must be obeyed in the expression of any chemical reaction. Accordingly, it would not be correct to write the reaction equation as:

$$N_2 + H_2 \rightarrow NH_3$$

as this does not properly account for all of the nitrogen and hydrogen atoms. Because there are two atoms of nitrogen in each molecule of N_2, then clearly each N_2 molecule must result in two molecules of NH_3. But merely writing:

$$N_2 + H_2 \rightarrow 2NH_3$$

is also incorrect, as this does not account for the additional hydrogen atoms in the two ammonia molecules. Because there are six hydrogen atoms in two molecules of NH_3, there must have been six atoms of hydrogen present at the start of the reaction. If the equation is then written as:

$$N_2 + 3H_2 \rightarrow 2NH_3$$

it can be seen that all of the atoms that were present before the reaction are also present after the reaction has taken place. The law of conservation of mass has been strictly obeyed, and the molar relationships between the atoms have been maintained. This maintenance of molar relationships when describing a chemical reaction, in adherence to the various laws of mass and proportion, is the essence of stoichiometry.

STOICHIOMETRY IN PHARMACOLOGY

Stoichiometry has a slightly different context in pharmacology, although the basic chemical principles underlying stoichiometry are equally important in biochemical systems. Pharmacology is the science of applying chemical compounds that do not typically occur in a biological system to assist the organism in overcoming an adverse condition such as disease or chronic pain. The efficacy of a pharmaceutical compound is measured against its effect on the biochemical system overall rather than its specific stoichiometry in the chemical sense. The concept of stoichiometry in biological systems is therefore often applied in terms such as the LD_{50}, the lethal dose for 50 percent of the test population.

Richard M. Renneboog, MSc

BIBLIOGRAPHY

Gribbin, John. *Science: A History, 1543–2001.* London: Lane, 2002. Print.

Johnson, Rebecca L. *Atomic Structure.* Minneapolis: Lerner, 2008. Print.

Mackay, K. M., R. A. Mackay, and W. Henderson. *Introduction to Modern Inorganic Chemistry.* 6th ed. Cheltenham: Nelson, 2002. Print.

Miessler, Gary L., Paul J. Fischer, and Donald A. Tarr. *Inorganic Chemistry.* 5th ed. Upper Saddle River: Prentice Hall, 2013. Print.

Myers, Richard. *The Basics of Chemistry.* Westport: Greenwood, 2003. Print.

Rang, H. P., et al. *Rang and Dale's Pharmacology.* 7th ed. New York: Elsevier, 2012. Print.

REACTION MECHANISMS

FIELDS OF STUDY

Organic Chemistry; Inorganic Chemistry; Biochemistry

SUMMARY

The concept of reaction mechanisms and its importance in chemistry-related fields is elaborated. A reaction mechanism is the process by which a reaction theoretically takes place and describes the reorganization of electrons and orbitals in the reactants as they interact to form the products.

PRINCIPAL TERMS

- **activated complex:** the various intermediate molecular structures that exist during a chemical reaction while the original chemical bonds are being broken and new bonds are being formed.
- **elementary reaction:** a chemical reaction that is completed in a single reaction step with only one transition state; often part of a multistep sequence of reactions that constitutes a mechanism.
- **reaction step:** a stage in a multistep reaction in which an elementary reaction converts either the reactants or a previously formed intermediate structure into either a new intermediate structure or the final products.
- **reactive intermediate:** a short-lived, highly reactive chemical species that is formed during an intermediate reaction step in a multistep reaction and typically cannot be isolated.
- **transition state:** an unstable structure formed during a chemical reaction at the peak of its potential energy that cannot be isolated and ultimately breaks down, either forming the products of the reaction or reverting back to the original reactants.

VISUALIZING CHEMICAL PROCESSES

A reaction mechanism is the hypothetical description of how a reaction occurs. It is based on observation and the logical application of the principles of modern atomic theory and molecular orbital theory. All reaction mechanisms begin with the observation of an overall reaction process, which is shown to occur between two reactants and to produce certain products. The purpose of a proposed reaction mechanism is to describe a logical means of transforming the reactants into the products that agrees with the theoretical principles governing molecular structures and their substituent functional groups as well as the characteristic behavior of atoms.

The purpose of a reaction mechanism is to describe the transformation of one compound into another by identifying individual electron movements and molecular orbital rearrangements that account for the observed changes that take place in the reaction. Reaction mechanisms consist of a sequence of elementary reactions, which in this context form individual reaction steps. Each individual reaction step must also be in agreement with the electron movements and orbital rearrangements that are allowed in theory.

Theory can be modified when empirical results call the existing theoretical principles into question, however. It has often been the case historically that a reaction that is expected to produce a certain

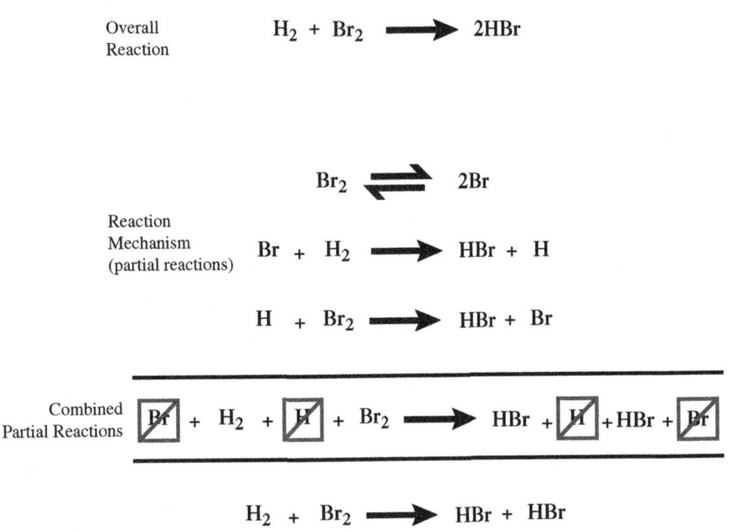

product by a certain mechanism instead produces a compound with the correct components but in an unexpected arrangement. In such cases, the observation provides new information about the actual reaction mechanism that has taken place, allowing the theoretical mechanism to be modified. The modified theoretical mechanism can then be more successfully applied to other reactions. For example, the reaction of triphenylmethyl halides with silver metal forms a particular dimer, which is a compound made from two similar monomers. For nearly seventy years after this reaction was identified, the product was assumed to be hexaphenylethane, but the compound did not behave the way hexaphenylethane should behave. Newer methods of analysis demonstrated that the structure of that particular dimer is not that of hexaphenylethane, and therefore the proposed mechanism for its formation in the reaction was wrong.

REACTIONS AND REACTION MECHANISMS

In any chemical reaction, atoms and molecules interact in certain ways, and any mechanism that describes the reaction is only one means of visualizing how their interaction may have transpired. Because reaction mechanisms are theoretical in nature, a given reaction mechanism can rarely if ever be proved to be the way in which the atoms and molecules have actually reacted. Even if a proposed mechanism accounts perfectly for the results of a reaction, it can only be said that those results were obtained because it is as though the atoms and molecules had interacted in that way.

Chemists visualize reaction mechanisms by drawing molecular structure diagrams and annotating them with standardized arrows that indicate the movement of electrons and orbitals. There are two main conventions used in drawing a proposed mechanism. The first is that the movement of two electrons from one orbital to another is shown by a curved arrow with a full, double-barbed arrowhead. The second is that the movement of individual electrons from one orbital to another is shown by a curved arrow with a half arrowhead. Thus, the formation of a bond between a chloride ion (Cl^-) and the *tert*-butyl carbonium ion, producing *tert*-butyl chloride, would be drawn as

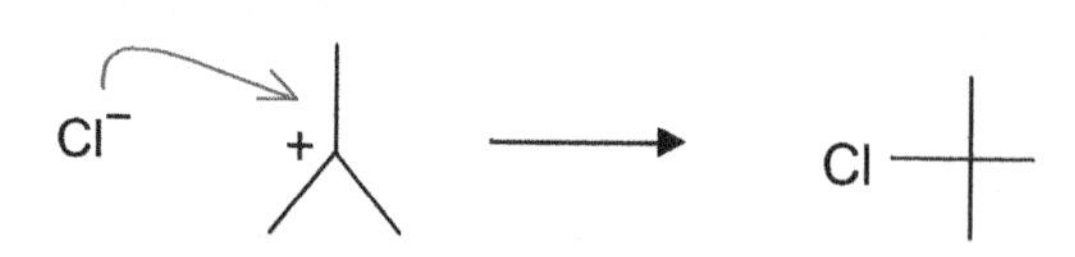

and the movement of the electrons in the diene and dienophile in a Diels-Alder reaction would be drawn as

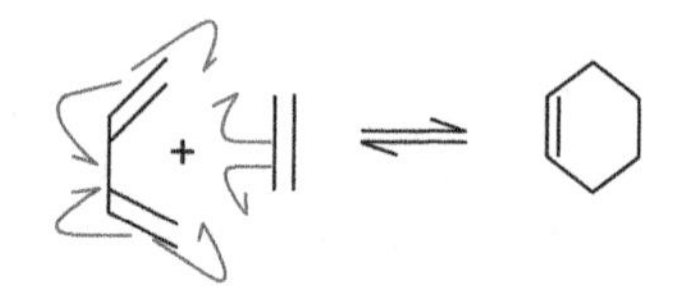

With these two simple conventions, a great deal of information about a particular reaction can be communicated in a very compact form.

An overall reaction equation can be written as follows:

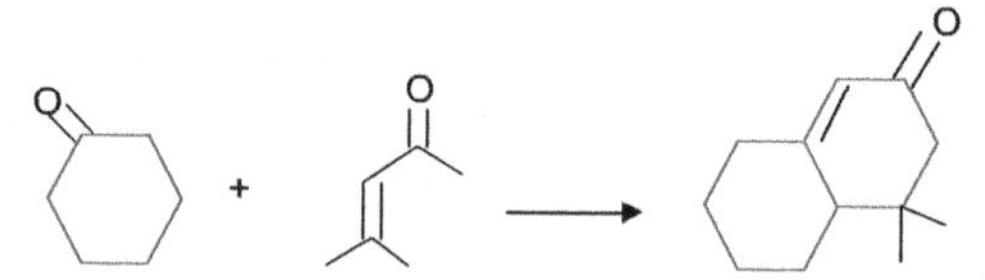

This equation represents a Robinson annulation reaction, which is a reaction that produces organic ring-shaped molecules. It identifies the initial reactants and the product but gives no information about the reaction conditions or any intermediary stages in the process. A mechanistic representation of just the first elementary reaction in the overall process would be drawn as

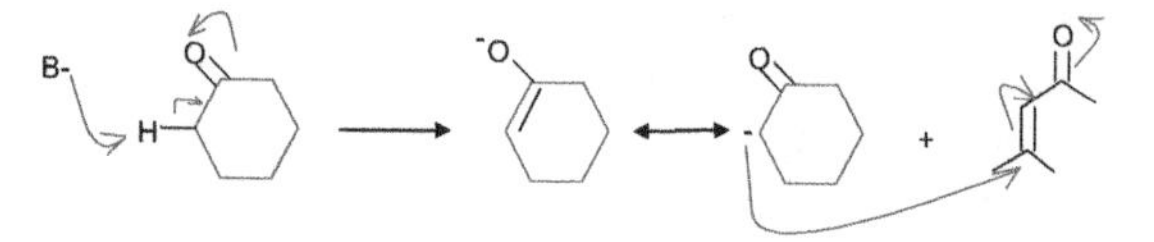

with several other elementary reactions following. For someone familiar with the conventions and the theoretical bases of the interactions, such diagrams very clearly communicate in a relatively small space concepts that would require several pages to describe in words.

When synthesizing complex molecules, such multistep procedures are the rule rather than the exception. The goal of each step in a multistep synthesis is to have complete control of the formation of the molecular structure of the product. A mechanistic reaction equation readily identifies all of the components

SAMPLE PROBLEM

The diagram below represents a reaction in which platinum metal (Pt) combines with molecular hydrogen (H_2) to form H_2Pt, which then reacts with 1-ethyl-2-methylcyclohexene to produce *cis*-1-ethyl-2-methylcyclohexane. Identify the catalyst and intermediates in the elementary steps of this reaction. What is the overall reaction equation?

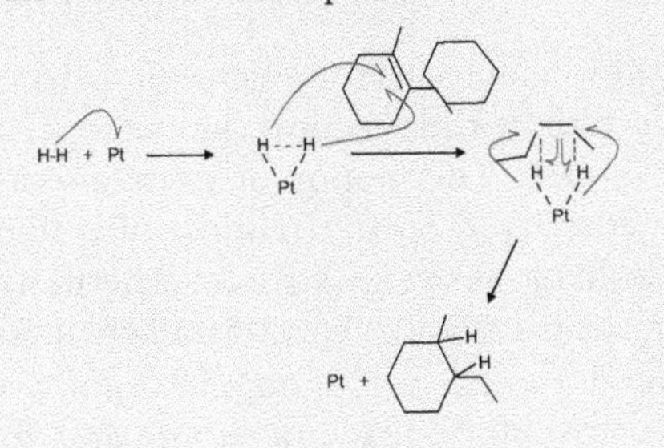

Answer:

Platinum metal is one of the reactants and is also recovered as a product. Since it is unchanged in the reaction, it is the catalyst for the addition of molecular hydrogen to the 1-ethyl-2-methylcyclohexene molecule. The H_2Pt structure is a stable intermediate that forms on the surface of the platinum metal. The structure involving the platinum-hydrogen complex and the cyclohexene molecule is the transition state, which exists at the point at which the pi bond of the cyclohexene breaks down and the two new C–H bonds form. The overall reaction is as follows:

1-ethyl-2-methylcyclohexene + $H_2 \rightarrow$

cis-1-ethyl-2-methylcyclohexane

of the reaction process, including transition states, reactive intermediates, catalysts (if present), and the activated complexes that may be involved in the overall reaction process.

MECHANISMS AND KINETICS

"Reaction kinetics" refers to the study of how reactions occur over time in relation to the amounts of reactants and products that are present in the reaction mixture. The amounts are expressed in terms of concentration, that is, the amount of a material that is present in a specific volume of the reaction mixture. The probable mechanism by which a reaction proceeds can often be discerned from the kinetic behavior of the reaction. Of particular interest in determining a reaction mechanism is the reaction step known as the rate-determining step. The overall rate of the reaction cannot be faster than the slowest reaction step in the mechanism. In cases where more than one mechanism can be reasonably proposed for a reaction, identification of the rate-determining step from the reaction kinetics can help identify which mechanism, if any, is more likely than the others to be accurate.

The presence of a catalyst alters the reaction kinetics by reducing the activation energy of the overall reaction, which is the energy that a chemical system must acquire in order for a reaction to proceed. It also changes the reaction mechanism because the catalyst is intimately involved in the reaction process along with the reactants, though it is not itself consumed in the reaction.

REACTION MECHANISMS IN BIOLOGICAL SYSTEMS

All biochemical reactions are mediated by enzymes. Due to the large size and complicated structures of enzymes, the mechanisms of enzyme-mediated reactions have a high level of complexity that is only found in biochemical environments. Essentially, the enzyme's active site encloses the substrate (reactant) in such a way that it forms a closed system in which the substrate interacts with its environment to effect a chemical change. Various functional groups in the enzyme's active site coordinate to both the substrate and any other molecule involved in the reaction. The individual steps of the overall reaction mechanism are therefore difficult to differentiate, since all steps take place in the same enzyme-substrate combination. Enzyme-mediated reactions exhibit characteristic kinetic behavior because of this intimate relationship.

Richard M. Renneboog, MSc

BIBLIOGRAPHY

Berg, Jeremy M., John L. Tymoczko, Gregory J. Gatto, and Lubert Strye. *Biochemistry.* 8th ed. New York: W. H. Freeman, 2015. Print.

Douglas, Bodie E., Darl H. McDaniel, and John J. Alexander. *Concepts and Models of Inorganic Chemistry.* 3rd ed. New York: Wiley, 1994. Print.

Miessler, Gary L., Paul J. Fischer, and Donald A. Tarr. *Inorganic Chemistry.* 5th ed. Upper Saddle River: Prentice Hall, 2013. Print.

Morrison, Robert Thornton, and Robert Neilson Boyd. *Organic Chemistry.* 7th ed. Englewood Cliffs, N.J.: Prentice Hall, 2003. Print.

Pine, Stanley H. *Organic Chemistry.* 5th ed. New York: McGraw, 1987. Print.

Sykes, Peter. *A Guidebook to Mechanism in Organic Chemistry.* 6th ed. London: Longman, 1986. Print.

Weeks, Daniel P. *Pushing Electrons: A Guide for Students of Organic Chemistry.* 4th ed. Belmont: Brooks, 2014. Print.

REACTION RATES

FIELDS OF STUDY

Physical Chemistry; Chemical Engineering; Biochemistry

SUMMARY

The study of chemical kinetics examines the behavior of chemical reaction systems over time. The rate of a chemical reaction depends on the concentration of the reactants and products and is affected by the presence and activity of materials that catalyze the particular process. This is of central importance in biological systems.

PRINCIPAL TERMS

- **chemical kinetics:** the branch of chemistry that studies the various factors that affect rates of chemical reactions.
- **closed system:** a physical or chemical reaction system defined by certain boundary conditions that prevent any components, reactants, or products from entering or exiting the system.
- **equilibrium:** the state that exists when the forward activity of a process is exactly equal to the reverse activity of that process.
- **rate-determining step:** in any multistep process, the step that proceeds at the slowest rate compared to the other steps in the process, thus determining the maximum rate at which the overall process takes place.
- **reaction mechanism:** the sequence of electron and orbital interactions that occurs during a chemical reaction as chemical bonds are broken, made, and rearranged.

VISUALIZING THE RATE OF REACTION

Reactions rates are typically studied as a reaction is carried out in a closed system. The higher the concentration of reactant materials in the system, the more rapidly the reaction can occur in a specific amount of time. The motion of atoms and molecules is often compared to the motion of a blindfolded person walking about in a room. The more people there are in the room, the more likely it is that the blindfolded person will bump into someone else. Similarly, the more reactant atoms and molecules there are in a reaction mixture, the more often they will bump into each other and react. In its simplest definition, chemical kinetics is counting the number of times collisions between atoms and molecules in a reaction mixture result in a reaction taking place over a specific amount of time.

The occurrence of reactions is governed by a number of factors, including the activation energy of the reaction, temperature, pressure, the geometry of the molecules, and the reduction-oxidation (redox) potential of the reacting components. Accordingly, different reactions occur at different rates. One could, for example, compare the rates at which 1-chloropropane and 1-bromopropane undergo the same substitution reaction under the same conditions and with the same reagent. Even though these are similar halogen elements, they exhibit different rates of reaction. Furthermore, the same element can exhibit measurably different rates of reaction if the reaction is carried out using two different isotopes, such as protium and deuterium. This process is termed the "isotope effect."

DETERMINING THE RATE OF A REACTION

Reaction rates are determined by direct observation and measurement. Typically, from the moment the reagents of a reaction are mixed, small samples of the live reaction mixture are extracted at intervals and analyzed. The disappearance of starting materials and the appearance of product materials over time shows the extent to which the reaction has progressed. Analytical methods such as high-performance liquid

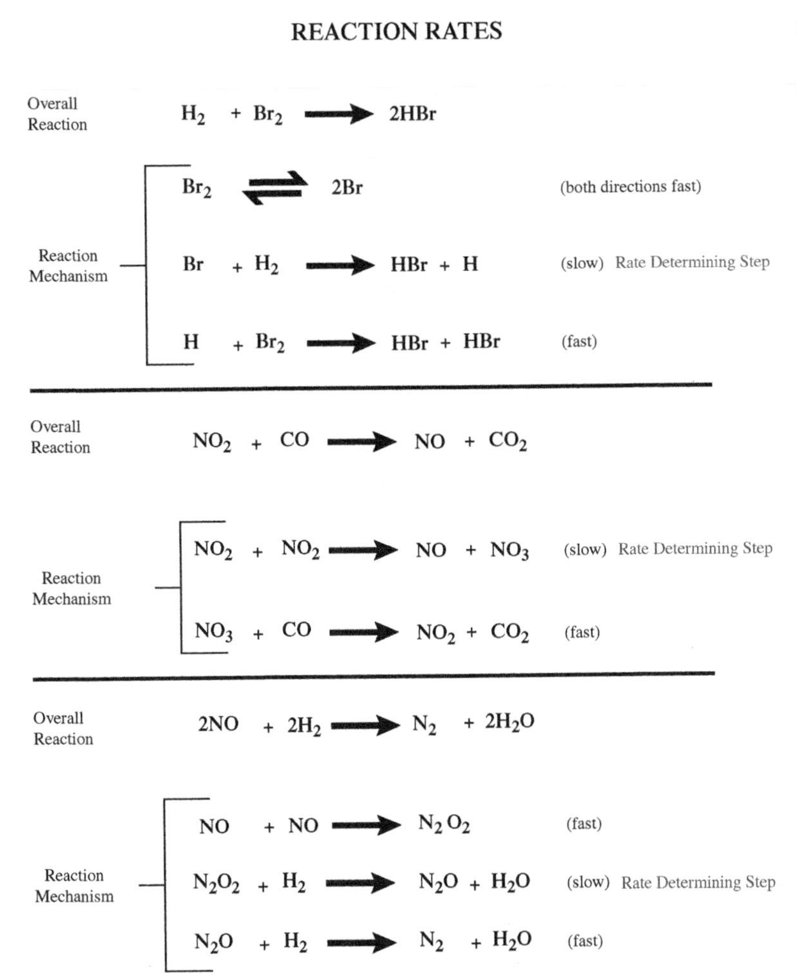

chromatography, gas chromatography, nuclear magnetic resonance spectrometry, and ultraviolet-visible spectrophotometry are routinely used to monitor the progress of a reaction. The measured changes of the absorbance peaks indicate the relative amounts of each component. Relating these to the known quantity of each component at the beginning of the reaction indicates the actual quantities of each. The measured results will always exhibit a mathematical relationship to the concentration of each individual component of the reaction mixture.

The rate of a particular reaction can provide insight into the mechanism by which the reaction proceeds. All reactions occur only at the rate of the slowest elementary step in the reaction mechanism. When more than one proposed mechanism is possible for an overall reaction, the rate of the reaction can demonstrate whether or not a catalyst is involved in the mechanism. The order of the reaction is also an important indicator of the probable mechanism. A reaction that exhibits first-order kinetics, for example, cannot occur by a mechanism that depends on just one single reactant because second-order reactions require the participation of two components in the rate-determining step.

The general form of the relationship for a first-order reaction, such as $A \rightarrow B$, is:

$$\text{rate} = k[A]$$

where k is the unique rate constant for that particular reaction and the rate is the measured rate of the reaction, specific to the conditions under which the reaction has been carried out. A reaction of the form $A + B \rightarrow C$ will have a rate equation of the form:

$$\text{rate} = k[A][B]$$

A reaction of the form $2A + B \rightarrow C$ has the kinetic expression:

$$\text{rate} = k[A]^2[B]$$

and so on for each individual reaction. The rate expression reflects the balanced chemical equation for the reaction, with the coefficients of each reactant becoming the exponent of its concentration term in the rate equation. Determining the rate equation for an unknown reaction can thus reveal information about the overall reaction process.

The rate expression for a complex reaction can quickly become too complex to be useful. In such cases, it is beneficial to determine the rate of reaction using a reaction mixture that has been saturated with so much of one of the components that its concentration effectively does not change throughout the course of the reaction. In this way, it is possible to force a more complex reaction to behave as though it were a first-order reaction. This method is particularly useful for determining the maximum rate of a reaction that involves a catalyst and especially for the study of enzyme-catalyzed reactions. Because the

SAMPLE PROBLEM

Given the following series of elementary steps in a reaction process, identify the rate-determining step and the corresponding rate law of the reaction.

$$2A \rightarrow A^* \text{ (fast)}$$

$$A^* + B \rightarrow A^*B \text{ (slow)}$$

$$A^*B + B \rightarrow C \text{ (fast)}$$

Answer:

The overall reaction can proceed only at the rate of the slowest elementary step in the process. Therefore, the rate of the reaction will be determined by the second elementary step, in which the activated complex A*B is formed:

$$A^* + B \rightarrow A^*B$$

The rate law can be written initially as

$$\text{rate}_2 = k_2[A^*][B]$$

because the rate of that particular step depends on the concentrations of both A* and B. But the concentration of A* is determined by the first elementary step of the process $2A \rightarrow A^*$. Since that step occurs more quickly than the next, the concentration of A* is determined by the rate of the first step, as

$$\text{rate}_1 = k'[A]^2$$

$$[A^*] = \frac{\left(k_1[A]^2\right)}{k'[A]^2}$$

Therefore, the overall rate of the reaction can be written as

$$\text{Rate} = k[A]^2[B]$$

Note that the rate of each step is unique to that step; thus, they may have very different values. Therefore, it is important to differentiate each rate from the others and to treat them in calculations as coefficients. The proportional values of any two rate constants will also be a constant, and it is common practice to indicate this by adding "primes" to a rate constant (k', k'', k''', and so on) as a method of keeping track of the corresponding proportional values.

rate of such reactions is quite variable (because the amount of enzyme-substrate activated complex that is present in the reaction mixture is inconsistent), saturating the reaction mixture ensures that the amount of enzyme-substrate complex is constant and at its maximum value throughout the course of the reaction.

EQUILIBRIUM AND RATE OF REACTION

For reactions that are readily reversible, equilibrium will be attained at some point, unless the products are somehow isolated from the mixture or more reactant materials are added. At equilibrium, the relative proportions of reactants and products are constant and do not change over time. The reaction does not cease, but rather is exactly balanced by the reverse reaction, and reactants are forming products at exactly the same rate that products are reverting back to reactants. In rate measurements, the relative concentrations of components change rapidly at the beginning of the reaction and trend to constant values as the system approaches equilibrium.

REACTION RATES IN BIOLOGICAL SYSTEMS

Understanding the rates of reactions in biological systems is crucial to the design and safe use of pharmaceuticals and other compounds. All materials are chemical in nature, and the ingestion of anything at all, from simple water to the most complex drugs, must pass through the metabolic system. The rate at which metabolic reactions process each substance determines the length of time that the material has an effect. It also determines the length of time required to prevent overdosing. A common application of this principle involves the consumption of alcohol, which is metabolized at a specific rate. Consuming alcohol faster than it can be metabolized results in overdosing, leading to impaired judgment, which can have fatal consequences, and even to alcohol poisoning, which may be equally deadly.

Richard M. Renneboog, MSc

BIBLIOGRAPHY

Berg, Jeremy M., John L. Tymoczko, Gregory J. Gatto, and Lubert Strye. *Biochemistry.* 8th ed. New York: W. H. Freeman, 2015. Print.

Laidler, Keith J. *Chemical Kinetics.* 3rd ed. New York: Harper, 1987. Print.

Lehninger, Albert L. *Biochemistry: The Molecular Basis of Cell Structure and Function.* 2nd ed. New York: Worth, 1975. Print.

Morrison, Robert Thornton, and Robert Neilson Boyd. *Organic Chemistry.* 7th ed. Englewood Cliffs, N.J.: Prentice Hall, 2003. Print.

Skoog, Douglas Arvid, Donald M. West, and F. James. Holler. *Skoog and West's Fundamentals of Analytical Chemistry.* 9th ed. Belmont (Calif.): Cengage Learning, 2014. Print.

REDOX REACTIONS

FIELDS OF STUDY

Inorganic Chemistry; Organic Chemistry; Biochemistry

SUMMARY

The concept of redox reactions and its importance in chemistry-related fields are elaborated. Oxidation and reduction are electron transfer processes that occur simultaneously in a chemical system; the amount of oxidation must exactly balance the amount of reduction.

PRINCIPAL TERMS

- **oxidation:** the loss of one or more electrons by an atom, ion, or molecule.
- **oxidation state:** a number that indicates the degree to which an atom or ion in a chemical compound has been oxidized or reduced.
- **oxidizing agent:** any atom, ion, or molecule that accepts one or more electrons from another atom, ion, or molecule in a reduction-oxidation (redox) reaction and is thus reduced in the process.
- **reducing agent:** any atom, ion, or molecule that donates one or more electrons to another atom, ion, or molecule in a reduction-oxidation (redox) reaction and is thus oxidized in the process.
- **reduction:** the gain of one or more electrons by an atom, ion, or molecule.

The Give and Take of Redox Reactions

The term "redox" is derived from reduction and oxidation, two inseparable processes of electron transfer that occur simultaneously in a great many chemical reactions. The process is similar in concept to paying for a purchase: the buyer must give money to the seller, and at the same time, the seller must accept the money from the buyer for the transaction to take place. In redox reactions, the "buyer" is the reducing agent that gives up a specific number of electrons, and the "seller" is the oxidizing agent that accepts that specific number of electrons.

In chemical terms, oxidation is the loss of electrons and reduction is the gain of electrons. This is perhaps easiest to keep in mind using a mnemonic such as "Let Every Orange Gorilla Eat Rice." There is

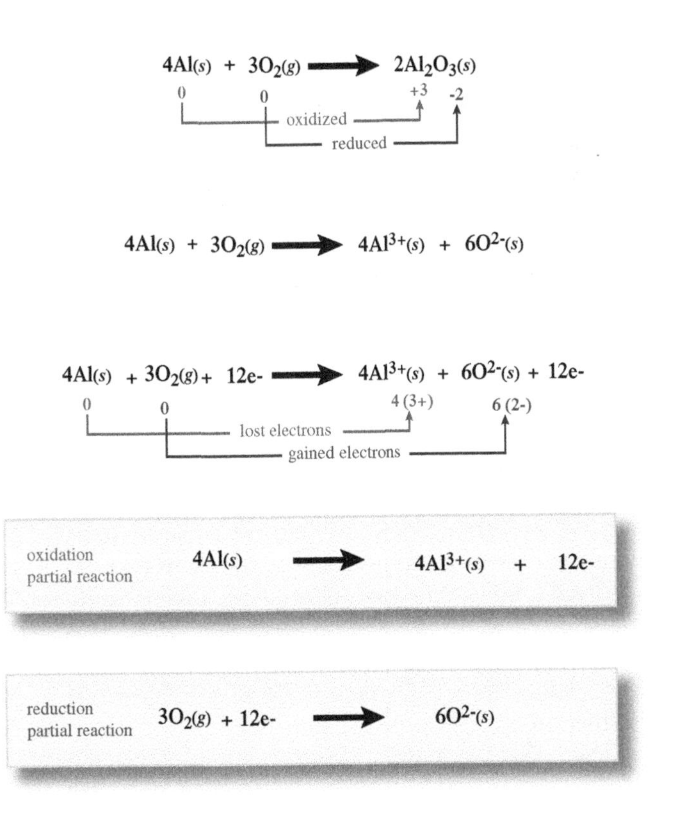

no set formula for how to remember the relationship, and any means of recalling the acronym LEO-GER (Loss of Electrons is Oxidation, Gain of Electrons is Reduction) is equally valid.

The most fundamental aspect of redox reactions is that, in the same way that mass must balance throughout the process of a chemical reaction, the electrons must also balance. This can be thought of as the law of conservation of charge, in that electrical charge can be neither created nor destroyed in a chemical reaction.

THE MEANING OF OXIDATION STATES

All chemical reactions take place at the level of the outermost electrons of the atoms involved in the interaction. Bonds are formed between atoms by the sharing of electrons, and a means of keeping track of the electrons is most helpful in understanding interactions at the atomic level. In many cases, it is the only way to determine the proper balance of a chemical reaction. The oxidation state concept is essentially a way for the chemist to quantify the relationship between two elements involved in a chemical reaction in which electron transfers take place.

The oxidation state of a single atom is essentially the number of electrons that particular atom has gained or given up control over in a chemical interaction. As a very simple example, a sodium atom that reacts with chlorine to produce sodium chloride (NaCl) has given up its one outermost electron to the chlorine atom. In so doing, the sodium atom has changed from the electrically neutral form of Na^0 to the positively charged sodium ion Na^+. At the same time, the chlorine atom has changed from the electrically neutral form of Cl^0 to the negatively charged chloride ion Cl^-. The oxidation state of the sodium atom has changed from 0 to +1, while the oxidation state of the chlorine atom has changed from 0 to −1. The sodium atom has been oxidized by the chlorine atom, just as the chlorine atom has been reduced by the sodium atom. Some confusion may arise from the terminology at this point, since the atom that was oxidized is labeled the reducing agent, and the atom that was reduced is labeled the oxidizing agent. This can be readily overcome, however, by recalling that Loss of Electrons is Oxidation, Gain of Electrons is Reduction.

The oxidation state of different atoms can seem overly large, ranging from −8 to +8, and one may easily mistake these values for the charge carried on the particular atom. It is important to remember that the oxidation state only represents the number of electrons that the atom has gained or lost control of in

SAMPLE PROBLEM

Balance the equation for the reaction between aluminum metal (Al) and sodium nitrite ($NaNO_2$) to produce ammonia (NH_3) and the anion AlO_2^- in basic solution.

Answer:

Write out the basic unbalanced equation:

$$Al^0 + NaNO_2 \rightarrow NH_3 + AlO_2^-$$

Next, determine the half reactions by assigning oxidation states and determining the electron transfers. Recall that the assigned oxidation states are −2 for each oxygen atom and +1 for each hydrogen atom.

Oxidation state of Al in AlO_2^-:

$$x + (-2 \times 2) = -1$$

$$x - 4 = -1$$

$$x = -1 + 4 = 3$$

Therefore, the oxidation state of Al in AlO_2^- is +3, and the half reaction for the aluminum is $Al^0 \rightarrow Al^{3+} + 3e^-$

Oxidation state of N in NO_2^-:

$$x + (-2 \times 2) = -1$$

The equation is solved in the same way as above. Therefore, the oxidation state of N in NO_2^- is +3.

Oxidation state of N in NH_3:

$$x + (1 \times 3) = 0$$

$$x + 3 = 0$$

$$x = 0 - 3 = -3$$

Therefore, the oxidation state of N in NH_3 is −3, and the half reaction for nitrogen is

$$N^{3+} + 6e^- \rightarrow N^{3-}$$

Continued on next page

SAMPLE PROBLEM CONTINUED

Balance the electron transfer:

$$2Al^0 \rightarrow 2Al^{3+} + 6e^-$$

Add together the balanced half reactions, canceling out the electrons:

$$N^{3+} + 2Al^0 \rightarrow 2Al^{3+} + N^{3-}$$

Next, substitute in the proper chemical formulas:

$$NO_2^- + 2Al^0 \rightarrow 2AlO_2^- + NH_3$$

Add in H_2O and OH^- as needed to complete the mass and charge balance:

$$NO_2^- + 2Al^0 + H_2O + OH^- \rightarrow 2AlO_2^- + NH_3$$

It is not necessary to add the sodium ions to the equation because they do not take part in the reaction and simply remain in solution.

its interaction with another atom. It may be helpful to also remember that the oxidation-state concept also applies to covalent compounds in which ions are not involved and electrons are shared between atoms to form chemical bonds. In such compounds, the oxidation states are determined by the theoretical gain or loss of electrons based on how electronegative each atom is.

ASSIGNING OXIDATION STATES

To facilitate understanding and application of the oxidation-state concept, a set of rules is used to assign the oxidation state of any particular atom.

- Rule 1: A hydrogen atom is always assigned an oxidation state of +1.
- Rule 2: An oxygen atom is always assigned an oxidation state of −2.
- Rule 3: A fluorine atom is always assigned an oxidation state of −1.
- Rule 4: Single-atom ions are always assigned an oxidation state that corresponds to their formal charge.

There are exceptions to these rules, but they are covered by other rules for assigning oxidation states. In organic compounds, overall oxidation states are assigned to a particular carbon atom according to the bonds it has formed.

- Rule 5: A carbon atom is assigned an oxidation state of −1 for each bond to a less electronegative atom, 0 for each bond to another carbon atom, and +1 for each bond to a more electronegative atom.
- Rule 6: To complete mass and charge balance, add H_2O and H^+ in acidic solution or H_2O and OH^- in basic solution as needed.

For example, perchloric acid has the chemical formula $HClO_4$. To balance the four oxygen atoms (each assigned an oxidation state of −2, for a total of −8) and the hydrogen atom (assigned an oxidation state of +1), the central chlorine atom must be assigned a state of +7. Since the chlorine atom is very electronegative and does not give up electrons to form a positively charged ion, it should be readily apparent that the oxidation state of +7 represents something entirely different from the atom's ionic charge.

HALF REACTIONS

To make the redox concept easier to comprehend, it is useful to break down a reaction to its component "half reactions." A half reaction isolates just the portion of the reaction corresponding to oxidation and reduction. For example, when magnesium metal, Mg, is dissolved in nitric acid, HNO_3, a redox reaction takes place in which aqueous magnesium nitrate, $Mg(NO_3)_2$, and hydrogen gas, H_2, are produced. The corresponding half reactions are as follows:

$$Mg^0 \rightarrow Mg^{2+} + 2e^-$$

$$H^{+1} + 1e^- \rightarrow H^0$$

These half reactions clearly indicate that the magnesium atom has given up two electrons and the hydrogen atoms have each accepted one electron. By balancing the number of electrons that are transferred and adding together the result, the balanced reaction can be determined. Here, there must be two molecules of HNO_3 to account for the two electrons accepted from the magnesium and the two atoms of hydrogen that are produced. Thus, the balanced equation is:

$$Mg + 2HNO_3 \rightarrow Mg(NO_3)_2 + H_2$$

Redox Reactions in Biological Systems

Redox reactions are exceedingly important in biological systems because they are the means by which all energy is transferred biochemically. Cellular respiration proceeds by the stepwise decomposition of glucose in the Krebs, or citric acid, cycle and the electron transport chain. Metabolism of the many different compounds that are consumed as food involves repeated stages of oxidation, mostly mediated by enzyme proteins, to extract energy and convert the materials into metabolites for elimination as waste.

Richard M. Renneboog, MSc

Bibliography

Berg, Jeremy M., John L. Tymoczko, Gregory J. Gatto, and Lubert Strye. *Biochemistry.* 8th ed. New York: W. H. Freeman, 2015. Print.

Douglas, Bodie E., Darl H. McDaniel, and John J. Alexander. *Concepts and Models of Inorganic Chemistry.* 3rd ed. New York: Wiley, 1994. Print.

Lafferty, Peter, and Julian Rowe, eds. *The Hutchinson Dictionary of Science.* 2nd ed. Oxford: Helicon, 1998. Print.

Lodish, Harvey, et al. *Molecular Cell Biology.* 7th ed. New York: Freeman, 2013. Print.

Miessler, Gary L., Paul J. Fischer, and Donald A. Tarr. *Inorganic Chemistry.* 5th ed. Upper Saddle River: Prentice Hall, 2013. Print.

Morrison, Robert Thornton, and Robert Neilson Boyd. *Organic Chemistry.* 7th ed. Englewood Cliffs, N.J.: Prentice Hall, 2003. Print.

Myers, Richard. *The Basics of Chemistry.* Westport: Greenwood, 2003. Print.

RNA/PROTEIN TRANSLATION

FIELDS OF STUDY

Biochemistry; Genetics; Molecular Biology

SUMMARY

The process by which RNA translates protein structures from the genetic code in the DNA molecule is described. The process is controlled by a simple series of three-nucleotide units called codons that determine which amino acids are to be assembled in which order to create every protein and enzyme in the entire biological system.

PRINCIPAL TERMS

- **anticodon:** a sequence of three nucleotide bases in transfer RNA (tRNA) that bonds to a complementary codon in messenger RNA (mRNA).
- **codon:** a sequence of three nucleotide bases that specifies a particular amino acid or control point in the process of protein synthesis.
- **gene expression:** the process by which RNA copies genes, which are specific segments of the DNA molecule, and uses the information to synthesize either proteins or other types of RNA.
- **peptide bond:** a covalent bond that links the carboxyl group of one amino acid to the amine group of another, enabling the formation of proteins and other polypeptides.
- **ribosome complex:** a structure consisting of ribosomal RNA (rRNA) and enzymes that decodes messenger RNA (mRNA) and coordinates the assembly of proteins from amino acids carried by transfer RNA (tRNA).

The Protein Assembly Line

Translation of the primary structure of a protein from the genetic code in the deoxyribonucleic acid (DNA) molecule is a kind of assembly-line manufacturing process that takes place in all biological systems. In essence, ribonucleic acid (RNA) reads the blueprint and assembly instructions from the DNA molecule, delivers the necessary parts to the "assembly line" in

RNA/PROTEIN TRANSLATION

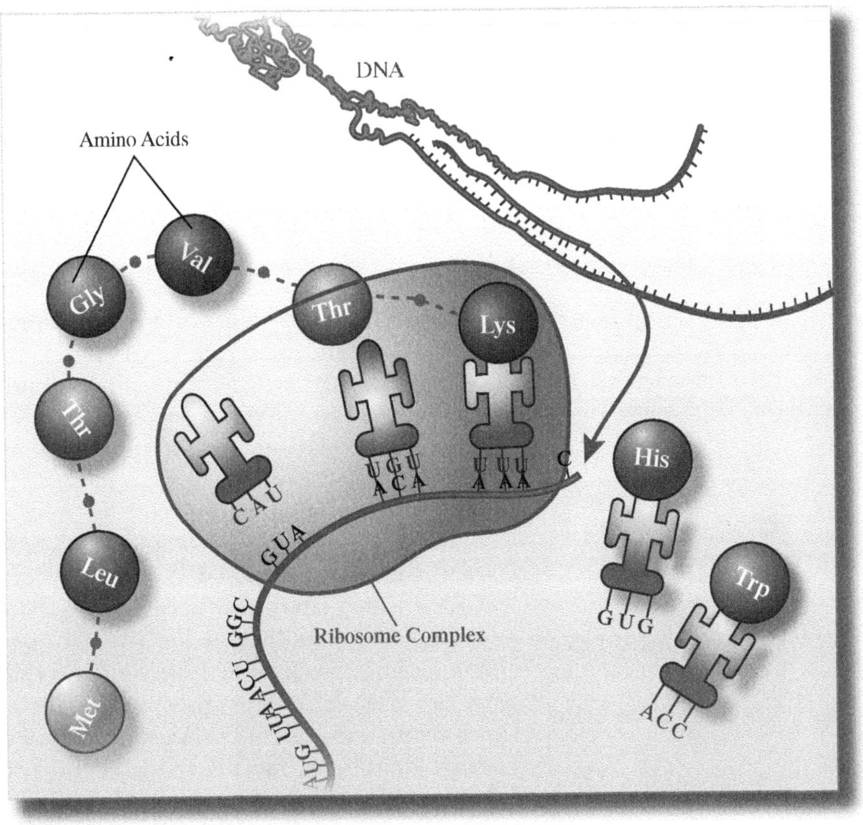

the ribosome complex, and assembles the parts into proteins according to the design specified by the DNA molecule.

The process is, of course, more complicated than that, as it is the result of many different enzyme-mediated chemical reactions that must occur in the proper sequence for each one of the many thousands of different proteins that are synthesized in the routine functioning of a living cell. The DNA molecule itself is composed of very few molecular components, but it is nevertheless subject to modification due to any number of causes. Each modification or perturbation of the normal structure of the DNA molecule has effects that are reflected in the cell's ability to produce the correct proteins for its proper functioning. A mutation of the DNA structure may prevent the production of a necessary protein, or it may cause the production of proteins that are at best unusable and at worst dangerous. Mutations that cause the normal process that halts protein synthesis to be disrupted can result in the formation of cancerous growths. Most often, however, the complexity of the process serves to prevent problems by simply shutting down the process when any aberration in the DNA structure is encountered.

Transcription and Translation

The structure of the DNA molecule consists of a very long backbone of deoxyribose sugar molecules alternating with phosphate groups. To each deoxyribose sugar molecule is attached one of four base molecules called nucleotides: adenine, cytosine, guanine, or thymine (commonly abbreviated A, C, G, and T). A second, equally long strand of DNA has a complementary structure to the first. When the two strands combine in the duplex DNA molecule to form its characteristic double-helix structure, the bases coordinate to each other, so that each instance of adenine or cytosine in one strand matches to thymine or guanine, respectively, in the other strand. The exact matching of the bases along the entire length of the two strands is required for the formation of the duplex DNA molecule.

To begin the transcription process—the process of copying a segment of DNA onto a segment of RNA—enzymes specific to the process attach to the DNA molecule and temporarily separate the two strands. This exposes the sequence of bases in the nucleotides of the DNA strand and begins the process of gene expression. Other enzymes called RNA polymerases then assemble free nucleotides to match the exposed sequence of DNA nucleotide bases. The strand that the RNA attaches to is called the template or noncoding strand, while the other strand, to which the template strand was originally attached, is called the coding strand. Because nucleotides always coordinate to their complementary bases, the sequence of the RNA strand will be identical to that of the DNA coding strand, with one difference: in RNA, the base

SAMPLE PROBLEM

Given the following sequence of nucleotides in a template strand of DNA, use an RNA codon table to identify the codons and the corresponding amino acids that will be assembled in sequence to form a new protein strand:

TTTCTTACCTTGTTGAGTGATTTTTACTTCAAACCGCGC

Answer:

First, convert the template strand to its complementary mRNA strand by replacing each nucleotide in the sequence with its RNA complement. Remember that U (uracil) is used in RNA instead of T (thymine), while the other nucleotides—C (cytosine), A (adenine), and G (guanine)—remain the same.

template DNA:

TTTCTTACCTTGTTGAGTGATTTTTACTTCAAACCG

mRNA: AAAGAAUGGAACAACUCACUAAAAAUGAAGUUUGGC

Divide the mRNA sequence into three-letter groups to isolate the codons:

AAA GAA UGG AAC AAC UCA CUA AAA AUG AAG UUU GGC

Identify the amino acid corresponding to each individual codon, using the standard genetic code chart:

- AAA: lysine (Lys)
- GAA: glutamic acid (Glu)
- UGG: tryptophan (Trp)
- AAC: asparagine (Asn)
- UCA: serine (Ser)
- CUA: leucine (Leu)
- AUG: methionine (Met)
- AAG: lysine (Lys)
- UUU: phenylalanine (Phe)
- GGC: glycine (Gly)

Now arrange the amino acids in their proper sequence according to the mRNA codons to determine the primary structure of the resulting polypeptide chain:

Lys-Glu-Trp-Asn-Asn-Ser-Leu-Lys-Met-Lys-Phe-Gly

Note that there are more combinations of the nucleotide bases than there are essential amino acids, and thus the same amino acid can be specified by different codons in the genetic code. In the above problem, both AAA and AAG code for the amino acid lysine.

uracil (U), rather than thymine, is used to complement adenine. Thus, a thymine base in the coding strand will coordinate to an adenine base in the template strand, which will in turn produce a uracil base in the RNA strand.

The formation of this new RNA strand is initiated and terminated by specific sequences of nucleotide bases on the DNA strand. When the "stop" sequence is encountered, the assembly process ends, and another enzyme separates the complementary RNA strand from the DNA strand, releasing it into the cytosol (intracellular fluid) of the cell as a molecule of messenger RNA (mRNA). At this point, transcription is complete and translation begins.

In the cytosol, the molecule of mRNA connects to ribosome complexes, which are themselves composed of another type of RNA called ribosomal RNA (rRNA). The binding of the mRNA to the rRNA occurs by matching nucleotide base sequences in the two strands. This exposes specific three-unit sequences of the mRNA termed codons, which identify the specific amino acid that is to be added onto the protein molecule being assembled at that location.

A third form of RNA, called transfer RNA (tRNA), transports particular amino acids to the site designated by the codon on the mRNA. Each tRNA molecule has a matching anticodon segment in its primary structure that attaches to the codon in the mRNA strand. In this way, amino acids are brought into the ribosome complex in the sequence that had been specified in the nucleotide sequence of the parent DNA molecule.

As each amino acid is brought into the ribosome complex by tRNA, the enzymes in the ribosome catalyze the formation of peptide bonds between the amino acids in the sequence. The resulting string of amino acids is called a polypeptide. After the polypeptide structure is released from the ribosome complex, it undergoes a process known as protein folding, through which it assumes the secondary structural features that characterize it as a protein or an enzyme. With the release of the new protein molecule, the process of translation is complete.

OVERLAPPING OF SEQUENCES IN MRNA

Each codon of the mRNA strand consists of three nucleotides, and merely shifting over one nucleotide in the sequence creates a new set of three nucleotides and therefore a new codon. For example, the DNA nucleotide series CCTACCTGG codes for the amino acids proline, threonine, and tryptophan. Shifting over two nucleotides creates the sequence TACCTGG, which codes for the amino acids tyrosine and leucine in an entirely different polypeptide chain. Viruses use this codon shifting to generate the various proteins required for their successful infection of a host cell.

THE GENETIC CODE AND THE HUMAN GENOME

The discovery of the molecular structure of DNA in 1943 enabled research that ultimately revealed the genetic code. Through numerous experimental studies that included the use of synthetic nucleotide sequences, researchers were eventually able to identify which groups of nucleotides were codons for which amino acids. Since the number of three-nucleotide combinations that can be formed with four nucleotides (sixty-four) is much greater than the twenty essential amino acids used in protein synthesis, there is significant redundancy in the genetic code, with several different codons specifying the same amino acid. The human genome was finally deciphered at the end of the twentieth century, and in February 2001, the journal *Nature* published its report of the first complete analysis of the human genome.

Richard M. Renneboog, MSc

BIBLIOGRAPHY

Berg, Jeremy M., John L. Tymoczko, Gregory J. Gatto, and Lubert Strye. *Biochemistry.* 8th ed. New York: W. H. Freeman, 2015. Print.

Dennis, Carina, Richard Gallagher, and Philip Campbell, eds. *The Human Genome.* Spec. issue of *Nature* 409.6822 (2001): 813–958. Print.

Lafferty, Peter, and Julian Rowe, eds. *The Hutchinson Dictionary of Science.* 2nd ed. Oxford: Helicon, 1998. Print.

Lehninger, Albert L. *Biochemistry: The Molecular Basis of Cell Structure and Function.* 2nd ed. New York: Worth, 1975. Print.

Lodish, Harvey, et al. *Molecular Cell Biology.* 7th ed. New York: Freeman, 2013. Print.

Reece, Jane B., et al. *Campbell Biology.* 10th ed. San Francisco: Cummings, 2013. Print.

Watson, James D. *The Double Helix.* New York: Atheneum, 1968. Print.

S

SALTS

FIELDS OF STUDY

Inorganic Chemistry; Physical Chemistry; Chemical Engineering

SUMMARY

The basic properties of salts are defined, and their behavior in solution is described. Salts are compounds composed of anions and cations, and they typically form crystalline solids with high melting points. The solubility of different salts varies considerably.

PRINCIPAL TERMS

- **anion:** any chemical species bearing a net negative electrical charge, which causes it to be drawn toward the positive pole, or anode, of an electrochemical cell.
- **cation:** any chemical species bearing a net positive electrical charge, which causes it to be drawn toward the negative pole, or cathode, of an electrochemical cell.
- **crystal:** a solid consisting of atoms, molecules, or ions arranged in a regular, periodic pattern in all directions, often resulting in a similarly regular macroscopic appearance.
- **dissociation:** the separation of a compound into simpler components.
- **electrolyte:** a material that ionizes in an appropriate solvent to produce an electrically conductive solution.

THE NATURE OF SALTS

Salts are compounds formed from the combination of an anion and a cation. They are the product of a neutralization reaction between an acidic compound and an alkaline, or basic, compound. Salts typically form crystal structures when in solid form. The reaction between the acid solution of hydrogen chloride (HCl) and the base ammonia (NH_3), for example, produces the salt ammonium chloride (NH_4Cl), a crystalline compound composed of ammonium cations (NH_4^+) and chloride anions (Cl^-). The cation in a salt is always derived from either a metal or a compound ion from a base such as ammonia. The anion is almost always that of an acid. Exceptions are known, especially in organometallic chemistry, in the form of complex ions that consist of a metal cation surrounded by several anionic ligands. For example, the complex anion $[Cr(CN)_6]^{3-}$ consists of a Cr^{3+} cation coordinated to six cyanide (CN^-) ions, resulting in a net charge on the complex of 3–.

Because of their ionic composition, salts are generally highly soluble in water and other polar solvents, via a process that involves the dissociation of component ions to form an electrolyte solution. While the anions and cations that make up a solid salt are strongly bound to each other by electrovalent (ionic) bonds, they remain entirely separate entities within the crystal structure, forming a lattice of individual ions rather than true molecules. Salts generally have high melting and boiling points because of the strength of the attraction between the anions and cations. Therefore, a true molecule of salt can exist only in isolation in the gas phase, or in a physical system that mimics the conditions of the gas phase. Because of the electrical charge that each salt ion carries as an electrolytic species in solution, salts serve an essential function in biochemical systems. In fact, all cell communication and energy processes, such as respiration, are electron-transfer processes involving ions derived from salts.

ATOMIC STRUCTURE AND SALT FORMATION

All salts are formed from ions, which are created by the gain or loss of electrons in the outermost, or valence, electron shell of an atom. All normal chemical reactions and interactions involve the valence electrons of atoms. The nucleus of the atom does not take part in normal chemical reactions, but the structure of the nucleus determines the elemental identity of the atom and its chemical behavior.

One significant aspect of that behavior is the thermodynamic stability of the overall atomic structure. This seems to be maximized when the outermost electron shell of the atom contains its full complement of electrons. This creates a significant driving force for most atoms to either gain or lose the number of valence-shell electrons necessary to achieve a full outermost electron shell. Each electron shell consists of a specific number of regions called orbitals that correspond to discrete energy levels for the electrons that may occupy a particular orbital. Individual orbitals are identified by a number and a letter; the number, properly called the principal quantum number, specifies the electron shell, while the letter indicates the shape of the orbital within the shell and the maximum number of electrons it can hold.

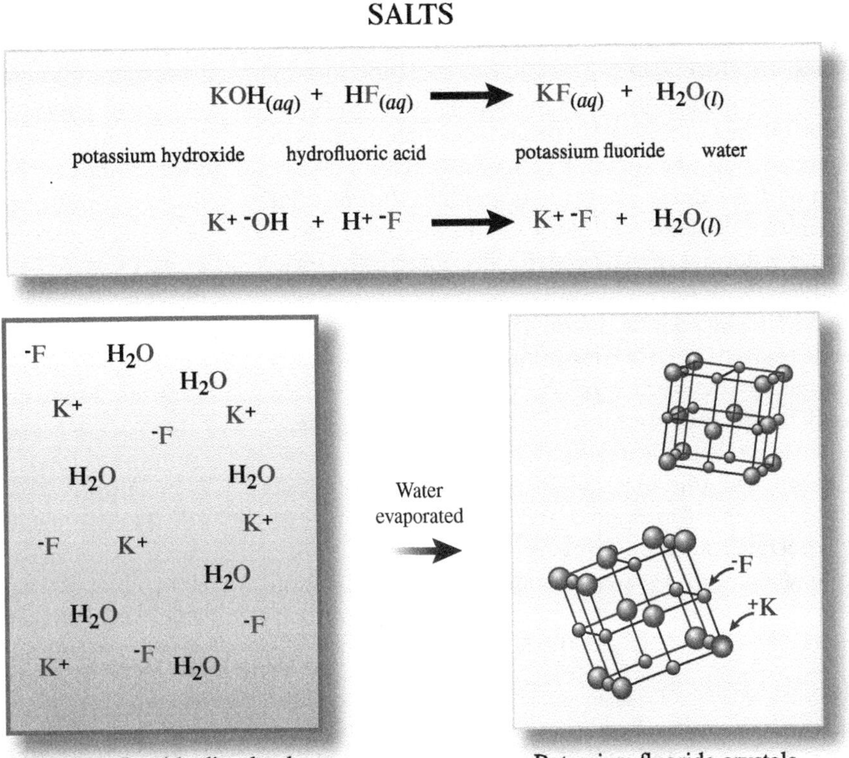

Potassium fluoride dissolved in water

Potassium fluoride crystals

Electrons enter the orbitals in the order 1*s*, 2*s*, 2*p*, 3*s*, 3*p*, 4*s*, 3*d*, 4*p*, 5*s*, 4*d*, 5*p*, 6*s*, 4*f*, 5*d*, 6*p*, 7*s* rather than in strict numerical order, due in part to the fact that the allowed energy levels of the orbitals become increasingly close together as their principal quantum number increases. In the transition metals, this tendency permits electrons to shift from one orbital to another in response to environmental influences, such as the types and numbers of ions or molecules that surround the particular atom in a compound. By giving up different numbers of valence electrons, transition metals are able to form a great variety of salts, while alkali and alkaline earth metals can typically only form a few. The number of available orbitals in transition metal atoms and ions also contributes to the ability of those elements to form large numbers of coordination compounds as complex ion salts.

The alkali, alkaline earth, and transition metals have their valence electrons in incompletely filled *s* and *d* orbitals such that giving up one, two, or three electrons results in a positively charged ion with an outermost electron shell that corresponds to that of the nearest lower noble-gas element. The opposite effect holds for the nonmetal elements (except the noble gases), which readily accept extra electrons into their valence electron shell in order to attain an outermost electron shell that corresponds to that of the nearest higher noble-gas element, gaining a negative charge in the process. Having a filled outermost electron shell gives anions and cations an electron distribution and stability similar to that of a noble gas, which partly explains why they remain separate entities in a salt crystal.

DISSOCIATION AND DISSOLUTION OF SALTS

The electron distribution that gives anions and cations stability in the solid phase also helps maintain the stability of those ions when they are separated from each other in a solution. Even though the anions and cations are separate chemical entities in the solid phase, the attractive force between their opposite electrical charges holds them tightly together in arrangements that satisfy various geometric conditions of atomic orbitals. Separation of the ions is the process of dissociation, and their dissociation energy is the energy required to separate the two oppositely charged ions in the gas phase. But gas-phase salt molecules are rather difficult to study, and this definition of dissociation energy is not practical in normal chemistry. Instead, the most commonly used dissociation energies of salts are typically stated in relation to the salts' behavior in water.

When placed into water, the highly polar water molecules begin to coordinate to the anions and cations at the surface of the crystal. The relatively weak bonds that are formed in this way cumulatively counteract the electrovalent bond between those ions and the solid material. When the actions of these two opposing forces on a particular ion are equal, the partially solvated ion is no longer bound to the salt crystal, and it is free to separate from the rest of the solid. As this occurs, the ion attracts more water molecules until it is completely surrounded and solvation is complete. The solvated ions maintain their electrical charge, but since equal numbers of each ion become dissolved, the resulting solution is electrically neutral. A fixed amount of water typically can dissolve only a specific amount of a particular salt because water molecules must surround each ion as it is solvated to maintain its solvated state. When the number of water molecules available for this has been exhausted, no more of the salt can dissolve, and the solution is said to be saturated.

Some ions require more water molecules than others to achieve solvation. Accordingly, each salt compound has a greater or lesser solubility in water. For example, 74.5 grams of calcium chloride ($CaCl_2$) will dissolve in 100 milliliters of room-temperature water, but the same amount of water will dissolve 209 grams of calcium iodide (CaI_2). The solubility of any particular salt is often expressed as its solubility product (K_{sp}), which is equal to the product of the concentrations of the constituent ions in a saturated solution, given in moles per liter. For a simple salt such as sodium chloride, the solubility product is expressed as:

$$K_{sp} = [Na^+][Cl^-]$$

A more complex salt such as trisodium phosphate (Na_3PO_4) dissociates into three sodium ions and one phosphate ion in solution. The solubility product expression for this salt is:

$$K_{sp} = [Na^+]^3[PO_4^{3-}]$$

Generally, the solubility product for any salt C_xA_y, where C = cation and A = anion, is:

$$K_{sp} = [C^+]x[A^-]y$$

The solubility product of any particular salt is a constant value, directly related to the temperature of the solution. If either ion concentration is increased, such as by the addition of another salt, the solubility product is exceeded and an appropriate amount of the solid salt will precipitate out of solution.

SAMPLE PROBLEM

When a solution of sodium iodide (NaI) is added to one of silver nitrate ($AgNO_3$), a cream-colored precipitate immediately forms. When the reaction is taken to completion, the precipitated salt can be filtered out and dried. Analysis of the filtered liquid shows that no silver ions or iodide ions are present. Write a balanced net ionic equation for the reaction and identify the type of bonding that has occurred.

Answer:

Since the reaction has apparently removed the silver and iodide ions from solution, the precipitated salt must be silver(I) iodide (AgI), formed according to the reaction equation

$$NaI(aq) + AgNO_3(aq) \rightarrow AgI(s) + NaNO_3(aq)$$

where "(*aq*)" indicates that the compound is in an aqueous solution and "(*s*)" indicates that it is in a solid state—in this case, the precipitate. The full ionic equation is

$$Na^+(aq) + I^-(aq) + Ag^+(aq) + NO_3^-(aq) \rightarrow Na^+(aq) + NO_3^-(aq) + AgI(s)$$

Because the Na^+ and NO_3^- ions remain unchanged in the reaction, they can be disregarded, giving the following net ionic equation:

$$Ag^+(aq) + I^-(aq) \rightarrow AgI(s)$$

The salt forms by the attraction of the positively charged silver cation and the negatively charged iodide anion. The bond formed is electrovalent, or ionic, in nature.

Conversely, adding more solvent or otherwise reducing the concentration of at least one of the ions will allow more of the salt to dissolve.

Types and Uses of Salts

The salt KH_2PO_4 is often called potassium diacid phosphate. The name indicates that the material is an acid salt. The $H_2PO_4^-$ ion in solution is able to release its two hydrogen atoms as protons (H^+) in the manner of an acid, according to the equilibrium equation:

$$H_2PO_4^- \rightleftharpoons H^+ + HPO_4^{2-} \rightleftharpoons 2H^+ + PO_4^{3-}$$

Similarly, calcium oxide (CaO) is described as a base salt because the oxide ion O^{2-} can act as a base and accept a proton from the water molecule, according to the equilibrium equation:

$$O^{2-} + H_2O \rightleftharpoons OH^- + OH^- \text{ (or } 2OH^-\text{)}$$

Salts serve numerous industrial purposes. Throughout much of human history, salts have been used as pigments because of the variety of colors associated with ions in different salts. In addition to being central to redox reactions, salts are essential to applications as diverse as battery technology, dealginating ponds with copper sulfate, and electroplating metal objects. In biochemical and medical research, certain salts, particularly those of organic acids, are used for the preparation of buffer solutions.

Richard M. Renneboog, MSc

Bibliography

Douglas, Bodie E., Darl H. McDaniel, and John J. Alexander. *Concepts and Models of Inorganic Chemistry.* 3rd ed. New York: Wiley, 1994. Print.

Haynes, W. H., PhD, David R. Lide, PhD, and Thomas J. Bruno, PhD, eds. *CRC Handbook of Chemistry and Physics,* 96th Edition. New York: CRC/Taylor & Francis Group, 2015. Print.

Holden, Alan, and Phylis Morrison. *Crystals and Crystal Growing.* New York: Doubleday, 1960. Print.

Jones, Mark Martin. *Chemistry and Society.* 5th ed. Philadelphia: Saunders College Pub., 1987. Print.

Miessler, Gary L., Paul J. Fischer, and Donald A. Tarr. *Inorganic Chemistry.* 5th ed. Upper Saddle River: Prentice Hall, 2013. Print.

Myers, Richard. *The Basics of Chemistry.* Westport: Greenwood, 2003. Print.

SPECTROSCOPY

FIELDS OF STUDY

Analytical Chemistry; Organic Chemistry; Biochemistry

SUMMARY

Spectroscopic analysis is the workhorse methodology of essentially all branches and fields of chemistry. Spectroscopy has largely replaced wet chemistry as an analytical technique. All spectroscopic methods of analysis involve measuring the interactions of matter with electromagnetic radiation.

PRINCIPAL TERMS

- **absorbance:** a measure of the amount of electromagnetic radiation of a particular frequency that a specific quantity of solid or solute matter absorbs; also called optical density.
- **chromophore:** the part of a molecule that absorbs specific wavelengths of electromagnetic radiation, often causing it to appear as a certain color.
- **composition:** the identities and relative proportions of different elements or components in a compound, mixture, or other material.
- **conjugation:** a system of alternating double and single bonds, characterized by the presence of a *p* orbital on three or more successive carbon atoms in a molecular structure, allowing their electrons to move freely through them.
- **crystallography:** the study of the properties and structures of crystals.
- **wavelength:** the distance between the same point on two successive waves, beyond which the shape of the wave begins to repeat.

THE BASICS OF SPECTROSCOPY

All matter interacts with electromagnetic radiation in some way. Electromagnetic radiation has both magnitude and direction and is a cyclic phenomenon with a specific period and wavelength. The effects of the interactions between matter and electromagnetic radiation can be measured and used for many different purposes.

How a certain type of matter interacts with electromagnetic radiation depends on the wavelengths, and therefore the energies, of the radiation in question. In atoms and molecules, electrons are bound in specific energy configurations called orbitals; they can transfer from one orbital to another only by absorbing or emitting energy that corresponds to a specific wavelength of electromagnetic radiation. Certain wavelengths of electromagnetic radiation can cause atoms and molecules to vibrate, rotate, or move in various other ways. The techniques used to observe and measure these interactions have an almost unlimited number of applications.

The science of spectroscopy began with the study of optics and the passage of sunlight through a triangular prism. The internal reflection and material properties of a triangular glass prism cause pure white sunlight passing through it to refract, spreading the visible light out according to wavelength into the familiar rainbow pattern of visible colors. Physicists noticed that sunlight that passed through another material medium, such as a candle flame, produced rainbow patterns in which segments were observed to be missing. As optical science developed, investigators found that the missing segments formed a distinct pattern that was characteristic of the medium that the sunlight had passed through. Eventually, researchers found that such interactions with light depend on wavelength and are characteristic of absorption, emission, scattering, diffraction, refraction, and rotation of the light. With modern technology, it is possible to fully identify and characterize the molecular structure of a compound from just a few molecules by observing how they interact with electromagnetic radiation. Application of the technology in the form of functional magnetic resonance imaging (fMRI) even allows medical researchers to observe the function of specific materials within the human body in real time.

SPECTROSCOPY

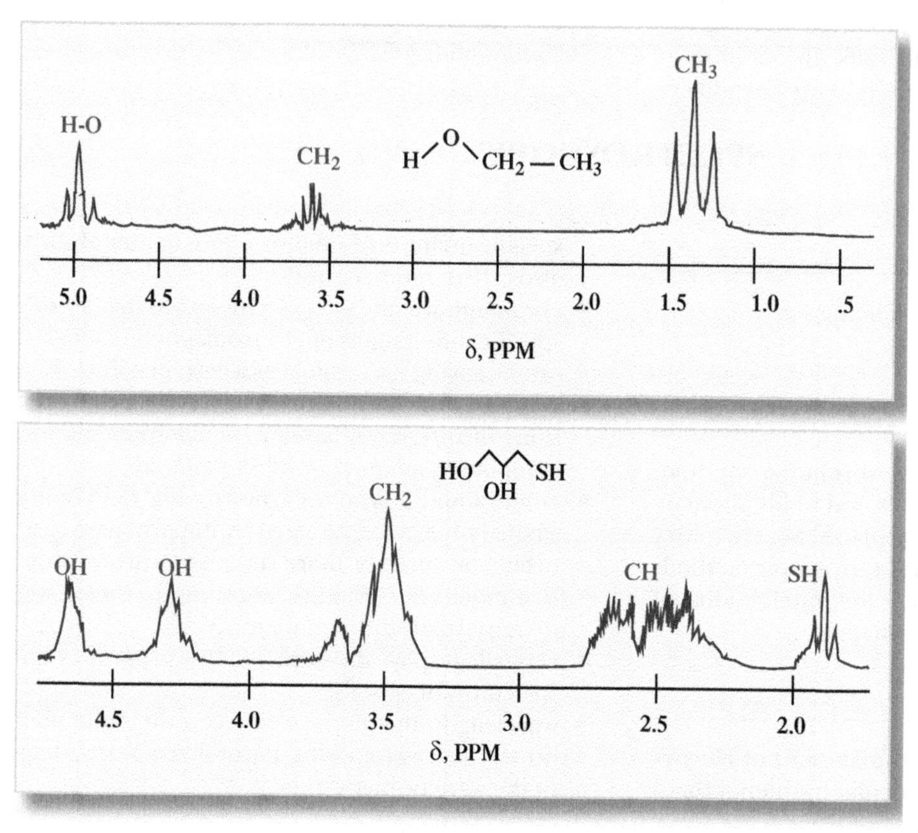

ABSORPTION OF LIGHT

The most common and versatile forms of spectroscopy are those that depend on the absorption of electromagnetic energy. Which technique to use depends on the wavelength and type of radiation involved, which can range from far infrared to x-ray wavelengths. Each form of spectroscopy relies on the particular absorbance of the material being studied—more specifically, of the chromophores in its constituent particles that absorb the energy. The data is then presented in a graph called an absorption spectrum, or simply a spectrum, which shows peaks or lines at the wavelengths or frequencies at which the energy was absorbed.

While the term "chromophore" was originally defined as the portion of a molecule that was responsible for the color of the compound, it is used in a broader sense to indicate the portion of a molecule that absorbs a particular wavelength of electromagnetic energy. In conjugated organic molecules, the chromophore is the conjugated system of alternating double (C=C) and single (C–C) carbon-carbon bonds created by overlapping *p* orbitals. Such chromophores typically absorb light in the ultraviolet-visible (UV/VIS) range, with the specific wavelengths absorbed depending on the extent of the conjugation. The electrons in the pi (ϖ) bonds (the secondary bonds in double bonds) absorb the energy, enabling them to jump between the orbital levels of the bond system. The form of spectroscopy that deals with wavelengths in the ultraviolet and visible range is known as UV/VIS spectroscopy.

The absorbance of a particular compound is proportional to the amount of the material that is doing the absorbing, typically in terms of the molar concentration. The higher the concentration of a particular compound in the test solution, the greater the amount of light that is absorbed. The process obeys the mathematics of the Beer-Lambert law, which relates absorbance to the concentration and thickness of the sample being observed, making this a useful property for determining the amount of material present.

Analyses of blood samples in medical service laboratories have long used UV/VIS spectroscopy on a large scale to test levels of cholesterol, triglycerides, glucose, and many other compounds that are normally present in blood. Forensic chemistry laboratories have also used this method when testing for the presence of toxic compounds or drugs in body serums. Because the absorbance method is so universal, it is routinely used in a wide array of industrial applications, including pharmaceutical manufacturing, DNA and enzyme analysis, ink and dye production, and environmental toxicology.

TYPES OF SPECTROSCOPY

The information provided by UV/VIS spectroscopy is very limited in regard to the molecular structure of compounds. Instead, molecular structures are routinely studied using mass spectrometry (to determine the molecular mass of the material), along with infrared (IR) spectroscopy and nuclear magnetic resonance (NMR) spectroscopy. The various bonds in a molecule vibrate in a number of modes, each of which has a particular associated energy. Individual bonds can vibrate linearly by stretching and relaxing along the bond axis. Bonds to the same atom, such as in a methylene group ($-CH_2$), can wag or rock by moving laterally with respect to each other, and they can vibrate with a scissoring motion relative to each other. Parts of molecules can rotate about any single bond, while double and triple bonds can twist but not rotate. In short, any type of motion that atoms within molecules can make has an associated energy. These energies fall within the IR range of wavelengths, making IR spectra useful for characterizing the functional groups in a molecule and for determining bond energies (also called bond strengths).

NMR is the workhorse of structural analysis, particularly for organic compounds. While an IR spectrum can identify the functional groups that are present in a molecular structure, the NMR spectrum characterizes the molecular structure itself, using the magnetic properties of atomic nuclei. The atomic nuclei of some elements spin, which generates an electromagnetic field and thereby enables them to be used in NMR procedures. The most common of these is proton NMR, or ^{1}H-NMR, which analyzes the behavior of hydrogen nuclei—essentially protons, as the nucleus of the most common isotope of hydrogen (hydrogen-1, or ^{1}H) contains one proton and zero neutrons.

Every nucleus is affected by the magnetic field of the other nuclei around it. In ^{1}H-NMR spectroscopy, the sample is dissolved in a solvent, and the resulting solution is placed in a strong magnetic field, spun, irradiated with a constant radio frequency, subjected to variations in the applied magnetic field, and scanned through a second range of radio frequencies. (Alternatively, the magnetic field can be kept constant and the radio frequency varied.) An atomic nucleus can spin either with or against the applied magnetic field, and changing, or "flipping," the spin requires energy input. Absorbance peaks are detected for a particular hydrogen nucleus when the radio frequencies match the resonance frequency of that nucleus—that is, the energy required to flip its spin. This higher-energy spin destabilizes the hydrogen atom, which then releases the same amount of energy. The electrons on that particular atom and adjacent atoms can shield or de-shield the nucleus from the applied magnetism and thereby affect how much energy is required to flip its spin. Thus, atoms are exposed to and respond to differing amounts of magnetism based on their positions in the molecule, which is how using NMR helps determine a molecular structure. For example, the hydrogen nuclei of a lone methyl group ($-CH_3$) next to a carbonyl group (C=O) produce a single sharp absorbance peak in a narrow range of frequencies relative to the single peak produced by a reference compound, typically tetramethylsilane (TMS), $Si(CH_3)_4$. The hydrogen nuclei of an ethyl group ($-CH_2CH_3$), however, produce a characteristic set of absorbance peaks in two different frequency ranges relative to TMS. One set consists of three symmetrically distributed peaks, called a triplet, and the other is a distinct set of four peaks, called a quartet.

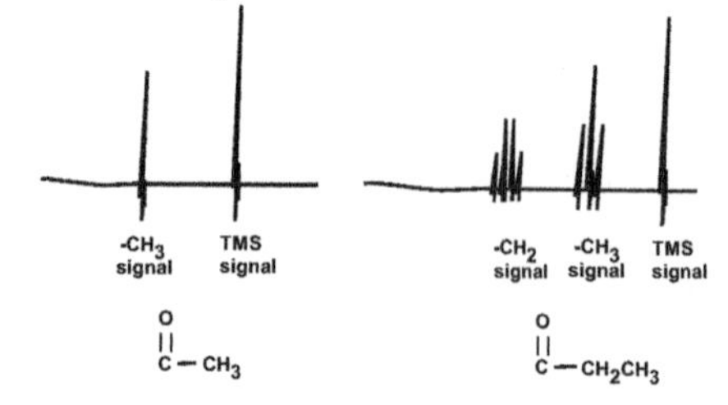

In the same way, every hydrogen nucleus in a molecular structure has its own corresponding signal in a ^{1}H-NMR spectrum. The values of the *x* axis of the spectrum are typically given in parts per million (ppm), which represents the difference between the resonance frequency of the sample and that of the reference (again, usually TMS), given in hertz, over the frequency to which the spectrometer is tuned, given in megahertz. The relationships between the different nuclei are demonstrated in the spectrum by various factors, including the position of the signal relative to that of the TMS signal, its intensity (height), its pattern, and the distance between peaks in the pattern, which is known as signal splitting. The position indicates the nucleus's immediate environment; signal splitting reveals which equivalent nearby nuclei are affecting a particular nucleus or set of nuclei. Mathematical integration of the area under each set of peaks reveals the relative ratio of the number of hydrogen nuclei that produced the peaks. In a characteristic ethyl group pattern, the ratio of the integral of the quartet to that of the triplet will be two to three, corresponding to two hydrogen nuclei in the $-CH_2$ group and three in the $-CH_3$ group.

While the ^{1}H-NMR spectrum reveals the relationship of the hydrogen atoms in the molecule—usually with sufficient detail to determine the molecule's complete structure—^{13}C-NMR reveals the relationship between the carbon-13 atoms in the molecule, ^{31}P-NMR looks at phosphorus-31, and so on. The process has been so well developed since its discovery that two- and three-dimensional spectra are routinely available. The ultimate extension of this technology is magnetic resonance imaging (MRI), used in diagnostic medicine. As might be imagined, reading spectra requires a great deal of practice and training in order to progress beyond the simplest of determinations.

SAMPLE PROBLEM

A compound containing four carbon atoms produces an IR spectrum in which the characteristic frequency of the carbonyl group (C=O) is observed. Its ^{1}H-NMR spectrum reveals a small quartet of peaks between 2 and 2.55 parts per million (ppm), a triplet between 1 and 1.12 ppm, and a singlet at 2.13 ppm relative to the TMS reference peak. Integration of the peaks gives a ratio of 2:3:3 (quartet:triplet:singlet). What is the identity of the compound?

Answer:

The quartet-triplet set is characteristic of an ethyl group, which, in combination with the carbonyl group, accounts for three of the four carbon atoms. The singlet peak is characteristic of a methyl group bonded to a carbonyl carbon, which accounts for the fourth carbon atom. The ratio of the quartet integral to the triplet integral is 2:3, as for an ethyl group, while the integral of the singlet is proportional to that of the triplet, indicating that it is associated with three hydrogen nuclei, as in a methyl group. Based on this information, the compound contains a carbonyl group, the carbon atom of which is bonded to a methyl group on one side and an ethyl group on the other. This makes the compound 2-butanone, or methyl ethyl ketone, which has the following structure:

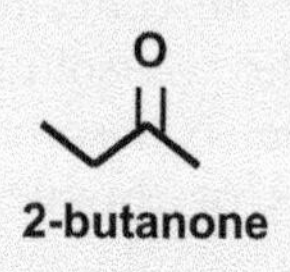

2-butanone

X-ray diffraction is another form of spectroscopy, used in crystallography to determine the three-dimensional arrangement of atoms in the molecules of a solid. The pattern of the diffracted x-rays is directly related to the arrangement of the atoms in the solid, as well as to the chemical composition of the material being examined.

Rotation of Light

Light can be plane-polarized (that is, made to vibrate in a single plane) very easily, as is commonly done with ordinary sunglasses. The light emitted from a typical LCD computer screen is polarized. If a polarizing filter is placed in front of the screen and slowly rotated, the amount of light passing through the filter will decrease until it reaches the point where no light passes through the filter. This principle is used in many spectroscopic devices to ensure that just a single wavelength is being used in an analysis. Polarimeters are devices used to measure optical activity, which is a property that exists in compounds when substituents are arranged asymmetrically about a central location such as a carbon atom. Optical isomers have the same molecular formula but different structures. Two optical isomers of a compound can have the exact same physical and chemical properties, but one will rotate the plane of plane-polarized light clockwise while the other rotates the plane by the same amount counterclockwise. Such isomers are known as enantiomers. The phenomenon was first demonstrated by French scientist Louis Pasteur (1822–95) in enantiomorphous crystals and was explained in principle by Dutch chemist Jacobus Henricus van't Hoff (1852–1911) and French chemist Joseph Achille le Bel (1847–1930) when they both independently proposed the tetrahedral arrangement of substituents about a carbon atom.

Richard M. Renneboog, MSc

Bibliography

Askeland, Donald R., Wendelin J. Wright, D. K. Bhattacharya, and Raj P. Chhabra. *The Science and Engineering of Materials.* Boston: Cengage Learning, 2016. Print.

Gribbin, John. *Science: A History, 1543–2001.* London: Lane, 2002. Print.

Hendrickson, James B., Donald J. Cram, and George S. Hammond. *Organic Chemistry.* 3rd ed. New York: McGraw, 1970. Print.

Lodish, Harvey, et al. *Molecular Cell Biology.* 7th ed. New York: Freeman, 2013. Print.

Morrison, Robert Thornton, and Robert Neilson Boyd. *Organic Chemistry.* 7th ed. Englewood Cliffs, N.J.: Prentice Hall, 2003. Print.

Silbey, Robert J., Robert A. Alberty, and Moungi G. Bawendi. *Physical Chemistry.* 5th ed. Hoboken: Wiley, 2012. Print.

Skoog, Douglas Arvid, Donald M. West, and F. James. Holler. *Skoog and West's Fundamentals of Analytical Chemistry.* 9th ed. Belmont (Calif.): Cengage Learning, 2014. Print.

Strobel, Howard A. and William R. Heineman. *Chemical Instrumentation: A Systematic Approach.* 3rd ed. New York: Wiley, 1989. Print.

SUBLIMATION

FIELDS OF STUDY

Physical Chemistry; Chemical Engineering; Environmental Chemistry

SUMMARY

The property of sublimation is described, and the thermodynamics of the process are discussed. Sublimation is the change of phase from solid to gas without passing through the intermediate liquid phase. Sublimation is a relatively common feature of volatile materials and plays a significant role in the natural water cycle.

PRINCIPAL TERMS

- **endothermic:** describes a process that requires the input of energy in the form of heat in order to proceed.
- **gas:** a state of matter in which material is fluid, has indefinite volume and shape, and is of variable density due to its ability to expand to fill any available space.

- **liquid:** a state of matter in which material is fluid, has definite volume but indefinite shape, and maintains a relatively constant density.
- **phase transition:** the change of matter from one state to another, such as from solid to liquid or liquid to gas, due to the transfer of thermal energy.
- **solid:** a state of matter in which material is nonfluid, has definite volume and shape, and maintains a near-constant density.

The Nature of Phase Transitions

Three distinct physical phases, or states of matter, are characteristic of most materials. They can exist as a solid, a liquid or a gas. The physical phase of any material is dependent on temperature, with a phase transition from solid to liquid taking place at the material's melting point and a similar transition from liquid to gas taking place at its boiling point. These temperatures are intensive characteristics of each material, meaning that they remain the same no matter how much of the material is present.

For pure compounds, the temperatures at which phase transitions take place are sharply defined, while for mixtures of compounds, the phase change generally takes place over a much broader range of temperatures. However, there are unique combinations of materials for which the melting point or boiling point is as sharply defined as for a pure compound. These combinations are said to be "eutectic" for solid-liquid phase changes and "azeotropic" for liquid-gas phase changes. Any phase change in a material that occurs by increasing the atomic or molecular energy of the material is an endothermic process, requiring the input of thermal energy from outside the thermodynamic system of the material.

The Process of Sublimation

Phase changes for most materials progress from solid through liquid to gas. There are many materials that decompose before reaching their melting or boiling point, and for these the progression of phase changes is presumed to follow the same pattern. The process of sublimation represents a break in the pattern, since a material undergoing sublimation passes directly from the solid phase to the gas phase without going through a liquid phase. It is possible that for at least some of these materials, the outermost layer of the solid phase, perhaps only one or two molecules in thickness, is in a liquid phase, but the difference is a technicality.

At any temperature, the atoms in a material are in a state of constant motion (even at the absolute zero of temperature). In gases and liquids, atoms and molecules physically move through space and regularly bump into each other; in solids, this degree of motion does not occur. In all three phases, however, atoms and the bonds between them vibrate in a variety of ways, with their kinetic energy proportional to the temperature. Accordingly, collisions occur in all three phases, and in solids below their melting point and liquids below their boiling point, collisions between molecules or atoms can impart sufficient energy to knock some particles free of the forces holding them together and drive them into a gas phase. When this happens with a liquid, it is called

Dry ice sublimation in water. Dry ice, which is solid carbon dioxide, will change from a solid state to the gaseous state at room temperature through the process of sublimation. The molecules of carbon dioxide leaving the dry ice are invisible to the naked eye. However, when the dry ice is placed in water it draws out water vapor with the carbon dioxide molecules, making the gas visible as fog. By Nevit, via Wikimedia Commons

SAMPLE PROBLEM

Identify which of the following are examples of a material undergoing sublimation:

- iodine at room temperature
- the wax of a lit candle
- dry ice at room temperature
- water ice heated to 100°C
- ice cubes left for a long time in a freezer
- naphthalene at room temperature

Answer:

Iodine at room temperature is a dark purple crystalline solid. Gaseous iodine forms as a purple haze about the solid. This is sublimation.

The wax of a lit candle is liquefied by the heat of the candle flame before it is drawn up into the candle wick, vaporized, and burned. This is not sublimation.

Dry ice is solid carbon dioxide at a temperature of about −79 °C (−112 °F). It goes directly into the gas phase as it warms. This is sublimation.

Water ice, when heated, first melts into liquid water. At 100 °C (212 °F), the liquid water boils and enters the gas phase. This is not sublimation.

Ice cubes left in a freezer for a long time will shrink in size, while ice crystals form on the inside of the freezer. This is caused by molecules of ice being converted into water vapor and then resolidifying on the walls of the freezer without passing through a liquid phase. This is sublimation.

Naphthalene is a compound from which moth balls and restroom deodorizer pucks are made. These do not melt at room temperature or dissolve in water. They undergo sublimation.

evaporation; when it happens with a solid, it is called sublimation.

SUBLIMATION AND VAPOR PRESSURE

When water is in a closed container, molecules are constantly escaping the liquid and becoming water vapor (evaporating); at the same time, molecules of water vapor are also reentering the liquid (condensing). The point at which the processes of evaporation and condensation in this closed system are happening at an equal rate is called equilibrium, and the pressure exerted by the gas from a liquid (such as water) or solid (such as water ice) in a closed system at equilibrium is called its "vapor pressure." Every element or compound has a vapor pressure, determined by the energetic motion of the atoms or molecules that constitute the material.

The stronger the intermolecular forces holding a material together, the lower the material's vapor pressure. Understandably, materials in which the intermolecular forces are weaker, and thus the vapor pressure is higher, are more prone to sublimation. Since the molecules being ejected from the material during sublimation must move against the pressure of the surrounding atmosphere, it requires more energy for sublimation to occur against higher pressures than against lower pressures.

Every material that has a vapor pressure in the solid phase, no matter how small it may be, can undergo sublimation. Water ice readily sublimates, and it is estimated that three-quarters of the snow and rain that falls in northern climates is returned to the atmosphere through sublimation and evaporation. Sublimation therefore plays an important role in the natural water cycle of the planet.

Richard M. Renneboog, MSc

BIBLIOGRAPHY

Crystal, David, ed. *The Cambridge Biographical Encyclopedia*. 2nd ed. New York: Cambridge UP, 1998. Print.

Jones, Mark Martin. *Chemistry and Society*. 5th ed. Philadelphia: Saunders College Pub., 1987. Print.

Myers, Richard. *The Basics of Chemistry*. Westport: Greenwood, 2003. Print.

Silbey, Robert J., Robert A. Alberty, and Moungi G. Bawendi. *Physical Chemistry*. 5th ed. Hoboken: Wiley, 2012. Print.

Wuts, Peter G. M., and Theodora W. Greene. *Greene's Protective Groups in Organic Synthesis*. 4th ed. Hoboken: Wiley-Interscience, 2007. Print.

SUBSTITUTION REACTION

FIELDS OF STUDY

Biochemistry; Genetics; Molecular Biology

SUMMARY

The processes of substitution reactions are described, and their value in synthesis is elaborated. Substitution reactions strictly adhere to stereochemical rules, permitting precise control of the molecular structure of the product.

PRINCIPAL TERMS

- **electrophilic:** describes a chemical species that tends to react with negatively charged or electron-rich species.
- **functional group:** a specific group of atoms with a characteristic structure and corresponding chemical behavior within a molecule.
- **intermediate:** a relatively stable but reactive chemical species that is formed during a chemical reaction and either reverts to the original reactants or continues in the reaction mechanism to form products.
- **leaving group:** an atom or group of atoms that is removed from a molecule during a chemical reaction, along with the electron pair that previously bonded it to the molecule.
- **nucleophilic:** describes a chemical species that tends to react with positively charged or electron-poor species.

The Basics of Substitution Reactions

Substitution reactions are easily recognized by the replacement of one substituent—an atom or functional group—in a reactant compound with a different group at the same position in the molecular structure. Substitution always replaces one component of a molecule with a different component not initially part of the molecule. The process must take place according to the same rules of chemical behavior that control the formation of a compound. When a compound undergoes a substitution reaction, the affected portion of the compound reacts in specific ways determined by the basic rules of chemical behavior. Two modes of substitution reactions are known: nucleophilic and electrophilic.

Nucleophilic Substitution

In nucleophilic substitution reactions, a nucleophile reacts with a substrate molecule (the molecule being acted upon) and replaces one of its existing substituents. The replaced substituent is known as the leaving group because it leaves the molecular structure when the substitution occurs. Substitution reactions require a substrate molecule with a suitable functional group, typically one that can become the leaving group.

A nucleophile is a chemical species that is attracted to, or reacts preferentially at, a species with a strong positive charge or a location with enhanced positive-charge character. A chemical bond between a carbon atom (C) and an iodine atom (I), for example, is polarized by the higher electronegativity (tendency to attract electrons) of the iodine atom. As a result, the carbon atom has a lower electron density and therefore more positive charge. In basic conditions, a hydroxide ion (OH^-), which has a high electron density and thus a negative charge, is able to associate with the carbon atom, as the opposite charges attract. The electron density of the hydroxide ion is greater than that of the iodine atom, which is sufficient to force the electrons in the C–I bond to shift toward the oxygen atom of the hydroxide ion and form a C–O bond. The reactants pass through a transition state as the bond to the oxygen atom strengthens and the bond to the iodine atom weakens. Since the four substituents bonded to a carbon atom in an organic molecule are arranged tetrahedrally about the carbon atom, the arrangement of the substituents is unique. Substitution reactions that occur in this manner are called S_N2 reactions (for "substitution nucleophilic bimolecular") and always conclude with an inversion of the original substituent arrangement about the carbon atom. This inversion is analogous to an umbrella being blown inside out by a strong wind. In a synthesis process, the inversion requirement is used to determine the stereochemistry of the product. The term "bimolecular" refers to the dependency of the rate of the reaction on the concentrations of both the nucleophile and the substrate.

A second mode of nucleophilic substitution occurs under acidic conditions. An organic alcohol (R–OH), for example, can be converted to an alkyl chloride by strong hydrochloric acid (HCl). Under the conditions of the reaction, the oxygen atom of the hydroxyl (–OH) group in the alcohol molecule acquires an additional proton in the form of a hydrogen cation (H^+), forming the intermediate structure $R–OH_2^+$. This species can release either the H^+, thus re-forming the original alcohol, or a neutral water molecule (H_2O), thus forming the corresponding carbonium ion R^+. A carbonium ion can do three things: rearrange its bonds and overall structure to form a more stable carbonium ion before reacting further, eject another H^+ to form a double bond (which would result in an elimination reaction), or gain a nucleophile. In this example, the chloride (Cl^-) ion of the hydrochloric acid acts as a nucleophile and forms a bond to the carbon atom of the carbonium ion to yield the alkyl chloride R–Cl. This type of substitution is termed S_N1, for "substitution nucleophilic unimolecular," meaning that the reaction rate depends only on the concentration of the substrate compound. The product of an S_N1 reaction is typically an equal mixture of molecules in which the substituents are inverted and molecules that retain their original arrangement , which reduces the efficiency of the substitution reaction by 50 percent, unless the orientation of the product molecules is not important.

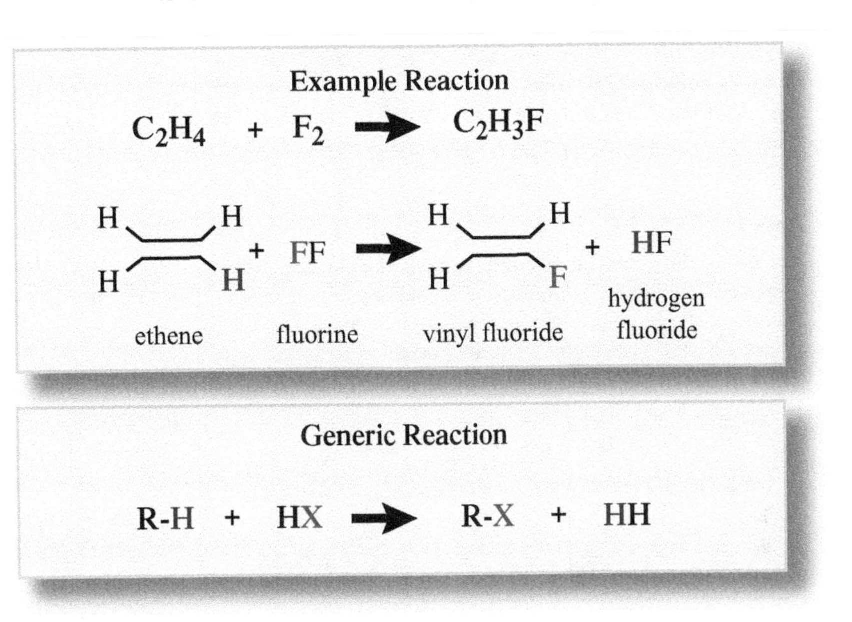

ELECTROPHILIC SUBSTITUTION REACTIONS

In electrophilic substitution reactions, the electrophile typically has a high degree of Lewis acidity, meaning that it has an affinity for negative charge and reacts readily with electron-rich chemical species. Electrophilic substitution reactions typically involve compounds with conjugated double-bond systems, especially the aromatic six-membered ring structure of the benzene molecule. Such compounds are notable for being otherwise unreactive. For example, in an electrophilic substitution reaction involving an aromatic compound such as toluene, which is a benzene molecule in which one of the hydrogen atoms has been replaced by a methyl group substituent ($-CH_3$), the electrophile coordinates to one of the pi bonds of the six-membered ring. Often a Lewis acid material, such as aluminum chloride ($AlCl_3$) or iron(III) chloride ($FeCl_3$), catalyzes such reactions, and the electrophilic species is added to one of the carbon atoms of the ring to form a charge-separated intermediate compound. Loss of the hydrogen atom or its relocation within the molecule affects the substitution reaction. When toluene reacts with "fuming sulfuric acid," a solution of sulfur trioxide (SO_3) in concentrated sulfuric acid (H_2SO_4), a molecule of sulfur trioxide is added to the toluene ring in such a way that one of its pi bonds gives up an electron, giving the toluene a positive charge, and the sulfur trioxide group acquires a negative charge by accepting the electron. Therefore, the carbon atom is bonded to both a hydrogen atom and the sulfur trioxide group. The hydrogen atom subsequently shifts, bonding to an oxygen atom of the sulfur trioxide group. The result is the formation of toluene sulfonic acid, in which a hydrogen atom has been substituted by the $-SO_3H$ group.

SAMPLE PROBLEM

In a reaction between 2-bromononane ($C_9H_{19}Br$) and potassium hydrogen sulfide (KHS), one of the products is the inorganic compound potassium bromide (KBr). Write an equation that identifies the substitution product.

Answer:

The Br^- in the potassium bromide product is from the 2-bromononane, and the K^+ is from the potassium hydrogen sulfide. Therefore, the bromine atom in the 2-bromononane has been replaced by the –HS group from the potassium hydrogen sulfide, and the reaction can be written as follows:

$$C_9H_{19}Br + KHS \rightarrow C_9H_{20}S + KBr$$

Alternatively, the reaction equation can also be written as

Br + KSH ⟶ HS + KBr

The substitution in the reaction has taken place through the replacement of Br^- by HS^-.

SUBSTITUTION REACTIONS IN BIOLOGICAL SYSTEMS

In both chemistry laboratories and chemical engineering applications, substitution reactions typically take place between the chemical compounds themselves. In biological systems, however, substitution reactions are moderated by enzymes that carry out specific modifications. The modification of various compounds through metabolism functions primarily to enhance the solubility of the compounds in the aqueous medium of body fluids and to improve their ability to pass through cell membranes. Bile acids are added to various compounds to facilitate their removal from the body by the kidneys. Enzyme actions bring about the substitution of the bile acid for some substituent on a metabolite compound. Pharmacologists work to understand this process and how it affects the activity and effectiveness of different drugs considered for use as medicines.

Richard M. Renneboog, MSc

BIBLIOGRAPHY

Douglas, Bodie E., Darl H. McDaniel, and John J. Alexander. *Concepts and Models of Inorganic Chemistry.* 3rd ed. New York: Wiley, 1994. Print.

Hendrickson, James B., Donald J. Cram, and George S. Hammond. *Organic Chemistry.* 3rd ed. New York: McGraw, 1970. Print.

Miessler, Gary L., Paul J. Fischer, and Donald A. Tarr. *Inorganic Chemistry.* 5th ed. Upper Saddle River: Prentice Hall, 2013. Print.

Morrison, Robert Thornton, and Robert Neilson Boyd. *Organic Chemistry.* 7th ed. Englewood Cliffs, N.J.: Prentice Hall, 2003. Print.

Wuts, Peter G. M., and Theodora W. Greene. *Greene's Protective Groups in Organic Synthesis.* 4th ed. Hoboken: Wiley-Interscience, 2007. Print.

SURFACE CHEMISTRY

FIELDS OF STUDY

Chemical Engineering; Physical Chemistry

SUMMARY

Surface chemistry is the study and application of the interaction between surfaces and chemical processes. Adsorption and chemisorption are central features of surface chemistry, and fundamental to catalytic processes involving two or more physical phases. The kinetics of surface chemistry is well studied, especially in regard to catalysis.

PRINCIPAL TERMS

- **alkene:** any organic compound that includes two carbon atoms connected by a double bond.
- **catalyst:** a chemical species that initiates or speeds up a chemical reaction but is not itself consumed in the reaction.
- **chemical kinetics:** the branch of chemistry that studies the various factors that affect rates of chemical reactions.
- **concerted reaction:** a reaction that proceeds from starting materials to end products in a single molecular process rather than via a stepwise mechanism.

- **enzyme:** a protein molecule that acts as a catalyst in biochemical reactions.
- **Gibbs free energy:** the energy in a thermodynamic system that is available to do work.
- **graphene:** an allotrope of carbon consisting of carbon atoms bonded together in a flat array of six-membered rings.
- **hydrogenation:** a chemical reaction in which two hydrogen atoms, usually in the form of molecular hydrogen (H2), are bonded to another molecule, almost always as a result of catalysis.
- **intermediate:** a relatively stable but reactive chemical species that is formed during a chemical reaction and either reverts back to the original reactants or continues in the reaction mechanism to form products.
- **polymerization:** a process in which small molecules bond together in a chain reaction to form a polymer, which is a much larger molecule composed of repeating structural units.
- **Rainey nickel:** a finely divided form of powdered nickel metal used to catalyze the hydrogenation of unsaturated functional groups in organic molecules.
- **rate-determining step:** in any multistep process, the step that proceeds at the slowest rate compared to the other steps in the process, thus determining the maximum rate at which the overall process takes place.
- **spontaneous reaction:** a chemical reaction that occurs without the input of energy from an outside source.

Adsorption, Absorption and Chemisorption

Surface chemistry is quite simply the study and application of chemical processes that occur at surfaces. Corrosion, catalysis and crystallization are historically the principal processes of interest, but the advances made in technology in recent years have expanded the field onto new areas of application. The processes that take place at a surface are not confined to solid surfaces. The interface between two immiscible liquid phases or between a gas and a liquid are also surfaces upon which different processes and reactions may take place rather than in the bulk matter of a phase. In all cases, however, it is essential that the materials involved come into contact at the surface. This is achieved in different ways, with the most common being adsorption and absorption. In adsorption, atoms and molecules of a material are attracted to the atoms and molecules that form the surface of a material. They adhere to the surface by the power of van der Waals attractive forces and by dipole-dipole attraction. A good analogy for this mechanism is the manner in which a toy balloon will adhere to a wall due to the build-up of a static electrical charge on the surface of the balloon. The balloon attaches itself to the wall without being fastened to it physically in any way and remains in place. In adsorption, a similar relationship exists between the atoms or molecules of the gas or liquid and the atoms or molecules of the surface. They are attached to the surface without being fastened to it by chemical bonds. This is a different process than absorption. In absorption, the atoms and molecules of the absorbed material do not interact chemically with the material that does the absorbing, generally, but this is also context-sensitive. A material may be absorbed by capillary action, as when a liquid is drawn into a paper towel. A material may also be absorbed by another material by diffusion through the surface and entry into the bulk

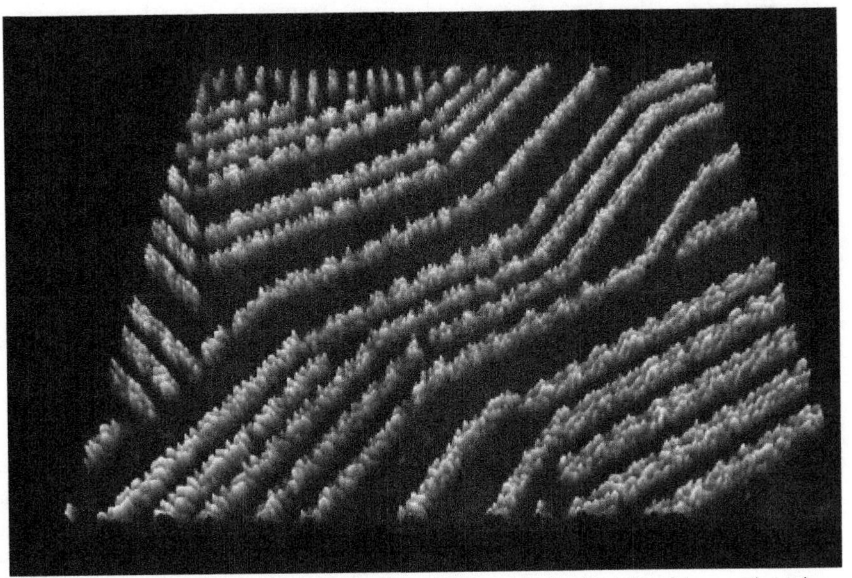

STM image of a quinacridone adsorbate. The self-assembled supramolecular chains of the organic semiconductor are adsorbed on a graphite surface. By Frank Trixler via Wikimedia Commons

matter of the absorbent phase. In these seemingly different processes, interaction between the materials is limited to attractive forces without the involvement of chemical bonding effects (except perhaps for hydrogen bonding to solvent molecules). The third major surface effect is chemisorption. Unlike adsorption and absorption, chemisorption is intimately involved with the formation of chemical bonds to the substrate surface in the processes that occur there. In chemisorption, the surface material forms chemical bonds to the other molecules that are there, effectively producing a reactive intermediate or catalyst that takes part in an overall chemical process. This is an important aspect of catalysis and plays an important role in determining the chemical kinetics of that process, typically by enabling the rate-determining step in the process. Adsorption is a spontaneous reaction, and as such has a negative value for the Gibbs free energy function:

$$\Delta G = \Delta H - T\Delta S$$

in which ΔG is the change in Gibbs free energy (the energy available in the system to perform useful work), ΔH is the change in the enthalpy of the system and ΔS is the change in the entropy of the system. Adsorption is also completely reversible, since no chemical bonds are involved. Chemisorption, on the other hand, is not a readily reversible process because it depends on the formation of chemical bonds. The only way that chemisorption can be reversed is through another bond formation process; that is, through another chemical reaction. Chemisorption is also spontaneous, but the nature of the bond formation that occurs in the process significantly raises the thermodynamics of the system. The heat of adsorption, example, is typically in the range of 20 to 40 kilojoules per mole, while the corresponding property for chemisorption is typically between 80 and 240 kilojoules per mole.

Other Surface Chemistry Processes

As mentioned above, advances in technology have significantly changed the scope of surface chemistry. Devices such as the atomic force microscope now enable surface scientists to examine surface properties directly and manipulate individual atoms to construct surfaces by design. New materials such as graphene provide substrate surfaces that can be chemically altered or augmented with nanoparticles or quantum dots, as desired. Such surface enhanced materials hold promise for great advances in many areas of application. The study of the chemical properties of the surfaces themselves and of the manner in which chemical processes occur on those surfaces is a vast and growing field. This is especially important in the study of catalysis and catalytic processes. While chemists have long known that the hydrogenation of an alkene can be catalyzed by a material such as Rainey nickel or that ethylene undergoes rapid polymerization in the presence of Ziegler-Natta catalysts, it has never been possible to examine the processes at surface level. Technology such as the atomic force microscope now makes such close examination a real possibility. In biochemistry, a great many reactions are catalyzed by an enzyme, a protein that has very specific surface structures that enable it to enter into an intimate combination with the appropriate substrate and so enable it to undergo a specific change. In this field as well new methods of examining the intimate surface details of the processes yield better understanding of their essential nature and applications in new ways.

Structure of Surfaces

A surface is essentially just a monolayer of atoms that forms a boundary between two phases. The arrangement of the atoms that form this layer is determined by the arrangement of the molecules within the bulk matter of the particular phase. For a solid surface, the atoms are arrayed in accord with the crystal lattice or stacking pattern of the material. For a material with a highly regular lattice, such as a crystal or a

SAMPLE PROBLEM

The rate of chemical reaction can be related directly to the rate at which a catalyst surface adsorbs a reacting material. What would be the effect on reaction rate if new reacting material is not adsorbed as fast as it reacts?

Answer:

If the rate of reaction is dependent on the material adsorbed on the catalyst surface, then as the material reacts the amount of adsorbed material would decrease over time. The rate at which product would be formed must therefore also decrease over time.

metal, the atoms at the surface also exhibit a highly regular arrangement. As the internal structure of the material becomes more disordered and amorphous, the arrangement of the atoms at the surface also becomes more disordered. One effect of this disorder is that the effective area of the surface also increases as the disorder increases. For a liquid surface, the degree of randomness of the atoms at the surface is maximized. The effects of the randomness of the distribution of atoms at a surface can be readily seen in two examples. In one case, the completely random surface exists between two liquid phases, one of which is a solution of a diamine and the other is a solution of a diacid chloride. Reaction occurs at this surface so readily that the surface itself can be removed from the interface as a continuous thread of nylon produced by the condensation reaction that occurs there. In the other case, the smooth surface of steel is observed to form rust (red iron oxide) at random locations. The pertinent feature of each of those locations is the presence of a dislocation or flaw in the regular arrangement of the surface atoms. In the absence of such surface irregularities, rust forms uniformly over the steel surface. The particular reaction that occurs in rust formation is an oxidation-reduction reaction between the neutral iron atoms and molecular oxygen (O_2) in the air. The valence electrons of the oxygen atoms interact with the corresponding atomic orbitals of the iron atoms, thus forming the required chemical bonds for chemisorption that can only be reversed by subsequent chemical reactions. In the function of a catalyst a similar effect occurs. The reacting materials generally form an intermediate complex with the catalyst material in such a way that the activation energy for their overall reaction is greatly reduced, often allowing them to react in the manner of a concerted reaction.

THE IMPORTANCE OF SURFACE CHEMISTRY

All chemical reactions take place at the level of the outermost or valence electrons of the atoms involved. This is exactly the level at which chemical and physical interactions take place on a surface, and at its finest definition the outermost valence electrons of an atom are effectively the surface of that atom. Understanding the processes that occur at a surface on the atomic scale is fundamental to understanding the nature of chemical reactions. On a somewhat larger scale, as with the manipulation of single atoms on a graphene surface, the development of new levels of understanding and application offer even more avenues of research and development that will lead to correspondingly new technologies.

Richard M. Renneboog M.Sc.

BIBLIOGRAPHY

Landolt, Dieter. *Corrosion and Surface Chemistry of Metals.* Boca Raton, FL: Taylor and Francis Group LLC, 2007. Print.

Somorjai, Gabor A., and Yimin Li. *Introduction to Surface Chemistry and Catalysis.* 2nd ed. New York: Wiley, 2010. Print.

Birdi, K.S. *Surface Chemistry Essentials.* Boca Raton, FL: CRC Press, 2014. Print.

McCafferty, E. *Surface Chemistry of Aqueous Corrosion Processes.* New York, NY: Springer, 2015. Print.

Nilsson, Anders, Lars G.M. Pettersson, and Jens K. Nørskov. *Chemical Bonding at Surfaces and Interfaces.* Jordan Hill, UK: Elsevier, 2008. Print.

Masel, Richard I. *Principles of Adsorption and Reaction on Solid Surfaces.* New York, NY: John Wiley & Sons, 1996. Print.

Kosmulski, Marek. *Chemical Properties of Material Surfaces.* New York, NY: Marcel Dekker Inc., 2001. Print.

Walas, Stanley M. *Reaction Kinetics for Chemical Engineers.* Boston, MA: Butterworths, 1989. Print.

SYNTHESIS

FIELDS OF STUDY

Organic Chemistry; Inorganic Chemistry; Chemical Engineering

SUMMARY

The process of synthesis and its importance in chemistry-related fields are described. Synthesis is the use of specific reactions to alter and elaborate a molecular composition and structure in order to obtain a desired compound.

PRINCIPAL TERMS

- **chemical reaction:** a process in which the molecules of two or more chemical species interact with each other in a way that causes the electrons in the bonds between atoms to be rearranged, resulting in changes to the chemical identities of the materials.
- **decomposition reaction:** a chemical reaction in which a single reactant breaks apart to create several products with smaller molecular structures.
- **law of conservation of mass:** the principle that all atoms present at the beginning of a process must also be present after the completion of the process; matter can be neither created nor destroyed, only changed from one form to another.
- **product:** a chemical species that is formed as a result of a chemical reaction.
- **reactant:** a chemical species that takes part in a chemical reaction.

The Value of Chemical Synthesis

Synthesis of a specific chemical compound is in principle the same as baking a cake. In both cases, specific materials are combined in the proper proportions and subjected to certain environmental conditions of temperature and pressure. Ideally, the materials combine in the desired manner to produce the desired result: in the case of a cake, something tasty to eat; in the case of chemical synthesis, the result may be a life-saving drug.

Chemical synthesis is a very valuable process economically, as it provides a vast array of useful materials that do not occur naturally. In biology, synthesis reactions are important to the maintenance of life, from the biosynthesis of glucose by green plants to the biosynthesis of DNA. Regardless of where they occur, all synthesis reactions obey the same basic rules of that govern chemical interactions, such as the law of the conservation of mass.

Historically, synthesis reactions were instrumental in the development of the science of chemistry from the pseudoscience of alchemy. The observation that combining various pure materials in the same way produced the same results and products led directly to the fundamental chemical laws of definite proportions and multiple proportions and, later, to the modern theory of atomic structure.

General Concepts of Synthesis

In a simple decomposition reaction, a reactant material, AB, breaks down into its component materials, A and B. At its simplest, synthesis, the opposite of decomposition, occurs by addition reactions in which combining material A with material B produces the product material AB:

$$A + B \rightarrow AB$$

Other simple addition combinations are also possible, such as A + 2B, 2A + 3B, and so on. Such combinations clearly illustrate the law of multiple proportions. In more complex syntheses, a variety of

SYNTHESIS

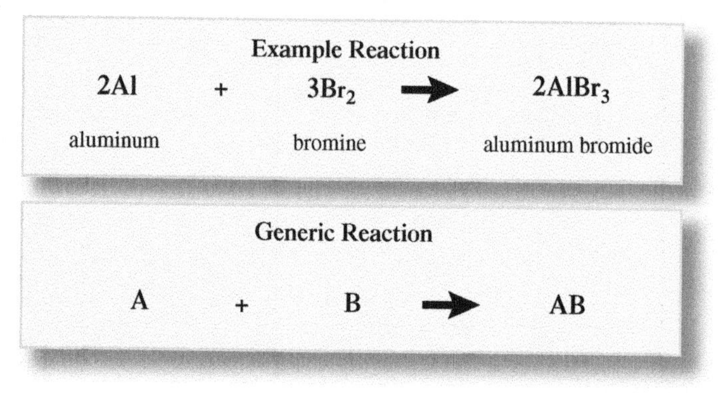

SAMPLE PROBLEM

Calcium carbonate ($CaCO_3$) can be synthesized by the reaction of calcium oxide (CaO) with carbon dioxide (CO_2). The organic compound tetrachloroethane ($Cl_2HC-CHCl_2$) can be prepared by the reaction of acetylene (C_2H_2) with chlorine gas (Cl_2). Write the balanced reaction equation for the two processes.

Answer:

CaO reaction: The chemical formulas for the reactants are CaO and CO_2. The chemical formula for the product is $CaCO_3$. Therefore, the basic reaction can be written initially as

$$CaO + CO_2 \rightarrow CaCO_3$$

Check the mass balance by counting the atoms on each side of the equation:

left-hand side: Ca, C, 3O

right-hand side: Ca, C, 3O

Since exactly the same number of each type of atom appears on both sides of the equation, the equation is balanced as written.

C_2H_2 reaction: The chemical formulas of the reactants are C_2H_2 and Cl_2. The chemical formula of the product is $C_2H_2Cl_4$. The basic reaction can be written initially as

$$C_2H_2 + Cl_2 \rightarrow C_2H_2Cl_4$$

Check the mass balance by counting the atoms on both sides of the equation:

left-hand side: 2C, 2H, 2Cl

right-hand side: 2C, 2H, 4Cl

There are two more chlorine atoms on the right than there are on the left. Therefore, a second unit of Cl_2 is required on the left-hand side, and the balanced reaction equation can be written as

$$C_2H_2 + 2Cl_2 \rightarrow C_2H_2Cl_4$$

reaction types may be used in a sequential or stepwise manner in order to maintain complete control over the direction of the synthesis and the products obtained at each step.

Organic compounds offer the greatest variety of possible structures, and synthesis reactions involving them are subject to very stringent rules determined by the molecular orbital interactions in each particular chemical reaction. As a result, undesirable side reactions are more likely to occur, consuming reactants without producing the desired products. Various functional groups can react in different ways, and it is common practice to use what is called a protecting group to prevent one part of a molecule from taking part in a reaction when it is not supposed to. The protecting group is added to a reactive functional group in order to convert it into a nonreactive group. After the reaction is complete, the protecting group can then be removed, regenerating the original functional group.

Typically, synthesis reactions are monitored by determining the yield of the desired product. An ideal synthesis reaction would produce a 100 percent yield, with all of the starting reactants converted entirely and only into the desired product. In practice, this is almost unheard of, especially in multistep synthesis procedures, in which the reduction of yield in each step is a cumulative effect. A five-step procedure, for example, may have an good yield of 80 percent in each individual step, but after the five steps have been completed, the overall yield of desired product would be only 0.80^5, or a mere 32 percent.

SYNTHESIS IN CHEMICAL ENGINEERING

Many important materials, especially various plastics and resins, are produced on an industrial scale. The very common plastics polyethylene and nylon are produced by simple addition chain reactions, in which individual molecules of the reactants bond to each other in a head-to-tail manner. The formation of polyethylene, for example, can be described by the following reaction equation:

$$nC_2H_4 \rightarrow -(CH_2CH_2)_n-$$

in which the coefficient n indicates an indeterminate number of ethylene molecules. The structure of nylon polymers is more complex, since nylon is derived from two different types of compounds. Other synthetic polymers have even more complex structures due to bond formation between different molecules. Such compounds are typically used as matrix resins in advanced composite materials from which the most advanced aircraft and other constructs are made.

Richard M. Renneboog, MSc

BIBLIOGRAPHY

Berg, Jeremy M., John L. Tymoczko, Gregory J. Gatto, and Lubert Strye. *Biochemistry.* 8th ed. New York: W. H. Freeman, 2015. Print.

Douglas, Bodie E., Darl H. McDaniel, and John J. Alexander. *Concepts and Models of Inorganic Chemistry.* 3rd ed. New York: Wiley, 1994. Print.

Hendrickson, James B., Donald J. Cram, and George S. Hammond. *Organic Chemistry.* 3rd ed. New York: McGraw, 1970. Print.

Jones, Mark Martin. *Chemistry and Society.* 5th ed. Philadelphia: Saunders College Pub., 1987. Print.

Miessler, Gary L., Paul J. Fischer, and Donald A. Tarr. *Inorganic Chemistry.* 5th ed. Upper Saddle River: Prentice Hall, 2013. Print.

Morrison, Robert Thornton, and Robert Neilson Boyd. *Organic Chemistry.* 7th ed. Englewood Cliffs, N.J.: Prentice Hall, 2003. Print.

Myers, Richard. *The Basics of Chemistry.* Westport: Greenwood, 2003. Print.

Wuts, Peter G. M., and Theodora W. Greene. *Greene's Protective Groups in Organic Synthesis.* 4th ed. Hoboken: Wiley-Interscience, 2007. Print.

T

THIOLS

FIELD OF STUDY

Organic Chemistry

SUMMARY

The characteristic properties and reactions of thiols are discussed. Thiols are theoretically unlimited in type. Their usefulness in organic synthesis reactions is limited, though they are significantly more reactive as nucleophiles than their oxygen analogs.

PRINCIPAL TERMS

- **functional group:** a specific group of atoms with a characteristic structure and corresponding chemical behavior within a molecule.
- **mercaptan:** an organic compound characterized by the presence of a sulfhydryl functional group (–SH).
- **odor:** the sensation created when molecules of a volatile chemical compound vaporize and bind to olfactory receptors in the nose.
- **organosulfur compound:** an organic compound containing one or more carbon-sulfur bonds.
- **redox reaction:** a reaction in which electrons are transferred from one atom, ion, or molecule (oxidation) to another (reduction).

The Nature of Thiols

Thiols, also known as mercaptans, are organosulfur compounds characterized by the presence of the sulfhydryl functional group (–SH), the sulfur analog of the hydroxyl functional group (–OH). Accordingly, the series of thiol compounds exactly parallels that of the alcohols, which feature a hydroxyl group bonded to a carbon atom, although the substitution of sulfur for oxygen radically alters the nature of the particular compound. For example, an alcohol compound responsible for the very pleasing scent of roses may become an eye-burning compound with a hideous odor when the hydroxyl group is replaced by the sulfhydryl group. The slightly pungent smell of ethanol is well known to most people; its sulfur analog, ethanethiol (also known as thioethanol or ethyl mercaptan), is equally well known as the odorant added in minute amounts to natural gas and propane to render any leakage obvious. The sulfur analog of the water molecule, hydrogen sulfide, is notorious as the cause of the odor of rotten eggs, while water is essentially odorless.

The difference in odor is attributable to the position of sulfur in period 3 of the periodic table of elements, the chart representing the known elements by atomic number and electron distribution; oxygen is in period 2, though both are in group 16. While both oxygen and sulfur nominally have six electrons in their outermost, or valence, electron shell, oxygen has its electrons in regions known as the $2s$ and $2p$ orbitals, while sulfur has its six valence electrons in the $3s$ and $3p$ orbitals. Being in period 3, sulfur also has the $3d$ orbitals available for electronic interactions. Accordingly, sulfur is able to exhibit a much broader range of chemical behaviors than is oxygen. For the same reasons, the sulfur atoms in thiols interact in very different ways with the scent receptors in the nasal passages, replacing the normal alcohol odors with unpleasant counterparts. Thiols occur naturally in biochemical systems in which compounds derived metabolically from the amino acid cysteine are produced.

Nomenclature of Thiols

Thiols have long been commonly known as mercaptans. That name derives from the Latin term *mercurium captans*, or "seizing mercury," which refers to the ease with which thiols form mercury-based compounds. Thiols are systematically named according to the parent compound from which they are derived. The methanol (CH_3OH) analog is called methanethiol (thiomethanol or methyl mercaptan, CH_3SH). The next in the series is ethanethiol (thioethanol or ethyl mercaptan, CH_3CH_2SH), then propanethiol ($CH_3CH_2CH_2SH$), and so on. The naming of isomeric structures is analogous to the naming of the normal alcohols.

The prefix "sulfanyl-" (when another group takes priority) or the suffix "-thiol" (when the sulfhydryl group takes priority) is used to indicate the sulfhydryl functional group as a substituent in larger molecules. When naming a thiol compound, some confusion may arise if the prefix "thio-" is used in place of "sulfanyl-." The sulfhydryl group can only be a substituent, but the sulfur atom can be incorporated into the framework of the molecular structure in place of an oxygen atom. Generally, "thio-" is used to indicate the presence of a sulfur atom, and in the International Union of Pure and Applied Chemistry (IUPAC) systematic naming convention, it is used to designate a sulfur atom that forms part of the framework molecular structure. Because of this ambiguity, the compounds 2-butanethiol and 2-thiobutane can be easily confused, though they have very different structures and properties:

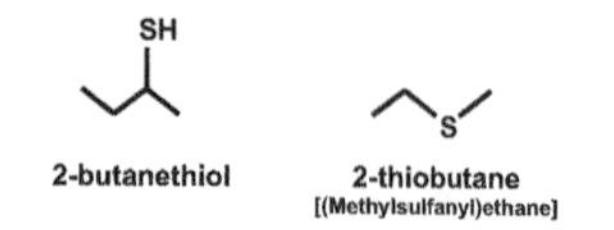

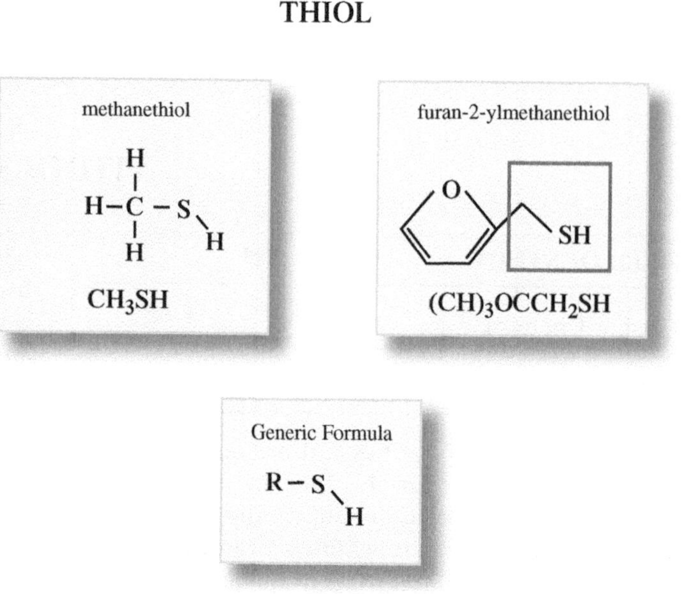

REACTIONS OF THIOLS

Thiols are reactive compounds that can produce a wide variety of compounds through both redox reactions (reduction-oxidation reactions) and non-redox reactions. Perhaps the most important reaction of thiols is the formation of disulfides, which can generally be formed under mild oxidation conditions. Disulfides play a significant role in the development of the three-dimensional structure of enzymes and proteins. The sulfhydryl substituents on different cysteine residues in the primary structure of the protein molecule can bond together by forming a disulfide linkage between the two sulfur atoms. This effectively locks other substituent groups and reactive functional groups into fixed spatial relationships relative to each other.

Oxidation of thiols under more vigorous conditions produces successive oxidation products by the addition of oxygen atoms to the sulfur atom as it is shifted into higher oxidation states. As part of a thiol, the sulfur atom is in the –2 oxidation state, analogous to an oxygen atom. Mild oxidation to form the disulfide brings the sulfur atom into the –1 oxidation state. Further oxidation first produces the corresponding sulfenic acid, characterized by the presence of an oxygen atom, and then adds a second oxygen atom to the sulfur atom to form the corresponding sulfinic acid derivative, RSO_2H (where R represents a generic alkane). Even further oxidation adds a third oxygen atom to produce the corresponding sulfonic acid compound, RSO_3H.

In the presence of elemental sulfur, thiols can form a variety of polysulfides in which a number of sulfur atoms are connected in a chain. This reaction plays a key role in the production of toughened rubber, which uses a process known as vulcanization. The presence of disulfides and similar cross-links between different rubber molecules in the polymer makes the rubber much stronger.

PRODUCTION OF THIOLS

A great many sulfur-containing compounds are produced by the reaction of compounds such as alkyl halides with elemental sulfur. Combustion of sulfur produces sulfur dioxide, for example, while bacterial action and environmental conditions conducive to chemical reduction naturally produce hydrogen sulfide. Vast quantities of elemental sulfur are recovered from the desulfurization of fossil fuels, providing a

SAMPLE PROBLEM

Given the chemical name 2-sulfhydrylcyclopentanone, draw the skeletal structure of the molecule.

Answer:

The name indicates that the material is a cyclopentane derivative with a carbonyl functional group ("-one), consisting of a carbon atom double bonded to an oxygen atom, at the first of five carbon atoms. There is a sulfhydryl group on the second carbon atom, adjacent to the carbonyl group. The structure of the compound is therefore

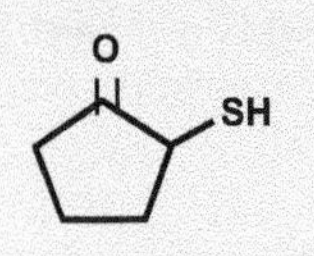

large supply for the conversion of other compounds. In addition, a number of small thiol compounds such as methanethiol and ethanethiol occur naturally and can be recovered through oil and gas processing. Mineral sources of sulfur are also common, since sulfur readily forms inorganic compounds with metals. Iron pyrite, a common iron ore popularly known as fool's gold, is composed partially of sulfur, and many other ores are also sulfides. The refining process typically releases the sulfur as sulfur dioxide, which is captured rather than released into the atmosphere.

Specific thiols are prepared by a nucleophilic displacement reaction with a sulfur compound. Sulfur compounds typically have greater reactivity as nucleophiles, or chemicals that tend to react with positively charged or electron-poor species, than their oxygen analogs, and the variety of compounds that sulfur can produce enables the use of a number of non-thiol sulfur compounds in nucleophilic reactions. One common method of producing thiols is to react an alkyl halide (an alkyl group bonded to a halogen) with sodium disulfide (Na_2S_2), followed by hydrolytic reduction of the sulfur-sulfur bond to form two sulfhydryl compounds.

Richard M. Renneboog, MSc

BIBLIOGRAPHY

Berg, Jeremy M., John L. Tymoczko, Gregory J. Gatto, and Lubert Strye. *Biochemistry.* 8th ed. New York: W. H. Freeman, 2015. Print.

Hendrickson, James B., Donald J. Cram, and George S. Hammond. *Organic Chemistry.* 3rd ed. New York: McGraw, 1970. Print.

Mann, J., et al. *Natural Products: Their Chemistry and Biological Significance.* New York: Wiley, 1994. Print.

Morrison, Robert Thornton, and Robert Neilson Boyd. *Organic Chemistry.* 7th ed. Englewood Cliffs, N.J.: Prentice Hall, 2003. Print.

Reece, Jane B., et al. *Campbell Biology.* 10th ed. San Francisco: Cummings, 2013. Print.

Robinson, Trevor. *The Organic Constituents of Higher Plants.* 6th ed. North Amherst: Cordus, 1991. Print.

Wuts, Peter G. M., and Theodora W. Greene. *Greene's Protective Groups in Organic Synthesis.* 4th ed. Hoboken: Wiley-Interscience, 2007. Print.

TOXINS, POISONS, AND VENOMS

FIELDS OF STUDY

Biochemistry

SUMMARY

The basic properties of toxins, poisons, and venoms are described, and their various methods of operation are discussed. Because of their molecular structures and modes of action, all three are inimical to the biochemistry of living things.

PRINCIPAL TERMS

- **cellular receptor:** a protein within a cell that specific signaling molecules bind to in order to elicit a specific response from the cell.
- **enzyme:** a protein molecule that acts as a catalyst in biochemical reactions.
- **neurotransmitter:** a chemical species that carries electrochemical signals across the synapses between neurons in the nervous system.
- **peptide:** an organic compound composed of a relatively small number of amino acid molecules,

held together by covalent bonds between a carbon atom of one molecule and a nitrogen atom of the next.

- **protein:** a biological polymer consisting of one or more long chains of amino acids linked by peptide bonds in a sequence specified by an organism's DNA.

The Nature of Toxins, Poisons, and Venoms

Biochemical systems such as living organisms are complex chemical factories in which tens of thousands of specific chemical reactions take place. Each reaction performs a function on either a specific material or a class of materials with similar structural features. Toxins, poisons, and venoms interfere with a system's proper functioning in ways that range from minor to fatal.

To understand how toxins, poisons, and venoms affect the chemical functions of the body, think of them as the various fluids used to keep an automobile running smoothly. Using windshield-wiper fluid in the radiator instead of a water–ethylene glycol solution would have only a minor effect on the efficiency of the cooling system, analogous to the automobile "organism" having a mild fever. On the other hand, using wiper fluid in place of lubricating oil would quickly result in the destruction of the engine, as the water would eat the metal and allow the engine's moving parts to grind against each other. Such "tissue damage" would quickly lead to the "death" of the automobile.

Toxins, poisons, and venoms are chemical compounds that have the same effects on living organisms as improper fluid replacements have on automobiles. They often come from other living organisms that produce these substances as part of their normal biochemical processes, either as metabolic waste products or as defensive or offensive mechanisms. The terms "toxin," "poison," and "venom" are often used interchangeably, but in fact they have distinct meanings. All three are indeed toxic to the targeted organism, and all may also be poisonous, but the source of the material determines the proper terminology.

Toxins

Toxins are bacterial waste products that act as poisons within the host organism. They may consist of normal biochemical materials that have been modified by metabolic processes, or they may be various proteins or peptides that interfere with the normal biochemical processes of the host. The human digestive tract, like the digestive tracts of all animals, is home to billions of bacterial flora that secrete compounds such as proteins and enzymes to break down food, making them essential to their host's health and normal digestive functions. When abnormal bacteria invade the host, they generally prove toxic to the endemic bacterial colonies—and ultimately to the host. In some cases, the endemic bacteria may be almost completely eradicated, leaving the host in a poor state of health until the invading bacteria can be eliminated and the proper strains regained.

VENOMOUS ANIMALS

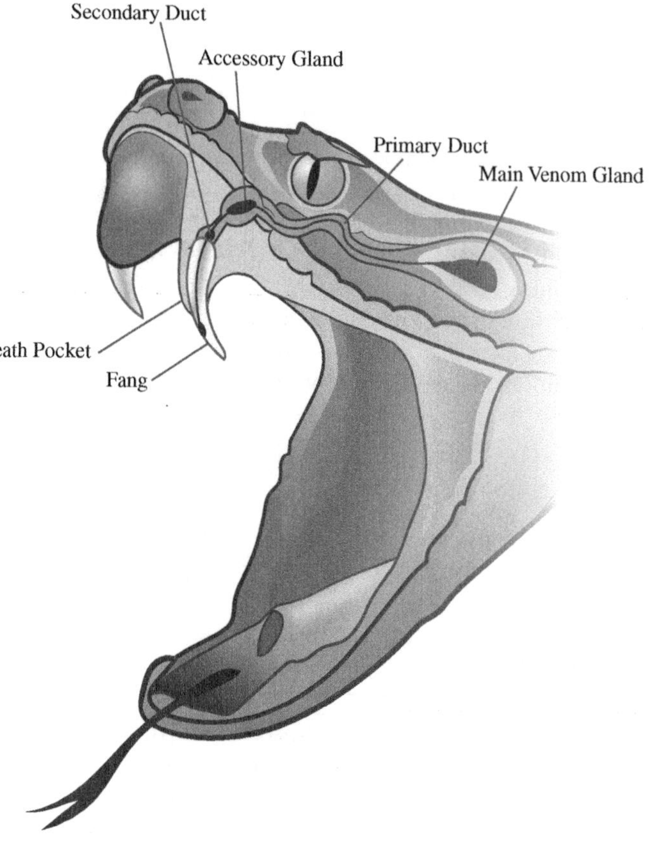

Venomous animals, such as rattlesnakes, have specialized anatomy for producing, storing, and delivering venom to their prey.

In other cases, recovery is not possible, as the toxins produced by the invading bacteria and viruses prove too damaging to the surrounding tissues.

The most damaging toxins are those against which the immune system of the host organism has little or no resistance. Bacteria such *E. coli* and the dinoflagellate algae responsible for "red tides" in the oceans produce toxins that affect the functioning of the nervous system of the host, and they may prove fatal if sufficiently severe. Viruses such as Ebola cause the host organism's own biochemical processes to produce compounds that actively destroy vital tissues, usually resulting in death.

POISONS

Poisons are chemical compounds and may be of animal, vegetable, or mineral origin. Given the precise nature of the biochemical systems within an organism, every chemical compound in existence can act as a poison at some point, whether by simply overloading normal biochemical pathways or by completely disrupting an organism's biochemical processes.

Human biochemistry includes and is dependent on the presence of many different elements of the periodic table. The vast majority of a living organism is organic in nature, meaning it is composed of principally carbon-based materials. Much smaller quantities of other elements, such as iron and selenium, serve as central atoms for various essential materials, including hemoglobin, cofactors in the active sites of enzymes, and electrolytes in cellular fluids. If a biochemical system has too little of these cofactors and micronutrients, it will not function properly; however, too much will effectively poison the system, interfering with the properly balanced functioning of the different biochemical pathways. Many biochemical cycles function in equilibrium with one another. When too much of one component is present, other components maintaining the equilibrium can become saturated, and the system cannot function as it should.

Certain mineral components are not native to human physiology, and their presence in the human body usurps the place of the chemical species that are supposed to be present, preventing the correct compounds and ions from performing their proper systemic functions. Hemoglobin, for example, transports oxygen in the red blood cells by reversibly coordinating the O_2 molecule. The presence of carbon monoxide (CO) or the cyanide ion (CN^-) usurps this function by binding irreversibly with the iron atom in hemoglobin, preventing it from transporting oxygen. Heavy metals such as lead, mercury, cadmium, and manganese poison the system in a similar way, though by different mechanisms. Lead and mercury poison the neurological system of an organism by interfering with the function of the neurotransmitter compounds that normally pass electrical signals from cell to cell, and they also are able to link to the protein molecules of cellular receptors and prevent cellular communication.

Poisons of animal origin are typically compounds secreted by various animals as defense mechanisms. In some cases, the poisons derive from the foods the animals consume. For example, caterpillars of the monarch butterfly feed exclusively on the milkweed plant, causing them carry a load of alkaloids that will poison and perhaps kill any birds that consumes them. Various toads are known to secrete compounds that are hallucinogenic when ingested, which might discourage other animals from eating them. Certain fish are known to secrete the compounds tetrodotoxin and saxitoxin, which interfere with the function of sodium-ion channels in membranes, effectively shutting down nerve function in a way that often results in death.

SAMPLE PROBLEM

Research the following three materials, identify each as either a toxin, a poison, or a venom, and tell how someone might be exposed to the material:

- urushiol
- muscarine
- crotamine

Answer:

Urushiol is the active component of poison ivy and is a toxin acquired through simple contact with the poison ivy plant. Muscarine is an active poison from the *Amanita muscaria* mushroom; it is acquired through ingestion of the mushroom and is often fatal. Crotamine is a polypeptide from the venom of certain snakes and is acquired through a bite from certain snakes.

All three materials have toxic effects. Once the affected person has incorporated the compound, the compound is commonly referred to as a toxin, even if it is in fact a poison or venom.

The largest group of poisons is organic compounds produced by plants, and the many pharmaceutical compounds designed to imitate them. These range from compounds as simple as oxalic acid (the bitter component of rhubarb) and nicotine (derived from tobacco and other alkaloid-producing plants) to complex molecules such as the aflatoxins (typically from molds found on peanuts) and ketophalloidin (from the death cap mushroom, *Amanita phalloides*). Alkaloids are a class of nitrogen-containing compounds that includes lysergic acid diethylamide (LSD), cocaine, caffeine, and theobromine (found in chocolate). Hundreds of thousands of such "natural product" compounds have been identified. The fact that essentially all natural compounds have physiological effects has led researchers to experiment with changing the molecular structures of such compounds in order to make them more potent and effective for pharmaceutical purposes. Thus, every pharmaceutical compound is a poison used to alleviate certain medical conditions, either by interfering with normal functions or by selectively poisoning an invading organism; painkillers are an example of the former, antibiotics of the latter. Every pharmaceutical compound is described with an LD_{50} dose, which is the dosage that would be lethal for 50 percent of individuals of a certain body weight given a predetermined amount of the compound. Simply stated, the higher the LD_{50} dose of a compound, the safer it is to use.

VENOMS

Venoms are unique compounds produced specifically by animals for the purpose of killing or severely incapacitating other animals. Snake venoms are the most familiar, but many other biting animals also produce venoms. Venoms are specialized proteins that work against the central nervous system or other tissues. In the worst cases, the venom kills tissue, requiring its immediate removal to save the life of the bite victim. In other cases, the venom works slowly to incapacitate the bite victim for a long period of time. Even a seemingly innocuous envenomating bite or sting, as from a small bee, can have serious consequences if the bite victim is allergic to the venom. Because venoms are small, specialized proteins, they have complex and poorly understood modes of action.

USES OF TOXINS, POISONS, AND VENOMS

The various properties and effects of toxins, poisons, and venoms can be applied to specific medical interventions. Botox, a preparation made from the botulinum toxin, is routinely used for cosmetic effects, but it is also effective in treating facial tics and other conditions of an impaired central nervous system. Toxins and venoms are used to prepare antitoxins and antivenins for front-line treatment of snake and other bites. Unfortunately, antivenins are generally creature specific, so to be of use, the antivenin specific to the venom in question must be available. Poisons of all kinds are studied intensively for pharmaceutical use. Though pharmaceuticals are not referred to as poisons, they all have poisonous thresholds for use.

Richard M. Renneboog, MSc

BIBLIOGRAPHY

Claus, Edward P., and Varro E. Tyler Jr. *Pharmacognosy*. Philadelphia: Lea, 1965. Print.

Jones, Mark Martin. *Chemistry and Society*. 5th ed. Philadelphia: Saunders College Pub., 1987. Print.

Kean, Sam. *The Disappearing Spoon: And Other True Tales of Madness, Love, and the History of the World from the Periodic Table of the Elements*. New York: Little, Brown, 2010. Print.

O'Neil, Maryadele J., ed. *The Merck Index: An Encyclopedia of Chemicals, Drugs, and Biologicals*. 15th ed. Cambridge: RSC, 2013. Print.

Rang, H. P., et al. *Rang and Dale's Pharmacology*. 7th ed. New York: Elsevier, 2012. Print.

Reece, Jane B., et al. *Campbell Biology*. 10th ed. San Francisco: Cummings, 2013. Print.

TRANSITION METALS

FIELDS OF STUDY

Inorganic Chemistry; Geochemistry; Metallurgy

SUMMARY

The basic properties and characteristics of the transition metals are presented. The transition metals are those elements whose valences are determined by the behavior of electrons in their *d* orbitals.

PRINCIPAL TERMS

- **d-block:** the portion of the periodic table containing the elements whose valence electrons are in their *d* orbitals.
- **element:** a form of matter consisting only of atoms of the same atomic number.
- **oxidation state:** a number that indicates the degree to which an atom or ion in a chemical compound has been oxidized or reduced.
- **periodic table:** the chart representing the known elements by atomic number and electron distribution.
- **valence electron:** an electron that occupies the outermost or valence shell of an atom and participates in chemical processes such as bond formation and ionization.

The Nature of the Transition Metals

The transition metals consist of the various elements in the d-block of the periodic table. The valence electrons of a transition metal atom are located in the atom's *d* orbitals, specific regions around the atom's nucleus. This arrangement of electrons is responsible for the multivalent nature of transition metal ions and for their ability to form numerous compounds with neutral organic molecules. The number of electrons allowed to occupy the *d* orbitals enables transition metal atoms to form compounds in which they take on different oxidation states. Many of the transition metals, including titanium, iron, gold, and mercury, are readily recognized and well known. Others, such as osmium and iridium, are scarce and thus much less familiar.

Atomic Structure of the Transition Metals

As with all elements, the atomic structure of transition metals consists of a central nucleus, which comprises positively charged protons and neutral neutrons, and negatively charged electrons, which are typically described as filling regions around the nucleus known as "orbitals." Types of orbitals include *s* orbitals, *p* orbitals, and *d* orbitals. The *d* orbitals first become relevant in atomic structure in the elements of the fourth period, or row, of the periodic table. Scandium is the first fourth-period element to be able to have electrons in its *d* orbital. Electrons are permitted to shift from orbital to orbital in response to environmental influences such as the types and numbers of ions or molecules that surround the particular atom in a compound. The transition metals are therefore readily able to take part in reduction-oxidation (redox) reactions, in which the oxidation states of the atoms involved change, due to the ease with which they donate or accept different numbers of valence electrons.

SAMPLE PROBLEM

Given the following transition metal compounds, identify the oxidation state of each metal atom and the Latin name corresponding to its elemental symbol:

- $Fe(OH)_3$
- $CuCl_2$
- $AuClO_4$

Answer:

The first compound is iron (III) hydroxide. The (III) indicates that the iron atom is in the +3 oxidation state, balancing the three hydroxide (–OH) groups, which each carry a negative charge. The symbol Fe derives from the Latin name for iron, *ferrum*.

The second compound is copper (II) chloride. The (II) indicates that the copper atom is in the +2 oxidation state, balancing the two chlorine ions, which each carry a negative charge. The symbol Cu derives from the Latin name for copper, *cuprum*.

The third compound is gold (I) perchlorate. The (I) indicates that the gold atom is in the +1 oxidation state, balancing the ClO_4 ionic compound, which carries a single negative charge. The symbol Au derives from the Latin name for gold, *aurum*.

TRANSITION METALS

78 Pt platinum 195

1 H hydrogen 1																	2 He helium 4
3 Li lithium 7	4 Be beryllium 9											5 B boron 11	6 C carbon 12	7 N nitrogen 14	8 O oxygen 16	9 F fluorine 19	10 Ne neon 20
11 Na sodium 28	12 Mg magnesium 29											13 Al aluminium 27	14 Si silicon 28	15 P phosphorus 31	16 S sulfur 32	17 Cl chlorine 35	18 Ar argon 40
19 K potassium 39	20 Ca calcium 40	21 Sc scandium 45	22 Ti titanium 48	23 Vi vanadium 51	24 Cr chromium 52	25 Mn manganese 55	26 Fe iron 56	27 Co cobalt 59	28 Ni nickel 59	29 Cu copper 64	30 Zn zinc 65	31 Ga gallium 70	32 Ge germanium 73	33 As arsenic 75	34 Se selenium 79	35 Br bromine 80	36 Kr krypton 84
37 Rb rubidium 85	38 Sr strontium 88	39 Y yttrium 89	40 Zr zirconium 91	41 I niobium 93	42 Mo molybdenum 96	43 Tc technetium 98	44 Ru ruthenium 101	45 Rh rhodium 103	46 Pd palladium 106	47 Ag silver 108	48 Cd cadmium 112	49 In indium 115	50 Sn tin 119	51 Sb antimony 122	52 Te tellerium 128	53 I iodine 127	54 Xe xenon 131
55 Cs caesium 133	56 Ba barium 137	*	72 Hf hafnium 178	73 Ta tantalum 181	74 W tungsten 184	75 Re rhenium 186	76 Os osmium 190	77 Ir iridium 192	78 Pt platinum 195	79 Au gold 197	80 Hg mercury 201	81 Tl thallium 204	82 Pb lead 207	83 Bi bismuth 209	84 Po polonium 209	85 At astatine 210	86 Rn radon 222
87 Fa francium 223	88 Ra radium 226	**	104 Rf rutherfordium 267	105 Db dubnium 268	106 Sg seaborgium 271	107 Bh bohrium 272	108 Hs hassium 270	109 Mt meitnerium 276	110 Ds darmstadtium 281	111 Rg roentgenium 280	112 Cn copernicum 285	113 Uut ununtrium 284	114 Fl flerovium 289	115 Uup ununpentium 288	116 Lv livermorium 293	117 Uus ununseptium 294	118 Uuo ununoctium 289

57 * La lanthanum 139	58 Ce cerium 140	59 Pr praseodymium 141	60 Nd neodymium 144	61 Pm promethium 145	62 Sm samarium 150	63 Eu europium 152	64 Gd gadolinium 157	65 Tb terbium 159	66 Dy dysprosium 163	67 Ho holmium 165	68 Er erbium 167	69 Tm thulium 168	70 Yb ytterbium 173	71 Lu lutetium 175
89 ** Ac actinium 227	90 Th thorium 232	91 Pa protactinium 231	92 U uranium 238	93 Np neptunium 237	94 Pu plutonium 244	95 Am americium 243	96 Cm curium 247	97 Bk berkelium 247	98 Cf californium 251	99 Es einsteinium 252	100 Fm fermium 257	101 Md mendelevium 258	102 No nobelium 259	103 Lr lawrendum 284

Transition Metal Compounds

The transition metals commonly form both monatomic and complex ions as well as coordination compounds in which a central transition metal atom is surrounded by a specific number of ligands, molecules or atoms that are often neutral or anionic. The bonds to the surrounding ligands are termed "coordination bonds" and are thought to form when an electron pair from the ligand enters a vacant orbital on the metal atom. Due to the strictly defined geometry of atomic orbitals, the ligands about the central metal atom can take on only select spatial arrangements, coordinating in structures that include the trigonal pyramid and trigonal planar arrangements of three ligands, the square planar and tetrahedral arrangements of four ligands, the trigonal bipyramid and square pyramid arrangements of five ligands, and the square bipyramid arrangement of six ligands. Higher coordination complexes are also known but are much less common. In addition, the same ligands can be shared by two metal atoms and thus form an extended structure, as in the chromate (CrO_4^{2-}) and dichromate ($Cr_2O_7^{2-}$) ions.

Transition Metals and Nomenclature

Historically, the study of metallurgy in medieval times marked the beginning of the practical science of chemistry. Because of the field's historical roots, many of the chemical symbols for elements derive from the medieval Latin names for the materials. In addition to the Latin root word, the name of a transition metal compound can reveal a great deal of information about the substance, including its oxidation state. For example, $Pb(NO_3)_2$ is lead (II) nitrate. The (II) indicates that the lead atom is in the +2 oxidation state. The chemical symbol for lead is Pb because it derives from *plumbum,* the Latin word for the element.

Extraction of Transition Metals

The transition metals are normally found in nature as ores or compounds and are typically extracted by mining and smelting operations that release the metals as relatively pure elements. It is also possible to find some of the transition metals as native metals—metals that exist in nature in their metallic form—rather than as ores or mineral compounds, but this is a rarity and requires very specific environmental conditions. Some of the transition metals, such as gold and platinum, are normally inert and are always found in their elemental form in nature rather than as compounds.

Richard M. Renneboog, MSc

Bibliography

Agricola, Georgius. *De re metallica.* Trans. Herbert Clark Hoover and Lou Henry Hoover. New York: Dover Publications, 1950. Print.

Gribbin, John. *Science: A History, 1543–2001.* London: Penguin, 2002. Print.

Johnson, Rebecca L. *Atomic Structure.* Minneapolis: Lerner, 2008. Print.

Kean, Sam. *The Disappearing Spoon: And Other True Tales of Madness, Love, and the History of the World from the Periodic Table of the Elements.* New York: Little, Brown, 2010. Print.

Mackay, K. M., R. A. Mackay, and W. Henderson. *Introduction to Modern Inorganic Chemistry.* 6th ed. Cheltenham: Nelson, 2002. Print.

Winter, Mark J. *The Orbitron: A Gallery of Atomic Orbitals and Molecular Orbitals.* University of Sheffield, n.d. Web.

TRIPLE POINT

FIELDS OF STUDY

Physical Chemistry

SUMMARY

The characteristics of the physical state known as the triple point are discussed. At the triple point, a pure material exists in solid, liquid, and gas states in thermodynamic equilibrium. The absolute, or Kelvin, temperature scale uses the triple point of water to set its primary reference point, absolute zero.

PRINCIPAL TERMS

- **critical pressure:** the pressure required to compress a particular gas into a liquid state at the critical temperature.
- **critical temperature:** the highest temperature at which a particular gas can be liquefied by pressure alone.
- **intensive properties:** the properties of a substance that do not depend on the amount of the substance present, such as density, hardness, and melting and boiling point.
- **kelvin:** the basic unit of the Kelvin temperature scale, equivalent in magnitude to one degree of the Celsius temperature scale; 0 kelvin is absolute zero, the lowest temperature theoretically possible, equal to −273.15 degrees Celsius.
- **thermodynamic equilibrium:** also known as thermal equilibrium, the condition in which a system and its surroundings are in the same energy and temperature state.

The Nature of the Triple Point

The word “triple” in triple point refers to the three standard states of matter: solid, liquid, and gas. At the triple point, an element or compound exists in all three states at once, in equilibrium with each other.

A triple point exists for every material that is capable of existing in all three states, not just for water. It is defined by a specific combination of temperature and pressure for a specific material. At the triple point, the material, in all three states, and its surroundings are in thermodynamic equilibrium. This is characterized by the intensive properties of the material, such as melting and boiling points, and does not depend on the extensive properties of the material, which are the properties that depend on the quantity of matter present, such as heat capacity and enthalpy.

SAMPLE PROBLEM

What happens to water at −30 °C and 1 atmosphere (atm) of pressure when the temperature is increased to 50 °C? When the pressure is reduced to 0.001 atm? When the temperature is decreased to −50 °C?

Answer:

Water ice forms at 0 °C and 1 atm of pressure by forming a crystal structure of water molecules. At different pressures and colder temperatures, the orientation of the water molecules changes, and several different forms of water ice are known. At −30 °C and 1 atm, water ice has a certain form.

When the temperature is raised to 50 °C, the ice absorbs energy from the surroundings, and the molecules become more energetic and move freely about as hot liquid. The vapor pressure of the liquid is also much higher at this temperature, and the transition of water molecules from liquid to gas increases due to the molecules' increased energy.

At −30 °C, reducing the pressure from 1 atm to 0.001 atm decreases the energy required for water molecules to sublimate from the solid phase to the gas phase, without passing through the liquid phase first, by a factor of one thousand. When the temperature is reduced to −50 °C, the energy barrier to sublimation increases, reducing the ability of water molecules to pass directly from the solid to the gas phase. Rearrangement of the water molecules in the ice structure may also occur, producing a harder, more crystalline form of ice than exists at −30 °C.

Conditions of the Triple Point

The conditions under which a material can be brought to its triple point are termed critical conditions. A gas can be forced into the liquid state by being placed under pressure. This can be observed with carbon dioxide in a glass cylinder such as the barrel of a syringe. When the syringe is plugged and the plunger is depressed, the clear, colorless carbon dioxide gas can be seen to condense into droplets of liquid carbon dioxide. The critical temperature of the compound is the highest temperature at which condensation can be achieved by pressure alone, while the critical pressure is the amount of pressure required to turn a gas into a liquid when the system is at its critical temperature.

Triple Point and the Absolute Temperature Scale

The realization that temperature plays a significant role in chemical and physical behavior marked a great leap forward for the science of chemistry. Means were sought to measure temperature changes, and many individuals developed their own temperature-measuring devices, although considerable time passed before the need for a standardized temperature scale was realized. A number of such scales are in use, including the Fahrenheit and Celsius scales, which use the freezing and boiling points of water as reference points, and the Rankine and Kelvin scales, which use absolute zero as their point of origin. The Fahrenheit and Rankine scales use degrees of the same magnitude; the Celsius and Kelvin scales also use a common degree. The Kelvin scale has become the international standard for measuring thermodynamic temperature changes.

The Celsius scale defines the freezing and boiling points of water as separated by 100 degrees. The triple point of water, at which water ice, liquid water, and water vapor are at thermodynamic equilibrium, is 0.01 degrees Celsius, or 273.16 kelvins. The Kelvin scale, developed by Lord Kelvin (1824–1907), was designed to be an absolute thermodynamic temperature scale, based on the linear relationship between temperature and volume of an ideal gas, with absolute zero defined as the point at which the volume of

TRIPLE POINT

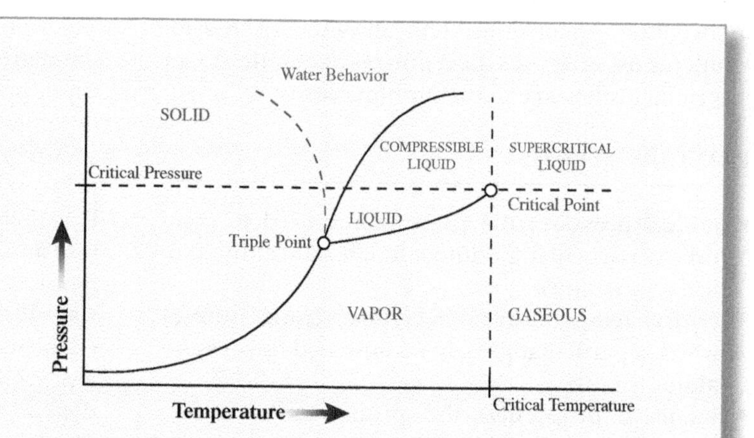

The triple point represents the temperature and pressure required for a substance to exist in 3 states (solid, liquid, and gaseous) at one time. Water behaves differently from most substances; as pressure increases it becomes liquid at lower temperatures.

the gas would become zero and all molecular movement would cease. In 1954, the international General Conference on Weights and Measures (*Conférence générale des poids et mesures*, or CGPM) defined the triple point of water as one of two fundamental fixed points on the Kelvin scale, the other being absolute zero. In 1967–68, the CGPM further established the magnitude of one kelvin as exactly 1/273.16 of the thermodynamic temperature of the triple point.

Richard M. Renneboog, MSc

BIBLIOGRAPHY

Askeland, Donald R., Wendelin J. Wright, D. K. Bhattacharya, and Raj P. Chhabra. *The Science and Engineering of Materials.* Boston: Cengage Learning, 2016. Print.

Daniels, Farrington, and Robert A. Alberty. *Physical Chemistry.* 3rd ed. New York: Wiley, 1966. Print.

Haynes, W. H., PhD, David R. Lide, PhD, and Thomas J. Bruno, PhD, eds. *CRC Handbook of Chemistry and Physics,* 96th Edition. New York: CRC/Taylor & Francis Group, 2015. Print.

Jones, Mark Martin. *Chemistry and Society.* 5th ed. Philadelphia: Saunders College Pub., 1987. Print.

Kean, Sam. *The Disappearing Spoon: And Other True Tales of Madness, Love, and the History of the World from the Periodic Table of the Elements.* New York: Little, Brown, 2010. Print.

Laidler, Keith J. *Chemical Kinetics.* 3rd ed. New York: Harper, 1987. Print.

Myers, Richard. *The Basics of Chemistry.* Westport: Greenwood, 2003. Print.

V

VALENCE BOND THEORY

FIELDS OF STUDY

Physical Chemistry

SUMMARY

Valence bond theory is a theoretical model to describe the shapes and interactions of atoms on the basis of shared pairs of electrons. Valence Shell Electron-Pair Repulsion (VSEPR) theory describes molecular structures as those that minimize the force of repulsion between electron pairs in the bonds and orbitals about a particular atom. Different atomic orbitals described by wave mechanics undergo hybridization to form an equal number of orbitals of equal energy and geometry.

PRINCIPAL TERMS

- **covalent bond:** a type of chemical bond in which electrons are shared between two adjacent atoms.
- **hybridize:** combine orbitals of similar quantum energy and different geometry to form an equal number of orbitals of equal quantum energy and identical geometry
- **Lewis structure:** a simplified representation of the bonds and unbonded electron pairs associated with the atoms of a molecule.
- **lone pair:** two valence electrons that share an orbital and are not involved in the formation of a chemical bond; also called a nonbonding pair.
- **repulsion:** an oppositional force that pushes two entities apart, such as the electrostatic repulsion between particles of like electrical charge.
- **stereochemistry:** the relative spatial arrangement of atoms and bonds in a molecular structure
- **structural formula:** a graphical representation of the arrangement of atoms and bonds within a molecule.
- **valence:** the maximum number of bonds an atom can form with other atoms based on its electron configuration; also called valency.
- **valence electron:** an electron that occupies the outermost or valence shell of an atom and participates in chemical processes such as bond formation and ionization.
- **valence shell:** the outermost energy level occupied by electrons in an atom.
- **VSEPR theory:** the Valence-Shell Electron-Pair Repulsion theory, which states that electron pairs in the valence shell of an atom will arrange themselves so that the electrostatic repulsion between them is minimized

ELECTRONS IN ATOMS

If light passes through a material before entering a prism, it is seen that some specific wavelengths are missing from the light emerging from the prism. These missing wavelengths always occur in specific patterns or series, such as the Balmer series and Rydberg series, associated with the movement of electrons within the atoms of the intervening material. In the modern theory of atomic structure, based on quantum mechanics, the energy associated with these specific wavelengths defines the region within the atom in which a particular electron can exist. Each of these atomic orbitals is described mathematically as corresponding to a specific shape and orientation in the atom, with the nucleus of the atom in the central position. The simplest orbital geometry is described as spherically symmetric about the nucleus, and is termed the *s* orbitals. The next orbital series is termed the *p* orbitals, consisting of three figure-eight, or dumbbell, shapes oriented at 90 degrees relative to each other. The five *d* orbitals are next, and have a more complex four-lobed shape. These are followed by the seven *f* orbitals, which have even more complex eight-lobed shapes. Each orbital can contain no more than two electrons, and the valence of the particular atom is determined by the number of electrons in

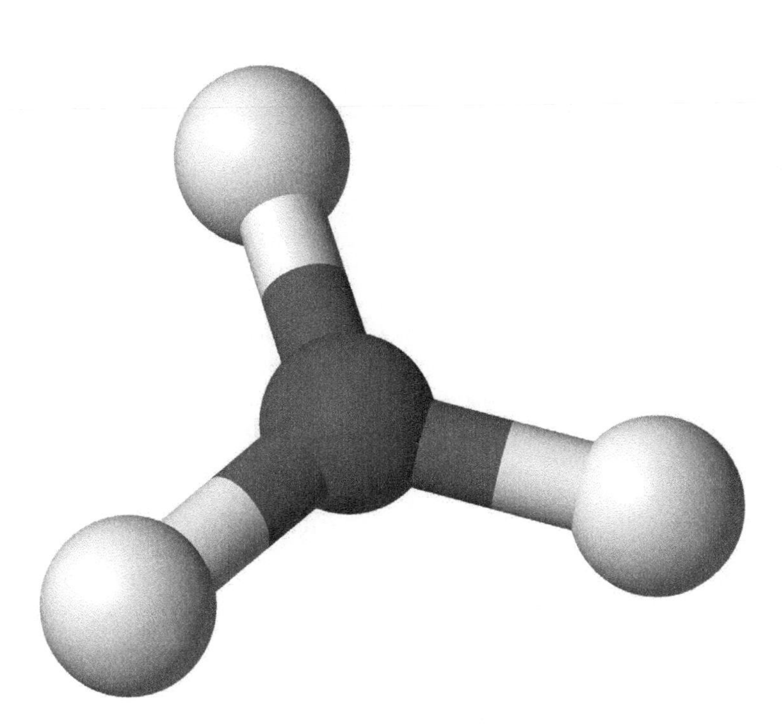

Trigonal-3D-balls. Licensed under Public Domain via Wikimedia Commons

its highest occupied series of atomic orbitals. The important feature of each orbital series is that its component orbitals correspond to a strict geometric arrangement about the nucleus. Yet spectroscopic measurements of molecules demonstrate that the arrangement of atoms within those molecules, or their stereochemistry, do not strictly adhere to the defined geometries of atomic orbitals. For example, a covalent bond between a carbon atom and a hydrogen atom in methane, CH_4, is formed when the valence electron of the hydrogen atom is shared in the valence shell of the carbon atom. The four C—H bonds that are formed are not at 90° relative to each other, but at an angle of about 109.5°. This is not predicted by the mathematics that describe the orbitals of an isolated atom, and requires that the one spherical *s* orbital and three dumbbell-shaped *p* orbitals combine at an average energy level, or hybridize, to form four sp^3 orbitals of equal energy. Filling all stable atomic orbitals with two shared or unshared electrons to form bonds is the basis of VSEPR theory, the Valence-Shell Electron Pair Repulsion theory. The basic premise of VSEPR theory is that the force of repulsion between the pairs of electrons in the bonds and the lone pairs of electrons about an atom drives the bonds to assume an orientation about the central atom in which the force of repulsion is minimized. In methane, and all other such carbon-centered systems, that orientation is achieved at the mutual angle of 109.5°. The structural formula and the distribution of electrons in molecules based on VSEPR rules are readily represented as Lewis structures or Lewis dot diagrams.

MULTIPLE BONDS AND RESONANCE

One of the shortcomings of valence bond theory is the manner in which it addresses compounds in which bonds between some atoms cannot be described adequately by VSEPR rules. These rules are efficient at describing single bonds between atoms on the basis of pairs of shared electrons in the outermost valence shells of those atoms, and calculated bond energies agree with observational measurements of those bond energies. For other compounds, however, there are significant differences between calculated and observed bond energies. The simple compound ethylene (C_2H_4), for example, has an observed carbon-carbon bond energy that does not agree with the value calculated on the basis of two single bonds between those two atoms. Ethyne, or acetylene, (C_2H_2) demonstrates an even greater discrepancy in that regard. Valence bond theory accommodates these discrepancies by positing the formation of a second type of single bond, and classifying their combinations as double or triple bonds. This second type of bond is theoretically formed by the sideways overlap of *p* orbitals that have not been hybridized. Double and triple bonds in VB theory are static bonds in the same sense as single bonds. A difficulty arises when there are systems in which such orbital combinations exist on more than two adjacent atoms. The classic case of this situation is the compound benzene (C_6H_6). Enthalpy calculations for benzene based on a system of three C—C single bonds alternating with three C==C double bonds do not agree with observed enthalpy measurements for the compound. If a diagram of the molecule as a hexagon is drawn, a second diagram can be drawn in which the bonds are shifted

SAMPLE PROBLEM

Calculate the total bond energy of benzene (C_6H_6) on the basis of multiple single bonds and compare it to the bond energy calculated with C=C double bonds using the following bond energy information:

C—C	352 kJ.mole^{-1}	(80.5 kcal.mole^{-1})
C—H	411 kJ.mole^{-1}	(98.2 kcal.mole^{-1})
C=C	607 kJ.mole^{-1}	(145 kcal.mole^{-1})

Answer:

The benzene molecule has six C—H single bonds, three C—C single bonds and three C=C double bonds. If the double bonds each consist of two single bonds, the total number of C—C single bonds is nine.

Therefore the total bond energy as single bonds would be

(9 X 352) + (6 X 411) = 5634 kJ.mole^{-1} (1313.7 kcal.mole^{-1})

As three C—C single bonds and three C=C double bonds, the total bond energy would be

(3 X 607) + (3 X 352) + (6 X 411) = 5343 kJ.mole^{-1} (1277 kcal.mole^{-1})

This result suggests that the C=C double bond is not composed of two C—C single bonds.

about the hexagon to their adjacent positions. The two structures these represent are exactly equivalent, and VB theory posits that the structure and bonding within the actual molecule resonates between these two structures so rapidly that they become indistinguishable from each other. The molecular structures of other compounds also demonstrate resonance. All such compounds are better described by molecular orbital (MO) theory.

THE IMPORTANCE OF VALENCE BOND THEORY

Valence bond theory provides a first step in gaining a sound understanding of the chemical nature of the world. Because all matter is believed to be composed of atoms, the universe and all that it contains is by definition chemical in nature. All of the properties and characteristics of matter and the changes that can be brought about in chemical reactions, including life itself, are therefore describable through the interaction of atoms via their valence shell electrons. While VB theory does not provide an ideal means of understanding how those chemical interactions occur, it does provide a satisfactory means of predicting the outcomes of the vast majority of chemical reactions.

Richard M. Renneboog, MSc

BIBLIOGRAPHY

Anslyn, Eric V., and Dennis A. Dougherty. *Modern Physical Organic Chemistry.* University Science Books, 2006. Print.

Brescia, Frank, John Arents, and Herbert Meislich. *Fundamentals of Chemistry.* 4th ed. New York, NY: Academic Press, 1980. Print.

Cooper, David L., ed. *Valence Bond Theory.* Elsevier, 2002. Print.

Gillespie, Ronald J., and István Hargittai. *The VSEPR Model of Molecular Geometry.* Dover, 2012. Print.

Kenkel, John. Basic *Chemistry Concepts and Exercises.* Boca Raton, FL: CRC Press, 2011. Print.

Mikulecky, Peter J., Michelle Rose Gilman and Kate Brutlag. *AP Chemistry for Dummies.* Hoboken: Wiley, 2009. Print.

Pfennig, Brian W. *Principles of Inorganic Chemistry.* Hoboken, NJ:Wiley, 2015. Print.

Shaik, Sason S., and Philippe C. Hiberty, A *Chemist's Guide to Valence Bond Theory.* Hoboken: Wiley, 2008. Print.

VALENCE SHELL

FIELDS OF STUDY

Organic Chemistry; Inorganic Chemistry; Physical Chemistry

SUMMARY

The characteristics of the valence electron shells of atoms are discussed. The valence shell holds the valence electrons of an atom. The number of valence electrons determines the chemical behavior of the atom and the number and type of chemical bonds it can form with other atoms.

PRINCIPAL TERMS

- **atomic model:** a theoretical representation of the structure and behavior of an atom based on the nature and behavior of its component particles.
- **electron:** a fundamental subatomic particle with a single negative electrical charge, found in a large, diffuse cloud around the nucleus.
- **electronegative:** describes an atom that tends to accept and retain electrons to form a negatively charged ion.
- **ionization energy:** the amount of energy required to remove an electron from an atom in a gaseous state.
- **octet rule:** the tendency of atoms when bonding to either accept or donate electrons in such a way that they end up with eight electrons in the outermost electron shell.
- **oxidation state:** a number that indicates the degree to which an atom or ion in a chemical compound has been oxidized or reduced.
- **unpaired electron:** a valence electron that occupies an orbital by itself and is not involved in the formation of a chemical bond.

The Modern Atomic Model

The electron was first identified as a charged particle by British physicist J. J. Thomson (1856–1940) in 1897, and the proton was discovered in 1917 by British physicist Ernest Rutherford (1871–1937). These discoveries revealed that atoms are not the indivisible absolutes that they were previously believed to be; rather, they are composite entities containing smaller, "subatomic" particles. The existence of the third essential subatomic particle, the neutron, was demonstrated by a third British physicist, James Chadwick (1891–1974), in 1932. With this last discovery, the observed behavior of the atom seemed to agree with the theoretical calculations of the developing field of quantum mechanics.

In simple terms, quantum mechanics describes the behavior of subatomic particles, such as electrons, on an infinitesimal scale. At the time, electrons were envisioned as being bound to the nucleus in the same way that the moon is bound to Earth, and terms such as "orbit" and "orbital" are still used to describe the relationship of electrons to the nucleus. However, it is now understood that electrons, like light, exhibit wave-particle duality; that is, they behave as both physical, massive (in the sense of having mass) particles and massless waves. This is perhaps the most difficult aspect of the modern atomic model to grasp, but it is fundamental to the behavior of electrons in atoms.

Early observations showed that electrons in an atom can have only very specific energies and that the transition of an electron from a lower energy level to a higher one can only occur when the electron acquires a specific amount of energy. Conversely, when the electron drops from a higher energy level to a lower one, it has to give up exactly that same amount of energy. The different energy levels that electrons can occupy in an atom are termed electron shells, and each one is identified by its principal quantum number (n), so that the first electron shell is the $n = 1$ shell, the second is the $n = 2$ shell, and so on. Each shell is divided into subshells that can hold only a specific number of one type of orbital (*s*, *p*, *d*, or *f*), and each orbital can contain no more than two electrons at any time.

In any neutral atom, the number of electrons is equal to the atomic number of the element, which is simply the number of protons in the nucleus; in other words, the numbers of electrons and protons are equal. The electrons are ordered in pairs in the orbitals of successive electron shells. The outermost shell of most of the noble gases (helium, neon, argon, krypton, xenon, and radon) contains a full complement of eight electrons; the exception is helium, which has only two electrons total. All other atoms have fewer than eight electrons in their outermost shells. The number of electrons in this shell determines the maximum number of bonds that a particular atom can form with other atoms. This is known as the valence of the atom, and the outermost (in most cases) electron shell is correspondingly called the valence shell—or, more accurately, its outermost electrons are known as valence electrons. (In most elements, valence electrons only occupy the outermost electron shell, but for the group of elements known as transition metals, they may also be found in inner shells.)

Valence Electrons and the Periodic Table

The number of electrons in the valence shell of an atom determines the chemical behavior of that atom. An atom is at an energy minimum, and therefore at its most stable, when the valence shell is completely filled with the eight electrons it is allowed to contain.

VALENCE SHELL

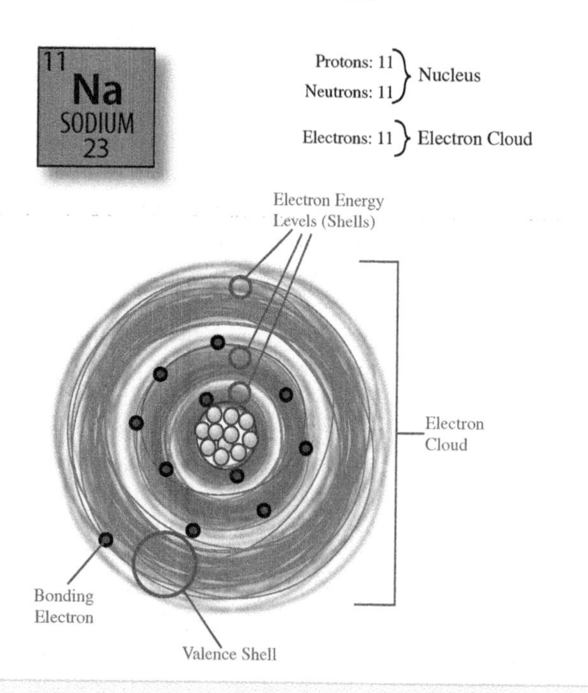

Electrons of an atom travel around the nucleus in the electron cloud. Most of the time electrons can be found at specific levels around the nucleus. In the above model, darker bands within the electron cloud, called electron shells, are locations where electrons are most likely to be found. The outermost shell is the valence shell.

This is the basis of the octet rule, and it is also why the noble gases are the most chemically inert, or nonreactive, elements. Elements that have six or seven electrons in their valence shell readily accept extra electrons to achieve the electron configuration of the next noble-gas element in the periodic table, while elements with only one or two electrons in their valence shell readily donate those electrons to achieve the electron configuration of the noble-gas element that precedes them. The more electrons that an atom has in its valence shell, the more electronegative it is said to be, meaning that it easily accepts extra electrons from other atoms. Conversely, atoms that easily give up their valence electrons are said to be electropositive. The energy required to remove an electron from the valence shell to produce a cation, or positively charged ion, is called the ionization energy. The energy necessary to remove the first electron from the neutral atom is called the first (or initial) ionization energy, the energy required to remove a second electron is the second ionization energy, and so on. Removing two electrons from the neutral atom at once requires energy equivalent to the sum of the first and second ionization energies.

The elements of the periodic table are arranged in vertical columns called groups and horizontal rows called periods. The periods are arranged in order of increasing principal quantum number, while the groups are arranged in order of their valence-shell electrons. The valence shell of the first two groups is the *s* subshell of the electron shell identified by the corresponding principal quantum number. (Although the letters *s*, *p*, *d*, and *f* technically identify the individual orbitals, each of which can hold only two electrons, they are often used to refer to the subshell containing the orbitals as well.) Group 1 elements (the lithium group) have just one electron in this orbital and very readily give it up to form the corresponding cation, which has a 1+ charge. When this happens, the valence shell of the ion has the same electron configuration as the noble-gas element immediately before it. For example, the electron distribution in the valence shell of the sodium cation (Na^+), which has the atomic number 11, is the same as that of the neon

SAMPLE PROBLEM

Write out the electron configuration of the element gallium (atomic number 31). Identify its group and period in the periodic table. Identify the subshell containing the valence electrons by its principal quantum number and orbital. How many valence electrons does it have? Can gallium form an ion with a 3+ charge?

Answer:

The order of electron subshells is

$$1s\ 2s\ 2p\ 3s\ 3p\ 3d\ 4s\ 4p\ 4d\ 4f\ 5s\ 5p\ 5d\ 5f$$

The *s* subshells can hold two electrons, the *p* subshells six, the *d* subshells ten, and the *f* subshells fourteen. There are thirty-one electrons in the gallium atom. The electron configuration is therefore

$$1s^2 2s^2 2p^6 3s^2 3p^6 3d^{10} 4s^2 4p^1$$

where the superscript number is the number of electrons occupying each subshell. The valence shell is the $n = 4$ shell, and it contains three valence electrons, in the 4*s* and 4*p* orbitals. Gallium should be able to form a 3+ ion, and it frequently does.

atom, which has the atomic number 10. The group 1 elements have the lowest ionization energy of all the elements. Group 2 elements (the beryllium group) have two electrons in their valence *s* orbital, which they can give up almost as easily to form cations with a 2+ charge.

The last six groups (13–18) have their valence electrons in the *p* subshell, which contains three *p* orbitals for a maximum possible total of six electrons. The elements in groups 13, 14, and 15 (the boron, carbon, and nitrogen groups) neither accept nor donate electrons readily to form ions. These elements most commonly form covalent bonds with other atoms by sharing unpaired electrons, which are single valence electrons that inhabit an orbital alone and are not part of a chemical bond. An unpaired electron can form a covalent bond by sharing the orbital of an unpaired electron on another atom (at which point, of course, they cease to be called unpaired electrons). The group 16 elements (the oxygen group) readily accept two electrons to form anions, or negatively charged ions, with a 2− charge. The group 17 elements (the halogens, also called the fluorine group) are the most electronegative of the elements and very quickly accept an electron from another atom to form a halide ion with a 1− charge. Finally, apart from helium, the noble-gas elements (group 18) have two *s* electrons and six *p* electrons, making a complete octet. In all cases, whether through the formation of ions or by sharing electrons covalently, an atom will achieve a complete valence-shell electron configuration.

In between these extremes are the transition metals, including the lanthanides and the actinides. The valence electrons of these elements are distributed between their *d* and *f* orbitals. These orbitals are close together in energy, much more numerous than *s* and *p* orbitals (the subshells contain five and seven orbitals, respectively), and do not conform to the octet rule.

OXIDATION STATES

A useful method for tracking valence electrons is to determine each atom's specific oxidation state. The oxidation state of an atom indicates the number of electrons it has either donated to or accepted from other atoms in order to form chemical bonds. This is especially useful in balancing redox (reduction-oxidation) reactions.

Richard M. Renneboog, MSc

BIBLIOGRAPHY

Douglas, Bodie E., Darl H. McDaniel, and John J. Alexander. *Concepts and Models of Inorganic Chemistry.* 3rd ed. New York: Wiley, 1994. Print.

Hendrickson, James B., Donald J. Cram, and George S. Hammond. *Organic Chemistry.* 3rd ed. New York: McGraw, 1970. Print.

Kean, Sam. *The Disappearing Spoon: And Other True Tales of Madness, Love, and the History of the World from the Periodic Table of the Elements.* New York: Little, Brown, 2010. Print.

Miessler, Gary L., Paul J. Fischer, and Donald A. Tarr. *Inorganic Chemistry.* 5th ed. Upper Saddle River: Prentice Hall, 2013. Print.

Morrison, Robert Thornton, and Robert Neilson Boyd. *Organic Chemistry.* 7th ed. Englewood Cliffs, N.J.: Prentice Hall, 2003. Print.

Myers, Richard. *The Basics of Chemistry.* Westport: Greenwood, 2003. Print.

Powell, P., and P. L. Timms. *The Chemistry of the Non-Metals.* London: Chapman, 1974. Print.

Wehr, M. Russell, James A. Richards Jr., and Thomas W. Adair III. *Physics of the Atom.* 4th ed. Reading: Addison-Wesley, 1984. Print.

Z

ZONE REFINING

FIELDS OF STUDY

Metallurgy; Crystallography

SUMMARY

Zone refining is a method of obtaining materials in a state of extremely high purity. The process depends in the migration of components between solid and liquid phases, in accord with the phase rule. In the process a narrow band of molten material is moved progressively along the length of a bar of material. A variation of the process is used to produce all silicon wafers used in the manufacture of computer chips and integrated circuits.

PRINCIPAL TERMS

- **alloy:** a mixture of a metal and at least one other element, often another metal; also known as a solid solution. .
- **composition:** the identities and relative proportions of different elements or components in a compound, mixture, or other material.
- **concentration:** the amount of a specific component present in a given volume of a mixture.
- **concentration gradient:** the gradual change in the concentration of solutes in a solution across a specific distance.
- **degrees of freedom, variance:** the smallest number of independent variables that must be specified to completely describe the state of a system.
- **element:** a form of matter consisting only of atoms of the same atomic number.
- **equilibrium constant:** a numerical value characteristic of a particular equilibrium reaction, defined as the ratio of the equilibrium concentration of the products to that of the reactants.
- **fusion:** change of state from solid to liquid.
- **intensive properties:** the properties of a substance that do not depend on the amount of the substance present, such as density, hardness, and melting and boiling point.
- **phase rule:** a simple mathematical relationship defining the variance of a system in terms of the number of phases in the system and their respective concentrations.

Intensive Properties and the Phase Rule

The zone refining principle was first reported in 1952, and quickly developed into a technique for producing ultra-pure materials. One of the intensive properties of a material in a solution or mixture is called the chemical potential. In a solution, the chemical potential of a component drives that particular material to migrate from a region of higher concentration toward one of lower concentration. This is most evident in the process of diffusion, and can be readily demonstrated by placing a drop of food coloring into a glass of water. With no stirring or other agitation to bring about mixing, the food coloring will be seen to disperse spontaneously, though slowly, throughout the entire volume of water. Not all solutions are liquids, however. An alloy of two or more metals is a solid solution that has been prepared by combining the metals in their liquid state and allowing the liquid solution to solidify. The composition of the alloy is defined by the concentration of each component. The term alloy can actually be applied to any material that consists of two or more components mixed together in an intimate combination, including compounds that are almost pure except for the presence of a small amount of contaminants. In all cases, each component acts as a contaminant of the other components, and the presence of an impurity normally reduces the melting point of the alloy in comparison to any of the components in their pure form. A pure element or compound will melt at a very specific temperature. But when two or more such materials are alloyed together the temperature of the alloy is reduced by an amount that depends on the relative concentrations of the components, and

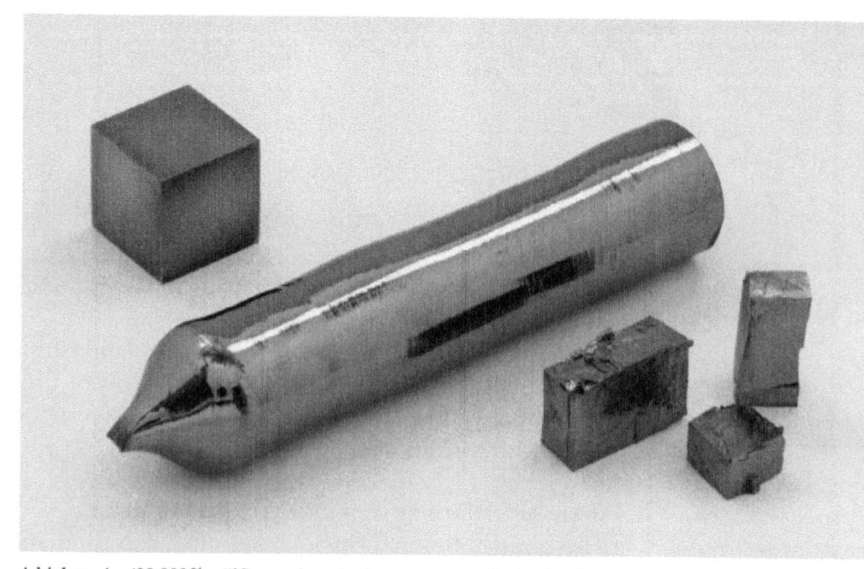

A high-purity (99.999% = 5N) tantalum single crystal, made by the floating zone process (cylindrical object in the center). By Alchemist-hp via Wikimedia Commons

melting, or fusion, occurs over a broadened range of temperatures. When melting of the alloy begins, the alloy does not melt uniformly and two phases are present rather than just one. The relative solubilities of the different components become effective, and this is the principle that makes the zone refining process capable of producing ultra-pure materials. The behavior of a system having more than one phase is defined in accordance with the phase rule, which describes the number of variables required for a complete description of the state of the system.

How Zone Refining Works

In the zone refining process a solid bar or rod of a material to be purified is heated by a heating coil so that just a narrow section of the material melts. The heating coil is moved slowly along the length of the bar so that the melted zone moves along with it. As the melt zone moves, the molten material solidifies again behind it and new material is melted ahead of it. In this way the condition of having three phases that are in equilibrium with each other is maintained. The concentration of impurities in the re-solidified region behind the melt zone is different from their concentration in the melt zone. Similarly, their concentration in the solid region ahead of the melt zone is also different. Thus, there are three concentration gradients determined by the relative concentrations of the impurities. The net result is that the chemical potential of the impurities in the bar drives them to move from one phase into the other progressively along the length of the bar. At the end of the process, the vast majority of impurities exist in the liquid phase at the end of the bar, while the material in the bar itself can have achieved a purity of 99.99999%. This level of purity has allowed very precise determination of the chemical and physical properties of each element and many compounds that can exist as a solid under conditions that allow the zone refining process to be carried out.

How the Phase Rule Works

The phase rule defines the number of independent variables, such as temperature and pressure, that must be used to completely describe that state of a

SAMPLE PROBLEM

Calculate the variance of a system having two components and three phases in equilibrium.

Answer:

The number of components, c, is 2.

The number of phases, p, is 3

The number of degrees of freedom, or variance, is calculated as

$$v = c - p + 2$$

$$= 2 - 3 + 2$$

$$= 1$$

Thus only one of temperature, pressure or concentration must be specified as a constant or fixed value in describing the properties and behavior of the system.

multiphase system. The calculation depends on the number of equilibrium conditions within the system, each of which has its own equilibrium constant. In addition, each phase (p) has its own concentration. These determine the variance (v), or degrees of freedom, of a system with a certain number of components (c). For a system in which only temperature, pressure and concentration are the independent variables, the variance of the system is calculated as:

$$v = c - p + 2$$

The variance of the system is related to the chemical potential of the components in each phase, and to the partial molal Gibbs free energy. These properties drive the migration of components from one phase to another in the zone refining process. The variance calculated in this way specified the number of variable factors that must be arbitrarily established in order for the system to be completely described.

THE IMPORTANCE OF ZONE REFINING

Zone refining provides access to materials in the highest known purity. This has been essential to determining the properties of many of the elements, and particularly of metallic elements. Chemists work with chemical reactions that are specific to the materials stated in a properly balanced chemical equation. The presence of impurities in the materials used has deleterious effects on the viability of those reactions when carried out, and ultimately interferes with understanding the manner in which the actual material behaves and how that knowledge can be applied in new and beneficial ways. The use of ultra-high purity metals such as gold in the preparation of metal-based drugs, for example, significantly reduces the possibility of any effects that might arise from the presence of trace contaminants. In a more universal application, a variation on zone refining is used to produce the large crystals of pure silicon that are used in the manufacture of computer chips and integrated circuits. In this process, a seed crystal of pure silicon is introduced to a molten mixture containing specific amounts of silicon, germanium and other elements required to induce the desired electronic properties. The crystal is turned slowly and withdrawn from the molten mass in such a way that a long, single crystal is extracted from the two-phase mixture. At the solid-liquid interface, as in zone refining, the material solidifying on the forming crystal is extremely pure. The crystal is subsequently sliced into wafers on which the miniscule electronic circuits of a computer chip will be etched. The process has been essential to the development of computer and electronic technology, and will be continue to be for many years to come.

Richard M. Renneboog, MSc

BIBLIOGRAPHY

Amos, S.W. *Principles of Transistor Circuits.* New York, NY: Elsevier, 2013. Print.

Atkins, Peter, and Julio de Paula. Atkins' Physical Chemistry. 10th ed. New York: Oxford University Press, 2014. Print.

Glicksman, Martin Eden. *Principles of Solidification. An Introduction to Modern Casting and Crystal Growth Concepts.* New York, NY: Springer, 2010. Print.

Gupta, K.M., and Nishu Gupta. *Advanced Electrical and Electronic Materials. Processes and Applications.* New York, NY: Wiley/Scrivener Publishing, 2015. Print.

Mühlbauer, Alfred. *History of Induction Heating & Melting.* Essen, GER: Vulkan-Verlag, 2008. Print.

Porter, David A., Kenneth E. Easterling, and Mohamed Y. Sherif. *Phase Transformations in Metals and Alloys.* 3rd ed. Boca Raton, FL: CRC Press, 2009. Print.

Walas, Stanley M. *Phase Equilibria in Chemical Engineering.* Stoneham, MA: Butterworth-Heinemann, 2013. Print.

Appendices

INTRODUCING THE PERIODIC TABLE

The periodic table is one of the first tools that students encounter in the study of chemistry. It is also the single most important depiction of the chemical elements, because each frame shows the principles underlying every one of the element's behaviors.

The periodic table organizes the different elements in periods according to the number of electron shells the elements have and in groups according to the configuration of their valence electrons. The information presented in the periodic table frame begins with the element symbol, the familiar one or two-letter abbreviation letter or letter pair used to represent an individual atom in chemical formulas. These symbols are uniquely assigned to each element, and are usually derived from the internationally recognized name of the element assigned by the International Union of Pure and Applied Chemistry (IUPAC). Generally, each symbol is taken from the universally recognized element name. For example, H stands for hydrogen, Cl for chlorine, and so on. Some symbols seem to defy this logic, however. The element symbol for tungsten is W, a letter which certainly does not appear anywhere in the word; the symbol comes from the element's earlier name, wolfram. Similarly, the symbols for copper (Cu) and iron (Fe) come from their original Latin names, cuprum and ferrum, respectively. Such symbols demonstrate a certain respect for the long tradition of science in the study of materials and the history of chemistry. Elements with very large atomic masses may be temporarily be assigned a three letter symbol, e.g. Uup, (for ununperntium or element number 115) until a common name is agreed upon.

Structure of Periodic Table

The periodic table presents the known elements ordered by the number of protons in the nucleus of an atom, in other words its atomic number, which is used to uniquely identify each element. The atomic number, as defined by the modern atomic theory, is the precise number of protons that are contained in the nucleus of any given atom of the element. Hydrogen, which is atomic number 1, has only a single proton in its nucleus; helium, atomic number 2, has two; and so on. This number generally appears above the element symbol in the periodic table.

Most elements exist in more than one isotope. The isotopes (nuclides) of an element have the same atomic number but differ in the number of neutrons. When it is important to distinguish between isotopes, the mass number is written as to the left of the element symbol and above the line and the atomic number written to the left and below the line. Thus hydrogen has three common isotopes

${}^{1}_{1}H$, also known as protium,
${}^{1}_{2}H$, deuterium, and
${}^{1}_{3}H$ or tritium, which is radioactive.

The atomic mass of each atom is almost exactly equal to the number of protons and neutrons in its nucleus and thus should be a whole number, as subatomic particles cannot be divided and still maintain their identity. Yet almost all atomic masses, which are displayed below the element symbols, are stated as decimals. This fractional value is due to the natural presence of isotopes of different elements, which are atoms that have the same number of protons in their nucleus but a different number of neutrons, thus changing the atomic mass of that particular atom. The atomic weight of an element is a proportional average of the atomic masses of the different isotopes, weighted according to the natural occurrence of each one. The element chlorine, for example, occurs naturally as two isotopes, one with a mass of 35 atomic mass units (u) and abundance of about 24 percent and the other with a mass of 37 u and abundance of 76 percent, giving it an atomic weight of 35.45. The same principle applies to all other naturally occurring elements.

Strictly speaking there are two other reasons why the atomic mass of an element is not simply an integral multiple of the atomic mass of hydrogen. The neutron has, in fact a slightly different mass than that of a proton plus the electron needed for overall neutrality. In addition, when ever an element is formed by nuclear fusion or by the radioactive decay of a heavier element; some matter is converted into energy according to Einstein's famous formula E=mc2. These effects are, however smaller then that of having several different isotopes, and will not be considered here.

According to modern atomic theory, electrons can possess only very specific amounts of energy, and only a maximum specified number of electrons can occupy each energy level, or electron shell, which corresponds to a defined area around the atomic nucleus. Each horizontal row, or period, represents the shell that contains the valence electrons of the elements within that period. The progression of the principal energy levels—that is the outermost electron shells—is typically shown on the extreme right-hand side of the periodic table, at the end of each period. Thus, the first period, containing hydrogen (H) and helium (He), corresponds to the first energy level, which can hold only two electrons; the second period, containing lithium (Li), beryllium (Be), boron (B) carbon (C), nitrogen (N), oxygen (O), fluorine (F), and neon (Ne), corresponds to the second level; and so on. At the bottom of the periodic table are two sets of elements, the lanthanides and the actinides, which would normally be part of the sixth and seventh periods but are instead shown at the bottom to save space.

The hydrogen atom, for which the quantum mechanical wave function can be found exactly, is taken as the model for electron in atoms. The nth energy level of hydrogen has 2n2 energy states. Thus for n=1 there are only two possible states, making up the 1s energy level. For n=2 there are 8 possible states, 2 in the *2s* level and 6 in the *2p*. For n=3, there are 18 possible states, 2 in the *2s*, 6 in the *2p* and 10 in the *2d*. For hydrogen, with only one electron all states with the same n have the same energy, but for many electron atoms, the *s* orbitals are owe in energy then the *p* orbitals, which are in turn lower in energy than the *d* orbitals a fact usually attributed to the fact that the *s* orbitals allow the electron to approach closer to the nucleus. For many electron atoms the orbitals fill up in the general order 1*s*, 2*s*, 2*p*, 3*s*, 3*p*, 4*s*, 3*d*, 4*p* , 5*s*, 4*f*, 4*d*, 5*p*, and so on.

The elements in the periodic table are also arranged in vertical columns, or groups. The elements in each group all have the same number of valence electrons and therefore are similarly reactive. The first column, for example, comprises the alkali metals: hydrogen, lithium, sodium (Na), potassium (K), rubidium (Rb), cesium (Cs), and francium (Fr). Each has just one electron in its outermost electron shell, which they all readily give up to form an ion with single positive charge. Accordingly, the alkali metals all form very similar types of compounds, undergo very similar reactions, and so on. Similarly, the group 2 elements—beryllium, magnesium (Mg), calcium (Ca), strontium (Sr), barium (Ba), and radium (Ra)—all have two valence electrons, which they readily give up to form ions with two positive charges.

As can be seen from this, the groups and periods of the periodic table contain essentially all of the information needed to understand the behavior of the chemical elements.

History of the Periodic Table

As the science of chemistry developed, many scientists recognized that certain elements exhibited similar chemical behaviors, and attempts were made to organize them in some way. Early attempts included De Chancourtois Telluric helix and Newlands law of octaves. However, because so many of the chemical elements were still unknown, these scientists had little success in establishing a meaningful system. As more materials came to be recognized as actual chemical elements, the periodicity of their behavior became more apparent. In the late nineteenth century, Russian chemist Dmitri I. Mendeleev (1834–1907) produced the first correctly ordered periodic table, by placing the elements in order of their atomic mass, starting a new period with each of the alkali metals and leaving gaps where elements seemed to be missing. Since the chemical properties of the elements suggested just a few interchanges of neighboring elements, Mendeleev made those and came up with a viable periodic table by 1869. Mendeleev considered the atom as truly indivisible and dismissed early reports of the first sub atomic particle, the electron.

Theoretical understanding of the observed periodicities came with four twentieth century discoveries: X-rays, the quantum mechanics of atomic structure, the existence of the neutron, and the discovery of intrinsic angular momentum or spin. Wilhelm Konrad Roentgen discovered x-rays in 1895, though there was some confusion about the nature of the rays for about a decade. By 1911, Rutherford had demonstrated that all the positive charge in the atom was concentrated in a tiny nucleus, and Neils Bohr had applied the quantum hypothesis to the allowed electron orbits in hydrogen. In 1913, Henry Moseley, a brilliant young Englishman who would soon lose his life in the battle for Gallipoli, used various metals as targets in x-ray tubes and found that the shortest wavelength produced decreased as one went along the periodic table, including the interchanges made by Mendeleev. By 1925 Erwin Schrodinger found a general equation for the behavior of very small particles. Although Schrödinger's equation could not be solved exactly for many electron atoms, approximate methods were soon found and it was clear that a picture of atoms as involving hydrogen-like orbitals was realistic. Each orbital was described by three quantum numbers: n (for the highest orbital, the row of the periodic table), an angular momentum quantum number which was 0 for *s* orbitals, 1 for *p* orbitals, 2 for *d* orbitals and so on, and a magnetic quantum number which described the behavior of the orbital in a magnetic field and could range from –l to +l. With the discovery of electron spin, also at about that time it became clear that at most two electrons could occupy any orbital. When the neutron was discovered a few years later the distinction between atomic mass and atomic number was complete and the few rearrangements made by Mendeleev justified

CHART OF THE NUCLIDES

The existence of most elements in several different isotopes is mainly a matter of nuclear physics, but has some implications for chemistry as well. The mass of different isotopes usually has only a small effect on the rates of chemical reactions, but deuterium (heavy hydrogen) does have twice the mass of ordinary hydrogen with the consequence that reactions taking place in heavy water may occur an order of magnitude slower than the same chemical reaction in aqueous solution. One result: heavy water, although chemically identical to ordinary drinking water, is in quantity a deadly poison, be cause of its effect on the rates of metabolic reactions.

Radioactive isotopes have an important role in medical and biological research as tracers. Allowing the exact sequence of molecular events to be determined. In radiocarbon dating one assumes that the production of !4C to occur at a constant rate due to cosmic rays and so the ratio of 12 C to !4C in once living plant matter, such as cotton or papyrus, could determine when the plant matter was grown.

Since organs like the thyroid gland absorb one element especially well. For the thyroid it is iodine, which is necessary to produce the hormone thyroxin. 131I is an isotope of iodine, which decays by beta particle emission with a half-life of 8 days. Such an isotope is ideal for therapeutic and diagnostic use since it is essentially gone from the patient's body after a few weeks.

A vital tool of the nuclear chemist is the chart of the nuclides, listing different properties of each atomic nucleus. The chart has as many columns as there are elements known and as many rows as there are isotopes of a given element. A plot of mass number against atomic number then shows all the possible transmutation of an element. An element that emits an alpha particle moves two squares to the right and four squares down in the table while a beta emitter moves one square to the right and one square up. The square corresponding to each isotope may be used to make information available on nuclear mass, decay modes, half lives, and so on. For nuclear medicine and archaeology the chart of the nuclides is indispensible.

THE PERIODIC TABLE AS A SOURCE OF ENTERTAINMENT

Chemists are a strange breed of human beings. They inhabit a world in which concepts like orbital hybridization, resonance, and electron spin are commonplace. The names of the 100 plus chemical elements read like the gazetteer to a foreign country. The periodic table has been printed on coffee cups, silk-screened onto men's ties, women's scarves and a host of teeand sweat-shirts. Perhaps it was inevitable that a night-club entertainer would set them to music. Look on the internet for Tom Lehrer's elements song (to the tune of Gilbert and Sullivan's Modern Major General) and several alternative visuals, and enjoy.

—Donald R. Franceschetti, Ph. D. and Richard M. Renneboog, MSc

BIBLIOGRAPHY

Douglas, Bodie E., Darl H. McDaniel, and John J. Alexander. *Concepts and Models of Inorganic Chemistry.* 3rd ed. New York: Wiley, 1994. Print.

Gribbin, John. *Science: A History, 1543–2001.* London: Lane, 2002. Print.

Haynes, W. H., PhD, David R. Lide, PhD, and Thomas J. Bruno, PhD, eds. *CRC Handbook of Chemistry and Physics,* 96th Edition. New York: CRC/Taylor & Francis Group, 2015. Print.

Jones, Mark Martin. *Chemistry and Society.* 5th ed. Philadelphia: Saunders College Pub., 1987. Print.

Kean, Sam. *The Disappearing Spoon: And Other True Tales of Madness, Love, and the History of the World from the Periodic Table of the Elements.* New York: Little, Brown, 2010. Print.

Mackay, K. M., R. A. Mackay, and W.Henderson. *Introduction to Modern Inorganic Chemistry.* 6th ed. Cheltenham: Nelson, 2002. Print.

Myers, Richard. *The Basics of Chemistry.* Westport: Greenwood, 2003. Print.

Powell, P., and P. L. Timms. *The Chemistry of the Non-Metals.* London: Chapman, 1974. Print.

PERIODIC TABLE OF THE ELEMENTS

Group→ ↓Period	1	2		3	4	5	6	7	8	9	10	11	12	13	14	15	16	17	18
	1 H																		2 He
	3 Li	4 Be												5 B	6 C	7 N	8 O	9 F	10 Ne
	11 Na	12 Mg												13 Al	14 Si	15 P	16 S	17 Cl	18 Ar
	19 K	20 Ca		21 Sc	22 Ti	23 V	24 Cr	25 Mn	26 Fe	27 Co	28 Ni	29 Cu	30 Zn	31 Ga	32 Ge	33 As	34 Se	35 Br	36 Kr
	37 Rb	38 Sr		39 Y	40 Zr	41 Nb	42 Mo	43 Tc	44 Ru	45 Rh	46 Pd	47 Ag	48 Cd	49 In	50 Sn	51 Sb	52 Te	53 I	54 Xe
	55 Cs	56 Ba	*	71 Lu	72 Hf	73 Ta	74 W	75 Re	76 Os	77 Ir	78 Pt	79 Au	80 Hg	81 Tl	82 Pb	83 Bi	84 Po	85 At	86 Rn
	87 Fr	88 Ra	†	103 Lr	104 Rf	105 Db	106 Sg	107 Bh	108 Hs	109 Mt	110 Ds	111 Rg	112 Cn	113 Uut	114 Fl	115 Uup	116 Lv	117 Uus	118 Uuo
			*	57 La	58 Ce	59 Pr	60 Nd	61 Pm	62 Sm	63 Eu	64 Gd	65 Tb	66 Dy	67 Ho	68 Er	69 Tm	70 Yb		
			†	89 Ac	90 Th	91 Pa	92 U	93 Np	94 Pu	95 Am	96 Cm	97 Bk	98 Cf	99 Es	100 Fm	101 Md	102 No		

TABLE OF THE ATOMIC WEIGHTS

The atomic weights, $A_r(E)$, of many elements are not invariant, but depend on the origin and treatment of the material. The standard atomic weight values and the uncertainties (in parentheses, following the last significant digit to which they are attributed) apply to elements from normal materials, where a normal material is a reasonably possible source for an element or its compounds in commerce, for industry or science; the material is not itself studied for some extraordinary anomaly and its isotopic composition has not been modified significantly in a geologically brief period (Wieser et al., 2011). The last significant figure of each tabulated value is considered reliable to ±1 except when a larger single digit uncertainty is inserted in parentheses following the atomic weight. For 12 of these elements, the standard atomic weight is given as an atomic-weight interval with the symbol [*a*, *b*] to denote the set of atomic-weight values in normal materials; thus, $a \leq A_r(E) \leq b$. The symbols *a* and *b* denote the lower and upper bounds of the interval [*a*, *b*], respectively. If a more accurate $A_r(E)$ value for a specific material is required, it should be determined. For 72 elements, $A_r(E)$ values and their evaluated uncertainties (in parentheses, following the last significant digit to which they are and attributed) are given. The footnotes to this table elaborate the types of variation that may occur for individual elements that may be larger than the listed uncertainties of values of $A_r(E)$ or may lie outside the values listed. Names of elements with atomic number 113, 115, 117, and 118 are provisional; they have been reported in the peer-reviewed, scientific literature, but they have not been named by IUPAC.

Alphabetic Order in English			Unabridged table		Abridged to five significant digits	
Element name	Symbol	Atomic number	Standard atomic weight	Footnotes	Standard atomic weight	Footnotes
actinium*	Ac	89				
aluminium (aluminum)	Al	13	26.981 5385(7)		26.982	
americium*	Am	95				
antimony	Sb	51	121.760(1)	g	121.76	g
argon	Ar	18	39.948(1)	g,r	39.948	g,r
arsenic	As	33	74.921 595(6)		74.922	
astatine*	At	85				
barium	Ba	56	137.327(7)		137.33	
berkelium*	Bk	97				
beryllium	Be	4	9.012 1831(5)		9.0122	
bismuth*	Bi	83	208.980 40(1)		208.98	
bohrium*	Bh	107				
boron	B	5	[10.806, 10.821]	m	[10.806, 10.821]	m
bromine	Br	35	[79.901, 79.907]		[79.901, 79.907]	
cadmium	Cd	48	112.414(4)	g	112.41	g
caesium (cesium)	Cs	55	132.905 451 96(6)		132.91	
calcium	Ca	20	40.078(4)	g	40.078(4)	g
californium*	Cf	98				
carbon	C	6	[12.0096, 12.0116]		[12.009, 12.012]	
cerium	Ce	58	140.116(1)	g	140.12	g
chlorine	Cl	17	[35.446, 35.457]	m	[35.446, 35.457]	m
chromium	Cr	24	51.9961(6)		51.996	
cobalt	Co	27	58.933 194(4)		58.933	

copernicium*	Cn	112				
copper	Cu	29	63.546(3)	r	63.546(3)	r
curium*	Cm	96				
darmstadtium*	Ds	110				
dubnium*	Db	105				
dysprosium	Dy	66	162.500(1)	g	162.50	g
einsteinium*	Es	99				
erbium	Er	68	167.259(3)	g	167.26	g
europium	Eu	63	151.964(1)	g	151.96	g
fermium*	Fm	100				
flerovium*	Fl	114				
fluorine	F	9	18.998 403 163(6)		18.998	
francium*	Fr	87				
gadolinium	Gd	64	157.25(3)	g	157.25(3)	g
gallium	Ga	31	69.723(1)		69.723	
germanium	Ge	32	72.630(8)		72.630(8)	
gold	Au	79	196.966 569(5)		196.97	
hafnium	Hf	72	178.49(2)		178.49(2)	
hassium*	Hs	108				
helium	He	2	4.002 602(2)	g r	4.0026	g r
holmium	Ho	67	164.930 33(2)		164.93	
hydrogen	H	1	[1.007 84, 1.008 11]	m	[1.0078, 1.0082]	m
indium	In	49	114.818(1)		114.82	
iodine	I	53	126.904 47(3)		126.90	
iridium	Ir	77	192.217(3)		192.22	
iron	Fe	26	55.845(2)		55.845(2)	

krypton	Kr	36	83.798(2)	g, m	83.798(2)	g, m
lanthanum	La	57	138.905 47(7)	g	138.91	g
lawrencium*	Lr	103				
lead	Pb	82	207.2(1)	g, r	207.2	g, r
lithium	Li	3	[6.938, 6.997]	m	[6.938, 6.997]	m
livermorium*	Lv	116				
lutetium	Lu	71	174.9668(1)	g	174.97	g
magnesium	Mg	12	[24.304, 24.307]		[24.304, 24.307]	
manganese	Mn	25	54.938 044(3)		54.938	
meitnerium*	Mt	109				
mendelevium*	Md	101				
mercury	Hg	80	200.592(3)		200.59	
molybdenum	Mo	42	95.95(1)	g	95.95	g
neodymium	Nd	60	144.242(3)	g	144.24	g
neon	Ne	10	20.1797(6)	g, m	20.180	g, m
neptunium*	Np	93				
nickel	Ni	28	58.6934(4)	r	58.693	r
niobium	Nb	41	92.906 37(2)		92.906	
nitrogen	N	7	[14.006 43, 14.007 28]		[14.006, 14.008]	
nobelium*	No	102				
osmium	Os	76	190.23(3)	g	190.23(3)	g
oxygen	O	8	[15.999 03, 15.999 77]		[15.999, 16.000]	
palladium	Pd	46	106.42(1)	g	106.42	g
phosphorus	P	15	30.973 761 998(5)		30.974	
platinum	Pt	78	195.084(9)		195.08	
plutonium*	Pu	94				

polonium*	Po	84				
potassium	K	19	39.0983(1)		39.098	
praseodymium	Pr	59	140.907 66(2)		140.91	
promethium*	Pm	61				
protactinium*	Pa	91	231.035 88(2)		231.04	
radium*	Ra	88				
radon*	Rn	86				
rhenium	Re	75	186.207(1)		186.21	
rhodium	Rh	45	102.905 50(2)		102.91	
roentgenium*	Rg	111				
rubidium	Rb	37	85.4678(3)	g	85.468	g
ruthenium	Ru	44	101.07(2)	g	101.07(2)	g
rutherfordium*	Rf	104				
samarium	Sm	62	150.36(2)	g	150.36(2)	g
scandium	Sc	21	44.955 908(5)		44.956	
seaborgium*	Sg	106				
selenium	Se	34	78.971(8)	r	78.971(8)	r
silicon	Si	14	[28.084, 28.086]		[28.084, 28.086]	
silver	Ag	47	107.8682(2)	g	107.87	g
sodium	Na	11	22.989 769 28 (2)		22.990	
strontium	Sr	38	87.62(1)	g, r	87.62	g, r
sulfur	S	16	[32.059, 32.076]		[32.059, 32.076]	
tantalum	Ta	73	180.947 88(2)		180.95	
technetium*	Tc	43				
tellurium	Te	52	127.60(3)	g	127.60(3)	g
terbium	Tb	65	158.925 35(2)		158.93	

thallium	Tl	81	[204.382, 204.385]		[204.38, 204.39]	
thorium*	Th	90	232.0377(4)		232.04	
thulium	Tm	69	168.934 22(2)		168.93	
tin	Sn	50	118.710(7)	g	118.71	g
titanium	Ti	22	47.867(1)		47.867	
tungsten	W	74	183.84(1)		183.84	
ununoctium*	Uuo	118				
ununpentium*	Uup	115				
ununseptium*	Uus	117				
ununtrium*	Uut	113				
uranium*	U	92	238.028 91(3)	g, m	238.03	g, m
vanadium	V	23	50.9415(1)		50.942	
xenon	Xe	54	131.293(6)	g, m	131.29	g, m
ytterbium	Yb	70	173.054(5)	g	173.05	g
yttrium	Y	39	88.905 84(2)		88.906	
zinc	Zn	30	65.38(2)	r	65.38(2)	r
zirconium	Zr	40	91.224(2)	g	91.224(2)	g

NOTE:

* Element has no stable isotopes. However, four such elements (Bi, Th, Pa, and U) do have a characteristic terrestrial isotopic composition, and for these elements, standard atomic weights are tabulated.

FOOTNOTES:

g Geological specimens are known in which the element has an isotopic composition outside the limits for normal material. The difference between the atomic weight of the element in such materials and that given in the table may exceed the stated uncertainty.

m Modified isotopic compositions may be found in commercially available material because the material has been subjected to an undisclosed or inadvertent isotopic fractionation. Substantial deviations in atomic weight of the element from that given in the table can occur.

r Range in isotopic composition of normal terrestrial material prevents a more precise Ar(E) being given; the tabulated Ar(E) value and uncertainty should be applicable to normal material.

Adapted from
M. Wang et al
The Ame2012 Atomic Mass Evaluation

NOBEL NOTES

This is the first of three articles, the second and third of which will appear in subsequent volumes: *Principles of Physics* and *Principles of Astronomy.* The overall premise is that discoveries found to be worthy of the most prestigious of scientific prizes generally have implications that will lead to other discoveries. In this first article we will be mainly concerned with the nature of matter, the nature of electromagnetism, the wave particle duality and the phenomena accessible in the laboratory without very expensive equipment. The second article discusses subatomic and subnuclear physics. The third article takes a look at our knowledge of the cosmos as a whole, and the possible futures for stars, galaxies and the universe.

Alfred Nobel, Swedish chemist, industrialist and pacifist, was born in 1833 and died in 1896. He was the inventor of dynamite, an explosive material with the energy of TNT (trinitrotoluene), but unlike TNT, one that could be manipulated and stored without the slightest danger of inadvertent detonation. While dynamite was a great boon to the mining and construction industries, it led to ever more powerful chemical bombs, making the human cost of warfare that much more terrible. An invention intended to save lives thus became responsible for millions of deaths in the wars of the twentieth century.

Nobel determined to do as much good as possible with the fortune he had gained from so lethal a discovery and therefore determined that the income from his fortune be awarded as prizes for achievements in five areas. The Nobel Prize for Peace was intended to honor those who had made the greatest contributions to peace between nations, with the prize recipient each year to be selected by a committee of five members of the Norwegian parliament. The Nobel Prize for Literature was to be selected each year by the Swedish Academy. The Nobel Prize for Physiology or Medicine would be selected by the Karolinska Institutet. The Royal Swedish Academy of Sciences selects the recipients the both Nobel Prize for Physics and the Nobel Prize for Chemistry. Nobel's will set up a foundation to fund the prizes in each of these areas. In 1968, the Nobel Prize in Economic Sciences was added in memory of Alfred Nobel.

Generally the prizes are awarded in Stockholm during December of each year. The Foundation's website, www.nobelprize.org, offers information about the history of the prizes as well details about prize winners in each of the five areas for each year since the inception of the prizes.

Under the terms of Nobel's will the prize can be split among at most three recipients, each of whom must be alive at the time the award is announced. In the sciences, at least, the prize is awarded to honor specific discoveries, more than the totality of an individual's work. Prize recipients receive a hand-calligraphed diploma written in Swedish, a gold medal of considerable worth, and a check derived from the income of the Nobel Foundation that year.

The first individual to win the Prize twice was Marie Curie, who shared the prize for physics in 1903 with Pierre Curie, her husband, and Henri Becquerel. She received the prize for chemistry by herself in 1911. Mme. Curie was the first member of a very exclusive club of two-time winners, which today includes Linus Pauling (for chemistry and peace), John Bardeen (for physics, twice) and Friedrich Sanger (for chemistry, twice). Receipt of the prize has had varying effects on the recipients' futures. For some, winning the prize led to the first university position they were able to hold as women in their own right. Some have used the occasion to change their fields of research. And some, like James D. Watson, deliberately sought the prize for its own sake.

Over its first decade, Nobel Prizes in Chemistry and Physics highlighted recent discoveries that were yet to be understood. It should be emphasized that as the twentieth century began, the existence of atoms in the solid and liquid states was open to debate; the mysterious cathode rays were considered by some to be disturbances in the aether rather than subatomic particles; and the structure of the atom, with its tiny positive nucleus, had yet to be revealed by Rutherford's scattering experiments. The very first Nobel Prize in Physics was awarded in 1901 to Konrad Roentgen for the discovery of x-rays. The Nobel Prize for Physics in 1903 was shared jointly by Pierre and Marie Curie and Henri Bequerel for their discoveries of radioactive elements, while the prize for 1906 went to Joseph John Thompson for showing that the cathode rays were indeed subatomic particles. The 1908 Nobel Prize for Chemistry went to Ernest Rutherford for his investigations of the products of radioactive decay and in 1911 the Prize for Chemistry went to the now widowed Marie Curie for the discovery of radium and polonium. While the awards thus far mentioned have been made for discoveries in physics, these important discoveries are now mentioned as part of every treatment of general chemistry.

The use of x-ray diffraction to determine the exact location of ions or molecule in a crystal was recognized by award of the physics prize to Max von Laue in 1914 and in 1915 to W. H. and W. L. Bragg (father and son). It soon became imperative for chemist and biochemists to grow a single crystal of their substance of interest. The discovery that x-ray diffraction could assist in determining molecular structures led to the work of Linus Pauling (Nobel Prize, 1954) on the nature of the chemical bond. The Nobel Prize in Physiology or Medicine 1962 went to J. D. Watson, F. H. C. Crick, and Maurice Wilkins for their work on the structure of DNA. As pointed out by J. D. Watson in his best selling book, *The Double Helix*, the x-ray crystallographer Rosalind Franklin was also heavily involved in the quest for DNA structure; however, she was deceased at the time of the award.

In 1962 Max Perutz and John C Kendrew received the Nobel Prize in Chemistry for work on the structure of globular proteins, particularly hemoglobin. Gerald Edelman and Rodney Porter shared the 1972 Nobel Prize in Physiology or Medicine for determining the structure of the antibody, again using x-ray diffraction.

The mid 1920's were heady days for physics. The wave particle duality was recognized by the Nobel Prizes for Physics awarded to Arthur H. Compton in 1917, and to Louis V. de Broglie in 1929. The 1922 Nobel Prize for Physics went to Niels Bohr for his model of the hydrogen atom, which worked perfectly in essence but could not be extended to many-electron atoms. That feat was accomplished by Werner Heisenberg (Nobel Prize in Physics, 1932) while Erwin Schrödinger and P.A.M. Dirac (Nobel Physics Prize, 1933) showed that the apparently different approaches of Heisenberg and Schröedinger were alternative formulations of the same mathematical description of nature. In 1943 O. Stern was honored for the discovery of electron spin made in the 1920's while in 1945 Wolfgang Pauli would receive the Physics Prize in part for the Pauli Principle which led to the shell model of the atom.

Winners of the Nobel awards were typically academics, who held university professorships or who were members of research institutes, but not always. The 1931 Nobel Prize in Chemistry was awarded to Irving Langmuir of the General Electric company in Schenectady, New York. Langmuir was one of the founders of chemical bonding theory and also one of the pioneers of surface science, a subject of tremendous commercial importance in the era of vacuum tube electronics. Over the years, very few prize winners had been directly connected with industry, but that may be changing as the cost of research grows. A much more recent example came in 1993 when the Nobel Prize for Chemistry went to Kary Mullis of Cetus Corporation for discovery of the Polymerase Chain Reaction.

WINNERS OF THE NOBEL PRIZE IN CHEMISTRY

2015	Tomas Lindahl, Paul Modrich, Aziz Sancar	"for mechanistic studies of DNA repair"
2014	Eric Betzig, Stefan W. Hell, William E. Moerner	"for the development of super-resolved fluorescence microscopy"
2013	Martin Karplus, Michael Levitt, Arieh Warshel	"for the development of multiscale models for complex chemical systems"
2012	Robert J. Lefkowitz, Brian K. Kobilka	"for studies of G-protein-coupled receptors"
2011	Dan Shechtman	"for the discovery of quasicrystals"
2010	Richard F. Heck, Ei-ichi Negishi, Akira Suzuki	"for palladium-catalyzed cross couplings in organic synthesis"
2009	Venkatraman Ramakrishnan, Thomas A. Steitz, Ada E. Yonath	"for studies of the structure and function of the ribosome"
2008	Osamu Shimomura, Martin Chalfie, Roger Y. Tsien	"for the discovery and development of the green fluorescent protein, GFP"
2007	Gerhard Ertl	"for his studies of chemical processes on solid surfaces"
2006	Roger D. Kornberg	"for his studies of the molecular basis of eukaryotic transcription"
2005	Yves Chauvin, Robert H. Grubbs, Richard R. Schrock	"for the development of the metathesis method in organic synthesis"
2004	Aaron Ciechanover, Avram Hershko, Irwin Rose	"for the discovery of ubiquitin-mediated protein degradation"
2003	Peter Agre	"for the discovery of water channels"
2003	Roderick MacKinnon	"for structural and mechanistic studies of ion channels"

2002	John B. Fenn, Koichi Tanaka,	"for their development of soft desorption ionisation methods for mass spectrometric analyses of biological macromolecules"
2002	Kurt Wüthrich	"for his development of nuclear magnetic resonance spectroscopy for determining the three-dimensional structure of biological macromolecules in solution"
2001	William S. Knowles, Ryoji Noyori	"for their work on chirally catalysed hydrogenation reactions"
2001	K. Barry Sharpless	"for his work on chirally catalysed oxidation reactions"
2000	Alan J. Heeger	"for the discovery and development of conductive polymers"
2000	Alan G. MacDiarmid, Hideki Shirakawa	"for the discovery and development of conductive polymers"
1999	Ahmed H. Zewail	"for his studies of the transition states of chemical reactions using femtosecond spectroscopy"
1998	Walter Kohn	"for his development of the density-functional theory"
1998	John A. Pople	"for his development of computational methods in quantum chemistry"
1997	Paul D. Boyer, John E. Walker	"for their elucidation of the enzymatic mechanism underlying the synthesis of adenosine triphosphate (ATP)"
1997	Jens C. Skou	"for the first discovery of an ion-transporting enzyme, Na+, K+ -ATPase"
1996	Robert F. Curl Jr., Sir Harold W. Kroto, Richard E. Smalley	"for their discovery of fullerenes"
1995	Paul J. Crutzen, Mario J. Molina, F. Sherwood Rowland	"for their work in atmospheric chemistry, particularly concerning the formation and decomposition of ozone"
1994	George A. Olah	"for his contribution to carbocation chemistry"
1993	Kary B. Mullis	"for his invention of the polymerase chain reaction (PCR) method"
1993	Michael Smith	"for his fundamental contributions to the establishment of oligonucleotide-based, site-directed mutagenesis and its development for protein studies"
1992	Rudolph A. Marcus	"for his contributions to the theory of electron transfer reactions in chemical systems"
1991	Richard R. Ernst	"for his contributions to the development of the methodology of high resolution nuclear magnetic resonance (NMR) spectroscopy"

1990	Elias James Corey	"for his development of the theory and methodology of organic synthesis"
1989	Sidney Altman, Thomas R. Cech	"for their discovery of catalytic properties of RNA"
1988	Johann Deisenhofer	"for the determination of the three-dimensional structure of a photosynthetic reaction centre"
1988	Robert Huber, Hartmut Michel	"for the determination of the three-dimensional structure of a photosynthetic reaction centre"
1987	Donald J. Cram, Jean-Marie Lehn, Charles J. Pedersen	"for their development and use of molecules with structure-specific interactions of high selectivity"
1986	Dudley R. Herschbach, Yuan T. Lee, John C. Polanyi	"for their contributions concerning the dynamics of chemical elementary processes"
1985	Herbert A. Hauptman, Jerome Karle	"for their outstanding achievements in the development of direct methods for the determination of crystal structures"
1984	Robert Bruce Merrifield	"for his development of methodology for chemical synthesis on a solid matrix"
1983	Henry Taube	"for his work on the mechanisms of electron transfer reactions, especially in metal complexes"
1982	Aaron Klug	"for his development of crystallographic electron microscopy and his structural elucidation of biologically important nucleic acid-protein complexes"
1981	Kenichi Fukui, Roald Hoffmann	"for their theories, developed independently, concerning the course of chemical reactions"
1980	Paul Berg	"for his fundamental studies of the biochemistry of nucleic acids, with particular regard to recombinant-DNA"
1980	Walter Gilbert, Frederick Sanger	"for their contributions concerning the determination of base sequences in nucleic acids"
1980		"for their contributions concerning the determination of base sequences in nucleic acids"
1979	Herbert C. Brown, Georg Wittig	"for their development of the use of boron- and phosphorus-containing compounds, respectively, into important reagents in organic synthesis"
1978	Peter D. Mitchell	"for his contribution to the understanding of biological energy transfer through the formulation of the chemiosmotic theory"
1977	Ilya Prigogine	"for his contributions to non-equilibrium thermodynamics, particularly the theory of dissipative structures"
1976	William N. Lipscomb	"for his studies on the structure of boranes illuminating problems of chemical bonding"
1975	John Warcup Cornforth	"for his work on the stereochemistry of enzyme-catalyzed reactions"

1975	Vladimir Prelog	"for his research into the stereochemistry of organic molecules and reactions"
1974	Paul J. Flory	"for his fundamental achievements, both theoretical and experimental, in the physical chemistry of the macromolecules"
1973	Ernst Otto Fischer, Geoffrey Wilkinson	"for their pioneering work, performed independently, on the chemistry of the organometallic, so called sandwich compounds"
1972	Christian B. Anfinsen	"for his work on ribonuclease, especially concerning the connection between the amino acid sequence and the biologically active conformation"
1972	Stanford Moore, William H. Stein	"for their contribution to the understanding of the connection between chemical structure and catalytic activity of the active centre of the ribonuclease molecule"
1971	Gerhard Herzberg	"for his contributions to the knowledge of electronic structure and geometry of molecules, particularly free radicals"
1970	Luis F. Leloir	"for his discovery of sugar nucleotides and their role in the biosynthesis of carbohydrates"
1969	Derek H. R. Barton	"for their contributions to the development of the concept of conformation and its application in chemistry"
1969	Odd Hassel	"for their contributions to the development of the concept of conformation and its application in chemistry"
1968	Lars Onsager	"for the discovery of the reciprocal relations bearing his name, which are fundamental for the thermodynamics of irreversible processes"
1967	Manfred Eigen, Ronald George Wreyford Norrish, George Porter	"for their studies of extremely fast chemical reactions, effected by disturbing the equlibrium by means of very short pulses of energy"
1966	Robert S. Mulliken	"for his fundamental work concerning chemical bonds and the electronic structure of molecules by the molecular orbital method"
1965	Robert Burns Woodward	"for his outstanding achievements in the art of organic synthesis"
1964	Dorothy Crowfoot Hodgkin	"for her determinations by X-ray techniques of the structures of important biochemical substances"
1963	Karl Ziegler, Giulio Natta	"for their discoveries in the field of the chemistry and technology of high polymers"
1962	Max Ferdinand Perutz, John Cowdery Kendrew	"for their studies of the structures of globular proteins"
1961	Melvin Calvin	"for his research on the carbon dioxide assimilation in plants"

1960	Willard Frank Libby	"for his method to use carbon-14 for age determination in archaeology, geology, geophysics, and other branches of science"
1959	Jaroslav Heyrovsky	"for his discovery and development of the polarographic methods of analysis"
1958	Frederick Sanger	"for his work on the structure of proteins, especially that of insulin"
1957	Lord (Alexander R.) Todd	"for his work on nucleotides and nucleotide co-enzymes"
1956	Sir Cyril Norman Hinshelwood, Nikolay Nikolaevich Semenov	"for their researches into the mechanism of chemical reactions"
1955	Vincent du Vigneaud	"for his work on biochemically important sulphur compounds, especially for the first synthesis of a polypeptide hormone"
1954	Linus Carl Pauling	"for his research into the nature of the chemical bond and its application to the elucidation of the structure of complex substances"
1953	Hermann Staudinger	"for his discoveries in the field of macromolecular chemistry"
1952	Archer John Porter Martin	"for their invention of partition chromatography"
1952	Richard Laurence Millington Synge	"for their invention of partition chromatography"
1951	Edwin Mattison McMillan, Glenn Theodore Seaborg	"for their discoveries in the chemistry of the transuranium elements"
1950	Otto Paul Hermann Diels, Kurt Alder	"for their discovery and development of the diene synthesis"
1949	William Francis Giauque	"for his contributions in the field of chemical thermodynamics, particularly concerning the behaviour of substances at extremely low temperatures"
1948	Arne Wilhelm Kaurin Tiselius	"for his research on electrophoresis and adsorption analysis, especially for his discoveries concerning the complex nature of the serum proteins"
1947	Sir Robert Robinson	"for his investigations on plant products of biological importance, especially the alkaloids"
1946	James Batcheller Sumner	"for his discovery that enzymes can be crystallized"
1946	John Howard Northrop, Wendell Meredith Stanley	"for their preparation of enzymes and virus proteins in a pure form"
1945	Artturi Ilmari Virtanen	"for his research and inventions in agricultural and nutrition chemistry, especially for his fodder preservation method"
1944	Otto Hahn	"for his discovery of the fission of heavy nuclei"

1943	George de Hevesy	"for his work on the use of isotopes as tracers in the study of chemical processes"
1939	Adolf Friedrich Johann Butenandt	"for his work on sex hormones"
1939	Leopold Ruzicka	"for his work on polymethylenes and higher terpenes"
1938	Richard Kuhn	"for his work on carotenoids and vitamins"
1937	Walter Norman Haworth	"for his investigations on carbohydrates and vitamin C"
1937	Paul Karrer	"for his investigations on carotenoids, flavins and vitamins A and B2"
1936	Petrus (Peter) Josephus Wilhelmus Debye	"for his contributions to our knowledge of molecular structure through his investigations on dipole moments and on the diffraction of X-rays and electrons in gases"
1935	Frédéric Joliot, Irène Joliot-Curie	"in recognition of their synthesis of new radioactive elements"
1934	Harold Clayton Urey	"for his discovery of heavy hydrogen"
1932	Irving Langmuir	"for his discoveries and investigations in surface chemistry"
1931	Carl Bosch, Friedrich Bergius	"in recognition of their contributions to the invention and development of chemical high pressure methods"
1930	Hans Fischer	"for his researches into the constitution of haemin and chlorophyll and especially for his synthesis of haemin"
1929	Arthur Harden, Hans Karl August Simon von Euler-Chelpin	"for their investigations on the fermentation of sugar and fermentative enzymes"
1928	Adolf Otto Reinhold Windaus	"for the services rendered through his research into the constitution of the sterols and their connection with the vitamins"
1927	Heinrich Otto Wieland	"for his investigations of the constitution of the bile acids and related substances"
1926	The (Theodor) Svedberg	"for his work on disperse systems"
1925	Richard Adolf Zsigmondy	"for his demonstration of the heterogenous nature of colloid solutions and for the methods he used, which have since become fundamental in modern colloid chemistry"
1923	Fritz Pregl	"for his invention of the method of micro-analysis of organic substances"
1922	Francis William Aston	"for his discovery, by means of his mass spectrograph, of isotopes, in a large number of non-radioactive elements, and for his enunciation of the whole-number rule"

1921	Frederick Soddy	"for his contributions to our knowledge of the chemistry of radioactive substances, and his investigations into the origin and nature of isotopes"
1920	Walther Hermann Nernst	"in recognition of his work in thermochemistry"
1918	Fritz Haber	"for the synthesis of ammonia from its elements"
1915	Richard Martin Willstätter	"for his researches on plant pigments, especially chlorophyll"
1914	Theodore William Richards	"in recognition of his accurate determinations of the atomic weight of a large number of chemical elements"
1913	Alfred Werner	"in recognition of his work on the linkage of atoms in molecules by which he has thrown new light on earlier investigations and opened up new fields of research especially in inorganic chemistry"
1912	Victor Grignard	"for the discovery of the so-called Grignard reagent, which in recent years has greatly advanced the progress of organic chemistry"
1912	Paul Sabatier	"for his method of hydrogenating organic compounds in the presence of finely disintegrated metals whereby the progress of organic chemistry has been greatly advanced in recent years"
1911	Marie Curie, née Sklodowska	"in recognition of her services to the advancement of chemistry by the discovery of the elements radium and polonium, by the isolation of radium and the study of the nature and compounds of this remarkable element"
1910	Otto Wallach	"in recognition of his services to organic chemistry and the chemical industry by his pioneer work in the field of alicyclic compounds"
1909	Wilhelm Ostwald	"in recognition of his work on catalysis and for his investigations into the fundamental principles governing chemical equilibria and rates of reaction"
1908	Ernest Rutherford	"for his investigations into the disintegration of the elements, and the chemistry of radioactive substances"
1907	Eduard Buchner	"for his biochemical researches and his discovery of cell-free fermentation"
1906	Henri Moissan	"in recognition of the great services rendered by him in his investigation and isolation of the element fluorine, and for the adoption in the service of science of the electric furnace called after him"
1905	Johann Friedrich Wilhelm Adolf von Baeyer	"in recognition of his services in the advancement of organic chemistry and the chemical industry, through his work on organic dyes and hydroaromatic compounds"
1904	Sir William Ramsay	"in recognition of his services in the discovery of the inert gaseous elements in air, and his determination of their place in the periodic system"

1903	Svante August Arrhenius	"in recognition of the extraordinary services he has rendered to the advancement of chemistry by his electrolytic theory of dissociation"
1902	Hermann Emil Fischer	"in recognition of the extraordinary services he has rendered by his work on sugar and purine syntheses"
1901	Jacobus Henricus van 't Hoff	"in recognition of the extraordinary services he has rendered by the discovery of the laws of chemical dynamics and osmotic pressure in solutions"

GLOSSARY

absorbance: a measure of the amount of electromagnetic radiation of a particular frequency that a specific quantity of solid or solute matter absorbs; also called optical density.

acid: a compound that can relinquish one or more hydrogen ions (Brønsted-Lowry acid-base theory) or that possesses vacant atomic orbitals to interact with electron-rich materials (Lewis acid-base theory).

acid derivative: a compound formed by modifying the molecular structure of an acid, such as an ester created by the reaction between a carboxylic acid and an alcohol.

activated complex: one of the various intermediate molecular structures that exist during a chemical reaction while the original chemical bonds are being broken and new bonds are being formed.

activation energy: the amount of energy a system requires for the formation of an activated complex from which a reaction can proceed.

acyl group: a functional group with the formula –RCO, where R is connected by a single bond to the carbon atom of the carbonyl (C=O) group.

acylation: a reaction process in which an acyl group is added to a compound.

adenosine triphosphate (ATP): a molecule consisting of adenine, ribose, and a triphosphate chain that is used to transfer the energy needed to carry out numerous cellular processes.

aerobic respiration: a form of cellular respiration that requires oxygen in order to generate energy from glucose.

alcohol: an organic compound in which a hydroxyl is the primary functional group and is bonded to a saturated carbon atom.

alkaline: describes a material that tends to increase the concentration of hydroxide ions in an aqueous solution, as well as conditions produced by the presence of bases.

alkene: any organic compound that includes two carbon atoms connected by a double bond.

alkoxyl: a functional group, also called an alkoxy group, consisting of an alkyl group bonded to an oxygen atom.

alkyl group: a functional group consisting of a hydrocarbon chain, usually with the general formula CnH2n+1.

alkylation: a combination reaction that results in the addition of an alkyl group to a molecule.

allotrope: one of two or more principal physical forms in which a single pure element occurs, due to differences in chemical bonding or the structural arrangement of the atoms; for example, diamond and graphite are two allotropes of carbon.

alloy: a mixture of a metal and at least one other element, often another metal; also known as a solid solution.

alpha particle: a particle consisting of two protons and two neutrons bound together, identical to a helium nucleus; typically produced in the process of alpha decay.

amino acid: an organic compound that contains both an amine and a carboxylic acid functional group and can, in some cases, combine with other amino acids to form proteins and other polypeptides.

amino group: a functional group containing a nitrogen atom bonded to two hydrogen atoms (–NH2).

ammonia: an inorganic compound consisting of a nitrogen atom bonded to three hydrogen atoms; exists as a colorless, pungent gas at room temperature.

amphoteric: describes a compound with the ability to act as either an acid or a base, depending on its environment and the other materials present.

amphoterism: the ability of a compound to act as either an acid or a base, depending on its environment and the other materials present.

anaerobic respiration: a form of cellular respiration that does not require oxygen in order to generate energy from glucose.

anion: any chemical species bearing a net negative electrical charge, which causes it to be drawn toward the positive pole, or anode, of an electrochemical cell.

anticodon: a sequence of three nucleotide bases in transfer RNA (tRNA) that bonds to a complementary codon in messenger RNA (mRNA).

aromatic hydrocarbon: a hydrocarbon in which the carbon atoms form a ring with alternating double and single bonds, distributed in such a way that all bonds are of equal length and strength; also called an arene.

aromaticity: a characteristic of certain ring-shaped molecules in which alternating double and single bonds are distributed in such a way that all bonds are of equal length and strength, giving the molecule greater stability than would otherwise be expected.

aryl group: a functional group derived from an aromatic ring.

Arrhenius equation: a mathematical function that relates the rate of a reaction to the energy required to initiate the reaction and the absolute temperature at which it is carried out.

atomic mass: the total mass of the protons, neutrons, and electrons in an individual atom.

atomic mass unit: a unit of mass defined as one-twelfth of the mass of one atom of carbon-12 (approximately 1.66 × 10–27 kilograms), equivalent to 1 gram per mole.

atomic model: a theoretical representation of the structure and behavior of an atom based on the nature and behavior of its component particles.

atomic weight: the average of the atomic masses of an element's isotopes, weighted according to their proportions in nature; often used interchangeably with "atomic mass."

atomic number: the number of protons in the nucleus of an atom, used to uniquely identify each element.

autocrine signaling: a type of cell signaling in which the signaling compound is produced within a cell and delivered to receptors on the outside of the same cell.

Avogadro's number (NA): 6.02214129 × 1023, often rounded to 6.022 × 1023; the number of particles (atoms or molecules) that constitute one mole of any element or compound, the mass of which in grams is numerically equal to the atomic or molecular weight of the material.

base: a compound that can relinquish one or more hydroxide ions (Brønsted-Lowry acid-base theory) or that possesses lone pairs of electrons that can interact with electron-poor materials (Lewis acid-base theory).

benzene: an organic compound with the molecular formula C_6H_6, consisting of a six-membered carbon ring with a hydrogen atom bonded to each carbon; in theory, the carbon-carbon bonds alternate between single and double bonds, but in fact they are all equal due to the electronic property of aromaticity.

beta particle: an electron or positron produced in the process of beta decay, in which a neutron of an unstable atom decomposes into a proton, and emitted from the nucleus at high speed.

biochemistry: the chemistry of living organisms and the processes incidental to and characteristic of life.

biomolecule: an organic molecule produced by a living organism.

bond energy: the amount of energy necessary to break the chemical bonds in a given molecule, measured in kilojoules per mole.

bonding: the formation of a link between two atoms as a result of the interaction of their valence electrons.

Boyle's law: the principle that states that the volume occupied by a gas varies in inverse proportion to the pressure.

Brønsted-Lowry acid-base theory: definitions of acids and bases developed separately in 1923 by Danish chemist Johannes Nicolaus Brønsted and English chemist Martin Lowry; defines an acid as any compound that can release a hydrogen ion and a base as any compound that can accept a hydrogen ion.

Brownian motion: the continuous, random motion of particles in a fluid medium, caused by impacts with the molecules that make up the medium.

calorie: a unit of energy, defined as the amount of energy required to raise the temperature of one gram of water by one degree Celsius; one calorie is equal to 4.184 joules.

carbohydrate: an organic compound containing hydroxyl (–OH) and carbonyl (C=O) groups, often with the general formula $Cx(H_2O)y$; includes sugars, starches, and celluloses.

carbon dating: a method of dating that uses the proportion of radioactive carbon-14 atoms remaining in organic material to determine how much time has elapsed since it was part of a living organism.

carbonyl group: a functional group consisting of a carbon atom double bonded to an oxygen atom.

carboxyl group: a functional group containing a carbon atom double bonded to an oxygen atom and single bonded to a hydroxyl group (–OH); has the formula CO2H, typically written –COOH.

carboxylic acid: an organic compound containing a carboxyl functional group and having the general formula RC(=O)OH.

catabolic reaction: a metabolic reaction in cells that breaks down large molecules into smaller ones, resulting in a release of energy.

catalyst: a chemical species that initiates or speeds up a chemical reaction but is not itself consumed in the reaction.

cation: any chemical species bearing a net positive electrical charge, which causes it to be drawn toward the negative pole, or cathode, of an electrochemical cell.

cell membrane: a biological membrane that forms a semipermeable barrier separating the interior of a cell from the exterior.

cellular receptor: a protein within a cell that specific signaling molecules bind to in order to elicit a specific response from the cell.

Charles's law: the principle that states that the volume occupied by a gas varies in direct proportion to the temperature.

chemical bond: a link between two atoms formed by the interaction of their valence electrons.

chemical formula: the combination of symbols and numerical coefficients that specifies the number and identity of the different atoms involved in a chemical reaction or molecular structure.

chemical kinetics: the branch of chemistry that studies the various factors that affect rates of chemical reactions.

chemical reaction: a process in which the molecules of two or more chemical species interact with each other in a way that causes the electrons in the bonds between atoms to be rearranged, resulting in changes to the chemical identities of the materials.

chromophore: the part of a molecule that absorbs specific wavelengths of electromagnetic radiation, often causing it to appear as a certain color.

closed system: a physical or chemical reaction system defined by certain boundary conditions that prevent any components, reactants, or products from entering or exiting the system.

codon: a sequence of three nucleotide bases that specifies a particular amino acid or control point in the process of protein synthesis.

coefficient: the number preceding a molecular, ionic, or atomic formula in a chemical equation that identifies the amount of each component present in the reaction being described.

cohesion: the tendency for like molecules of a substance to stick together due to their shape and electronic structure.

combination reaction: a chemical reaction in which two or more reactants combine to form a single product.

combustion: a reaction between a fuel and an oxidizing agent that results in new chemical compounds and the release of heat; most often takes place between organic material and molecular oxygen, in which case the products include carbon dioxide and water.

complementary strand: one of the two strands of nucleotides that make up a DNA molecule, with each nucleotide in one strand corresponding to the position of its complementary nucleotide (cytosine for guanine, adenine for thymine, and vice versa) in the other.

composition: the identities and relative proportions of different elements or components in a compound, mixture, or other material.

compound: a chemically unique material whose molecules consist of several atoms of two or more different elements.

compressive stress: a force that acts to push on or compress a material.

concentration: the amount of a specific component present in a given volume of a mixture.

concentration gradient: the gradual change in the concentration of solutes in a solution across a specific distance.

concerted reaction: a reaction that proceeds from starting materials to end products in a single molecular process rather than via a stepwise mechanism.

conductivity: the ability of a material to transfer heat (thermal conductivity) or electricity (electrical conductivity) from one point to another.

conjugate acid: the material formed from a base when it accepts a proton (H+) from an acid, thus gaining a unit of positive charge.

conjugate base: the material formed from an acid when it donates a proton (H+) to a base, thus losing a unit of positive charge.

conjugate pair: an acid and the conjugate base that is formed when it donates a proton, or a base and the conjugate acid that is formed when it accepts a proton.

conjugation: the overlap of *p* orbitals between three or more successive carbon atoms in a molecular structure, creating a system of alternating double and single bonds through which electrons can move freely.

covalent bond: a type of chemical bond in which electrons are shared between two adjacent atoms.

critical pressure: the pressure required to compress a particular gas into a liquid state at the critical temperature.

critical temperature: the highest temperature at which a particular gas can be liquefied by pressure alone.

crystal: a solid consisting of atoms, molecules, or ions arranged in a regular, periodic pattern in all directions, often resulting in a similarly regular macroscopic appearance.

crystallography: the study of the properties and structures of crystals.

cyclic: describes a compound whose molecular structure forms a closed ring.

cyclic delocalization: a property of some ring molecules, such as benzene, in which the overlap of orbitals between the atoms that make up the ring allows their electrons to move freely about the molecule.

cyclohexane: a saturated hydrocarbon composed of six methylene bridges ($-CH_2-$) bonded in a six-membered ring structure; has the molecular formula C_6H_{12}.

cyclopentane: a saturated hydrocarbon composed of five methylene bridges ($-CH_2-$) bonded in a five-membered ring structure; has the molecular formula C_5H_{10}.

cyanocarbon: an organic compound containing multiple cyano groups, which consist of a carbon atom triple bonded to a nitrogen atom.

d-block: the portion of the periodic table containing the elements whose valence electrons are in their *d* orbitals.

decomposition reaction: a chemical reaction in which a single reactant breaks apart to create several products with smaller molecular structures.

deformation: any permanent change in the shape of an object as a result of the application of force or a change in temperature.

dehydration reaction: a chemical reaction in which hydrogen and oxygen atoms are removed from the reactants and combine to form water.

density: the amount of a material contained within a particular space, usually expressed as mass per unit volume.

deoxyribonucleic acid (DNA): a large molecule formed by two complementary strands of nucleotides that encodes the genetic information of all living organisms.

derivative: a compound that is obtained by subjecting a similar parent compound to one or more chemical reactions that target certain functional groups, leaving the basic molecular structure unaltered.

deuterium: an isotope of hydrogen that contains one neutron and one proton; occurs naturally in about 1 in 6,500 hydrogen atoms.

diatomic: describes a molecule or ion consisting of two atoms that are chemically bonded to each other.

Diels-Alder reaction: a reaction in which an activated conjugated diene compound reacts with a suitable dienophile to form a six-membered ring structure; named for Otto Diels and Kurt Alder.

diene: an organic compound that contains two carbon-carbon double bonds (C=C) in its molecular structure.

dienophile: an organic compound containing a carbon-carbon double bond (C=C) that reacts preferentially with a suitable diene in a Diels-Alder reaction.

diffusion: the process by which different particles, such as atoms and molecules, gradually become intermingled due to random motion caused by thermal energy.

dipole: the separation of positive and negative charges within a single molecule due to electron density being relatively high in one part of the molecule and relatively low in another.

dipole-dipole: describes a type of interaction, either attraction or repulsion, between two molecular dipoles, which are molecules that are polarized due to the greater concentration of electrons in one area.

dissociation: the separation of a compound into simpler components.

dissociation constant: a characteristic value representing the extent to which a compound dissociates into component ions in a certain solvent and under specific conditions.

double bond: a type of chemical bond in which two adjacent atoms are connected by four bonding electrons rather than two.

double displacement: a substitution reaction in which the atoms of two elements exchange places in their respective compounds.

ductility: the ability of a solid material to be deformed by the application of tensile (pulling) force, such as bending, without breaking or fracturing.

electrical charge: a property of subatomic particles that causes them to exert a force on each other, either attractive (if their charges are of opposite signs) or repulsive (if they are of the same sign); by convention, a proton is assigned a charge of 1+ and an electron is assigned a charge of 1−.

electrolysis: the passage of an electric current through a solution or molten material to induce a chemical reaction, resulting in a reduced chemical species at the negative terminal of the electrolytic cell and an oxidized species at the positive terminal.

electrolyte: a material that ionizes in an appropriate solvent to produce an electrically conductive solution.

electron: a fundamental subatomic particle with a single negative electrical charge, found in a large, diffuse cloud around the nucleus.

electron configuration: the order and arrangement of electrons within the orbitals of an atom or molecule.

electron shell: a region surrounding the nucleus of an atom that contains one or more orbitals capable of holding a specific maximum number of electrons.

electron spectroscopy: an analytical method of studying atoms and molecules based on the absorption or emission of electromagnetic radiation by their electrons.

electron transport chain: a series of oxidation and reduction (redox) reactions in which the electrons released by oxidation are transferred from one molecule to the next, ultimately enabling the production of ATP.

electronegative: describes an atom that tends to accept and retain electrons to form a negatively charged ion.

electrophilic: describes a chemical species that tends to react with negatively charged or electron-rich species.

electrophilic addition: an addition reaction in which an electrophile, typically a positively charged chemical species, is bonded to a nucleophile, a molecule with a free pair of electrons that can be easily donated; results in the breaking of a multiple bond to form two single bonds.

electropositive: describes an atom that tends to lose electrons to form a positively charged ion.

electrovalent bond: an alternate term for an ionic bond, which is a type of chemical bond formed by mutual attraction between two ions of opposite charges.

element: a form of matter consisting only of atoms of the same atomic number.

element name: the official name by which each element is known, according to the standards of the International Union of Pure and Applied Chemistry (IUPAC).

element symbol: a one- or two-letter abbreviation uniquely assigned to each element, usually derived from the internationally recognized name of the element.

elementary reaction: a chemical reaction that is completed in a single reaction step with only one transition state; often part of a multistep sequence of reactions that constitutes a mechanism.

empirical formula: a chemical formula that indicates the relative proportion of atoms of each element present in one molecule of a substance, which may not be equal to the total number of atoms present.

endergonic: synonym for endoergic; describes a reaction process that requires the input of energy in the form of work in order to proceed.

endocrine signaling: a type of cell signaling in which the signaling compound is produced in one location in the body and transported to a receptor site some distance away.

endosome: an intracellular compartment that sorts and transports material taken into a cell via endocytosis.

endothermic: describes a process that requires the input of energy in the form of heat in order to proceed.

enthalpy: the total heat content within a thermodynamic system, defined as internal energy plus the product of pressure and volume; also, the change in heat content associated with a chemical process.

enzyme: a protein molecule that acts as a catalyst in biochemical reactions.

equilibrium: the state that exists when the forward activity of a process is exactly equal to the reverse activity of that process.

equilibrium constant: a numerical value characteristic of a particular equilibrium reaction, defined as the ratio of the equilibrium concentration of the products to that of the reactants.

ester: a class of compounds characterized by a carbonyl group bonded to an alkoxy group, formed by the condensation of an acid and an alcohol.

ether: an organic compound consisting of an oxygen atom bonded to two alkyl or aryl groups.

exergonic: synonym for exoergic; describes a reaction process that can occur spontaneously and releases energy in the form of work.

exponential decay: a process of decomposition in which the amount of non-decayed material decreases at a rate proportional to the current amount of material present rather than the original amount.

f-block: the portion of the periodic table containing the elements whose valence electrons are in their *f* orbitals, namely the lanthanides and the actinides.

fluorination: the addition or substitution of fluorine atoms as substituents in compounds.

fracture: a dislocation in the internal structure of an object that causes it to break into two or more pieces.

freezing point: the temperature at which a liquid undergoes a phase change to become a solid.

functional group: a specific group of atoms with a characteristic structure and corresponding chemical behavior within a molecule.

fundamental particle: one of the smaller, indivisible particles that make up a larger, composite particle; commonly used to refer to electrons, protons, and neutrons, although these are themselves composed of various actual fundamental particles, such as quarks, leptons, and certain types of bosons.

gamma ray: ionizing electromagnetic radiation of higher energy and shorter wavelength than x-rays, typically produced by the nucleus of an atom undergoing radioactive decay.

gas: a state of matter in which material is fluid, has indefinite volume and shape, and is of variable density due to its ability to expand to fill any available space.

geminal: describes the relationship between two functional groups, usually similar or identical, bonded to the same (typically carbon) atom in the same molecule.

gene expression: the process by which RNA copies genes, which are specific segments of the DNA molecule, and uses the information to synthesize either proteins or other types of RNA.

Gibbs free energy: the energy in a thermodynamic system that is available to do work.

half-life: the length of time required for one-half of a given amount of material to decompose or be consumed through a continuous decay process.

halide: a binary compound consisting of a halogen element (fluorine, chlorine, bromine, iodine, or astatine) bonded to a non-halogen element or organic group; alternatively, an anion of a halogen element.

halogens: the elements in group 17 of the periodic table (fluorine, chlorine, bromine, iodine, and astatine), all of which are highly electronegative and tend to accept one electron to form a negative ion.

heat energy: the kinetic energy of the random motion of a system's component particles that contributes to the temperature of that system.

homogeneous mixture: a physical combination of different materials that has a generally uniform distribution of composition, and therefore properties, throughout its mass; called a solution when liquid and an alloy when solid metal.

hybrid structure: a representation of molecular structure that averages a number of possible molecular structures that are equivalent in terms of the arrangement of orbitals and electrons within them.

hydrocarbon: an organic compound composed solely of carbon and hydrogen atoms.

hydrochloric acid: a corrosive aqueous solution of hydrogen chloride, present in the common digestive fluid of the stomach.

hydrogen bond: a weak type of chemical bond formed by the attraction of a hydrogen atom to an electronegative atom—an atom with a strong tendency to attract electrons—in the same or another molecule.

hydrogen ion: a hydrogen atom that has lost its one electron, represented by the symbol H+.

hydrogenation: a chemical reaction in which two hydrogen atoms, usually in the form of molecular hydrogen (H2), are bonded to another molecule, almost always as a result of catalysis.

hydrolysis: the cleavage of a chemical bond caused by the presence of water.

hydronium ion: a polyatomic ion with the formula H3O+, formed by the addition of the hydrogen cation (H+) to a molecule of water; also called oxonium (IUPAC preference) or hydroxonium.

hydroxide: an anion consisting of one oxygen atom and one hydrogen atom, represented as OH−; also, an ionic compound in which the OH− ion is bonded to another element or group.

hydroxyl group: a primary functional group consisting of an oxygen atom covalently bonded to a single hydrogen atom.

hypertonic: describes a solution with a greater concentration of solutes than the solution to which it is being compared; in biology, a solution with a greater solute concentration than the cytoplasm of a cell.

hypotonic: describes a solution with a lower concentration of solutes than the solution to which it is being compared; in biology, a solution with a lower solute concentration than the cytoplasm of a cell.

ideal gas law: the principle that states that the product of the pressure and volume of a gas is directly proportional to the product of the absolute temperature and the number of moles of the gas, expressed by the equation $PV = nRT$.

intensive properties: the properties of a substance that do not depend on the amount of the substance present, such as density, hardness, and melting and boiling point.

intermediate: a relatively stable but reactive chemical species that is formed during a chemical reaction and either reverts back to the original reactants or continues in the reaction mechanism to form products.

ion: an atom, molecule, or neutral radical that has either lost or gained electrons and is therefore electrically charged.

ionic bond: a type of chemical bond formed by mutual attraction between two ions of opposite charges.

ionization: the process by which an atom or molecule loses or gains one or more electrons to acquire a net positive or negative electrical charge.

ionization energy: the amount of energy required to remove an electron from an atom in a gaseous state.

isomer: one of two or more chemical species that have the same molecular formula but different molecular structures.

isomeric: describes chemical species that have the same molecular formula but different molecular structures.

isotonic: describes a solution with the same concentration of solutes as the solution to which it is being compared; in biology, a solution with the same solute concentration as the cytoplasm of a cell.

isotope: an atom of a specific element that contains the usual number of protons in its nucleus but a different number of neutrons.

IUPAC: the International Union of Pure and Applied Chemistry, an organization that establishes international standards and practices for chemistry.

juxtacrine signaling: a type of cell signaling in which the signaling compound is produced within a cell and delivered to receptors in an adjacent cell via physical contact.

kelvin: the basic unit of the Kelvin temperature scale, equivalent in magnitude to one degree of the Celsius temperature scale; 0 kelvin is absolute zero, the lowest temperature theoretically possible, equal to –273.15 degrees Celsius.

Krebs (citric acid) cycle: a cyclic series of biochemical reactions that completes the conversion of glucose into carbon dioxide, water, and adenosine triphosphate (ATP), consuming and then regenerating citric acid in the process.

law of conservation of mass: the principle that states that all atoms present at the beginning of a process must also be present after the completion of the process; matter can be neither created nor destroyed, only changed from one form to another.

law of definite proportions: the principle that states that the proportions by mass of the elements that combine to form a specific compound are definite and do not vary.

law of multiple proportions: the principle that states that the atoms that make up any particular compound are present in simple, whole-number ratios.

leaving group: an atom or group of atoms that is removed from a molecule during a chemical reaction, along with the electron pair that previously bonded it to the molecule.

Lewis acid-base theory: definitions of acids and bases developed in 1923 by American chemist Gilbert N. Lewis; defines an acid as any chemical species that can accept an electron pair and a base as any chemical species that can donate an electron pair.

Lewis structure: a simplified representation of the bonds and unbonded electron pairs associated with the atoms of a molecule.

lipid: a type of biomolecule that is soluble in organic nonpolar solvents and generally insoluble in water; includes fats, waxes, and the major components of organic oils.

liquid: a state of matter in which material is fluid, has definite volume but indefinite shape, and maintains a relatively constant density.

logarithm: the exponent, or power, to which a specific base number must be raised to produce a given value; commonly abbreviated "log."

lone pair: two valence electrons that share an orbital and are not involved in the formation of a chemical bond; also called a nonbonding pair.

macromolecule: a very large molecule; most often refers to polymers but can also refer to single molecules with extended, non-polymeric structures.

main group elements: the elements in groups 1 (excluding hydrogen), 2, and 13–18 of the periodic table; alternatively, all elements (excluding hydrogen) that are neither lanthanides, actinides, nor transition metals.

malleability: the ability of a solid material to be deformed by the application of compressive (pushing) force, such as hammering, without breaking or fracturing.

man-made element: an element or isotope that does not occur naturally but is synthesized in high-energy particle accelerators by bombarding other elements with streams of nuclear particles.

mass: an intrinsic property of matter that determines the extent to which it can be acted on by a force.

matter: any substance that occupies physical space and is composed of atoms or the particles that make up atoms.

mercaptan: an organic compound characterized by the presence of a sulfhydryl functional group (–SH).

metallic bond: a type of chemical bond formed by the sharing of delocalized electrons between a number of metal atoms.

moiety: a specific portion of a molecular structure.

mole: the amount of any pure substance that contains as many elementary units (approximately 6.022 × 1023) as there are atoms in twelve grams of the isotope carbon-12.

molecular formula: a chemical formula that indicates how many atoms of each element are present in one molecule of a substance.

molecule: the basic unit of a compound, composed of two or more atoms connected by chemical bonds.

monatomic: describes a molecule or ion consisting of only one atom.

monomer: a molecule capable of bonding to other molecules to form a polymer.

multiple bond: a bond formed by two atoms sharing two or more electron pairs; includes double bonds and triple bonds.

multivalent: describes an atom that has the ability to accept or donate more than one valence electron and thus can exist in more than one oxidation state.

neurotransmitter: a chemical species that carries electrochemical signals across the synapses between neurons in the nervous system.

neutral: describes a chemical solution that has a pH of 7 and thus is neither acidic nor basic.

neutralization: a chemical reaction between an acid and a base that results in the formation of a salt, usually accompanied by water.

neutron: a fundamental subatomic particle in the atomic nucleus that is electrically neutral and about equal in mass to the mass of one proton.

noble gases: a group of gases, including helium, neon, and argon, that occur naturally only as monatomic materials and do not normally form compounds.

nomenclature: a system of specific names or terms and the rules for devising or applying them; in chemistry, refers mainly to the system of names for chemical compounds as established by the International Union of Pure and Applied Chemistry (IUPAC).

nucleic acid: a biopolymer consisting of many different nucleotides bonded together; includes both DNA and RNA.

nucleophilic: describes a chemical species that tends to react with positively charged or electron-poor species.

nucleotide: the basic structural component of DNA and RNA, consisting of a ribose (in RNA) or deoxyribose (in DNA) sugar molecule bonded to a phosphate group and one of five nucleobases: cytosine, adenine, guanine, thymine (DNA only), or uracil (RNA only).

nucleus: the central core of an atom, consisting of specific numbers of protons and neutrons and accounting for at least 99.98 percent of the atomic mass.

octet rule: the tendency of atoms when bonding to either accept or donate electrons in such a way that they end up with eight electrons in the outermost electron shell.

odor: the sensation created when molecules of a volatile chemical compound vaporize and bind to olfactory receptors in the nose.

orbital: a specific region of space about the nucleus of an atom in which electrons of a given energy level are most likely to be found.

organic acid: an acid derived from an organic compound.

organic chemistry: the study of the chemical identities, behaviors, and reactions of carbon-based compounds and materials.

organic compound: generally, a compound containing one or more carbon atoms, although some carbon-containing compounds are considered inorganic.

organophosphorus compound: an organic compound containing one or more carbon-phosphorus bonds.

organosulfur compound: an organic compound containing one or more carbon-sulfur bonds.

osmosis: the passage of solvent molecules through a semipermeable membrane from a region of low solute concentration to one of higher concentration; also the primary mechanism by which water moves through cell walls.

osmotic pressure: the pressure that would have to be applied to a solution to prevent the flow of solvent through a semipermeable membrane.

oxidation: the loss of one or more electrons by an atom, ion, or molecule.

oxidation state: a number that indicates the degree to which an atom or ion in a chemical compound has been oxidized or reduced.

oxide: a compound formed by the reaction of any element with oxygen, such as carbon dioxide, carbon monoxide, iron oxide, or diphosphorus pentoxide.

oxidizing agent: any atom, ion, or molecule that accepts one or more electrons from another atom, ion, or molecule in a reduction-oxidation (redox) reaction and is thus reduced in the process.

oxoanion: an ion consisting of one or more central atoms bonded to a number of oxygen atoms and bearing a net negative electrical charge.

paracrine signaling: a type of cell signaling in which the signaling compound is produced in one location in the body and delivered to receptors in a nearby cell.

parent chain: the longest continuous hydrocarbon chain in an organic compound, used as the basis for its unique name identification.

partial charge: a term used to indicate a degree of charge separation in bonded atoms due to the different electronegativities of the atoms at either end of the bond

passive transport: the passage of materials through a membrane with no input of energy required.

peptide: an organic compound composed of a relatively small number of amino acid molecules, held together by covalent bonds between a carbon atom of one molecule and a nitrogen atom of the next.

peptide bond: a covalent bond that links the carboxyl group of one amino acid to the amine group of another, enabling the formation of proteins and other polypeptides.

periodic table: the chart representing the known elements by atomic number and electron distribution.

periodicity: the tendency of elements with similar electron distributions in their valence shells to exhibit similar chemical properties, such as ionization energy, atomic radius, and electronegativity.

pH: a numerical value that represents the acidity or basicity of a solution, with 0 being the most acidic, 14 being the most basic, and 7 being neutral.

pH indicator: a compound that changes color according to the pH of a solution, or a device that measures the pH of a solution electronically.

phagocytosis: a type of endocytosis in which solid particles are taken into the cytoplasm of a cell through the cell membrane.

phase transition: the change of matter from one state to another, such as from solid to liquid or liquid to gas, due to the transfer of thermal energy.

phenyl group: a cyclic functional group with the formula $-C_6H_5$, similar to a benzene molecule but with one fewer hydrogen atom.

phosphane: a class of compounds with the general formula PnHn+2; also, an alternative name for the compound phosphine (PH_3).

pi bond: a covalent chemical bond formed when parallel *p* orbitals of two adjacent atoms overlap in a side-by-side manner to form two molecular orbitals.

pinocytosis: a type of endocytosis in which extracellular fluid and any substances it may contain are taken into the cytoplasm of a cell through the cell membrane.

plasticity: the ability of a material to undergo deformation without breaking.

plutonium: atomic number 94, an extremely toxic and dangerously radioactive element of the actinide group that has several known isotopes.

polar: describes a molecule or functional group in which there is a difference in the distribution of electronic charge, causing one part of the molecule or group to be relatively electrically positive and another part to be relatively electrically negative.

polarity: a characteristic of a molecule or functional group in which there is a difference in the distribution of electronic charge, causing one part of the molecule or group to be relatively electrically positive and another part to be relatively electrically negative.

polyatomic: describes a molecule or ion consisting of multiple atoms that are chemically bonded to each other.

polyether: an organic compound characterized by the presence of multiple ether linkages (C–O–C) in its molecular structure.

polymer: a large molecule formed by the concatenation of many individual smaller molecules, known as monomers.

polymerase chain reaction: a laboratory method in which a very small amount of DNA can be replicated thousands or even millions of times, using free nucleotides and an enzyme called DNA polymerase.

polymerization: a process in which small molecules bond together in a chain reaction to form a polymer, which is a much larger molecule composed of repeating structural units.

polyol: an organic compound containing multiple hydroxyl groups.

positron: the antiparticle of the electron, with identical physical properties but the opposite (positive) electrical charge.

potential energy: the energy contained in an object due to its position, composition, or arrangement that is capable of being translated into kinetic energy or the performance of work; for example, the energy contained in the bonds between atoms, which can be used to fuel a chemical reaction.

precipitant: a substance that, when added to a solution, causes a component of the solution to become solid and separate from the liquid.

primary: describes an organic compound in which one of the hydrogen atoms bonded to a central atom is replaced by another atom or group of atoms, called a substituent.

probability density: in reference to electrons, the probability of finding a particular electron in a given region of space within an atom or molecule; also called electron density.

product: a chemical species that is formed as a result of a chemical reaction.

proportion: the quantity of a unit component relative to the others in the same system.

protein: a biological polymer consisting of one or more long chains of amino acids linked by peptide bonds in a sequence specified by an organism's DNA.

protium: the essential form of hydrogen, containing one proton and no neutron; the most common form of matter in the known universe.

proton: a fundamental subatomic particle with a single positive electrical charge, found in the atomic nucleus.

proton acceptor: a compound or part of a chemical compound that has the ability to accept a proton (H+) from a suitably acidic material in a chemical reaction.

proton donor: a compound or part of a chemical compound that has the ability to relinquish a proton (H+) to a suitably basic material in a chemical reaction.

protonation: the addition of a proton, in the form of a hydrogen cation (H+), to an atom, ion, or molecule.

pyrophoricity: a property of some solids and liquids that causes them to spontaneously combust when exposed to air.

quantum number: one of four numbers that describe the energy level, orbital shape, orbital orientation, and spin of an electron within an atom.

R (generic placeholder): a symbol used primarily in organic chemistry to represent a hydrocarbon side chain or other unspecified group of atoms in a molecule; can be used specifically for an alkyl group, with Ar used to represent an aryl group.

radioactive decay: the loss of particles from the nucleus of an unstable atom in the form of ionizing radiation.

radioactivity: the emission of subatomic particles due to the spontaneous decay of an unstable atomic nucleus, the process ending with the formation of a stable atomic nucleus of lower mass.

radioisotope: any radioactive isotope of an element that undergoes spontaneous nuclear fission until a stable, nonradioactive isotope is formed.

rare earth elements: a group of chemical elements in the periodic table that includes the fifteen lanthanide elements (lanthanum, cerium, praseodymium, neodymium, promethium, samarium, europium, gadolinium, terbium, dysprosium, holmium, erbium, thulium, ytterbium, and lutetium) as well as scandium and yttrium.

rate-determining step: in any multistep process, the step that proceeds at the slowest rate compared to the other steps in the process, thus determining the maximum rate at which the overall process takes place.

reactant: a chemical species that takes part in a chemical reaction.

reaction mechanism: the sequence of electron and orbital interactions that occurs during a chemical reaction as chemical bonds are broken, made, and rearranged.

reaction rate: how much of a particular reaction or reaction step occurs per unit time.

reaction step: a stage in a multistep reaction in which an elementary reaction converts either the reactants or a previously formed intermediate structure into either a new intermediate structure or the final products.

reactive intermediate: a short-lived, highly reactive chemical species that is formed during an intermediate reaction step in a multistep reaction and typically cannot be isolated.

reactivity: the propensity of a chemical species to undergo a reaction under applied conditions.

receptor: a molecule or molecular structure, typically an enzyme or other protein, that interacts only with specific compounds, often triggering a biochemical response in the cells to which the receptor is attached.

redox reaction: a reaction in which electrons are transferred from one atom, ion, or molecule (oxidation) to another (reduction).

reducing agent: any atom, ion, or molecule that donates one or more electrons to another atom, ion, or molecule in a reduction-oxidation (redox) reaction and is thus oxidized in the process.

reduction: the gain of one or more electrons by an atom, ion, or molecule.

repulsion: an oppositional force that pushes two entities apart, such as the electrostatic repulsion between particles of like electrical charge.

resonance: a method of graphically representing a molecule with both single and multiple covalent bonds whose valence electrons are not associated with one particular atom or bond; two or more diagrams, or resonance structures, depict each possible arrangement of single and multiple bonds, while the true structure of the molecule is somewhere between the different resonance structures, an intermediate form known as the resonance hybrid.

reverse osmosis: the application of pressure to a solution in order to overcome the osmotic pressure of a semipermeable membrane and force water to pass through it in the direction opposite to normal osmotic flow.

ribonucleic acid (RNA): a category of large molecules, typically consisting of a single strand of nucleotides, that perform various functions in cells, including the transcription of DNA molecules and the transfer of specific genetic information for protein synthesis.

ribosome complex: a structure consisting of ribosomal RNA (rRNA) and enzymes that decodes messenger RNA (mRNA) and coordinates the assembly of proteins from amino acids carried by transfer RNA (tRNA).

RNA polymerase: the enzyme responsible for initiating gene transcription in order to assemble and replicate strands of RNA.

salt: an ionic compound produced by the reaction of an acid and a base, formed either by combining electron-rich and electron-poor species or by replacing the hydrogen cation ($H+$) of the acid with another cation from the base.

saturated: describes an organic compound in which carbon atoms are attached to other atoms via single bonds only, allowing the compound to contain the maximum possible number of hydrogen atoms.

secondary: describes an organic compound in which two of the hydrogen atoms bonded to a central atom are replaced by other atoms or groups of atoms, called substituents.

sedimentation: the gradual accumulation of solid particles as they settle out of suspension in a fluid.

semipermeable membrane: a membrane that allows the passage of a material, such as water or another solvent, from one side to the other while preventing the passage of other materials, such as dissolved salts or another solute.

semi-structural formula: a chemical formula that does not show the complete molecular structure of a compound but gives sufficient information to differentiate it from its isomers.

shared pair: the two electrons shared between two atoms in a normal covalent bond.

side chain: a group of atoms that branches off from the main chain, or backbone, of an organic molecule.

single bond: a type of chemical bond in which two adjacent atoms are connected by a single pair of electrons via the direct overlap of their atomic orbitals.

single displacement: a substitution reaction in which atoms of one element replace atoms of another element in a compound.

solid: a state of matter in which material is non-fluid, has definite volume and shape, and maintains a near-constant density.

solubility: the ability of a particular substance, or solute, to dissolve in a particular solvent at a given temperature and pressure.

solute: any material that is dissolved in a liquid or fluid medium, usually water.

solvent: any fluid, most commonly water, that dissolves other materials.

spontaneous reaction: a chemical reaction that occurs without the input of energy from an outside source.

standard temperature and pressure (STP): standardized conditions under which experiments are conducted in order to maintain consistency between results; established as 273.15 kelvins and 105 pascals by the International Union of Pure and Applied Chemistry (IUPAC), though other organizations use different standards.

stereochemistry: the relative spatial arrangement of atoms and bonds in a molecular structure

stoichiometry: the relative quantities of substances that make up compounds or participate in chemical reactions, or the branch of chemistry that deals with these relative quantities.

structural formula: a graphical representation of the arrangement of atoms and bonds within a molecule.

substitution reaction: a chemical reaction in which one component of a compound is replaced by a different atom or group of atoms without altering the basic structure of the molecule.

supernatant: the liquid or fluid that remains after a substance is precipitated from a solution.

synthesis reaction: a chemical reaction in which two or more reactants combine to form a single, more complex product; also, a reaction executed with the goal of creating a specific product or products.

synthetic: produced by artificial means or manipulation rather than by naturally occurring processes.

tensile stress: a force that acts to pull or stretch a material.

tertiary: describes an organic compound in which three of the hydrogen atoms bonded to a central atom are replaced by other atoms or groups of atoms, called substituents.

thermodynamic equilibrium: also known as thermal equilibrium, the condition in which a system and its surroundings are in the same energy and temperature state.

thermodynamics: the branch of physics that deals with the relationships between energy, heat, and work within a physical system.

transcription factor: a protein that binds to DNA in order to initiate, regulate, or block gene transcription.

transition metals: the elements of the periodic table that have valence electrons in their *d* orbitals.

transition state: an unstable structure formed during a chemical reaction at the peak of its potential energy that cannot be isolated and ultimately breaks down, either forming the products of the reaction or reverting back to the original reactants.

translation: the overall process of RNA-mediated protein synthesis, entailing transcription of genetic information from the DNA molecule by messenger RNA (mRNA), assembly of the corresponding amino acids by transfer RNA (tRNA), and formation of the protein molecule in ribosomes by ribosomal RNA (rRNA).

transmitter: a biochemical compound produced to trigger a specific response at a corresponding receptor site.

transuranium elements: the elements in the periodic table that have an atomic number greater than that of uranium (92), all of which are unstable and radioactive.

triple bond: a type of chemical bond in which two adjacent atoms are connected by six bonding electrons rather than two.

tritium: an unstable radioisotope of hydrogen that contains one proton and two neutrons.

trivalent: describes an atom that has the ability to exist in three different oxidation states and to form compounds accordingly.

unpaired electron: a valence electron that occupies an orbital by itself and is not involved in the formation of a chemical bond.

unsaturated: describes an organic compound in which carbon atoms are attached to other atoms via double or triple bonds, preventing the compound from containing the maximum possible number of hydrogen atoms.

unstable: describes a chemical species with a structure or composition that is prone to spontaneous decomposition.

valence: the maximum number of bonds an atom can form with other atoms based on its electron configuration; also called valency.

valence electron: an electron that occupies the outermost or valence shell of an atom and participates in chemical processes such as bond formation and ionization.

valence shell: the outermost energy level occupied by electrons in an atom.

vesicle: a small bubble within a cell, surrounded by a double layer of lipids.

vicinal: describes the relationship between two functional groups bonded to adjacent (typically carbon) atoms in the same molecule.

VSEPR theory: the valence-shell electron-pair repulsion theory, which states that electron pairs in the valence shell of an atom will arrange themselves so that the electrostatic repulsion between them is minimized.

water soluble: describes a compound that can be dissolved by water.

wavelength: the distance between the same point on two successive waves, beyond which the shape of the wave begins to repeat.

BIBLIOGRAPHY

Abbott, David. *The Biographical Dictionary of Scientists.* New York: P. Bedrick, 1984. Print.

Acheson, R.M. *An Introduction to the Chemistry of Heterocyclic Compounds.* 3rd ed. New York: Wiley, 2008. Print.

Adleman, Leonard M. *Science*, New Series, Vol. 266, No. 5187 (Nov. 11, 1994), 1021-1024.

Agricola, Georgius. *De re metallica.* Trans. Herbert Clark Hoover and Lou Henry Hoover. New York: Dover, 1950. Print.

Aldersey-Williams, Hugh. *The Most Beautiful Molecule: The Discovery of the Buckyball.* New York: John Wiley, 1995.

Amos, Martyn. *Theoretical and Experimental DNA Computation.* Natural Computing Series. Springer (2010)

Amos, S.W. *Principles of Transistor Circuits.* New York, NY: Elsevier, 2013. Print.

Anslyn, Eric V. and Dougherty, Dennis A. M*odern Physical Organic Chemistry.* University Science Books, 2006. Print.

Arnaut, Luís, Sebastião Formosinho, and Hugh Burrows. *Chemical Kinetics: From Molecular Structure to Chemical Reactivity.* Oxford: Elsevier, 2007. Print.

Askeland, Donald R., Wendelin J. Wright, D. K. Bhattacharya, and Raj P. Chhabra. *The Science and Engineering of Materials.* Boston: Cengage Learning, 2016. Print.

Atkins, Peter, and Julio de Paula. *Atkins' Physical Chemistry.* 10th ed. New York: Oxford University Press, 2014. Print.

Bachrach, Steven M. *Computational Organic Chemistry.* Hoboken: Wiley, 2007. Print.

Bailey, James E., and David F. Ollis. *Biochemical Engineering Fundamentals.* 2nd ed. New York: McGraw, 1988. Print.

Barrett, Jack. *Structure and Bonding.* Cambridge, UK: The Royal Society of Chemistry, 2001. Print.

Bartlett, Neil. *The Oxidation of Oxygen and Related Chemistry: Selected Papers of Neil Bartlett.* Singapore: World Scientific, 2001. Print.

Basu, Chhandak, ed. *PCR Primer Design (Methods in Molecular Biology).* Humana Press, 2015.

Bayse, Craig A., and Julia L Brumaghim. *Biochalcogen Chemistry. The Biological Chemistry of Sulfur, Selenium and Tellurium.* American Chemical Society, 2015. Print.

Begon, Michael, Colin R. Townsend, and John L. Harper. *Ecology: From Individuals to Ecosystems.* 4th ed. Malden: Blackwell, 2006. Print.

Bell, Jerry A. *Chemistry: A Project of the American Chemical Society.* New York: Freeman, 2005. Print.

Berg, Jeremy M., John L. Tymoczko, Gregory J. Gatto, and Lubert Strye. *Biochemistry.* 8th ed. New York: W. H. Freeman, 2015. Print.

Bhat, H.L. *Introduction to Crystal Growth. Principles and Practice.* Boca Raton, FL: CRC Press, 2015. Print.

Biasonni, Roberto, and Alessandro Raso, eds. *Quantitative Real-Time PCR: Methods and Protocols (Methods in Molecular Biology).* Humana Press, 2014..

Birdi, K.S. *Surface Chemistry Essentials.* Boca Raton, FL: CRC Press, 2014. Print.

Boreskov, G.K. *Heterogenous Catalysis.* New York, NY: Nova Science Publishers, 2003. Print.

Bottani, Eduardo J., and Juan Tascón, M.D., eds. *Adsorption by Carbons.* Jordan Hill, UK: Elsevier, 2008. Print.

Boyd, Richard N. *An Introduction to Nuclear Astrophysics.* Chicago, IL: University of Chicago Press, 2008. Print.

Bréchignac, Catherine, Philippe Houdya, and Marcel Lahmani, eds. *Nanomaterials and Nanochemistry.* New York, NY: Springer-Verlag, 2007. Print

Brescia, Frank, John Arents, and Herbert Meislich. *Fundamentals of Chemistry.* 4th ed. New York, NY: Academic Press, 1980. Print.

Brey, Wallace S. *Physical Chemistry and Its Biological Applications.* New York, NY: Academic Press, 1978. Print.

Brock, W. H. *The Chemical Tree: A History of Chemistry.* New York: Norton, 2000. Print.

Bustin, Stephen A., ed. *The PCR Revolution: Basic Technologies and Applications.* Cambridge University Press, 2014.

Byrne, Michael. "The Black Holes We See in Space Might Already Be White Holes." *Motherboard.* Vice Media, 21 July 2015. Web. 30 Apr. 2015.

Cain, Fraser. "Never a Star: Did Supermassive Black Holes Form Directly?" *Universe Today.* Author, 7 Sept. 2007. 30 Apr. 2015.

Campbell, F.C., ed. *Phase Diagrams. Understanding the Basics Materials.* Park, OH: ASM International, 2012. Print.

Campbell, Neil A. *Biology.* Menlo Park: Benjamin, 1987. Print.

Capper, *Peter, and Michael Mauk, eds. Liquid Phase Epitaxy of Electronic, Optical and Optoelectronic Materials.* Hoboken, NJ: Wiley, 2007. Print.

Carey, Francis A., and Richard J. Sundberg. *Advanced Organic Chemistry: Part A—Structure and Mechanisms.* 5th ed. New York: Springer, 2007. Print

Carter, C. Barry, and Grant M. Norton. *Ceramic Materials Science and Engineering.* New York: Springer Science + Business Media, 2007. Print.

Claus, Edward P., and Varro E. Tyler Jr. *Pharmacognosy.* Philadelphia: Lea, 1965. Print.

Clugson, Michael, and Rosalind Flemming. *Advanced Chemistry.* New York: Oxford UP, 2000. Print.

Cobb, Allan B. *The Basics of Nonmetals.* New York, NY: Rosen Publishing, 2014. Print.

Cobb, Cathy, and Harold Goldwhite. *Creations of Fire: Chemistry's Lively History from Alchemy to the Atomic Age.* New York: Plenum, 1995. Print.

Cooper, David L., ed. *Valence Bond Theory.* Elsevier, 2002. Print.

Corbridge, D.E.C. *Phosphorus. Chemistry, Biochemistry and Technology.* 6th ed. Boca Raton, FL: CRC Press, 2013. Print.

Costanzo, Linda S. *Physiology: Cases and Problems.* 4th ed. Baltimore: Lippincott, 2012. Print.

Crowe, Jonathan, and Tony Bradshaw. *Chemistry for the Biosciences: The Essential Concepts.* 2nd ed. Oxford: Oxford UP, 2010. Print.

Crystal, David, ed. *The Cambridge Biographical Encyclopedia.* 2nd ed. New York: Cambridge UP, 1998. Print.

Daniels, Farrington, and Robert A. Alberty. *Physical Chemistry.* 3rd ed. New York: Wiley, 1966. Print.

Davison, Sydney George, and K.W. Sulston, *Green-Function Theory of Chemisorption.* New York: Springer, 2006. Print.

Delgado, Juan L., et al. "Buckyballs," *Polyarenes II.* Topics in Current Chemistry: Volume 350, 2014, pp. 1-64.

Dennis, Carina, Richard Gallagher, and Philip Campbell, eds. *The Human Genome.* Spec. issue of *Nature* 409.6822 (2001): 813–958. Print.

Deutch, Yvonne, ed. *Science against Crime.* London: Marshall, 1982. Print.

Doering, Robert and Yoshio Nishi, eds. *Handbook of Semiconductor Manufacturing Technology.* 2nd ed. Boca Raton, FL: CRC Press, 2008. Print.

Douglas, Bodie E., Darl H. McDaniel, and John J. Alexander. *Concepts and Models of Inorganic Chemistry.* 3rd ed. New York: Wiley, 1994. Print.

Dubois, Jean-Marie and Belin-Ferré, *Esther Complex Metallic Alloys. Fundamentals and Applications.* New York, NY: John Wiley & Sons, 2010. Print.

Dumé, Belle. "Bountiful buckyballs resolve interstellar mystery" *physicsworld.* Web. 20 Aug. 2015. <http://physicsworld.com/cws/article/news/2015/jul/16/bountiful-buckyballs-resolve-interstellar-mystery>

Dyson, Freeman. *Origins of Life.* 2nd ed. New York: Cambridge UP, 1999. Print.

Emsley, John. *Nature's Building Blocks: An A–Z Guide to the Elements.* New ed. New York: Oxford UP, 2011. Print.

Eu, Byung Chan and Al-Ghoul, Mazen. *Chemical Thermodynamics with examples for Non-equilibrium Processes.* Hackensack, NJ: World Scientific, 2010. Print.

Favre, Henry, and Warren H. Powell. *Nomenclature of Organic Chemistry.* Cambridge: Royal Soc. of Chemistry, 2014. Print.

Fenichell, Stephen. *Plastic: The Making of a Synthetic Century.* New York: Harper, 2009. Print.

Fleming, Ian. *Molecular Orbitals and Organic Chemical Reactions.* Hoboken: Wiley, 2010. Print.

Fontana, W., and L. Buss. ""The Arrival of the Fittest": Toward a Theory of Biological Organization." *Bulletin of Mathematical Biology* 56.1 (1994): 1-64. Web.

GaBany, R. Jay. "A Singular Place." *Cosmotography.* Author, n.d. Web. 30 Apr. 2015.

Gillespie, Ronald J., and István Hargittai. *The VSEPR Model of Molecular Geometry.* Dover, 2012. Print.

Glicksman, Martin Eden. *Principles of Solidification. An Introduction to Modern Casting and Crystal Growth Concepts.* New York, NY: Springer, 2010. Print.

Gooch, Jan W. *Encyclopedic Dictionary of Polymers.* New York: Springer, 2007. Print.

Graedel, T. E. "The Pieces of the Periodic Table Nature's Building Blocks: An A-Z Guide to the Elements, by John Emsley. *World of the Elements,* Elements of the World, by Hans-Jurgen Quadbeck-Seeger." *Journal of Industrial Ecology* 13.1 (2009): 154-55. Web.

Greene, Theodore W., and Peter G. M. Wuts. *Greene's Protective Groups in Organic Synthesis.* 4th ed. New York: Wiley, 2007. Print.

Greenwood, N.N. and Earnshaw, A. *Chemistry of the Elements.* 2nd ed. Burlington, MA: Elsevier Butterworth-Heinemann, 2005. Print.

Gregersen, Erik, ed. T*he Britannica Guide to the Atom.* New York: Britannica, 2011. Print.

Gribbin, John. *Science: A History, 1543–2001.* London: Lane, 2002. Print.

Gupta, K.M., and Nishu Gupta. *Advanced Electrical and Electronic Materials. Processes and Applications.* New York, NY: Wiley/Scrivener Publishing, 2015. Print.

Hagen, Jens. I*ndustrial Catalysis. A Practical Approach.* 3rd ed. Weinheim, GER: Wiley-VCH Verlag, 2015. Print.

Halka, Monika, and Brian Nordstrom. *Halogens and Noble Gases.* New York, NY: Facts on File Inc., 2010. Print.

Hall, Judy. *The Crystal Bible. A Definitive Guide to Crystals.* Cincinnati, OH: Walking Stick Press, 2003. Print.

Hall, Shannon. "How Do Black Holes Get Super Massive?" *Universe Today.* Fraser Cain, 13 Aug. 2013. Web. 30 Apr. 2015.

Hamid, Arabnia R. and Quoc Nam Tran. *Emerging Trends in Computational Biology, Bioinformatics, and Systems Biology: Algorithms and Software Tools.* Emerging Trends in Computer Science and Applied Computing. Morgan Kaufmann (2015).

Haynes, W. H., PhD, David R. Lide, PhD, and Thomas J. Bruno, PhD, eds. *CRC Handbook of Chemistry and Physics,* 96th Edition. New York: CRC/Taylor & Francis Group, 2015. Print.

Hendrickson, James B., Donald J. Cram, and George S. Hammond. *Organic Chemistry.* 3rd ed. New York: McGraw, 1970. Print.

Herbert, R. B. *The Biosynthesis of Secondary Metabolites.* 2nd ed. London: Chapman & Hall, 1989. Print.

Herbst, Eric and van Dishoeck, Ewine F. "Complex Organic Interstellar Molecules" *The Annual Review of Astronomy and Astrophysics 2009.* 47:427-480. Retrieved from <http://www.annualreviews.org>

Herman, Marian A., Wolfgang Richter and Helmut Sitter. *Epitaxy. Physical Principles and Technical Implementation.* New York, NY: Springer-Verlag, 2004. Print.

Hillert, Mats. *Phase Equilibria, Phase Diagrams and Phase Transformations. Their Thermodynamic Basis.* 2nd ed. New York, NY: Cambridge University Press, 2007. Print.

Hoffmann, Roald. *The Same and Not the Same.* New York: Columbia UP, 1995. Print.

Holden, Alan, and Phylis Morrison. *Crystals and Crystal Growing.* New York: Doubleday, 1960. Print.

House, James E., and Kathleen A. House. *Descriptive Inorganic Chemistry.* Waltham, MA: Academic Press, 2016. Print.

Hudgins, Douglas M. "Interstellar Polycyclic Aromatic Compounds and Astrophysics." *The Astrophysics & Astrochemistry Laboratory.* NASA Ames Research Center, n.d. Web.

Hudson, John B. *Surface Science. An Introduction* New York, NY: John Wiley & Sons, 1998. Print.

Ibach, Harald. *Physics of Surfaces and Interfaces.* New York, NY: Springer, 2006. Print.

Ignatova, Zoya, and Israel Martinez-Perez. *DNA Computing Models.* Springer (2010).

Ihde, Aaron J. *The Development of Modern Chemistry.* New York: Harper & Row, 1964. Print.

Jacobs, Adam. *Understanding Organic Reaction Mechanisms.* Cambridge: Cambridge UP, 1997. Print.

Johnson, Rebecca L. *Atomic Structure.* Minneapolis: Lerner, 2008. Print.

Jones, Mark Martin. *Chemistry and Society.* 5th ed. Philadelphia: Saunders College Pub., 1987. Print.

Jorgensen, William L., and Lionel Salem. T*he Organic Chemist's Book of Orbitals.* New York: Academic, 1973. Print.

Kaufman, Myron. *Principles of Thermodynamics.* New York, NY: Marcel Dekker Inc., 2002. Print.

Kean, Sam. *The Disappearing Spoon: And Other True Tales of Madness, Love, and the History of the World from the Periodic Table of the Elements.* New York: Little, Brown, 2010. Print.

Kenkel, John. *Basic Chemistry Concepts and Exercises.* Boca Raton, FL: CRC Press, 2011. Print.

Kittel, Charles. *Introduction to Solid State Physics.* 8th ed. Hoboken, NJ: Wiley, 2005. Print.

Knystautas, Émile, ed. *Engineering Thin Films and Nanostructures With Ion Beams.* Boca Raton, FL: Taylor & Francis/CRC Press, 2005. Print.

Kolasinski, Kurt W. *Surface Science. Foundations of Catalysis and Nanoscience.* 3rd ed. New York, NY: John Wiley & Sons, 2012. Print.

Kosmulski, Marek. *Chemical Properties of Material Surfaces.* New York, NY: Marcel Dekker Inc., 2001. Print.

Kragh, Helge. *Niels Bohr and the Quantum Atom: The Bohr Model of Atomic Structure, 1913–1925*. Oxford: Oxford UP, 2012. Print.

Kröger, Ferdinand A. T*he Chemistry of Imperfect Crystals*. Amsterdam: North-Holland Publ., 1974. Print.

Kucera, Jane. *Reverse Osmosis: Design, Processes, and Applications for Engineers*. Hoboken: Wiley, 2010. Print.

Kuech, Thomas F., ed. *Handbook of Crystal Growth III Thin Films and Epitaxy: Basic Techniques, and Materials, Processes and Technology*. 2nd ed. Waltham, MA: Elsevier, 2015. Print.

Kutney, Gerald. Sulfur. *History, Technology, Applications & Industry*. Toronto, ON: ChemTec Publishing, 2007. Print.

Kwok, Sun. *Physics and Chemistry of the Interstellar Medium*. Herndon, VA: University Science Books, 2007. Print.

Lafferty, Peter, and Julian Rowe, eds. *The Hutchinson Dictionary of Science*. 2nd ed. Oxford: Helicon, 1998. Print.

Laidler, Keith J. *Chemical Kinetics*. 3rd ed. New York: Harper, 1987. Print.

Landolt, Dieter. C*orrosion and Surface Chemistry of Metals*. Boca Raton, FL: Taylor and Francis Group LLC, 2007. Print.

Lane, Nick. *Oxygen: The Molecule That Made the World*. Oxford: Oxford UP, 2009. Print.

Lehninger, Albert L. *Biochemistry: The Molecular Basis of Cell Structure and Function*. 2nd ed. New York: Worth, 1975. Print.

Leubner, Ingo H. *Precision Crystallization. Theory and Practice of Controlling Crystal Size*. Boca Raton, FL: CRC Press, 2010. Print.

Lewis, Richard J., and Gessner G. Hawley. *Hawley's Condensed Chemical Dictionary*. 15th ed. New York: Wiley, 2007. Print.

Lew, Kristi. *Chemical Reactions*. New York: Infobase, 2008. Print.

Lodish, Harvey, et al. *Molecular Cell Biology*. 7th ed. New York: Freeman, 2013. Print.

Lo, Y. M. Dennis, Rossa W. K. Chiu, and K. C. Allen Allen Chan, eds. *Clinical Applications of PCR*. 2nd ed. Totowa, NJ: Humana, 2006. Print.

Lüth, Hans. *Solid Surfaces, Interfaces and Thin Films*. 6th ed. New York, NY: Springer, 2015. Print.

Mackay, K. M., R. A. Mackay, and W. Henderson. *Introduction to Modern Inorganic Chemistry*. 6th ed. Cheltenham: Nelson, 2002. Print.

Madou, Marc J. and Morrison, S. Roy. *Chemical Sensing with Solid State Devices*. San Diego, CA: Academic Press, 1989. Print.

Mann, J., et al. *Natural Products: Their Chemistry and Biological Significance*. New York: Wiley, 1994. Print.

Marshall, John. "Introduction: Black Holes." *New Scientist*. Reed Business Information, 6 Jan. 2010. Web. 30 Apr. 2015.

Masel, Richard I. *Principles of Adsorption and Reaction on Solid Surfaces*. New York, NY: John Wiley & Sons, 1996. Print.

Masterton, William L., Cecile N. Hurley, and Edward Neth. *Chemistry: Principles and Reactions*. 7th ed. Belmont: Brooks, 2012. Print.

Matson, Michael L., and Alvin W. Orbaek. *Inorganic Chemistry for Dummies*. Hoboken: Wiley, 2013. Print.

McCafferty, E. *Surface Chemistry of Aqueous Corrosion Processes*. New York, NY: Springer, 2015. Print.

Miessler, Gary L., Paul J. Fischer, and Donald A. Tarr. *Inorganic Chemistry*. 5th ed. Upper Saddle River: Prentice Hall, 2013. Print.

Mikulecky, Peter J., Michelle Rose Gilman and Kate Brutlag. *AP Chemistry for Dummies*. Hoboken: Wiley, 2009. Print.

Moiseyev, Valentin N. *Titanium Alloys. Russian Aircraft and Aerospace Applications*. Boca Raton, FL: CRC Press, 2006. Print.

Moller, Simon. *PCR (The Basics)*. Taylor & Francis, 2006.

Morrison, Robert Thornton, and Robert Neilson Boyd. *Organic Chemistry*. 7th ed. Englewood Cliffs, N.J.: Prentice Hall, 2003. Print.

Mott, Nevill F., and Ronald Wilfried. *Gurney. Electronic Processes in Ionic Crystals*. New York, NY: Dover, 1964. Print

Mühlbauer, Alfred. *History of Induction Heating & Melting*. Essen, GER: Vulkan-Verlag, 2008. Print.

Murata, Satoshi, and Satoshi Kobayashi (eds). *DNA Computing and Molecular Programming: 20th International Conference*, Kyoto, Japan, September 22-26, 2014. Springer (2014).

Myers, Richard. *The Basics of Chemistry*. Westport: Greenwood, 2003. Print.

Nastasi, M., and Mayer, J.W. *Ion Implantation and Synthesis of Materials*. New York, NY: Springer-Verlag, 2006. Print.

National Research Council. *National Science Education Standards*. S.I.: National Academy, 1996. Print

Next Generation Science Standards: For States, by States. Washington, D.C.: National Academies, 2013. Print.

Ngo, Christian and Marcel H. Van de Voorde, *Nanotechnology in a Nutshell.* Atlantis Press, 2014, pp. 71-84.

Nicolis, G., and I. Prigogine. *Exploring Complexity: An Introduction.* New York: W.H. Freeman, 1989. Print.

Nilsson, Anders, Lars G.M. Pettersson, and Jens K. Nørskov. *Chemical Bonding at Surfaces and Interfaces.* Jordan Hill, UK: Elsevier, 2008. Print.

Nishinauga, Tatau, ed. *Handbook of Crystal Growth. Fundamentals: Thermodynamics, Kinetics, and Transport and Stability.* 2nd ed. Waltham, MA: Elsevier, 2015. Print.

Nolan, Tania, and Stephen A. Bustin, eds. *PCR Technology: Current Innovations.* CRC Press, 2013.

O'Neil, Maryadele J., ed. T*he Merck Index: An Encyclopedia of Chemicals, Drugs, and Biologicals.* 15th ed. Cambridge: RSC, 2013. Print.

Orton, John. *Molecular Beam Epitaxy. A Short History.* New York, NY: Oxford UP, 2015. Print.

Paun, Gheorghe, Grzegorz Rozenber, and Arto Salmaa. *DNA Computer: New Computing Paradigms.* Texts in Theoretical Computer Science. Springer (2006).

Pelczar, Michael J., Jr., E. C. S. Chan, and Noel R. Krieg. *Microbiology: Concepts and Applications.* New York: McGraw, 1993. Print.

Pfennig, Brian W. *Principles of Inorganic Chemistry.* Hoboken, NJ:Wiley, 2015. Print.

Pine, Stanley H. *Organic Chemistry.* 5th ed. New York: McGraw, 1987. Print.

Pohl, Udo W. *Epitaxy of Semiconductors: Introduction to Physical Principles.* New York, NY: Springer-Verlag, 2013. Print.

Poole, Colin. *Gas Chromatography.* Waltham, MA: Elsevier, 2012. Print.

Porter, David A., Kenneth E. Easterling, and Mohamed Y. Sherif. *Phase Transformations in Metals and Alloys.* 3rd ed. Boca Raton, FL: CRC Press, 2009. Print.

Pöttgen, Rainer, and Dirk Johrendt. *Intermetallics: Synthesis, Structure, Function.* Boston, MA: Walter de Gruyter GmbH, 2014. Print.

Powell, P., and P. L. Timms. *The Chemistry of the Non-Metals.* London: Chapman, 1974. Print.

Prigogine, I., and Isabelle Stengers. *Order out of Chaos: Man's New Dialogue with Nature.* Toronto: Bantam, 1984. Print.

Rang, H. P., et al. *Rang and Dale's Pharmacology.* 7th ed. New York: Elsevier, 2012. Print.

Rauk, Arvi. *Orbital Interaction Theory of Organic Chemistry.* 2nd ed. Hoboken, NJ: John Wiley & Sons, 2001. Print.

Raymond, Kenneth W. *General, Organic, and Biological Chemistry: An Integrated Approach*. 4th ed. Hoboken: Wiley, 2014. Print.

Redd, Nola Taylor. "Black Holes: Facts, Theory, and Definition." *Space.com.* Purch, 9 Apr. 2015. Web. 30 Apr. 2015.

Reece, Jane B., et al. *Campbell Biology.* 10th ed. San Francisco: Cummings, 2013. Print.

Report of the Committee of Ten on Secondary School Studies; with the Reports of the Conferences Arranged by the Committee. New York: Pub. for the National Education Association, by the American Book, 1894. Print.

Rieger, Philip H. *Electrochemistry.* 2nd ed. New York, NY: Chapman & Hall, 1994. Print.

Rimini, Emanuele. *Ion Implantation: Basics to Device Fabrication.* Boston, MA: Kluwer Academic Publishing, 1995. Print.

Riordan, Michael, and Lillian Hoddeson. *Crystal Fire: The Birth of the Information Age.* New York: Norton, 1997. Print.

Robbers, James E., Marilyn K. Speedie, and Varro E. Tyler. *Pharmacognosy and Pharmacobiotechnology.* Rev. ed. Baltimore: Williams, 1996. Print.

Roberts, Michael, Michael Jonathan Reiss, and Grace Monger. *Advanced Biology.* Cheltenham: Nelson, 2000. Print.

Robinson, Trevor. *The Organic Constituents of Higher Plants.* 6th ed. North Amherst: Cordus, 1991. Print.

Rodgers, Glen E. *Descriptive Inorganic, Coordination, & Solid-State Chemistry.* Toronto, ON: Nelson Education, 2011. Print.

Ross, Julian R.H. *Heterogeneous Catalysis Fundamentals and Applications.* Boston, MA: Elsevier. Print.

Rudolph, Peter. ed. *Bulk Crystal Growth. Basic Techniques, and Growth Mechanisms and Dynamics.* Waltham, MA: Elsevier, 2015. Print.

Ryan, Sean G. and Andrew J. Norton. *Stellar Evolution and Nucleosynthesis.* New York, NY: Cambridge UP, 2010. Print.

Sankararaman, Sethuraman. *Pericyclic Reactions.* Weinheim: Wiley, 2005. Print.

Sauthoff, Gerhard. *Intermetallics.* New York, NY: VCH, 1995. Print.

Schramm, David N. T*he Big Bang and Other Explosions in Nuclear & Particle Astrophysics.* River Edge, NJ: World Scientific Publishing, 1996. Print.

Sears, William M. Jr., *Helium: The Disappearing Element.* New York, NY: Springer, 2015. Print.

Shaik, Sason S., and Philippe C. Hiberty, *A Chemist's Guide to Valence Bond Theory.* Hoboken: Wiley, 2008. Print.

Shanbogh, Pradeep and Nalini G. Sundaram. "Fullerenes Revisited." *Resonance* (February 2015): pp. 123-135.

Shasha, Dennis E., and Cathy Lazere. *Natural Computing: DNA, Quantum Bits, and the Future of Smart Machines.* W.W. Norton & Company (2010).

Shaviv, Giora. *The Synthesis of the Elements.* New York, NY: Springer, 2012. Print.

Silbey, Robert J., Robert A. Alberty, and Moungi G. Bawendi. *Physical Chemistry.* 5th ed. Hoboken: Wiley, 2012. Print.

Skoog, Douglas Arvid, Donald M. West, and F. James. Holler. *Skoog and West's Fundamentals of Analytical Chemistry.* 9th ed. Belmont (Calif.): Cengage Learning, 2014. Print.

Smith, Michael B., and Jerry March. *March's Advanced Organic Chemistry: Reactions, Mechanisms, and Structure.* 7th ed. Hoboken: Wiley, 2013. Print.

Somorjai, Gabor A., and Yimin Li. *Introduction to Surface Chemistry and Catalysis.* 2nd ed. New York: Wiley, 2010. Print.

Stevens, T. S., and W. E. Watts. *Selected Molecular Rearrangements.* New York: Van Nostrand, 1973. Print.

Stringfellow, Gerald B. *Organometallic Vapor-Phase Epitaxy: Theory and Practice.* Boston, MA: Academic Press, 1989. Print.

Strobel, Howard A. and William R. Heineman. *Chemical Instrumentation: A Systematic Approach.* 3rd ed. New York: Wiley, 1989. Print.

Stryer, Lubert, Jeremy M. Berg, and John L. Tymoczko. *Biochemistry.* 7th ed. New York: Freeman, 2012. Print.

Stuart A. Kauffman, "Whispers from Carnot: The Origins of Order and Principles of Adaptation in Complex Non Equilibrium Systems, in Complexity; Metaphors, Models and Reality, ed. By G.A. Cowan, D. Pines, D. Meltzer, *SFI Studies in the Sciences of Complexity,* Vol 19, (Addison-Wesley,1994, New York).

Stuart A. Kaufmann, *The Origins of Order: Self Organization and Selection in Evolution,* (Oxford, New York, 1993)

Suggs, J. Wm. *Organic Chemistry.* Hauppauge, NY: Barron's Educational Series, 2002. Print.

Sunagawa, Ichiro. C*rystals. Growth, Morphology and Perfection.* New York, NY: Cambridge University Press, 2005. Print.

S. W. Wilson, "The Genetic Algorithm and Simulated Evolution" in *Artificial Life,* ed. by C. G. Langton SFI Studies in the Sciences of Complexity, Vol 19, (Addison-Wesley, 1989, New York).

Sykes, Peter. *A Guidebook to Mechanism in Organic Chemistry.* 6th ed. London: Longman, 1986. Print.

Thakur, Vijay Kumar and Manju Kumari Thakur, eds. *Chemical Functionalization of Carbon Nanomaterials: Chemistry and Applications.* Boca Raton, FL: Taylor & Francis Group, 2016.

The Britannica Guide to the 100 Most Influential Scientists. Introd. John Gribbin. London: Constable, 2008. Print.

"The Human Genome." *Nature* 409.6822 (2001): 813–958. Print.

Thro, Ellen. *Genetic Engineering: Shaping the Material of Life.* New York: Facts On File, 1995. Print.

Tielens, A.G.G.M. *The Physics and Chemistry of the Interstellar Medium.* New York, NY: Cambridge University Press, 2005. Print.

Tomanek, David. *Guide Through the Nanocarbon Jungle: Buckyballs, Nanotubes, Graphene, and Beyond.* Morgan and Claypool (2014).

Toy, Arthur D.F. *The Chemistry of Phosphorus.* New York, NY: Pergamon Press, 1975. Print.

van Pelt-Verkuil, Elizabeth, Alex van Belkum, and John P. Hays. *Principals and Technical Aspects of PCR Amplification.* Springer, 2010.

Verschuur, Gerrit. *The Invisible Universe: The Story of Radio Astronomy.* New York, NY: Springer, 2015. Print.

Vieira, Ernest R. *Elementary Food Science.* 4th ed. Gaithersburg: Aspen, 1999. Print.

Walas, Stanley M. Phase *Equilibria in Chemical Engineering.* Stoneham, MA: Butterworth-Heinemann, 2013. Print.

Walas, Stanley M. *Reaction Kinetics for Chemical Engineers.* Boston, MA: Butterworths, 1989. Print.

Wang, Zhaocai, Jian Tan, Dongmei Huang, Yingchao Ren, and Zuwen Ji. "A Biological Algorithm to Solve the Assignment Problem Based on DNA Molecules Computation." *Applied Mathematics and Computation.* Volume 244 (2014), pp. 183-190.

Watson, James D. *The Double Helix.* New York: Atheneum, 1968. Print.

Weeks, Daniel P. *Pushing Electrons: A Guide for Students of Organic Chemistry.* 4th ed. Belmont: Brooks, 2014. Print.

Wehr, M. Russell, James A. Richards Jr., and Thomas W. Adair III. *Physics of the Atom.* 4th ed. Reading: Addison-Wesley, 1984. Print

Wiberg, Egon and Wiberg, Nils. *Inorganic Chemistry.* San Diego, CA: Academic Press, 2001. Print.

Widmann, D., H. Mader, and H. Friedrich. *Technology of Integrated Circuits.* New York, NY: Springer-Verlag, 2000. Print.

Wink, Donald J., Sharon Fetzer-Gislason, and Sheila McNicholas. *The Practice of Chemistry.* New York: Freeman, 2004. Print.

Winter, Arthur. *Organic Chemistry 1 for Dummies.* Hoboken: Wiley, 2014. Print.

Winter, Mark J. *The Orbitron: A Gallery of Atomic Orbitals and Molecular Orbitals.* University of Sheffield, n.d. Web.

Wixom, Robert L., and Charles W. Gehrke. *Chromatography: A Science of Discovery.* Hobboken: Wiley, 2010. Print.

Wright, Margaret Robson. *An Introduction to Aqueous Electrolyte Solutions.* Hoboken: Wiley, 2007. Print.

Wuts, Peter G. M., and Theodora W. Greene. *Greene's Protective Groups in Organic Synthesis.* 4th ed. Hoboken: Wiley-Interscience, 2007. Print.

Yamada, Koichi M.T. and Winnewissner, Gisbert, eds. *Interstellar Molecules. Their Laboratory and Interstellar Habitat.* New York, NY: Springer, 2011. Print.

Zemaitis, Joseph F., Diane M. Clark, Marshall Rafal, and Noel C. Scrivner. *Handbook of Aqueous Electrolyte Thermodynamics.* New York: Wiley-Interscience, 2010. Print.

Zumdahl, Steven S., and Susan Zumdahl. *Chemistry.* 9th ed. Belmont: Brooks Coles, 2013. Print.

GENERAL INDEX

Q

R

X

Y

Z